APPLIED AND INDUSTRIAL MATHEMATICS IN ITALY III

Series on Advances in Mathematics for Applied Sciences – Vol. 82

APPLIED AND INDUSTRIAL MATHEMATICS IN ITALY III

Selected Contributions from the 9th SIMAI Conference

Rome, Italy **15 – 19 September 2008**

Edited by

Enrico De Bernardis
INSEAN, Roma, Italy

Renato Spigler
Università Roma Tre, Roma, Italy

Vanda Valente
IAC-CNR, Roma, Italy

World Scientific

NEW JERSEY • LONDON • SINGAPORE • BEIJING • SHANGHAI • HONG KONG • TAIPEI • CHENNAI

Published by

World Scientific Publishing Co. Pte. Ltd.

5 Toh Tuck Link, Singapore 596224

USA office: 27 Warren Street, Suite 401-402, Hackensack, NJ 07601

UK office: 57 Shelton Street, Covent Garden, London WC2H 9HE

Library of Congress Cataloging-in-Publication Data

SIMAI Conference (9th : 2008 : Rome, Italy)

Applied and industrial mathematics in Italy III : Selected Contributions from the 9th SIMAI Conference, Rome, Italy, 15–19 September 2008 / edited by Enrico De Bernardis, Renato Spigler & Vanda Valente.

p. cm. -- (Series on advances in mathematics for applied sciences ; v. 82)

Includes bibliographical references.

ISBN-13: 978-981-4280-29-7 (hardcover : alk. paper)

ISBN-10: 981-4280-29-1 (hardcover : alk. paper)

1. Engineering mathematics--Industrial applications--Congresses.
2. Manufacturing processes--Mathematical models--Congresses. I. De Bernardis, Enrico. II. Spigler, Renato, 1947. III. Valente, Vanda. IV. Title.

TA329.S56 2009
620.001'51--dc22

2009031468

British Library Cataloguing-in-Publication Data

A catalogue record for this book is available from the British Library.

Printed in Singapore by World Scientific Printers

PREFACE

In recent years, research activities in Italy concerning applied and industrial mathematics have been conducted by many scientists with different experience, from university, research institutions and industry. They addressed a number of specific subjects. Besides classical topics — such as fluid dynamics, elasticity, continuous mechanics — new problems have been considered, with application to a variety of situations and phenomena such as, for example, traffic modelling, hemodynamics, biofilms; moreover, a number of research projects of interest for the industry have also been developed.

The state-of-the-art of research in applied and industrial mathematics in Italy can be assessed from the large number of contributions presented at the IX Congress of Italian Society for Applied and Industrial Mathematics (SIMAI), held in Rome in September 2008. In this volume 50 papers are proposed, which were selected among all those presented at the Congress.

In order to properly select these papers from a large number of submitted contributions, the Editors of this volume were supported by the expertise of all members of SIMAI Scientific Committee: Ubaldo Barberis, Nicola Bellomo, Luca Formaggia, Giorgio Fotia, Alessandro Iafrati, Roberto Natalini, Mario Primicerio, Luigia Puccio and Alessandro Speranza. We would like to thank all of them, as well as all the referees involved in the selection procedure.

The Editors

Roma, July 20, 2009

CONTENTS

Multichannel Wavelet Scheme for Color Image Processing

S. Agreste* and A. Vocaturo†

Department of Mathematics
Messina University, Italy
**sagreste@unime.it*
†avocaturo@unime.it

Every day massive quantities of two-dimensional documents are produced, stored and transmitted. Digital color images are the most prominent type of data in this category. To process these objects a variety of powerful and sophisticated wavelet-based schemes have been developed and implemented.
Considering that digital color images are vector objects with three components, they can be suitably processed in the context of a multichannel Multi Resolution Analysis.
In this work we focus on a new vector approach to multichannel wavelet analysis of color images based on several classes of *full rank* filters, showing the results of a wide experimentation.

Keywords: Multichannel wavelet, digital image processing

1. Introduction

The bulk of information necessary to represent digital objects grows day by day, such as the number of pixel of digital camera or videocamera. So it has become necessary and essential to create image files with manageable and transmittable size. Compression algorithms are used in the standards such as JPEG and MPEG to reduce the number of bits required for representing an image or a video sequence. The wavelet transform has emerged as a cutting edge technology, within the field of image compression. Wavelet based coding provides substantial improvements in picture quality at higher compression ratios. Over the past few years a variety of powerful and sophisticated wavelet-based schemes for image compression have been developed and implemented.

Historically, the concept of wavelets is originated from the study of time-frequency signal analysis, wave propagation, and sampling theory. In 1982,

Jean Morlet[1,2] first introduced the idea of wavelets as a family of functions constructed by using translation and dilation of a *mother wavelet* for the analysis of nonstationary signals. Today wavelet analysis is an axciting method for solving difficult problems in several disciplines with applications as diverse as data compression, image processing, pattern recognition (Ref. 3,4). Since the early 1990s, multiwavelets have been introduced as generalization of wavelet functions (Ref. 5,6). Multiwavelets with multiplicity r are vector-values functions. A wavelet functions are the scalar case, when the multiplicity $r = 1$.

Unlike the scalar case, some extra degrees of freedom are allowed, which can be used to construct functions with several desirable properties, combining, for example, orthogonality, symmetry, short support and vanishing moments. All these *good properties* are needed for efficiently processing two-dimensional signals, hence multiwavelets are more powerful than wavelet in image processing. In general the application of the multiwavelet decomposition/reconstruction scheme requires two additional steps respect to the scalar case. The first consists in finding from a given set of input data $\{y_1, y_2, ..., y_m\}$, r sequences of initial coefficients $\left\{c_i^{k,0}\right\}_{k \in Z}$, $i = 1, ..., r$, needed by the analysis phase. This step is called *pre-filtering* of the data, since it can be seen as the application of a filter to the initial data. The second step is called *post-filtering*. It consists in finding the output data from a vector of r entries obtained from the synthesis phase.

The multichannel wavelet has been introduced to analysis multichannel signal in Ref. 7,8. In this case the pre and postfiltering steps are not required. This function allows to process a signal with r *vector-valued* signal by means only one decomposition/reconstruction or better that processes the signal as a multichannel signal with possibly intricate correlations between some of these channels. A good example of multichannel signal is a color digital image. In fact because color image has at least three components in according to the color model representation. The color pixels are vectors. In RGB system, each point $\mathbf{c}$ can be interpreted as a vector $\mathbf{c} = [c_R, c_G, c_B]^T$. The components of $\mathbf{c}$ are the RGB components of a color image at a point (Ref. 9).

In this paper we are going to present an innovative denoising and compression algorithm based on multichannel wavelet. The work is organized as follows. We start by presenting the multichannel wavelet theory from Multichannel MultiResolution Analysis point of view and then we present the denoising and compression algorithms. In the last part we expose some experimental results obtained. We have looked over the results with PSNR

(Peak Signal to Noise Ratio) and WPSNR (Weight Peak Signal to Noise Ratio) values.

2. Multichannel MultiResolution Analysis and Multichannel Wavelet

In order to model and analyze a multichannel signal we introduce the *Multichannel MultiResolution Analysis* (MMRA). This analysis is based on the concept of *matrix refinable functions* known from the study of *full rank* stationary vector subdivision scheme.[10]

We consider the MMRA like a natural extension of the well-known *multiresolution analysis* to vector valued functions. It is necessary to observe that here we are not considering an MRA with multiplicity, that is, a scalar MRA generated by a finite number of functions instead of a single one. We are interested in a MMRA for the space:

$$L_2(\mathbb{R})^{Z_r} = \left\{ \boldsymbol{f} : \mathbb{R} \to \mathbb{R}^r : \|\boldsymbol{f}\|_2 = \left(\sum_{j \in \mathbb{Z}_r} \int_{\mathbb{R}} |f_j(x)|^2 \, dx \right)^{\frac{1}{2}} < \infty \right\},$$

of square integrable vector fields.

A MMRA is defined by a *nested* sequence of closed subspace $V_0 \subset V_1 \subset \ldots \subset L_2(\mathbb{R})^{\mathbb{Z}_r}$ with the properties that:

- they are *shift invariant*
  $\boldsymbol{f} \in V_k \Leftrightarrow \boldsymbol{f}(\cdot + j) \in V_k, \quad j \in Z;$
- they are *scaled* versions of each other
  $\boldsymbol{f} \in V_0 \Leftrightarrow \boldsymbol{f}(2^k \cdot) \in V_k;$
- the are generated by *stable* integer translates of certain vector fields:
  $V_0 = span\{\boldsymbol{f}_j : j \in \mathbb{Z}_r\}$, where the translates of vector fields $\boldsymbol{f}_j, j \in \mathbb{Z}_r$, forms a *Riesz basis* in $L_2(\mathbb{R})^r$ that is,

$$\left\| \sum_{j \in \mathbb{Z}_r} \sum_{k \in \mathbb{Z}} c_{jk} \boldsymbol{f}_j(\cdot - k) \right\|_2 \approx \left(\sum_{j \in \mathbb{Z}_r} \sum_{k \in \mathbb{Z}} |c_{jk}|^2 \right)^{\frac{1}{2}}$$

  Let $\boldsymbol{F} \in L_2(\mathbb{R})^{\mathbb{Z}_r \times \mathbb{Z}_r}$ be the matrix value function with rows $\boldsymbol{f}_j^T$, $j \in \mathbb{Z}_r$ and $\boldsymbol{c}(k)$, $k \in \mathbb{R}^{\mathbb{Z}_r}$ be the vector of entries c_{jk}, $j \in \mathbb{Z}_r$, this condition can be written as $|\boldsymbol{F} * \boldsymbol{c}|_2 \sim |\boldsymbol{c}|_2$.

Any MMRA is generated by a *matrix refinable* function $\boldsymbol{F} \in L_2(\mathbb{R})^{\mathbb{Z}_r \times \mathbb{Z}_r}$, that is, a function for which there exists a finitely supported *mask*

$$\mathbf{A} = \left(\mathbf{A}(k) \in \mathbb{R}^{\mathbb{Z}_r \times \mathbb{Z}_r} : k \in \mathbb{Z}\right) \in \ell_{00}^{\mathbb{Z}_r \times \mathbb{Z}_r}$$

such that

$$\boldsymbol{F} = (\boldsymbol{F} * \mathbf{A})\,(2\cdot) := \sum_{k \in \mathbb{Z}} \boldsymbol{F}(2 \cdot -k)\mathbf{A}(k). \tag{1}$$

where $*$ symbol represents the *convolution* operator between matrix valued function $\boldsymbol{F}$ and matrix sequences $\boldsymbol{A}$.

Equation 1 is called *matrix refinement equation.* Like in the classical scalar MRA, refinable functions also form the building blocks for the MMRA. The matrix-valued function $\boldsymbol{F} \in L_2(\mathbb{R})^{\mathbb{Z}_r \times \mathbb{Z}_r}$ which generates the MMRA, is called the *scaling function* of the MMRA.

To introduce the relative multichannel wavelet it is necessary to suppose that exists a *wavelet* function for any MMRA generated by an *orthogonal* matrix function, that is, by a matrix function $\boldsymbol{F} \in L_2(\mathbb{R})$ such that:

$$\langle \boldsymbol{F}, \boldsymbol{F}(\cdot - j) \rangle = \delta_{0j}, \quad j \in \mathbb{Z}$$

where $\langle \cdot, \cdot \rangle$ represents the skew symmetric bilinear form

$$\langle \cdot, \cdot \rangle : L_2(\mathbb{R})^{\mathbb{Z}_m \times \mathbb{Z}_k} \times L_2(\mathbb{R})^{\mathbb{Z}_m \times \mathbb{Z}_l} \longrightarrow \mathbb{R}^{\mathbb{Z}_k \times \mathbb{Z}_l}$$

defined as

$$\langle F, G \rangle := \int_{\mathbb{R}} \boldsymbol{F}^T(x)\boldsymbol{G}(x)dx.$$

Since an MMRA consists of a nested sequence of spaces

$$... \subset V_{j-1} \subset V_j \subset V_{j+1} \subset ... \qquad , j \in \mathbb{N}$$

it is possible to define the relative orthogonal complements:

$$W_j := V_{j+1} \ominus V_j, \text{ i.e. } V_{j+1} = V_j \oplus W_j, \quad j \in \mathbb{N}_0$$

Let $\boldsymbol{F} \in V_j$ and let $\mathbf{G} \in W_j$, the orthogonality condition has to be understood in the sense that

$$\langle \boldsymbol{F}, \boldsymbol{G} \rangle = \mathbf{0}.$$

A function $\boldsymbol{G} \in V_1$ is called multichannel *wavelet* for the MMRA if

$$W_j = \left\{\mathbf{G} * \mathbf{c}(2^j \cdot) : \mathbf{c} \in \ell_2(\mathbb{Z})^{\mathbb{Z}_r}\right\}, \quad j \in \mathbb{N}_0,$$

and it is called an *orthonormal wavelet* if $\mathbf{G}$ is moreover orthonormal, i.e., if

$$\langle \mathbf{G}, \mathbf{G}(\cdot - j) \rangle = \delta_{0j}\mathbf{I}, \quad j \in \mathbb{N}_0$$

Suppose that $\boldsymbol{F}$ is an orthonormal $\boldsymbol{A}$-refinable matrix function, i.e. $\boldsymbol{A}^T\boldsymbol{A} = 2\boldsymbol{I}$, then there exists a bi-infinite matrix $B = [\boldsymbol{B}(j-2k) : j, k \in \mathbb{Z}]$ where

$$\mathbf{B} = \big(\mathbf{B}(k) \in \mathbb{R}^{\mathbb{Z}_r \times \mathbb{Z}_r} : k \in \mathbb{Z}\big) \in \ell_{00}^{\mathbb{Z}_r \times \mathbb{Z}_r}$$

such that $B^T A = 0$ and $B^T B = 2I$. Moreover this matrix B satisfies $AA^T + BB^T = 2I$. This matrix B is the key to the wavelet construction, in fact

$$\boldsymbol{G} := \boldsymbol{F} * \boldsymbol{B}(2\cdot) \in V_1.$$

To introduce a *fast wavelet transform* it is necessary to define, for any finitely supported matrix valued sequence $\boldsymbol{A} \in \ell_{00}^{\mathbb{Z}\times\mathbb{Z}_r}$, the *subdivision operator* $S_{\boldsymbol{A}}$ and the *decimation operator* $S^*_{\boldsymbol{A}}$:

***subdivision operator*:**

$$S_{\boldsymbol{A}}\boldsymbol{c} = \sum_{k\in\mathbb{Z}} \boldsymbol{A}(\cdot - 2k)\boldsymbol{c}(k) \quad \boldsymbol{c} \in \ell_2(\mathbb{Z})^{\mathbb{Z}_r}$$

***decimation operator*:**

$$S^*_{\boldsymbol{A}}\boldsymbol{c} = \sum_{j\in\mathbb{Z}} \boldsymbol{A}^T(j - 2\cdot)\boldsymbol{c}(j), \quad \boldsymbol{c} \in \ell_2(\mathbb{Z})^{\mathbb{Z}_r}$$

Then we can define the decomposition part of the pyramid scheme by setting

$$\boldsymbol{c}_{j-1} = \frac{1}{2}S^*_{\boldsymbol{A}}\boldsymbol{c}_j \quad \text{and} \quad \boldsymbol{d}_{j-1} = \frac{1}{2}S_{\boldsymbol{B}}\boldsymbol{c}_j \tag{2}$$

and the reconstruction part as

$$\boldsymbol{c}_j = S_{\boldsymbol{A}}\boldsymbol{c}_{j-1} + S_{\boldsymbol{B}}\boldsymbol{d}_{j-1} \tag{3}$$

3. Experimental Results

One of the most attractive features of multichannel wavelets is their effectiveness in the context of the signal compression and denoising. For this reason, we have experimented with the developed *full rank* filters in image compression and denoising based on *wavelet shrinkage*. Wavelet shrinkage, developed by *Donoho et al*[11,12] , selects the wavelet coefficients with significant energy by means *thresholding*. This technique will zero out many small coefficients, which results in efficient representation. There are two kinds of thresholding: *hard* and *soft* (Ref.13). Let I be the digital color image. The outputs entries $I^{hard}_{l,j,k}$ and $I^{soft}_{l,j,k}$ of the thresholding techniques with threshold δ are:

$$I^{hard}_{l,j,k} = \begin{cases} I_{l,j,k}, & if\ |I_{l,j,k}| > \delta \\ 0, & otherwise \end{cases}$$

$$I^{soft}_{l,j,k} = \begin{cases} sign(I_{l,j,k})(|I_{l,j,k}| - \delta), & if\ |I_{l,j,k}| > \delta \\ 0, & otherwise \end{cases}$$

Thresholding is lossy algorithm because the original signal cannot be reconstructed exactly starting from the processed signal.

We have developed denoising and compression algorithms based on wavelet shrinkage. In particular, we have applied the thresholding technique to multichannel wavelet coefficients. The performance of the multichannel wavelet scheme has been tested by means PSNR (Peak Signal to Noise Ratio) and WPSNR (Weight Peak Signal to Noise Ratio) values on the classical test images: *Lena*, *Baboon*, *Peppers*, *Lake*, *F16* and *House*. Instead, to value the compression score we have used the *energy ratio* in percentage defined by:

$$E = \frac{\left\|\hat{I}\right\|_2^2 \times 100}{\|I\|_2^2} \tag{4}$$

where I and $\hat{I}$ are respectively the original and compressed images.

Table 1. PSNR values of test images relative to the denoising processing by means soft (center) and hard (right) thresholding

Test image	**PSNR**		
	Noise	Den. ST	Den. HT
Lena	23.24	29.16	28.40
Peppers	22.23	28.86	27.62
Baboon	20.89	21.53	21.88
F16	20.47	27.43	26.26
Lake	21.42	26.10	26.03
House	21.41	27.37	26.72

In Table 1 and Table 2 there are explained the respectively PSNR and WPSNR values relative to the denoising technique. In the "*Noise*" column, we have inserted the PSNR (WPSNR) values relative to the original image and the image corrupted by noise. The "*Den. ST*"' and "*Den. HT*"' columns represent the PSNR (WPSNR) values relative to the original image and the

Table 2. WPSNR values of test images relative to the denoising processing by means soft (center) and hard (right) thresholding

Test image	WPSNR		
	Noise	Den. ST	Den. HT
Lena	38.10	38.27	38.55
Peppers	37.87	38.59	38.05
Baboon	33.57	36.53	35.10
F16	36.06	36.12	36.20
Lake	35.28	37.05	36.74
House	36.97	37.08	37.25

Table 3. PSNR corresponding to different value of compression putting to zero from 70% to 90% of wavelet coefficients by means hard thresholding

Test image	PSNR			
	70%	80%	85%	90%
Lena	35.7	21.1	17.1	14.0
Baboon	27.2	21.8	17.5	13.4
Peppers	39.0	30.2	25.0	20.3

image corrupted by noise after the denoising operation by means soft (*Den. ST*) and hard (*Den. HT*) thresholding. Fig. 1 represents an example of denoising operation by means hard and soft thresholding.

Table 4. WPSNR corresponding to different value of compression putting to zero from 70% to 90% of wavelet coefficients by means hard thresholding

Test image	WPSNR			
	70%	80%	85%	90%
Lena	51.3	29.3	25.3	22.1
Baboon	46.6	34.5	26.5	21.0
Peppers	51.1	40.1	34.5	29.2

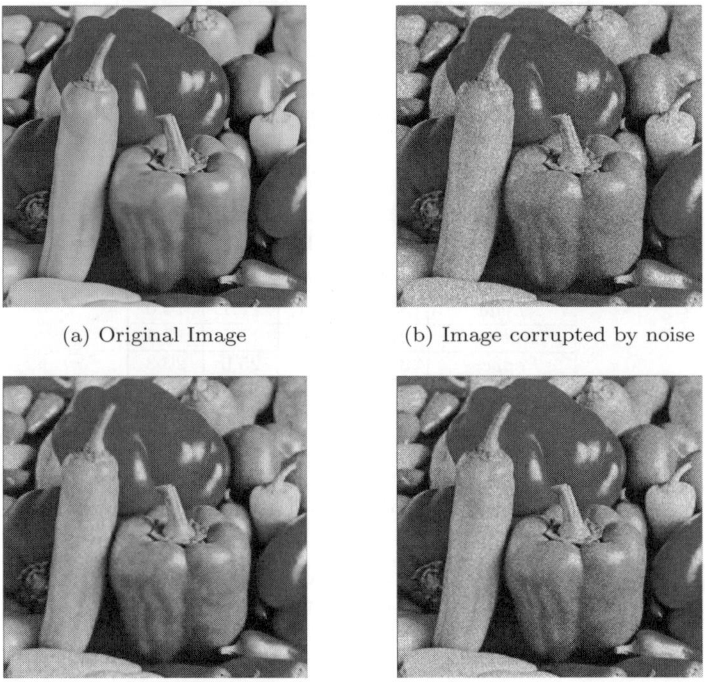

(a) Original Image

(b) Image corrupted by noise

(c) Image corrupted by noise after soft thresholding

(d) Image corrupted by noise after hard thresholding

Fig. 1. Example of denoising operation by means hard and soft thresholding

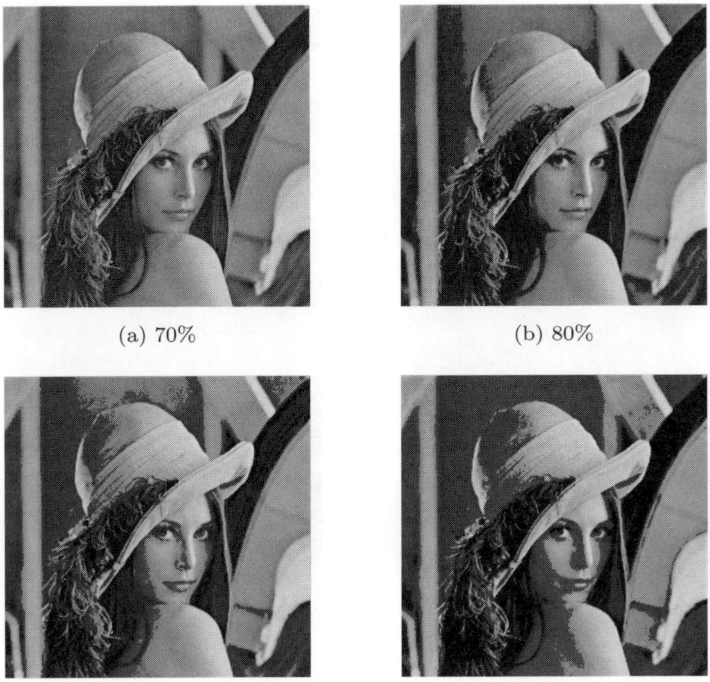

(a) 70% (b) 80%

(c) 85% (d) 90%

Fig. 2. Example of compression *Lena* image by means hard thresholding putting to zero 70% (a), 80% (b), 85% (c) and 90% (d) of wavelet coefficients

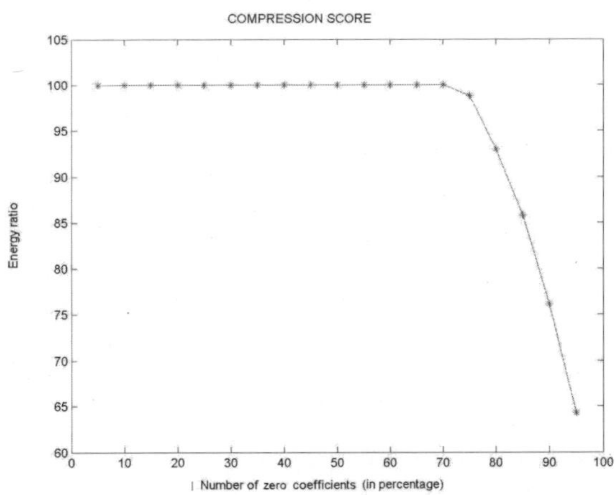

Fig. 3. Plot of energy ratio in percentage relative of *Lena* image

(a) 70%

(b) 80%

(c) 85%

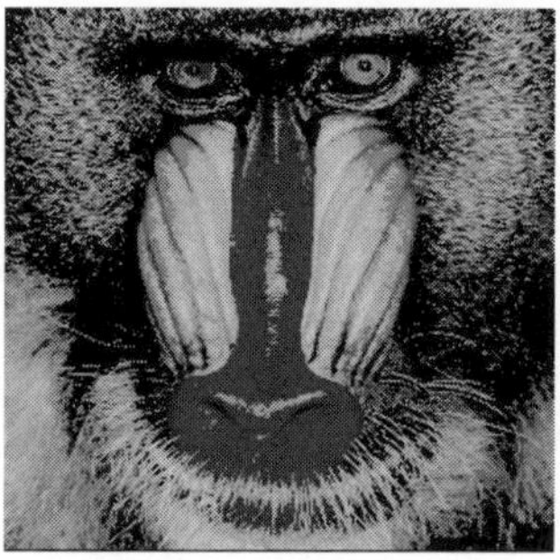

(d) 90%

Fig. 4. Example of compression of *Baboon* image by means hard thresholding putting to zero 70% (a), 80% (b), 85% (c) and 90% (d) of wavelet coefficients

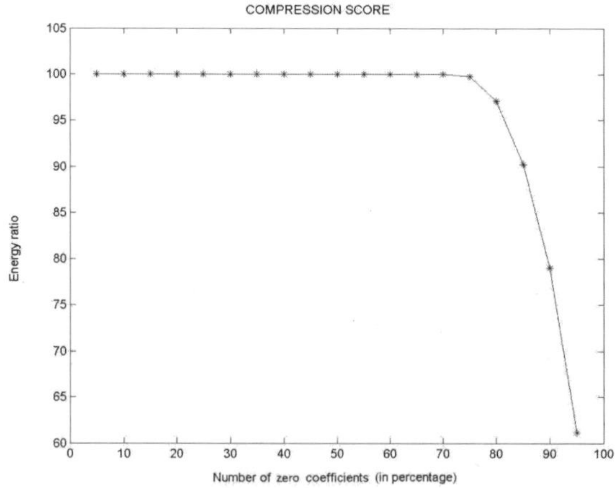

Fig. 5. Plot of energy ratio in percentage relative of *Baboon* image

In Table 3 and Table 4 the PSNR and WPSNR values are presented corresponding to different values of compression respectively for Lena, Baboon and Peppers images. It can be observed that we obtain a good quality of compressed images putting to zero until 80% of coefficients (see Fig. 2 and Fig. 4). In Fig. 3 and Fig. 5 we have plotted the compression score of *Lena* and *Baboon* images corresponding to different values of thresholding obtained annihilating from 5% to 95% wavelet coefficients.

4. Conclusion

In this paper we investigated about the multichannel wavelets. We presented an overview about the mathematical theory showing a matrix wavelet approach which has the potential to form a convenient method for the analysis of vector-valued signals. Then we applied this vectorial approach on multichannel signal such as digital color images showing the results of a wide experimentation. *Lena*, *Baboon*, *Peppers*, *Lake*, *F16* and *House* images have been processed. PSNR and WPSNR values are illustrated relative to denoising and compression technique.
Next step is to build and test other multichannel wavelet based on several classes of *full rank* filters and to built an useful tool and a suitable software library for vectorial signal processing.

Acknowledgements

This work was supported by: PRA Projects of University of Messina and GNCS of INdAM.

References

1. J. Morlet, G. Arens, E. Fourgeau and D. Glard, *Geophysics* **47**, 203 (1982).
2. J. Morlet, G. Arens, E. Fourgeau and D. Giard, *Geophysics* **47**, 222 (1982).
3. L. Debnath, *Wavelet Transforms and Their Applications* (Birkhauser, 2002).
4. A. Cohen, *Numerical Analysis of Wavelet Methods* (Elsevier, 2003).
5. T. N. T. Goodman and S. L. Lee, *Transactions of the American Mathematical Society* **342**, 307 (1994).
6. G. Plonka and V. Strela, From wavelets to multiwavelets, in *Mathematical Methods for Curves and Surfaces II*, eds. M. Dahlem, T. Lyche and L. Shumaker (Vanderbilt University Press, 1998) pp. 1–25.
7. S. Bacchelli, M. Cotronei and T. Sauer, *Adv. Appl. Math.* **29**, 581 (2002).
8. S. Bacchelli, M. Cotronei and T.Sauer, *BIT 42* **2**, 231 (1998).
9. R. Gonzalez and R. Woods, *Digital Image Processing*, second edn. (Prentice Hall, 2001).

10. C. Conti, M. Cotronei and T. Sauer, Full rank interpolatory vector subdivision schemes, in *Curve and Surface Fitting: Avignon 2006*, eds. A. Cohen, J. L. Merrien and L. S. Eds (Nashboro Press, Nashville, 2007).
11. D. L. Donoho and I. M. Johnstone, *Biometrika* **81**, 425 (1994).
12. D. L. Donoho and I. M. Johnstone, *Ann. Statist* **26**, 879 (1998).
13. G. Strang and T. Nguyen, *Wavelets and Filter Banks* (Wellesley-Cambridge Press,U.S., 1996).

A Space-time Galerkin BEM for 2D Exterior Wave Propagation Problems

Alessandra Aimi[♮], Mauro Diligenti, Ilario Mazzieri, Stefano Panizzi

Department of Mathematics, Università di Parma, Italy
[♮]*E-mail: alessandra.aimi@unipr.it*

Chiara Guardasoni[*]

Department of Mathematics, Università degli Studi di Milano, Italy
[*]*E-mail: chiara.guardasoni@unimi.it*

Here we consider wave propagation for 2D Dirichlet or Neumann exterior problems reformulated in terms of boundary integral equations with retarded potential. Starting from a natural energy identity, a space-time weak formulation for the boundary integral equations is proposed and discussed. Some numerical results and applications are presented to show the effectiveness of the introduced approach.

Keywords: wave propagation, boundary integral equations, Galerkin BEM

1. Introduction

Time-dependent problems that are frequently modelled by hyperbolic partial differential equations can be dealt with boundary integral equations (BIEs)[4,9,10] . The boundary element techniques applicable to time dependent problems are classified as follows[8] : *transform method* based on the Laplace transform or the Fourier transform; *direct method* based on the formulation using the time dependent fundamental solution of the hyperbolic partial differential equation and the construction of the BIEs via representation formulas and jump relations, where the use of single layer and double layer potentials follows in large part the known formalism related to elliptic problems; and *time-stepping method* based on the finite difference approximation of time derivatives. Domain-type numerical methods, such as finite element method and finite difference method, require a discretization of not just the surface of the domain of interest, as it is with the boundary element method (BEM), but its interior as well, thereby increasing the modelling

requirements and the size of the problem for comparable accuracy and creating difficulties due to wave reflections at artificial boundaries for infinite or semi-infinite domains. On the other hand, the BEM discretizes only the surface of the domain of interest and takes automatically into account the radiation condition at infinity for exterior problems.
The purpose of this concise paper is to present a direct space-time Galerkin method for the discretization of the *retarded potential* BIEs applied to solve Dirichlet or Neumann wave propagation in two dimensional exterior problems. Special attention is devoted to a formulation based on a natural energy identity that leads to a space-time weak formulation of the corresponding BIEs[2] . In the last section some examples demonstrate the computational efficiency associated with this approach and serve to validate the weak formulation proposed.

2. Two dimensional wave equation

2.1. *Dirichlet problem*

Here we will consider a Dirichlet problem for the wave equation exterior to an open arc $\Gamma \subset \mathbb{R}^2$:

$$u_{tt} - \triangle u = 0, \qquad \mathbf{x} \in \mathbb{R}^2 \setminus \Gamma, t \in (0,T) \tag{1}$$

$$u(\mathbf{x},0) = u_t(\mathbf{x},0) = 0, \qquad \mathbf{x} \in \mathbb{R}^2 \setminus \Gamma, \tag{2}$$

$$u(\mathbf{x},t) = g(\mathbf{x},t), \qquad (\mathbf{x},t) \in \Sigma_T := \Gamma \times [0,T]\,. \tag{3}$$

In this case, the boundary datum $g(\mathbf{x},t)$ represents the value of the excitation field over Γ. Having set $r = \|\mathbf{x} - \boldsymbol{\xi}\|_2$, we can consider the single-layer representation of the solution of (1)-(3):

$$u(\mathbf{x},t) = \int_\Gamma \int_0^t G(r,t-\tau)\varphi(\boldsymbol{\xi},\tau)\, d\tau\, d\gamma_\xi, \qquad \mathbf{x} \in \mathbb{R}^2 \setminus \Gamma, t \in (0,T), \tag{4}$$

where $\varphi = \left[\dfrac{\partial u}{\partial \mathbf{n}}\right]$ is the jump of the normal derivative of u along Γ and

$$G(r,t-\tau) = \frac{1}{2\pi} \frac{H[t-\tau-r]}{[(t-\tau)^2 - r^2]^{\frac{1}{2}}} \tag{5}$$

is the forward fundamental solution of the two dimensional wave operator, with $H[\cdot]$ the Heaviside function. With a limiting process for $\mathbf{x}$ tending to Γ and using the assigned Dirichlet boundary condition we obtain the space-time BIE

$$\int_\Gamma \int_0^t G(r,t-\tau)\varphi(\boldsymbol{\xi},\tau)\, d\tau\, d\gamma_\xi = g(\mathbf{x},t), \qquad \mathbf{x} \in \Gamma,\ t \in (0,T), \tag{6}$$

which can be written with the compact notation

$$V\varphi = g. \tag{7}$$

A usual way to introduce a weak formulation for (7) is to project the BIE using $L^2(\Sigma_T)$ scalar product, but, as it has been proved[3] , this weak formulation gives rise to instability phenomena in the discretization phase. An alternative approach is suggested by the well-known conservation law satisfied by the (real-valued) solutions to the d'Alambert equation. Integrating with respect to space-time in $\mathbb{R}^2 \times (0,T)$ and taking into account that u and u_t vanish for $t=0$, we get the energy identity:

$$\mathcal{E}(T,u) := \frac{1}{2}\int_{\mathbb{R}^2}\left(\frac{\partial u(\mathbf{x},T)}{\partial t}^2 + |\nabla u(\mathbf{x},T)|^2\right)d\mathbf{x} = \int_{\Sigma_T}\frac{\partial u}{\partial t}(\mathbf{x},t)\left[\frac{\partial u}{\partial \mathbf{n}}\right](\mathbf{x},t)dtd\gamma_x.$$

In a previous paper[2] we have introduced a natural space-time weak formulation for (7) related to the energy of the wave equation:

given $g \in H^1_{\{0\}}(\Sigma_T)$, find $\varphi \in L^2(\Sigma_T)$ such that

$$a_{\mathcal{E}}(\varphi,\psi) = < g_t, \psi >_{L^2(\Sigma_T)}, \quad \forall \psi \in L^2(\Sigma_T), \tag{8}$$

where the bilinear form $a_{\mathcal{E}}(\varphi,\psi) : L^2(\Sigma_T) \times L^2(\Sigma_T) \to \mathbb{R}$ is defined by

$$a_{\mathcal{E}}(\varphi,\psi) = < (V\varphi)_t, \psi >_{L^2(\Sigma_T)} = \int_\Gamma \int_0^T (V\varphi)_t\,(\mathbf{x},t)\,\psi(\mathbf{x},t)\,dt\,d\gamma_x. \tag{9}$$

Now, we come to the crucial question of the coerciveness of the energy functional, made more precise by the following Proposition[2] :

Proposition 2.1. *There exists a sequence $(\varphi_n)_{n\in\mathbb{N}}$ of non vanishing functions in $C_0^\infty(\Sigma_T)$, such that, for any $s \in \mathbb{R}$ we have*

$$\lim_{n\to\infty} \frac{a_{\mathcal{E}}(\varphi_n,\varphi_n)}{\|\varphi_n\|_s^2} = 0, \tag{10}$$

where $\|\cdot\|_s$ stands for the norm in $H^s(\mathbb{R}^2)$.

The main consequence of Proposition 2.1 is that, in general, the quadratic form $a_{\mathcal{E}}(\varphi,\varphi)$ cannot be coercive with respect to any Sobolev norm. Therefore, in order to obtain satisfactory *a priori* bounds on the *energetic* Galerkin approximated solutions of problem (7), we need to complement the information coming from the quadratic form $a_{\mathcal{E}}(\varphi,\varphi)$ with some other arising from alternative arguments. In this respect, we recall two possible strategies[2] :
1) to obtain a constraint on the oscillations in the space variable of the

approximating solutions. In fact, as we have shown[2] , under a suitable constraint, the energetic norm is coercive with respect to the $L^2(\Sigma_T)$ norm.
2) to modify (9) adding another bilinear form which takes into account also the skew-symmetric part of the operator $\varphi \mapsto (V\varphi)_t$.

2.2. *Neumann problem*

Now, we introduce also a 2D Neumann model problem, for which we will give numerical results in the last section. Let us consider the scattering problem by a crack in an unbounded elastic isotropic medium. Let $\Omega = \mathbb{R}^2 \setminus \Gamma$ be this medium and Γ the crack represented by an open arc. Let Γ^- and Γ^+ denote the lower and upper faces of the crack and $\mathbf{n}$ the normal unit vector to Γ, oriented from Γ^- to Γ^+. As usual, the total displacement field can be represented as the sum of the incident field (the wave propagating without the crack) and the scattered field. In a 3D elastic isotropic medium, there are three plane waves propagating in a fixed direction: the P wave, the SH wave and the SV wave. The 2D antiplane problem corresponds to an incident SH wave, when all quantities are independent of the third component z (in particular, the crack has to be invariant with respect to z). The scattered wave satisfies the following Neumannn problem for the wave operator (without loss of generality we will consider a dimensionless problem which can be obtained after an appropriate scaling of the units):

$$u_{tt} - \triangle u = 0, \qquad \mathbf{x} \in \mathbb{R}^2 \setminus \Gamma, t \in (0,T) \tag{11}$$

$$u(\mathbf{x},0) = u_t(\mathbf{x},0) = 0, \qquad \mathbf{x} \in \mathbb{R}^2 \setminus \Gamma \tag{12}$$

$$\frac{\partial u}{\partial \mathbf{n}}(\mathbf{x},t) = g(\mathbf{x},t), \qquad (\mathbf{x},t) \in \Sigma_T := \Gamma \times [0,T]\,. \tag{13}$$

In (11)-(13) the unknown function u stands for the third component of the displacement field and g is the datum, which is the opposite of the normal derivative of the incident wave along Γ, i.e. $g = -\frac{\partial u^I}{\partial \mathbf{n}}$.
Let us consider the double-layer representation of the solution of (11)-(13):

$$u(\mathbf{x},t) = \int_\Gamma \int_0^t \frac{\partial G(r,t-\tau)}{\partial \mathbf{n}_\xi}\, \phi(\boldsymbol{\xi},\tau)\, d\tau\, d\gamma_\xi\,, \quad \mathbf{x} \in \mathbb{R}^2 \setminus \Gamma, t \in (0,T), \tag{14}$$

where G is given in (5) and $\phi = [u]$ is the jump of u along Γ. Taking the normal derivative with respect to $\mathbf{x}$ of (14), with a limiting process for $\mathbf{x}$ tending to Γ and using the assigned Neumann boundary condition, we obtained the space-time hypersingular BIE

$$\int_\Gamma \int_0^t \frac{\partial^2 G(r,t-\tau)}{\partial\, \mathbf{n}_x \partial \mathbf{n}_\xi}\, \phi(\boldsymbol{\xi},\tau)\, d\tau\, d\gamma_\xi = g(\mathbf{x},t)\,, \quad \mathbf{x} \in \Gamma, t \in (0,T), \tag{15}$$

which can be written with the compact notation

$$D\phi = g. \tag{16}$$

The energetic weak problem related to (16) will be of the form:

$$\tilde{a}_{\mathcal{E}}(\phi,\eta) =< g, \eta_t >_{L^2(\Sigma_T)},$$

where

$$\tilde{a}_{\mathcal{E}}(\phi,\eta) :=< D\phi\,,\,\eta_t >_{L^2(\Sigma_T)}= \int_\Gamma \int_0^T (D\phi)(\mathbf{x},t)\,\eta_t(\mathbf{x},t)\,dt\,,\,d\gamma_x$$

and η is a suitable test function belonging to the same functional space of ϕ. The hypersingular integral operator D can be equivalently expressed in the following way:

$$D\phi(\mathbf{x},t) = \int_\Gamma \frac{\partial^2 r}{\partial \mathbf{n}_x \partial \mathbf{n}_\xi} \int_0^t G(r,t-\tau)\left[\phi_t(\xi,\tau) + \frac{\phi(\boldsymbol{\xi},\tau)}{t-\tau+r}\right] d\tau\, d\gamma_\xi +$$

$$\int_\Gamma \frac{\partial r}{\partial \mathbf{n}_x}\frac{\partial r}{\partial \mathbf{n}_\xi} \int_0^t G(r,t-\tau)\left[\phi_{tt}(\boldsymbol{\xi},\tau) + \frac{2\,\phi_t(\boldsymbol{\xi},\tau)}{t-\tau+r} + \frac{3\,\phi(\boldsymbol{\xi},\tau)}{(t-\tau+r)^2}\right] d\tau d\gamma_\xi. \tag{17}$$

3. Galerkin BEM discretization

We introduce the discretization for a crack of length L, constituted by M elements $\{e_1,\cdots,e_M\}$ with $e_i \cap e_j = \emptyset$, $\cup_{j=1}^M \overline{e}_j = \overline{\Gamma}$, $length(e_i) \leq h$. The functional background compels one to choose spatially shape functions belonging to $L^2(\Gamma)$ for Dirichlet problems and to $H_0^1(\Gamma)$ for Neumann problems. Hence we use standard piecewise polynomial boundary element functions $w_j(x)$, $j = 1,\cdots,M_h$ suitably defined in relation to the introduced mesh over Γ. For time discretization we consider an uniform decomposition of the time interval $[0,T]$ with time step $\Delta t = T/N_{\Delta t}$, $N_{\Delta t} \in \mathbb{N}^+$, generated by the $N_{\Delta t}+1$ instants

$$t_k = k\,\Delta\,t, \quad k = 0,\cdots,N_{\Delta t}$$

and we choose temporally piecewise constant shape functions for Dirichlet problems and piecewise linear shape functions for Neumann problems, although higher degree shape functions can be used. Note that, for this particular choice, our shape functions, denoted with $v_k(t)$, will be defined as

$$v_k(t) = H[t-t_k] - H[t-t_{k+1}]\,, \qquad k = 0,\cdots,N_{\Delta t}-1$$

for Dirichlet problems or as

$$v_k(t) = R(t - t_k) - 2\,R(t - t_{k+1}) + R(t - t_{k+2}),$$

where $R(t - t_k) = \frac{t-t_k}{\Delta t}\, H[t - t_k]$ is the ramp function, for Neumann problems. Hence, the unknown approximate solution of the problem at hand will be expressed as

$$\sum_{k=0}^{N_{\Delta t}-1} \sum_{j=1}^{M_h} \alpha_j^{(k)}\, w_j(\mathbf{x})\, v_k(t). \tag{18}$$

The discretization coming from energetic weak formulation produces the linear system

$$A_{\mathcal{E}}\, x_{\mathcal{E}} = b_{\mathcal{E}}. \tag{19}$$

In the case of a Dirichlet problem, the matrix elements, after double analytic integration in the time variables, are of the form

$$\sum_{\alpha,\beta=0}^{1} (-1)^{\alpha+\beta} \int_\Gamma w_i(\mathbf{x}) \int_\Gamma \mathcal{B}(r, t_{h+\alpha}, t_{k+\beta}) w_j(\boldsymbol{\xi})\, d\gamma_\xi\, d\gamma_x, \tag{20}$$

where, having set $\Delta_k^h = t_h - t_k$,

$$\mathcal{B}(r, t_h, t_k) = -\frac{1}{2\pi}\, H[\Delta_k^h - r] \left[\log\left(\Delta_k^h + \sqrt{(\Delta_k^h)^2 - r^2}\right) - \log r\right].$$

In the case of a Neumann problem, the matrix elements, after double analytic integration in the time variables, are of the form

$$\sum_{\alpha,\beta,\delta=0}^{1} (-1)^{\alpha+\beta+\delta} \int_\Gamma w_i(\mathbf{x}) \int_\Gamma \mathcal{C}(r, t_{h+\alpha}, t_{k+\beta+\delta})\, w_j(\boldsymbol{\xi}) d\gamma_\xi\, d\gamma_x, \tag{21}$$

where

$$\mathcal{C}(r, t_h, t_k) = \frac{H[\Delta_k^h - r]}{2\pi\,\Delta t} \Big[\frac{\mathbf{r}\cdot\mathbf{n}_x\ \mathbf{r}\cdot\mathbf{n}_\xi}{r^2}\ \frac{\Delta_k^h \sqrt{(\Delta_k^h)^2 - r^2}}{r^2} +$$
$$\frac{(\mathbf{n}_x\cdot\mathbf{n}_\xi)}{2} \Big[\log(\Delta_k^h + \sqrt{(\Delta_k^h)^2 - r^2}) - \log r - \frac{\Delta_k^h \sqrt{(\Delta_k^h)^2 - r^2}}{r^2} \Big]\Big].$$

Weakly singular, singular and hypersingular double integrals in space variables are efficiently calculated with numerical quadrature schemes[1] widely used for BIEs related to elliptic problems, after a careful subdivision of the integration domain due to the presence of the Heaviside function.

Anyway, the above elements depend on the difference $t_h - t_k$ and in particular they vanish if $t_h \le t_k$. Hence, matrix $A_{\mathcal{E}}$ has a block lower triangular

Toeplitz structure. Each block has dimension M_h. If we indicate with $A_{\mathcal{E}}^{(\ell)}$ the block obtained when $t_h - t_k = (\ell+1)\,\Delta t$, $\ell = 0, \ldots, N_{\Delta t} - 1$, the linear system can be written as

$$\begin{pmatrix} A_{\mathcal{E}}^{(0)} & 0 & 0 & \cdots & 0 \\ A_{\mathcal{E}}^{(1)} & A_{\mathcal{E}}^{(0)} & 0 & \cdots & 0 \\ A_{\mathcal{E}}^{(2)} & A_{\mathcal{E}}^{(1)} & A_{\mathcal{E}}^{(0)} & \cdots & 0 \\ \cdots & \cdots & \cdots & \cdots & 0 \\ A_{\mathcal{E}}^{(N_{\Delta t}-1)} & A_{\mathcal{E}}^{(N_{\Delta t}-2)} & \cdots & A_{\mathcal{E}}^{(1)} & A_{\mathcal{E}}^{(0)} \end{pmatrix} \begin{pmatrix} \alpha^{(0)} \\ \alpha^{(1)} \\ \alpha^{(2)} \\ \vdots \\ \alpha^{(N_{\Delta t}-1)} \end{pmatrix} = \begin{pmatrix} b_{\mathcal{E}}^{(0)} \\ b_{\mathcal{E}}^{(1)} \\ b_{\mathcal{E}}^{(2)} \\ \vdots \\ b_{\mathcal{E}}^{(N_{\Delta t}-1)} \end{pmatrix}, \tag{22}$$

where: $\alpha^{(\ell)} = \left(\alpha_j^{(\ell)}\right)$ and $b_{\mathcal{E}}^{(\ell)} = \left(b_{\mathcal{E},j}^{(\ell)}\right)$, $j = 1, \ldots, M_h$. The solution of (22) is obtained with a block forward substitution, i.e. at every time instant $t_\ell = (\ell+1)\,\Delta t$ we solve a reduced linear system of the type:

$$A_{\mathcal{E}}^{(0)} \alpha^{(\ell)} = b_{\mathcal{E}}^{(\ell)} - (A_{\mathcal{E}}^{(1)} \alpha^{(\ell-1)} + \cdots + A_{\mathcal{E}}^{(\ell)} \alpha^{(0)}). \tag{23}$$

Procedure (23) is a time-marching technique, where the only matrix to be inverted is the positive definite $A_{\mathcal{E}}^{(0)}$ diagonal block, while all the other blocks are used to update at every time step the right-hand side. Owing to this procedure we can construct and store only the blocks $A_{\mathcal{E}}^{(0)}, \cdots, A_{\mathcal{E}}^{(N_{\Delta t}-1)}$ with a considerable reduction of computational cost and memory requirement.

4. Numerical results

In the following, we will present some numerical results obtained for 2D exterior problems starting from the proposed energetic weak formulation. We consider (1) – (3) with $\Gamma = \{(x,0), x \in [0,1]\}$ and Dirichlet boundary datum shown in figure 1, and we fix the observation time interval $[0,2]$.

$$g(x,t) = \begin{cases} \sin 2\pi(t-0.1) & \text{if } 0.1 < t < 0.6\,, \\ 0 & \text{elsewhere}\,. \end{cases}$$

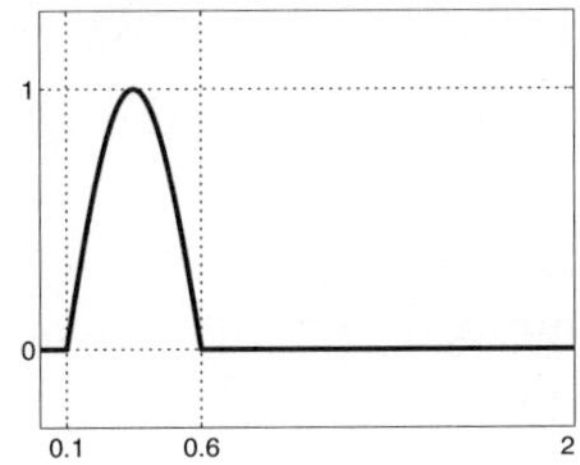

Fig. 1. Dirichlet boundary condition

As uniform temporal and spatial discretization steps we use $\Delta t = 0.05$ and $h = 0.05$ respectively and we adopt spatial constant shape and test functions.

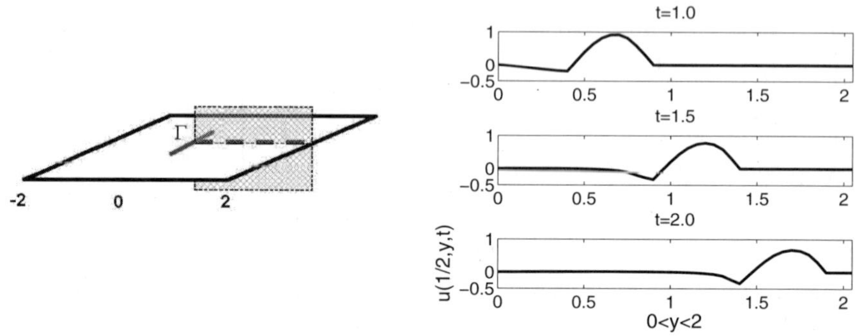

Fig. 2. Section of the solution $u(x, y, t)$ in $x = \frac{1}{2}$ in some time instants.

In figure 2 we present a section of the solution $u(x, y, t)$ of the problem $(1) - (3)$, in $x = 0.5$, for $t \in (0, 2)$. As one can observe, the wave travelling away from the boundary Γ assumes a structure similar to the Dirichlet boundary datum but with diminishing intensity.

As second example, we consider a Dirichlet problem exterior to cracks with a corner, that are a progressive modification of the rectilinear one $\Gamma_0 = \{(x, 0) : x \in [0, 1]\}$: in particular, the middle point of the crack has increasing ordinate $y_i = 2^{-(4-i)}$, $i = 1, \cdots, 3$, in order to obtain cracks Γ_i as shown in figure 3. The Dirichlet boundary condition is

$$g(x, t) = H[t]f(t)\,, \quad \text{where} \quad f(t) = \begin{cases} \sin^2(\frac{\omega t}{2}) & \text{if } \; 0 \le t \le \frac{\pi}{\omega} \\ 1 & \text{if } \; t \ge \frac{\pi}{\omega} \end{cases}$$

and observation time interval is $[0, 2]$. As uniform temporal discretization we use $\Delta\, t = 0.0125$ and cracks are uniformly subdivided into 40 boundary elements where we adopt constant shape and test functions.

In figure 3 we show the time histories of the approximate solution in the middle point of the 20th element of the mesh introduced on Γ_i, for $i = 0, \cdots, 3$. As one can note, the higher is the ordinate y_i, the greater is the difference between time histories on Γ_i and time history on Γ_0 at the beginning of the simulation, while asymptotic behavior of the approximate solutions seems to become similar for all the cracks.

Now, we present numerical results related to two dimensional Neumann exterior problems (11)-(13), for $\Gamma = \{(x, 0) : x \in [0, 1]\}$. The incident wave

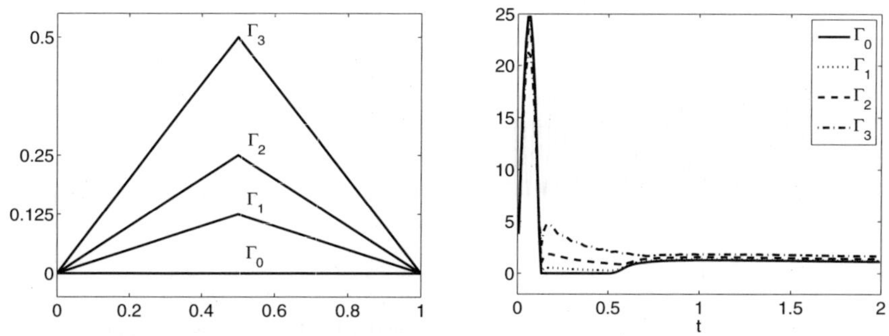

Fig. 3. Cracks Γ_i, $i = 0, \cdots, 3$ and time history of the solution φ near the corner for different cracks.

$u^I(\mathbf{x}, t)$ is a plane wave propagating in direction k with unitary amplitude:

$$u^I(\mathbf{x}, t) = f(t - k \cdot \mathbf{x}) \qquad \text{with} \qquad k = (\cos\theta, \sin\theta).$$

Hence, the Neumann datum on Γ in (11) will be:

$$g(\mathbf{x}, t) = -\frac{\partial}{\partial \mathbf{n}_x} f(t - k \cdot \mathbf{x})\Big|_{\Gamma}. \tag{24}$$

We show the results obtained for two different functions f, that have been chosen for the known asymptotic behavior of the solution, which allows us to validate the approximate solution.

Plane harmonic wave. Because of the initial conditions in (11), we consider a wave which becomes harmonic after a fixed time[7] :

$$f(t) = \begin{cases} 0 & \text{if} \quad t < 0, \\ \frac{1}{2}(1 - \cos\omega t) & \text{if} \quad 0 \le t \le \frac{\pi}{\omega}, \\ \sin\left(\frac{\omega t}{2}\right) & \text{if} \quad t \ge \frac{\pi}{\omega}, \end{cases} \tag{25}$$

where ω represents the frequency. In this case the solution has to become harmonic too, with the same period as the incident wave, i.e. $P = 4\pi/\omega$. The fixed circular frequency $\omega = 8\pi$ is such that the wave length $\lambda = 2\pi/\omega$ is equal to a quarter the crack length.

We choose a uniform decomposition of the space interval $[0, 1]$ in 20 subintervals ($h = 0.05$) and we decompose the observation time interval $(0, 10)$ in 400 equal parts ($\Delta t = 0.025$). For this numerical simulation we choose spatial linear shape and test functions. In figure 4 we show the time harmonic behavior for $\theta = \pi/3$ of the crack opening displacement (COD) ϕ at $x = 0.4$. Note that the COD is zero till the time instant $t^* = x\cos\theta = x/2$, since the incident wave doesn't invest the whole crack simultaneously. In

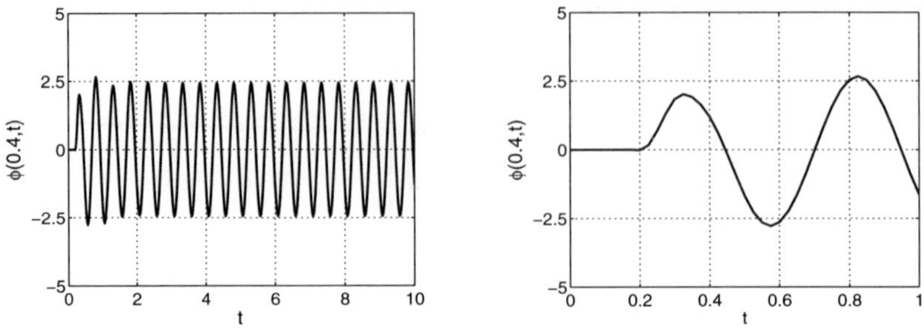

Fig. 4. Density $\phi(x,t)$ calculated in $x = 0.4$ for $\theta = \pi/3$, with a zoom.

order to verify that the period of ϕ coincides with the period of the incident wave, we show in figure 5 the approximate solution ϕ on Γ in time instants separated by multiples of the time period.

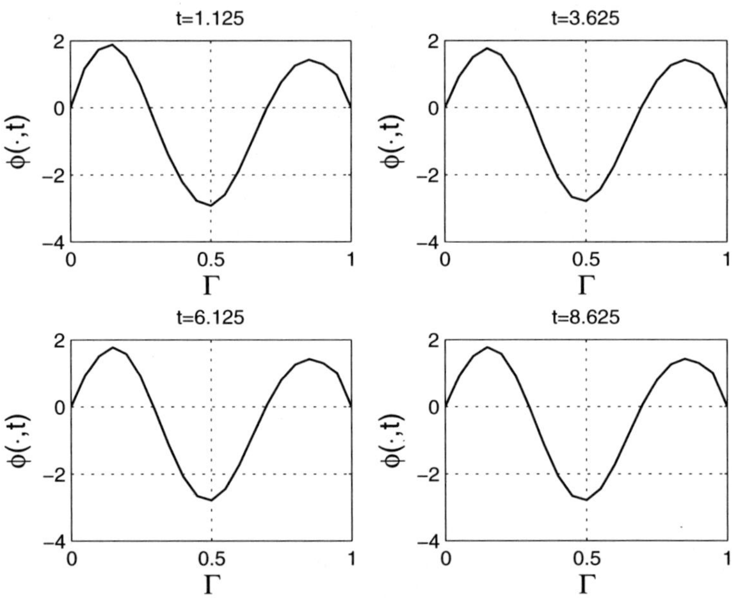

Fig. 5. Solution $\phi(x,t)$ in different time instants.

Plane linear wave. Let us consider another example always[7] , where the Neumann boundary conditions comes from this choice:

$$f(t) = t\,H[t].$$

In this case, the Neumann datum (24) tends to the constant value $g_\theta = \sin\theta$

when t tends to infinity. The solution u tends to the solution of the static problem

$$-\triangle u_\infty = 0 \quad \text{in} \quad \mathbb{R}^2 \setminus \Gamma, \qquad\qquad \frac{\partial u_\infty}{\partial \mathbf{n}} = g_\theta \quad \text{on } \Gamma, \tag{26}$$

and the associated jump $\phi_\theta^\infty(x) = [u_\infty]$ across Γ can be computed explicitly:

$$\phi_\theta^\infty(x) = \sin\theta\sqrt{x(1-x)}\ .$$

Then we can compare the solution $\phi(x,t)$ with the static solution $\phi_\theta^\infty(x)$. Let us consider the crack Γ decomposed in 20 subintervals ($h = 0.05$), the final observation time $T = 10$, the time step $\Delta t = 0.025$ and spatial linear shape and test functions. Also in this case, the numerical solution obtained for large times and shown in figure 6 for $\theta = \pi/4$ is in perfect agreement with the corresponding one reported in Bécache[7] . Note that points of Γ, symmetric with respect to $x = 0.5$ behaves in different ways at the beginning of the simulation, since the incident wave does not invest simultaneously the crack; then they assume a symmetric behavior for sufficiently large times. Finally in figure 7 we present the approximate COD

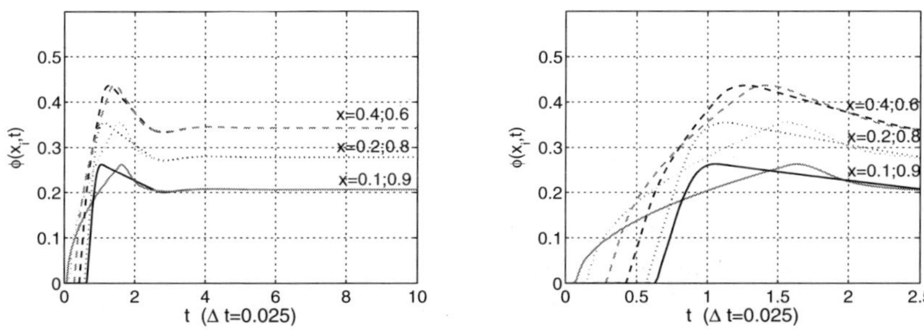

Fig. 6. Density $\phi(x,t)$ evaluated in some points of Γ for $\theta = \pi/4$.

for $T = 4, 5, 10$ together with the analytical solution of the corresponding static problem: the four curves overlap each other.

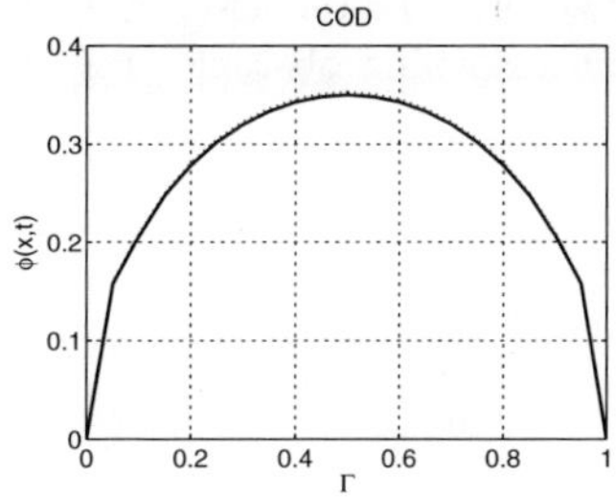

Fig. 7. COD at $t = 4, 5, 10$ compared with static solution for $\theta = \pi/4$.

Acknowledgements

Work done within a research project financed by the Italian Ministry of University and of Research (MIUR) prot.2007JL35WY_002.

References

1. A. Aimi, M. Diligenti and G. Monegato. New Numerical Integration Schemes for Applications of Galerkin BEM to 2D Problems, *Int. J. Numer. Meth. Engng.* **40** (1997) pp. 1977–1999.
2. A. Aimi, M. Diligenti, C. Guardasoni, I. Mazzieri and S. Panizzi. An energy approach to space-time Galerkin BEM for wave propagation problems. Tech. Report, 2008/487 Department of Mathematics, Parma University, Italy.
3. A. Aimi and M. Diligenti. A new space-time energetic formulation for wave propagation analysis in layered media by BEMs, *Int. J. Numer. Methods Engrg.* **75** (2008) pp. 1102–1132.
4. H. Antes. A boundary element procedure for transient wave propagations in two-dimensional isotropic media. *Finite Elem. Anal. Des.* **1** (1985) pp. 313–322.
5. A. Bachelot, L. Bounhoure and A. Pujols. Time dependent integral method for Maxwell's system. In: Mathematical and numerical aspects of wave propagation, *SIAM, Philadelphia*, (1995) pp. 151–159.
6. A. Bamberger and T. Ha Duong. Formulation variationnelle pour le calcul de la diffraction d'une onde acoustique par une surface rigide. *Math. Methods Appl. Sci.* **8** (1986) pp. 598–608.
7. E. Bécache. A variational boundary integral equation method for an elastodynamic antiplane crack, *Int. J. Numer. Methods Engng.*, **36** (1993) pp. 969–984.
8. M. Costabel. Time-dependent problems with the boundary integral method. In: E. Stein, R. de Borst, T.J.R. Hughes, editors, *Encyclopedia of Computational Mechanics*, John Wiley, 2004.
9. A. Frangi. Elastodynamics by BEM: a new direct formulation. *Int. J. Num. Meth. Engng.* **45** (1999) pp. 721–740.
10. L. Gaul and M. Schanz. A comparative study of three boundary element approaches to calculate the transient response of viscoelastic solids with unbounded domains, *Comput. Methods Appl. Mech. Eng.* **179** (1-2) (1999) pp. 111–123.
11. T. Ha Duong, B. Ludwig and I. Terrasse. A Galerkin BEM for transient acoustic scattering by an absorbing obstacle. *Int. J. Num. Meth. Engng.* **57** (2003) pp. 1845–1882.

Application of Laplace Transformation to the Solution of Particular Systems of ODE's for Nuclear Engineering

Franco Almerico

Intelligent Systems S.r.l.
Via Garibaldi 16, 20035 Lissone (MI), Italy
f.almerico@gmail.com

A methodology applying the Laplaces transformation to the solution of particular systems of Ordinary Differential Equations (ODEs) is presented that can be used in the calculation of inventory of radioactive nuclides and fission products in the core of a nuclear power reactor. A number of simplifying assumptions regarding fission products generation by thermal and fast fission of Uranium and Plutonium isotopes are introduced, but a detailed description of the decay process, including branching of decay chains, is used. The proposed mathematical modeling is indeed general and could be applied to other classes of problem involving the solution of systems of chained ODEs.

Keywords: Reactor isotopes inventory, radioactive decay, Laplace transformation, Ordinary Differential Equations, recursive formulas, source term.

1. Introduction

One of the issues for the design and licensing of nuclear power plants is the evaluation of radiation doses environment and plant personnel in case of accidental radioactive release, in particular radioactive Iodine and Noble Gases isotopes doses because they pose a serious hazard to public and personnel health. The present article describes a methodology to calculate the inventory of radioactive iodine and noble gas fission products in the core of a Light Water Reactor (LWR) and their subsequent release to the containment system after a Loss of Coolant Accident (LOCA). The proposed model represents an extension of that described in TID-14844: "Calculation of Distance Factors for Power and Test Reactor Sites". (1) In spite of a number of simplifying assumptions regarding fission products generation by thermal and fast fission of ^{235}U, ^{238}U, ^{239}Pu, and ^{241}Pu are introduced, a detailed description of the decay process, including branching of decay chains, is used. In this regard the model can be considered as an intermediate step between the very simple steady state model of TID-14844 and the more complicated and complete mathematical models on which depletion codes, like the ORIGEN code(2), are based. The results of the above model can then be used as input to models describing the in-plant contamination transport, in order to determine the time-dependent releases of iodine, krypton, and xenon to the environment, and radiation doses to plant personnel, after the hypothesized LOCA. The main objective of the model is to obtain a conservative estimate of the iodine and noble gas fission products reactor core inventory and their subsequent release to the containment system after a LOCA, with the goal of achieving a mathematical description suitable for an efficient, flexible and easy to use computer implementation through an analytical or numerical algorithm.

2. Description and Solution of the Analytical Model

2.1. *General Equations*

The distribution of fission products throughout a reactor core is a very complicated function of the initial composition, power history, and time-space-energy dependent distribution of the neutron flux. Therefore a detailed mathematical representation of the processes describing the changes in reactor composition and the production of fission products cannot be readily carried out by simple analytic procedures.(3,4,5) In safety analysis and environmental hazard assessment, estimates of the reactor core inventory can be obtained by using the following one-group rate balance equations:

$$\frac{dNi(t)}{dt} = \sum_j \gamma_{j\to i}\sigma_{f,j}\Phi N_j(t) + \sum_k g_{k\to i}\sigma_k\Phi N_k(t) + \sum_l h_{l\to i}\lambda_l N_l(t) - \sigma_{a,i}\Phi N_i(t) - \lambda_i N_i(t) \quad (1)$$

where

i, j, k, l = nuclide indices.

$N_i(t)$, $N_j(t)$, $N_k(t)$, $N_l(t)$ = core abundances of nuclides i, j, k, l, respectively [atom].

Φ = average total flux of neutrons in thermal equilibrium; this is taken as a constant [neutrons/cm2-s].

$\sigma_{f,j}, \sigma_{a,i}, \sigma_k$ = effective microscopic cross-sections for: fission of nuclide j, neutron absorption by nuclide i, neutron absorption by nuclide k, respectively [cm2], so that when the σ's are multiplied by the total thermal flux, Φ, the proper reaction rates are obtained.

$\gamma_{j\to i}$ = independent fission yield of nuclide i from the fission of nuclide j.

$g_{k\to i}$ = number of nuclide i produced per unit neutron absorption by nuclide k.

$h_{l\to i}$ = number of nuclides i produced per unit decay of nuclide l.

λ_i, λ_l = radioactive decay constants of nuclides i and l, respectively [1/s].

The first term on the RHS represents the formation rate of nuclide i from the fission of nuclide j at time t. The second term represents the production rate of nuclide i from all the neutron absorbing reactions of nuclide k at time t. The third term represents the production rate of nuclide i from radioactive decay of all the precursors of nuclide i at time t. The fourth and the fifth terms represent the rate of disappearance of nuclide i by neutron absorption and radioactive decay respectively at time t.
Noting that $\sigma_{f,j}\Phi N_j(t)$ represents the fissioning rate of nuclide j at time t, we can write:

$$\sigma_{f,j}\Phi N_j(t) = \sum_j \gamma_{j\to i}\frac{P_j(t)}{\epsilon_j} \tag{2}$$

where:
$P_j(t)$ = power generated at time t by fission of nuclide j.
ϵ_j = average recoverable energy from one fission of nuclide j.

Substituting Eq. (2) into (1) yields:

$$\begin{aligned}\frac{dNi(t)}{dt} &= \sum_j \gamma_{j\to i}\frac{P_J(t)}{\epsilon_j} + \sum_k g_{k\to i}\sigma_k\Phi N_k(t) + \sum_l h_{l\to i}\lambda_l N_l(t) \\ &\quad - \sigma_{a,i}\Phi N_i(t) - \lambda_i N_i(t)\end{aligned} \tag{3}$$

When absorption terms are negligible (i.e. $g_{k\to i} = 0$ and $\sigma_{a,i} = 0$), the above equation simplifies to:

$$\frac{dNi(t)}{dt} = \sum_j \gamma_{j\to i}\frac{P_j(t)}{\epsilon_j} + \sum_l h_{l\to i}\lambda_l N_l(t) - \lambda_i N_i(t) \tag{4}$$

2.1.1. *Treatment of the Coupling Term*

To treat the coupling term

$$\sum_l h_{l\to i}\lambda_l N_l(t) \tag{5}$$

appearing in Eqs. (3) and (4), four decay modes: β^+, β^-, electron capture (EC), and isomeric transition have been retained on the basis of their presence in the decay chains containing iodine and noble gases isotopes: α-decay and neutron emission have not been considered. Under these conditions the only coupling by decay is between the equations representing the time evolution of nuclides having the same mass number. A careful analysis of the decay chains of iodine and noble gases shows that all the decay chains of interest can be regrouped into chains with branching and without branching.

Decay Chains Without Branching. Decay chains without branching can be represented by the linear sequence shown below:

$$\begin{array}{ccccccc} X_1 \rightarrow & X_2 \rightarrow \dots & \rightarrow X_{i-1} & \rightarrow X_i \rightarrow & X_{i+1} \rightarrow \dots & \rightarrow X_M(stable) \\ \uparrow & \uparrow & \uparrow & \uparrow & \uparrow & \uparrow \\ \gamma_1 & \gamma_2 & \gamma_{i-1} & \gamma_i & \gamma_{i+1} & \gamma_M \end{array}$$

The sequence can include any of the four decay modes introduced as long as the chain remains unbranched, that is each nuclide has one daughter only. The indexation of the nuclides in the chain obeys to the following rule: for a given atomic number, the metastable nuclide, if present, precedes the ground state. The *M-th* nuclide represents the last considered nuclide of the chain, that is the stable one. The same scheme can also represent the unbranched β^+ or EC decay section of a given decay chain; the stable nuclide is now in the *M'-th* position, in the sequence. Therefore a decay chain encompassing both β^- and β^+ or EC decay modes is divided into two separate decay chains leading to the same stable nucleus from different sides of the nuclide valley.

Decay Chains with Branching. To define the sequencing of nuclides in a branched decay chain, the coupling between the nuclides of the chain is first considered. A given nuclide can be at the most produced by the decay of three precursors. This would happen, for example, to the ground state of the nuclide of mass number N and atomic number Z when it is produced by decay of its isomer and by decay of the isomeric precursors of atomic number Z-1 for a β^- decay chain (or Z+1 for a β^+ or EC decay chain). Therefore each nuclide in a decay chain of specified mass number can be considered as potentially produced by its three precursors, with the following branching fractions:

α_{i-1} = branching fraction of the decay of nuclide *i-1* to nuclide *i*.
β_{i-2} = branching fraction of the decay of nuclide *i-2* to nuclide *i*.
δ_{i-3} = branching fraction of the decay of nuclide *i-3* to nuclide *i*.

Mathematical Representation. The coupling term becomes:

$$\sum_l h_{l\to i}\lambda_l N_l(t) = \alpha_{i-1}\lambda_{i-1}N_{i-1}(t) + \beta_{i-2}\lambda_{i-2}N_{i-2}(t) + \delta_{i-3}\lambda_{i-3}N_{i-3}(t) \tag{6}$$

with $\delta_{i-3} = 0$ for $\imath \leq 3$, $\beta_{i-2} = 0$ for $\imath \leq 2$, and $\alpha_{i-1} = 0$ for $i = 1$, to properly account for the inexistence of some or all precursors for the first three members of the chain.

For unbranched chain, the above expression simplifies to:

$$\sum_l h_{l\to i}\lambda_l N_l(t) = \alpha_{i-1}\lambda_{i-1}N_{i-1}(t) \tag{7}$$

with $\alpha_{i-1} = 0$ for $i = 0$.

2.2. *Solution of the Reactor Inventory Equations*

Substituting Eq. (5) or (6) into Eq. (3) or (4) yields the system of equations describing the formation of fission products in the reactor core. Because the reactor core is assumed to be uncontaminated by fission products at the beginning of reactor operation, the initial condition is given by: $N_i(t=0) = 0$.

The method of solution is illustrated by looking at Eqs. (3) and (4), after substitution of the coupling term given by Eq. (5), i.e.:

$$\frac{dNi(t)}{dt} = \sum_j \gamma_{j\to i}\frac{P_j(t)}{\epsilon_j} + \alpha_{i-1}\lambda_{i-1}N_{i-1}(t) + \beta_{i-2}\lambda_{i-2}N_{i-2}(t) + \delta_{i-3}\lambda_{i-3}N_{i-3}(t) - (\lambda_i + \sigma_{a,i}\Phi)N_i(t) \tag{8}$$

for i=1,. . ., M,

where the second term on the RHS of Eq. (3) has been neglected. M is the number of nuclides in a chain of specified mass number. The system of equations represented by Eq. (8) is an inhomogeneous system of linear ordinary differential equations, with constant coefficients, in the unknowns $N_1, ..., N_i, ..., N_M$, with vanishing initial conditions. In this work, the Laplace's transformation (6) has been chosen as a method of solution of Eqs. (8). Its main advantage is that it allows expressing the solutions of the system by means of recursive formulas, with very little algebraic effort. With respect to a numerical solution, the benefits are: exactness of results, ease of programming, and speed of computation.

2.2.1. *Solution by Laplace Transform*

The solution proceeds as follows. Taking the Laplace's transform of Eq. (8), and isolating the Laplace's transform of $N_i(t)$ we get:

$$\begin{aligned}\overline{N}_i(s) = (\sum_j \gamma_{j\to i}\frac{P_j(t)}{\epsilon_j}\frac{1}{s} + N_i(0) + \alpha_{i-1}\lambda_{i-1}\overline{N}_{i-1}(s) \\ + \beta_{i-2}\lambda_{i-2}\overline{N}_{i-2}(s) + \delta_{i-3}\lambda_{i-3}\overline{N}_{i-3}(s))\frac{1}{(s+(\lambda_i+\sigma_{a,i}\Phi))}\end{aligned} \tag{9}$$

where the $\overline{N}_i(s)$'s represent the Laplace's transforms of the time-dependent solutions, the $N_i(0)$'s the initial conditions, and s is the parameter of the transformation. The above expression constitutes a recursive formula which allows to express the solution of index i in terms of the *(i-1)-th*, the *(i-2)-th*, and the *(i-3)-th* solutions.

Of particular importance is the situation when $\sigma_{a,i} = 0$, a condition generally realized except for a few nuclides (in particular: ^{135}Xe).
Eq. (9) then becomes:

$$\begin{aligned}\overline{N}_i(s) = (\sum_j \gamma_{j\to i}\frac{P_j(t)}{\epsilon_j}\frac{1}{s} + N_i(0) + \alpha_{i-1}\lambda_{i-1}\overline{N}_{i-1}(s) \\ + \beta_{i-2}\lambda_{i-2}\overline{N}_{i-2}(s) + \delta_{i-3}\lambda_{i-3}\overline{N}_{i-3}(s))\frac{1}{(s+\lambda_i)}\end{aligned} \tag{10}$$

For Eq. (10), we can write:

$$\overline{N}_i(s) = \sum_{j=0}^{m}\sum_{k=1}^{\mu_j-1}\frac{A_i^{j,k}}{(s+\lambda_j)^k} \tag{11}$$

where:

$-\lambda_j$ = poles of $\overline{N}_i(s)$; it is to be noted that $\lambda_0 = 0$;
m = number of distinct poles $-\lambda_j$
μ_j = multiplicity of pole $-\lambda_j$
$A_i^{j,k}$ = constants to be evaluated.

It is worthwhile to notice that, given the structure of Eq. (10) the poles $-\lambda_j$ are just the opposite of the decay constants. So they can be multiple if some of the nuclides have the same half-life. This case has been given a special treatment in Sect. 2.3, so that we can assume that the decay constants have distinct numerical values. For this case the transforms take the form:

$$\overline{N}_i(s) = \sum_{j=0}^{i} \frac{A_i^j}{(s+\lambda_j)} \tag{12}$$

The corresponding inverse Laplace's transforms are:

$$N_i(t) = \sum_{j=0}^{i} A_i^j e^{-\lambda_j(t)} \tag{13}$$

which yield the desired time-dependent solutions.

2.2.2. *Evaluation of the Precoefficients A*

The precoefficients A_i^j can be found as residues of the transforms $\overline{N}_i(s)$ in the neighborhood of the poles $-\lambda_j$, by taking the limit: $A_i^j = lim_{s\rightarrow -\lambda_j}\overline{N}_i(s)(s+\lambda_j)$ for $j = 0, 1, ..., i$.

The results are:

$$A_i^0 = (\sum_j \gamma_{j\rightarrow i}\frac{P_j}{\epsilon_j} + \alpha_{i-1}\lambda_{i-1}A_{i-1}^0 + \beta_{i-2}\lambda_{i-2}A_{i-2}^0 + \delta_{i-3}\lambda_{i-3}A_{i-3}^0)\frac{1}{\lambda_i} \tag{14}$$

$$A_i^1 = (\alpha_{i-1}\lambda_{i-1}A_{i-1}^1 + \beta_{i-2}\lambda_{i-2}A_{i-2}^1 + \delta_{i-3}\lambda_{i-3}A_{i-3}^1)\frac{1}{\lambda_i - \lambda_1} \tag{15}$$

...

$$A_i^m = (\alpha_{i-1}\lambda_{i-1}A_{i-1}^m + \beta_{i-2}\lambda_{i-2}A_{i-2}^m + \delta_{i-3}\lambda_{i-3}A_{i-3}^m)\frac{1}{\lambda_i - \lambda_m} \tag{16}$$

...

$$A_i^i = N_i(0) - (A_i^0 + A_i^1 + ... + A_i^{i-1}) \tag{17}$$

Expressions (14), (15), (16), (17) make possible the evaluation of the precoefficients of the $i - th$solution once the precoefficients of the $(i-3) - th$, $(i-2) - th$ and $(i-1) - th$ solutions are known. Therefore we can obtain in sequence all the solutions up to the last one. We only need to memorize the coefficients A_i^j of the three preceding solutions and calculate the time dependent solution according to Eq. (13).

2.2.3. *Equilibrium Approximation*

Usually, the first few members of most decay chains are very short-lived; their half-lives have been given the same value of 10^{-10} s (10) because of physical limitations in their measurement. From the analytical point of view, this introduces the complication of having to deal with the presence of multiple poles in expanding Eq. (12). To overcome the difficulty, and since these nuclides reach equilibrium in a time period insignificantly small with respect to the generation times considered in this study, an equilibrium approximation has been used when $\lambda_i > 10^5$ [1/s]: the first non short-lived nuclide appearing in the chain is considered to be produced directly by fission with a yield equal to its cumulative yield.

2.3. *Modeling of the Radioactive Release from the Reactor Core*

2.3.1. *Background*

Presently the calculation of fission product release from the reactor core, in the event of a LOCA, is conducted with a great degree of conservatism. The guidelines to follow in determining the radioactive release under LOCA conditions are given in Regulatory Guides 1.3 and 1.4 for BWR's and PWR's, respectively (7,8); they are:

Halogens (iodine isotopes): (1) a fraction of 0.5 of the initial core inventory is instantaneously released to the primary system at the beginning of the accident. (2) a fraction of 0.5 of this release, or 0.25 of the initial core inventory, is instantaneously released to the primary containment building. Moreover a fraction of 0.91 of the iodine released is assumed to be in elemental form, a fraction of 0.05 in particulate form and 0.04 in organic form (methyl iodide).

noble gases (krypton and xenon isotopes). all of the the initial core inventory is instantaneously released to the primary containment structure. These values have been shown to be very conservative and are believed to be typical of a core meltdown condition. (13)

2.3.2. *Description of the Model*

In this work the NRC recommendations have been implemented. Moreover, the delayed releases due to decay of the precursors, have been incorporated. Therefore the total release is the result of an immediate release, occurring at the onset of the accident, and of a time-dependent delayed release taking place over the whole accident duration.

Prompt Release. Since decay of the very short-lived nuclides can be considered as an instantaneous process, their decay products are assumed to contribute to the initial core inventory; therefore their eventual impact is lumped into the immediate release. Under these conditions, designating by $N_{i,c}(0)$, the core inventory of the $i-th$ member of a specified decay chain, one can write:

$$N_{1,c}(0) = 0, \ldots, N_{k,c}(0) = 0 \tag{18}$$

$$N_{k+1,c}(0) = N_{k+1}(\tau_0) + \alpha_k \sum_{l=1}^{k} N'_{l,c}(\tau_0) \tag{19}$$

$$N_{k+2,c}(0) = N_{k+2}(\tau_0) + \beta_k \sum_{l=1}^{k} N'_{l,c}(\tau_0) \tag{20}$$

$$N_{k+3,c}(0) = N_{k+3}(\tau_0) + \delta_k \sum_{l=1}^{k} N'_{l,c}(\tau_0) \tag{21}$$

$$N_{k+4,c}(0) = N_{k+4}(\tau_0) \tag{22}$$

$$\ldots$$

$$N_{M,c}(0) = N_M(\tau_0), \tag{23}$$

where

k = index of the last short-lived precursor

$N_j(\tau_0)$ = reactor inventory of nuclide j after a reactor operating time, τ_0 ; it is obtained from Sect. 2.

$N'_{l,c}(\tau_0)$ is evaluated with the following recursive formula:

$$N'_{l,c}(\tau_0) = N_l(\tau_0) + \alpha_{l-1} N'_{l-1,c}(\tau_0) + \beta_{l-2} N'_{l-2,c}(\tau_0) + \delta_{l-3} N'_{l-3,c}(\tau_0) \tag{24}$$

The mathematical expression for the prompt release is now given by:

$$N_{i,j}(0) = f_{i,j} F_i N_{i,c}(0) \tag{25}$$

where:

i = nuclide index in the specified decay chain

$N_{i,j}(0)$ = amount of the $i-th$ nuclide in the decay chain instantaneously released to the containment building in physical form j [atom]

$N_{i,c}(0)$ = amount of nuclide i in the core, taking the fast transients into account [atom]; it is obtained from Eqs. (18-23) and (24)

F_i = release fraction of nuclide i from the core to the containment building

$f_{i,j}$ = fraction of nuclide i which is in the $j-th$ physical form (particulate, elemental, organic).

Delayed Release. The mathematical model for the delayed release starts from the core inventory equations modified to take into account the fast decay transients and the instantaneous release.

The initial conditions are given by the amount of each nuclide remaining in the core after the prompt release or:

$$N^*_{i,c}(0) = (1 - F_{1i})N_{i,c}(0) \tag{26}$$

where:

$N^*_{i,c}(0)$ = core inventory of nuclide i right after the prompt release [atom].

$N_{i,c}(0)$ = is defined in Eqs. (18-23)

F_{1i} = fraction of nuclide i released from the core to the primary system.

The system of equations describing the rate of change of the fission product inventory in the core is now:

$$\begin{aligned}\frac{dN^*_{i,c}(t)}{dt} &= -\lambda_i N^*_{i,c} + (\alpha_{i-1}\lambda_{i-1}N^*_{i-1,c} \\ &\quad + \beta_{i-2}\lambda_{i-2}N^*_{i-2,c} + \delta_{i-3}\lambda_{i-3}N^*_{i-3,c})(1 - R_i)\end{aligned} \tag{27}$$

where:

R_i = delayed release fraction from the core to the primary system for nuclide i; it is assumed: $R_i = 0.5$ for iodine; 1.0 for noble gases; 0 for all others.

The term: $(\alpha_{i-1}\lambda_{i-1}N^*_{i-1,c} + \beta_{i-2}\lambda_{i-2}N^*_{i-2,c} + \delta_{i-3}\lambda_{i-3}N^*_{i-3,c})(1 - R_i)$ represents the formation rate of nuclide i from the decay of its precursors, assuming that decay of precursors and release of then produced nuclide i are two simultaneous processes. The delayed release source term is now given by:

$$S_i(t) = (\alpha_{i-1}\lambda_{i-1}N^*_{i-1,c} + \beta_{i-2}\lambda_{i-2}N^*_{i-2,c} + \delta_{i-3}\lambda_{i-3}N^*_{i-3,c})F_i \tag{28}$$

This expression represents the delayed release rate of nuclide i into the containment system, in unit of [atom/s].

Solution of the Delayed Release Equations. Eqs. (27) constitute a system of linear first order ODE's to be integrated applying the initial conditions (26). The method chosen is identical to the one presented in Sect. 2. So the transformed solution and the time-dependent solution can be expressed as:

$$\overline{N}^*_{i,c}(s) = \sum_{j=1}^{i} \frac{A_i^j}{s * \lambda_j} \quad and \quad N^*_{i,c}(t) = \sum_{j=1}^{i} A_i^j e^{-\lambda_j t} \tag{29}$$

with:

$$A_i^1 = (\alpha_{i-1}\lambda_{i-1}A_{i-1}^1 + \beta_{i-2}\lambda_{i-2}A_{i-2}^1 + \delta_{i-3}\lambda_{i-3}A_{i-3}^1)\frac{(1-R_i)}{\lambda_i - \lambda_1} \tag{30}$$

$$\cdots$$

$$A_i^j = (\alpha_{i-1}\lambda_{i-1}A_{i-1}^j + \beta_{i-2}\lambda_{i-2}A_{i-2}^j + \delta_{i-3}\lambda_{i-3}A_{i-3}^j)\frac{(1-R_i)}{\lambda_i - \lambda_j} \tag{31}$$

$$\cdots$$

$$A_i^i = N^*_{i,c}(0) - (A_i^1 + A_i^2 + ... + A_i^{i-1}) \tag{32}$$

The time dependent delayed release rate for nuclide i , expressed in unit of [atom], and the integrated release, expressed in unit of [atom], can be written as follows:can be written as:

$$S_i(t) = \sum_{j=1}^{i-1} S_i^j e^{-\lambda_j t} \quad and \quad I_i(t) = \int_0^t S_i(t^{'})dt^{'} = \sum_{j=1}^{i-1} \frac{S_i^j}{\lambda_j}(1 - e^{-\lambda_j t}) \tag{33}$$

3. Computer Implementation and Discussion of Results

The solution procedures shown in the preceeding Sections have been computer implemented in two programs, named SOURCE and CORE, written in FORTRAN V language (9,10). The implementation upgraded an old version of the model and of the computer codes developed for Control Room Habitability and HVAC Systems Performance Evaluation (14). The results of computation of radioactive fission products inventory in a sample problem for a PWR, with a power of 3565 MWth, specific power of 30 [MW/metric ton of U], lifetime averaged neutron flux of 2.9 x 10^{13} [neutron/cm^2s], fuel enrichment of 3.3% in ^{235}U, recoverable energy from thermal fission in ^{235}U of 200 MeV and operation time of 3 years, applied to a set of 41 decay chains producing iodine or noble gases isotopes, for a total of 250 nuclides, are obtained in few minutes on a Intel Pentium Windows XP Workstation.

The results of the sample problem have been compared to those obtained by the ORIGEN code (2) and to those given in the TID-14844 report(1). The comparison is shown in Fig. 1, 2 and 3 for Krypton (^{83m}Kr, ^{85m}Kr, ^{85}Kr, ^{87}Kr, ^{88}Kr, ^{89}Kr), Iodine (^{131}I, ^{132}I, ^{133}I, ^{134}I, ^{135}I) and Xenon (^{131m}Xe, ^{133m}Xe, ^{133}Xe, ^{135m}Xe, ^{135}Xe, ^{138}Xe) isotopes respectively and show a reasonable agreement with the results of the ORIGEN code and of TID-14844 report.

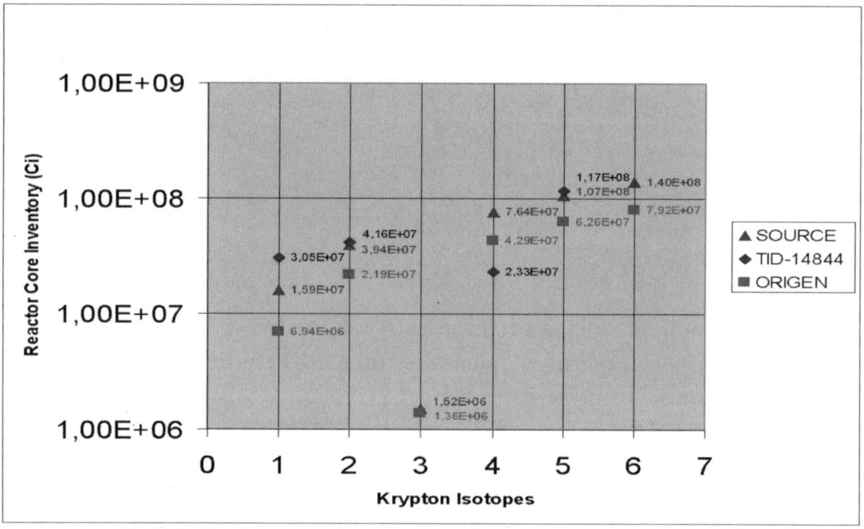

Fig. 1. Results of model for Krypton

4. Conclusions

A model to calculate the inventory of radioactive iodine and noble gas fission products in the core of a LWR and their subsequent release to the reactor containment system, after a LOCA, has been presented and computer implemented. The solution method based on Laplace's transform has been shown to be computationally fast and precise, thus the model can be used to readily provide a conservative evaluation of the input term required

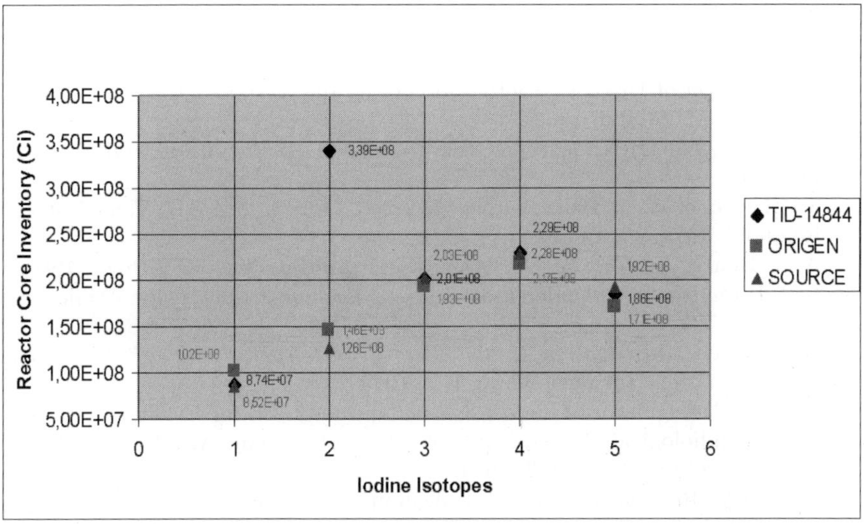

Fig. 2. Results of model for Iodine

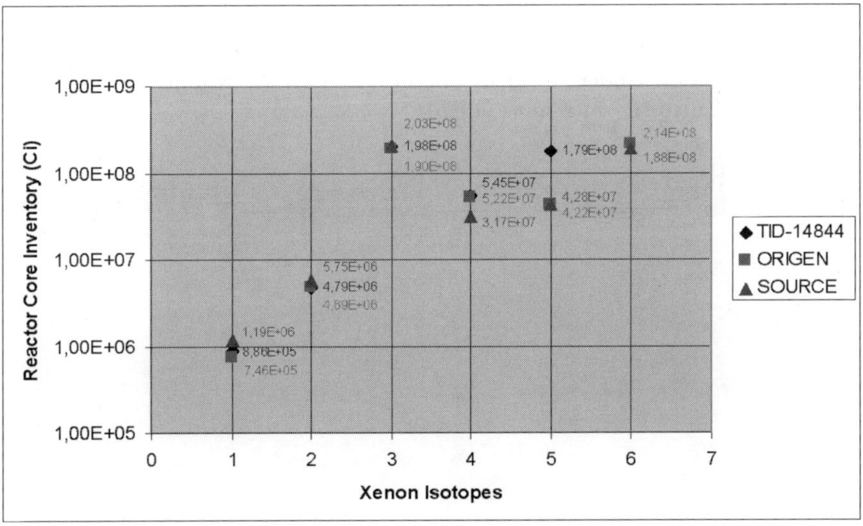

Fig. 3. Results of model for Xenon

for radiological assessment of control room habitability under accident conditions.

References

1. J. J. DiNunno, F. D. Anderson, R. E. Baker, F. C. Waterfield, TID-14844 "Calculation of Distance Factors for Power and Test Reactor Sites", Washington, D. C. (March 1962).
2. M. J. Bell, The ORNL Isotope Generation and Depletion Code," ORNL-4628, May 1973.
3. S. Glasstone, G. I. Bell, *Nuclear Reactor Theory*, 2nd ed., Van Nostrand Reinhold, 1970.
4. A. F. Henry, *Nuclear Reactor Analysis*, II printing, The MIT Press, 1980.
5. L. J. Hamilton, J. J. Duderstadt, *Nuclear Reactor Analysis*, John Wiley and Sons, 1975.
6. G. Doetsch, *Introduction to the Theory and Applications of the Laplace Transformation*, Springer Verlag Ed., 1970.
7. U.S. NRC, Regulatory Guide 1.3, Assumptions Used for Evaluating the Potential Radiological Consequences of a Loss of Coolant Accident for Boiling Water Reactors, Revision 2, 1974.
8. U.S. NRC, Regulatory Guide 1.4, Assumptions Used for Evaluating the Potential Radiological Consequences of a Loss of Coolant Accident for Pressurized Water Reactors, Revision 2, 1974.
9. F. L. Friedmann, E. B. Koffmann, *Problem Solving and Structured Programming in FORTRAN*, II edition, Addison-Wesley, 1981.
10. Microsoft, FORTRAN Power Station User and Programmers Manual, Microsoft Corporation, Seattle, Washington, 2001.
11. F. Almerico and A.J. Machiels, Performance Evaluation of Control Room HVAC Systems Under Accident Conditions (Part 2: Exhibits), UILU-ENG-86-5302, University of Illinois (1986)

Mathematical Modeling of Coordinate Measuring Machine (CMM) Probe Head and Sensor Calibration by Least Squares Optimization and Numerical Solution of the Resulting System of Non-linear Algebraic Equations

Franco Almerico

Intelligent Systems S.r.l.
Via Garibaldi 16, 20035 Lissone (MI), Italy
f.almerico@gmail.com

A methodology for mathematical modeling of Computerized Measuring Machine (CMM) probe head and touching sensor calibration is presented, aimed at the calculation of 3D sensor calibration parameters by means of multivariable Least Squares error optimization. The resulting system of non-linear algebraic equations is solved by an iterative numerical method procedure and the obtained calibration parameters are then used in the programming and automation of the CMM measuring process. In order to validate calibration results, sensors parameters are compared with those obtained by other calibration software and methods. Applications of the method can be several, mainly in Mechanical Production sector.

Keywords: Coordinate Measuring Machine, CMM, 3D measurement error, probe qualification, least squares optimization, partial differential equations, systems of non-linear algebraic equations, non-linear Jacobi iteration.

1. Introduction

The present article describes a methodology to model the Calibration of a Coordinate Measuring Machine (CMM) Probe Head and contact Sensor through a mathematical method based on Least Squares Optimization of the combined CM end Probe head measurement error. The model developed and the optimization method result in a system of non-linear algebraic equations for which a numerical solution is envisaged and the 3D calibration error parameters are derived very efficiently. The Engineering Problem to cope with is the identification of the factors influencing measuring error of a CMM probe head and connected contact sensor. In order to achieve that goal it is necessary to develop a methodology to model the physico-geometric features and the qualification process' errors of the combination of a probe head and of its contact sensor, that is the process adopted to calculate the qualification parameters that, once applied to the measurement process, allow obtaining correct coordinates measures. Another aspect of the engineering problem is that, due to the very complex calculations, the methodology and the mathematical model have to be computer implemented and model in a fast and precise computer code within a flexible and complete procedure in order to allow their use in real industrial processes.

2. Objectives

Giving the above engineering problem requisites the objective of the work presented in this article was:

(1) to develope a tool that allowed obtaining an estimate of the error deriving from the physico-geometric imperfections of a CMM probe head and contact sensor;
(2) to define suitable correction parameters of each configuration of probe head/contact sensor, to be applied in the measuring process to account for the above error and to set up a calculation procedure;
(3) to obtain mathematical description apt to be efficiently computer implemented, through a reliable and robust analytical or numerical algorithm;
(4) to easily apply and integrate the obtained code in a software package for CMM measuring process management and control, with characteristics of generality, flexibility, ease of use, speed of computation and possibly facilitating process improvement and simplification;
(5) to establish the acceptability of model and of the solution method by validation of results with comparison to other models/SW packages.

3. Mathematical Modeling

A complete mathematical modeling of a CMM, such to allow precise prediction and control of time-dependent machine trajectory, is a difficult task depending on many factors, often not easily predictable. A 3D CAD modeling of the whole CMM can cope with geometry representation needs but not with a full errors representation. A simpler approach to construct a full CMM mathematical model addressed to the evaluation of the measurement errors is that of deriving mathematical models of suitable subparts of a CMM, that allow then to evaluate the full CMM measurement error by relying upon other CMM parts' internal error controls. This introduces a need for breaking a CMM in submodels.

3.1. *CMM Modeling*

A CMM can have different configurations, from 3 to 9/10 motion axes and indexed or continuous rotating tables, operated by a governing unit with a Numerical Control (NC) or Direct Computerized Control (DCC). These machines can also be operated manually, with a joystick or keyboard, or from computer or keyboard. The typical structure of a CMM is illustrated in Fig. 1.

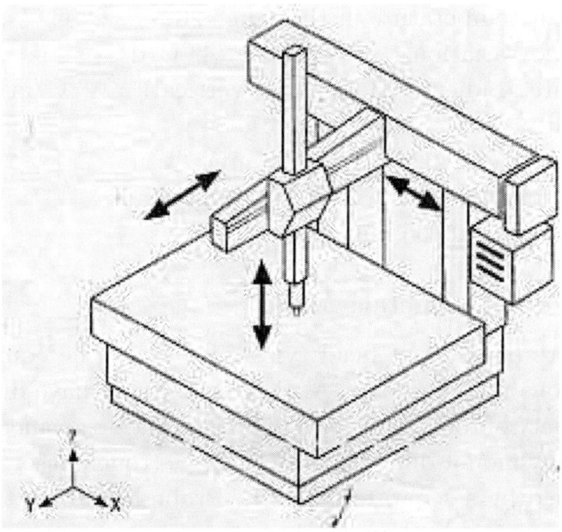

Fig. 1. Typical Structure of a CMM

The position of the probe head's mounting thread could be fully defined by its axes coordinates with respect to CMM axes origin, or to any other absolute or relative origin, but the value of axes coordinates are affected by various errors.

3.1.1. *CMM Measurement Errors*

CMM measurement errors may at least include:

- axis linearity error

- axis rectilinearity and orthogonality errors;
- axes volumetric error;
- temperature deformation error;
- encoders and measurement system errors;
- axes motion control system errors.

These errors are controlled by NC logic/firmware and parameters.

3.1.2. *CMM Operational Data*

In order to ensure the specified precision (accuracy) and repeatability CMM must operate within the following operational data ranges:

- Operating temperature range form 10 to 35° C
- Environment temperature in the range 20 +/− 2° C
- Temperature variation < 1° C/h or < 4° C/d
- Temperature gradients (Horizontal/Vertical) < 1° C/m
- Repeatability < 0.02 mm
- Precision (X,Y,Z) 0.02 + L/50.000 mm
- Precision (volume) 0.03 + L/30.000 mm
- Resolution 0.05 / 0.005 mm

3.2. *Probe Head Geometric Model*

Probe heads are of different head types. They may be equipped with indexed/continuous rotating axes (A, B) control, may have different mountings, and a variety of tools to fit with different objects geometry and shapes. Probe heads may mount different sensors types (touch, laser, optical,..) and may be governed by a separate NC unit. Probe head orientation could be fully defined by giving A, B angles, but these angles are affected by various errors that have to be taken into account in the measurements.

3.2.1. *Probe Head Errors*

Probe head errors may at least include:

- probe head mounting (rototraslation) error
- probe head axes angles error
- probe head axes rotation centers error
- temperature deformation error.

All these errors are dealt with by the probe head and CMM software or by design/manufacturing (errors may be guaranteed to be < 0.0025 mm).

3.3. *Sensor Physico-Geometric Model*

A touch trigger sensor is a stylus with a small radius ball on the tip and works as a switch that close/open an electric circuit when deflection of the stylus is greater than a threshold angle. It is integrated in a limit force sensor. Touch trigger sensors may encompass many combinations of extension bars, according to objects geometry and type of measurement, but the number of bars should be minimized in order to reduce the elastic deflection that otherwise can be non neglegible. Any combination of probe head A, B angle orientation, extension bars and sensor stylus can be named a probe and it is unique. Theorethycally it could be fully defined by the values of the following four parameters: A, B, total bar+stylus length and ruby ball radius. But this is not enough, and well see why in the next subsections.

3.3.1. *Probe (Head + Sensor) Errors*

Probe errors may at least include:

- probe head mounting error
- probe head orientation error
- extension bar + stylus length error
- rectilinearity error
- ruby balls radius, sphericity, position error
- sensor triggering angle (pretravel error)

all these errors may be globally estimated through the qualification process.

3.4. *Qualification Process*

The qualification process is performed by touching N points of a calibrated sphere with a probe. The process is repeated for each probe used, with same sphere, allowing the determination of "probe" global error, average radius and 3D displacement of the ruby ball center with respect to a CMM controlled point, that is of the qualification parameters. The qualification sphere, also called Universal Datum Sphere, is a calibrated, certified geometry special steel sphere, used as test object in "probes" qualification. Radius and sphericity are certified but the position may be unknown.

3.4.1. *Qualification Process Modeling*

The coordinates of the ruby balls center, when it is in contact with a perfectly spherical datum sphere as in Fig. 2, are linked by the following equation:

$$(x - x_{cq})^2 + (y - y_{cq})^2 + (z - z_{cq})^2 = (R_T + R_Q)^2 \tag{1}$$

where:

$R_T + R_Q = R =$ sensor's centers sphere
$R_T =$ ruby ball nominal radius
$R_Q =$ datum sphere nominal radius.

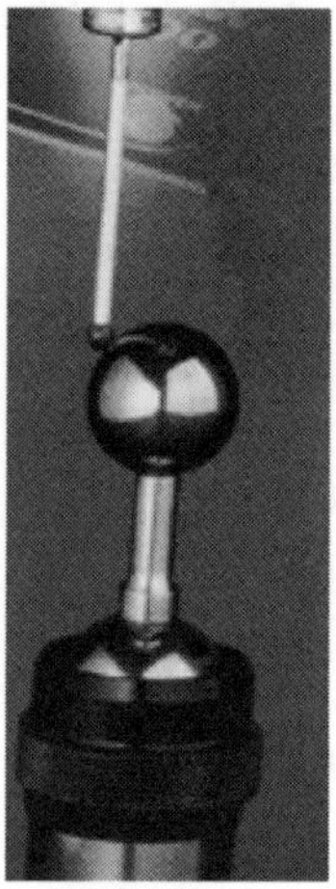

Fig. 2. Sensor's qualification process

In the perfect sphericity case, only four distinct points would be needed to determine radius R and center C of sensor's centers sphere, with R_T unknown. The equation would be

$$(x - x_{cq})^2 + (y - y_{cq})^2 + (z - z_{cq})^2 = R^2 \tag{2}$$

according to Fig. 3.

Actually, because of the sphericity error and the sensor triggering error, we can only approximate the sensor's centers sphere, thus needing more than four points. As the datum sphere sphericity error is much less than the ruby balls sphericity error and is also less than the CMM resolution and error, we can assume the datum sphere be perfectly spherical, with center in point $C(x'_{cq}, y'_{cq}, z'_{cq})$ and radius R_Q. The radius of the ruby ball's centers

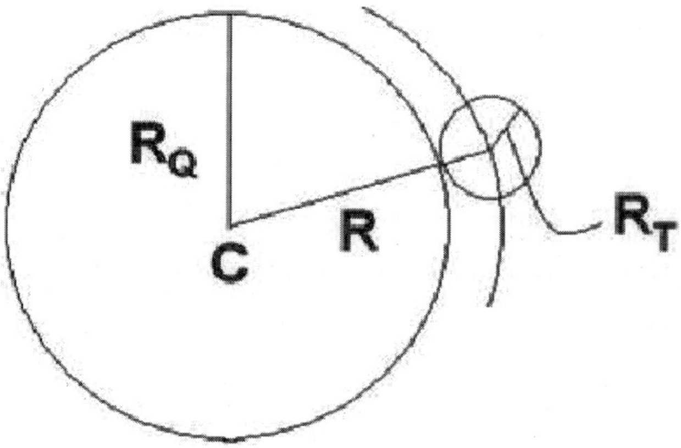

Fig. 3. Ideal geometry for sensor's qualification process

sphere R' and the datum sphere radius R_Q are linked by the relationship $R' = R'_T + R_Q$, where R'_T is the ruby ball average radius.

If we measure the datum sphere in N distinct points $P_i(x'_i, y'_i, z'_i)$, we have a situation like that shown in Fig. 4, where:

$$(x'_i - x'_{cq})^2 + (y'_i - y'_{cq})^2 + (z'_i - z'_{cq})^2 = R'^2_i \qquad (3)$$

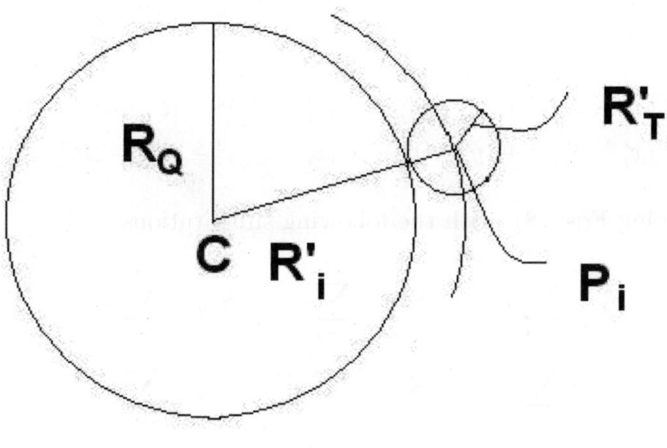

Fig. 4. Real geometry for sensor's qualification process

We can calculate the quadratic error W as given by:

$$W = \sum_{i=1}^{N} (R_i'^2 - R'^2)^2. \tag{4}$$

3.4.2. *Qualification Parameters Optimization*

The quadratic error W can also be written as follows:

$$\begin{aligned} W = \sum_{i=1}^{N} [(x_i' - x_{cq}')^2 + (y_i' - y_{cq}')^2 + (z_i' - z_{cq}')^2 \\ - (x_i' - x_{cq}')^2 + (y_i' - y_{cq}')^2 + (z_i' - z_{cq}')^2]^2 \end{aligned} \tag{5}$$

or as:

$$W = \sum_{i=1}^{N} [(x_i' - x_{cq}')^2 + (y_i' - y_{cq}')^2 + (z_i' - z_{cq}')^2 - R'^2]^2 \tag{6}$$

which can be minimized by taking:

$$\frac{\partial W}{\partial x_{cq}'} = 0 \qquad \frac{\partial W}{\partial y_{cq}'} = 0 \qquad \frac{\partial W}{\partial z_{cq}'} = 0 \qquad \frac{\partial W}{\partial R'} = 0 \tag{7}$$

Which in turn gives the following expressions:

$$\begin{cases} \sum_{i=1}^{N} 2[(x_i' - x_{cq}')^2 + (y_i' - y_{cq}')^2 + (z_i' - z_{cq}')^2 - R'^2][-2x_i' + 2x_{cq}'] = 0 \\ \sum_{i=1}^{N} 2[(x_i' - x_{cq}')^2 + (y_i' - y_{cq}')^2 + (z_i' - z_{cq}')^2 - R'^2][-2y_i' + 2y_{cq}'] = 0 \\ \sum_{i=1}^{N} 2[(x_i' - x_{cq}')^2 + (y_i' - y_{cq}')^2 + (z_i' - z_{cq}')^2 - R'^2][-2z_i' + 2z_{cq}'] = 0 \\ \sum_{i=1}^{N} 2[(x_i' - x_{cq}')^2 + (y_i' - y_{cq}')^2 + (z_i' - z_{cq}')^2 - R'^2][-2R'] = 0 \end{cases} \tag{8}$$

Developing Eqs. (8), with the following substitutions:

$$\begin{aligned} S_x &= \sum_{i=1}^{N} x_i' & S_{x^2} &= \sum_{i=1}^{N} x_i'^2 & S_{xy} &= \sum_{i=1}^{N} x_i' y_i' \\ S_y &= \sum_{i=1}^{N} y_i' & S_{y^2} &= \sum_{i=1}^{N} y_i'^2 & S_{xz} &= \sum_{i=1}^{N} x_i' z_i' \\ S_z &= \sum_{i=1}^{N} z_i' & S_{z^2} &= \sum_{i=1}^{N} z_i'^2 & S_{yz} &= \sum_{i=1}^{N} y_i' z_i' \end{aligned} \tag{9}$$

and

$$
\begin{aligned}
S_{x^3} &= \sum_{i=1}^{N} x_i'^3 & S_{y^2} &= \sum_{i=1}^{N} y_i'^3 & S_{z^3} &= \sum_{i=1}^{N} z_i'^3 \\
S_{x^2y} &= \sum_{i=1}^{N} x'^2 y_i' & S_{xy^2} &= \sum_{i=1}^{N} x_i' y_i'^2 & S_{yz^2} &= \sum_{i=1}^{N} y_i' z_i'^2 \\
S_{x^2z} &= \sum_{i=1}^{N} x_i'^2 z_i' & S_{xz^2} &= \sum_{i=1}^{N} x_i' z_i'^2 & S_{y^2z} &= \sum_{i=1}^{N} y_i'^2 z_i'
\end{aligned}
\tag{10}
$$

we obtain a system of four third degree non-linear algebraic equations in the unknowns $x'_{cq}, y'_{cq}, z'_{cq}, R'$ whose solution provides the spatial coordinates of the center of the ruby balls' centers sphere and its radius.

4. Numerical Solution of Qualification Equations

The above system of equations, Eqs. (10), is of the following type:

$$
\begin{cases}
f_1(x'_{cq}, y'_{cq}, z'_{cq}, R') = 0 \\
f_2(x'_{cq}, y'_{cq}, z'_{cq}, R') = 0 \\
f_3(x'_{cq}, y'_{cq}, z'_{cq}, R') = 0 \\
f_4(x'_{cq}, y'_{cq}, z'_{cq}, R') = 0
\end{cases}
\tag{11}
$$

or:

$$F(\mathbf{x}) = 0 \quad f(\mathbf{x}) \in \mathbb{R}_{\ltimes} \tag{12}$$

with

$$\mathbf{x} = (x'_{cq}, y'_{cq}, z'_{cq}, R') \quad \mathbf{x} \in \mathbb{R}_{\ltimes} \tag{13}$$

and $n = 4$.

The system of equations (11) can be solved with the non-linear Jacobi iteration method by searching the solutions in $\mathbb{R}_{\ltimes}$ of the equation:

$$\mathbf{x} = \mathbf{G}(\mathbf{x}) \tag{14}$$

which turns out to be a solution of

$$\mathbf{F}(\mathbf{x}) = \mathbf{0} \tag{15}$$

By posing

$$\mathbf{G}(\mathbf{x}) = \mathbf{x} - \mathbf{MF}(\mathbf{x}) \tag{16}$$

with

$$\mathbf{M} = \mathbf{diag}(\mathbf{m_{11}}, \mathbf{m_{22}}, \ldots, \mathbf{m_{nn}}) \tag{17}$$

we get the following iteration formula:

$$x_j^{(i+1)} = x_j^{(i)} - m_{jj} f_j(x^{(i)}) \tag{18}$$

for $j = 1, 2, \ldots, n$, where we posed

$$m_{jj} = \frac{1}{\left(\frac{\partial f_j}{\partial x_j}\right)_{x=x^i}} \tag{19}$$

with

$$x_j^{i+1} = x_j^i - \frac{1}{\left(\frac{\partial f_j}{\partial x_j}\right)_{x=x^i}} f_j(x^i) \tag{20}$$

for $j = 1, 2, ..., n$.

The method converges if the first approximation of the solution $\mathbf{x}^{(0)}$ is not too far from the solution $\mathbf{x}$ and if it is satisfied the criterion:

$$\|G'(x)\|_\infty < 1 \tag{21}$$

where

$$\|G'(x)\|_\infty = max_{i=1,\ldots,n} \sum_{j=1}^{n} |g_{ij}| \tag{22}$$

and

$$\|G'(x)\|_\infty \tag{23}$$

is the uniform norm of the Jacobian Matrix.

Developing the partial derivatives of $G(x)$ the condition can also be written as:

$$\|G'(x)\|_\infty = max_{i=1,\ldots,n} \sum_{j=1}^{n} \left\{ \frac{-\frac{\partial f_i(x)}{\partial x_j}\frac{\partial f_i(x)}{\partial x_i} + f_i(x)\frac{\partial^2 f_i(x)}{\partial x_i \partial x_j}}{\left(\frac{\partial f_j(x)}{\partial x_i}\right)^2} + \frac{\left|f_i(x)\frac{\partial^2 f_i(x)}{\partial x_i^2}\right|}{\left(\frac{\partial f_i(x)}{\partial x_i}\right)^2} \right\} \leq 1. \tag{24}$$

The iteration method is then interrupted when the following condition is verified:

$$\|x^{(r)} - x^{(r-1)}\| \leq \epsilon \tag{25}$$

here $\|x\| = \sqrt{\sum_{i=1}^{n} x_i^2}$ is the Euclidean norm of x.

5. Computer Implementation and Results

The mathematical method has been computer implemented and a thorough test of the numerical algorithm has been performed resulting very fast and apt to be included into a CMM software package to be used in production industry.

The iteration method converges at the solution in very few (in most cases from 6 to 10) iterations. The values of R_T have been compared with those produced with another certified software and are in agreement within an error tolerance < 0.05 mm. Coordinates measurements have also been compared obtaining spatial errors < 0.07 mm.

The mathematical modeling of probes and qualification process proved to be very flexible and general and has been introduced in software modules of a CMM application package for 3D coordinate measuring in mechanical manufacturing industry.

6. Conclusions

A modeling of the qualification process of the probes of a CMM has been produced that allows estimate the error associated with a probe for CMM measurement. A mathematical model has been constructed that allows to describe all the different configurations of probe head and contact sensors, to represent the qualifcation process, and to calculate 3D corrections to probes geometry.

Model has been implemented and showed to be accurate, computationally fast, and allows to describe all different configurations of probe head and contact sensors and for these reasons has been integrated into a CMM software package used by a manufacturing industry in everyday work.

Acknowledgements

I like to thank a lot my son, Davide, for the help he gave me in the preparation of the presentation of this work to the 9th SIMAI 2008 Congress and the testing of the CMM qualification software.

References

1. P.G. Ranky, Programming Industrial Robots in FMS, Robotica, 2, pp. 87-92.
2. Fan You Chen, Gripping mechanisms for industrial robots.- An overview, Mechanisms and Machine Theory, 1982, 17(5), pp. 299-311.
3. Yoram Koren, Computer Control of Manufacturing Systems, The Mc Graw Hill International Editions, 1986.
4. Igor Aleksander, Computer Control of Manufacturing Systems, The Mc Graw Hill International Editions, 1986.
5. G. Dahlquist, A. Björck, Numerical Methods, PrenticeHall, NJ, 1974
6. F. Fontanella, A. Pasquali, Calcolo Numerico - Metodi e algoritmi, Pitagora Editrice, Bologna, 1984
7. J.M. Ortega, W.C. Rheinboldt, Iterative Solution of Nonlinear Equations in Several Variables, SIAM, Classics in Applied Mathematics, Philadelphia, 2000
8. B. Siciliano, L. Sciavicco, L. Villani, G. Oriolo, Robotica Modellistica, pianificazione e controllo, The Mc Graw Hill Companies, 2008
9. L. Sciavicco, B. Siciliano, Robotica industriale Modellistica e controllo di manipolatori, The Mc Graw Hill Companies, 2000

Weighted Traffic Equilibrium Problem with Delay in Non-pivot Hilbert Spaces

Annamaria Barbagallo*, Stéphane Pia†
Dipartimento di Matematica e Informatica, Facoltà di Scienze MM.FF.NN.
Università degli Studi di Catania, Italy
* *barbagallo@dmi.unict.it*
† *pia@dmi.unict.it*

In the paper, the retarded weighted traffic equilibrium problem in non-pivot Hilbert spaces is introduced and the equivalence between the equilibrium condition and a weighted variational inequality is proved. Moreover, the variational formulation allows to apply the variational inequalities theory to obtain some existence and regularity results.

Keywords: Non-pivot Hilbert spaces, retarded weighted traffic equilibrium problem, weighted variational inequalities, existence and regularity results.

1. Introduction

The aim of the paper is to present the retarded weighted traffic equilibrium problem. The importance of delay effects is well-known in the theory of population dynamics or in physics phenomena. In mathematical modeling, the importance of delay effects in population dynamics has led several authors to propose a rich variety of delay differential equations [11]. For a brief survey on this kind of differential equations the interested reader can refer to [9]. In economic markets models the importance of delay derives from the fact that, even if the propagation of information through the network could be considered in certain cases instantaneous, producers and consumers usually take their decisions after they get all the data available. Mathematical modeling of economic markets by network theory has a long history [5,12]. Recently, delay effects have been introduced in traffic network modeling and developed within the variational framework (see [7,13] and [3]). The information through the network travels at finite speed, therefore the users take a certain time before adjusting their path choices and reaching an equilibrium state. As a consequence, it is reasonable to think that demand requirements at time t are satisfied only at time $t + h$, after a delay $h > 0$. In reality, the delay should depend on the variable t, but, for the sake of simplicity, it is assumed that it has a constant positive value.

In [8], S. Giuffrè and S. Pia introduced the weighted traffic equilibrium problem and showed some existence and regularity result for the equilibrium solution. The model introduced in [8] is an extension of the dynamic traffic model presented in [6]. The extension develops in two directions. The first one is related to the operator involved in the description of the equilibrium and the second one is related to the spaces used. To motivate the first enhancement, we can highlight that a very important difficulty in the dynamic traffic equilibrium problem is related to the real time cost determination of the flow over the links in the transportation network. More exactly, we want to investigate about the distribution of the traffic flow over routes connecting the same origin/destination pair in order to have the optimal distribution of the flow in the transportation network. For this reason, it is fundamental to know what is the traffic density over each route. The collection of this information could be very costly and moreover it is very difficult and almost impossible to aggregate data on a real time basis. The idea developed by the SENSEable City Laboratory at MIT is that, in the case of an urban traffic network,this information can be roughly collected using mobile devices connection data. As clarified in [14,15], it is possible to compute these data in order to evaluate the traffic repartition over a monitored area. This usage of mobile devices has inspired the introduction of weighted to define a bilinear form able to take into account the different densities over flows. The authors introduce also a second class of weights with the intention to extends the existence domain and in a certain sense to remove artificial congestions, that means congestions solved in a bigger functional space.

The paper is organized as follows. In Section 2 we recall some important notions about the dual representation of an Hilbert space, and we introduce non-pivot Hilbert spaces. Moreover, we present the weighted variational inequalities. In Section 3, we introduce the retarded weighted traffic equilibrium problem and we show the equivalence between the retarded weighted variational inequality which expresses the equilibrium and the retarded weighted Wardrop's condition. In Section 4 and Section 5, we apply general existence and regularity results to show analogous results for the retarded weighted traffic equilibrium solution with delay effects.

2. Basic Concepts

Every time we work with a Hilbert space V, it is necessary to decide if we identify the topological dual space $V^* = \mathcal{L}(V, \mathbb{R})$ with V or not. It is in the common habit to make the identification, one of the reasons is that, in this case, the vectors of the polar of a set in V are in V. In some cases the identification does not make sense. An example is given by $V = L^2(\mathbb{R}, (1+|x|)) \subset L^2(\mathbb{R})$ (dense subspace of $L^2(\mathbb{R})$) endowed with the inner product:

$$(u, v)_V = \int_{\mathbb{R}} (1 + |x|)u(x)v(x)dx.$$

An element $\varphi \in L^2(\mathbb{R})^*$ is also an element of V^*, but if we identify φ to an element $f \in L^2(\mathbb{R})$, this function does not define a linear form on V and the expression $\varphi(v) = \langle f, v \rangle_V$ does not have meaning on V (for more details see [1]).
Let us consider a Hilbert space V and its topological dual V^*. In the following, the space V is not identified to V^* but it is known that there exists an isometry $J : V \to V^*$ such that for all $x \in V$, $J(x) = \text{grad}\left(\frac{\|x\|^2}{2}\right)$, J is linear and it is called the duality mapping.

Now we introduce some basic objects which will be used in our model. Let Ω be an open subset of $\mathbb{R}$, let $a : \Omega \to \mathbb{R}^+ \setminus \{0\}$ be a continuous and strictly positive function called "weight" and let $s : \Omega \to \mathbb{R}^+ \setminus \{0\}$ be a continuous and strictly positive function called "real time traffic density" (RTTD). The bilinear form, defined into $C_0(\Omega) \times C_0(\Omega)$ (continuous function with compact support on Ω) by

$$\langle f, g \rangle_{a,s} = \int_{\Omega} f(\omega)g(\omega)a(\omega)s(\omega)d\omega, \quad \forall f, g \in C_0(\Omega), \tag{1}$$

represents clearly an inner product. We remark that if a is a weight, $a^{-1} = 1/a$ is also a weight.

Definition 2.1. We call $L^2(\Omega, a, s)$ a completion of $C_0(\Omega)$ with respect to the inner product (1).

At this point, we are able to introduce the infinite-dimensional non-pivot Hilbert spaces. Let us consider two vector $a, s \in \mathbb{R}^n$, and let a_i, s_i be the components, for $i = 1, \ldots, n$. Denoting by $V_i = L^2(\Omega, \mathbb{R}, a_i, s_i)$ and $V_i^* = L^2(\Omega, \mathbb{R}, a_i^{-1}, s_i)$, the space $V = \prod_{i=1}^n V_i$ is a Hilbert space with respect to the inner product

$$(F,G)_V = (F,G)_{\mathbf{a},\mathbf{s}} = \sum_{i=1}^{m} \int_\Omega F_i(\omega)G_i(\omega)a_i(\omega)s_i(\omega)d\omega, \quad \forall F,G \in V,$$

and the space $V^* = \prod_{i=1}^n V_i^*$ is a Hilbert space with respect to the inner product

$$(F,G)_{V^*} = (F,G)_{\mathbf{a}^{-1},\mathbf{s}} = \sum_{i=1}^{m} \int_\Omega \frac{F_i(\omega)G_i(\omega)s_i(\omega)}{a_i(\omega)} d\omega, \quad \forall F,G \in V^*.$$

Moreover, the following bilinear form, defined into $V^* \times V$ by

$$\langle f,x\rangle_{V^*\times V} = \langle f,x\rangle_{\mathbf{s}} = \sum_{i=1}^{m} \int_\Omega f_i(\omega)x_i(\omega)s_i(\omega)d\omega, \quad \forall f \in V^*,\ \forall x \in V,$$

represents a duality between V and V^* and the duality mapping is given by $J(F) = (a_1F_1,\ldots,a_nF_n)$ (for details see [8]).

Let us now recall some important definitions on the operator defined on a subset of the non-pivot Hilbert space V. An operator $C : S \to V^*$ is said to be

- *pseudomonotone in the sense of Karamardian* (shortly *K-pseudomonotone*) on S if for all $x_1, x_2 \in S$

$$\langle C(x_2), x_1 - x_2\rangle_{\mathbf{s}} \geq 0 \Longrightarrow \langle C(x_1), x_1 - x_2\rangle_{\mathbf{s}} \geq 0, \quad \forall x_1,x_2 \in S;$$

- *strictly pseudomonotone* on S if

$$\langle C(x_2), x_1 - x_2\rangle_{\mathbf{s}} \geq 0 \Longrightarrow \langle C(x_1), x_1 - x_2\rangle_{\mathbf{s}} > 0,\ \forall x_1,x_2 \in S,\ x_1 \neq x_2.$$

Let $\mathbb{K}$ be a convex subset of V. An operator $C : \mathbb{K} \to V^*$ is said to be

- *hemicontinuous in the sense of Fan* (shortly *F-hemicontinuous*) if for any $x \in \mathbb{K}$, the function

$$\mathbb{K} \ni \xi \to \langle C(\xi), \xi - x\rangle_{\mathbf{s}}$$

is lower semi-continuous on $\mathbb{K}$;

- *lower hemicontinuous along line segments* if for any $x,y \in \mathbb{K}$, the function

$$\mathbb{K} \ni \xi \to \langle C(\xi), x - y\rangle_{\mathbf{s}}$$

is lower semi-continuous on the line segment $[x,y]$.

Let us introduce weighted variational inequalities defined into a non-pivot Hilbert space V.

Definition 2.2. Let $\mathbb{K}$ be a nonempty, convex and closed subset of V and let $C : \mathbb{K} \to V^*$ be a vector-function. The weighted variational inequality is the problem to find a vector $x \in \mathbb{K}$, such that

$$\langle C(x), y - x \rangle_{\mathbf{s}} \geq 0, \quad \forall y \in \mathbb{K}. \tag{2}$$

Let us set

$$K(t) = \{f(t) \in \mathbb{R}^m : \ f \in K\},$$

since $K(t)$ is a closed and convex set, we can introduce the finite-dimensional variational inequality associated with (2)

$$\langle C(t, x(t)), y(t) - x(t) \rangle_{\mathbf{s}(t)} \geq 0, \quad \forall y(t) \in K(t), \text{ a.e. in } \Omega. \tag{3}$$

We remark that x is solution to (2) $\Leftrightarrow$ $x(t)$ is solution to (3) for almost every $t \in \Omega$ (see [4]).

3. Retarded Weighted Traffic Equilibrium Problem

We suppose now for an easier reading that $\Omega =]0, T[$ and for $h > 0$ we define $\Omega_h =]0, T + h[$. We consider a variant case of the model described in [6] and [13].

Let us introduce a network $\mathcal{N}$, which is represented by a graph $G = [N, L]$, where N is the set of nodes (i.e. cross-roads, airports, railway stations) and L is the set of directed links between the nodes. Let a denote a link of the network connecting a pair of nodes and let r be a path consisting of a finite sequence of links which connect an Origin-Destination (O/D) pair of nodes. In the network there are n links and m paths. Let W denote the set of O/D pairs with typical O/D pair w_j, $|W| = l$ and $m > l$. The set of paths connecting the O/D pair w_j is represented by $\mathcal{R}_j$ and the entire set of paths in the network by $\mathcal{R}$. Let Ω be an open subset of $\mathbb{R}$, let $\mathbf{a} = \{a_1, \ldots, a_m\}$ and $\mathbf{a}^{-1} = \{a_1^{-1}, \ldots, a_m^{-1}\}$ be two families of weights such that for each $1 \leq r \leq m$, $a_r \in \mathcal{C}(\Omega, \mathbb{R}^+ \setminus \{0\})$. We introduce also the family called real time density $\mathbf{s} = \{s_1, \ldots, s_m\}$ such that for each $1 \leq r \leq m$, $a_i \in \mathcal{C}(\Omega, \mathbb{R}^+ \setminus \{0\})$. We associate to each path r, $r = 1, 2, \ldots, m$ the components a_r and s_r of the weights $\mathbf{a}$ and $\mathbf{s}$, respectively. By means of these components, we define the spaces $V = L^2(\Omega, \mathbb{R}, \mathbf{a}, \mathbf{s})$ and $V^* = L^2(\Omega, \mathbb{R}, \mathbf{a}^{-1}, \mathbf{s})$, as introduced in Section 2.

Let $F \in L^2(\Omega, \mathbb{R}, \mathbf{a}, \mathbf{s})$ denote the path flow vector-function. Let $\lambda, \mu \in L^2(\Omega, \mathbb{R}^m, \mathbf{a}, \mathbf{s})$ be the capacity constraints functions, such that $\lambda \leq \mu$. Let

$\rho_j \in L = \bigcap_{r=1}^{m} L^2(\Omega, \mathbb{R}, a_r, s_r)$ represent the travel demand associated with the users travelling between O/D pair w_j and let $\rho = (\rho_1, \dots, \rho_j, \dots, \rho_l)^T$ be the total demand vector-function. Furthermore, we consider the cost trajectory $C : \Omega \times L^2(\Omega, \mathbb{R}^m, \mathbf{a}, \mathbf{s}) \to L^2(\Omega, \mathbb{R}^m, \mathbf{a}, \mathbf{s})$.

Now we want to introduce delay effects in the weighted traffic equilibrium model. We suppose that the traffic demand at time t is satisfied after a delay $h > 0$. Therefore, the set of all the retarded feasible flows is given by

$$\mathbb{K}_h = \{F(t+h) \in L^2(\Omega, \mathbb{R}, \mathbf{a}, \mathbf{s}) : \ \lambda(t+h) \leq F(t+h) \leq \mu(t+h), \quad \text{a.e. in } \Omega, \\ \Phi F(t+h) = \rho(t), \quad \text{a.e. in } \Omega\}, \tag{4}$$

where Φ is the O/D pairs-paths incidence matrix, whose typical entry ϕ_{jr} is 1 if path r connects the pair w_j and 0 otherwise.

Definition 3.1. A flow $H \in L^2(\Omega, \mathbb{R}, \mathbf{a}, \mathbf{s})$ is said to be a retarded equilibrium flow if

$$H \in \mathbb{K}_h : \int_\Omega \langle C(t, H(t+h)), F(t+h) - H(t+h) \rangle_{\mathbf{s}(t)} dt \geq 0, \ \forall F \in \mathbb{K}_h. \tag{5}$$

We remark that weighted variational inequality (5) is equivalent to the pointed retarded weighted variational inequality

$$H \in \mathbb{K}_h : \langle C(t, H(t+h)), F(t+h) - H(t+h) \rangle_{s(t)} \geq 0, \forall F(t) \in \mathbb{K}_h(t), \ \text{a.e. in } \Omega,$$

where

$$\mathbb{K}_h(t) = \{F(t+h) \in \mathbb{R}^m : \lambda(t+h) \leq F(t+h) \leq \mu(t+h), \ \Phi F(t+h) = \rho(t)\}.$$

It is possible to prove the equivalence between condition (5) and what we will call a **weighted retarded Wardrop condition** (6). More precisely we have:

Theorem 3.1. *A flow $H \in \mathbb{K}_h$ is a retarded equilibrium flow in the sense of (5) if and only if $\forall w \in \mathcal{W}$, $\forall q, m \in \mathcal{R}(w)$, a.e. in Ω:*

$$s_q(t) C_q(t, H(t+h)) < s_m(t) C_m(t, H(t+h)) \\ \Longrightarrow H_q(t+h) = \mu_q(t+h) \text{ or } H_m(t+h) = \lambda_m(t+h). \tag{6}$$

Proof. Let us assume that (6) holds. For every $w \in \mathcal{W}$, we consider the following sets

$$A = \{q \in \mathcal{R}(w) : \ H_q(t+h) < \mu_q(t+h), \ \text{a.e. in } \Omega\}$$
$$B = \{m \in \mathcal{R}(w) : \ H_m(t+h) > \lambda_m(t+h), \ \text{a.e. in } \Omega\}$$

From (6) it follows

$$s_q(t)C_q(t, H(t+h)) \geq s_m(t)C_m(t, H(t+h)), \ \forall q \in A, \ \forall m \in B, \ \text{a.e. in } \Omega.$$

Then, there exists a function $\gamma_w(t) : \Omega \to \mathbb{R}$ such that a.e. in Ω

$$\inf_{q \in A} s_q(t)C_q(t, H(t+h)) \geq \gamma_w(t) \geq \sup_{m \in B} s_m(t)C_m(t, H(t+h)).$$

Let $F \in K_h$ be arbitrary. For every $r \in \mathcal{R}(w)$ such that $s_r(t)C_r(t, H(t+h)) < \gamma_w(t)$, a.e. in Ω_h, it results $r \notin A$, that is $H_r(t+h) = \mu_r(t+h)$, a.e. in Ω_h. This implies $F_r(t+h) - H_r(t+h) \leq 0$, a.e. in Ω_h and then

$$(s_r(t)C_r(t, H(t+h)) - \gamma_w(t))(F_r(t+h) - H_r(t+h)) \geq 0, \ \text{a.e. in } \Omega.$$

Likewise for every $r \in \mathcal{R}(w)$ such that $s_r(t)C_r(t, H(t+h)) > \gamma_w(t)$ a.e. in Ω, it results $r \notin B$ and

$$(s_r(t)C_r(t, H(t+h)) - \gamma_w(t))(F_r(t+h) - H_r(t+h)) \geq 0, \ \text{a.e. in } \Omega.$$

It results

$$\begin{aligned}\sum_{r=1}^{n} s_r(t)C_r(t, H(t+h))(F_r(t+h) - H_r(t+h)) \\ \geq \gamma_w(t) \sum_{r=1}^{n} (F_r(t+h) - H_r(t+h)) \\ = \gamma_w(t)(\rho_w(t) - \rho_w(t)) = 0\end{aligned}$$

and finally summing up $\forall w \in \mathcal{W}$ we get the result integrating on Ω.

Now, suppose that (6) does not hold. Then there exists $w \in \mathcal{W}$, $q, m \in \mathcal{R}(w)$ and a set $E \subseteq \Omega$ with positive measure such that a.e. in E it results

$$\begin{aligned}s_q(t)C_q(t, H(t+h)) < s_m(t)C_m(t, H(t+h)) \\ \Longrightarrow H_q(t+h) < \mu_q(t+h), \ \text{or } H_m(t+h) > \lambda_m(t+h).\end{aligned}$$

For $t \in E$, we set

$$\delta(t+h) = \min\{\mu_q(t+h) - H_q(t+h), H_m(t+h) - \lambda_m(t+h)\}.$$

It results $\delta(t+h) > 0$ a.e. in E. We construct $F : \Omega \to \mathbb{R}$ in the following way:

$$F_q(t+h) = \begin{cases} H_q(t+h) + \delta(t+h), & \text{a.e. in } E, \\ H_q(t+h), & \text{a.e. in } \Omega \setminus E, \end{cases}$$

$$F_m(t+h) = \begin{cases} H_m(t+h) - \delta(t+h), & \text{a.e. in } E, \\ H_m(t+h), & \text{a.e. in } \Omega \setminus E, \end{cases}$$

$$F_r(t+h) = H_r(t+h), \quad \text{for } r \neq q, m, \ \text{a.e. in } \Omega.$$

Since $F \in \mathbb{K}_h$, then one obtains

$$\int_\Omega \sum_{r=1}^{n} C_r(t, H(t+h))(F_r(\omega+h) - H_r(t+h))s_r(\omega)dt$$
$$= \int_E \delta(t)[s_q(t)C_q(t, H(t+h)) - s_m(t)C_m(t, H(t+h))]dt < 0.$$

Thus H is not an equilibrium. □

4. Existence of Equilibria

In this Section, we obtain an existence result for the retarded weighted model, we can state the following theorem:

Theorem 4.1. *Each one of the following conditions is a sufficient condition for the existence of solutions for problem (5):*

i) $\forall H, F \in \mathbb{K}_h$ *we have*

$$\int_\Omega \langle C(t, H(t+h)), F(t+h) - H(t+h)\rangle_{s(t)} dt \geq 0$$
$$\Longrightarrow \int_\Omega \langle C(t, F(t+h)), F(t+h) - H(t+h)\rangle_{s(t)} dt \geq 0,$$

and $\forall F, G \in \mathbb{K}_h$ *the function:*

$$H \to \int_\Omega \langle C(t, H(t+h)), F(t+h) - G(t+h)\rangle_{s(t)} dt$$

is weakly upper semicontinuous on the segment $[F, G]$*;*

ii) $\forall F \in \mathbb{K}_h$ *the function:*

$$H \to \int_\Omega \langle C(t, H(t+h)), F(t+h) - H(t+h)\rangle_{s(t)} dt$$

is weakly upper semicontinuous.

Proof. We remark that $\mathbb{K}_h$ is closed , convex and bounded, hence weakly compact. Setting $t + h = y$, from (5) we get the following problem:

Find $H \in \mathbb{K}^h$ such that

$$\int_{\Omega_h} \langle C(y-h, H(y)), F(y) - H(y)\rangle_{s(y-h)} dy, \ \forall F \in \mathbb{K}^h \tag{7}$$

where

$$\mathbb{K}^h := \{F \in V^h : \ \lambda(y) \leq F(y) \leq \mu(y), \quad \text{a.e. in } \Omega_h,$$
$$\Phi F(y) = \rho(y-h), \quad \text{a.e in } \Omega_h\}$$

where $V^h = \prod_{i=1}^n L^2(\Omega_h, \mathbb{R}, a_i, s_i)$. We denote by C_h and s_h the mappings such that:

$$C_h(y, H(y)) = C(y-h, H(y)), \quad \forall y \in \Omega,$$
$$\mathbf{s}_h(y) = \mathbf{s}(y-h), \quad \forall y \in \Omega.$$

Then, (7) can be written

$$H \in \mathbb{K}^h : \int_\Omega \langle C_h(y, H(y)), F(y) - H(y)\rangle_{\mathbf{s}_h(y)} dy \geq 0, \quad \forall F \in \mathbb{K}^h. \qquad (8)$$

Therefore, we can apply Corollary 5.1 of [6] and give sufficient condition for the existence of a solution to (8). But if C satisfies condition (i) on $\mathbb{K}_h$, namely $\forall H, F \in \mathbb{K}_h$ one has

$$\int_\Omega \langle C(t, H(t+h)), F(t+h) - H(t+h)\rangle_{\mathbf{s}(t)} dt \geq 0$$
$$\Longrightarrow \int_\Omega \langle C(t, F(t+h)), H(t+h) - F(t+h)\rangle_{\mathbf{s}(t)} dt \leq 0$$

is pseudomonotone which implies the pseudomonotony of C_h on $\mathbb{K}^h$. If C satisfies the second part of condition (i) on $\mathbb{K}_h$, namely $\forall F \in \mathbb{K}_h$ the function

$$H \to \int_\Omega \langle C(t, H(t+h)), F(t+h) - G(t+h)\rangle_{\mathbf{s}(t)} dt$$

is upper semicontinuous on the segment $[F, G]$, which implies the semicontinuity of C_h on $[F, G]$. And if C satisfies condition (ii) on $\mathbb{K}_h$, namely $\forall F \in \mathbb{K}_h$ the function

$$H \to \int_\Omega \langle C(t, H(t+h)), F(t+h) - H(t+h)\rangle_{\mathbf{s}(t)} dt,$$

is upper semicontinuous, which implies the hemicontinuity of C_h on $\mathbb{K}^h$. Therefore we get the claim. □

5. Regularity of Equilibria

Now, let us present a regularity result for the retarded weighted traffic equilibrium problem. In order to show the regularity result for the retarded weighted traffic equilibrium problem, we recall the well-known property of set convergence due to C. Kuratowski [10], that is a generalization of the classical Hausdorff definition of a metric for the space of closed subsets of a (compact) metric space.

More precisely, let (X, d) be a metric space and let $\{K_n\}_{n\in\mathbb{N}}$ be a sequence of subsets of X. Recall that

$$d - \underline{\lim}_n K_n = \{x \in X : \exists \{x_n\}_{n\in\mathbb{N}} \text{ eventually in } K_n \text{ such that } x_n \xrightarrow{d} x\},$$
$$d - \overline{\lim}_n K_n = \{x \in X : \exists \{x_n\}_{n\in\mathbb{N}} \text{ frequently in } K_n \text{ such that } x_n \xrightarrow{d} x\},$$

where *eventually* means that there exists $\delta \in \mathbb{N}$ such that $x_n \in K_n$ for any $n \geq \delta$, and *frequently* means that there exists an infinite subset $N \subseteq \mathbb{N}$ such that $x_n \subset K_n$ for any $n \in N$. Finally we can remind the Kuratowski convergence of sets.

Definition 5.1. We say that $\{K_n\}_{n\in\mathbb{N}}$ converges to some subset $K \subseteq X$ in Kuratwoski's sense, and we briefly write $K_n \to K$, if $d - \underline{\lim}_n K_n = d - \overline{\lim}_n K_n = K$. Thus, in order to verify that $K_n \to K$, it suffices to check that

(K1) $d - \overline{\lim}_n K_n \subseteq K$, i.e. for any subsequence $\{x_n\}_{n\in\mathbb{N}}$ converging to x in X, such that x_n lies in K_n for all $n \in \mathbb{N}$, then the limit x belongs to K;

(K2) $K \subset d - \underline{\lim}_n K_n$, i.e. for any $x \in K$, there exists a sequence $\{x_n\}_{n\in\mathbb{N}}$ converging to x in X such that x_n lies in K_n for all $n \in \mathbb{N}$.

Remark 5.1. Under assumption of continuity for the functions λ, μ and ρ, the set of feasible flows as in (4) fulfils conditions (K1) and (K2) (see for example the proof of Theorem 3.2 in [2]).

We can now show the continuity result for the solution of the retarded weighted traffic equilibrium problem associated with a strictly pseudo-monotone operator, uniformly with respect to $t \in \Omega$, namely $\forall x_1(t), x_2(t) \in \mathbb{K}_h(t)$, a.e. in Ω

$$\langle C(t, x_2(t)), x_1(t) - x_2(t)\rangle_{\mathbf{s}(t)} \geq 0 \implies \langle C(t, x_1(t)), x_1(t) - x_2(t)\rangle_{\mathbf{s}(t)} > 0,$$

applying Theorem 4.2 in [4] and Remark 5.1.

Theorem 5.1. *Let h be a positive number and let $L^2(\Omega, \mathbb{R}, \boldsymbol{a}, \boldsymbol{s})$ and L be as specified in Section 3. Let $\lambda, \mu \in C(\Omega, \mathbb{R}^m_+) \cap L^2(\Omega, \mathbb{R}, \boldsymbol{a}, \boldsymbol{s})$ and let $\rho \in C(\Omega, \mathbb{R}^l_+) \cap L$ be vector-functions. Let $C : \Omega \times \mathbb{K}_h \to V^*$ be a continuous function so that $C(t, \cdot)$ is strictly pseudo-monotone uniformly with respect to $t \in \Omega$. Then the solution map of the retarded weighted traffic equilibrium problem is continuous on Ω.*

This result is very useful for numerical results. In other words the continuity of the solution permits to approximate the solution of the evolutionary problem with an interpolation of scale functions [4].

References

1. J-P. Aubin. *Analyse Fonctionelle appliquée.* Editions PUF, 1987.
2. A. Barbagallo, Regularity results for time-dependent variational and quasi-variational inequalities and computational procedures, *Math. Models Methods Appl. Sci.*, **17** (2007), pp. 277–304.
3. A. Barbagallo, On the regularity of retarded equilibria in time-dependent traffic equilibrium problems, to appear on *Nonlinear Anal., Theory Methods Appl.*.
4. A. Barbagallo and S. Pia, Weighted variational inequalities in non-pivot Hilbert spaces with applications, to appear on *Compt. Optim. Appl.*
5. A.A. Cournot, *Researches into the Mathematical Principles of the Theory of Wealth*, English Translation, MacMillan, London, 1897.
6. P. Daniele, A. Maugeri and W. Oettli, Time-Dependent Traffic Equilibria, *J. Optim. Theory Appl.*, **103** (1999), pp. 543-555.
7. T.L. Friesz, D. Bernstein, T.E. Smith, R.L. Tobin and B.W. Wie, A variational inequality formulation of the dynamic network user equilibrium problem, *Oper. Res.* **41** (1993), pp. 179–191.
8. S.Giuffré and S.Pia, Weighted Traffic Equilibrium in Non-Pivot Hilbert Spaces, *to appear on Nonlinear Anal., Theory Methods Appl.*
9. J. Hale. *Functional Differential Equations.* Springer, New York, 1971.
10. C. Kuratowski, Les functions semi-continues dans l'espace des ensembles fermés, *Fund. Math.*, **18** (1932), pp. 148–159.
11. J.D. Murray. *Mathematical Biology.* Springer, Berlin, 1993.
12. A.C. Pigou. *The Economics of Welfare.* MacMillan, London, 1920.
13. F. Raciti. Equilibrium in time-dependent traffic networks with delay. In F. Giannessi, A. Maugeri and P. Pardalos, editors, *Equilibrium Problems: Nonsmooth Optimization and Variational Inequality Models*, pp. 247–253, Kluwer, 2001.
14. C. Ratti, R.M. Pulselli, S. Williams and D. Frenchman. Mobile landscapes: Using location data from cell-phones for urban analysis. *Environment and Planning B: Planning and Design*, **33** (2006), pp. 727-748.
15. C. Ratti, A. Sevtsuk, S. Huang, R. Pailer, *Mobile Landscapes: Graz in Real Time.* Proceedings of the 3rd Symposium on LBS & TeleCartography, 28–30 November, Vienna, Austria (2005).

THERMODYNAMICS OF TYPE-II HIGH T_c SUPERCONDUCTORS

F. BARRILE AND L. RESTUCCIA*

Dipartimento di Matematica, Facoltà di Scienze MM.FF.NN.
Università degli Studi di Messina, Italy
** lrest@dipmat.unime.it*

In this paper we construct a geometric model for the thermodynamics of a vortex field in type-II high-T_c superconductors, within a nonconventional model based on the extended irreversible thermodynamics with internal variables. For this purpose we derive the transformation induced by the process and the dynamical system for a *simple material element* of the vortex lattice. Furthermore, we obtain the expression of the entropy function, the necessary conditions for its existence and the entropy 1-form, as a starting point to investigate a thermodynamical phase space and the state laws.

Keywords: Type-II high Tc superconductors; Non equilibrium Thermodynamics; Internal variables; Geometric models.

1. Introduction

In this paper, in the framework of extended irreversible thermodynamics with internal variables, we derive a geometric model for a vortex field in a type-II high T_c superconductor placed in an applied magnetic field ([1], [2]). The obtained results represent a starting point to investigate the extended-thermodynamics state space and to derive suitable laws of state. A type-II superconductor behaves in a different way than a type-I superconductor. In fact a type-I superconductor, when the intensity of the external applied magnetic field has intensity less than a critical value H_c, expels the magnetic flux from the material and it is in the Meissner state. On the contrary, for a type-II superconductor there exist two critical fields for these materials. For applied magnetic field intensities less than a lower critical field H_{c_1}, the superconductor is in the Meissner state. Applied magnetic field intensities fields greater than the an upper critical field H_{c_2} destroy the superconductivity. On the range $H_{c_1} \leq H \leq H_{c_2}$, the superconductor is in the mixed (or vortex) state. It is well known, above the lower critical field H_{c_1}, the

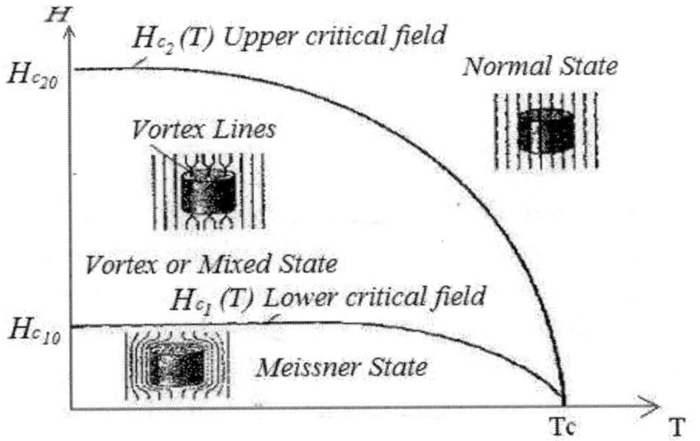

Fig. 1. The H - T phase diagrams for superconductor.

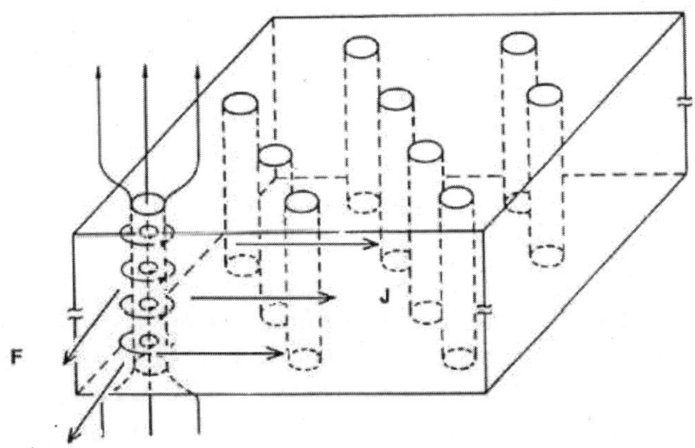

Fig. 2. Structure of array vortices (**F** - Lorentz force, **J** - supercurrent).

magnetic flux penetrates a type II superconductor on a λ_o London depth (see Figs. 1 and 2). Now, it can be observed that when the superconductor is limited by a plane, if the applied magnetic field is parallel to it, the magnetic flux is expelled from the material (Meissner effect). On the contrary, if the applied magnetic field is perpendicular to such limiting plane, the magnetic flux penetrates the superconductor along a finite number of lines, called *Abrikosov vortices* (or flux lines, flux tubes, fluxons), each carrying a quantum of magnetic flux. These tiny vortices of supercurrent tend to arrange themselves in a triangular (also quadratic or hexagonal) flux-line lattice. Because of the fact that each vortex line has a sign (has definite vorticity), lines of the opposite signs annihilate themselves. So, the density of the vortex field can vary. Inside each vortex there is a core in the normal phase, then outside there is the superconducting phase. In such situation the supercurrent flows around each vortex line at an average distance of 100 Å (on the range of the λ_o London depth). The imperfections of the crystal lattice (like dislocations, defect points , grain boundaries, etc) pin the vortex lines. The vortex array has some mechanical properties, coming mainly from the elastic properties of the superconducting body. However, since the vortices are created by the applied magnetic field and as the supercurrent flows around of each vortex, Lorentz force interactions exist among them (see Fig. 2). Those interactions create an additional mechanical field inside the type-II superconductor. If the Lorentz force between the vortices is greater than the pinning force the vortex lattice has an elastic behavior when the external applied magnetic field is near the lower critical limit value H_{c_1}. But it transfers itself smoothly like a "fluid" when the external applied magnetic field is near the upper critical limit value H_{c_2}. The "fluidity" of the vortex array is also observed when the density of the supercurrent is above its critical value and/or the temperature is sufficiently high. It is assumed that the pinning force is negligible and the vortices are soft and parallel to each other. The vortex motion (creep) is accompanied by an energy dissipation. Such motion is damped by a force proportional to the vortex velocity. Hence, the vortex field has viscous-elastic properties. Type-II high-T_c superconductor materials find applications in several fundamental technological sectors: in the field of optoelectronics; in the field of electronic sensors; in the processes of fabrication of microwave devices; in the realization of SQUID sensors in a biomedical frame; in the field of electromagnetic screens; in applied computer science; in the technology for integrated circuits VLSI (very large scale integration); in the energy transport sector, creating magnetic levitation and suspension. In [3]-[7] a thermodynamical model was developed for describing the vortex field in type-II high-T_c superconductors. A method to linearize the obtained field equations was applied and the magneto-mechanical wave propagation problem was considered for vortex lattice both in the solid and fluid states. As an example the dispersion problem in YBCO-ceramics was studied. In this paper we construct a thermodynamical model with internal variables for a vortex field in type-II high-T_c superconductors choosing a new vector of state. A geometrization technique is applied to obtain a geometric model for the thermodynamics of a simple material element for this vortex lattice. In particular, we derive the dynamical system, the morphism defined on the fibre bundle of the processes, the transformation induced by the processes, the expression of the entropy function, the necessary conditions for its existence and the entropy 1-form. Such entropy 1-form is the starting point to introduce and investigate an extended thermodynamical phase space (see [8]) and to derive the state laws (see [9]). In [10], [11] and [12] geometrical models for complex media were derived in the same geometrized framework by one of the authors (LR).

2. Equations governing the description of viscous-elastic field of vortices in the type-II superconductors

In [3]-[7], in the framework of extended thermodynamics with internal variables, a nonconventional model for the viscous-elastic field of vortices in type-II superconductors was developed. It was assumed that *the characteristic volume* of the body is sufficiently large for averaging all the physical quantities taken into consideration. The dimensions of such volume are much greater than the London penetration depth λ_o and the coherence length ξ (i.e. the radius of a vortex line). Only *depinned (soft) vortices*, averaged in the characteristic volume are considered. Any creation or annihilation of the vortices is omitted. *The mass density* ρ of the vortex field is the density of the vortices related to the molar volume of the material. ρ has dimensions kgm^{-3}. *The interaction between vortices* is due only to the Lorentz force. *The energy dissipation* occurs only because of the viscosity η of the vortex field caused by ohmic-like resistance of the normal state inside the vortex core. The relaxation feature of *the thermal field* is not taken into account, because of very low temperatures of the considered material. Only *small deformations* of the vortex field are considered and they are described by the linear strain tensor $\varepsilon_{ij} = \frac{1}{2}(u_{i,j} + u_{j,i})$. We indicate by u_i the components of the displacement vector of a vortex field and by $v_i = \dot{u}_i$. The symbol " $\dot{}$ " indicates the material derivative. In this paper we consider the magnetization of the vortex field, that was not taken into consideration in [3]-[7]. We choose as internal variables the local density of Cooper's superconducting electrons n^S and the supercurrent density j_i^S, satisfying the first London equation. n^S is defined by $n^S = \psi\psi^* = |\psi|^2$, where ψ denotes the wave function describing the probability density of superelectrons (ψ^* is the complex conjugate to ψ). The evolution equation for ψ is a suitable Schrödinger's equation. The vector of state (i.e. the set of independent variables) is the following: $C = \left\{\varepsilon_{ij}, \dot{\varepsilon_{ij}}, \mathcal{E}_i, B_i, n^S, T, j_i^S, q_i, n^S{}_{,i}, T_{,i}\right\}$, where $\dot{\varepsilon}_{ij}$ indicates the viscoelastic character of the vortex field, $\mathcal{E}_i$ is the electromotive intensity in the comoving frame, B_j is the magnetic induction, e is the internal energy density, T is the absolute temperature and q_i is the heat flux. The gradients $T_{,i}, n^S{}_{,i}$ concern possible nonlocal effects in the material. The behavior of the vortex lattice is governed by three groups of fundamental laws. The first group concerns *the mass density balance*, *the momentum balance* and *the density of the internal energy balance*, respectively

$$\dot{\rho} + \rho v_{k,k} = 0, \tag{1}$$

$$\rho \dot{v_k} - \sigma_{ik,i} - \epsilon_{kij}\left(j_i^N + j_i^S\right) B_j - M_i B_{i,k} - f_k = 0, \tag{2}$$

$$\rho \dot{e} - \sigma_{ik} v_{k,i} + q_{k,k} - \left(j_i^N + j_i^S\right) \mathcal{E}_i - \rho B_k \dot{\mathcal{M}}_k - \rho r = 0, \tag{3}$$

where $\mathbf{M}$ is the magnetization vector, given by $\mathbf{M} = \rho \boldsymbol{\mathcal{M}}$, σ_{ik} is the viscoelastic symmetric stress tensor, j_i^N is the normal current which satisfies Ohm's law, f_k is a given external body force and r is the heat source distribution. In the following the given external body force f_k and the heat source distribution r are neglected. The second group of laws deals with the electromagnetic field and concerns Maxwell's equations, the constitutive equations and *London's first equation* [2], in the form:

$$\epsilon_{ijk} E_{k,j} = -\frac{\partial B_i}{\partial t}, \quad B_{k,k} = 0, \quad \epsilon_{ijk} H_{k,j} - \frac{\partial D_i}{\partial t} = j_i^S + j_i^N, \quad E_{k,k} = 0, \tag{4}$$

$$B_i = \mu_0 H_i + M_i, \quad D_k = \epsilon_0 E_k \quad \frac{\partial j_k^S}{\partial t} = \frac{1}{\Lambda}\left(E_k + \epsilon_{kls} v_l B_s\right), \quad \Lambda = \mu_0 \lambda_0^2. \tag{5}$$

In (5) μ_0 is the permeability and ϵ_0 is the permittivity of vacuum. In $(5)_3$ the convection current has been disregarded respect to the conduction current. In $(5)_4$ the free charge density has been neglected. Moreover, the following relation holds true

$$\mathcal{E}_i = E_i + \epsilon_{ijk} v_j B_k. \tag{6}$$

The third group of fundamental laws concerns *the time evolution of fluxes and internal variables*

$$\dot{q}_k - Q_k(C) = 0, \qquad \dot{j}_k^S - J_k^S(C) = 0, \qquad \rho \dot{n}^S + j_{k,k}^S = N^S(C). \tag{7}$$

Equ. $(7)_3$ may be treated as the balance equation of superelectrons. All the admissible solutions of the proposed evolution equations should be restricted by the following *entropy inequality*:

$$\rho\dot{S} + \Phi^S_{k,k} - \frac{\rho r}{T} \geq 0, \tag{8}$$

where S denotes the entropy per unit mass and $\mathbf{\Phi}^S$ is the entropy flux associated with the fields of the set C given by

$$\mathbf{\Phi}^S = \frac{1}{T}\mathbf{q} + \mathbf{k}, \tag{9}$$

with $\mathbf{k}$ an additional term called *extra entropy flux density.* The use of the second law of thermodynamics in the form of the entropy inequality (8) gives us a possibility to determine all the constitutive functions that in our case form the set

$$\mathcal{Z} = \left\{\sigma_{ik}, M_i, e, j^N_k, J^S_k, Q_k, S, \Phi^S_k, \mu^{n^S}\right\}. \tag{10}$$

In (10) μ^{n^S} is the potential of the superelectrons. In [3] and [7] the entropy inequality (8) was analyzed by Liu's theorem [13], and constitutive relations were obtained, using isotropic polynomial representations of proper constitutive functions satisfying the objectivity and material frame indifference principles (see Smith's theorem [14]). In particular, mechanical phenomenological properties of the vortex lattice field versus H ($H_{c_1} < H < H_{c_2}$) where studied and proper constitutive laws for the stress tensor σ_{ij} were derived in the viscous-elastic case, where $\sigma_{ij} = \sigma_{ij}(C_1)$, with $C_1 = \{\varepsilon_{ij},\ \dot{\varepsilon}_{ij},\ B_i\}$ [3], and in the thermo-visco-elastic case where, $\sigma_{ij} = \sigma_{ij}(C_2)$, with $C_2 = \{\varepsilon_{ij},\ \dot{\varepsilon}_{ij},\ T,\ B_i\}$ [7]. In this last case, in the linear approximation, in the expression for σ_{ik} the temperature T is replaced by the relative temperature $\Theta = \frac{T-T_c}{T_c}$. Its values should not exceed the critical one T_c. Hence, it follows the definition of Θ:

$$\Theta = \frac{T - T_c}{T_c}, \quad 0 < T < T_c, \quad \longrightarrow \quad \Theta < 0,$$

$$\Theta = 0 \quad if \quad T = T_c, \quad \Theta = -1 \quad if \quad T = 0.$$

In this way the relative temperature within the superconducting phase is always negative. In [3] and [7] two parameters determining an actual state of vortices were introduced

$$\alpha = \left(\frac{H_{c2} - H}{H_{c2} - H_{c1}}\right), \qquad \alpha = \begin{cases} 0\ if\ H = H_{c2} \\ 1\ if\ H = H_{c1} \end{cases},$$

$$\beta = \left(\frac{H-H_{c1}}{H_{c2}-H_{c1}}\right), \qquad \beta = \begin{cases} 0\ if\ \ H = H_{c1} \\ 1\ if\ H = H_{c2}, \end{cases}$$

$$\alpha + \beta = \begin{cases} 0 \qquad if\ H = H_{c1}\ or\ H = H_{c2} \\ f(H)\ if \qquad H_{c1} < H < H_{c2}. \end{cases}$$

Using the formalism $\sigma_{ij} = \tau_{ij} + \frac{1}{3}\sigma_0\delta_{ij}$, where $\sigma_0 = \sigma_{kk}$, and writing $\sigma_0 = \sigma_0^{lattice} + \sigma_0^{fluid}$ and $\tau_{ij} = \tau_{ij}^{lattice} + \tau_{ij}^{fluid}$, with the help of the parameters α and β, in the case of thermo-visco-elastic it was obtained :

$$\sigma_0^{lattice} = \alpha(2\mu + 3\lambda)\varepsilon_{kk} - 3\lambda^T\Theta, \quad \sigma_0^{fluid} = -3\beta p,$$

$$\tau_{ij}^{lattice} = 2\mu\alpha\varepsilon_{ij} - \tfrac{2}{3}\mu\alpha\varepsilon_{kk}\delta_{ij} + 2\mu_L\alpha\dot{\varepsilon}_{ij} - \tfrac{2}{3}\mu_L\alpha\dot{\varepsilon}_{kk}\delta_{ij}, \quad \tau_{ij}^{fluid} = \beta D\dot{\varepsilon}_{ij},$$

and finally

$$\sigma_{ij} = \left[\left(\frac{1}{3}\alpha K - \frac{2}{3}\alpha G\right)\varepsilon_{kk} - \frac{2}{3}\alpha\eta\dot{\varepsilon}_{kk} - \lambda^T\Theta - \beta p\right]\delta_{ij}$$

$$+2\alpha G\varepsilon_{ij} + 2(\alpha+\beta)\eta\dot{\varepsilon}_{ij},$$

where the material-like coefficients responsible for the thermomechanical properties of the vortex field can be called as follows: λ *and* μ – Lamé constants of the lattice, λ^T – thermoelastic constant of the lattice, λ_L *and* μ_L – viscoelastic constants of the lattice, p – pressure of the lattice in the fluid state. Furthermore, $K = 2\mu + 3\lambda$ is the lattice elastic bulk modulus, $G = \mu$ is the lattice shear modulus, $\mu_L = \eta$ and $D = 2\eta$ (with η the viscosity coefficient of the lattice).

3. A geometric model for type-II high-T_c superconductors

Now, taking into account the model and the results presented in the previous section, we construct a geometric thermodynamical model for the description of the vortex field in type-II high-Tc superconductors following [15]- [21]. We introduce the transformation induced by the process and we derive the dynamical system for a *simple material element* of the vortex lattice. Then, we obtain the expressions for the entropy function and the entropy 1-form. Thus, we consider a material element and we define the state space at time t as the set B_t of all state variables which "fit" the configuration of the element at time t. B_t is assumed to have the structure of a finite dimensional manifold. The "total state space" is the disjoint union $\mathcal{B} = \bigcup_t \{t\} \times B_t$ with a given natural structure of fibre bundle over

$\mathbb{R}$, where time flows (see [15]-[17]). If the instantaneous state space B_t does not vary in time, then $\mathcal{B}$ has the topology of the Cartesian product: $\mathcal{B} \simeq \mathbb{R} \times B$. Moreover, we consider an abstract space of processes, i.e. a set Π of functions $P^i_t : [0,t] \to \mathcal{G}$, where $[0,t]$ is any time interval and $t \in \mathbb{R}$ is the so called duration of the process. The space $\mathcal{G}$ is a suitable target space defined by the problem under consideration and i is a label ranging in an unspecified index set for all allowed processes. For the given state space B we suppose that the set Π is such that the following statements hold:
i) $\exists D : \Pi \to \mathcal{P}(B)$, where $\mathcal{P}(B)$ is the set of all subsets of B. D is the *domain function* and $D^i_t \equiv D(P^i_t)$ is called the domain of the i-th process (of duration t);
ii) $\exists R : \Pi \to \mathcal{P}(B)$; R is the *range function* and $R^i_t \equiv R(P^i_t)$ is called the range of the i-th process (of duration t);
iii) Considering the restrictions $P^i_t = P^i_t \mid_{[0,\tau]}$ $(\tau \leq t)$ new processes are obtained (*restricted processes*) and they satisfy the following statements: $\forall \tau < t \; D(P^i_t) \subseteq D(P^i_\tau)$. Incidentally, this implies that $\bigcap_{\tau=0}^{t} D(P^i_\tau) = D(P^i_t)$, where t is the maximal duration.
Then, a continuous function χ is defined such that

$$\chi : \mathbb{R} \times \Pi \to C^0(B,B); \qquad (t, P^i_t) \to \rho^i_t, \tag{11}$$

where $\rho^i_t : b \in D^i_t \subseteq B \to \rho^i_t(b) = b_t \in R^i_t \subseteq B,$ so that for any instant of time t and for any process $P^i_t \in \Pi$ a continuous mapping, ρ^i_t, called *transformation induced by the process*, is generated. ρ^i_t gives point by point a correspondence between the initial state b and the final state $\rho^i_t(b) = b_t$. Now, we can define the following function of time:

$$\lambda^i_b(\tau) = \begin{cases} b & if\ \tau = 0 \qquad with\ \ b \in D^i_t \\ \rho^i_t(b) & if\ \tau \in]0,t] \end{cases} \tag{12}$$

such that the transformation induced is given by $\delta : \mathbb{R} \longrightarrow \mathbb{R} \times B$

$$\delta : \tau \in \mathbb{R} \longrightarrow \delta(\tau) = (\tau, \lambda^i_b(\tau)) \in \mathbb{R} \times B. \tag{13}$$

With these positions the transformation is interpreted as a curve δ in the union of all state spaces such that it intersects the instantaneous state space just once. In this geometric model we assume that the state variables are

$$\left(F_{ij}, \dot{F_{ij}}, e, D_i, B_i, n^S, j^S_i, q_i, T_{,i}, n^S{}_{,i}\right).$$

The full state space is then $B = Lin(\mathfrak{V}) \oplus Lin(\mathfrak{V}) \oplus \mathbb{R} \oplus \mathfrak{V} \oplus \mathfrak{V} \oplus \mathbb{R} \oplus_{i=1}^{4} \mathfrak{V}$, where $\mathfrak{V} \simeq \mathbb{R}^3$. The process P^i_t is described by the following functions

$$P^i_t(\tau) = [\mathbf{L}(\tau), \dot{\mathbf{L}}(\tau), h(\tau), \boldsymbol{\mathcal{H}}(\tau), \boldsymbol{\Xi}(\tau), G^S(\tau), \boldsymbol{\mathcal{J}}^S(\tau), \boldsymbol{\mathcal{Q}}(\tau), \boldsymbol{\Theta}(\tau), \mathcal{N}^S(\tau)] \in \mathcal{G},$$

where

$$\mathcal{H}_i = \epsilon_{ijk}H_{k,j} - (j_i^N + j_i^S),\ \Xi_i = -\epsilon_{ijk}E_{k,j},\ h = (j_i^N + j_i^S)\mathcal{E}_i - q_{i,i} + \rho B_i \dot{\mathcal{M}}_i,$$

$$G^S = -j_{k,k}^S + N^S$$

and $\mathcal{G}$ is given by $\mathcal{G} = Lin(\mathfrak{V}) \oplus Lin(\mathfrak{V}) \oplus \mathbb{R} \oplus \mathfrak{V} \oplus \mathfrak{V} \oplus \mathbb{R} \oplus_{i=1}^{4} \mathfrak{V}$. Moreover, the constitutive functions $\boldsymbol{\sigma}$, $\boldsymbol{\mathcal{M}}$, T, $\mathbf{J}^N$, $\mathbf{J}^S$ and $\mathbf{Q}$ are defined in the following way

$$\boldsymbol{\sigma} : \mathbb{R} \times B \longrightarrow Sym(\mathfrak{V}), \quad \boldsymbol{\mathcal{M}} : \mathbb{R} \times B \longrightarrow \mathfrak{V},$$

$$T : \mathbb{R} \times B \longrightarrow \mathbb{R}^{++}, \mathbf{J^N} : \mathbb{R} \times B \longrightarrow \mathfrak{V},\ \ \mathbf{J^S} : \mathbb{R} \times B \longrightarrow \mathfrak{V},\ \ \mathbf{Q} : \mathbb{R} \times B \longrightarrow \mathfrak{V}.$$

We assume that for each pair (P_t^i, b) we have the following dynamical system

$$\begin{cases} \dot{\mathbf{F}} = \mathbf{L}(\tau)\mathbf{F}(\tau) \\ \ddot{\mathbf{F}} = \dot{\mathbf{L}}(\tau)\mathbf{F}(\tau) + \mathbf{L}(\tau)\dot{\mathbf{F}}(\tau) \\ \rho \dot{e} = \boldsymbol{\sigma}(\delta) \cdot \mathbf{L}(\tau) + h(\tau) \\ \dot{\mathbf{D}} = \boldsymbol{\mathcal{H}}(\tau) \\ \dot{\mathbf{B}} = \boldsymbol{\Xi}(\tau) \\ \rho \dot{n}^S = G^S(\tau) \\ \dot{\mathbf{j}}^S = \boldsymbol{\mathcal{J}}^S(\delta) \\ \dot{\mathbf{q}} = \boldsymbol{\mathcal{Q}}(\delta) \\ \dot{\nabla T} = \boldsymbol{\Theta}(\tau) \\ \dot{\nabla n^S} = \mathcal{N}^S(\tau). \end{cases} \tag{14}$$

Such dynamical system determines a linear morphism G defined on fibre bundle of process in the following way
$G : (\mathbf{F}, \dot{\mathbf{F}}, e, \mathbf{D}, \mathbf{B}, n^S, \mathbf{j}^S, \mathbf{q}, \nabla T, \nabla n^S, \mathbf{L}, \dot{\mathbf{L}}, h, \boldsymbol{\mathcal{H}}, \boldsymbol{\Xi}, G^S, \boldsymbol{\mathcal{J}}^S, \boldsymbol{\mathcal{Q}}, \boldsymbol{\Theta}, \mathcal{N}^S) \rightarrow$ $(\mathbf{F}, \dot{\mathbf{F}}, e, \mathbf{D}, \mathbf{B}, n^S, \mathbf{j}^S, \mathbf{q}, \nabla T, \nabla n^S, \dot{\mathbf{F}}, \ddot{\mathbf{F}}, \dot{e}, \dot{\mathbf{D}}, \dot{\mathbf{B}}, \dot{n}^S, \dot{\mathbf{j}}^S, \dot{\mathbf{q}}, \dot{\nabla T}, \dot{\nabla n^S})$,
which in matrix form is expressed by:

$$\left(\mathbf{F}, \dot{\mathbf{F}}, e, \mathbf{D}, \mathbf{B}, n^S, \mathbf{j}^S, \mathbf{q}, \nabla T, \nabla n^S, \dot{\mathbf{F}}, \ddot{\mathbf{F}}, \dot{e}, \dot{\mathbf{D}}, \dot{\mathbf{B}}, \dot{n}^S, \dot{\mathbf{j}}^S, \dot{\mathbf{q}}, \dot{\nabla T}, \dot{\nabla n^S}\right)^T =$$

$$\begin{pmatrix} \mathbf{I} & \mathbf{0} \\ \mathbf{0} & \mathbf{A} \end{pmatrix} \left(\mathbf{F}, \dot{\mathbf{F}}, e, \mathbf{D}, \mathbf{B}, n^S, \mathbf{j}^S, \mathbf{q}, \nabla T, \nabla n^S, \mathbf{L}, \dot{\mathbf{L}}, \mathrm{h}, \boldsymbol{\mathcal{H}}, \boldsymbol{\Xi}, G^S, \boldsymbol{\mathcal{J}}^S, \boldsymbol{\mathcal{Q}}, \boldsymbol{\Theta}, \mathcal{N}^S\right)^T,$$

with

$$\mathbf{A} = \begin{pmatrix} \mathbf{F} & 0 & 0 & 0 & 0 & 0 & 0 & 0 & 0 & 0 \\ \dot{\mathbf{F}} & \mathbf{F} & 0 & 0 & 0 & 0 & 0 & 0 & 0 & 0 \\ \frac{\sigma}{\rho} & 0 & \frac{1}{\rho} & 0 & 0 & 0 & 0 & 0 & 0 & 0 \\ 0 & 0 & 0 & 1 & 0 & 0 & 0 & 0 & 0 & 0 \\ 0 & 0 & 0 & 0 & 1 & 0 & 0 & 0 & 0 & 0 \\ 0 & 0 & 0 & 0 & 0 & 1 & 0 & 0 & 0 & 0 \\ 0 & 0 & 0 & 0 & 0 & 0 & 1 & 0 & 0 & 0 \\ 0 & 0 & 0 & 0 & 0 & 0 & 0 & 1 & 0 & 0 \\ 0 & 0 & 0 & 0 & 0 & 0 & 0 & 0 & 1 & 0 \\ 0 & 0 & 0 & 0 & 0 & 0 & 0 & 0 & 0 & 1 \end{pmatrix} \tag{15}$$

Finally, $\rho_t^i(b)$ is given by

$$\rho_t^i(b) = \begin{cases} \mathbf{F} & = \int_0^t \left[\mathbf{L}(\tau)\mathbf{F}(\tau)\right] dt \\ \dot{\mathbf{F}} & = \int_0^t \left[\dot{\mathbf{L}}(\tau)\mathbf{F}(\tau) + \mathbf{L}(\tau)\dot{\mathbf{F}}(\tau)\right] dt \\ e & = \int_0^t \frac{1}{\rho}\left[\boldsymbol{\sigma}(\delta)\cdot\mathbf{L}(\tau) + h(\tau)\right] dt \\ \mathbf{D} & = \int_0^t \left[\boldsymbol{\mathcal{H}}(\tau)\right] dt \\ \mathbf{B} & = \int_0^t \left[\boldsymbol{\Xi}(\tau)\right] dt \\ n^S & = \int_0^t \left[G^S(\tau)\right] dt \\ \mathbf{j}^S & = \int_0^t \left[\mathbf{J}^S(\delta)\right] dt \\ \mathbf{q} & = \int_0^t \left[\mathbf{Q}(\delta)\right] dt \\ \nabla T & = \int_0^t \left[\boldsymbol{\Theta}(\tau)\right] dt \\ \nabla n^S & = \int_0^t \left[\mathcal{N}^S(\tau)\right] dt \end{cases} \tag{16}$$

By using the system of differential equations (14), following standard procedures (see [15]-[17] and [21]) in this geometrical structure we are able to introduce an "entropy function", which is related to a transformation between the initial and the final states b and $\rho_t^i(b) = b_t$, respectively, in the following way:

$$s(\rho_t^i, b, t) = -\int_0^t \frac{1}{\rho}\nabla\cdot\mathbf{\Phi}^S d\tau, \tag{17}$$

where $\mathbf{\Phi}^S$ is defined according to equation (9).

Using the internal energy balance and the relation $\boldsymbol{\mathcal{E}} = (\mathbf{E} + \mathbf{v}\wedge\mathbf{B})$, we obtain the following expression for $\nabla\cdot\mathbf{q}$:

$$\nabla\cdot\mathbf{q} = -\rho\dot{e} + \boldsymbol{\sigma}\cdot(\dot{\mathbf{F}}\mathbf{F}^{-1}) + (\mathbf{j}^N + \mathbf{j}^S)\cdot\boldsymbol{\mathcal{E}} + \mathbf{B}\cdot\dot{\boldsymbol{\mathcal{M}}} \tag{18}$$

$$= -\rho\dot{e} + \boldsymbol{\sigma}\cdot(\dot{\mathbf{F}}\mathbf{F}^{-1}) + (\nabla\wedge\frac{\mathbf{B}}{\mu_0} - \dot{\mathbf{D}})\cdot\boldsymbol{\mathcal{E}} + \mathbf{B}\cdot\dot{\mathbf{M}} - \frac{\dot{\rho}}{\rho}\mathbf{B}\cdot\mathbf{M}.$$

Then, using (9) we get

$$s = \int_0^t -\frac{1}{\rho T}\nabla\cdot\mathbf{q}d\tau + \int_0^t \frac{1}{\rho T^2}\mathbf{q}\cdot\nabla T d\tau - \int_0^t \frac{1}{\rho}\nabla\cdot\mathbf{k}d\tau, \tag{19}$$

so that the final expression for $s(\rho_t^i, b, t)$ is

$$s(\rho_t^i, b, t) = \int_\delta \Omega, \tag{20}$$

$$\text{where}\quad \Omega = \frac{1}{T}de - \frac{1}{\rho T}(\boldsymbol{\sigma}\mathbf{F}^{-T})\cdot d\mathbf{F} + \frac{1}{\rho T}\boldsymbol{\mathcal{E}}\cdot d\mathbf{D} - \frac{1}{\rho T}\mathbf{B}\cdot d\mathbf{B}$$

$$-\frac{1}{\rho}\left[\frac{1}{T^2}\mathbf{q}\cdot\nabla T - \frac{1}{T}\nabla\wedge\frac{\boldsymbol{B}}{\mu_0}\cdot\boldsymbol{\mathcal{E}} + \frac{\mu_0}{T}\mathbf{B}\cdot\dot{\mathbf{H}} + \frac{\dot{\rho}}{\rho T}\mathbf{B}\cdot\mathbf{M} - \nabla\cdot\mathbf{k}\right]d\tau. \tag{21}$$

In equ. (21) we have used the relation $\boldsymbol{\sigma}\cdot(\dot{\mathbf{F}}\mathbf{F}^{-1}) = (\boldsymbol{\sigma}\mathbf{F}^{-T})\cdot\dot{\mathbf{F}}$, with $\mathbf{F}^{-T} = (\mathbf{F}^{-1})^T$ (T denoting matrix transposition). Thus, the entropy function is now calculated as an integral along a path into the space $\mathbb{R}\times B$ of all thermodynamic variables together with the independent time variable (see [15]-[17]) and Ω is a 1-form in $\mathbb{R}\times B$ called the entropy 1-form. In components the entropy 1-form Ω becomes:

$$\Omega = \omega_\mu d\mathfrak{q}^\mu + \omega_0 dt = \omega_A d\mathfrak{q}^A, \quad (A = 1, 2, ..., 11) \tag{22}$$

$$\text{where}\qquad \mathfrak{q}^A = (\mathbf{F}, \dot{\mathbf{F}}, e, \mathbf{D}, \mathbf{B}, n^S, \mathbf{j}^S, \mathbf{q}, \nabla T, \nabla n^S, \tau)$$

$$\text{and}\qquad \boldsymbol{\omega}_A = \Bigg(-\frac{1}{\rho T}\boldsymbol{\sigma}\mathbf{F}^{-T}, 0, \frac{1}{T}, \frac{1}{\rho T}\boldsymbol{\mathcal{E}}, -\frac{1}{\rho T}\mathbf{B}, 0, 0, 0, 0, 0,$$

$$\frac{1}{\rho T}\left(\frac{1}{T}\mathbf{q}\cdot\nabla T - \nabla\wedge\frac{\boldsymbol{B}}{\mu_0}\cdot\boldsymbol{\mathcal{E}} + \mu_0\mathbf{B}\cdot\dot{\mathbf{H}} + \frac{\dot{\rho}}{\rho}\mathbf{B}\cdot\mathbf{M}\right) - \frac{1}{\rho}\nabla\cdot\mathbf{k}\Bigg).$$

Thus, by external differentiation,the following 2-form is obtained:

$$d\Omega = \frac{1}{2}A_{\mu\lambda}d\mathfrak{q}^\mu \wedge d\mathfrak{q}^\lambda + E_\lambda dt \wedge d\mathfrak{q}^\lambda, \tag{23}$$

where $A_{\mu\lambda} = \partial_\mu\omega_\lambda - \partial_\lambda\omega_\mu$ and $E_\lambda = \partial_0\omega_\lambda - \partial_\lambda\omega_0$.

Applying the closure conditions for the entropy 1-form, the necessary conditions for the existence of the entropy function during the processes under consideration are

$$\partial_{\mathbf{B}}\left(-\frac{1}{\rho T}\boldsymbol{\sigma}\mathbf{F}^{-T}\right) = \partial_{\mathbf{F}}\left(-\frac{1}{\rho T}\mathbf{B}\right), \quad \partial_e\left(-\frac{1}{\rho T}\boldsymbol{\sigma}\mathbf{F}^{-T}\right) = \partial_{\mathbf{F}}\left(\frac{1}{T}\right),$$

$$\partial_e\left(-\frac{1}{\rho T}\mathbf{B}\right) = \partial_{\mathbf{B}}\left(\frac{1}{T}\right), \quad \partial_{\mathbf{D}}\left(-\frac{1}{\rho T}\mathbf{T}\mathbf{F}^{-T}\right) = \partial_{\mathbf{F}}\left(\frac{1}{\rho T}\boldsymbol{\mathcal{E}}\right),$$

$$\partial_{\mathbf{D}}\left(-\frac{1}{\rho T}\mathbf{B}\right)=\partial_{\mathbf{B}}\left(\frac{1}{\rho T}\boldsymbol{\mathcal{E}}\right),\quad \partial_{\mathbf{D}}\left(\frac{1}{T}\right)=\partial_{e}\left(\frac{1}{\rho T}\boldsymbol{\mathcal{E}}\right),$$

$$\frac{\partial\omega_A}{\partial\mathbf{q}^B}=0\qquad (A=1,3,4,5,11;\ B=2,6,...,10),$$

$$\partial_{\mathbf{t}}\left(-\tfrac{1}{\rho T}\mathbf{T}\mathbf{F}^{-T}\right)=$$
$$\partial_{\mathbf{F}}\left[\tfrac{1}{\rho T}\left(\tfrac{1}{T}\mathbf{q}\cdot\nabla T-\nabla\wedge\tfrac{\boldsymbol{B}}{\mu_0}\cdot\boldsymbol{\mathcal{E}}+\mu_0\mathbf{B}\cdot\dot{\mathbf{H}}+\tfrac{\dot\rho}{\rho}\mathbf{B}\cdot\mathbf{M}\right)-\tfrac{1}{\rho}\nabla\cdot\mathbf{k}\right],$$
$$\partial_{\mathbf{t}}\left(-\tfrac{1}{\rho T}\mathbf{B}\right)=\partial_{\mathbf{B}}\left[\tfrac{1}{\rho T}\left(\tfrac{1}{T}\mathbf{q}\cdot\nabla T-\nabla\wedge\tfrac{\boldsymbol{B}}{\mu_0}\cdot\boldsymbol{\mathcal{E}}+\mu_0\mathbf{B}\cdot\dot{\mathbf{H}}+\tfrac{\dot\rho}{\rho}\mathbf{B}\cdot\mathbf{M}\right)-\tfrac{1}{\rho}\nabla\cdot\mathbf{k}\right],$$
$$\partial_{\mathbf{t}}\left(\frac{1}{T}\right)=\partial_{e}\left[\frac{1}{\rho T}\left(\frac{1}{T}\mathbf{q}\cdot\nabla T-\nabla\wedge\frac{\boldsymbol{B}}{\mu_0}\cdot\boldsymbol{\mathcal{E}}+\mu_0\mathbf{B}\cdot\dot{\mathbf{H}}+\frac{\dot\rho}{\rho}\mathbf{B}\cdot\mathbf{M}\right)-\frac{1}{\rho}\nabla\cdot\mathbf{k}\right],$$
$$\partial_{\mathbf{t}}\left(\frac{1}{\rho T}\boldsymbol{\mathcal{E}}\right)=\partial_{\mathbf{D}}\left[\frac{1}{\rho T}\left(\frac{1}{T}\mathbf{q}\cdot\nabla T-\nabla\wedge\frac{\boldsymbol{B}}{\mu_0}\cdot\boldsymbol{\mathcal{E}}+\mu_0\mathbf{B}\cdot\dot{\mathbf{H}}+\frac{\dot\rho}{\rho}\mathbf{B}\cdot\mathbf{M}\right)-\frac{1}{\rho}\nabla\cdot\mathbf{k}\right].$$

Moreover, if the entropy 1-form is closed and its coefficients are regular, this form is exact and the existence of an upper-potential S, satisfying relation $S(b_t)-S(b)\geq s$, is ensured for all $P_t^i\in\Pi$, with $b_t=\rho_t^i(b)$ [21]. Such 1-form is the stating point to introduce a thermodynamical phase space in a suitable way [8]. Furthermore, from the conditions obtained for the existence of the entropy function, the thermodynamical state laws can be investigated following [9].

References

1. T.P. Orlando and K.A. Delin, *Foundations of applied superconductivity* (Addison-Wesley Publishing Company, Reading, Massachusetts, 1991).
2. M. Cyrot and D. Pavuna, *Introduction to superconductivity and high-T_c materials* (World Scientific, Singapore, 1992).
3. B. Maruszewski and L. Restuccia, *Mechanics of a vortex lattice in superconductors. Phenomenological approach*, in *Trends in Continuum Physics*, eds. B.T. Maruszewski, W. Muschik and A. Radowicz (World Scientific, Singapore, 1999), pp. 220-230.
4. L. Restuccia and B. Maruszewski, *On a thermodynamical model for type II high Tc superconductors. Theory and applications*, in *Mathematical Models And Methods For Smart Materials, Series on Advances in Mathematics for Applied Sciences*, eds. M. Fabrizio, B. Lazzari and A. Morro, Vol.62 (World Scientific, 2002), pp. 283-296.
5. L. Restuccia and B. Maruszewski, *Vortex lattice waves in a high-T_c superconductor*, in *THERMAL STRESSES'99*, eds. J.J. Skrzypek and R. B. Hetnarski (Cracow, University of Technology, Cracow, Poland, 1999), pp. 89-92.

6. L. Restuccia and B. Maruszewski, *Magnetoacoustic waves in a vortex fluid in high-Tc superconductors*, in *THERMAL STRESSES '2001* (Osaka University, Osaka, Japan, 2001), pp.3-8.
7. B. Maruszewski, *Atti Accademia Peloritana dei Pericolanti di Messina* **LXXXVI**, Suppl. 1, DOI: 10.1478/C1S0801014 (2008).
8. S. Preston, J.Vargo, *Atti Accademia Peloritana dei Pericolanti di Messina* **LXXXVI**, Suppl. 1, DOI: 10.1478/C1S0801019 (2008).
9. M. Dolfin, S. Preston, L. Restuccia, Geometry of the entropy form (to be published).
10. M. Dolfin, M. Francaviglia and L. Restuccia, *Technische Mechanik* **24**, 137 (2004).
11. M. Francaviglia, L. Restuccia and P. Rogolino, *Journal of Non-Equilibrium Thermodynamics* **29**, 221 (2004).
12. D. German, L. Restuccia, *Thermodynamics of piezoelectric media with dislocations*, in *Applied and Industrial Mathematics in Italy II, Series on Advances in Mathematics for Applied Sciences*, eds. V. Cutello, G. Fotia and L. Puccio, Vol.75, (World Scientific, 2007).
13. I-Shih Liu, *Arch.Rat.Mech.Anal.* **46**,131 (1972) .
14. G.F. Smith , *Int.J.Engng.Sci.* **9**,899 (1971) .
15. M. Dolfin, M. Francaviglia and P. Rogolino, *J. Non-Equilib. Thermodyn.* **23**, 250 (1998).
16. M. Dolfin, M. Francaviglia, and P. Rogolino, *Periodica Polytechnica Serie Mech. Eng.* **43**, 29(1999).
17. V. Ciancio, M. Dolfin, M. Francaviglia, P. Rogolino, *J. Non-Equilib. Thermodyn.* **26**, 255 (2001).
18. W. Noll, *Arch. Rat. Mech. Anal.* **2**, 197 (1958).
19. B. D. Coleman and M. E. Gurtin, *J. Chem. Phys.* **47**, 597 (1967).
20. W. Noll, *Arch. Rat. Mech. Anal.* **48**, 1 (1972).
21. B. D. Coleman and D. R. Owen, *Arch. Rat. Mech. Anal.* **54**, 1 (1974).

EXISTENCE AND UNIQUENESS FOR A THREE DIMENSIONAL MODEL OF FERROMAGNETISM

V. BERTI* and M. FABRIZIO†

Dipartimento di Matematica, Università degli Studi di Bologna, P.zza di Porta S. Donato 5, I-40126, Bologna, Italy
* *berti@dm.unibo.it*
† *fabrizio@dm.unibo.it*

C. GIORGI

Dipartimento di Matematica, Università degli Studi di Brescia, Via Valotti 9, I-25133 Brescia, Italy
giorgi@ing.unibs.it

A three-dimensional model describing the behavior of a ferromagnetic material is proposed. Since the passage between paramagnetic and ferromagnetic state at the Curie temperature is a second order phase transition, the phenomenon is set in the context of the Ginzburg-Landau theory, identifying the order parameter with the magnetization $\mathbf{M}$. Accordingly, $\mathbf{M}$ satisfies a vectorial Ginzburg-Landau equation, interpreted as a balance law of the internal order structure. Thermoelectromagnetic effects are included by coupling Maxwell's equation and the appropriate representation of the heat equation. The resulting differential system is completed with initial and boundary conditions and its well-posedness is proved.

Keywords: Ferromagnetic materials, phase transitions, well-posedness.

1. Introduction

The model we propose describes the paramagnetic-ferromagnetic transition induced by variations of the temperature. The phenomenon of ferromagnetism occurs in some metals like iron, nickel, cobalt, when an external magnetic field yields a large magnetization inside the material, due to the alignment of the spin magnetic moments. Below a critical value θ_c of the temperature, called Curie temperature, even if the external magnetic field is removed, the spin magnetic moments stay aligned. However, when the temperature overcomes the threshold value θ_c, the material returns to the paramagnetic state, since the residual alignment of the spins disappears. The passage from the paramagnetic to the ferromagnetic state can be interpreted as a second order phase transition according to Landau's terminology (see Refs. 2,7), since no latent heat is released or absorbed. Therefore, we set the phenomenon into the general framework of the Ginzburg-Landau theory. The first step is the identification of a suitable order parameter. Indeed, each phase transition can be interpreted as a passage from a less ordered structure to a more ordered one and vice versa. In this context, the order parameter is the variable measuring the internal order of the material.

In ferromagnets, the internal order is given by the alignment of the spin magnetic moments. Accordingly, the order parameter can be identified with the magnetization vector, which means that the internal order is due both to the number of the oriented spin and to their direction. The vectorial structure of the order parameter leads to a three-dimensional model for ferromagnetism. Like other phase transitions models, the magnetization satisfies a Ginzburg-Landau equation which can be interpreted as a balance law of the internal order (see Ref. 4). Moreover, since the transition is induced by variations of the temperature we couple an appropriate representation of the heat equation deduced from the energy balance and consistent with the principles of thermodynamics. Finally, the evolution equations are completed with Maxwell's equations describing the variations of the electromagnetic field inside the ferromagnet. For the resulting differential system, endowed with boundary and initial conditions, existence and uniqueness of solutions have been proved (see Ref. 1).

2. Evolution equations

We denote by $\Omega \subset \mathbb{R}^3$ the domain occupied by the ferromagnetic material. Maxwell's equations rule the variations of the electromagnetic field inside Ω

$$\nabla \times \mathbf{E} = -\dot{\mathbf{B}}, \qquad \nabla \times \mathbf{H} = \dot{\mathbf{D}} + \mathbf{J}, \tag{1}$$

$$\nabla \cdot \mathbf{B} = 0, \qquad \nabla \cdot \mathbf{D} = \rho_e, \tag{2}$$

where $\mathbf{E}, \mathbf{H}, \mathbf{D}, \mathbf{B}, \mathbf{J}, \rho_e$ are respectively the electric field, the magnetic field, the electric displacement, the magnetic induction, the current density and the free charge density. We assume the electromagnetic isotropy of the material and the following constitutive equations

$$\mathbf{D} = \varepsilon \mathbf{E}, \qquad \mathbf{B} = \mu \mathbf{H} + \mathbf{M}, \qquad \mathbf{J} = \sigma \mathbf{E}, \tag{3}$$

where ε, μ, σ are respectively the dielectric constant, the magnetic permeability, the conductivity, and $\mathbf{M}$ is the magnetization vector. Moreover, as usual in ferromagnetic models, we will neglect the displacement current $\varepsilon\dot{\mathbf{E}}$. Accordingly, from (1)–(3) we deduce

$$\mu\dot{\mathbf{H}} + \dot{\mathbf{M}} = -\frac{1}{\sigma}\nabla \times \nabla \times \mathbf{H}, \qquad \nabla \cdot (\mu\mathbf{H} + \mathbf{M}) = 0. \tag{4}$$

In order to model the paramagnetic-ferromagnetic transition at the Curie temperature, we have to specify the law relating the variation of the magnetization to the magnetic field. Like other macroscopic models for phase transitions, this relation can be deduced by a balance law of the order inside the material. In particular, the order of the internal structure in a ferromagnetic material is due both to the position and to the direction of the magnetic spins. This leads to the choice of the magnetization vector as the variable which measures "amount of order" of the internal structure.

Two different approaches are allowed. A possible choice[1] consists in identifying the order parameter of the transition with the components of the magnetization $\mathbf{M}$. A different model[4] is based on the decomposition

$$\mathbf{M} = \varphi\mathbf{m}, \qquad \varphi \geq 0, \qquad |\mathbf{m}| = 1. \tag{5}$$

In the first case the phase transition is characterized by a vector valued order parameter. Thus, we define the internal structure order as the vector-valued measure whose representation is

$$\mathbf{K}^i(\mathcal{A}) = \int_{\mathcal{A}} \rho\mathbf{k}\,dx, \qquad \mathcal{A} \subset \Omega,$$

where ρ is the mass density. The vector $\mathbf{k}$, called specific internal structure order, accounts for the internal order of the spins determined by their orientation. Moreover, we assume a representation of the external structure order vector in the form

$$\mathbf{K}^e(\mathcal{A}) = \int_{\partial\mathcal{A}} \mathbf{P}\mathbf{n}_{\mathcal{A}}\,ds + \int_{\mathcal{A}} \rho\boldsymbol{\sigma}\,dx, \qquad \mathcal{A} \subset \Omega,$$

where $\partial\mathcal{A}$ denotes the boundary of $\mathcal{A}$ and $\mathbf{n}_{\mathcal{A}}$ is its unit outward normal vector. Here $\mathbf{P}$ is a second order tensor such that $\mathbf{P}\mathbf{n}_{\mathcal{A}}$ provides the specific flux of the structure order through $\partial\mathcal{A}$ and $\boldsymbol{\sigma}$ is the structure order supply. In particular, the tensor $\mathbf{P}(x,t)$ describes the distribution of the internal order in a neighborhood of x at the instant t, while $\boldsymbol{\sigma}$ represents a source of internal order inside the domain Ω. In the sequel, we assume $\boldsymbol{\sigma} = \mathbf{0}$. Moreover, we suppose that the mass density is constant and for simplicity we let $\rho = 1$.

The structure order balance on every sub-body $\mathcal{A} \subset \Omega$ reads

$$\mathbf{K}^i(\mathcal{A}) = \mathbf{K}^e(\mathcal{A}),$$

from which we obtain a.e. in Ω the local equation

$$\mathbf{k} = \nabla \cdot \mathbf{P}. \tag{6}$$

The previous equation gives the relation between the magnetization and the magnetic field, provided that constitutive equations for $\mathbf{k}$ and $\mathbf{P}$ are chosen. We assume

$$\mathbf{k} = \gamma\dot{\mathbf{M}} + \theta_c F'(|\mathbf{M}|) + \theta G'(|\mathbf{M}|) - \mathbf{H} \tag{7}$$

$$\mathbf{P} = \nu\nabla\mathbf{M}, \tag{8}$$

where γ, ν are positive constants and $\theta > 0$ is the absolute temperature.

Since the paramagnetic-ferromagnetic transition is a typical second order phase transition, a possible choice of the functions F, G is (see Ref. 3)

$$F(|\mathbf{M}|) = \frac{1}{4}|\mathbf{M}|^4 - \frac{1}{2}|\mathbf{M}|^2, \qquad G(|\mathbf{M}|) = \frac{1}{2}|\mathbf{M}|^2. \tag{9}$$

With this assumption, the function $W = \theta_c F + \theta G$ admits the minimum $\mathbf{M} = 0$ when $\theta \geq \theta_c$. Conversely, below the Curie temperature, the minimum is attained when the modulus of the magnetization is

$$M = \frac{\theta_c - \theta}{\theta_c} \in (0, 1).$$

As we will see later, W represents the part of the free energy depending on the magnetization. Therefore, the minimum M of the free energy is such that $M \to 0$ continuously as $\theta \to \theta_c$, which is the typical situation occurring in second-order phase transitions.

It is worth noting that the dependence of F, G on $|\mathbf{M}|$ implies that these functions characterize the transition in isotropic ferromagnets, when the magnetization has no preferred direction at equilibrium.

Substitution of (7)-(9) into (6) leads to the Ginzburg-Landau equation

$$\gamma\dot{\mathbf{M}} = \nu\Delta\mathbf{M} - \theta_c(|\mathbf{M}|^2 - 1)\mathbf{M} - \theta\mathbf{M} + \mathbf{H}. \tag{10}$$

The model based on the decomposition (5), requires the introduction of two distinct evolution equations for the modulus φ and the direction $\mathbf{m}$ of the magnetization. The variable φ, identified with the order parameter, can be interpreted as a phase field, and it satisfies the scalar Ginzburg-Landau equation

$$\gamma\dot{\varphi} = \nu\Delta\varphi - \theta_c F'(\varphi) - \theta G'(\varphi) + \mathbf{H} \cdot \mathbf{m}. \tag{11}$$

Such equation can be deduced by a balance law on the internal structure, written in the local form as

$$k = \nabla \cdot \mathbf{p},$$

where, like in the previous model

$$k = \gamma\dot{\varphi} + \theta_c F'(\varphi) + \theta G'(\varphi) - \mathbf{H} \cdot \mathbf{m} \qquad \mathbf{p} = \nu\nabla\varphi.$$

Equation (11) needs to be coupled with an evolution equation for the versor $\mathbf{m}$. For example, in the classical Landau-Lifshitz model,[6] $\mathbf{m}$ satisfies

$$\dot{\mathbf{m}} = -\tau\mathbf{m} \times \mathbf{H} - \lambda\mathbf{m} \times (\mathbf{m} \times \mathbf{H}), \qquad \tau, \lambda > 0, \tag{12}$$

where the term $-\tau\mathbf{m} \times \mathbf{H}$ allows to explain the precession of $\mathbf{m}$ around $\mathbf{H}$, and $-\lambda\mathbf{m} \times (\mathbf{m} \times \mathbf{H})$ accounts for dissipation.

Henceforth, we consider the model based on equations (4), (10). Similar arguments can be used to deal with the case described by (4), (11), (12).

In order to provide a coherent model able to include both thermal and electromagnetic effects, it is essential to deduce the appropriate representation of the heat equation. As known, the thermal balance law is expressed by the equation

$$h = -\nabla \cdot \mathbf{q} + r,$$

where h is the rate at which heat is absorbed, $\mathbf{q}$ is the heat flux vector and r is the heat source. Moreover, denoting by e the internal energy, from the first law of thermodynamics, we deduce

$$\dot{e} = \mathcal{P}^i + h,$$

where $\mathcal{P}^i$ is the internal power. Hence

$$\dot{e} = \mathcal{P}^i - \nabla \cdot \mathbf{q} + r. \tag{13}$$

The latter provides the heat equation governing the evolution of the temperature, once we assume constitutive equations for e, $\mathbf{q}$ and give the representation of the internal power $\mathcal{P}^i$.

By following the approach proposed in Ref. 4, equation (6) is regarded as a field equation able to yield a power (and then an energy variation) $\mathcal{P}^i_{\mathbf{M}}$, related the internal structure order of the material. Therefore, by keeping the electromagnetic effects into account, the internal power is given by the sum

$$\mathcal{P}^i = \mathcal{P}^i_{\mathbf{M}} + \mathcal{P}^i_{\text{el}},$$

where $\mathcal{P}^i_{\rm el}$ is defined as

$$\mathcal{P}^i_{\rm el} = \dot{\mathbf{D}}\cdot\mathbf{E} + \dot{\mathbf{B}}\cdot\mathbf{H} + \mathbf{J}\cdot\mathbf{E} = \varepsilon\dot{\mathbf{E}}\cdot\mathbf{E} + \mu\dot{\mathbf{H}}\cdot\mathbf{H} + \dot{\mathbf{M}}\cdot\mathbf{H} + \sigma|\mathbf{E}|^2.$$

The representation of the internal power $\mathcal{P}^i_{\mathbf{M}}$, is obtained by multiplying (6) by $\dot{\mathbf{M}}$, i.e.

$$\mathbf{k}\cdot\dot{\mathbf{M}} + \nu\nabla\mathbf{M}\cdot\nabla\dot{\mathbf{M}} = \nu\nabla\cdot(\nabla\mathbf{M}^T\dot{\mathbf{M}}),$$

where the superscript T denotes the transpose of a tensor. The previous identity leads to the identification of the internal and external power as

$$\mathcal{P}^i_{\mathbf{M}} = \mathbf{k}\cdot\dot{\mathbf{M}} + \nu\nabla\mathbf{M}\cdot\nabla\dot{\mathbf{M}}, \qquad \mathcal{P}^e_{\mathbf{M}} = \nu\nabla\cdot(\nabla\mathbf{M}^T\dot{\mathbf{M}}).$$

Therefore, (7) and (9) imply

$$\mathcal{P}^i_{\mathbf{M}} = \gamma|\dot{\mathbf{M}}|^2 + \nu\nabla\mathbf{M}\cdot\nabla\dot{\mathbf{M}} + [\theta_c(|\mathbf{M}|^2-1)+\theta]\mathbf{M}\cdot\dot{\mathbf{M}} - \mathbf{H}\cdot\dot{\mathbf{M}}.$$

Now, we choose the following constitutive equations for the internal energy and the heat flux

$$e = c(\theta) + \frac{1}{2}\varepsilon|\mathbf{E}|^2 + \frac{1}{2}\mu|\mathbf{H}|^2 + \frac{\theta_c}{4}(|\mathbf{M}|^2-1)^2 + \frac{\nu}{2}|\nabla\mathbf{M}|^2 \tag{14}$$

$$\mathbf{q} = -k(\theta)\nabla\theta, \qquad k(\theta) > 0. \tag{15}$$

Substitution into (13) leads to the evolution equation for the temperature

$$c'(\theta)\dot{\theta} - \sigma|\mathbf{E}|^2 - \gamma|\dot{\mathbf{M}}|^2 - \theta\mathbf{M}\cdot\dot{\mathbf{M}} = \nabla\cdot(k(\theta)\nabla\theta) + r. \tag{16}$$

Let us prove the consistence of this model with the second law of thermodynamics, written by means of the Clausius-Duhem inequality as

$$\dot{\eta} \geq -\nabla\cdot\left(\frac{\mathbf{q}}{\theta}\right) + \frac{r}{\theta},$$

where η is the entropy density. By introducing the free energy $\psi = e - \theta\eta$ and using (13), from the previous inequality we obtain

$$\dot{\psi} + \eta\dot{\theta} + \frac{\mathbf{q}}{\theta}\cdot\nabla\theta - \mathcal{P}^i \leq 0.$$

Substituting the expression of the internal power we deduce

$$\begin{aligned}&\dot{\psi} + \eta\dot{\theta} + \frac{\mathbf{q}}{\theta}\cdot\nabla\theta - \gamma|\dot{\mathbf{M}}|^2 - \nu\nabla\mathbf{M}\cdot\nabla\dot{\mathbf{M}}\\ &-[\theta_c(|\mathbf{M}|^2-1)+\theta]\mathbf{M}\cdot\dot{\mathbf{M}} - \varepsilon\dot{\mathbf{E}}\cdot\mathbf{E} - \mu\dot{\mathbf{H}}\cdot\mathbf{H} - \sigma|\mathbf{E}|^2 \leq 0.\end{aligned} \tag{17}$$

This inequality leads to the following representations

$$\psi = \frac{\mu}{2}|\mathbf{H}|^2 + \frac{\varepsilon}{2}|\mathbf{E}|^2 + \frac{\nu}{2}|\nabla\mathbf{M}|^2 + \frac{\theta_c}{4}(|\mathbf{M}|^2-1)^2 + \frac{\theta}{2}|\mathbf{M}|^2 + \alpha(\theta) \tag{18}$$

$$\eta = -\frac{\partial\psi}{\partial\theta} = -\frac{1}{2}|\mathbf{M}|^2 - \alpha'(\theta), \tag{19}$$

where α satisfies the equation

$$\alpha(\theta) - \theta\alpha'(\theta) = c(\theta).$$

Substitution of (15), (18), (19) into (17) provides the reduced inequality

$$-\frac{k(\theta)}{\theta}|\nabla\theta|^2 - \sigma|\mathbf{E}|^2 - \gamma|\dot{\mathbf{M}}|^2 \leq 0,$$

which holds along any process and proves the thermodynamical consistence of the model.

In order to specialize the differential equation describing the evolution of the temperature, we suppose that heat conductivity and specific heat depend on θ according the polynomial laws

$$k(\theta) = k_0 + k_1\theta, \qquad c(\theta) = c_1\theta + \frac{c_2}{2}\theta^2, \qquad\qquad k_0, k_1, c_1, c_2 > 0.$$

In addition, we restrict our attention to processes for which the fields $\mathbf{E}$, $\dot{\mathbf{M}}$, $\nabla\theta$ are small enough so that the quadratic terms $|\mathbf{E}|^2$, $|\dot{\mathbf{M}}|^2$, $|\nabla\theta|^2$ are negligible with respect to other contributions in (16). After dividing by θ, the energy balance (16) reduces to

$$c_1\partial_t(\ln\theta) + c_2\dot{\theta} - \mathbf{M}\cdot\dot{\mathbf{M}} = k_0\Delta(\ln\theta) + k_1\Delta\theta + \hat{r}$$

and for the sake of simplicity, we assume that $\hat{r} = r/\theta$ is a known function of x, t.

3. Well posedness of the problem

In the previous section we have deduced the differential equations ruling the evolution of the ferromagnetic material, namely

$$\gamma\dot{\mathbf{M}} = \nu\Delta\mathbf{M} - \theta_c(|\mathbf{M}|^2 - 1)\mathbf{M} - \theta\mathbf{M} + \mathbf{H} \tag{20}$$

$$c_1\partial_t(\ln\theta) + c_2\dot{\theta} - \mathbf{M}\cdot\dot{\mathbf{M}} = k_0\Delta(\ln\theta) + k_1\Delta\theta + \hat{r} \tag{21}$$

$$\mu\dot{\mathbf{H}} + \dot{\mathbf{M}} = -\frac{1}{\sigma}\nabla\times\nabla\times\mathbf{H} \tag{22}$$

$$\nabla\cdot(\mu\mathbf{H} + \mathbf{M}) = 0. \tag{23}$$

This system is completed by the initial and boundary conditions

$$\mathbf{M}(x,0) = \mathbf{M}_0(x), \quad \theta(x,0) = \theta_0(x), \quad \mathbf{H}(x,0) = \mathbf{H}_0(x), \quad x \in \Omega \tag{24}$$

$$\nabla\mathbf{M}\,\mathbf{n} = \mathbf{0}, \quad \nabla\theta\cdot\mathbf{n} = 0, \quad (\nabla\times\mathbf{H})\times\mathbf{n} = 0, \quad \text{on } \partial\Omega. \tag{25}$$

Concerning the boundary conditions, we have assumed that the magnetization satisfies the Neumann boundary condition typical of phase transitions.

The same homogeneous Neumann condition is required for the temperature. Finally, we observe that, in the quasi-steady approximation, Maxwell's equation $(1)_2$ reduces to

$$\sigma\mathbf{E} = \nabla \times \mathbf{H},$$

so that $(25)_3$ implies the continuity of the tangential component of the electric field across $\partial\Omega$.

In view of $(1)_1$ and (23), if we impose the following conditions on the initial data

$$(\mu\mathbf{H}_0 + \mathbf{M}_0) \cdot \mathbf{n}|_{\partial\Omega} = 0, \qquad \nabla \cdot (\mu\mathbf{H}_0 + \mathbf{M}_0) = 0, \tag{26}$$

then at any subsequent $t > 0$ (see Ref. 5)

$$(\mu\mathbf{H} + \mathbf{M}) \cdot \mathbf{n}|_{\partial\Omega} = 0, \qquad \text{and} \qquad \nabla \cdot (\mu\mathbf{H} + \mathbf{M}) = 0 \ \text{ a.e. in } \Omega.$$

hold.

Definition 3.1. A triplet $(\mathbf{M}, \theta, \mathbf{H})$ such that

$$\begin{aligned}
&\mathbf{M} \in L^2(0, T, H^2(\Omega)) \cap H^1(0, T, L^2(\Omega)) \\
&\theta \in L^2(0, T, H^1(\Omega)), \quad \theta > 0, \quad \ln\theta \in L^2(0, T, H^1(\Omega)) \\
&c_1 \ln\theta + c_2\theta \in H^1(0, T, H^1(\Omega)') \\
&\mathbf{H} \in L^2(0, T, H^1(\Omega)) \cap H^1(0, T, H^1(\Omega)')
\end{aligned}$$

satisfying (24)-(25) almost everywhere, is a weak solution of (20)-(25) if

$$\begin{aligned}
&\gamma\dot{\mathbf{M}} + \nu\Delta\mathbf{M} + [\theta_c(|\mathbf{M}|^2 - 1) + \theta]\mathbf{M} - \mathbf{H} = 0, \qquad \text{a.e. in } \Omega \\
&\int_\Omega \Big[c_1\partial_t(\ln\theta)\omega + c_2\dot{\theta}\omega - \mathbf{M}\cdot\dot{\mathbf{M}}\omega + k_0\nabla(\ln\theta)\cdot\nabla\omega \\
&\qquad + k_1\nabla\theta\cdot\nabla\omega - \hat{r}\omega\Big]\,dx = 0 \\
&\int_\Omega \left(\mu\dot{\mathbf{H}}\cdot\mathbf{w} + \dot{\mathbf{M}}\cdot\mathbf{w} + \frac{1}{\sigma}\nabla\times\mathbf{H}\cdot\nabla\times\mathbf{w}\right) dx = 0 \\
&\mu\nabla\cdot\mathbf{H} + \nabla\cdot\mathbf{M} = 0, \qquad \text{a.e. in } \Omega
\end{aligned}$$

for any $\omega, \mathbf{w} \in H^1(\Omega)$ and a.e. $t \in (0, T)$.

Existence of weak solutions is proved by introducing a suitable approximation of the logarithmic nonlinearities. More precisely, we denote by $\ln_\varepsilon$ the Yosida approximation of the logarithm function defined as

$$\ln_\varepsilon \tau = \frac{\tau - \rho_\varepsilon(\tau)}{\varepsilon}, \qquad \tau \in \mathbb{R},$$

where $\rho_\varepsilon(\tau)$ is the unique solution of equation

$$\rho_\varepsilon(\tau) + \varepsilon \ln \rho_\varepsilon(\tau) = \tau.$$

Then we consider the approximated problem obtained by replacing into (21) the logarithm function with its regularization $\ln_\varepsilon$.

Definition 3.2. A triplet $(\mathbf{M}_\varepsilon, \theta_\varepsilon, \mathbf{H}_\varepsilon)$ such that

$$\begin{aligned}
&\mathbf{M}_\varepsilon \in L^2(0,T,H^2(\Omega)) \cap H^1(0,T,L^2(\Omega))\\
&\theta_\varepsilon \in L^2(0,T,H^1(\Omega)), \quad \ln_\varepsilon \theta_\varepsilon \in L^2(0,T,H^1(\Omega))\\
&c_1 \ln_\varepsilon \theta_\varepsilon + c_2\theta_\varepsilon \in H^1(0,T,H^1(\Omega)')\\
&\mathbf{H}_\varepsilon \in L^2(0,T,H^1(\Omega)) \cap H^1(0,T,H^1(\Omega)')
\end{aligned}$$

and satisfying (24)-(25) almost everywhere, is a weak solution of the approximated problem if

$$\dot{\mathbf{M}}_\varepsilon + \nu\Delta\mathbf{M}_\varepsilon + [\theta_c(|\mathbf{M}_\varepsilon|^2 - 1) + \theta]\mathbf{M}_\varepsilon - \mathbf{H}_\varepsilon = 0, \qquad \text{a.e. in } \Omega \quad (27)$$

$$\int_\Omega \Big[c_1\partial_t(\ln_\varepsilon \theta_\varepsilon)\omega + c_2\dot{\theta}_\varepsilon\omega - \mathbf{M}_\varepsilon \cdot \dot{\mathbf{M}}_\varepsilon\omega + k_0\nabla(\ln_\varepsilon \theta_\varepsilon)\cdot\nabla\omega + k_1\nabla\theta_\varepsilon\cdot\nabla\omega - \hat{r}\omega\Big]dx = 0 \quad (28)$$

$$\int_\Omega \left(\mu\dot{\mathbf{H}}_\varepsilon \cdot \mathbf{w} + \dot{\mathbf{M}}_\varepsilon \cdot \mathbf{w} + \frac{1}{\sigma}\nabla\times\mathbf{H}_\varepsilon\cdot\nabla\times\mathbf{w}\right)dx = 0 \quad (29)$$

$$\mu\nabla\cdot\mathbf{H}_\varepsilon + \nabla\cdot\mathbf{M}_\varepsilon = 0 \qquad \text{a.e. in } \Omega \quad (30)$$

for any $\omega, \mathbf{w} \in H^1(\Omega)$ and a.e. $t \in (0,T)$.

Lemma 3.1. *Let $\hat{r} \in L^2(0,T,H^1(\Omega)')$ and $\mathbf{M}_0 \in H^1(\Omega)$, $\theta_0, \mathbf{H}_0 \in L^2(\Omega)$, such that (26) hold. For every $\varepsilon > 0$ and $T > 0$, there exists a solution $(\mathbf{M}_\varepsilon, \theta_\varepsilon, \mathbf{H}_\varepsilon)$ in the sense of Definition 3.2.*

Proof. The proof is based on the Galerkin procedure. The advantage of introducing a family of approximating problems stands in the properties of the Yosida regularization.

The details of the proof can be found in Ref. 1. □

Existence of solutions to (20)-(25) is proved, by showing that the solutions $(\mathbf{M}_\varepsilon, \theta_\varepsilon, \mathbf{H}_\varepsilon)$ of the approximated problem converge to a solution $(\mathbf{M}, \theta, \mathbf{H})$ of the original problem as $\varepsilon \to 0$.

In the sequel we denote by $\|\cdot\|$ the usual $L^2(\Omega)-$norm.

Theorem 3.1 (Existence). *Let* $\hat{r} \in L^2(0, T, H^1(\Omega)')$ *and* $\mathbf{M}_0 \in H^1(\Omega)$, $\theta_0, \mathbf{H}_0 \in L^2(\Omega)$, *such that (26) hold. For every* $T > 0$, *problem (20)-(25) admits a solution* $(\mathbf{M}, \theta, \mathbf{H})$ *in the sense of Definition 3.1.*

Proof. We test (27) with $\dot{\mathbf{M}}_\varepsilon$, (28) with θ_ε, (29) with $\mathbf{H}_\varepsilon$. Adding the resulting equation and integrating over $\Omega \times (0, t)$, we obtain

$$\frac{1}{2}\left[\nu\|\nabla \mathbf{M}_\varepsilon\|^2 + c_1 \int_\Omega I_\varepsilon(\theta_\varepsilon)dx + c_2\|\theta_\varepsilon\|^2 + \mu\|\mathbf{H}_\varepsilon\|^2 + \frac{\theta_c}{2}\|\mathbf{M}_\varepsilon\|_{L^4}^4\right]$$
$$+ \int_0^t \left[\gamma\|\dot{\mathbf{M}}_\varepsilon\|^2 + k_1\|\nabla\theta_\varepsilon\|^2 + \frac{1}{\sigma}\|\nabla \times \mathbf{H}_\varepsilon\|^2 + k_0 \int_\Omega \ln_\varepsilon'(\theta_\varepsilon)|\nabla\theta_\varepsilon|^2 dx\right] d\tau$$
$$\leq C_0 + \frac{c_1}{2}\int_\Omega |I_\varepsilon(\theta_0)|dx, \tag{31}$$

where $C_0 > 0$ depends on the norms of the initial data and

$$I_\varepsilon(\tau) = \int_0^\tau s \ln_\varepsilon'(s)ds = \tau \ln_\varepsilon(\tau) - \int_0^\tau \ln_\varepsilon(s)ds, \qquad \tau \in \mathbb{R}.$$

In order to prove an estimate independent of ε, we exploit the inequality

$$|\ln_\varepsilon \tau| \leq |\ln \tau|, \qquad \tau > 0$$

which leads to

$$|I_\varepsilon(\theta_0)| \leq \theta_0 \ln_\varepsilon \theta_0 + \int_0^{\theta_0} |\ln_\varepsilon s|ds \leq 2\theta_0^2.$$

Thus, from (31) we deduce

$$\|\mathbf{M}_\varepsilon\|_{H^1}^2 + \|\theta_\varepsilon\|^2 + \|\mathbf{H}_\varepsilon\|^2 + \int_0^t \left[\|\dot{\mathbf{M}}_\varepsilon\|^2 + \|\nabla\theta_\varepsilon\|^2 + \|\nabla \times \mathbf{H}_\varepsilon\|^2\right] d\tau \leq C_0. \tag{32}$$

Now let us test (28) with $c_1 \ln_\varepsilon(\theta_\varepsilon) + c_2\theta_\varepsilon$ thus obtaining

$$\frac{1}{2}\frac{d}{dt}\|c_1 \ln_\varepsilon \theta_\varepsilon + c_2\theta_\varepsilon\|^2 + \frac{k_0 c_1}{2}\|\nabla(\ln_\varepsilon \theta_\varepsilon)\|^2 + k_1 c_2\|\nabla\theta_\varepsilon\|^2$$
$$\leq \lambda\left(\|c_1 \ln_\varepsilon \theta_\varepsilon + c_2\theta_\varepsilon\|^2 + \|\mathbf{M}_\varepsilon\|_{H^1}^2\|\dot{\mathbf{M}}_\varepsilon\|^2 + \|\hat{r}\|_{H^1(\Omega)'}^2 + \|\nabla\theta_\varepsilon\|^2\right),$$

where λ is a positive constant. By recalling (32), Gronwall's inequality implies

$$\|c_1 \ln_\varepsilon \theta_\varepsilon + c_2\theta_\varepsilon\|^2 + \int_0^t \|\nabla(\ln_\varepsilon \theta_\varepsilon)\|^2 d\tau \leq C_0$$

and comparison with (28) yields

$$\int_0^t \|c_1\partial_t \ln_\varepsilon \theta_\varepsilon + c_2\dot{\theta}_\varepsilon\|_{H^1(\Omega)'}^2 dt \leq C_0.$$

The previous a priori estimates allow to pass to the limit as $\varepsilon \to 0$ into (27)-(30) and to obtain a solution $(\mathbf{M}, \theta, \mathbf{H})$ of (20)-(25), satisfying the inequality

$$\|\mathbf{M}\|_{H^1}^2 + \|\theta\|^2 + \|\mathbf{H}\|^2 + \int_0^t \left(\|\dot{\mathbf{M}}\|^2 + \|\nabla\theta\|^2 + \|\nabla \times \mathbf{H}\|^2\right) d\tau \le C_0. \quad \square$$

Theorem 3.2 (Uniqueness). *Let* $(\mathbf{M}_0, \theta_0, \mathbf{H}_0) \in H^1(\Omega) \times L^2(\Omega) \times L^2(\Omega)$ *satisfying (26) and* $\hat{r} \in L^2(0, T, H^1(\Omega)')$. *Then problem (20)-(24) admits a unique solution* $(\mathbf{M}, \theta, \mathbf{H})$.

Proof. Let $(\mathbf{M}_1, \theta_1, \mathbf{H}_1)$, $(\mathbf{M}_2, \theta_2, \mathbf{H}_2)$ be two solutions, with the same initial data and sources. We introduce the differences

$$\mathbf{M} = \mathbf{M}_1 - \mathbf{M}_2, \qquad \theta = \theta_1 - \theta_2, \qquad \xi = \ln\theta_1 - \ln\theta_2, \qquad \mathbf{H} = \mathbf{H}_1 - \mathbf{H}_2,$$

which solve the differential problem

$$\begin{aligned} \gamma\dot{\mathbf{M}} &= \nu\Delta\mathbf{M} - \theta_c(|\mathbf{M}_1|^2 - 1)\mathbf{M}_1 + \theta_c(|\mathbf{M}_2|^2 - 1)\mathbf{M}_2 \\ &\quad -\theta_1\mathbf{M}_1 + \theta_2\mathbf{M}_2 + \mathbf{H} && (33) \\ c_1\dot{\xi} + c_2\dot{\theta} &= \mathbf{M}_1 \cdot \dot{\mathbf{M}}_1 - \mathbf{M}_2 \cdot \dot{\mathbf{M}}_2 + k_0\Delta\xi + k_1\Delta\theta && (34) \\ \mu\dot{\mathbf{H}} + \dot{\mathbf{M}} &= -\frac{1}{\sigma}\nabla \times \nabla \times \mathbf{H}. && (35) \end{aligned}$$

By integrating (34) and (35) over $(0, t)$, we obtain

$$c_1\xi + c_2\theta = \frac{1}{2}(|\mathbf{M}_1|^2 - |\mathbf{M}_2|^2) + \int_0^t (k_0\Delta\xi + k_1\Delta\theta)\, d\tau = 0 \qquad (36)$$

$$\mu\mathbf{H} + \mathbf{M} + \frac{1}{\sigma}\int_0^t \nabla \times \nabla \times \mathbf{H} d\tau = 0. \qquad (37)$$

Let us multiply (33) by $\mathbf{M}$, (36) by $k_0\xi + k_1\theta$ and (37) by $\mathbf{H}$. Adding the resulting equations and integrating over Ω, we deduce

$$\begin{aligned} &\frac{1}{2}\frac{d}{dt}\left[\gamma\|\mathbf{M}\|^2 + \frac{1}{\sigma}\left\|\int_0^t \nabla \times \mathbf{H} d\tau\right\|^2 + \left\|\int_0^t (k_0\nabla\xi + k_1\nabla\theta)\, d\tau\right\|^2\right] \\ &+\nu\|\nabla\mathbf{M}\|^2 + \mu\|\mathbf{H}\|^2 + \int_\Omega (c_1\xi + c_2\theta)(k_0\xi + k_1\theta) dx \le J_1 + J_2, \qquad (38) \end{aligned}$$

where

$$J_1 = \int_\Omega \left[-\theta_c(|\mathbf{M}_1|^2 - 1)\mathbf{M}_1 + \theta_c(|\mathbf{M}_2|^2 - 1)\mathbf{M}_2 - \theta_1\mathbf{M}_1 + \theta_2\mathbf{M}_2\right] \cdot \mathbf{M} dx$$

$$J_2 = \frac{1}{2}\int_\Omega \mathbf{M} \cdot (\mathbf{M}_1 + \mathbf{M}_2)(k_0\xi + k_1\theta) dx.$$

By means of Young's, Hölder's inequalities and the embedding inequalities

$$\|\mathbf{w}\|_{L^p} \leq C\|\mathbf{w}\|_{H^1}, \qquad p = 1, ..., 6, \ \mathbf{w} \in H^1(\Omega),$$
$$\|\mathbf{w}\|_{L^\infty} \leq C\|\mathbf{w}\|_{H^2}, \qquad \mathbf{w} \in H^2(\Omega),$$

the integrals J_1, J_2 can be estimated as

$$J_1 + J_2 \leq \frac{\nu}{2}\|\nabla\mathbf{M}\|^2 + \frac{\mu}{2}\|\mathbf{H}\|^2 + \frac{1}{2}(k_0 c_1\|\xi\|^2 + k_1 c_2\|\theta\|^2) + \zeta(t)\|\mathbf{M}\|^2,$$

where

$$\zeta(t) = C(1 + \|\theta_1(t)\|^2_{H^1} + \|\mathbf{M}_1(t)\|^2_{H^2} + \|\mathbf{M}_2(t)\|^2_{H^2}).$$

Substitution into (38) and Gronwall's inequality prove $\mathbf{M} = \mathbf{0}$, $\theta = 0$, $\mathbf{H} = \mathbf{0}$. □

References

1. V. Berti, M. Fabrizio and C. Giorgi, *J. Math. Anal. Appl.* **335**, 661 (2009).
2. M. Brokate, J. Sprekels *Hysteresis and phase transitions* (Springer, New York, 1996).
3. M. Fabrizio, *Internat. J. Engrg. Sci.* **44**, 529 (2006).
4. M. Fabrizio, C. Giorgi, A. Morro, *Math. Methods Appl. Sci.* **31**, 627 (2008).
5. M. Fabrizio, A. Morro, *Electromagnetism of Continuous Media* (Oxford University Press, 2003).
6. L.D. Landau, E.M. Lifshitz, *Phys. Z. Sowietunion* **8**, 153 (1935).
7. L.D. Landau, E.M. Lifshitz, L.P. Pitaevskii, *Electrodynamics of continuous media* (Pergamon Press, Oxford, 1984).

A GSS METHOD FOR OBLIQUE ℓ_1 PROCRUSTES PROBLEMS

C. BOGANI

On leave from School of Mathematics, University of Birmingham, Birmingham, B15 2TT, UK

M. G. GASPARO*, A. PAPINI**

Dipartimento di Energetica "S. Stecco" Università di Firenze, Firenze, 50134, Italia
**E-mail: mariagrazia.gasparo@unifi.it*
***E-mail: alessandra.papini@unifi.it*

We propose a Generating Search Set method for solving the oblique ℓ_1 Procrustes problem. Implementative details, algorithmic choices and theoretical properties of the method are discussed. The results of some numerical experiments are reported.

Keywords: Piecewise smooth optimization, nonlinear constraints, direct search methods.

1. Introduction

Given $C \in \mathbb{R}^{m\times n}$ and $B \in \mathbb{R}^{m\times p}$, the oblique ℓ_1 Procrustes problem is

$$\min_{Q\in\mathcal{OB}(n,p)} \|CQ - B\|_{\ell_1}, \tag{1}$$

where $\mathcal{OB}(n,p) = \{Q \in \mathbb{R}^{n\times p} : \operatorname{diag}(Q^TQ) = (1,1,\ldots,1)^T\}$, and $\|A\|_{\ell_1} = \sum_{j=1}^{r} \|a_j\|_1$, for $A = [a_1 \ldots a_r]$. Problem (1) is a variant of the orthogonal Procrustes problem (where the constraint is $Q^TQ = I$), which is considered whenever more degrees of freedom are needed, as, e.g., in the Procrustes Analysis, a well known technique used in many applications (factor analysis [8], statistical shape analysis [5], etc.). Classically, Procrustes problems are defined by means of the Frobenius norm. The use of the ℓ_1 norm was recently advocated because of its robustness with respect to outliers ([10], [11], [4]). An appealing feature of (1) is that both the objective function and the constraint have a separable structure with respect to the columns of Q, so that the problem splits into p problems in $\mathbb{R}^n$ of the form

$$\min_{x \in \mathcal{S}} \|Cx - b\|_1, \tag{2}$$

with $b \in \mathbb{R}^m$ and $\mathcal{S} = \{x \in \mathbb{R}^n : x^T x = 1\}$. The objective function in (2) is not everywhere continuously differentiable, and this lack of smoothness has to be properly treated. Some ways to solve (2) are known, basically founded on different smooth reformulations. For example, we can introduce two slack vectors $u, v \in \mathbb{R}^m$, and obtain the following reformulation

$$\begin{array}{rl} & \min_{u,v,x} \sum_{j=1}^{m} (u_j + v_j) \\ \text{s.t.} & Cx - u + v = b, \\ & u, v \geq 0\,,\;\; x^T x = 1, \end{array} \tag{3}$$

which can be solved by standard nonlinear programming methods. Alternatively, in [11], the objective function is approximated by $\phi_\alpha(x) = (Cx - b)^T \tanh(\alpha(Cx - b))$ for some large α, and then the projected gradient flow method is used to minimize $\phi_\alpha(x)$ over $\mathcal{S}$, by solving the IVP

$$\begin{array}{l} \dot{x}(t) = P_{\mathcal{S}}(-\nabla \phi_\alpha(x(t))), \quad t \in [0, T], \\ x(0) \in \mathcal{S} \;\text{ given}, \end{array} \tag{4}$$

where $P_{\mathcal{S}}$ is the orthogonal projection on $\mathcal{S}$, and T is taken large enough to reach the equilibrium. This approach avoids increasing the dimension of the problem, but the choice of the smoothing parameter α is not trivial.

Here we propose a different approach: (2) is maintained in its original form, and solved by a Generating Set Search (`GSS`) method. `GSS` methods are a class of direct search methods [6], that we briefly describe for a generic objective function f. Given the iterate $x^{(k)}$, a step-length Δ_k, and a finite set of search directions $\Gamma_k = \{d_1^{(k)}, \ldots, d_p^{(k)}\}$, the core of a `GSS` iteration consists in sampling f at the trial points $y_j = x^{(k)} + \Delta_k d_j^{(k)}$, $j = 1, \ldots, p$. If a suitable decrease condition holds for some y_r, the iteration is *successful*, the new iterate is $x^{(k+1)} = y_r$, and a new step-length $\Delta_{k+1} \geq \Delta_k$ is chosen; otherwise, the iteration is *unsuccessful*, the new iterate is $x^{(k+1)} = x^{(k)}$, and a step-length $\Delta_{k+1} < \Delta_k$ is selected. A commonly used condition to declare success is the *sufficient decrease* condition $f(y_r) \leq f(x^{(k)}) - \rho(\Delta_k)$, where $\rho : \mathbb{R}^+ \to \mathbb{R}^+$ is a continuous, increasing *forcing function* such that $\rho(t) = o(t)$ for $t \to 0$. In the constrained case, all iterates must belong to the feasible set Ω; hence, infeasible trial points must be properly treated. In the `GSS` method we propose for solving (2), the feasibility requirement is satisfied by projecting onto $\mathcal{S}$ the points $y = x^{(k)} + \Delta_k d$, with $d \in \Gamma_k$, so that the trial points actually used are of form $z = P_{\mathcal{S}}(y) = \frac{y}{\|y\|_2}$ [a].

The definition of the search direction set Γ_k is crucial for the convergence properties. In particular, for smooth unconstrained problems, the existence of stationary limit points for the sequence $\{x^{(k)}\}$ is proved, under mild assumptions, if at each iteration Γ_k *positively spans* $\mathbb{R}^n$, that is if any $d \in \mathbb{R}^n$ is a nonnegative linear combination of the elements of Γ_k [6]. In fact this property ensures that the sampling around $x^{(k)}$ is rich enough to infer the local behaviour of f, and eventually decide whether a limit point $\hat{x}$ is stationary. However, this does not work for constrained and/or nonsmooth problems. In these cases, one way to ensure the existence of stationary limit points consists in exploiting some extra structure, if any. For example, in the linearly constrained case, where Ω is a polyhedron, the behaviour of f near $x^{(k)}$ can be captured, if Γ_k takes into account the geometry of Ω around $x^{(k)}$. When $x^{(k)}$ is safely far from the boundary $\partial\Omega$, Γ_k can be defined as in the unconstrained case. Otherwise, Γ_k must *generate* the cone of feasible directions at x, for all x close to $x^{(k)}$; that is any direction in these cones must be a nonnegative linear combination of the elements of Γ_k [6]. An approach of the same flavour can be followed to minimize $f(x) = \|Cx - b\|_1$, which is locally Lipschitz continuous over $\mathbb{R}^n$, and has a highly structured set of nondifferentiability $\mathcal{H} = \cup_{j=1}^m \{x \in \mathbb{R}^n : c_j^T x = b_j\}$, where $c_1^T, \ldots, c_m^T$ are the rows of C [2]. The idea is that when $x^{(k)}$ is far from $\mathcal{H}$, we can choose Γ_k as in the smooth case. Otherwise, there is a set of hyperplanes $c_j^T x = b_j$, $j \in \{1, \ldots, m\}$ near $x^{(k)}$, and Γ_k must generate any polyhedral cone associated to any subset of such hyperplanes.

The main contribution of the paper is a natural, though nontrivial, extension of previous ideas to handle the nonlinear equality constraint $x \in \mathcal{S}$, which yields a new and promising GSS algorithm for solving the ℓ_1 Procrustes problem. Different approaches to treat nonlinear constraints with GSS methods have been proposed in the literature. We refer to [6] for a discussion on this issue and its related difficulties, in the smooth case. The paper is organized as follows. In Section 2, we discuss how to define and construct a proper set of search directions Γ_k, and present the *Projected GSS* algorithm to solve (2). In Section 3, we study the convergence properties of the method. In Section 4, we report on some numerical experiments.

[a] In the literature, some different ways to treat infeasible points have been proposed ([1], [6], [7]). Such points are either suitably projected onto Ω, or, if the *barrier approach* is adopted (i.e., the value $+\infty$ is assigned to f outside Ω), simply discarded.

2. The method

Let us introduce some notations. The tangent space at $x \in \mathcal{S}$ is

$$T(x) = \{d \in \mathbb{R}^n : x^T d = 0\}. \tag{5}$$

For a given $\eta \geq 0$, $J_\eta(x) = \{j : |c_j^T x - b_j| \leq \eta \|c_j\|_2\}$ is the set of the indices of the nondifferentiability hyperplanes close to x within η (η-active indices at x). The cardinality of $J_\eta(x)$ is denoted by $\ell_\eta(x)$. In particular, $J_0(x)$, for short $J(x)$, is the set of active indices at x, and $\ell(x) = \ell_0(x)$ its cardinality. If $J_\eta(x) \neq \emptyset$, say $J_\eta(x) = \{j_1, \ldots, j_{\ell_\eta(x)}\}$, the tangent space $T(x)$ is partitioned into a family of polyhedral cones $\mathcal{F}_\eta(x) = \left\{\mathcal{T}_{s,\eta}(x),\ s = (s_1, \ldots, s_{\ell_\eta(x)}),\ s_i = \pm 1,\ i = 1, \ldots, \ell_\eta(x)\right\}$, where $\mathcal{T}_{s,\eta}(x) = \mathcal{K}_{s,\eta}(x) \cap T(x)$, and

$$\mathcal{K}_{s,\eta}(x) = \{d \in \mathbb{R}^n : s_i c_{j_i}^T d \leq 0,\ i = 1, \ldots, \ell_\eta(x)\} \tag{6}$$

is the polyhedral cone associated with the normals of the η-active hyperplanes, and with the vector of signs s. The cones $\mathcal{T}_{s,\eta}(x)$ are said *nondegenerate* if $M_\eta(x) = [c_{j_1} \ldots c_{j_{\ell_\eta(x)}}\ x]$ has full column rank, and *degenerate* otherwise. The notations $\mathcal{F}_0(x)$, $\mathcal{T}_{s,0}(x)$, $\mathcal{K}_{s,0}(x)$, and $M_0(x)$ will be shortened to $\mathcal{F}(x)$, $\mathcal{T}_s(x)$, $\mathcal{K}_s(x)$, and $M(x)$, respectively.

Let us turn to the definition of Γ_k. To treat the constraint, we choose $\Gamma_k \subset T(x^{(k)})$. Let us first suppose $x^{(k)}$ far from $\mathcal{H}$, so that $J_\eta(x^{(k)}) = \emptyset$. Then, Γ_k must contain a positive basis $\mathcal{B}^+(T(x^{(k)}))$ for $T(x^{(k)})$ [b]. We recall that a positive basis of a subspace X can be defined, for example, by

$$\mathcal{B}^+(X) = \mathcal{B}(X) \cup -\mathcal{B}(X), \tag{7}$$

where $\mathcal{B}(X)$ is any basis of X. A basis for $T(x^{(k)})$ can be easily obtained, e.g. from the QR decomposition $Q^T x^{(k)} = e_1$ [c]. Indeed, if $Q = [q_1 \ldots q_n]$, it is well known that $\{q_2, \ldots, q_n\}$ spans the null space of $[x^{(k)}]^T$, that is $T(x^{(k)})$. Now, let us suppose $J_\eta(x^{(k)}) \neq \emptyset$. In this case we can handle the presence of the nondifferentiability hyperplanes near $x^{(k)}$, by imposing that Γ_k includes a set of generators for each cone $\mathcal{T}_{s,\eta}(x^{(k)})$ [d]. Recalling that $\mathcal{T}_{s,\eta}(x^{(k)})$ is defined by the inequalities in (6) and the equality in (5), next

---

[b] A set $\mathcal{B}^+(X) \subset X$ is a positive basis for a subspace $X \subseteq \mathbb{R}^n$, if each vector in X is a nonnegative linear combination of the vectors in $\mathcal{B}^+(X)$, and no proper subset of $\mathcal{B}^+(X)$ positively spans X. If X has dimension p, the cardinality of $\mathcal{B}^+(X)$ is at most $2p$ [9].

[c] Here, and in the following, $\{e_1, \ldots, e_n\}$ is the canonical basis of $\mathbb{R}^n$.

[d] A set $\mathcal{G}(\mathcal{T}) \subset \mathcal{T}$ is a set of generators for the polyhedral cone $\mathcal{T}$, if each element of $\mathcal{T}$ is a nonnegative linear combination of the elements of $\mathcal{G}(\mathcal{T})$; furthermore, $\mathcal{G}(\mathcal{T})$ is said minimal, if no proper subset of $\mathcal{G}(\mathcal{T})$ generates the whole $\mathcal{T}$.

Theorem 2.1 and Corollary 2.1 can be exploited in the nondegenerate case to compute the generators.

Theorem 2.1 ([9, Theorem 5.1]). *Let* $M = [c_1 \dots c_\ell \, y_1 \dots y_r] \in \mathbb{R}^{n \times q}$, *with* $q = \ell + r \leq n$, *be a full column rank matrix, and let* $Z = [z_1 \dots z_n] \in \mathbb{R}^{n \times n}$ *be an invertible matrix, such that* $Z^T M = [e_1 \dots e_q]$. *Then the set*

$$\mathcal{G}(\mathcal{T}) \equiv \{-z_1, \dots, -z_\ell\} \cup \{Zu, \ u \in \mathcal{B}^+(\text{span}\{e_{q+1}, \dots, e_n\})\}$$

is a minimal set of generators for the cone

$$\mathcal{T} = \{d \in \mathbb{R}^n : c_i^T d \leq 0, \ i = 1, \dots, \ell, \text{ and } y_i^T d = 0, \ i = 1, \dots, r\}. \tag{8}$$

The following corollary says that, under the assumptions of Theorem 2.1, given the generators of $\mathcal{T}$, the generators of any other cone defined by changing the signs of the inequalities in (8), are obtained for free.

Corollary 2.1. *Under the assumptions and notations of Theorem 2.1, a set of generators for any cone*

$$\mathcal{T}_s = \left\{d \in \mathbb{R}^n : s_i c_i^T d \leq 0, \ s_i = \pm 1, i = 1, \dots, \ell, \text{ and } y_i^T d = 0, i = 1, \dots, r\right\}$$

is included in

$$\Gamma \equiv \{z_1, \dots, z_\ell, -z_1, \dots, -z_\ell\} \cup \{Zu, \ u \in \mathcal{B}^+(\text{span}\{e_{q+1}, \dots, e_n\})\}. \tag{9}$$

Proof. Let $M_s = [s_1 c_1 \dots s_\ell c_\ell \ y_1 \dots y_r] = M D_s$, where D_s is the $q \times q$ diagonal matrix $D_s = \text{diag}(s_1, \dots, s_\ell, 1, \dots, 1)$. Then it can be easily verified that a set of generators for $\mathcal{T}_s$, included in Γ, is given by

$$\mathcal{G}(\mathcal{T}_s) \equiv \{-s_i z_i, \ i = 1, \dots, \ell\} \cup \{Zu, \ u \in \mathcal{B}^+(\text{span}\{e_{q+1}, \dots, e_n\})\}.$$

Indeed, it is sufficient to apply Theorem 2.1 to the matrix M_s, and observe that the matrix $Z_s = [s_1 z_1 \dots s_\ell z_\ell \ z_{\ell+1} \dots z_n] = Z \, \text{diag}\{D_s, I_{n-q}\}$ is such that $Z_s u = Zu$ for all $u \in \mathcal{B}^+(\text{span}\{e_{q+1}, \dots, e_n\})$, and $Z_s^T M_s = \text{diag}\{D_s, I_{n-q}\}[e_1 \dots e_q] D_s = [e_1 \dots e_q]$. □

We recall now another corollary of Theorem 2.1, which will be useful for the analysis of convergence developed in the next section.

Corollary 2.2 ([9, Corollary 5.2]). *Under the assumptions and notations of Theorem 2.1, the set* Γ *in (9) contains a set of generators for any cone defined by the equalities and by any subset of the inequalities in (8).*

Given M as in Theorem 2.1, the matrix $W = M(M^TM)^{-1} = [w_1 \dots w_q]$, and any basis $\mathcal{B}(N(M^T)) = \{v_1, \dots, v_{n-q}\}$ for the null space of M^T, it is easily verified that the matrix $Z = [w_1 \dots w_q \ v_1 \dots v_{n-q}]$ satisfies the requirements of Theorem 2.1. Turning to the problem at hand, if $M = M_\eta(x^{(k)})$ has full column rank, we can apply Corollary 2.1 to deduce that

$$\Gamma_\eta(x^{(k)}) = \{w_1, \dots, w_{\ell_\eta(x^{(k)})}, -w_1, \dots, -w_{\ell_\eta(x^{(k)})}\} \cup \mathcal{B}^+(N(M^T)) \qquad (10)$$

contains the generators for any cone in the family $\mathcal{F}_\eta(x^{(k)})$. If $\ell_\eta(x^{(k)}) = 0$, $M_\eta(x^{(k)})$ is the vector $x^{(k)}$ and (10) reduces to $\Gamma_\eta(x^{(k)}) = \mathcal{B}^+(T(x^{(k)}))$.

Remark 2.1. Corollary 2.2 says that the set $\Gamma_\eta(x^{(k)})$ in (10) includes also a positive basis for the tangent space $T(x^{(k)})$, as well as a set of generators for every cone in $T(x^{(k)})$, defined by any subset of the inequalities in (6).

The matrix W and a basis $\mathcal{B}(N(M^T))$ can be easily obtained by the QR factorization or the SVD of M. The latter is our preference, since it is the best numerical tool to evaluate the rank of a matrix. Let $U \in \mathbb{R}^{n\times n}$ and $V \in \mathbb{R}^{q\times q}$ be orthogonal matrices such that $M = U\Sigma V^T$, with $\Sigma = [\bar{\Sigma}\ 0]^T$, $\bar{\Sigma} = \mathrm{diag}(\sigma_1, \dots, \sigma_q)$. Let $U = [u_1 \dots u_n]$. In the nondegenerate case, $\sigma_1, \dots, \sigma_q$ are positive and $N(M^T) = \mathrm{span}\{u_{q+1}, \dots, u_n\}$; then $W = -[u_1 \dots u_q]\bar{\Sigma}^{-1}V^T$ and $\mathcal{B}^+(N(M^T))$ can be obtained by (7).

The computation of Γ_k is not so simple when $M_\eta(x^{(k)})$ is column rank deficient, because a set of generators needs to be traced for all of the $2^{\ell_\eta(x^{(k)})}$ cones $\mathcal{T}_{s,\eta}(x^{(k)})$. We are aware that the cost can be prohibitive. Nevertheless, some computations can be saved by carefully organizing the algorithm, and exploiting the fact that $\mathcal{T}_{-s,\eta}(x^{(k)}) = -\mathcal{T}_{s,\eta}(x^{(k)})$ for any s. In the algorithm PGSS below, for a given s, we compute a set of generators $\mathcal{G}(T_{s,\eta}(x^{(k)})) = \{d_1, \dots, d_p\}$ of $\mathcal{T}_{s,\eta}(x^{(k)})$, and test at once the trial points corresponding to $\pm d_i$, for $i = 1, \dots, p$. If an acceptable point is found, the computation of the remaining generators is skipped and the iteration concluded; otherwise, another vector s is checked.

Algorithm PGSS (Projected GSS)

Data: $x^{(0)} \in \mathcal{S}$, $\Delta_0 > 0$, $\eta > 0$, $\theta_{max} \in (0,1)$, $\lambda_{max} > 1$.
For $k = 0, 1, \dots$ until convergence **do**
  Determine $J_\eta(x^{(k)})$, set ℓ equal to its cardinality, and create $M = M_\eta(x^{(k)})$.
  If $\mathrm{rank}(M) = \ell + 1$ **then**
    Create $\Gamma_k = \Gamma_\eta(x^{(k)})$ using (10).
    If $\|Cz - b\|_1 \le \|Cx^{(k)} - b\|_1 - \rho(\Delta_k)$ for some $z = P_{\mathcal{S}}(x^{(k)} + \Delta_k d)$ and $d \in \Gamma_k$ **then** set $flag = 1$ **else** set $flag = 0$.
  else

Define the matrix $S = [s^{(1)} \dots s^{(2^{\ell-1})}]$ whose columns are all sign vectors in $\mathbb{R}^\ell$ with first component equal to $+1$. Set $flag = 0$ and $t = 0$.

While $flag = 0$ and $t < 2^{\ell-1}$ **do**

Set $t = t + 1$ and $s = s^{(t)}$. Compute a set of generators $\mathcal{G}(\mathcal{T}_{s,\eta}(x^{(k)}))$.

If $\|Cz - b\|_1 \leq \|Cx^{(k)} - b\|_1 - \rho(\Delta_k)$ for some $z = P_{\mathcal{S}}(x^{(k)} + \Delta_k d)$ and $d \in \mathcal{G}(\mathcal{T}_{s\eta}(x^{(k)})) \cup -\mathcal{G}(\mathcal{T}_{s,\eta}(x^{(k)}))$ **then** set $flag = 1$.

End while

End if

If $flag = 0$ **then** set $x^{(k+1)} = x^{(k)}$ and $\Delta_{k+1} = \theta_k \Delta_k$ with $\theta_k \in (0, \theta_{max}]$ **else** set $x^{(k+1)} = z$ and $\Delta_{k+1} = \lambda_k \Delta_k$ with $\lambda_k \in [1, \lambda_{max}]$.

End do

3. Convergence analysis

Convergence properties of GSS methods have been studied in a number of papers. Among them, it is worth to cite the review paper [6], for an excellent overview on the state of the art of derivative-free optimization in the smooth case, and the paper [1], where a simple analysis clarifies the relationship between the choice of the search directions and the optimality properties for nonsmooth problems. In both papers, it is proved the existence of a subsequence converging to a point with some stationarity properties, related to the problem. In our case, $\hat{x} \in \mathcal{S}$ is said a stationary point if

$$f'(\hat{x}; d) = \lim_{t \to 0^+} \frac{f(\hat{x} + td) - f(\hat{x})}{t} \geq 0 \quad \forall d \in T(\hat{x}), \tag{11}$$

where $f(x) = \|Cx - b\|_1$. We notice that the directional derivative $f'(x; d)$ is defined for all x and d, because f is convex. If $x \in \mathbb{R}^n \backslash \mathcal{H}$, it is easily seen that $f'(x; d) = g(x)^T d$, $\forall d \in \mathbb{R}^n$, with $g(x) = \sum_{j=1}^m \operatorname{sign}(c_j^T x - b_j) c_j$. Instead, if $x \in \mathcal{H}$, it can be proved [2] that

$$f'(x; d) = g_s(x)^T d, \quad \forall d \in \mathcal{K}_s(x), \tag{12}$$

where, assuming for simplicity $J(x) = \{1, \dots, \ell(x)\}$,

$$g_s(x) = -\sum_{j \in J(x)} s_j c_j + \sum_{j \notin J(x)} \operatorname{sign}(c_j^T x - b_j) c_j.$$

The convergence analysis focuses on the subsequence $\{x^{(k)}\}_{k \in \mathcal{U}}$ of the unsuccessful iterates, where no improved point was produced, i.e.

$$f(P_{\mathcal{S}}(x^{(k)} + \Delta_k d^{(k)})) > f(x^{(k)}) - \rho(\Delta_k), \quad \forall d^{(k)} \in \Gamma_k, \ k \in \mathcal{U}, \tag{13}$$

and is based on taking a limit in (13) to obtain nonnegativity results for some directional derivatives of f at a limit point $\hat{x}$ of $\{x^{(k)}\}_{k \in \mathcal{U}}$.

Theorem 3.1. *Let $\{x^{(k)}\}$ be the sequence of iterates produced by the PGSS algorithm. Assume that $\beta_L, \beta_U > 0$ exist such that*

$$\beta_L \leq ||d^{(k)}|| \leq \beta_U, \ \forall d^{(k)} \in \Gamma_k, \tag{14}$$

and that the forcing function $\rho : \mathbb{R}^+ \to \mathbb{R}^+$ is continuous, increasing, and satisfies $\rho(t) = o(t)$. Let $K \subseteq \mathcal{U}$ be an infinite subset of indices such that $\{x^{(k)}\}_{k\in K} \to \hat{x}$ and $\{d^{(k)}\}_{k\in K} \to \hat{d}$, with $d^{(k)} \in \Gamma_k$. Then $f'(\hat{x}; \hat{d}) \geq 0$.

Proof. Since $K \subseteq \mathcal{U}$, inequality (13) holds for all $k \in K$, and then

$$\frac{f(x^{(k)} + \Delta_k \hat{d}) - f(x^{(k)})}{\Delta_k} > \frac{f(x^{(k)} + \Delta_k \hat{d}) - f(P_{\mathcal{S}}(x^{(k)} + \Delta_k d^{(k)}))}{\Delta_k} - \frac{\rho(\Delta_k)}{\Delta_k}. \tag{15}$$

Let us show that the right hand side of (15) vanishes with $k \to \infty$. Indeed, we easily see that $P_{\mathcal{S}}(x + \Delta d) = \frac{x+\Delta d}{\sqrt{1+\Delta^2 \|d\|^2}} = x + \Delta d + O(\Delta^2 \|d\|^2)$, $\forall x \in \mathcal{S}$ and $d \in T(x)$. This relation, together with the local Lipschitz continuity of f, the assumptions on ρ, and (14), yields

$$\frac{f(x^{(k)} + \Delta_k d^{(k)}) - f(P_{\mathcal{S}}(x^{(k)} + \Delta_k d^{(k)}))}{\Delta_k} - \frac{\rho(\Delta_k)}{\Delta_k} = o(\Delta_k). \tag{16}$$

Hence the claim is easily proved, because the assumptions on ρ imply $\lim_{k\to\infty} \Delta_k = 0$ (see [3, Lemma 3.1], [6, sec. 3.7]). Now, since for convex functions the usual directional derivative is equal to the (Clarke) generalized directional derivative, we have:

$$f'(\hat{x}; \hat{d}) = \limsup_{y\to\hat{x},\, t\to 0+} \frac{f(y + t\hat{d}) - f(y)}{t} \geq \limsup_{k\in K,\, k\to\infty} \frac{f(x^{(k)} + \Delta_k \hat{d}) - f(x^{(k)})}{\Delta_k}.$$

Then, the thesis follows from (15) and (16). □

The stationarity condition (11) will be in the end proved if we can apply Theorem 3.1 to a sufficiently rich set of directions. Let $J(\hat{x}) = \emptyset$, and let $\mathcal{B}^+(T(\hat{x})) = \{d_1, \ldots, d_p\}$ be a positive basis for $T(\hat{x})$. Then, it is sufficient to ensure $f'(\hat{x}; d_j) \geq 0$, for $j = 1, \ldots, p$. Indeed, for any $d \in T(\hat{x})$, we have $d = \sum_{j=1}^{p} \alpha_j d_j$, with $\alpha_1, \ldots, \alpha_p \geq 0$, and $f'(\hat{x}; d) = \sum_{j=1}^{p} \alpha_j f'(\hat{x}; d_j) \geq 0$. When $J(\hat{x}) \neq \emptyset$, $T(\hat{x})$ is partitioned into the family $\mathcal{F}(\hat{x})$ of polyhedral cones, and we have to repeat the previous argument for each cone $\mathcal{T}_s(\hat{x})$. More precisely, for any $d \in T(\hat{x})$ there exists s such that $d \in \mathcal{T}_s(\hat{x})$. Given the set of generators $\mathcal{G}(\mathcal{T}_s(\hat{x})) = \{d_1, \ldots, d_{p_s}\}$, we have $d = \sum_{j=1}^{p_s} \alpha_j d_j$, with $\alpha_1, \ldots, \alpha_{p_s} \geq 0$. Then, if $f'(\hat{x}; d_j) \geq 0$, for $j = 1, \ldots, p_s$, (12) yields:

$$f'(\hat{x}; d) = g_s(\hat{x})^T \sum_{j=1}^{p_s} \alpha_j d_j = \sum_{j=1}^{p_s} \alpha_j g_s(\hat{x})^T d_j = \sum_{j=1}^{p_s} \alpha_j f'(\hat{x}; d_j) \geq 0.$$

To summarize, $\hat{x}$ *is stationary for problem (2) if and only if* $f'(\hat{x}; d) \geq 0$ *for all* $d \in \Gamma(\hat{x})$, where

$$\Gamma(\hat{x}) = \begin{cases} \mathcal{B}^+(T(\hat{x})) & \text{for } J(\hat{x}) = \emptyset \\ \bigcup_{\mathcal{T}_s \in \mathcal{F}(\hat{x})} \mathcal{G}(\mathcal{T}_s) & \text{for } J(\hat{x}) \neq \emptyset. \end{cases}$$

Hence, we can ensure the existence of stationary limit points of $\{x^{(k)}\}$, provided that the following assumption is satisfied.

Assumption 3.1. *Given a limit point* $\hat{x}$ *of the sequence* $\{x^{(k)}\}$, *and the sequence* $\{\Gamma_k\}$, *there exist an infinite set of indices* $K \subseteq \mathcal{U}$ *and a sequence of subsets* $\hat{\Gamma}_k \subseteq \Gamma_k$, *such that* $\{\hat{\Gamma}_k\}_{k \in K} \to \hat{\Gamma} \supseteq \Gamma(\hat{x})$.

Theorem 3.2. *Let* $\{x^{(k)}\}$ *be the sequence of iterates produced by the* PGSS *algorithm. Assume that (14) holds, and that the forcing function* ρ *satisfies the assumptions of Theorem 3.1. Then, if Assumption 3.1 holds, each limit point* $\hat{x}$ *of the subsequence* $\{x^{(k)}\}_{k \in \mathcal{U}}$ *is a stationary point for problem (2).*

Proof. The existence of at least a limit point $\hat{x}$ of $\{x^{(k)}\}_{k \in \mathcal{U}}$ is ensured by the compactness of $\mathcal{S}$. Then the thesis is immediate, because Assumption 3.1 and Theorem 3.1 imply that $f'(\hat{x}; d) \geq 0$, for all $d \in \Gamma(\hat{x})$. □

Next theorem shows that Assumption 3.1 holds for the PGSS algorithm in the nondegenerate case. The degenerate case is still an open problem.

Theorem 3.3. *Let* $\hat{x}$ *be a limit point of the sequence* $\{x^{(k)}\}$ *produced by the algorithm* PGSS, *with* $J(\hat{x}) = \{j_1, \ldots, j_{\ell(\hat{x})}\}$. *Assume that, when a positive basis of a subspace is needed, it is obtained by applying (7) to an orthonormal basis of the subspace. Then, if* $M(\hat{x}) = [c_{j_1} \ldots c_{j_{\ell(\hat{x})}}\ \hat{x}]$ *has full column rank, there exist* $\eta > 0$, $\hat{\Gamma} \supseteq \Gamma(\hat{x})$, *and an infinite set of indices* K, *such that* $\Gamma_k = \Gamma_\eta(x^{(k)})$ *for all* $k \in K$ *(see (10)), and* $\{\Gamma_k\}_{k \in K} \to \hat{\Gamma}$.

Proof. Let $\{x^{(k)}\}_{k \in \bar{K}}$ be a subsequence converging to $\hat{x}$. For sufficiently large values of $k \in \bar{K}$, the cones in $\mathcal{F}_\eta(x^{(k)})$ are nondegenerate if η is sufficiently small, and $J(\hat{x}) \subseteq J_\eta(x^{(k)})$. Therefore, there exist $L \geq \ell(\hat{x})$ and $K' \subseteq \bar{K}$ such that, $\forall k \in K'$, $J_\eta(x^{(k)}) = \{j_1, \ldots, j_{\ell(\hat{x})}, j_{\ell(\hat{x})+1}, \ldots, j_L\}$, and $\mathrm{rank}(M_k) = L+1$, where $M_k = M_\eta(x^{(k)}) = [c_{j_1} \ldots c_{j_L}\ x^{(k)}]$. Then, if $L > 0$,

$$\Gamma_k = \Gamma_\eta(x^{(k)}) = \{w_1^{(k)}, \ldots, w_L^{(k)}, -w_1^{(k)}, \ldots, -w_L^{(k)}\} \cup \mathcal{B}^+(N(M_k^T)),$$

where $w_j^{(k)}$ is the j-th column of $W_k = M_k(M_k^T M_k)^{-1}$. Since $\{M_k\}_{k \in K'} \to \hat{M} \equiv [c_{j_1} \ldots c_{j_L}\ \hat{x}]$, it also holds $\{W_k\}_{k \in K'} \to \hat{W} \equiv \hat{M}(\hat{M}^T \hat{M})^{-1}$. In particular, this implies that $\{w_j^{(k)}\}_{k\in K'} \to \hat{w}_j$, for $j = 1, \ldots, L$, where $\hat{w}_j$ is the j-th column of $\hat{W}$. Given an orthonormal basis $\{v_1^{(k)}, \ldots, v_p^{(k)}\}$ of $N(M_k^T)$, let $\mathcal{B}^+(N(M_k^T)) = \{v_1^{(k)}, \ldots, v_p^{(k)}, -v_1^{(k)}, \ldots, -v_p^{(k)}\}$, with $p{=}n{-}L{-}1$. By compactness arguments, there exists $K \subseteq K'$ such that $\{v_1^{(k)}, \ldots, v_p^{(k)}\}_{k\in K} \to \{\hat{v}_1, \ldots, \hat{v}_p\}$. Then $\{\Gamma_k\}_{k\in K} \to \hat{\Gamma}$, with $\hat{\Gamma} = \{\hat{w}_1, \ldots, \hat{w}_L, -\hat{w}_1, \ldots, -\hat{w}_L\} \cup \{\hat{v}_1, \ldots, \hat{v}_p, -\hat{v}_1, \ldots, -\hat{v}_p\}$. Using the continuity of the nullspace and the orthonormality of the bases, we easily see that $\{\hat{v}_1, \ldots, \hat{v}_p\}$ is an orthonormal basis for $N(\hat{M}^T)$, and $\{\hat{v}_1, \ldots, \hat{v}_p, -\hat{v}_1, \ldots, -\hat{v}_p\}$ is a positive basis for $N(\hat{M}^T)$. Then, by Corollary 2.1, $\hat{\Gamma}$ includes a set of generators for any cone

$$\mathcal{T}_{L,s} = \left\{d \in \mathbb{R}^n : s_i c_{j_i}^T d \le 0,\ s_i = \pm 1, i = 1, \ldots, L, \text{ and } \hat{x}^T d = 0\right\}.$$

Moreover, recalling Corollary 2.2 and Remark 2.1, we deduce that $\hat{\Gamma}$ contains a positive basis for $T(\hat{x})$, and, if $\ell(\hat{x}) > 0$, a set of generators for any cone $\mathcal{T}_s \in \mathcal{F}(\hat{x})$. This concludes the proof for $L > 0$. When $L = 0$, we have $M_k = [x^{(k)}]$ and $\Gamma_k = \mathcal{B}^+(T(x^{(k)})) = \mathcal{B}^+(N(M_k^T))$ for all $k \in K'$. Using the continuity of the tangent space, and the same compactness arguments as above, it exists a subset $K \subseteq K'$ such that $\{\Gamma_k\}_{k\in K} \to \hat{\Gamma} = \mathcal{B}^+(T(\hat{x}))$. □

4. Numerical experiments

The effectiveness of the `PGSS` algorithm was verified by solving in `MATLAB 7.0.4` some test problems (see [10] and [11]) on a processor PENTIUM 4 (2.80 GHz). The results in this section were obtained using $\eta = 10^{-3}$, $\Delta_0 = 0.5$, and $\rho(t) = 10^{-4}t^2$. Further, we used $\theta_k = 0.25$ and $\lambda_k = 1$ for each k, and the stopping criterion $\Delta_k < 10^{-7}(1 + \|x^{(k)}\|_\infty)$. If needed, we computed a set of generators for a degenerate cone by the `cddlib` code by Fukuda[e]. The search directions were normalized in the euclidean norm.

Example 4.1. Here we consider the problem (1), where $B, C \in \mathbb{R}^{24\times 4}$ arise in the classical *24 psychological tests problem* of Holzinger and Harman (see [8, Table 10.2 and Table 12.3]). For this problem, we compared the performance of the `PGSS` algorithm with the method described in [11], briefly `TW` method, using several values for the smoothing parameter α. The integrator `ode15s` from the `MATLAB ODE suite`, with tolerances equal to 10^{-7}, was used for (4). The integration was stopped when the decrease in the objective function $\|Cx - b\|_1$ within one step was less than 10^{-7}. In a first set of experiments we used two "reasonable" starting guesses:

[e]`http://www.ifor.math.ethz.ch/~fukuda/cdd_home/cdd.html`

Table 1. Results for the 24 psychological tests problem

	$f^*[Q^{(0)}=I]$	CPU	$f^*[Q^{(0)}=Q_{ls}]$	CPU
PGSS	7.923359	8.86	7.923587	8.12
TW, $\alpha = 10^3$	7.925103	138.52	7.925020	60.34
TW, $\alpha = 10^4$	7.924266	163.98	7.925549	75.42
TW, $\alpha = 10^5$	7.949950	154.77	7.925483	69.12
TW, $\alpha = 10^6$	8.031399	151.25	7.997582	67.55

$Q^{(0)} = I$ and $Q^{(0)} = Q_{ls}$, where Q_{ls} is the solution of the least squares problem $\min_Q \|CQ - B\|_2$, normalized so that $Q_{ls} \in \mathcal{OB}(4,4)$. In Table 1, we report the value $f^* = \|CQ^* - B\|_{\ell_1}$, where Q^* is the computed solution, and the CPU time (in centiseconds). The table shows that the PGSS algorithm reached lower values f^* than TW, with lower computational effort. In a second set of experiments, we used 100 randomly generated initial guesses, normalized to fit in $\mathcal{OB}(4,4)$. The hystograms in the figure refer to the computed values for f^*. PGSS produced $f^* \in [7.9, 8.0]$ in almost the 90% of the runs, while the results of TW strongly depend on α: for $\alpha = 10^3$, in almost 50% of the cases TW reached values slightly lower than PGSS; for $\alpha = 10^6$ its behavior largely deteriorated. We also solved reformulation (3) of the problem by the fmincon function of the MATLAB Optimization Toolbox. The results were similar to those of PGSS, with CPU times from 6 to 10 times higher.

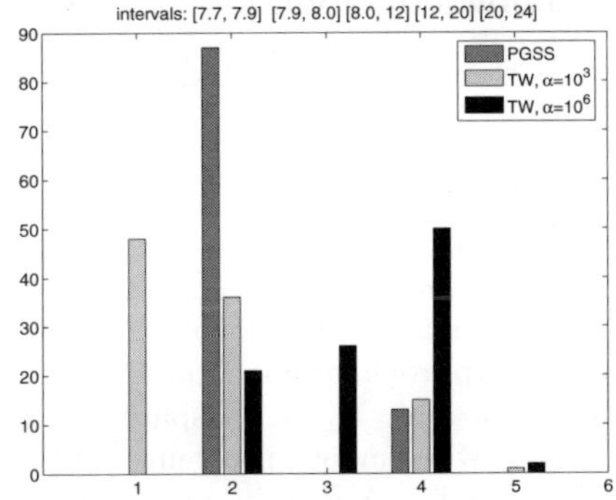

Example 4.2. This example arises in the context of shape analysis [5]. We are given 30 specimens of handwritten digit "3", each of them described by 13 special points (landmarks), whose coordinates can be collected in a 13×2 matrix [f]. In the following, we will denote B_i the matrix associated to the i-th digit "3". In the experiments we considered the oblique Procrustes problems, where the matrix $C \in \mathbb{R}^{13\times 2}$ describes the "average digit 3" (that is each element of C is the arithmetic mean of the corresponding elements of $B_1, \ldots, B_{30}$), and the target matrix B is B_i, for $i = 1, \ldots, 30$. Here we comment the results obtained

[f]See http://www.maths.nottingham.ac.uk/personal/ild/bookdata/digit3.dat.

with $Q^{(0)} = I$ (similar results were obtained with $Q^{(0)} = Q_{ls}$). In 12 cases, we had to use `cddlib` because of degeneracy. This was not excessively time consuming, due to the very small dimension of the problem ($n = p = 2$): indeed, the mean CPU time was 1.26cs for these degenerate problems, and 0.87cs for the non-degenerate ones. The average number of function evaluations over all 30 problems was 82, and the average number of matrix factorizations was 22. In the figure we show the target (solid line), the average digit "3" (dotted line), and the digit "3" corresponding to CQ^* (dashdot line), for four different targets.

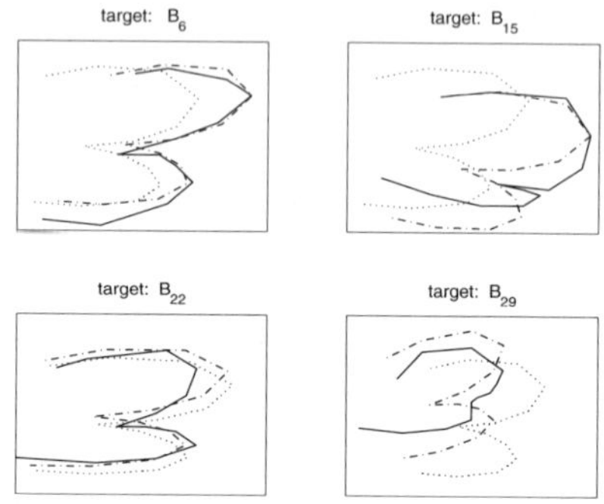

Acknowledgements

This work was supported by Dip. Energetica "S. Stecco", and MIUR.

References

1. C. Audet, J. E. Dennis Jr., Analysis of Generalized Pattern Searches, *SIAM J. Optim.*, **13** (2003), pp. 889–903.
2. C. Bogani, M. G. Gasparo, A. Papini, A Pattern Search Method for Discrete L_1–Approximation, *J. of Opt. Theory and Appl.*, **134** (2007), pp. 47–59.
3. C. Bogani, M. G. Gasparo, A. Papini, Generating set search methods for piecewise smooth problems, *SIAM J. Optim.*, **20** (2009), pp. 321–335.
4. S. Boyd, L. Vanderberghe, *Convex Optimization*, Cambridge University Press, Cambridge, UK, 2006.
5. I. L. Dryden, K. V. Mardia, *Statistical Shape Analysis*, Wiley, Chichester, 1998.
6. T. G. Kolda, R. M. Lewis, V. Torczon, Optimization by direct search: new perspectives on some classical and modern methods, *SIAM Rev.*, **45** (2003), pp. 385–482.
7. S. Lucidi, M. Sciandrone, P. Tseng, Objective-derivative-free methods for constrained optimization, *Math. Program. Ser. A*, **92** (2002), pp. 37–59.
8. S. Mulaik, *The Foundations of Factor Analysis*, McGraw-Hill, NY, 1972.
9. C. J. Price, I. D. Coope, Frames and Grids in unconstrained and linearly constrained optimization: a nonsmooth approach, *SIAM J. Optim.*, **14** (2003), pp. 415–438.
10. N. T. Trendafilov, On the ℓ_1 Procrustes problem, *Future Generation Computer Systems*, **19** (2003), pp. 1177–1186.
11. N. T. Trendafilov, G.A. Watson, The ℓ_1 oblique Procrustes problem, *Statistics and Computing*, **14** (2004), pp. 39–51.

Snake Segmentation of Multiple Sclerosis Lesions for Assisted Diagnosis by Cluster Analysis-Based Techniques

Lilla Bonanno*, Pietro Lanzafame, Alessandro Celona, Silvia Marino, Barbara Spanó, Placido Bramanti

IRCCS Centro Neurolesi "Bonino Pulejo"
Via Provinciale Palermo Ctr. Casazza 98124 Messina, Italy
**E-mail: lilla.bonanno@centroneurolesi.it*
www.centroneurolesi.it

Luigia Puccio

Dipartimento di Matematica, Facoltà di Scienze MM.FF.NN.
Università degli Studi di Messina, Italy
E-mail: gina@dipmat.unime.it
www.unime.it

Magnetic Resonance Imaging (MRI), allowing in-vivo detection of lesions, is today a crucial tool for diagnosis of Multiple Sclerosis (MS). Although the detection of lesions are not sufficient for a diagnosis of MS because of similarity with patterns detected in other neurological diseases, taking into account different radiological informations, MRI findings can often yield a high degree of confidence. We used a snake based procedure for segmentation of lesion then proposing a method based on Cluster Analysis to support clinicians in the diagnosis of MS. By identifying a minimum set of significant descriptors, our algorithm can help neurologist and neuroimaging expert to distinguish MS plaques from other kinds of lesions.

Keywords: Magnetic Resonance Imaging, Multiple Sclerosis, Snake, Cluster Analysis.

1. Introduction

Multiple Sclerosis (MS) is an autoimmune condition in which the immune system attacks the Central Nervous System (CNS).[19] Magnetic Resonance Imaging (MRI) has become the most sensitive paraclinical test in diagnosis, assessment of disease evolution and treatment of the effects in MS (fig. 1(a)). We used a snake based procedure for segmentation of lesion then proposing a method based on Cluster Analysis to support clinicians in the diagnosis of MS. By identifying a minimum set of significant descriptors, our algorithm can help neuroimaging expert to distinguish MS plaques from other kinds of lesions. MRI is used as a prognostic tool at the first presentation of symptoms, suspicious of brain demyelination. Multiple hyperintense lesions on T2-weighted sequences are the characteristic MR appearance of MS. The majority of lesions are small, although, they can occasionally measure several centimeters in diameter. MS lesions are usually small with intermediate high signal intensity with less severe degree of inflammation. MS lesions tend to have an ovoid configuration with the major axis perpendicular to the ventricular borders. Most lesions, especially in the early stages of the disease, are evidend on conventional MRI but diffuse irregular hyperintensities have also been demonstrated in the later stages of the disease. These areas with poorly defined borders, are usually seen around the ventricles and called dirty appearing white matter (DAWM).

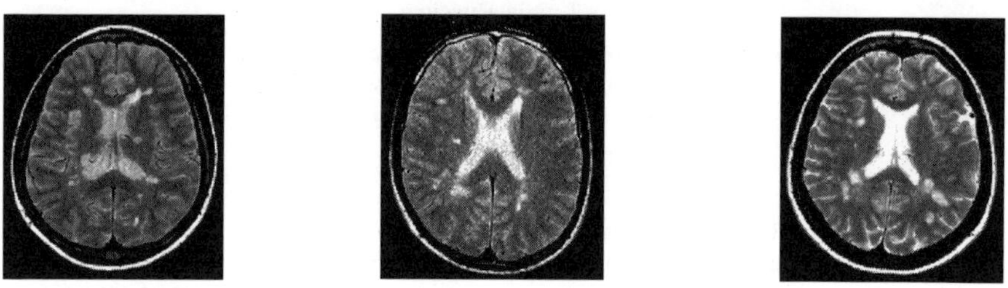

(a) (b) (c)

Fig. 1. Axial T2-weighted images of patients with MS (a), Cadasil (b) and DM1 (c) demonstrate multiple hyperintense lesions with periventricular predominance.

Cerebral autosomal dominant arteriopathy with subcortical infarcts and leukoencephalopathy (CADASIL) is an autosomal dominant condition causing recurrent subcortical strokes, often complicated by a subcortical dementia. MR imaging plays a central role in the diagnosis and evaluation of patients with CADASIL (fig. 1(b)). The radiologic appearance of rarefaction of the white matter is also seen as corresponding high signal intensity on T2-weighted MR imaging diagnosis. In particular CADASIL may be misdiagnosed as MS especially in early stage.[16] In T2-weighted magnetic resonance imaging (MRI) scans in both conditions may show diffuse periventricular white matter hyperintensities. Myotonic dystrophy type 1 (dystrophia myotonica, DM1) (fig. 1(c)), the most common form of adult muscular distrophy, is a progressive autosomal dominant disorder.[10,13] A number of neuroimaging studies have evaluated the relevance of brain involvement in DM1. In particular, MRI studies have demonstrated that focal white matter (WM) lesions can be found frequently in the brains of DM1 patients.[1,4,9,15] On T2 images, we detected small, hyperintense abnormalities in the WM of patients: patient showed punctuate, periventricular and subcortical WM lesions.

2. Objective of work

The individualization of an object in a digital image is one of the most common problems in image analysis. Among the various types of images, we reviewed, with special attention medical ones, which are often affected by disturbances since stemming the observation by digital acquisition, which can cause difficulties in applying techniques for recovery of the contour of the object. This operation is difficult both for the great variability of the forms that objects can have and for the image quality. We have considered the analytical and geometric parameters from spline approximating curves aiming to evaluate the automatic acquisition of the coordinates of these curves through the use of snakes and to find the analytical parameters describing the curve in order to allow comparison with other images of the same type. To provide a more accurate description of biological profiles, a method was developed based on the best approximation of the data using the cubic spline functions. In particular we are interested in analyzing the brain regions focusing the main problem in the necessity to recognize the contour. In applications like this (recovery of contour), a crucial phase is the collection of data. To do so 3 types of methods may be used:

- Manuals methods;
- Iteractive methods assisted by the computer;
- Automatic methods that use some analyzers of image.

2.1. *Deformable Models: Snake*

The deformable models are curves or surfaces defined inside an image that can move thanks to the influence of internal forces. These internal forces are defined inside the curve or the same surface and then there are also the external forces, calculated through same data of the image. The internal forces are considered to maintain the model regularity during the deflection. On the contrary, the external forces are defined to move towards an object of the contour or to another area of the image. In this way, the deformable models offer robustness to some disorders of the image and to some holes of the contour. Moreover these deformable models allow to integrate the elements of contour in a coherent and substantial mathematical description; for this reason they force the contours extracted to be regular and also to consider other informations in advance about the object. The deformable model is known as snake. It showed a good behaviour compared to data.[11] The snakes are planar deformable contours, useful in many problems about the analysis of the image. Then, the snake may be seen as a deformable curve or as an active contour. In a snake different forces act: *Elasticity, Stiffness, Viscosity, External force, Trades of gradient scalar, Min delta/max delta, Iterations of GVF, Iterations contours, Noise parameter, Blured Gaussian, Sigma Gaussian, Image of contrast.*

For all the examples we dimensioned the image to $200 * 150$ pixels. Subsequently, these curves were included in a program of Matlab produced by D. Tomazevic, C. Xu J. L. Prince, and freely available Web[21] *sdemo*, to extract the parameters of the snake and compare these parameters obtained from the snake with those obtained with a manual measurement. Opening the files of the curve with the program *sdemo* we obtain the following figure (fig. 2). We have considered the *gradient* and then left *sigma* to zero. Then instead of selecting the standard vector field, which calculates the standard external force vector field using gradient, we selected the *Gradient Vector Field* (GVF) which calculates the gradient vector field external force and *Normalize* to normalize vectors in vector field. In this case it is necessary to give a value to *Mu*, the parameter for the *Gradient Vector Field* (GVF) (which we have left to 0.1) and the number of iterations to calculate *Gradient Vector Field* (GVF) where we put 80. Manually we have traced the initial snake of the figure of brain lesion and automatically the snake is traced in the blured image and in the potential field with snake (GVF) (fig.3).

3. Analysis of geometrical parameters and results

For the analysis of geometrical parameters we saved the snake coordinates, to be analyzed by the SnakeIter Matlab routine:

$$xlswrite('snake.xls', x,' coordinated',' A1');$$

$$xlswrite('snake.xls', y,' coordinated',' B1');$$

that allows to store the coordinates of the final snake, in a file excel, in two columns and precisely in column A1 there will be the coordinates for x

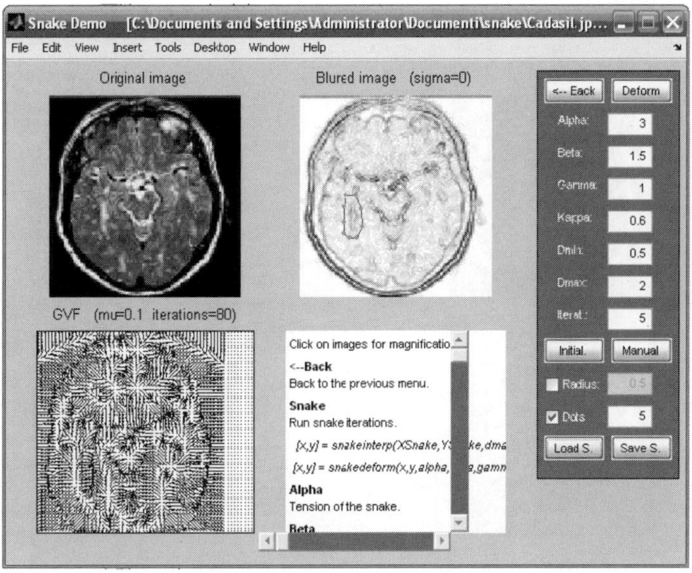

Fig. 2. Initialization of the GVF Snake.

and in column B1 the coordinates of y. As already said, we used a statistical program, STATISTICA 6.0, to underline the differences among various curves and that allow to group them according to the technique considered. The purpose of cluster analysis is to identify a smaller number of groups such that the elements that belong to a group are, in some sense, more similar between them and not to the elements belonging to other groups. The fundamental point of departure is the definition of a similarity measure or distance between objects (that is, between the lines of the matrix of data). Another fundamental point is the rule that allow to form some groups. According to the type of data, we may obtain different sizes. For quantitative data we can have measures of distance, on the contrary for qualitative data we may have measures of association. Among the measures of distances we have taken in consideration the Euclidean metric or Euclidean distance that is the distance between two points that measure, for example, the Euclidean distance between two points $P = (p_1, p_2, ..., p_n)$ and $Q = (q_1, q_2, ..., q_n)$ in $\mathbf{R}^n$ is defined as (1):

$$d(P, Q) = \|P - Q\|_2 = \left(\sum_{i=1}^{n} (p_i - q_i)^2 \right)^{1/2} . \qquad (1)$$

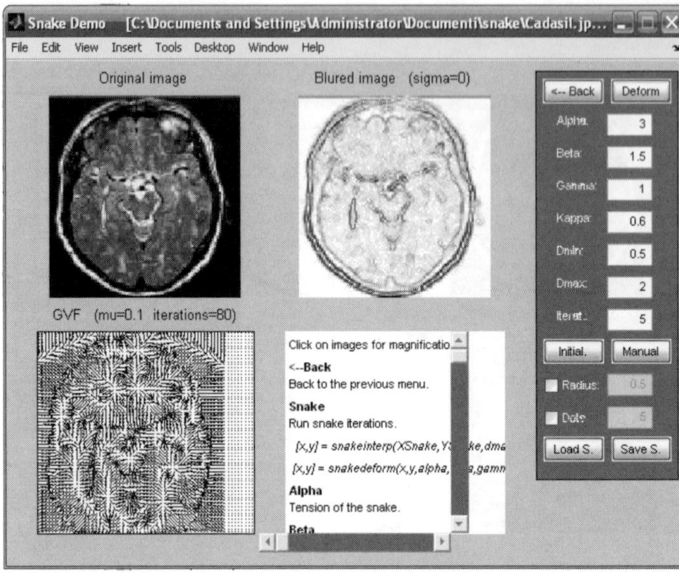

Fig. 3. Final configuration of the GVF snake.

Another measure that we have considered is the correlation is 1-Pearson r. The whole process is based on the definition of criteria for allocation of units to cluster (or a small cluster to a larger one). There are several possible criteria, and consequently, differents aggregative algorithms.

Single Linkage that defines the distance between two clusters as the minimum distance between them (2):

$$d_{(A,B)C} = min\,(d_{AC}, d_{BC})\,. \tag{2}$$

Complete Linkage or the criterion of complete bond, that defines the distance between two clusters as the maximum distance between them (3):

$$d_{(A,B)C} = max\,(d_{AC}, d_{BC})\,, \tag{3}$$

where A, B and C denote the clusters.

By applying these techniques to the curves, we can obtain the dendograms, that is a graph representations that display the level of aggregation of units or clusters according to growing ordinates. According to tree diagram obtained we divided the curves into different groups.

Various techniques have been applied:

3.1. *Cluster analysis with Euclidean distances*

In the case of Euclidean distance, we used the criterion Complete Linkage. In table (1) you can see that the profiles have been grouped into 17 groups using this tecnique. Another curves are isolated, this means that they have different features.

Table 1. Complete Linkage with Euclidean Distances.

Groups	Lesions
1	*pz6_29asm*
2	*pz1_24sm*
3	*pz1_20csm*
4	*pz1_26sm, pz1_20bsm*
5	*pz1_20asm*
6	*pz5_31asteinert, pz2_29steinert, pz4_19bcadasil*
7	*pz1_25asm, pz7_18steinert, pz1_31steinert, pz5_30sm, pz4_22asteinert, pz4_19acadasil*
8	*pz5_31asm, pz1_40sm, pz1_29sm, pz7_16steinert, pz1_30asm, pz2_25steinert, pz2_29sm, pz10_32steinert, pz4_32sm, pz1_30bsteinert, pz2_32asm, pz1_33sm, pz5_26sm, pz2_27cadasil*
9	*pz5_34asm, pz1_34csm, pz1_30asteinert, pz1_30asteinert, pz7_37steinert, pz1_34dsm, pz1_37asm, pz1_35sm, pz1_38sm, pz4_37cadasil*
10	*pz4_22cadasil, pz1_31cadasil*
11	*pz1_30csm*
12	*pz2_30steinert*
13	*pz5_31bsteinert, pz5_20steinert, pz1_23asm, pz4_22bsteinert, pz1_21sm, pz5_23sm, pz1_30dsteinert*
14	*pz1_23bsm, pz4_29sm, pz1_30sm, pz1_31sm, pz7_15steinert, pz1_37bsm, pz1_34bsm, pz6_29bsm, pz2_28steinert, pz1_30csteinert, pz5_31bsm, pz5_34bsm, pz1_32cadasil*
15	*pz5_28steinert, pz4_28steinert*
16	*pz4_38sm*
17	*pz1_25bsm, pz2_32bsm, pz1_34asm, pz4_23cadasil, pz3_31sm, pz1_28sm, pz1_25csm, pz6_31sm, pz3_42cadasil, pz1_28cadasil*

3.2. *Cluster analysis with 1-Pearson r*

With the technique of 1-Pearson r we made different tests to distinguish a Complete Linkage from a Single Linkage.

3.3. *Complete Linkage with 1-Pearson r*

With this technique the profiles have been grouped into 11 groups. In this case the curves have not been isolated and this means that, they can be grouped in the same class, although they have different forms.

3.4. *Single Linkage with 1-Pearson r*

With this technique the profiles have been grouped into 19 groups. In this case another curves are isolated.

4. Analysis of parameters and results

For the parameters analysis we used the method of discreet approximation to least squares, adopted for the reconstruction of flat curves with the help of cubic spline, parametric and periodical functions. For the determination of certain parameters we have applied to data come from the snakes the algorithm described in Forme.[7] This algorithm is divided into two phases. In the first phase set data, derived from snake, are normalized and pre-processed, to get a new file, containing a constant number of coordinates of points equidistant for each curve. These coordinates, which constitute the input of the second phase, are drawn from spline approximating data, obtained with the method of least squares with constraints of periodicity because a closed curve must be rebuild. In the second phase, from groups of data that make up the output of the first phase are derived by applying again the method of least squares to the spline approximating the contours of the objects under consideration. In this case we obtain some analytical equations concerning the same space of functions and so comparable. For each curve set parameters descriptors form derives from the analytical equations of spline and its derivatives. In the experimental part we did the cluster analysis taking into consideration all the lesions that were in the same brain region (periventricular one). These curves were normalized to give rise to comparable items so that they are described in the same area of reference.

4.1. *Complete Linkage with 1-Pearson r*

In case the 1-Pearson r we used the criterion Complete Linkage. As it can be seen from the tree diagram (fig. 4) profiles have been grouped into 5 groups thanks to this technique. Some curves are isolated.

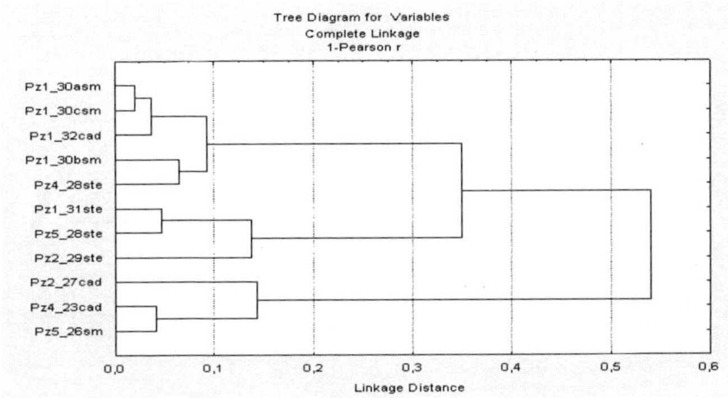

Fig. 4. Complete Linkage with 1-Pearson r.

4.2. *Single Linkage with 1-Pearson r*

In this case, as it can be seen from dendogram (fig. 5), the curves have unique features.

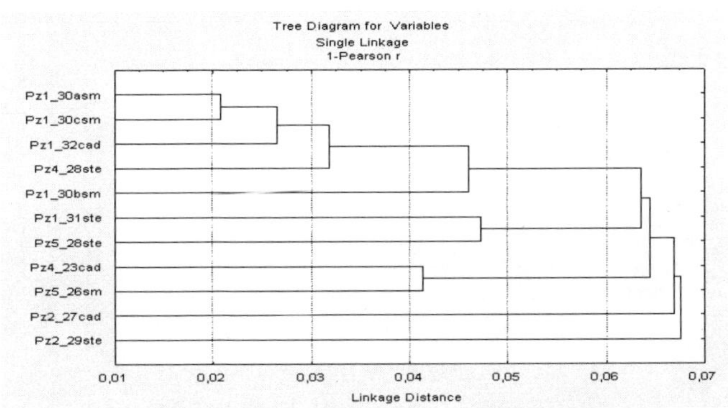

Fig. 5. Single Linkage with 1-Pearson r.

4.3. *Complete Linkage with 1-Pearson r*

In this case, (fig. 6) we have four groups, only two curves are isolated.

4.4. *Single Linkage with Euclidean distances*

In this case, (fig. 7) we have three groups, only two curves are isolated. Only two curves have features similar to other.

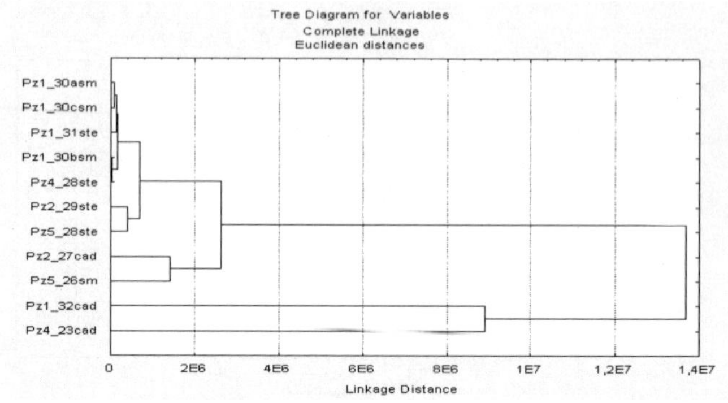

Fig. 6. Complete Linkage with Euclidean Distances.

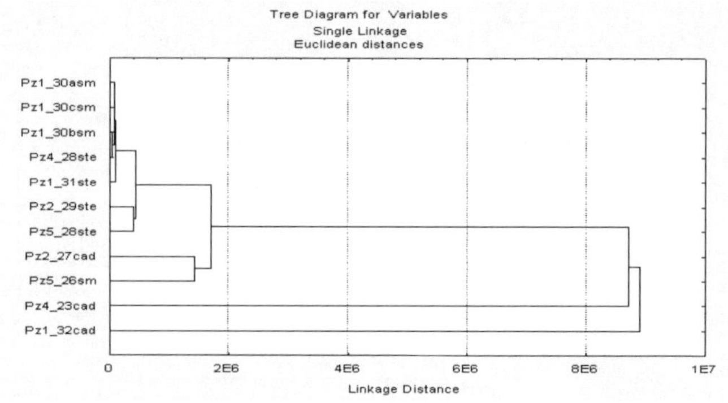

Fig. 7. Complete Linkage with Euclidean Distances.

In the case of analytical parameters we considered the Ward's method. This method is aimed at minimizing the intra-groups variance, so it can only be used for quantitative variables. By applying this technique to the curves we obtain the following dendogram (fig. 8):

4.5. *Ward's method with 1-Pearson r*

We get six groups (fig. 8) but there are two curves isolated.

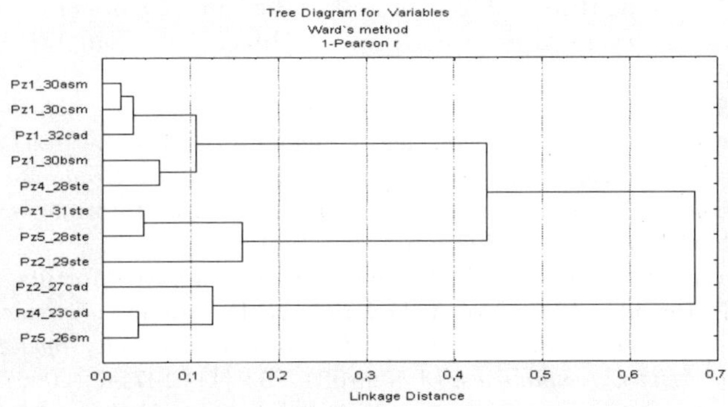

Fig. 8. Ward's method with 1-Pearson r.

5. Conclusion

From the results obtained, we have seen that several analysed curves have similar features to be grouped in the same class. In addition, we have observed that the use of snake, may facilitate the automatic acquisition of the coordinates about the considered profiles. Moreover, with the technique of cluster analysis, as seen, in the same class are different diseases and we have false positives. To resolve this problem, we should strengthen the method and then add other conditions in order to better differentiate the diseases. In conclusion, we can say however that the problem to determine a minimum set of parameters that describe the form remains open and the search is in full development in this area.

References

1. G. Bachmann, M.S. Damian, M. Koch, G. Schilling, B. Fach, S. Stoppler, *The clinical and genetic correlates of MRI findings in myotonic dystrophy*, Neuroradiology 38:629-635 (1996).
2. L. Bonanno, *Funzioni Snake e Cluster Analysis per lo studio di ultrastrutture sub-cellulari*, Tesi di Laurea Magistrale in Matematica, Università di Messina (2007).
3. V. Cavallari, G. Cenacchi, A. Armigliato, *Atti workshop di analisi d'immagine e morfologia quantitativa*, (Taormina, 12-13 Ottobre 1992).
4. B. Censori, L. Provinciali, M. Danni, L. Chiaramoni, M. Maricotti, N. Foschi et al., *Brain involvement in myotonic dystrophy: MRI features and their relationship to clinical and cognitive conditions*, Acta Neurol Scand 90:211-217 (1994).
5. T. F. Cootes, G. J. Edwards and C. J. Taylor, *Active appearance model*, Proc. European Conf. Comp. Vis., pp.484-498 (1998).

6. T. F. Cootes, A. Hill, C. J. Taylor and J. Haslam, *Use of active shape models for locating structures in medical images*, Imag. Vis. Computing J., vol. 12, no. 6, pp. 355-366 (1994).
7. F. De Stefano, *FORME: Un pacchetto interattivo per la morfologia quantitativa*, Tesi di Laurea in Informatica, Università di Messina (1998).
8. D. Fritsch, S. Pizer, L. Yu, V. Johnson and E. Chaney, *Segmentation of medical image objects using deformable shape loci*, Proc. Information Processing in Medical Imaging IPMI'97, pp. 127-140 (1997).
9. A. Giorgio, M. T. Dotti, M. Battaglini, S. Marino, M. Mortilla, M. L. Stromillo, P. Placido, A. Orrico, A. Federico, N. De Stefano, *Cortical damage in brains of patients with adult-form of myotonic dystrophy type 1 and no or minimal MRI abnormalities*, (J Neurol.); 253 (11):1471-7 (2006).
10. P.S. Harper, *Myotonic dystrophy*, Philadelphia PA: WB Saunders (2002).
11. M. Kass, A. Witkin and D. Terzopoulos, *Snakes: Active Contour Models*, International Journal of Computer Vision, 321-331 (1988).
12. C. Li, J. Liu and M. Fox, *Segmentation of edge preserving gradient vector flow: an approach toward automatically initializing and splitting of snakes*, (CVPR), pp.162-167 (2005).
13. A. Mankodi, C.A. Thornton, *Myotonic syndromes*, Curr Opin Neurol 15:545-552 (2002).
14. T. McInerney and D. Terzopoulos, *Deformable Models in Medical Image Analysis: A Survey*, Medical Image Analysis, 1(2) (1996).
15. Y. Miaux, J. Chiras, B. Eymard, M.C. Lauriot-Prevost, H. Radvanyi, N. Martin-Duverneuil et al., *Cranial MRI findings in myotonic dystrophy*, Neuroradiology 39:166-170 (1997).
16. S. O'Riordan, A.M. Nor, M. Hutchinson, *CADASIL imitating multiple sclerosis: the importance of MRI markers*, Mult Scler (2002);8:430-432.
17. N. Paragios, R. Deriche, *Geodesic active regions and level set methods for supervised texture segmentation*, (IJCV), 46(3): 223-247 (2002).
18. N. Paragios, O. Mellina-Gottardo, V. Ramesh, *Gradient Vector Flow Geometric Active Contours*, (IEEE T-PAMI), 26(3):402-407 (2004).
19. M.A. Sahraian, A. Shakouri Rad, M. Motamedi, H. Pakdaman and E.W. Radue, *Magnetic Resonance Imaging abnormalities in multiple sclerosis: a review* Iran. J. Radiol., (Summer 2007).
20. C. Xu and J. Prince, *Snakes, shapes and gradient vector flow*, (IEEE T-IP), 7(3):359-369 (1998).
21. http://iacl.ece.jhu.edu/projects/gvf/snakedemo/

Geometric Probabilities for Three-dimensional Tessellations and Raster Classifications

Vittoria Bonanzinga, Loredana Sorrenti

DIMET, Facoltà di Ingegneria
Università degli Studi "Mediterranea" di Reggio Calabria, Italy
vittoria.bonanzinga@unirc.it
loredana.sorrenti@unirc.it

In this paper we study the statistics of intersections of a sphere with tiles in regular three-dimensional tessellations consisting by cubes with cubic obstacles. We also show that the obtained results are useful in the determination of a classification rule for raster conversions in GIS and GRASS GIS.

Keywords: Geometric probability, stochastic geometry, random sets, random convex sets, GIS, GRASS GIS.

1. Introduction.

The intersection of one spatial features with another is a fundamental geographic operation. A particular instance of intersection occurs when one of the features is a tile in a regular tessellation, and the other is a line, an area or a volume. Several examples follow: calculation of distance between two points referenced in the United Kingdom National Grid, determination of the number of United States Geological Survey, quadrangle maps sheets required to cover a specific watershed, rasterization of a vector landcover map. In such instances, the probability that a feature intersects one of the tiles in a regular tessellation is of interest. There is a growing literature on the applications in Geographic Information Science involving the analysis of data that are stored or measured in regular tessellation on some projected portion of the earth surface.[2,5] In this paper we consider the common GIS application of converting paper land cover maps to a digital raster data base. A variety of encoding rules exist. One of these chooses the class of polygons occupying the central point of each raster cell. There is of course a tradeoff between precision and data volume; corse cell resolution leads to smaller and more tractable data sets at the cost of information loss. To what degree can this loss be estimated in advance of the procedure? For example: what is the probability that a land cover patch of a specific size is captured by raster encoding at a specific cell resolution? This is a specific case of a more general problem: the probability that a spatial feature will intersect at least one central point in a regular tessellation. Solutions for these sorts of problems appear in recent GIS literature,[6] but their mathematical underpinnings are identifiable in texts of geometric probability.[7,10–12] In fact integral geometry provides instruments and methods to solve problems of this kind and actually there exists an active research area working in this direction.[1,4] The power of such solutions is to characterize computational complexity of GIS algorithms and thereby estimate their efficiency.

In general, Earth and its features are located and evolve in 3D space and time. However, for most applications a projection of geospatial data to a flat plane is sufficient; therefore two-dimensional representation of geographical features (with data georeferenced by their horizontal coordinates) is the most common. GIS provides the most comprehensive support for 2D data. Recently a new Geographic Information System, commonly referred to as GRASS GIS,[9] is developing; it is used for data management processing, graphics production, spatial modelling. Recent versions of GRASS GIS include a 3D raster model for volume data.

In view of these new recent developments the study of the statistics of intersections in the three-dimensional case becomes a powerful instrument for eventual encoding rules for 3D raster data model.

In the second section we point out the problem in the two-dimensional case, we illustrate the solution given by Gooldchild and Shortridge[6] and we show that such solution is a special case of a more general result of integral geometry, given by Duma and Stoka.[4]

In the third section we extend the results of[4] to the three-dimensional case. As an application, the solution determined[6] for 2D raster conversion is extended for the analogue problem in 3D case.

2. Buffon's type solutions and 2D raster classification applications for circles.

When converting vector class data to raster, a classification rule must be used to assign a class value to each raster cell. One possibility is to use the class of the polygon occupied by the central point of the cell, a point-counting method of area measurement.[8] A concern with such an approach is that small polygons may disappear in the conversion process because they fail to overlap any cell center point. Then it is fundamental to answer to the following.

Question 2.1. What is the probability that a polygon is missed during conversion to raster?

The required probability is clearly a function of raster cell size and polygon shape and area. In the following,[4] the probability that a circle with radius r intersects one of the squares of a quadratic lattice $\mathfrak{R}(A, a)$ with side A, with quadratic obstacles with sides $2a$ (see figure 1), is computed.

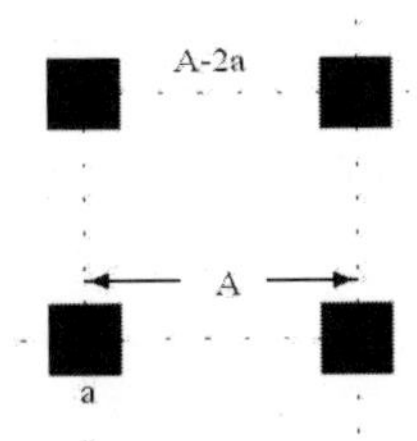

Fig. 1. Fundamental tile of $\mathfrak{R}(A, a)$.

Theorem 2.1. *The probability that a random circle of constant radius r, uniformly distributed in a bounded region of the plane intersects the lattice $\mathfrak{R}(A, a)$ is:*

$$\frac{4a^2+8ar+\pi r^2}{A^2}, \qquad \text{if } r \leq \frac{A-2a}{2}, \tag{1}$$

$$1 - \frac{(A-2a)^2-(A-2a)\sqrt{4r^2-(A-2a)^2}-2\beta r^2}{A^2}, \quad \text{if } \frac{A-2a}{2} < r < \frac{A-2a}{\sqrt{2}}, \tag{2}$$

$$1 \qquad \text{if } r \geq \frac{A-2a}{\sqrt{2}}, \tag{3}$$

where $\beta = \frac{\pi}{2} - 2\arccos(\frac{A-2a}{2r})$.

When $a = 0$ the quadratic obstacles are points and $\mathfrak{R}(A, a)$ represents a grid point mesh (see figure 2).

Corollary 2.1. *The probability that a circle with radius r overlaps a square grid mesh with spacing A is*

$$p = \frac{\pi r^2}{A^2}, \text{ if } r \leq \frac{A}{2}, \tag{4}$$

$$p = \frac{\pi r^2}{A^2} - \frac{4r^2}{A^2}\arccos\frac{A}{2r} + \frac{1}{A}\sqrt{4r^2 - A^2}, \text{ if } \frac{A}{2} < r < \frac{A\sqrt{2}}{2}, \tag{5}$$

$$p = 1, \text{ if } r \geq \frac{A\sqrt{2}}{2}. \tag{6}$$

Formulas (4), (5), (6) coincide with the results given by Goodchild and Shortridge in.[6]

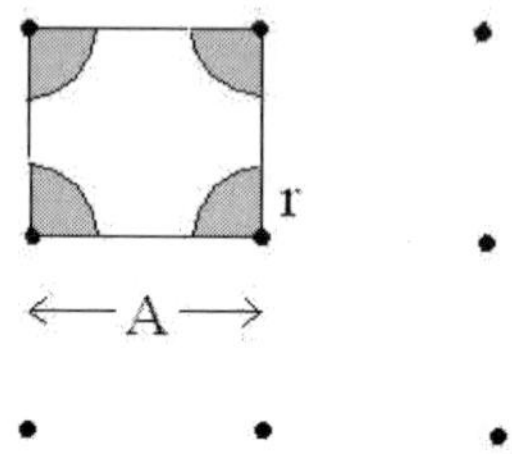

Fig. 2. Geometric representation of the probability that a circle with radius r intersects a gridpoint mesh.

3. Buffon's type solutions and 3D raster classification applications for spheres.

In this section we first determine the probability of intersection of a sphere with the sides of cubic tessellations with cubic obstacles, using techniques of integral geometry. Finally we apply the obtained results to 3D raster classifications.

Duma and Stoka[4] compute the probability that a test body T does not intersect a lattice consisting by square with quadratic obstacles, in the Euclidean plane E_2, in several cases, when T is a circle or a line segment. In this section we consider the analogue problem in the Euclidean space E_3. Let be given, in the euclidian space E_3, a system of perpendicular aces, a lattice $\mathfrak{R}(L, a)$ consisting by cubes C with edges $2a$ having as symmetry center the points $M_{h,k,l} = (hL, kL, lL)$, $h, k, l \in \mathbb{Z}$ and the edges parallel to the coordinate aces, as in figure 3. Let T be a test body with centroid P and oriented axis of rotation $\mathbf{d}$. We want to determine the probability p_T that the test body T does not intersect the lattice $\mathfrak{R}(L, a)$ when T is a sphere with constant radius r placed with random position in the lattice. In order to do this we consider the points $M_{0,0,0}$, $M_{L,0,0}$, $M_{0,L,0}$, $M_{L,L,0}$, $M_{0,0,L}$, $M_{L,0,L}$, $M_{0,L,L}$, $M_{L,L,L}$, and the cube $\mathfrak{C}_0$ with vertexes these points. We denote by $\mathcal{M}$ the set of the spheres having the barycentre in the cube $\mathfrak{C}_0$ and by $\mathcal{N}$ the set of test body completely lying within $\mathfrak{C}_0$ but not intersecting the eight cubes C with symmetry center the points $M_{0,0,0}$, $M_{L,0,0}$, $M_{0,L,0}$,

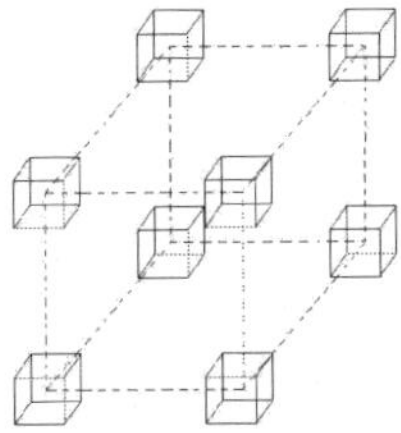

Fig. 3.

$M_{L,L,0}$, $M_{0,0,L}$, $M_{L,0,L}$, $M_{0,L,L}$, $M_{L,L,L}$. We have:

$$p_T = \frac{\mu(\mathcal{M})}{\mu(\mathcal{N})} \tag{7}$$

where μ is the Lebesgue measure. The measures $\mu(\mathcal{M})$ and $\mu(\mathcal{N})$ are computed using the elementary Kinematic measure of the euclidian space E_3:[10]

$$dK = dx \wedge dy \wedge dz \wedge d\Omega \wedge d\phi \tag{8}$$

where x, y, z is the center of the sphere in a rectangular coordinate system, $d\Omega$ the element of the solid angle

$$d\Omega = |\sin\theta| d\psi \wedge d\theta \tag{9}$$

and ϕ the angle of rotation, $0 \leq \phi \leq 2\pi$, $0 \leq \psi \leq 2\pi$, $0 \leq \theta \leq \frac{\pi}{2}$.

Theorem 3.1. *If $L > 2a$, the probability p_S that a sphere S with random position and constant radius r, uniformly distributed in a limited region of the space E_3, does not intersect the lattice $\mathfrak{R}(L,a)$ is:*

$$\begin{aligned} p_S = 1 &- \frac{4\pi a r^2 + 20a^2 r + 16a^3}{L^3} - \frac{2\pi r^3}{3L^3} \\ &- \frac{4a}{L} + \frac{12a^2}{L^2}, \qquad \text{if } r \leq \frac{L-2a}{2}; \end{aligned} \tag{10}$$

$$\begin{aligned} p_S = 1 &- \frac{8a^2}{L^3} + \frac{12a^2}{L^2} - \frac{6a}{L} \\ &- \left(\frac{1}{L} + \frac{2a^2}{L^3} - \frac{4a}{L^2}\right)\sqrt{4r^2 - (L-2a)^2} \\ &- 2\beta r\left(\frac{1}{L^2} - \frac{2a}{L^3}\right), \qquad \text{if } \frac{L-2a}{2} < r < \frac{L-2a}{\sqrt{2}}; \end{aligned} \tag{11}$$

$$p_S = 0, \qquad \text{if } r \geq \frac{L-2a}{\sqrt{2}}, \tag{12}$$

where $\beta = \frac{\pi}{2} - 2\arccos\left(\frac{L-2a}{2r}\right)$.

Proof. We denote by $\mathfrak{C}_1$ the solid (lying within $\mathfrak{C}_0$), with the following property: a point P is inside $\mathfrak{C}_1$ if and only if the sphere S, with centre P is completely contained within $\mathfrak{C}_o$ and does not intersect any cube in $\mathfrak{C}_o$. We have:

$$\begin{aligned}\mu(\mathcal{M}) &= \int_0^{2\pi} d\phi \int_0^{\pi} d\psi \int_0^{\frac{\pi}{2}} \sin\theta d\theta \int_{\{(x,y,z)\in\mathfrak{C}_o\}} dxdydz \\ &= 2\pi^2 \cdot (\, Volume\ of\ \mathfrak{C}_o) = 2\pi^2 L^3.\end{aligned} \tag{13}$$

$$\begin{aligned}\mu(\mathcal{N}) &= \int_0^{2\pi} d\phi \int_0^{\pi} d\psi \int_0^{\frac{\pi}{2}} \sin\theta d\theta \int_{\{(x,y,z)\in\mathfrak{C}_1\}} dxdydz \\ &= 2\pi^2 \cdot (\, Volume\ of\ \mathfrak{C}_1).\end{aligned} \tag{14}$$

If S is a solid figure we denote by V_S the volume of S.
By substituting (13) and (14) in (7) we obtain:

$$p_S = \frac{V_{\mathfrak{C}_1}}{L^3}. \tag{15}$$

First Case. Suppose $r \leq \frac{L-2a}{2}$. In order to compute the volume of $\mathfrak{C}_1$ we note that $\mathfrak{C}_1$ is the union of three solid figures: a solid S_1 with base the plain surface $\mathfrak{B}_1$ represented in figure 4 and height a, a solid S_2 consisting of a cube with edge $L - 2a$ minus four parellelepipeds with edges a, a, r, minus two cylinders with radius r and height a, minus a semisphere with radius r, and a solid S_3 equal to S_1. Then we have:

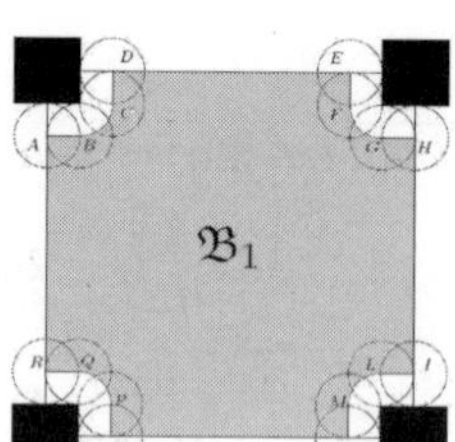

Fig. 4.

$$\begin{aligned} V_{S_1} &= (Area\mathfrak{B}_1) \cdot a \\ &= [4a(L-2a-2r)+(L-2a)^2-\pi r^2] \cdot a \\ &= -4a^3-8a^2r+aL^2-\pi ar^2. \end{aligned} \tag{16}$$

$$\begin{aligned} V_{S_2} &= (L-2a)^3-4ar(a)-2\pi r^2 a-\frac{1}{2}\cdot\frac{4}{3}r^3 \\ &= -8a^3+12a^2L-4a^2r-6aL^2-2\pi ar^2+L^3-\frac{2\pi r^3}{3}. \end{aligned} \tag{17}$$

$$\begin{aligned} V_{\mathfrak{C}_1} &= 2\cdot V_{S_1}+V_{S_2} \\ &= -16a^3+12a^2L-20a^2r-4aL^2-4\pi ar^2+L^3-\frac{2\pi r^3}{3}. \end{aligned} \tag{18}$$

Then by substituting formula (18) in (15) we obtain probability (10).

Second Case. Suppose $\frac{L-2a}{2} < r < \frac{L-2a}{\sqrt{2}}$. We first compute the angle $\beta = P'\hat{Q}'S'$ represented in figure 5.

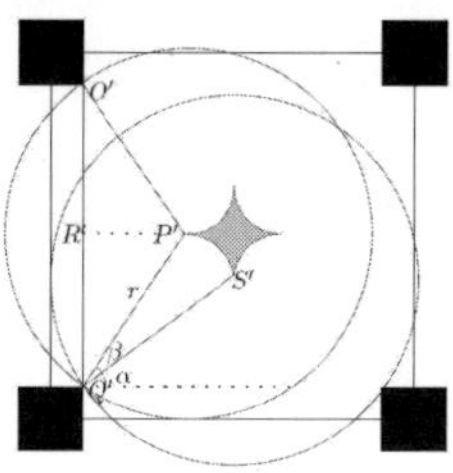

Fig. 5.

We note that

$$\frac{L-2\alpha}{2} = r\cos\alpha,$$

where $\alpha = O'P'Q'$. It follows that

$$\alpha = \arccos\frac{L-2a}{2r}.$$

Then

$$\beta = \frac{\pi}{2}-2\alpha = \frac{\pi}{2}-2\arccos\frac{L-2a}{2r}.$$

In this case $\mathfrak{C}_1$ is a cube with edge $L-2a$ minus four prisms with base the triangle $O'P'Q'$ (see figure 5), and height $L-2a$, minus $\frac{2\beta}{\pi}$ of a cylinder with height $L-2a$ and base with radius r. We have:

$$P'R' = r\sin\alpha = r\sqrt{1-\frac{(L-2a)^2}{4r^2}} = \frac{1}{2}\sqrt{4r^2-(L-2a)^2},$$

and

$$\begin{aligned} Area(O'P'Q') &= (L-2a)\cdot\frac{1}{2}\sqrt{4r^2-(L-2a)^2}\cdot\frac{1}{2} \\ &= \frac{1}{4}(L-2a)\sqrt{4r^2-(L-2a)^2}. \end{aligned}$$

Then

$$\begin{aligned} V_{\mathfrak{C}_1} &= (L-2a)^3-(L-2a)^2\sqrt{4r^2-(L-2a)^2} \\ &-\frac{2\beta}{\pi}\left(\pi r^2\cdot(L-2a)\right). \end{aligned} \tag{19}$$

Then by substituting formulas (19) in (15) we obtain probability (11).

Third case. Suppose $r \geq \frac{L-2a}{2r}$. In this case each sphere with radius r does not intersect one of the cubes C of $\mathfrak{R}(L,a)$. Then $p_S = 0$. □

Now we consider the lattice $\mathfrak{R}'(L,a)$ obtained by $\mathcal{R}(L,a)$ adding the plane portions delimited by the following segments: $\{(x,kL,lL) : x \in [hL+a,(h+1)L-a]\}$, $\{(hL,y,lL) : y \in [hL+a,(h+1)L-a]\}$, $\{(hL,kL,z) : z \in [hL+a,(h+1)L-a]\}$,$h,k,l \in \mathbb{Z}$.

Theorem 3.2. *If $L > 4a$, the probability p_S that a sphere S with radius r does not intersect one of the edges of $\mathfrak{R}'(L,a)$ is the following:*

$$\begin{aligned} p_S &= 1-\frac{16a^3-4a^2r+4ar^2(\pi-5)+4r^3(4-\pi)}{L^3} \\ &\quad -\frac{2\pi r^3}{3L^3}-\frac{8ar-12a^2-8r^2}{L^2}-\frac{4a+2r}{L}, && \text{if } r < a \end{aligned} \tag{20}$$

$$p_S = 1-\frac{2\pi a^2}{3L^3}-\frac{6r}{L}+\frac{12r^2}{L^2}-\frac{8r^3}{L^3}, \qquad \text{if } a \leq r < \frac{L}{2} \tag{21}$$

$$p_S = 0, \qquad \text{if } r \geq \frac{L}{2}. \tag{22}$$

Proof. We denote by $\mathfrak{C}_2$ the solid (lying within $\mathfrak{C}_0$) with the following property: a point P is inside $\mathfrak{C}_2$ if and only if the sphere S, with center P is completely contained within $\mathfrak{C}_0$ and does not intersect any cube in $\mathfrak{C}_0$ and any plane portion delimited by the segments $\{(x,kL,lL) : x \in [hL+a,(h+1)L-a]\}$, $\{(hL,y,lL) : y \in [hL+a,(h+1)L-a]\}$, $\{(hL,kL,z) : z \in [hL+a,(h+1)L-a]\}$,$h,k,l \in \mathbb{Z}$. Arguing as in the proof of Theorem 3.1 we obtain

$$\mu(\mathcal{M}) = 2\pi^2 L^3, \tag{23}$$

$$\mu(\mathcal{N}) = 2\pi^2\cdot(\mathit{Volume\ of\ }\mathfrak{C}_2). \tag{24}$$

By substituting (23) and (24) in (7) we obtain:

$$p_S = \frac{V_{\mathfrak{C}_2}}{L^3}. \tag{25}$$

First case. Suppose that $r < a$. In order to compute the volume of $\mathfrak{C}_2$ we note that $\mathfrak{C}_2$ is the union of three solid figures: a solid S_1 with height $a - r$ and base the plain surface $\mathfrak{B}'_1$ of figure 6, a solid S_2 consisting of a cube with edge $L - 2a$ minus four rectangular parallelepipeds with edges a, r, $a - r$, two cylinders with radius r and height $a - r$, minus one semisphere with radius r, a solid S_3 equal to S_1. We note that, since $a < r$ then $a - r > 0$

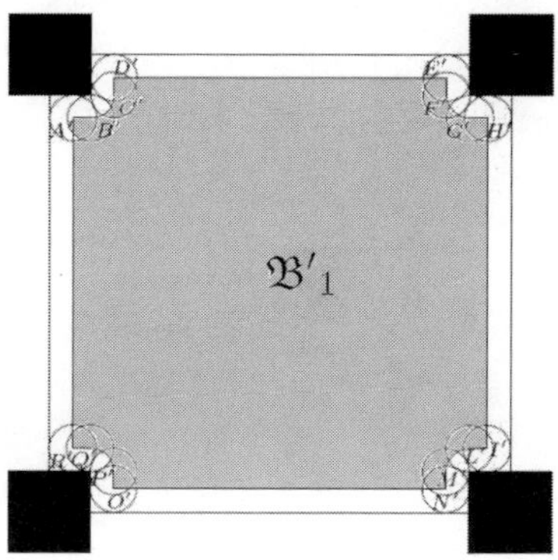

Fig. 6.

and since $L > 4a$, then $L - 2a - 2r > 0$. We have:

$$\begin{aligned} V_{S_1} &= Area(\mathfrak{B}'_1) \cdot (a - r) \\ &= [4(a - r)(L - 2a - 2r) + (L - 2a)^2 - \pi r^2](a - r) \\ &= -4a^3 + 4a^2 r + aL^2 - 4aLr + ar^2(8 - \pi) - L^2 r \\ &+4Lr^2 + r^2(\pi - 8). \end{aligned} \tag{26}$$

$$\begin{aligned} V_{S_2} &= (L - 2a)^3 - 4ar(a - r) - 2\pi r^2(a - r) - \frac{2}{3}\pi r^3 \\ &= -8a^3 + 12a^2 L - 4a^2 r - 6aL^2 + 2ar^2(2 - \pi) + L^3 \\ &+2\pi r^3 - \frac{2}{3}\pi r^3. \end{aligned} \tag{27}$$

$$\begin{aligned} V_{\mathfrak{C}_2} &= 2 \cdot V_{S_1} + V_{S_2} \\ &= -16a^3 + 12a^2 L + 4a^2 r - 4aL^2 - 8aLr + 4ar^2(5 - \pi) \\ &+L^3 - 2L^2 r + 8Lr^2 + 4r^3(\pi - 4) - \frac{2}{3}\pi r^3. \end{aligned} \tag{28}$$

Then by substituting formula (28) in (25) we obtain probability (19).
Second Case. Suppose $a \leq r < \frac{L}{2}$. In this case $\mathfrak{C}_1$ is a cube with edge

$L-2r$ minus one semisphere with radius a (see figure 7 for the plane section of this solid figure). Then:

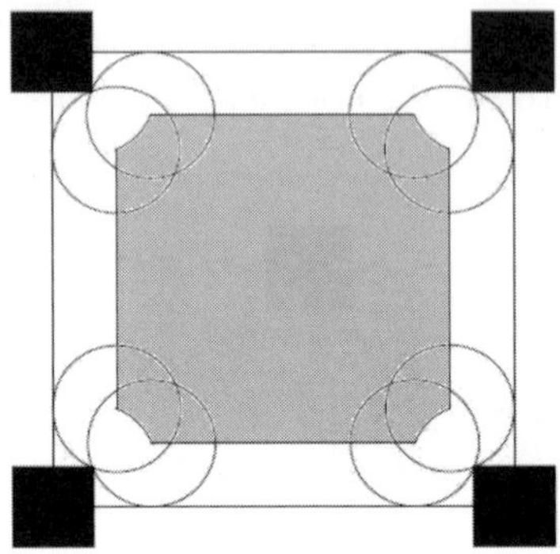

Fig. 7.

$$V_{\mathfrak{C}_2} = (L - 2r)^3 - \frac{2}{3}\pi a^2. \tag{29}$$

Then by substituting formula (29) in (25) we obtain probability (20).
Third Case. Suppose $r \geq \frac{L}{2}$. In this case the sphere S intersects one of the cubes of $\mathfrak{R}'(L, a)$. Then $p_S = 0$. □

Remark 3.1. *If $a = 0$, then $\mathfrak{R}'(L, a)$ is a lattice whose fundamental tile is a cube with edge L. In this case, it follows from Theorem 3.2 that:*

$$p_S = 1 - \frac{6r}{L} + \frac{12r^2}{L^2} - \frac{8r^3}{L^3}. \tag{30}$$

Then the probability of intersection for a sphere with diameter $2r < L$ is

$$p'_S = \frac{6r}{L} - \frac{12r^2}{L^2} + \frac{8r^3}{L^3}. \tag{31}$$

Remark 3.2. *A. Duma and M. Stoka*[3] *determined the following formula for computing the probability of intersection of an arbitrary convex test body K with constant width k, with a lattices of planes whose fundamental cell is a rectangular parallelepiped, with sides a, b, c:*

$$p_K^{(3)} = \frac{k(ab + bc + ca)}{abc} - \frac{k^2(a + b + c)}{abc} + \frac{k^3}{abc}, \tag{32}$$

which coincides with the formula obtained by M. Stoka for a sphere with diameter k.[13] *If $k = 2r$ and $a = b = c = L$ formula (32) coincides with formula (31).*

As an application of the results of this section, we are able to answer to the following.

Question 3.1. What is the probability that a sphere is missed during a 3D conversion to raster?

When $a = 0$ the cubic obstacles are points and $\mathfrak{R}(L, a)$ represents a 3D grid point mesh.

Corollary 3.1. *The probability that a sphere with radius r overlaps a cubic grid mesh with spacing L is*

$$p = \frac{2\pi r^3}{L^3},\ \textit{if}\ r \le \frac{L}{2}, \tag{33}$$

$$p = \frac{1}{L}\sqrt{4r^2 - L^2} + 2\beta\frac{r^2}{L^2},\ \textit{if}\ \frac{L}{2} < r < \frac{L\sqrt{2}}{2}, \tag{34}$$

$$p = 1\ \textit{if}\ r \ge \frac{L - 2a}{\sqrt{2}}, \tag{35}$$

where $\beta = \frac{\pi}{2} - 2\arccos\frac{L}{2r}$.

Proof. The assertion follows from Theorem 3.1. □

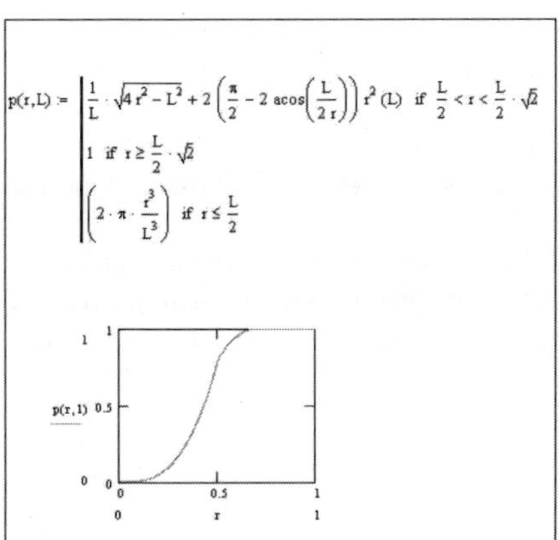

Fig. 8. Graph of the probability of intersection of a sphere with a 3D mesh of spacing 1.

The graph in figure 8 illustrates the implementation of formulas (33), (34), (35) and the probability of intersection for spheres with radius ranging from 0 to 1 on a grid mesh with spacing 1.

References

1. V. Bonanzinga and L. Sorrenti, Buffon type problems with multiple intersections for rectangle in the euclidean plane, *Pub. Inst. Stat. Univ. Paris*, **52/3** (2008), pp. 61–69.
2. B. Boots, Spatial Tessellations, in *Geographical Information Systems: Principles, Techniques Iussues*, 2nd Edition, edited by P. A. Longley, M. F. Goodchild, D. J. Maguire, D. W. Rhind (New York: Wiley & Sons), (1998), pp. 503–526.
3. A. Duma and M. Stoka, Geometric probabilities for convex bodies of revolution in the euclidean space E_3, *Rend. Circ. Mat. Palermo*, serie II-Suppl. **65** (2000), pp. 109–115.
4. A. Duma and M. Stoka, Geometric probabilities for quadratic lattices with quadratic obstacles, *Ann. I.S.U.P.*, **48/1–2**, (2004), pp. 19–42.
5. M. F. Gooldchild, A hierarchical spatial data structure for global geographic information systems, *CVGIP: Graphical Models and Image Processing*, **54/1**, (1992), pp. 31–44.
6. M. F. Gooldchild and A. M. Shortridge, Geometric probability and GIS: some applications for the statistics of intersections, *International Journal of Geographical Information Science*, **16/3**, (2002), pp. 227–243.
7. D. A. Klain and G. C. Rota, Introduction to Geometric Probability, Cambridge University Press, (1997).
8. D. H. Maling, Measurement from Maps: Principles and Methods of Cartometry, Oxford Pergamon Press, (1989).
9. H. Mitasova and M. Neteler, Open source GIS: a GRASS GIS approach, 2nd Edition, Kluwer Academic Publishers, (2004).
10. L. A. Santaló, Integral Geometry and Geometric Probability, Addison Wesley, Mass,(1976).
11. H. Solomon, Geometric Probability, *CBMS-NSF Regional Conference Series in Applied Mathematics*, Society for Industrial and Applied Mathematics, Philadelphia, **28**, (1978).
12. M. Stoka, Probabilitá e Geometria, Herbita (1982).
13. M. Stoka, Sur quelques problèmes de probabilités géométriques pour des réseaux dans l'espace euclidien E_n, *Pub. Inst. Stat. Univ. Paris*, **34/3**, (1989).

A DYNAMIC CONTACT PROBLEM BETWEEN TWO THERMOELASTIC BEAMS

G. BONFANTI

Dipartimento di Matematica, Università degli Studi di Brescia, Brescia, Italy
E-mail: bonfanti@ing.unibs.it
dm.ing.unibs.it/bonfanti/

M. G. NASO

Dipartimento di Matematica, Università degli Studi di Brescia, Brescia, Italy
E-mail: naso@ing.unibs.it
dm.ing.unibs.it/naso/

We deal with a system modelling the vibrations of two thermoelastic beams in dynamic unilateral contact across a joint. Under suitable mechanical and thermal boundary conditions, the evolution problem is shown to possess an energy decaying exponentially to zero, as time goes to infinity.

Keywords: Thermoelastic beam, Signorini condition, contact, asymptotic behavior.

1. Introduction.

We consider a model for the vibrations of two uniform thermoelastic beams that are in unilateral contact across a mechanical joint with clearance (see Fig. 1). Denoting by $u = u(x,t) : (0,l_0) \times (0,T) \to \mathbb{R}$ and $v = v(x,t) : (l_0,l) \times (0,T) \to \mathbb{R}$ the vertical displacements, and by $\theta = \theta(x,t) : (0,l_0) \times (0,T) \to \mathbb{R}$ and $\varphi = \varphi(x,t) : (l_0,l) \times (0,T) \to \mathbb{R}$ the thermal moments of the beams, we deal with the following evolution system

$$\begin{aligned}
&u_{tt}(x,t) + k_1\, u_{xxxx}(x,t) + m\,\theta_{xx}(x,t) = 0 \text{ in } (0,l_0) \times (0,T),\\
&\theta_t(x,t) - \tau_1\,\theta_{xx}(x,t) - m\,u_{xxt}(x,t) = 0 \quad \text{in } (0,l_0) \times (0,T),\\
&v_{tt}(x,t) + k_2\, v_{xxxx}(x,t) + m\,\varphi_{xx}(x,t) = 0 \text{ in } (l_0,l) \times (0,T),\\
&\varphi_t(x,t) - \tau_2\,\varphi_{xx}(x,t) - m\,v_{xxt}(x,t) = 0 \quad \text{in } (l_0,l) \times (0,T).
\end{aligned} \tag{1}$$

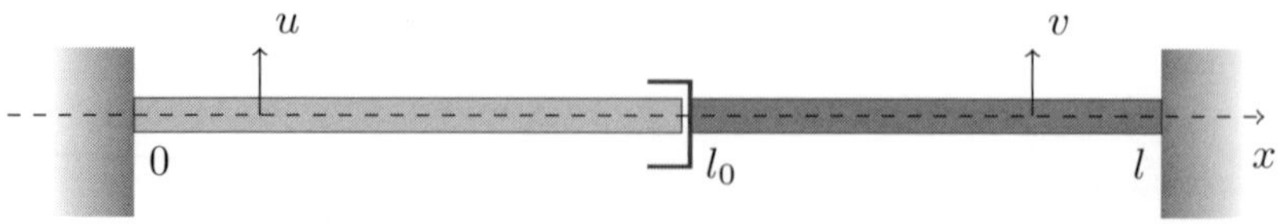

Fig. 1. The two thermoelastic beams in unilateral contact

Here and in what follows, the subscripts x and t indicate partial derivatives. Coefficients k_i, τ_i $(i = 1, 2)$, and m satisfy $k_i, \tau_i > 0$ and $m \neq 0$. Without loss of generality we may assume that $m > 0$. We add to the system (1) the initial conditions

$$\begin{aligned} &u(x,0) = u_0(x),\ u_t(x,0) = u_1(x),\ \theta(x,0) = \theta_0(x),\ \text{ in } (0, l_0), \\ &v(x,0) = v_0(x),\ v_t(x,0) = v_1(x),\ \varphi(x,0) = \varphi_0(x),\ \text{in } (l_0, l), \end{aligned} \tag{2}$$

for some assigned functions $u_0, u_1, \theta_0 : (0, l_0) \to \mathbb{R}$ and $v_0, v_1, \varphi_0 : (l_0, l) \to \mathbb{R}$, and we supplement it with the boundary conditions at $x = 0$ and $x = l$

$$\begin{aligned} &u(0,t) = 0,\ u_x(0,t) = 0,\ \theta_x(0,t) = 0,\ \text{in } (0,T), \\ &v(l,t) = 0,\ \ v_x(l,t) = 0,\ \ \varphi_x(l,t) = 0,\ \text{in } (0,T). \end{aligned} \tag{3}$$

For a detailed derivation of the modeling of thermoelastic beams, we refer, e.g., to Ref. 1. Moreover, we model the joint at $x = l_0$ with the Signorini non penetration condition (see, e.g., Refs. 2–4). We assume that (see Fig. 2) the joint with gap g can be *asymmetrical*, and the gap $g = g_1 + g_2$, where the positive constants g_1 and g_2 are respectively the *upper clearance* and the *lower one*, when the system is at rest.

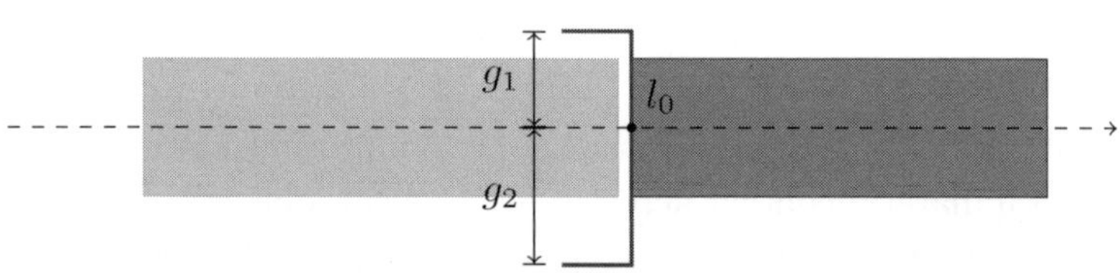

Fig. 2. The joint at $x = l_0$ with clearance $g = g_1 + g_2$

The right end of the left beam is required to be within the clearance of the left end of the right beam. Thus,

$$v(l_0, t) - g_2 \leq u(l_0, t) \leq v(l_0, t) + g_1, \quad t \in (0, T).$$

When strict inequalities hold, namely

$$v(l_0, t) - g_2 < u(l_0, t) < v(l_0, t) + g_1,$$

the ends are *free*, and

$$\sigma_1(l_0,t) = \sigma_2(l_0,t) \equiv 0,$$

where

$$\sigma_1(l_0,t) = -k_1\, u_{xxx}(l_0,t) - m\,\theta_x(l_0,t),$$

$$\sigma_2(l_0,t) = -k_2\, v_{xxx}(l_0,t) - m\,\varphi_x(l_0,t),$$

denote the shear stress of the left beam and of the right one, respectively. When the ends are *in contact*, the stresses are equal, thus,

$$\sigma_1(l_0,t) = \sigma_2(l_0,t) =: \sigma(l_0,t).$$

Moreover, when the contact takes place

- at the lower end, then

$$u(l_0,t) = v(l_0,t) - g_2 \quad \text{and} \quad \sigma(l_0,t) \geq 0;$$

- at the upper end, then

$$u(l_0,t) = v(l_0,t) + g_1 \quad \text{and} \quad \sigma(l_0,t) \leq 0.$$

By introducing the subdifferential of the indicator function χ_v,

$$\chi_v(\phi) = \begin{cases} 0, & \text{if } v - g_2 \leq \phi \leq v + g_1, \\ +\infty, & \text{otherwise}, \end{cases}$$

denoted by $\partial\chi_v$

$$\partial\chi_v(\phi) = \begin{cases} (-\infty, 0], & \text{if } \phi = v - g_2, \\ 0, & \text{if } v - g_2 < \phi < v + g_1, \\ [0, +\infty) & \text{if } \phi = v + g_1, \end{cases}$$

the *contact conditions* may be summarized in the following form:

$$v(l_0,t) - g_2 \leq u(l_0,t) \leq v(l_0,t) + g_1, \tag{4}$$

$$\sigma_1(l_0,t) = \sigma_2(l_0,t) =: \sigma(l_0,t), \tag{5}$$

$$-\,\sigma(l_0,t) \in \partial\chi_{v(l_0,t)}(u(l_0,t)), \tag{6}$$

in $(0,T)$. Moreover, we suppose that

$$k_1\, u_{xx}(l_0,t) = k_2\, v_{xx}(l_0,t), \quad \text{in } (0,T), \tag{7}$$

and

$$\theta(l_0, t) = \varphi(l_0, t), \quad \text{in } (0, T), \tag{8}$$

and hence that the ends, evaluated at $x = l_0$, exert equal moments on each other. Finally, at $x = l_0$ we assume the following mechanical transmission conditions (see, e.g., Ref. 5)

$$u_x(l_0, t) = v_x(l_0, t), \quad \text{in } (0, T), \tag{9}$$

and the thermal ones

$$\tau_1 \, \theta_x(l_0, t) = \tau_2 \, \varphi_x(l_0, t), \quad \text{in } (0, T). \tag{10}$$

The aim of the present paper is to establish a global in time existence result to problem (1)–(10) and analyze its longtime behavior. In particular, we prove that the system possesses an energy decaying exponentially to zero, as time goes to infinity. This investigation is a subsequent continuation of the study begun in Ref. 6 where an analogous contact problem was considered, under different thermal conditions on the joint point. More precisely, in Ref. 6 homogeneous Dirichlet boundary conditions were prescribed on the temperature variables at $x = l_0$, namely $\theta(l_0, t) = 0$ and $\varphi(l_0, t) = 0$ in $(0, T)$. In the present contribution, we aim to generalize such a condition assuming (8). On the other hand, by (10), we are supposing that there is a continuity in the heat flux at the joint point. Related to the new set of boundary conditions, further analytical difficulties arise, expecially in the investigation of the asymptotic behavior of the solutions. The exponential decay of the energy associated to the system is achieved under a suitable choice of the initial data θ_0 and φ_0 (see (13) below). An analogous analysis will be carried out in the forthcoming paper[7] where the system (1)–(10) will be considered for different choices of the coefficients characterizing the materials and for subsequent different boundary conditions on the joint point. As for the literature on related subjects, we observe that the dynamics of contact problems have been extensively investigated both from the modellistic and the analytical point of view (see, e.g., Refs. 3,4,8–15). Concerning the study of the asymptotic behavior of the solutions, we recall, e.g., Refs. 16–19. For a more extended overview, we address the reader to Ref. 6.

Finally, we briefly sketch the plan of the paper. In Sec. 2 we make precise the functional setting and the notation. In Sec. 3 we establish the existence of a weak solution to the problem (1)–(10) (Theorem 3.1) by a regularization-a priori estimates-passage to the limit procedure. In Sec. 4, we focus on the proof of the exponential stability of a solution to the problem (1)–(10) as time goes to infinity (see Theorem 4.2). First, we work in a approximate framework: we find the exponential decay for the approximate solution by introducing a suitable Lyapunov functional and by using the multiplier method. Then, by weak lower semicontinuity arguments, we achieve the exponential decay for a solution to the original problem.

2. Functional setting and notation.

To give a *variational formulation* of our problem, we introduce the following spaces

$$\mathcal{V}_1 = \left\{ w \in H^2(0,l_0): \quad w(0)=0, \quad w_x(0)=0 \right\},$$

$$\mathcal{V}_2 = \left\{ w \in H^2(l_0,l): \quad w(l)=0, \quad w_x(l)=0 \right\},$$

$$\mathcal{H}_1 = H^1(0,l_0),$$

$$\mathcal{H}_2 = H^1(l_0,l),$$

$$\mathcal{M} = \left\{ (\vartheta_1,\vartheta_2) \in L^2(0,l_0) \times L^2(l_0,l): \quad \int_0^{l_0} \vartheta_1(x)\,dx + \int_{l_0}^{l} \vartheta_2(x)\,dx = 0 \right\},$$

$$\mathcal{H} = \{ (\vartheta_1,\vartheta_2) \in \mathcal{H}_1 \times \mathcal{H}_2: \quad \vartheta_1(l_0) = \vartheta_2(l_0) \},$$

$$\mathcal{K} = \{ (w_1,w_2) \in \mathcal{V}_1 \times \mathcal{V}_2: \quad w_2(l_0) - g_2 \le w_1(l_0) \le w_2(l_0) + g_1 \quad \text{and} \quad w_{1x}(l_0) = w_{2x}(l_0) \}.$$

Moreover, concerning the initial data, we assume that

$$(u_0,v_0) \in \mathcal{K}, \tag{11}$$

$$(u_1,v_1) \in L^2(0,l_0) \times L^2(l_0,l), \tag{12}$$

$$(\theta_0,\varphi_0) \in \mathcal{M}. \tag{13}$$

We may now specify the variational problem we are dealing with by introducing the following definition of weak solution to the problem (1)–(10).

Definition 2.1. Let $u_0, v_0, u_1, v_1, \theta_0$, and φ_0 be given as in (11)–(13). A quadruple (u, θ, v, φ) is a *weak solution* to the problem (1)–(10) when

$$(u, v) \in W^{1,\infty}\left(0, T; L^2(0, l_0) \times L^2(l_0, l)\right) \cap L^\infty(0, T; \mathcal{K}), \tag{14}$$

$$(\theta, \varphi) \in L^\infty\left(0, T; \mathcal{M}\right) \cap L^2(0, T; \mathcal{H}), \tag{15}$$

$$u(x, 0) = u_0(x), \quad \text{in } (0, l_0),$$

$$v(x, 0) = v_0(x), \quad \text{in } (l_0, l),$$

and satisfies the relations

$$\begin{aligned}
&\int_0^T \int_0^{l_0} \left[-u_t(w-u)_t + k_1 u_{xx}(w-u)_{xx} - m\theta_x(w-u)_x\right] dx\, dt \\
&\quad + \int_0^T \int_{l_0}^{l} \left[-v_t(z-v)_t + k_2 v_{xx}(z-v)_{xx} - m\varphi_x(z-v)_x\right] dx\, dt \\
&\qquad \geq \int_0^{l_0} u_1(w(\cdot,0) - u_0) dx + \int_{l_0}^{l} v_1(z(\cdot,0) - v_0) dx,
\end{aligned} \tag{16}$$

for all $(w, z) \in W^{1,1}(0, T; L^2(0, l_0) \times L^2(l_0, l)) \cap L^2(0, T; \mathcal{K})$ such that $w(\cdot, T) = u(\cdot, T)$ and $z(\cdot, T) = v(\cdot, T)$, and

$$\begin{aligned}
&\int_0^T \int_0^{l_0} \left(-\theta\psi_t + \tau_1\theta_x\psi_x + m u_{xx}\psi_t\right) dx\, dt \\
&\quad + \int_0^T \int_{l_0}^{l} \left(-\varphi\eta_t + \tau_2\varphi_x\eta_x + m v_{xx}\eta_t\right) dx\, dt \\
&= \int_0^{l_0} (\theta_0 - m u_{0xx})\psi(\cdot, 0)\, dx + \int_{l_0}^{l} (\varphi_0 - m v_{0xx})\eta(\cdot, 0)\, dx,
\end{aligned} \tag{17}$$

for all $(\psi, \eta) \in W^{1,1}(0, T; L^2(0, l_0) \times L^2(l_0, l)) \cap L^2(0, T; \mathcal{H})$ such that $\psi(\cdot, T) = 0$ and $\eta(\cdot, T) = 0$.

3. Global existence result.

To prove the global existence of a weak solution to problem (1)–(10), we list in that follows the steps as developed in Ref. 6.

First, we approximate problem (1)–(10) by a *penalization procedure* obtained by regularizing the Signorini contact condition with a normal compliance condition (see, e.g., Refs. 3,4,18,20).

Let us introduce the families of initial data $\{u_0^\varepsilon\}_{\varepsilon>0}$, $\{u_1^\varepsilon\}_{\varepsilon>0}$, $\{v_0^\varepsilon\}_{\varepsilon>0}$, $\{v_1^\varepsilon\}_{\varepsilon>0}$, $\{\theta_0^\varepsilon\}_{\varepsilon>0}$, and $\{\varphi_0^\varepsilon\}_{\varepsilon>0}$ satisfying

$$(u_0^\varepsilon, v_0^\varepsilon) \in \left[H^4(0,l_0) \times H^4(l_0,l)\right] \cap \mathcal{K}, \tag{18}$$

$$(u_1^\varepsilon, v_1^\varepsilon) \in \mathcal{V}_1 \times \mathcal{V}_2, \tag{19}$$

$$(\theta_0^\varepsilon, \varphi_0^\varepsilon) \in \left[H^2(0,l_0) \times H^2(l_0,l)\right] \cap \mathcal{M}. \tag{20}$$

For any $\varepsilon > 0$ let us consider the following system

$$\begin{aligned}
&u_{tt}^\varepsilon(x,t) + k_1\, u_{xxxx}^\varepsilon(x,t) + m\,\theta_{xx}^\varepsilon(x,t) = 0 \text{ in } (0,l_0)\times(0,T),\\
&\theta_t^\varepsilon(x,t) - \tau_1\,\theta_{xx}^\varepsilon(x,t) - m\,u_{xxt}^\varepsilon(x,t) = 0 \quad \text{in } (0,l_0)\times(0,T),\\
&v_{tt}^\varepsilon(x,t) + k_2\, v_{xxxx}^\varepsilon(x,t) + m\,\varphi_{xx}^\varepsilon(x,t) = 0 \text{ in } (l_0,l)\times(0,T),\\
&\varphi_t^\varepsilon(x,t) - \tau_2\,\varphi_{xx}^\varepsilon(x,t) - m\,v_{xxt}^\varepsilon(x,t) = 0 \quad \text{in } (l_0,l)\times(0,T),
\end{aligned} \tag{21}$$

together with the initial conditions

$$\begin{aligned}
&u^\varepsilon(x,0) = u_0^\varepsilon(x),\ u_t^\varepsilon(x,0) = u_1^\varepsilon(x),\ \theta^\varepsilon(x,0) = \theta_0^\varepsilon(x), \text{ in } [0,l_0],\\
&v^\varepsilon(x,0) = v_0^\varepsilon(x),\ v_t^\varepsilon(x,0) = v_1^\varepsilon(x),\ \varphi^\varepsilon(x,0) = \varphi_0^\varepsilon(x), \text{ in } [l_0,l],
\end{aligned} \tag{22}$$

and the boundary conditions at $x = 0$ and $x = l$

$$\begin{aligned}
&u^\varepsilon(0,t) = 0,\ u_x^\varepsilon(0,t) = 0,\ \theta_x^\varepsilon(0,t) = 0, \text{ in } [0,T],\\
&v^\varepsilon(l,t) = 0,\ v_x^\varepsilon(l,t) = 0,\ \varphi_x^\varepsilon(l,t) = 0, \text{ in } [0,T].
\end{aligned} \tag{23}$$

Let the joint be described by

$$k_1\, u_{xx}^\varepsilon(l_0,t) = k_2\, v_{xx}^\varepsilon(l_0,t), \quad u_x^\varepsilon(l_0,t) = v_x^\varepsilon(l_0,t), \qquad \text{in } [0,T], \tag{24}$$

$$\theta^\varepsilon(l_0,t) = \varphi^\varepsilon(l_0,t), \quad \tau_1\,\theta_x^\varepsilon(l_0,t) = \tau_2\,\varphi_x^\varepsilon(l_0,t), \qquad \text{in } [0,T]. \tag{25}$$

and

$$\sigma_1^\varepsilon(l_0,t) = \sigma_2^\varepsilon(l_0,t) := \sigma^\varepsilon(t), \quad \text{in } [0,T], \tag{26}$$

where

$$\sigma_1^\varepsilon(l_0,t) = -\,k_1\, u_{xxx}^\varepsilon(l_0,t) - m\,\theta_x^\varepsilon(l_0,t), \tag{27}$$

$$\sigma_2^\varepsilon(l_0,t) = -\,k_2\, v_{xxx}^\varepsilon(l_0,t) - m\,\varphi_x^\varepsilon(l_0,t), \tag{28}$$

and

$$\begin{aligned}
\sigma^\varepsilon(t) = &-\frac{1}{\varepsilon}\left\{[u^\varepsilon(l_0,t) - v^\varepsilon(l_0,t) - g_1]^+ - [v^\varepsilon(l_0,t) - u^\varepsilon(l_0,t) - g_2]^+\right\}\\
&- \varepsilon u_t^\varepsilon(l_0,t) + \varepsilon v_t^\varepsilon(l_0,t).
\end{aligned} \tag{29}$$

Here and in the sequel, $[f]^+ := \max\{f, 0\}$ denotes the positive part of f.

By means of the Faedo-Galerkin method, we can prove the following result establishing the well-posedness of the regularized problem (21)–(29). For more details we address the reader to an analogous result proposed in Ref. [6, Proposition 3.2].

Proposition 3.1. *Let $u_0^\varepsilon, v_0^\varepsilon, u_1^\varepsilon, v_1^\varepsilon, \theta_0^\varepsilon, \varphi_0^\varepsilon$ be given as in* (18)–(20) *and compatible with the boundary conditions* (23)–(29) *for $t = 0$. Then, there exists a unique quadruple $(u^\varepsilon, \theta^\varepsilon, v^\varepsilon, \varphi^\varepsilon)$ such that*

$$
\begin{aligned}
(u^\varepsilon, v^\varepsilon) \in W^{2,\infty}&\left(0, T; L^2(0, l_0) \times L^2(l_0, l)\right) \\
&\cap W^{1,\infty}\left(0, T; H^2(0, l_0) \times H^2(l_0, l)\right) \\
&\qquad \cap L^\infty\left(0, T; H^4(0, l_0) \times H^4(l_0, l)\right), \quad (30)
\end{aligned}
$$

$$
(\theta^\varepsilon, \varphi^\varepsilon) \in W^{1,\infty}(0, T; \mathcal{M}) \cap L^\infty(0, T; H^2(0, l_0) \times H^2(l_0, l)), \quad (31)
$$

fulfilling (21)–(29).

Subsequently, we aim to consider a sequence of approximate solutions (provided by Proposition 3.1) and analyze their convergence (as $\varepsilon \to 0$) to a weak solution to the system (1)–(10). To this goal, from now on, we let ε vary, say, in $(0, 1)$ and, concerning the approximating initial data, we assume that

$$
\begin{aligned}
(u_0^\varepsilon, v_0^\varepsilon) &\to (u_0, v_0) \quad \text{in} \quad H^2(0, l_0) \times H^2(l_0, l), && (32)\\
(u_1^\varepsilon, v_1^\varepsilon) &\to (u_1, v_1) \quad \text{in} \quad L^2(0, l_0) \times L^2(l_0, l), && (33)\\
(\theta_0^\varepsilon, \varphi_0^\varepsilon) &\to (\theta_0, \varphi_0) \quad \text{in} \quad L^2(0, l_0) \times L^2(l_0, l). && (34)
\end{aligned}
$$

Next, arguing as in Ref. [6, Section 3.2], we can derive suitable a priori estimates on the quadruple $(u^\varepsilon, \theta^\varepsilon, v^\varepsilon, \varphi^\varepsilon)$ and we pass to the limit with respect to the approximating parameter showing that a sequence of approximate solutions converges to a solution to the original problem. Then, we obtain the following result

Theorem 3.1. *Under the assumptions* (11)–(13)*, there exists a weak solution to problem* (1)–(10) *(in the sense of Definition 2.1).*

4. Exponential decay.

Let (u, θ, v, φ) be a weak solution to the problem (1)–(10) provided by Theorem 3.1. Denoting by

$$E(t, u, \theta, v, \varphi) := \frac{1}{2} \int_0^{l_0} \left[|u_t(x,t)|^2 + k_1 |u_{xx}(x,t)|^2 + |\theta(x,t)|^2 \right] dx$$

$$+ \frac{1}{2} \int_{l_0}^{l} \left[|v_t(x,t)|^2 + k_2 |v_{xx}(x,t)|^2 + |\varphi(x,t)|^2 \right] dx$$

the *energy associated to the system*, we aim to establish that $E(t)$ decays exponentially as $t \to +\infty$ (see Theorem 4.2).

As a first step, we prove that the energy

$$\mathcal{E}^\varepsilon(t) := E(t, u^\varepsilon, \theta^\varepsilon, v^\varepsilon, \varphi^\varepsilon)$$

$$+ \frac{1}{2\varepsilon} \left\{ \left| [u^\varepsilon(l_0,t) - v^\varepsilon(l_0,t) - g_1]^+ \right|^2 + \left| [v^\varepsilon(l_0,t) - u^\varepsilon(l_0,t) - g_2]^+ \right|^2 \right\},$$

associated to the penalized system (21)–(29) decays exponentially.

Theorem 4.1. *Let $(u^\varepsilon, \theta^\varepsilon, v^\varepsilon, \varphi^\varepsilon)$ be the quadruple provided by Proposition 3.1. Then there exist two positive constants M and γ, independent of ε and t, such that*

$$\mathcal{E}^\varepsilon(t) \leq M\, \mathcal{E}^\varepsilon(0)\, e^{-\gamma t}, \quad t \geq 0. \tag{35}$$

Sketch of the proof. We set

$$\mathcal{L}^\varepsilon(t) := N\mathcal{E}^\varepsilon(t) + \frac{1}{8} \left[\mathcal{I}_1^{u,\theta}(t) + \mathcal{I}_1^{v,\varphi}(t) + \frac{\varepsilon}{2} |u^\varepsilon(l_0,t) - v^\varepsilon(l_0,t)|^2 \right]$$

$$+ \delta \left[\mathcal{I}_2^{u,\theta}(t) + \mathcal{I}_2^{v,\varphi}(t) \right] + \frac{1}{m} \left[\mathcal{I}_3^{u,\theta}(t) + \mathcal{I}_3^{v,\varphi}(t) \right], \tag{36}$$

where the positive constants N and δ will be fixed later. The auxiliary functionals are defined by

$$\mathcal{I}_1^{u,\theta}(t) := \int_0^{l_0} u^\varepsilon(x,t)\, u_t^\varepsilon(x,t)\, dx,$$

$$\mathcal{I}_1^{v,\varphi}(t) := \int_{l_0}^{l} v^\varepsilon(x,t)\, v_t^\varepsilon(x,t)\, dx,$$

$$\mathcal{I}_2^{u,\theta}(t) := -\int_0^{l_0} q(x)\, u_x^\varepsilon(x,t)\, u_t^\varepsilon(x,t)\, dx,$$

$$\mathcal{I}_2^{v,\varphi}(t) := -\int_{l_0}^{l} q(x)\, v_x^\varepsilon(x,t)\, v_t^\varepsilon(x,t)\, dx,$$

$$\mathcal{I}_3^{u,\theta}(t) := \int_0^{l_0} \left[\int_0^x \theta^\varepsilon(\xi,t)\, d\xi\right] q(x)\, u_t^\varepsilon(x,t)\, dx,$$

$$\mathcal{I}_3^{v,\varphi}(t) := -\int_{l_0}^{l} \left[\int_x^l \varphi^\varepsilon(\xi,t)\, d\xi\right] q(x)\, v_t^\varepsilon(x,t)\, dx,$$

where $q(x) = x - l_0$, with $x \in [0,l]$. Using the Young and Poincaré inequalities, and the Sobolev embedding theorems, we can prove

$$\sum_{i=1}^{3} |\mathcal{I}_i^{u,\theta}(t)| + \sum_{i=1}^{3} |\mathcal{I}_i^{v,\varphi}(t)| + \varepsilon\, |u^\varepsilon(l_0,t) - v^\varepsilon(l_0,t)|^2 \leq C\, E^\varepsilon(t),$$

for a positive constant C independent of ε and t. Then, if N is large enough, there exist two positive constants C_1 and C_2, independent of ε and t, such that

$$C_1\, \mathcal{E}^\varepsilon(t) \leq \mathcal{L}^\varepsilon(t) \leq C_2\, \mathcal{E}^\varepsilon(t). \tag{37}$$

Moreover, choosing δ small enough, we infer that there exists a positive constant C_0, independent of ε and t, such that

$$\frac{d}{dt}\mathcal{L}^\varepsilon(t) \leq -C_0 \mathcal{E}^\varepsilon(t) \leq -\frac{C_0}{C_2}\mathcal{L}^\varepsilon(t).$$

Hence

$$\mathcal{L}^\varepsilon(t) \leq \mathcal{L}^\varepsilon(0)\, e^{-\frac{C_0}{C_2}t}$$

gives

$$\mathcal{E}^\varepsilon(t) \leq \frac{C_2}{C_1}\mathcal{E}^\varepsilon(0)\, e^{-\frac{C_0}{C_2}t}, \quad t \geq 0,$$

that is (35), with $M = C_2/C_1$ and $\gamma = C_0/C_2$. □

Finally, thanks to weak lower semicontinuity arguments, the main result follows

Theorem 4.2. *Let* (u, θ, v, φ) *be a weak solution to the problem* (1)–(10) *provided by Theorem 3.1. Then, there exist two positive constants* M *and* γ, *independent of* t, *such that*

$$E(t, u, \theta, v, \varphi) \leq M\, E(0, u, \theta, v, \varphi)\, e^{-\gamma t}, \quad t \geq 0. \tag{38}$$

References

1. J. Lagnese and J.-L. Lions, *Modelling analysis and control of thin plates*, Recherches en Mathématiques Appliquées [Research in Applied Mathematics], Vol. 6 (Masson, Paris, 1988).
2. G. Duvaut and J.-L. Lions, *Inequalities in mechanics and physics* (Springer-Verlag, Berlin, 1976).
3. K. L. Kuttler, A. Park, M. Shillor and W. Zhang, *Math. Comput. Modelling* **34**, 365 (2001).
4. K. L. Kuttler and M. Shillor, *Dyn. Contin. Discrete Impuls. Syst. Ser. B Appl. Algorithms* **8**, 93 (2001).
5. T. F. Ma and H. Portillo Oquendo, *Bound. Value Probl.* , Art. ID 75107, 14 (2006).
6. G. Bonfanti, J. E. Muñoz Rivera and M. G. Naso, *J. Math. Anal. Appl.* **345**, 186 (2008).
7. G. Bonfanti, M. Fabrizio, J. E. Muñoz Rivera and M. G. Naso, On the energy decay for a thermoelastic contact problem involving heat transfer, Preprint.
8. H. Antes and P. D. Panagiotopoulos, *The boundary integral approach to static and dynamic contact problems*, International Series of Numerical Mathematics, Vol. 108 (Birkh˙Equality and inequality methods.
9. K. T. Andrews, M. Shillor and S. Wright, *J. Elasticity* **42**, 1 (1996).
10. M. I. M. Copetti, *M2AN Math. Model. Numer. Anal.* **38**, 691 (2004).
11. C. M. Elliott and Q. Tang, *Nonlinear Anal.* **23**, 883 (1994).
12. K. L. Kuttler and M. Shillor, *Nonlinear World* **2**, 355 (1995).
13. P. Shi and M. Shillor, *European J. Appl. Math.* **1**, 371 (1990).
14. M. Shillor, M. Sofonea and J. J. Telega, *Models and analysis of quasistatic contact*, Lect. Notes Phys., Vol. 655 (Springer, 2004).
15. M. Stavroulaki and G. E. Stavroulakis, *Int. J. Appl. Math. Comput. Sci.* **12**, 115 (2002).
16. H. Gao and J. E. Muñoz Rivera, *J. Differential Equations* **186**, 52 (2002).
17. J. E. Muñoz Rivera and M. de Lacerda Oliveira, *IMA J. Appl. Math.* **58**, 71 (1997).
18. J. E. Muñoz Rivera and S. Jiang, *J. Math. Anal. Appl.* **217**, 423 (1998).
19. M. Nakao and J. E. Muñoz Rivera, *J. Math. Anal. Appl.* **264**, 522 (2001).
20. J. A. C. Martins and J. T. Oden, *Comput. Methods Appl. Mech. Engrg.* **40**, 327 (1983).

AN ITERATIVE THRESHOLDING ALGORITHM FOR THE NEURAL CURRENT IMAGING

G. BRETTI

Università Campus Bio-Medico di Roma
Via A. del Portillo 21, 00128 Roma, Italy
and Dip. Me.Mo.Mat., Università di Roma "La Sapienza"
Via A. Scarpa 16, 00161 Roma, Italy
E-mail: gab.bretti@gmail.com

F. PITOLLI

Dip. Me.Mo.Mat., Università di Roma "La Sapienza"
Via A. Scarpa 16, 00161 Roma, Italy
E-mail: pitolli@dmmm.uniroma1.it

Neural current imaging aims at analyzing the functionality of the human brain through the localization of those regions where the neural current flows. The reconstruction of an electric current distribution from its magnetic field measured in the outer space, gives rise to a highly ill-posed and ill-conditioned inverse problem. We use a joint sparsity constraint as a regularization term and we propose an efficient iterative thresholding algorithm to recover the current distribution. Some numerical tests are also displayed.

Keywords: Electric current imaging, Magnetoencephalograpy, Inverse problem, Sparsity constraint, Iterative thresholding, Multiscale basis.

1. Introduction

Bioelectric current imaging aims at analyzing the functionality of the human internal organs through the localization of those regions where their electric activity arises. In particular, magnetoencephalographic (MEG) neuroimaging aims at detecting the brain active regions through the measurements of the tiny magnetic field generated outside the brain by neural currents (see,[4] and references therein).

Since the neuromagnetic field can be very weak and affected by high noise, the measurements are performed by very sensitive magnetometers based on Superconducting QUantum Interference Devices (SQUIDs). Nevertheless, a sophisticated data analysis is needed.

MEG measurements do not directly provide a neural current image and we need a model linking the current distribution to the external magnetic field. Actually, electric current imaging requires to solve a linear inverse problem which usually does not have a unique solution[7,12] so that further constraints on the solution have to be added. Regularization techniques which use a quadratic constraint, give good results when the quantities under observation are equally distributed in time or space.[7] However, the current distribution we want to reconstruct usually has a *sparse representation* so that can be represented has a linear combination of *few* localized basic currents.[5] To promote sparsity in the reconstruction of scalar quantities, a regularization technique based on *non quadratic constraints* was introduced in [3] and the solution of the inverse problem was approximated by a convergent thresholded Landweber algorithm. A generalization to the vector case by means of a *joint sparsity constraint* was proposed in [9] and joint sparsity was used in [8] to regularize the MEG inverse problem. Following,[10] in this paper we introduce an efficient algorithm to numerically solve the MEG inverse problem and we show that the resulting thresholded iteration is convergent. Numerical tests show that joint sparsity outperforms Tikhonov regularization and soft thresholding,[6] especially in the case of highly noisy data.
The paper is organized as follows. The inverse MEG problem is briefly recalled in Section 2. Then, in Section 3 we introduce the joint sparsity constraint as a regularization term. A convergent iterative thresholding algorithm which solves the MEG inverse problem is outlined in Section 4. Finally, some numerical tests are displayed in Section 5.

2. The MEG Inverse Problem

Let us model the head as a set of different regions, say V_k, $k = 0, \dots, K$, with constant conductivity σ_k, such that

$$V_k \subset V_{k+1}, \qquad k = 0, \dots, K-1. \tag{1}$$

The regions V_k, $k = 0, \dots, K-1$, represent different anatomical parts, i.e. the brain, the cerebrospinal fluid, the meninges, the skull, etc. The neural currents $\vec{\mathbf{J}}$ are confined to the brain, represented by the innermost region V_0. The external magnetic field $\vec{\mathbf{B}}$ is linked with the current $\vec{\mathbf{J}}$ by the *Biot-*

Savart law[12]

$$\vec{\mathbf{B}}(\vec{\mathbf{r}}) = \vec{\mathbf{B}}_\infty(\vec{\mathbf{r}}) - \\ - \frac{\mu_0}{4\pi}\sum_{k=0}^{K}(\sigma_{k+1}-\sigma_k)\int_{\partial V_k}\frac{\Phi(\vec{\mathbf{r}}')\,\vec{\mathbf{e}}_k(\vec{\mathbf{r}}')\times(\vec{\mathbf{r}}-\vec{\mathbf{r}}')}{|\vec{\mathbf{r}}-\vec{\mathbf{r}}'|^3}\,d\big(\partial V_k(\vec{\mathbf{r}}')\big)\,, \tag{2}$$

where $\mathbf{B}_\infty$ is the magnetic field in an infinite homogeneous medium with magnetic permeability μ_0, i.e.

$$\vec{\mathbf{B}}_\infty(\vec{\mathbf{r}}) = \frac{\mu_0}{4\pi}\int_{V_0}\vec{\mathbf{J}}(\vec{\mathbf{r}}\,')\times\frac{\vec{\mathbf{r}}-\vec{\mathbf{r}}\,'}{|\vec{\mathbf{r}}-\vec{\mathbf{r}}\,'|^3}\,d\vec{\mathbf{r}}'\,, \tag{3}$$

Φ is the electric potential on the surfaces ∂V_k and $\vec{\mathbf{e}}_k$ is the unit normal w.r.t. ∂V_k, the surface between V_k and V_{k+1}. Note that σ_{K+1} is set equal to 0.

Let $\vec{\mathbf{q}}_l$, $l = 1, \ldots, N$, be the site coordinates of the magnetometers, which are located on a spherical surface Ω centered at the origin with $\delta := dist(V_K, \partial\Omega) > 0$. Usually, just the normal component w.r.t. Ω is measured, i.e.

$$B_r(\vec{\mathbf{q}}_l) := \vec{\mathbf{B}}(\vec{\mathbf{q}}_l)\cdot\vec{\mathbf{e}}_r(\vec{\mathbf{q}}_l), \quad l = 1, \ldots, N, \tag{4}$$

where $\vec{\mathbf{e}}_r(\vec{\mathbf{q}}_l)$ is the unit normal w.r.t. Ω in $\vec{\mathbf{q}}_l$.

Here, we assume that the regions V_k are concentric spheres centered at the origin with increasing radius. Then, recalling that for any three vectors in $\mathbf{R}^3$ it holds $v\times w\cdot z = -z\times w\cdot v$, one has:

$$\mathcal{B}(\vec{\mathbf{J}}, \vec{\mathbf{q}}_l) := B_r(\vec{\mathbf{q}}_l) = \frac{\mu_0}{4\pi}\int_{V_0}\frac{\vec{\mathbf{e}}_r(\vec{\mathbf{q}}_l)\times(\vec{\mathbf{r}}\,'-\vec{\mathbf{q}}_l)}{|\vec{\mathbf{r}}\,'-\vec{\mathbf{q}}_l|^3}\cdot\vec{\mathbf{J}}(\vec{\mathbf{r}}\,')\,d\vec{\mathbf{r}}\,'\,. \tag{5}$$

We remark that, due to the spherical symmetry, the magnetic field generated by the electric potential does not contribute to B_r.

Note that the normal component of $\vec{\mathbf{B}}$ on a surface external to the brain uniquely determines the magnetic field out of the head as soon as the current flux on the scalp is known. In fact, this is the case since the current flux is null.

The MEG inverse problem aims at reconstructing the current distribution $\vec{\mathbf{J}}$ starting from the normal component of the magnetic field measured in $\vec{\mathbf{q}}_l$. In order to identify the current sources from the measurements $\mathbf{M} = (m_1, \ldots, m_N)$ we have to minimize the discrepancy

$$\Delta(\vec{\mathbf{J}}) := \left\|\mathbf{G}(\vec{\mathbf{J}}) - \mathbf{M}\right\|^2_{\mathbf{R}^N}\,, \tag{6}$$

where $\mathbf{G}(\vec{\mathbf{J}}) = \big(\mathcal{B}(\vec{\mathbf{J}}, \vec{\mathbf{q}}_1), ..., \mathcal{B}(\vec{\mathbf{J}}, \vec{\mathbf{q}}_N)\big)$.

Unfortunately, this is a strongly ill-posed problem since there exist silent currents that do not produce any magnetic field in the outer space, so that non unique solutions can be expected. For these reasons the minimization of the discrepancy might not be feasible and some regularization technique is required, see [7,11].

3. A Joint Sparsity Constraint for the MEG Inverse Problem

Let us assume that the current density $\vec{\mathbf{J}} = (J_1, J_2, J_3) \in L_2(V_0; \mathbf{R}^3)$ can be *sparsely* represented by a suitable dictionary $\mathcal{D} := (\psi_\lambda)_{\lambda\in\mathbf{\Lambda}}$, i.e.

$$J_\ell \approx \sum_{\lambda\in\mathbf{\Lambda}_S} j_\lambda^\ell \psi_\lambda\,, \qquad j_\lambda^\ell = \langle J_\ell, \psi_\lambda\rangle\,, \qquad \ell = 1,2,3, \tag{7}$$

where $\mathbf{\Lambda}_S \subset \mathbf{\Lambda}$ is the set of the *few* significant coefficients j_λ^ℓ.[5] As a dictionary we choose a stable multiscale basis $(\psi_\lambda)_{\lambda\in\mathbf{\Lambda}} \in L_2(V_0, \mathbf{R})$, for instance, a wavelet basis[2] or a frame system.[1]

By using the decomposition (7), the discrepancy (6) can be written as

$$\Delta(\vec{\mathbf{j}}) = \left\|T\vec{\mathbf{j}} - \mathbf{M}\right\|_{\mathbf{R}^N}^2, \tag{8}$$

where $\vec{\mathbf{j}} := (j_\lambda^\ell)_{\lambda\in\Lambda,\ell=1,2,3}$ and the operator $T : \ell_2(\Lambda, \mathbf{R}^3) \to \mathbf{R}^N$ has entries given by

$$(T\vec{\mathbf{j}})_l = \sum_{\ell=1}^{3}\sum_{\lambda\in\mathbf{\Lambda}} j_\lambda^\ell \frac{\mu_0}{4\pi}\int_{V_0}\left(\frac{\vec{\mathbf{e}}_r(\vec{\mathbf{q}}_l)\times(\vec{\mathbf{r}}\,'-\vec{\mathbf{q}}_l)}{|\vec{\mathbf{r}}\,'-\vec{\mathbf{q}}_l|^3}\right)_\ell \psi_\lambda(\vec{\mathbf{r}}\,')\,d\vec{\mathbf{r}}\,'. \tag{9}$$

Thus, the MEG inverse problem can be set as follows. *Given a set of magnetic field measurements* $\mathbf{M}$, *determine the decomposition coefficients* $\vec{\mathbf{j}}$ *that minimize the functional*

$$\mathcal{J}_{\mathbf{\Psi}}(\vec{\mathbf{j}}) := \mathbf{\Delta}(\vec{\mathbf{j}}) + \mathbf{\Psi}_{\mathcal{D}}(\vec{\mathbf{j}}), \tag{10}$$

where $\mathcal{D}$ *is a dictionary such that the solution of the minimum problem is sparsely represented, and* $\mathbf{\Psi}_{\mathcal{D}}$ *is the* joint sparsity measure

$$\mathbf{\Psi}_{\mathcal{D}}(\vec{\mathbf{j}}, v) := \sum_{\lambda\in\mathbf{\Lambda}} v_\lambda\,\|\vec{\mathbf{j}}_\lambda\|_p + \sum_{\lambda\in\mathbf{\Lambda}} \omega_\lambda\|\vec{\mathbf{j}}_\lambda\|_2^2 + \sum_{\lambda\in\mathbf{\Lambda}} \theta_\lambda(\rho_\lambda - v_\lambda)^2, \quad p \geq 1. \tag{11}$$

introduced in[9]. Here, $(\theta_\lambda)_{\lambda\in\mathbf{\Lambda}}$, $(\rho_\lambda)_{\lambda\in\mathbf{\Lambda}}$, $(\omega_\lambda)_{\lambda\in\mathbf{\Lambda}}$ are positive sequences and $\|\cdot\|_p$ denotes the usual p-norm for vectors in $\mathbf{R}^3$.

Thus, the MEG inverse problem with the joint sparsity constraint consists in minimizing the functional

$$\mathcal{J}^{(p)}_{\theta,\omega,\rho}(\vec{\mathbf{j}}, v) = \sum_{l=1}^{N} \left|(T\vec{\mathbf{j}})_l - m_l\right|^2 + \sum_{\lambda\in\mathbf{\Lambda}} v_\lambda \, \|\vec{\mathbf{j}}_\lambda\|_p + \sum_{\lambda\in\mathbf{\Lambda}} \omega_\lambda \|\vec{\mathbf{j}}_\lambda\|_2^2 + \sum_{\lambda\in\mathbf{\Lambda}} \theta_\lambda (\rho_\lambda - v_\lambda)^2 \tag{12}$$

jointly with respect to both $\vec{\mathbf{j}}$ and v, restricted to $v_\lambda \geq 0$. The minimization of $\mathcal{J}^{(p)}_{\theta,\omega,\rho}$ promotes that all entries of the vector $\vec{\mathbf{j}}_\lambda$ have the *same sparsity* pattern. Note that v serves as an indicator of large values of $\|\vec{\mathbf{j}}_\lambda\|_p$ and $0 \leq v_\lambda \leq \rho_\lambda$, $\lambda \in \mathbf{\Lambda}$, at the minimum. The quadratic term $\sum_{\lambda\in\mathbf{\Lambda}} \omega_\lambda \|\vec{\mathbf{j}}_\lambda\|_2^2$ makes the overall functional convex, depending on a suitable choice of the sequence $(\omega_\lambda)_{\lambda\in\mathbf{\Lambda}}$ (see [9] for the details).

We remark that when $\theta_\lambda = 0$ and $\omega_\lambda = \alpha$ (a fixed constant) for all $\lambda \in \Lambda$, we obtain the usual Tikhonov regularization since $(v_\lambda)_{\lambda\in\Lambda} = 0$ in this case.

4. An Iterative Thresholding Algorithm for the MEG Inverse Problem

The minimizer $(\vec{\mathbf{j}}^*, v^*)$ of the functional $\mathcal{J}^{(p)}_{\theta,\rho,\omega}$ can be approximated by the following iterative thresholding algorithm deduced from [10].

Algorithm JS

$$
\begin{cases}
\textit{Let } \gamma \textit{ be a suitable relaxation parameter} \\[2ex]
\textit{Choose the positive sequences } \quad (\theta_\lambda)_{\lambda\in\mathbf{\Lambda}},\ (\rho_\lambda)_{\lambda\in\mathbf{\Lambda}},\ (\omega_\lambda)_{\lambda\in\mathbf{\Lambda}} \\[2ex]
\textit{Choose } \quad \vec{\mathbf{j}}^{(0)} = 0 \\[2ex]
\textit{For } \quad 0 \le k \le K \quad \textit{do} \\[2ex]
\quad \textit{Let } \ \nu_\lambda^{(0)} = \rho_\lambda \ \textit{and} \ \ \vec{\mathbf{z}}_\lambda^{(0)} = \vec{\mathbf{j}}_\lambda^{(k)} + \gamma\left(T^*(M - T\vec{\mathbf{j}}^{(k)})\right)_\lambda, \quad \lambda \in \mathbf{\Lambda} \\[2ex]
\quad \textit{For } \quad 0 \le r \le R \quad \textit{do} \\[2ex]
\quad\quad \vec{\mathbf{z}}_\lambda^{(r+1)} = S^{(p)}_{\nu_\lambda^{(r)}}\left(\vec{\mathbf{z}}_\lambda\right), \quad \lambda \in \Lambda \\[2ex]
\nu_\lambda^{(r+1)} = \begin{cases} \rho_\lambda - \frac{1}{2\theta_\lambda(1+\omega_\lambda)}\|\vec{\mathbf{z}}_\lambda^{(r+1)}\|_p & \textit{if } \|\vec{\mathbf{z}}_\lambda^{(r+1)}\|_p < 2\theta_\lambda(1+\omega_\lambda)\rho_\lambda \\ 0 & \textit{otherwise} \end{cases}, \quad \lambda \in \mathbf{\Lambda} \\[2ex]
\textit{Approximate } \ \vec{\mathbf{j}}_\lambda^{(k+1)} \approx \dfrac{\vec{\mathbf{z}}_\lambda^{(R+1)}}{1+\omega_\lambda}, \qquad \lambda \in \mathbf{\Lambda} \\
\textit{Compute } \ v_\lambda^{(K+1)} = \begin{cases} \rho_\lambda - \frac{1}{2\theta_\lambda}\|\vec{\mathbf{j}}_\lambda^{(K+1)}\|_p & \textit{if } \|\vec{\mathbf{j}}_\lambda^{(K+1)}\|_p < 2\theta_\lambda\rho_\lambda \\ 0 & \textit{otherwise} \end{cases}, \quad \lambda \in \mathbf{\Lambda}.
\end{cases}
$$

We remark that the values $(v_\lambda^{(K+1)})_{\lambda\in\mathbf{\Lambda}}$ are not always needed.

The operator $S_\nu^{(p)}(\vec{\mathbf{z}}) = \left(S_\nu^{(p)}(\vec{\mathbf{z}})\right)_{\ell=1,2,3}$ is a thresholding operator whose explicit expression for $p = 1, 2$ is (cf. [9]):

$$
\left(S_\nu^{(1)}(\vec{\mathbf{z}})\right)_\ell = s_\nu^{(1)}(z_\ell) := \begin{cases} z_\ell - sign(z_\ell)\frac{\nu}{2} & \text{if } \ |z_\ell| > \frac{\nu}{2}, \\[2ex] 0 & \text{otherwise,} \end{cases} \tag{13}
$$

$$S_\nu^{(2)}(\vec{\mathbf{z}}) := \begin{cases} \frac{\|\vec{\mathbf{z}}\|_2 - \nu/2}{\|\vec{\mathbf{z}}\|_2}\,\vec{\mathbf{z}} & \text{if} \quad \|\vec{\mathbf{z}}\|_2 > \frac{\nu}{2}, \\ 0 & \text{otherwise.} \end{cases} \tag{14}$$

Theorem 4.1. *Let $p = 1, 2, \infty$, and assume*

$$\inf_{\lambda \in \mathbf{\Lambda}} \theta_\lambda (s_{\min} + \omega_\lambda) > \frac{\kappa_p}{4},$$

where $s_{\min}$ is the minimum of the spectrum of T^*T (note that in this case $s_{\min} = 0$), with $\kappa_p = 3$ for $p = 1$ and $\kappa_p = 1$ for $p = 2, \infty$. Then, the Algorithm JS converges strongly to the unique pair $(\vec{\mathbf{j}}^*, v^*)$ minimizing the functional $\mathcal{J}^{(p)}_{\theta,\rho,\omega}$. Moreover, the following error estimate holds:

$$\|\vec{\mathbf{j}}^{(k)} - \vec{\mathbf{j}}^*\|_2 \leq \beta^k \, \|\vec{\mathbf{j}}^{(0)} - \vec{\mathbf{j}}^*\|_2, \tag{15}$$

where

$$\beta := \sup_{\lambda \in \mathbf{\Lambda}} \frac{4\theta_\lambda(1 - s_{\min})}{4\theta_\lambda(1 + \omega_\lambda) - \kappa_p} < 1.$$

Proof. From [10, Lemma 2.1] it follows that $\mathcal{J}^{(p)}_{\theta,\rho,\omega}$ is strictly convex and has a unique minimizer. Then, the claim follows from Prop. 3.4 and Th. 4.3 in [10]. □

In order to implement the algorithm we need to evaluate $T^*T\,\vec{\mathbf{j}}$, whose explicit expression is given by (cf. [8])

$$\left(T^*T\vec{\mathbf{j}}\right)_{\lambda,\ell} = \sum_{\mu \in \mathbf{\Lambda}} \sum_{m=1}^{3} \left(\sum_{l=1}^{N} (A_{\ell,l}\psi_\lambda)(A_{m,l}\psi_\mu) \right) j_\mu^m, \qquad \lambda \in \mathbf{\Lambda}, \quad \ell = 1,2,3, \tag{16}$$

where the operator $A_{\ell,l} : L_2(V_0; \mathbf{R}) \to \mathbf{R}$ is defined as

$$A_{\ell,l}\, f := \frac{\mu_0}{4\pi} \int_{V_0} \left(\frac{\vec{\mathbf{e}}_z(\vec{\mathbf{q}}_l) \times (\vec{\mathbf{r}}\,' - \vec{\mathbf{q}}_l)}{|\vec{\mathbf{r}}\,' - \vec{\mathbf{q}}_l|^3} \right)_\ell f(\vec{\mathbf{r}}\,')\, d\vec{\mathbf{r}}\,'. \tag{17}$$

Let $\mathcal{M}$ be the matrix whose entries are the coordinates of T^*T in the multiscale basis $(\psi_\lambda)_{\lambda \in \mathbf{\Lambda}}$, i.e.

$$\mathcal{M}_{(\lambda,\ell),(\mu,m)} := \sum_{l=1}^{N} (A_{\ell,l}\psi_\lambda)(A_{m,l}\psi_\mu), \qquad \lambda, \mu \in \mathbf{\Lambda}, \quad \ell, m = 1,2,3. \tag{18}$$

From (16) it follows

$$\left(T^*T\vec{\mathbf{j}}\right)_{\lambda,\ell} = \sum_{\mu \in \mathbf{\Lambda}} \sum_{m=1}^{3} j_\mu^m \, \mathcal{M}_{(\lambda,\ell),(\mu,m)}, \qquad \lambda \in \mathbf{\Lambda}, \quad \ell = 1,2,3. \tag{19}$$

Since $\mathcal{M}$ is a bi-infinite matrix, in order to implement an efficient procedure to compute $T^*T\,\vec{\mathbf{j}}^{(k)}$ we need the amplitude of the entries of $\mathcal{M}$ to decay fast when λ and μ increase. Let us choose as multiscale basis a compactly supported wavelet basis with $\Omega_\lambda := \text{supp}\,(\psi_\lambda) \sim 2^{-|\lambda|}$, where $|\lambda|$ denote the spatial resolution scale of ψ_λ. Moreover, we assume that the basis functions have d vanishing moments, a prescribed smoothness, and fast decay, i.e. $|\psi_\lambda| \leq C2^{-3/2|\lambda|}$. It can be shown that $\mathcal{M}$ has compressibility properties w.r.t. such a basis (cf. [8]), so that $T^*T\,\vec{\mathbf{j}}^{(k)}$ can be evaluated efficiently.

5. Numerical Tests

Under the assumption that the magnetometers are very close to the skull, the MEG setting reduces to a bidimensional problem in which the current flows in the (x, y)-plane and the magnetic field is measured by a magnetometer array located at a given height. Although in this case the inverse problem has a unique solution, it can be ill-conditioned for the presence of high noise.

In the numerical tests the brain activity is modeled by a horizontal bidimensional current dipole located in the plane $\Pi_0 = \{x, y \in \mathbf{R}, z = 0\}$ and the magnetic field is sampled by 400 magnetometers located on a regular horizontal grid at height $\delta = 1$, see Fig. 1. Note that we use adimensional measure units in the tests.

We choose as a multiscale basis the Daubechies orthonormal wavelets with $d = 4$ vanishing moments and discretize the plane Π_0 with 32 pixels for each dimension. Finally, 2 multiscale levels are used for the current decomposition. At each multiscale level, the thresholding parameter $(\rho_\lambda)_{\lambda\in\boldsymbol{\Lambda}}$ is chosen equal to τ times the mean of the wavelet coefficients $\vec{\mathbf{j}}^{(1)}$.

In Fig. 2 the current distribution reconstructed after 100 iterations of Algorithm `JS` with $\tau = 1$ is shown. As for the other parameters, we have set $\theta_\lambda = 10^{-4}$ and $\omega_\lambda = 10^{-3}$ for each $\lambda \in \Lambda$. In the figure, the current image obtained by joint-thresholding, i.e. Algorithm JS with $p = 2$ and $R > 0$, (left) is compared with the current image obtained by uncoupled soft-thresholding, i.e. Algorithm JS with $p = 1$ and $R = 0$, (right).

In Fig. 4 the current distribution reconstructed after 100 iterations of Algorithm `JS` is displayed for $\tau = 20$ in the case of high white Gaussian noise with linear signal to noise ratio equal to 0.1 is added to the magnetic field (see Fig. 3).

In case of measurements with high noise, Tikhonov regularization is not able to give an accurate current image: when the regularization parameter

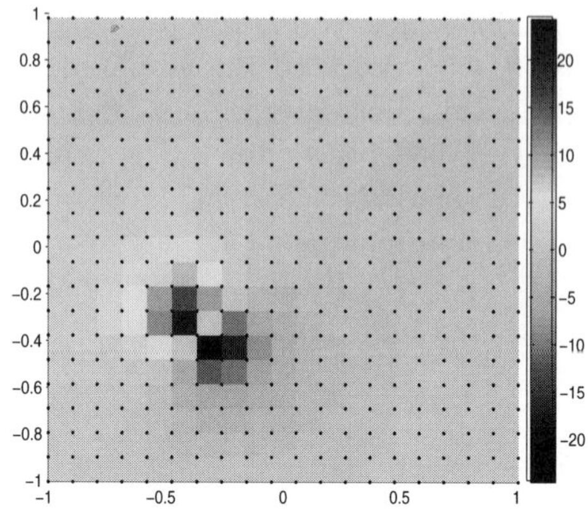

Fig. 1. The magnetic field produced by a current dipole located in (-0.4,-0.4,0). The black points represent the magnetometer sites.

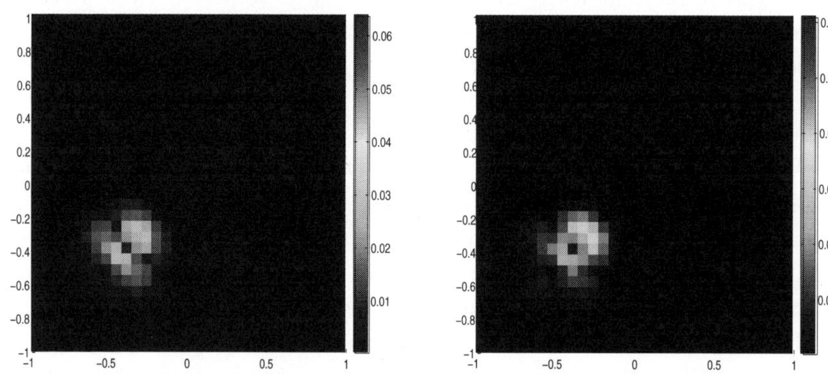

Fig. 2. The current intensity reconstructed starting from the magnetic field displayed in Fig. 1 by using joint-thresholding (left) and uncoupled soft-thresholding (right).

is chosen equal to 0 the current is not reconstructed at all (see Fig. 5, left), while a regularization parameter greater than 0 gives a blurred image (see Fig. 5, right, where the regularization parameter is chosen by means of the discrepancy principle and is equal to 356).

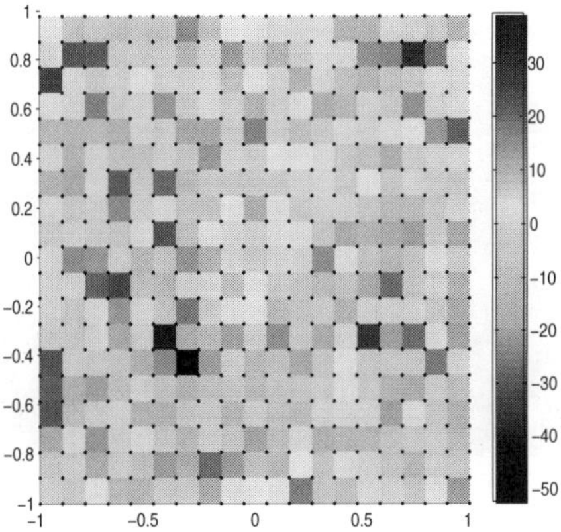

Fig. 3. The noisy magnetic field produced by a current dipole located in (-0.4,-0.4,0). The black points represent the magnetometer sites.

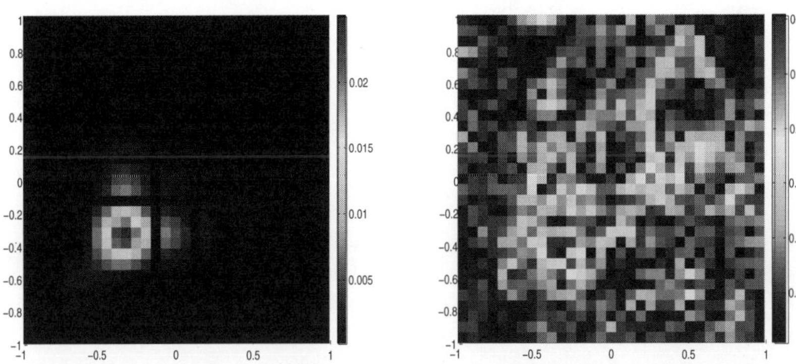

Fig. 4. The current intensity reconstructed starting from the magnetic field displayed in Fig. 3 by using joint-thresholding (left) and uncoupled soft-thresholding (right).

The ℓ_1-norm

$$\|(\vec{\mathbf{j}}_\lambda)_{\lambda\in\Lambda}\|_{\ell_1} := \sum_{\lambda\in\mathbf{\Lambda}} \|\vec{\mathbf{j}}_\lambda\|_{\mathbf{R}^2} \tag{20}$$

as a function of the thresholding parameter τ is displayed in Fig. 6 for noiseless and noisy data.

The numerical tests show that the proposed algorithm outperforms both

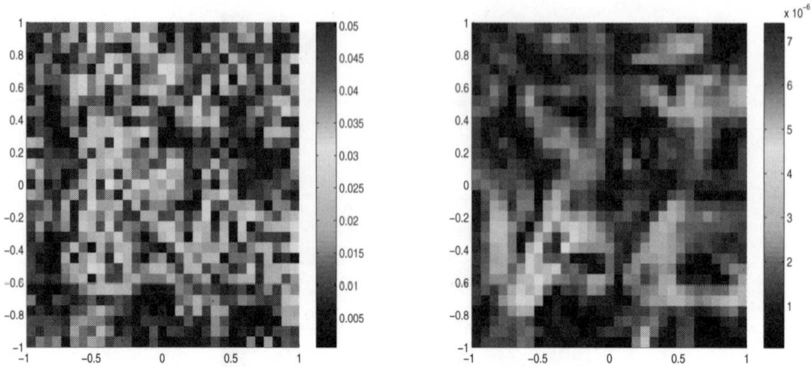

Fig. 5. The current distribution reconstructed by using the Tikhonov regularization starting from the noisy magnetic field displayed in Fig. 3. Two different values of the regularization parameter have been used: 0 (left) and 356 (right).

Tikhonov regularization and uncoupled iterative thresholding, especially in case of high noise.

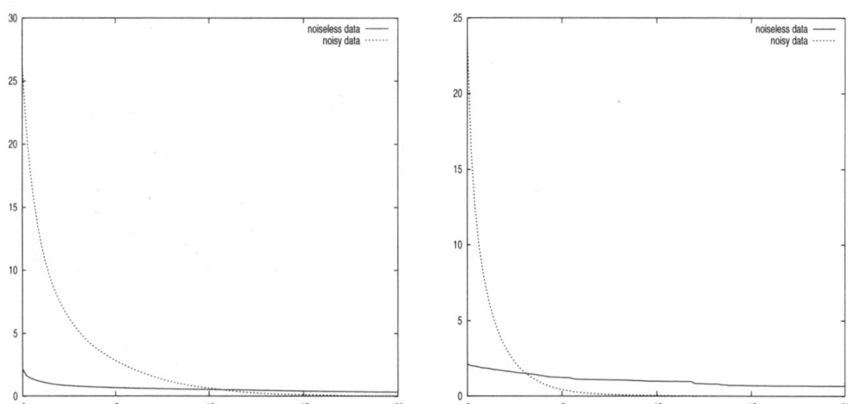

Fig. 6. The ℓ_1-norm of the reconstructed current as a function of the thresholding parameter τ. The left graph refers to the joint-thresholding case while the right one refers to soft-thresholding.

References

1. O. Christensen, *An Introduction to Frames and Riesz Bases*, Birkhäuser, 2003.
2. I. Daubechies, Ten Lectures on Wavelets, *SIAM*, 1992.
3. I. Daubechies, M. Defrise & C. De Mol, An iterative thresholding algorithm for linear inverse problems with a sparsity constraint, *Comm. Pure Appl. Math.* **57** (2004), 1413–1457.
4. C. Del Gratta, V. Pizzella, F. Tecchio & L. Romani, Magnetoencephalography - a noninvasive brain imaging method with 1 ms time resolution, *Rep. Prog. Phys.* **64**, (2001), 1759–1814.
5. D.L. Donoho, Superresolution via Sparsity Constraints, *SIAM J. Math. Anal.* **23** (1992), 1309–1331.
6. D.L. Donoho, De-noising by soft-thresholding, *IEEE Trans. Inform. Theory* **41** (1995), 613–627.
7. H.W. Engl, M. Hanke & A. Neubauer, *Regularization of inverse problems*, Kluwer, Dordrecht, 2000.
8. M. Fornasier & F. Pitolli, Adaptive iterative thresholding algorithms for magnetoencephalography (MEG), *J. Comput. Appl. Math.*, **221**, 2008, 386–395.
9. M. Fornasier & H. Rauhut, Recovery algorithms for vector valued data with joint sparsity constraints, *SIAM J. Numer. Anal.*, **46**, 2008, 577–613.
10. M. Fornasier & H. Rauhut, Iterative thresholding algorithms, *Appl. Comput. Harmon. Anal.*, **25**, 2008, 187–200.
11. R. Kress, L. Kühn & R. Potthast, Reconstruction of a current distribution from its magnetic field, *Inverse Problems 18* (2002), 1127–1146.
12. J. Sarvas, Basic mathematical and electromagnetic concepts of the biomagnetic inverse problem, *Phys. Med. Biol. 32* (1987), 11–22.

NO SAMPLING LINEAR SAMPLING FOR 3D INVERSE SCATTERING PROBLEMS

M. BRIGNONE

Dipartimento di Matematica, Università degli Studi di Genova, INDAM, and CNR-INFM LAMIA,
Genova, I-16146, Italy
E-mail: brignone@dima.unige.it

R. ARAMINI

Dipartimento di Ingegneria e Scienza dell'Informazione, Università degli Studi di Trento,
Povo di Trento, I-38100, Italy
E-mail: aramini@science.unitn.it

G. BOZZA

Dipartimento di Ingegneria Biofisica ed Elettronica, Università degli Studi di Genova,
Genova, I-16145, Italy
E-mail: bozza@dibe.unige.it

M. PIANA

Dipartimento di Informatica, Università degli Studi di Verona, and CNR-INFM LAMIA,
Verona, I-37134, Italy
E-mail: piana@sci.univr.it

In Linear Sampling the profile of an unknown object can be visualized from far-field scattering data by plotting the norm of the Tikhonov regularized solution of a set of discretized far-field equations sampled over a computational grid. Here we present a new 3D formulation of the Linear Sampling, in which the set of far-field equations for all sampling points and all polarizations is replaced by a single functional equation, whose Tikhonov regularized solution can be analytically determined. Such an implementation is computationally much faster than the traditional one and does not reduce the quality of the reconstructions.

Keywords: Inverse scattering, linear sampling method, regularization theory.

1. Introduction

The linear sampling method[1–3] is a well-known numerical procedure for the visualization of an unknown 2D or 3D object from measurements of far-field scattering data. It is based on a *general theorem*,[4] which is concerned with a one-parameter family of linear integral equations of the first kind, the parameter being the sampling point z in physical space: for each z, the corresponding equation is known as the *far-field equation.* The integral kernel of the far-field equation is independent of z and is formed by the *far-field pattern* of the scattered field for all incidence and observation directions, while the right-hand side is a z-dependent and analytically known function. The general theorem states the existence, for each z in $\mathbb{R}^2$ or $\mathbb{R}^3$, of an approximate solution to the far-field equation, such that its L^2-norm tends to blow up when z approaches the boundary of the scatterer from inside, and can be made arbitrarily large when z is outside. Inspired by this theorem, the linear sampling method exploits a suitable monotonic map of this L^2-norm as indicator function of the support of the scatterer.

The main features of the method are its robustness and generality: in fact, it provides satisfactory reconstructions of penetrable or impenetrable scatterers in a large variety of electromagnetic or acoustic scattering conditions, with almost no a priori information.

In two recent papers, a new formulation of the method has been proposed, enhancing the rapidity of the visualization process. In the 2D case,[5] the traditional one-parameter family of far-field equations is replaced by a single linear functional equation set in L^2-based Hilbert spaces. In this approach, called *no-sampling linear sampling*, a single regularization procedure is applied to this functional equation and therefore the computation performed by the method is notably faster. In Ref. 6 the no-sampling linear sampling has been recently extended to the 3D case, thus resulting in an extremely fast algorithm: objects that are reconstructed in around half an hour by traditional linear sampling, are reconstructed with comparable accuracy by this fully no-sampling procedure in around one minute.

However, the formulation proposed in Ref. 6 is strictly related to a particular zero-order discretization of the integral appearing in the far-field equation, according to which the unit sphere is approximated by a (not necessarily uniform) parallel/meridian-shaped mesh naturally induced by the spherical coordinates. The aim of this contribution is then to overcome such limitation, by allowing for a completely general set of incidence and observation directions, represented as a set of points on the unit sphere: in this case, the simplest possible mesh is triangular and can be obtained from the points themselves by using a standard algorithm. Moreover, a triangular mesh naturally suggests using a first-order discretization of the integral, rather than a zero-order one. As a numerical validation, we firstly test our formulation against the well-known "teapot" scattering data used in Ref. 3, where the incidence and observation directions generate a uniform triangular mesh: the result is that the teapot is reconstructed with comparable accuracy, but in a much shorter time. Then, we also revisit an example already discussed in Ref. 6: the incidence and observation directions, originally conceived for a parallel/meridian-shaped mesh, are now reorganized to form a non-uniform triangular mesh. In this case, both the reconstruction quality and the computational time are very similar to those of Ref. 6.

2. The linear sampling method

In the framework of three-dimensional and time-harmonic electromagnetic scattering,[3,7] we consider the direct problem of determining the total electric field $\vec{E} = \vec{E}(\vec{x})$ that solves the system

$$\text{curl curl } \vec{E}(\vec{x}) - k^2 N(\vec{x})\vec{E}(\vec{x}) = 0, \qquad \vec{x} \in \mathbb{R}^3 \setminus D_{\vec{J}}, \quad (1)$$

$$\vec{E}(\vec{x}) = \vec{E}^s(\vec{x}) + \vec{E}^i(\vec{x}), \qquad \vec{x} \in \mathbb{R}^3, \quad (2)$$

$$\lim_{|\vec{x}|\to\infty} \left(\text{curl } \vec{E}^s(\vec{x}) \times \vec{x} - \mathrm{i}k|\vec{x}|\vec{E}^s(\vec{x})\right) = 0, \quad (3)$$

where k is the wavenumber in the homogeneous and lossless background medium, $N = N(\vec{x})$ is the 3×3 symmetric matrix representing the (possibly anisotropic) refraction index of the propagation media in all $\mathbb{R}^3$ (with N equal to the identity matrix outside the scatterer), $D_{\vec{J}}$ is the support of the current density source $\vec{J}$, while $\vec{E}^s = \vec{E}^s(\vec{x})$ and $\vec{E}^i = \vec{E}^i(\vec{x})$ are the scattered and the incident electric field respectively; limit (3) represents the Silver-Müller radiation condition, which holds uniformly in all directions $\hat{x} = \vec{x}/|\vec{x}|$.

In the following, the electric incident field $\vec{E}^i$ will be chosen as a plane wave propagating along the direction $\hat{d}$ and polarized along $\vec{p} \in \mathbb{R}^3$ (so that $\vec{p} \cdot \hat{d} = 0$), i.e.

$$\vec{E}^i(\vec{x}) = \vec{p}\, e^{\mathrm{i}k\vec{x}\cdot\hat{d}}, \qquad \vec{x} \in \mathbb{R}^3. \quad (4)$$

The Stratton-Chu formula[7] implies that the scattered field $\vec{E}^s$ is characterized by the asymptotic behavior

$$\vec{E}^s(\vec{x}) = \frac{e^{\mathrm{i}kr}}{r} \left\{ \vec{E}_\infty(\hat{x}; \hat{d}, \vec{p}) + \mathcal{O}\left(\frac{1}{r}\right) \right\} \qquad \text{as } r = |\vec{x}| \to \infty; \quad (5)$$

in the previous equality, $\vec{E}_\infty(\cdot;\hat{d},\vec{p})$ denotes the *far-field pattern* of $\vec{E}^s$: it belongs to the vector space of tangential fields $L^2_t(\Omega) := \{\vec{f}(\cdot) \in (L^2(\Omega))^3 \mid \vec{f}(\hat{x}) \cdot \vec{\nu}(\hat{x}) = 0 \; \forall \hat{x} \in \Omega\}$, where $\vec{\nu}(\hat{x})$ is the unit vector normal to $\Omega := \{\vec{x} \in \mathbb{R}^3, \; |\vec{x}| = 1\}$ in $\hat{x}$ and $\vec{f}(\hat{x}) \cdot \vec{\nu}(\hat{x})$ is the usual scalar product in $\mathbb{C}^3$ between $\vec{f}(\hat{x})$ and $\vec{\nu}(\hat{x})$.

The inverse problem we are interested in is that of determining the location and shape of the unknown scatterer by using a set of measurements of the far-field pattern of $\vec{E}^s$ for a finite number of incidence and observation directions. To this end, we first consider the *far-field operator*

$$F : L^2_t(\Omega) \ni \vec{g}(\cdot) \mapsto \int_\Omega \vec{E}_\infty(\hat{x};\hat{d},\vec{g}(\hat{d}))ds(\hat{d}) \in L^2_t(\Omega). \tag{6}$$

Then, for each point $\vec{z} \in \mathbb{R}^3$ and each polarization $\hat{q} \in \Omega$, we introduce the *far-field equation*

$$(F\vec{g}_{\vec{z},\hat{q}}(\cdot))\,(\hat{x}) = \vec{E}_{e,\infty}(\hat{x};\vec{z},\hat{q}), \qquad \hat{x} \in \Omega, \tag{7}$$

in the unknown $\vec{g}_{\vec{z},\hat{q}}(\cdot) \in L^2_t(\Omega)$, where $\vec{E}_{e,\infty}(\hat{x};\vec{z},\hat{q})$ is the far-field pattern of an elementary dipole located in $\vec{z}$ and oriented along the unit vector $\hat{q}$, i.e.[3]

$$\vec{E}_{e,\infty}(\hat{x};\vec{z},\hat{q}) := \frac{\mathrm{i}k}{4\pi}(\hat{x} \times \hat{q}) \times \hat{x}\; e^{-\mathrm{i}k\hat{x}\cdot\vec{z}}. \tag{8}$$

The general theorem at the basis of the linear sampling method states the existence of an approximate solution to Eq. (7) whose L^2-norm blows up to infinity when $\vec{z}$ approaches the boundary of the scatterer from inside and stays arbitrarily large outside: so the idea arises of using a suitable monotonic map of such a norm as indicator function of the support of the scatterer. However, the compactness of the far-field operator F makes solving the far-field equation (7) an ill-posed problem.

We now observe that real experiments can only provide a finite number of measurements, typically affected by noise; this means that, if $\{\hat{d}_{\ell_{\hat{d}}} \in \Omega : \ell_{\hat{d}} = 1,\ldots,L_{\hat{d}}\}$ and $\{\hat{x}_{\ell_{\hat{x}}} \in \Omega : \ell_{\hat{x}} = 1,\ldots,L_{\hat{x}}\}$ are the incidence and observation directions respectively, then only a discrete and noisy version of the far-field pattern

$$\vec{E}^H_\infty(\hat{x}_{\ell_{\hat{x}}};\hat{d}_{\ell_{\hat{d}}},\hat{p}) := \vec{E}_\infty(\hat{x}_{\ell_{\hat{x}}};\hat{d}_{\ell_{\hat{d}}},\hat{p}) + \vec{H}(\ell_{\hat{x}};\ell_{\hat{d}},\hat{p}) \tag{9}$$

(for $\ell_{\hat{d}} = 1,\ldots,L_{\hat{d}}$ and $\ell_{\hat{x}} = 1,\ldots,L_{\hat{x}}$) is at disposal; here $\vec{H}(\ell_{\hat{x}};\ell_{\hat{d}},\hat{p})$ denotes the amount of noise affecting each measured value of the far-field pattern. Moreover, we shall assume to know a bound h on the noise level (see the next section for more details).

A discretization of the far-field equation is then required, and the first step is to project the far-field patterns onto a basis. For any incidence direction $\hat{d}_{\ell_{\hat{d}}}$, let us consider two unit vectors $\hat{\xi}_1(\hat{d}_{\ell_{\hat{d}}})$ and $\hat{\xi}_2(\hat{d}_{\ell_{\hat{d}}})$ spanning the tangent plane to Ω in $\hat{d}_{\ell_{\hat{d}}}$. Since $\vec{g}_{\vec{z},\hat{q}}(\cdot) \in L^2_t(\Omega)$, we have

$$\vec{g}_{\vec{z},\hat{q}}(\hat{d}_{\ell_{\hat{d}}}) = g^1_{\vec{z},\hat{q}}(\hat{d}_{\ell_{\hat{d}}})\hat{\xi}_1(\hat{d}_{\ell_{\hat{d}}}) + g^2_{\vec{z},\hat{q}}(\hat{d}_{\ell_{\hat{d}}})\hat{\xi}_2(\hat{d}_{\ell_{\hat{d}}}), \qquad \ell_{\hat{d}} = 1, \ldots, L_{\hat{d}}, \tag{10}$$

where, for each $i = 1, 2$, we put $g^i_{\vec{z},\hat{q}}(\hat{d}_{\ell_{\hat{d}}}) := \vec{g}_{\vec{z},\hat{q}}(\hat{d}_{\ell_{\hat{d}}}) \cdot \hat{\xi}_i(\hat{d}_{\ell_{\hat{d}}})$.

Next, we consider a triangular meshing of the unit sphere Ω with vertices defined by the $L_{\hat{d}}$ incidence directions: it follows from Euler's formula that the corresponding number of triangles forming the mesh is $T_{\hat{d}} = 2L_{\hat{d}} - 4$. Then, following Ref. 6, the discretized version of the far-field equation (7) can be written as

$$\sum_{\ell_{\hat{d}}=1}^{L_{\hat{d}}} [g^1_{\vec{z},\hat{q}}(\hat{d}_{\ell_{\hat{d}}})\vec{E}^H_\infty(\hat{x}_{\ell_{\hat{x}}}; \hat{d}_{\ell_{\hat{d}}}, \hat{\xi}_1(\hat{d}_{\ell_{\hat{d}}}))$$
$$+ g^2_{\vec{z},\hat{q}}(\hat{d}_{\ell_{\hat{d}}})\vec{E}^H_\infty(\hat{x}_{\ell_{\hat{x}}}; \hat{d}_{\ell_{\hat{d}}}, \hat{\xi}_2(\hat{d}_{\ell_{\hat{d}}}))]\,\omega_{\ell_{\hat{d}}} = \vec{E}_{e,\infty}(\hat{x}_{\ell_{\hat{x}}}; \vec{z}, \hat{q}), \tag{11}$$

where the weights $\omega_{\ell_{\hat{d}}}$ are related to the quadrature formulas used for discretizing the integrals over the triangular mesh. Here we use a linear approximation of the integrand function over each triangle T_n, with $n = 1, \ldots, T_{\hat{d}}$, by interpolating the values taken by this function at the vertices of T_n: then, if $J_{\ell_{\hat{d}}} := \{n \in \{1, \ldots, T_{\hat{d}}\} : \hat{d}_{\ell_{\hat{d}}} \in T_n\}$, we obtain

$$\omega_{\ell_{\hat{d}}} = \frac{1}{3} \sum_{n \in J_{\ell_{\hat{d}}}} A_{T_n}, \tag{12}$$

where A_{T_n} denotes the measure of the area of the n-th triangle T_n.

For any observation direction $\hat{x}_{\ell_{\hat{x}}}$ and any polarization $\hat{q}$, the vector equation (11) can be decomposed in two scalar equations: then, we can collect the resulting $2L_{\hat{x}} \times 2L_{\hat{d}}$ scalar equations in a more compact form, by means of the following linear system

$$\mathbf{F}^h \mathbf{G}_{\vec{z},\hat{q}} = \mathbf{E}_{e,\infty}(\vec{z}, \hat{q}), \tag{13}$$

where $\mathbf{F}^h$ is the $2L_{\hat{x}} \times 2L_{\hat{d}}$ matrix, independent of both $\vec{z}$ and $\hat{q}$, storing the information provided by the far-field measurements, $\mathbf{G}_{\vec{z},\hat{q}}$ is the unknown column vector and $\mathbf{E}_{e,\infty}(\vec{z}, \hat{q})$ is obtained by discretizing expression (8). Here the bound h on the noise level is used as a superscript to distinguish $\mathbf{F}^h$ from the corresponding noise-free version $\mathbf{F}$.

3. The no-sampling approach

If $A_i > 0$ for any $i = 1, 2, 3$, let $T = \bigotimes_{i=1}^{3}[-A_i, A_i]$ be a domain containing the scatterer and let $\mathcal{B} := T \times \Omega$. The new mathematical set-up requires the introduction of the Hilbert space $[L^2(\mathcal{B})]^{2L}$, for a generic $L \in \mathbb{N}$, equipped with the scalar product

$$(\mathbf{f}(\cdot), \mathbf{g}(\cdot))_{2,L} := \int_{\mathcal{B}} (\mathbf{f}(\vec{z}, \hat{q}), \mathbf{g}(\vec{z}, \hat{q}))_L \, d\vec{z} \, ds(\hat{q}), \tag{14}$$

where $(\cdot\,, \cdot)_L$ denotes the weighted scalar product of $\mathbb{C}^{2L}$ induced by the discretization, i.e. defined, for all $\mathbf{w}^1 := \left\{w_t^1\right\}_{t=1}^{2L}$, $\mathbf{w}^2 := \left\{w_t^2\right\}_{t=1}^{2L} \in \mathbb{C}^{2L}$, as

$$(\mathbf{w}^1, \mathbf{w}^2)_L := \sum_{\ell=1}^{L} \omega_\ell \left[w_\ell^1 \, \overline{w}_\ell^2 + w_{L+\ell}^1 \, \overline{w}_{L+\ell}^2\right], \tag{15}$$

with ω_ℓ given by relation (12), while the bar means complex conjugation. Moreover, we shall denote by $\|\cdot\|_{2,L}$ the norm induced by the scalar product (14).

We can now introduce the linear operator

$$\mathsf{F}^h : \left[L^2(\mathcal{B})\right]^{2L_{\hat{d}}} \ni \mathbf{G}(\cdot) \mapsto \left\{\sum_{t=1}^{2L_{\hat{d}}} (\mathbf{F}^h)_{st} \, G_t(\cdot)\right\}_{s=1}^{2L_{\hat{x}}} \in \left[L^2(\mathcal{B})\right]^{2L_{\hat{x}}}, \tag{16}$$

where $(\mathbf{F}^h)_{st}$ are the elements of the matrix $\mathbf{F}^h$ and $\mathbf{G}(\cdot) := \{G_t(\cdot)\}_{t=1}^{2L_{\hat{d}}}$.

The definition of the operator F^h allows collecting the infinitely many algebraic systems (13) into a single functional equation, written in $[L^2(\mathcal{B})]^{2L_{\hat{x}}}$ for the unknown $\mathbf{G}(\cdot) \in [L^2(\mathcal{B})]^{2L_{\hat{d}}}$ as:

$$[\mathsf{F}^h \mathbf{G}(\cdot)](\cdot) = \mathbf{E}_{e,\infty}(\cdot), \tag{17}$$

where $\mathbf{E}_{e,\infty}(\cdot)$ is the element of $[L^2(\mathcal{B})]^{2L_{\hat{x}}}$ obtained from $\mathbf{E}_{e,\infty}(\vec{z}, \hat{q})$ by simply regarding the sampling pair $(\vec{z}, \hat{q})$ as an independent variable on $\mathcal{B}$.

It can be proved that the inverse problem of determining the generalized solution of the functional equation (17) is well-posed;[8] nonetheless, it remains ill-conditioned, thus requiring in any case a regularization procedure. If compared with the traditional pointwise implementation, the novelty of the no-sampling approach consists in the fact that now the regularization of Eq. (17) is performed independently of both $\vec{z}$ and $\hat{q}$: therefore, only one value of the regularization parameter needs to be determined. In Ref. 6 we have shown that, if $r^h := \operatorname{rank} \mathbf{F}^h$ and $\{\sigma_p^h, \mathbf{u}_p^h, \mathbf{v}_p^h\}_{p=1}^{r^h}$ is the (weighted) singular system of $\mathbf{F}^h$, the Tikhonov regularized solution of Eq. (17) is given by

$$\mathbf{G}_\alpha(\cdot) = \sum_{p=0}^{r^h-1} \frac{\sigma_p^h}{(\sigma_p^h)^2 + \alpha} \left(\mathbf{E}_{e,\infty}(\cdot), \mathbf{v}_p^h\right)_{L_{\hat{x}}} \mathbf{u}_p^h. \tag{18}$$

Then, the generalized discrepancy principle[9] allows fixing the optimal value α^* for the regularization parameter α as the zero of the generalized discrepancy function

$$\rho(\alpha) := \left\| \left[\mathsf{F}^h \mathbf{G}_\alpha(\cdot)\right](\cdot) - \mathbf{E}_{e,\infty}(\cdot)\right\|_{2,L_{\hat{x}}}^2 - h^2 \left\|\mathbf{G}_\alpha(\cdot)\right\|_{2,L_{\hat{d}}}^2, \tag{19}$$

where the bound h on the noise level can be identified with the maximum of the (weighted) singular values of the matrix $\mathbf{F}^h - \mathbf{F}$. The analytical expression of $\rho(\alpha)$ given in Ref. 6 allows a very fast computation of its zero α^*.

Taking inspiration from the general theorem at the basis of the traditional linear sampling method, the 3D reconstruction of the unknown scatterer can be obtained by plotting a C–level surface of the indicator function $1/\Psi$, where

$$\begin{aligned}\Psi(\vec{z}) &:= \int_\Omega \|\mathbf{G}_{\alpha^*}(\vec{z},\hat{q})\|_{L_{\hat{x}}}^2 \, ds(\hat{q}) \\ &= \frac{4\pi}{3} \sum_{j=1}^{3} \sum_{p=0}^{r^h-1} \frac{(\sigma_p^h)^2}{[(\sigma_p^h)^2 + \alpha^*]^2} \left| \left(\mathbf{E}_{e,\infty}(\vec{z},\hat{e}_j), \mathbf{v}_p^h\right)_{L_{\hat{x}}} \right|^2 \quad \forall \vec{z} \in T, \end{aligned} \tag{20}$$

being $\{\hat{e}_j \,:\, j = 1,2,3\}$ the canonical basis of $\mathbb{R}^3$.

An almost automatic recipe to fix the value of C for the optimal visualization of the scatterer is described in Ref. 6. We choose

$$C := \int_0^1 \frac{1}{\Psi(\gamma(t))} \, dt, \tag{21}$$

where $\gamma : [0,1] \to \mathbb{R}^3$ is a plane curve obtained by applying an active contour technique to the 2D map describing the restriction of the indicator function to a finite plane region containing a slice of the scatterer.

4. Numerical results

In this section we describe two numerical simulations. The first example is concerned with the reconstruction of the perfectly conducting scatterer shown in Fig. 1(a) and contained in a cubic domain $T = [-0.4\,\mathrm{m}\,,\, 0.4\,\mathrm{m}] \times [-0.4\,\mathrm{m}\,,\, 0.4\,\mathrm{m}] \times [-0.2\,\mathrm{m}\,,\, 0.4\,\mathrm{m}]$. The direct far-field data, computed by using CESC solver, are just those used in Ref. 3; moreover we corrupt each value by 7% Gaussian noise. The wavenumber is $k = 56\,\mathrm{m}^{-1}$ and the 252 incidence/observation directions are uniformly distributed on the unit sphere as shown in Fig. 1(b). In order to determine the threshold value C for the indicator function, we follow the scheme described above. More precisely, if we refer $\mathbb{R}^3$ to the usual Cartesian coordinate system (x_1, x_2, x_3), we restrict the indicator function $1/\Psi$ to the plane of Cartesian equation $x_3 = 0.05\,\mathrm{m}$ and fix the value C by applying an active contour technique to this restricted visualization map. The corresponding C-level surface of $1/\Psi$ provides the reconstruction of the scatterer, which is presented in Fig. 1(c). The regularization algorithm described by relations (18) and (19) is performed in around 10 s, while the overall visualization procedure is completed in less than 3 minutes.

For the second example we consider the non-connected and dielectric scatterer presented in Fig. 1(d), characterized by constant $\epsilon_r = 2.0$ and $\sigma = 0.0\,\mathrm{S\,m}^{-1}$ for all its connected components, and contained in the bounded cubic region $T = [-1.5\,\mathrm{m}, 1.5\,\mathrm{m}]^3$. The far-field data are computed by using a stabilized biconjugate-gradient fast Fourier transform method of moments code[10] and each value is corrupted by 7% Gaussian noise. We choose a wavenumber $k = 6\,\mathrm{m}^{-1}$ and consider 146 incidence and observation directions distributed on the unit sphere so to form the non-uniform triangular mesh shown in Fig. 1(e). The scatterer is firstly cut by the plane in $\mathbb{R}^3$ of Cartesian equation $x_2 = 0.9\,\mathrm{m}$, then the usual deformable model is applied to the corresponding visualization map: the result is the reconstruction shown in Fig. 1(f). Owing to the smaller number of incidence and observation directions with respect to the previous example, the overall visualization procedure is now completed in less than 30 s, where only around 1 s is taken by the regularization algorithm.

5. Conclusions

In this contribution we have presented a no-sampling implementation of the linear sampling method for 3D electromagnetic inverse scattering problems, by adopting a first-order discretization of the far-field operator for any triangular mesh over the unit sphere. As a result, the formulation proposed in Ref. 6 is generalized in order to allow for any possible choice of the incidence and observation directions, without impairing the computational efficiency of the method.

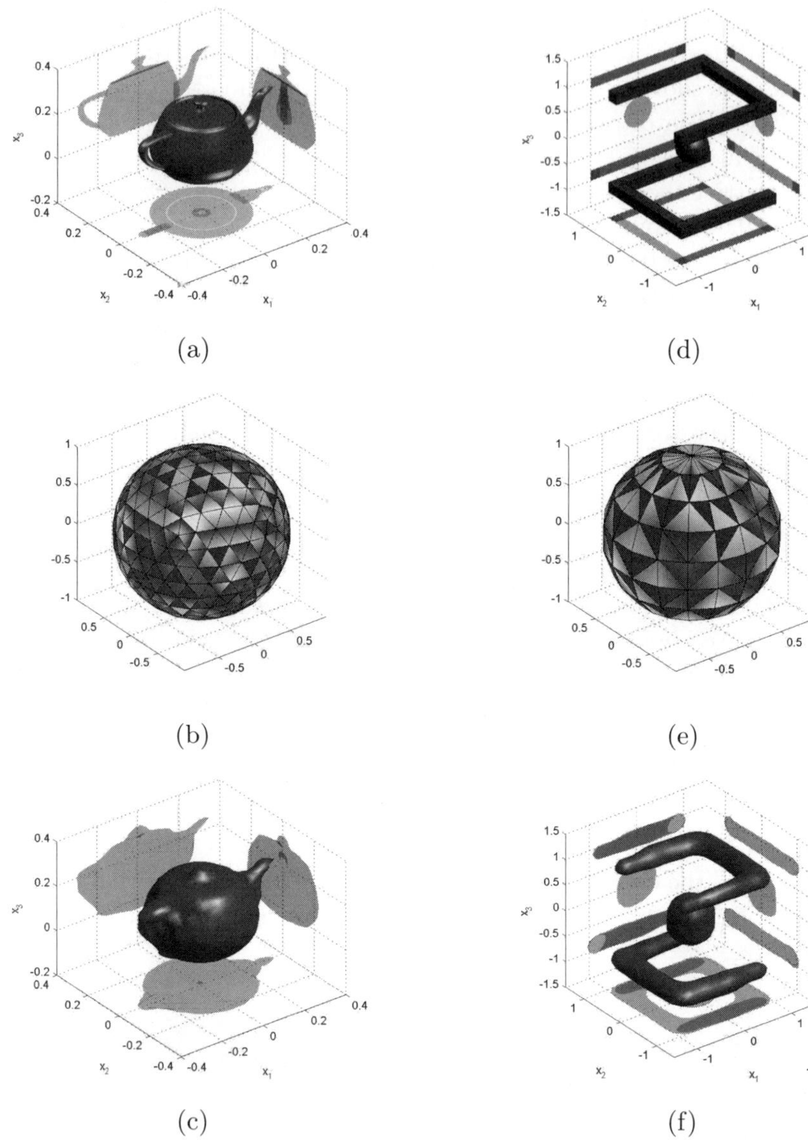

Fig. 1. (a) Exact geometry of the perfectly conducting teapot; (b) uniform mesh formed on the unit sphere by the 252 views chosen for the reconstruction of the teapot; (c) no-sampling linear sampling reconstruction of the teapot. (d) Exact geometry of the dielectric object characterized by constant $\epsilon_r = 2.0$ and $\sigma = 0.0\,\mathrm{S\,m^{-1}}$; (e) non-uniform mesh formed on the unit sphere by the 144 views chosen for the reconstruction of the dielectric object; (f) no-sampling linear sampling reconstruction of the dielectric object.

Acknowledgements

We would like to thank Prof. Houssem Haddar for providing the far-field data of the first numerical example, as well as Cristina Campi and Giovanni Giorgi for helpful discussions during the preparation of the manuscript.

References

1. D. Colton and A. Kirsch, *Inverse Problems* **12**, 383 (1996).
2. D. Colton, M. Piana and R. Potthast, *Inverse Problems* **13**, 1477 (1997).
3. D. Colton, H. Haddar and M. Piana, *Inverse Problems* **19**, S105 (2003).
4. F. Cakoni and D. Colton, *Georg. Math. J.* **10**, 911 (2003).
5. R. Aramini, M. Brignone and M. Piana, *Inverse Problems* **22**, 2237 (2006).
6. M. Brignone, G. Bozza, R. Aramini, M. Pastorino and M. Piana, *Inverse Problems* **25**, 015014 (2009).
7. D. Colton and R. Kress, *Inverse Acoustic and Electromagnetic Scattering Theory*, 2nd edition (Springer, Berlin, 1998).
8. R. Aramini, On some open problems in the implementation of the linear sampling method, PhD thesis, Dipartimento di Matematica, Università degli Studi di Trento, (Italy, 2007), pp. x + 238.
9. A. N. Tikhonov, A. V. Goncharsky, V. V. Stepanov and A. G. Yagola, *Numerical Methods for the Solution of Ill-Posed Problems* (Kluwer, Dordrecht, 1995).
10. Q. Z. Zhong, H. L. Qing, C. Xiao, E. Ward, G. Ybarra and W. T. Joines, *IEEE Trans. Biomed. Eng.* **50**, 1180 (2003).

TIME SERIES ANALYSIS OF DATA FROM STRESS ECG

C. CAMMAROTA

Dipartimento di Matematica
'La Sapienza' Università di Roma
p.le A. Moro 2, 00185 Roma, Italy
E-mail: cammar@mat.uniroma1.it
www.mat.uniroma1.it/people/cammarota

The heartbeat time series of the electrocardiogram recorded during stress test is non stationary, showing both decreasing and increasing trends and time variability of the variance. The analysis of the extrema is used to investigate the durations of accelerations and decelerations of the residuals obtained subtracting the trend. The time mean of these durations is used as a statistic to test the hypothesis that the residuals are independent and identically distributed (i.i.d.) variables. In this hypothesis the expectation of the statistic is 3/2; the rejection region of the test is computed by numerical simulation. Data analysis performed over the heartbeat series of 14 healthy subjects shows that the mean is significantly greater 3/2 and different in stress and recovery.

Keywords: Time series; Extrema; Stress ECG.

1. Introduction

The electrocardiogram (ECG) is the recording on the body surface of the electrical activity generated by the heart. In 1903 Einthoven introduced a method of recording the signal (the magnitude is of the order of mV) the labeling of the various waves and investigated a variety of cardiac abnormalities. The main peak is called R wave, corresponding to the contraction of ventricles (systole) (Figure 1). The RR interval is the time between two consecutive R peaks and it is inversely proportional to the instantaneous heart rate. The RR time series shows a variability reminiscent of complex systems, stochastic or deterministic chaotic, that is known as Heart Rate Variability (HRV). The HRV is mainly due to the autonomic control, to respiratory interaction,to humoral regulation. Spectral analysis[1] is the main tool used to investigate the control of the autonomic system on the heart rate. It is believed that the parasympathetic control is related to high frequency (HR) spectral components of RR sequence (0.15 Hz - 0.4 Hz) and the sympathetic one is related to low frequency (LF) components (0.04 Hz - 0.15 Hz). These results are supported by experiments in humans performed with pharmacological blockade or stimulation such as the tilt test.

It is also conjectured that sympathetic stimulation produces acceleration of heart rate, and parasympathetic one produces deceleration. It is known that these controls act on the scale of a few seconds, reflecting in the dependence of a small number of consecutive elements of the RR time series. In order to investigate this dependence the RR series is coded into a sequence of binary symbols and the frequency of suitable short words is computed (symbolic analysis). This method provides a characterization of the autonomic control similar to the one of spectral analysis in ECG data recorded during tilt test.[2] Binary words of fixed length can be used to code periods of acceleration and deceleration in RR series of 24 hours (Holter recording). Data analysis of normal subjects shows a significant difference between positive and negative accelerations,[3] suggesting that sympathetic and parasympathetic controls are not symmetric.

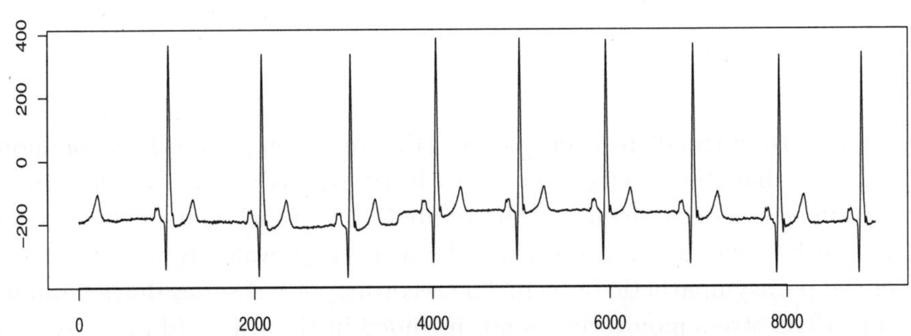

Fig. 1. The electrocardiogram of a normal subject; the sharp peaks (R) correspond to systole.

The ECG stress test is performed to evaluate the presence of myocardial ischaemia. In normal subjects during effort myocardial blood flow increases three or four times compared to rest, as a consequence of increased oxygen request. The RR series extracted from the ECG recorded during a stress test shows: a decreasing trend (stress phase), a global minimum (acme) and

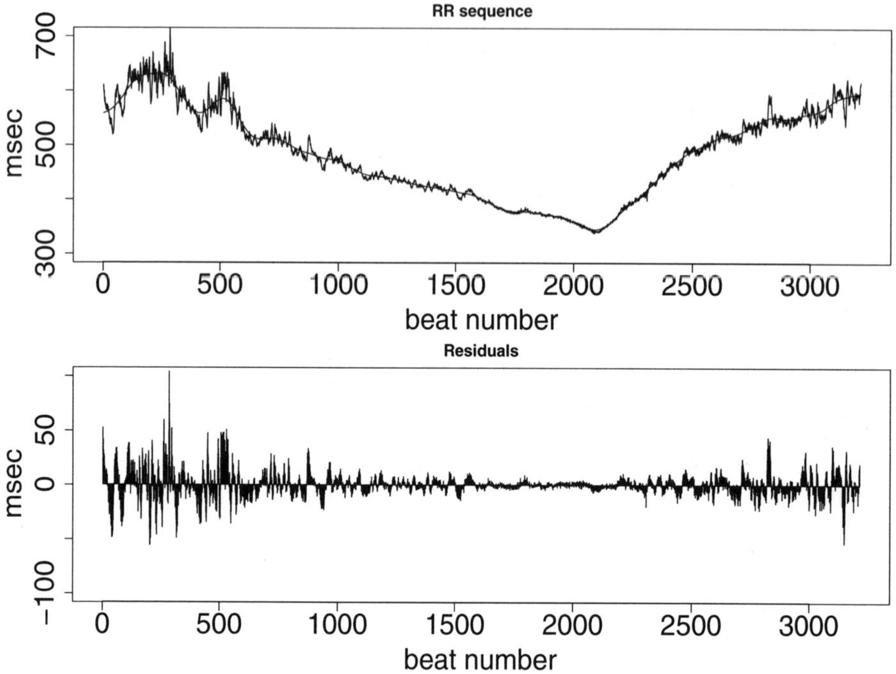

Fig. 2. First panel: RR series extracted from the ECG recorded during stress test of a normal subject. Second panel: Residuals of the RR series after detrending.

an increasing trend (recovery phase) (Figure 2, top panel). It is commonly believed that during stress, in which heart rate is increasing, the heart is prevalently under influence of the sympathetic branch and during recovery, in which heart rate is decreasing, the heart is prevalently under influence of the parasympathetic branch. The recording of RR series during ambulatory ECG stress monitoring is not included in the standard protocols. The analysis of this type of series is new in the HRV literature,[4] and can provide new insights into the neuroautonomic control of heart rate. In that paper the "analysis of extrema" was proposed as a method of investigation of non stationary series. The sequence of the lengths of monotonicity intervals was extracted, its properties were investigated and the time mean of the lengths was used as a statistic in the data analysis. This method was applied to the RR series of 14 healthy subjects, providing a clear evidence of the control.

In the present contribution we improve the results obtained in[4] in three directions. First, in the analysis of extrema, after a summary of known results, we give a new one concerning the variance of the statistic. Second we find the rejection region of a test of serial independence, simulating series of independent and identically distributed (i.i.d.) random variables (r.v.). Third we improve data analysis, performing the analysis of the variability after detrending the series, in order to compare the results previously obtained without detrending.

2. Analysis of extrema

The analysis of extrema is based on the idea that a relevant information for a time series is contained in the extrema (local maxima and minima). This idea is also used in the " Empirical Mode Decomposition" (EMD),[5] a method of analysis of continuous non stationary signals. Mathematical properties of the extrema of the intrinsic mode functions of EMD are non trivial; in the case of white noise some results have been obtained by using numerical simulations.

The analogous of white noise in time series is a sequence of i.i.d. variables. In this case some results on the extrema can be obtained.[4] We summarize them below.

We model the observed time series $x_1, ..., x_n$ as the realization of a sequence of r. v. $X_1, ..., X_n$, with joint continuous distribution P. We consider the subsets of R^3 : $A = \{x_1 < x_2,\ x_2 > x_3\}$ corresponding to maximum condition and $B = \{x_1 > x_2,\ x_2 < x_3\}$ corresponding to minimum condition and the event $E = A \cup B$, corresponding to an extreme (or turning point).

For the sequence $X_i' = (X_{i-1}, X_i, X_{i+1}) \in R^3,\ i = 2, ..., n-1$, we consider the sequence of r.v. $T_i,\ i = 1, ...$ defined as the occurrence times of the event E and the sequence of r.v. defined as the recurrence times

$$H_i = T_{i+1} - T_i, \quad i = 1, ...$$

We refer to the H_i as to the "lengths of monotonicity intervals". We focus on the properties of the H_i's under the assumption that the variables X_i are i.i.d. with continuous distribution. In these assumptions the sequence $X_i',\ i \geq 2$ is also stationary. Hence the H_i's form a stationary sequence under the conditional probability $P(\quad | X_2' \in E)$.[6]

The first result concerns the conditional mean of the H_i's.

Theorem 2.1. *For the i.i.d. sequence $X_1, X_2, ...$ of continuous variables, the variable "length of the monotonicity interval" has a conditional mean*

$$E(H_1 \mid X_2' \in E) = \frac{1}{P(X_2' \in E)} = \frac{3}{2} \tag{1}$$

The second result concerns the conditional distribution of H_1.

Theorem 2.2. *For the i.i.d. sequence $X_1, X_2, ...$ of continuous variables, the variable "length of the monotonicity interval" has a discrete density*

$$P(H_1 = s \mid X_2' \in E) = 3\left[\frac{1}{(s+1)!} - 2\frac{1}{(s+2)!} + \frac{1}{(s+3)!}\right], \quad s \geq 1 \quad (2)$$

We now give a new result concerning the conditional variance of the length of the monotonicity intervals.

Theorem 2.3. *For the i.i.d. sequence $X_1, X_2, ...$ of continuous variables, the variable "length of the monotonicity interval" has a conditional variance*

$$\text{Var}(H_1 \mid X_2' \in E) = 3\left[2e - \frac{63}{12}\right] \quad (3)$$

Proof.

We compute the variance using in short notations

$$Var(H) = E(H^2) - E(H)^2$$

By the distribution 2.2 we have

$$E(H^2) = 3\sum_{s=1}^{\infty} s^2 \left[\frac{1}{(s+1)!} - 2\frac{1}{(s+2)!} + \frac{1}{(s+3)!}\right]$$

We compute separately the three contributions. After some easy calculations one gets

$$\frac{s^2}{(s+1)!} = \frac{(s+1-1)^2}{(s+1)!} = \frac{1}{(s-1)!} - \frac{1}{s!} + \frac{1}{(s+1)!}$$

$$\frac{s^2}{(s+2)!} = \frac{(s+2-2)^2}{(s+2)!} = \frac{1}{s!} - \frac{3}{(s+1)!} + \frac{4}{(s+2)!}$$

$$\frac{s^2}{(s+3)!} = \frac{(s+3-3)^2}{(s+3)!} = \frac{1}{(s+1)!} - \frac{5}{(s+2)!} + \frac{9}{(s+3)!}$$

Collecting the three summands we get

$$s^2\left[\frac{1}{(s+1)!} - 2\frac{1}{(s+2)!} + \frac{1}{(s+3)!}\right]$$

$$= \frac{1}{(s-1)!} - \frac{3}{s!} + \frac{8}{(s+1)!} - \frac{13}{(s+2)!} + \frac{9}{(s+3)!}$$

This quantity can be decomposed into 4 telescopic series and a residual term:

$$\frac{1}{(s-1)!} - \frac{1}{s!} - \frac{2}{s!} + \frac{2}{(s+1)!} + \frac{6}{(s+1)!} - \frac{6}{(s+2)!} - \frac{7}{(s+2)!} + \frac{7}{(s+3)!} + \frac{2}{(s+3)!}$$

The sums of the four series take the values 1, -2, 3, -7/6 and the residual is

$$2 \sum_{s=1}^{\infty} \frac{1}{(s+3)!} = 2(e - \frac{8}{3})$$

Finally we have obtained

$$E(H^2) = 3\left(2e + 2 - \frac{39}{6}\right)$$

Using that $E(H) = 3/2$ we get the result. □

3. A test of serial independence

For a time series $X_1, X_2, \ldots, X_n$ we consider the lengths of the monotonicity intervals $H_1, H_2, \ldots, H_N$, and the statistic

$$D_n = \frac{1}{N} \sum_{j=1}^{N} H_j$$

i.e. the time mean of the variables $H_1, H_2, \ldots, H_N$. Note that N is random and depends on n. Given n, the distribution of D_n does not depend on the distribution of the X_i if these are independent. By using the ergodic theorem one can see that the statistic D_n converges almost everywhere:

$$\lim_{n \to \infty} D_n = \frac{3}{2}, \quad \text{a.e.}$$

The distribution of D_n depends only on n and can be computed simulating on a computer a large number of i.i.d. sequences. This simulation and the data analysis is performed using the statistical software R.[7] In table 1 two examples of quantiles useful for applications are reported. The lengths of data series $n = 500$ and $n = 1000$ are typical of the present application. These ideas allow to perform a test of serial independence: if the observed value of D_n falls outside an interval centered in 3/2 the i.i.d. hypothesis is rejected. The main interest in this test with respect to other tests is: it does not assume a specific distribution for the data; the statistic has a clear meaning related to acceleration of heart rate.

Table 1. Rounded values of the quantiles of the simulate distribution for the statistic D_n, the time mean length of monotonicity intervals, in case of two sequences with n= 500 and n=1000 data.

probability	1%	2.5%	5%	95%	97.5%	99%
quantile n= 500	1.42	1.43	1.44	1.56	1.58	1.59
quantile n=1000	1.43	1.44	1.45	1.55	1.56	1.57

4. Data analysis

In the data analysis of [4] the analysis of the extrema was performed on the RR time series. Here we perform the same type of analysis on the sequence of residuals obtained after detrending the RR series. The trend is estimated using a cubic spline (see Figure 2, first panel) and the residuals are computed (second panel). The residuals are non stationary in variance, but this has a weak effect on the statistic D_n, since the time variation of the variance is very slow if compared to the length of an interval of acceleration that is typically of a few units.

A group of 14 healthy subjects underwent to stress test in the Laboratory of Cardiology of University of Rome 'La Sapienza'. The ECG was recorded with PC-ECG 1200 (Norav Medical Ltd.), which provides output digital signal with an amplitude resolution of 2.441 microV and 500 Hz sampling frequency. The 50 Hz power-line interference and voluntary muscular activity were removed by using a discrete wavelet transform filter. An automated method was used for R peaks detection from the V5 lead. The RR sequence was extracted. All the cases show a profile of the RR series similar to the one of Figure 2(first panel). The filtered series shows a unique global point of minimum (acme); the original series restricted to beat numbers smaller (larger) than acme is called 'stress (recovery)phase' respectively. The analysis of the extrema was applied to the residuals (Figure 2, second panel) for both phases. The results are reported in table 2.

5. Conclusions

Statistical analysis of the data has given the following results.

Table 2. Rounded values of the statistic D_n, mean length of monotonicity intervals of the 14 cases. First column: case number; second column: stress phase; third column: recovery phase.

case	stress	recovery	case	stress	recovery
1	1.80	1.89	8	1.75	2.02
2	1.77	1.81	9	1.92	1.92
3	1.76	2.60	10	2.22	2.26
4	1.94	2.12	11	1.78	2.34
5	2.00	2.04	12	1.68	1.80
6	1.73	2.23	13	1.77	2.03
7	2.13	2.28	14	1.56	1.46

(i) In all the cases, with exception of case 14, the statistic D_n has values greater than 1.59 (the 0.99 quantile for $n = 500$) both for stress and recovery.

(ii) The values in stress are significantly greater than in recovery, as confirmed by a standard t-test.

Result (i) means that the hypothesis of independence for the residuals has to be rejected. More precisely it suggests that the control system acts prolonging the durations of acceleration and deceleration. In case 14 the hypothesis is not rejected: the residuals can be considered an i.i.d. sequence and this suggests a reduced neuroautonomic control. Therefore the action of the neuroautonomic control reflects not only in the trend of the RR series but also in the residuals.

Result (ii) suggests that during exercise the control of the sympathetic and parasympathetic branch of the neuroautonomic system are of different intensity and that they could be quantified by the durations of acceleration and deceleration in the residuals of the RR series.

Acknowledgments.

We thank M. Curione (Clinical Science Department of 'La Sapienza' University of Rome) for collecting data and for valuable discussions.

References

1. A. Malliani, M. Pagani, F. Lombardi and S. Cerutti, *Circulation* **84**, 482 (1991).
2. C. Cammarota and E. Rogora, *Physical Review E, Statistical, nonlinear, and soft matter physics* **74**, 042903 (2006).
3. C. Cammarota and E. Rogora, *Chaos, Solitons & Fractals* **32**, 1649 (2007).

4. C. Cammarota and M. Curione, *Mathematical medicine and biology* **25**, 87 (2008).
5. N. Huang, Z.Shen, S. Long, M.C.Wu, H. Shih, Q. Zheng, N-C.Yen, C. Tung and H. Liu, *Proc.Roy.Soc. London* **A454**, 903 (1998).
6. L. Breiman, *Probability* (SIAM, 1993).
7. R Development Core Team (2005), A language and environment for statistical computing. Url: http://www.R-project.org.

Transfer Across Scale Irregular Domains

Raffaela Capitanelli

Dipartimento di Metodi e Modelli Matematici per le Scienze Applicate,
Università degli Studi di Roma "La Sapienza",
Via A. Scarpa 16, 00161 Roma, Italy
raffaela.capitanelli@uniroma1.it

We prove existence, uniqueness and regularity results for a mixed Dirichlet-Robin problem on scale irregular domains. These results are comparable with the numerical study on the Laplacian transfer across fractal surfaces.

Keywords: Mixed Dirichlet-Robin problems, Fractals, Irregular interfaces.

1. Introduction.

Brownian motion, percolation, diffusion, aggregation, roughening and other random processes naturally give rise to fractal geometries. When the properties of these processes are due more to the hierarchy of their geometry than the random character of this hierarchy, it is possible to understand the properties of these random object by the study of the properties of deterministic fractals with the same fractal dimension. For example, in Refs. 6, 7 and 18, it is shown that the response of a random electrode of given fractal dimension is very close to that of a deterministic electrode with the same fractal dimension.

The aim of the paper is to study the same phenomenon from a mathematical point of view. More precisely, we consider the current flowing through an electrochemical cell as shown in Fig. 1, where the working electrode is a scale irregular Koch curve. This problem can be formally stated as

$$\begin{cases} -\Delta u = f & \text{in } \Omega^{(\xi)} \\ u = 0 & \text{on } \Gamma_0 \\ \frac{\partial u}{\partial \nu} = 0 & \text{on } \Gamma_1 \\ \frac{\partial u}{\partial \nu} + cu = d & \text{on } \Gamma_2 \\ \frac{\partial u}{\partial \nu} = 0 & \text{on } \Gamma_3 \end{cases} \tag{1}$$

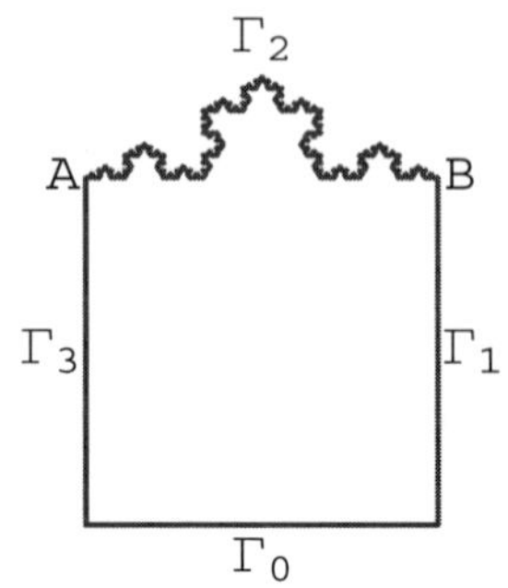

Fig. 1. An electrochemical cell.

where $\Omega^{(\xi)}$ is the set bounded by $\Gamma_0 = \{(x,y) \in \mathbb{R}^2 : 0 < x < 1, y = -1\}$, $\Gamma_1 = \{(x,y) \in \mathbb{R}^2 : x = 1, -1 < y < 0\}$, $\Gamma_3 = \{(x,y) \in \mathbb{R}^2 : x = 0, -1 < y < 0\}$, and Γ_2 is $\overset{\circ}{K}{}^{(\xi)} = K^{(\xi)} \setminus \{P_1, P_2\}$ ($K^{(\xi)}$ denotes the irregular scale Koch curve, $P_1 = (0,0)$ and $P_2 = (1,0)$); moreover, f is a given function in $L^2(\Omega^{(\xi)})$, $c \geqslant 0$ and d are constant.

We prove existence and uniqueness of the variational solution of (1) in Theorem 3.1.

In order to have a better insight of the problem, we consider the so-called *prefractal* problem, that is, the same problem where Γ_2 instead of being the scale irregular Koch curve $K^{(\xi)}$ is the n-th prefractal curves approximating $K^{(\xi)}$: we obtain existence and uniqueness results in Theorem 4.1 and regularity results in Theorem 4.2. In particular, these results fit the numerical study of the Laplacian transfer across fractal surfaces given in Ref. 6. We remark that, in Ref. 3, we have obtained existence, uniqueness and regularity results for the same mixed Dirichlet-Robin problem when the working electrode Γ_2 is both a Koch-type curve and the n-th prefractal curve approximating the Koch-type curve. Moreover, in Ref. 4, we have proved the strong convergence of suitable extensions of solutions of prefractal problems to the solution of the corresponding fractal problem. These type of results are useful for the numerical approximation of the fractal problem itself: our guess is that similar asymptotic results hold also for scale irregular interfaces.

The layout of the paper is as follows. In the second section, we recall the definitions and the properties of the scale irregular Koch curves. In the third section, we prove existence and uniqueness results when the interface Γ_2 is a scale irregular Koch curve. In section 4, we consider the problem when the interface is the n-th prefractal curves approximating the scale irregular Koch curve and we prove existence, uniqueness and regularity results.

2. Scale irregular Koch curves

In this section, we recall the definition of scale irregular Koch curves (see Refs. 1 and 16). We remark that these fractals do not have any exact self-similarity even if they are spatially homogeneous: they provide a good example for environment dependent fractals.

The scale irregularity comes from the fact that at each prefractal steps the family of contractive similitudes to apply is to be chosen from a finite set of families: the jumps from one family to another are meant to represent the influence of the environment on the morphogenesis of the fractal structure.

Let $A = \{1, 2\}$: for $a \in A$ let $2 < \ell_a < 4$, and, for each $a \in A$, let

$$\Psi^{(a)} = \{\psi_1^{(a)}, \dots, \psi_4^{(a)}\} \tag{2}$$

be the family of contractive similitudes $\psi_i^{(a)} : \mathbb{C} \to \mathbb{C}$, $i = 1, \dots, 4$, with contraction factor ${\ell_a}^{-1}$:

$$\psi_1^{(a)}(z) = \frac{z}{\ell_a}, \qquad \psi_2^{(a)}(z) = \frac{z}{\ell_a} e^{i\theta(\ell_a)} + \frac{1}{\ell_a},$$

$$\psi_3^{(a)}(z) = \frac{z}{\ell_a} e^{-i\theta(\ell_a)} + \frac{1}{2} + i\sqrt{\frac{1}{\ell_a} - \frac{1}{4}}, \qquad \psi_4^{(a)}(z) = \frac{z-1}{\ell_a} + 1,$$

where

$$\theta(\ell_a) = \arcsin\left(\frac{\sqrt{\ell_a(4-\ell_a)}}{2}\right). \tag{3}$$

Let $\Xi = A^{\mathbb{N}}$; we call $\xi \in \Xi$ an environment. We define a left shift S on Ξ such that if $\xi = (\xi_1, \xi_2, \xi_3, \dots)$, then $S\xi = (\xi_2, \xi_3, \dots)$. For $B \subset \mathbb{R}^2$ set

$$\Phi^{(a)}(B) = \bigcup_{i=1}^{4} \psi_i^{(a)}(B)$$

and

$$\Phi_n^{(\xi)}(B) = \Phi^{(\xi_1)} \circ \dots \circ \Phi^{(\xi_n)}(B).$$

The fractal $K^{(\xi)}$ associated with the environment sequence ξ is defined by

$$K^{(\xi)} = \overline{\bigcup_{n=1}^{+\infty} \Phi_n^{(\xi)}(\Gamma)}$$

where $\Gamma = \{P_1, P_2\}$ with $P_1 = (0,0)$ and $P_2 = (1,0)$. This set is not in general self-similar, but the family $\{K^{(\xi)}, \xi \in \Xi\}$ satisfies the following relation

$$K^{(\xi)} = \Phi^{(\xi_1)}(K^{(S\xi)}).$$

Moreover, they are spatially homogeneous: in fact, the volume measure $\mu^{(\xi)}$ on $K^{(\xi)}$ satisfies a sort of homogenous condition. Before to describing this measure, we introduce some notations. For $\xi \in \Xi$, we define the word space

$$W = W^{(\xi)} = \{(w_1, w_2, ...) : 1 \leqslant w_i \leqslant 4\}$$

and, for $w \in W$, we set $w|n = (w_1, ..., w_n)$ and $\psi_{w|n} = \psi_{w_1}^{(\xi_1)} \circ \cdots \circ \psi_{w_n}^{(\xi_n)}$.

The volume measure $\mu^{(\xi)}$ is the unique Radon measure on $K^{(\xi)}$ such that

$$\mu^{(\xi)}(\psi_{w|n}(K^{(S^n\xi)})) = \frac{1}{4^n}$$

for all $w \in W$.

The fractal set $K^{(\xi)}$ and the volume measure $\mu^{(\xi)}$ depend on the oscillations in the environment sequence ξ (see Ref. 1). We denote by $h_a^{(\xi)}(n)$ the frequency of the occurrence of a in the finite sequence $\xi|n$, $n \geqslant 1$:

$$h_j^{(\xi)}(n) = \frac{1}{n} \sum_{i=1}^{n} \mathbf{1}_{\{\xi_i = a\}}, \ a = 1, 2.$$

Let p_a be a probability distribution on A, and suppose that ξ satisfies

$$h_a^{(\xi)}(n) \to p_a, \ \ n \to +\infty,$$

where $0 \leq p_a \leq 1, \ \ p_1 + p_2 = 1$ and

$$|h_a^{(\xi)}(n) - p_a| \leq \frac{1}{n}, \ a = 1, 2, \ (n \geq 1),$$

that is, we consider the case of the fastest convergence of the occurrence factors.

Under these conditions, the measure $\mu^{(\xi)}$ has the property that there exists two positive constants C_1, C_2, such that,

$$C_1 r^{d^{(\xi)}} \leq \mu^{(\xi)}(B(P, r) \cap K^{\xi}) \leq C_2 r^{d^{(\xi)}}, \quad \forall P \in K^{(\xi)}, \tag{4}$$

with

$$d^{(\xi)} = \frac{\ln 4}{p_1 \ln \ell_1 + p_2 \ln \ell_2}, \tag{5}$$

where $B(P, r)$ denotes the Euclidean ball with center in P and radius $0 < r \leq 1$ (see Refs. 15 and 16). According to Jonsson and Wallin (see Ref. 10), we say that $K^{(\xi)}$ is a d-set with respect to the Hausdorff measure $\mathcal{H}^d$, with $d = d^{(\xi)}$.

3. Scale irregular fractal case

In this section, we prove the existence and the uniqueness of the variational solution of the fractal problem when the interface Γ_2 is a scale irregular Koch curve (for analogous results when the interface Γ_2 is a Koch-type curves, see Ref. 3). The main tools are some trace on $d-$sets and the related trace spaces, the Besov spaces $B^{p,q}_{\alpha}$ (see Refs. 11, 19–21). Moreover, we use a suitable Green's formula on fractal domains and the space $B^{2,2}_{\beta,0}$, the "fractal" analogue of the Lions-Magenes space $H^{\frac{1}{2}}_{0,0}$ (see Ref. 14) (for the Green's formula on fractal domains and the definition of the spaces $B^{2,2}_{\beta,0}$, see Ref. 12).

Theorem 3.1. *For any $f \in L^2(\Omega^{(\xi)})$, there exists one and only one solution u of the following problem*

$$\begin{cases} \text{find } u \in V(\Omega^{(\xi)}) := \{u \in H^1(\Omega^{(\xi)}) : u = 0 \text{ on } \Gamma_0\} \quad s.t. \\ \int_{\Omega^{(\xi)}} \nabla u \nabla v \, dxdy + c \int_{K^{(\xi)}} u \, v \, d\mu^{(\xi)} = \int_{\Omega^{(\xi)}} f \, v \, dxdy + d \int_{K^{(\xi)}} v \, d\mu^{(\xi)} \\ \forall \, v \in V(\Omega). \end{cases} \tag{6}$$

Such u solves the following problem

$$\begin{cases} -\Delta u = f & \text{in } L^2(\Omega^{(\xi)}) \\ u = 0 & \text{in } H^{\frac{1}{2}}(\Gamma_0) \\ \frac{\partial u}{\partial \nu} = 0 & \text{in } (H^{\frac{1}{2}}_{0,0}(\Gamma_1))' \\ \frac{\partial u}{\partial \nu} + cu = d & \text{in } (B^{2,2}_{\beta,0}(K^{(\xi)}))' \\ \frac{\partial u}{\partial \nu} = 0 & \text{in } (H^{\frac{1}{2}}_{0,0}(\Gamma_3))', \end{cases} \tag{7}$$

with $\beta = \frac{d^{(\xi)}}{2}$.

Proof. The proof follows by applying Lax-Milgram Theorem to the bilinear form

$$a(u,v) = \int_{\Omega^{(\xi)}} \nabla u \, \nabla v \, dxdy + c \int_{K^{(\xi)}} u \, v \, d\mu^{(\xi)} \tag{8}$$

and by using the trace Theorems cited before. By usual duality results, we achieve the proof (for a characterization of the dual of Besov spaces, see Refs. 11 and 21). □

4. Prefractal case

We study the problem (1) when the interface Γ_2 is the n-th prefractal curves approximating the scale irregular Koch curve in order to understand the behavior of the solutions of the same problem on the fractal structure in a better way. Let K_0 be the line segment of unit length having as endpoints $P_1 = (0,0)$ and $P_2 = (1,0)$. We set, for each n in $\mathbb{N}$, $K_n^{(\xi)} = \Phi_n^{(\xi)}(K_0)$: $K_n^{(\xi)}$ is the so-called n-th prefractal curve (see Fig. 2 and Fig. 3).

Fig. 2. Examples of prefractal curves.

Fig. 3. Examples of prefractal curves.

Then, we study the problem (1) when the interface $\Gamma_2 = \overset{\circ}{K}{}_n^{(\xi)} = K_n^{(\xi)} \setminus$

$\{P_1, P_2\}$. Obviously, in this case, we consider $\Omega_n^{(\xi)}$, that is, the set bounded by Γ_0 Γ_1, Γ_2, Γ_3; moreover, we put $c = c_n \geqslant 0$ and $d = d_n$, that is, we consider the case in which also the coefficients c and d depend on the index of the prefractal curves. We remark that in this paper, the constants c_n and d_n are not determined since we work with n fixed; in the asymptotic analysis, on the contrary, these constants will be chosen in a suitable way (see Ref. 4; for similar results, see Ref. 13). In the following, we fix $\xi \in \Xi$.

Theorem 4.1. *For any $f \in L^2(\Omega_n^{(\xi)})$, for every n in $\mathbb{N}$, there exists one and only one solution u of the following problem*

$$\begin{cases} \text{find } u_n \in V(\Omega_n^{(\xi)}) := \{u_n \in H^1(\Omega_n^{(\xi)}) : \ u_n = 0 \text{ on } \Gamma_0\} \quad \text{such that} \\ \int_{\Omega_n^{(\xi)}} \nabla u_n \nabla v \, dxdy + c_n \int_{K_n^{(\xi)}} u_n \, v \, ds = \\ \qquad\qquad\qquad \int_{\Omega_n^{(\xi)}} f \, v \, dxdy + d_n \int_{K_n^{(\xi)}} v \, ds \\ \forall \, v \in V(\Omega_n^{(\xi)}). \end{cases} \tag{9}$$

Such u solves

$$\begin{cases} -\Delta u_n = f & \text{in } L^2(\Omega_n^{(\xi)}) \\ u_n = 0 & \text{in } H^{\frac{1}{2}}(\Gamma_0) \\ \frac{\partial u_n}{\partial \nu} = 0 & \text{in } (H^{\frac{1}{2}}_{0,0}(\Gamma_1))' \\ \frac{\partial u_n}{\partial \nu} + c_n u_n = d_n & \text{in } (H^{\frac{1}{2}}_{0,0}(K_n^{(\xi)}))' \\ \frac{\partial u_n}{\partial \nu} = 0 & \text{in } (H^{\frac{1}{2}}_{0,0}(\Gamma_3))'. \end{cases} \tag{10}$$

Proof. The thesis follows by applying Lax-Milgram Theorem to the bilinear form

$$a_n(u_n, v) = \int_{\Omega_n^{(\xi)}} \nabla \, u_n \nabla \, v \, dx \, dy + c_n \int_{K_n^{(\xi)}} u_n \, v \, ds \,, \tag{11}$$

where ds denotes the arc length Lebesgue measure on polygonal curve $K_n^{(\xi)}$. Then, by using trace Theorems on polygonal curves (see Refs. 2, 8, 17), for any $f \in L^2(\Omega_n^{(\xi)})$, we obtain the existence and the uniqueness of the weak solution u_n. We conclude the proof by usual duality arguments. □

In the following, we prove global regularity results in terms of ordinary fractional Sobolev spaces. Since the coefficients are constants, in order to determine global regularity, we have to study two crucial elements: the irregular boundary and the boundary mixed Dirichlet-Robin conditions.

Theorem 4.2. *Let u_n be the weak solution of the prefractal problem (9). Then, set* $\ell^* = \min(\ell_1, \ell_2)$,

$$u_n \in H^s(\Omega_n^{(\xi)}), \textit{ with } \quad s < 1 + \frac{\pi}{\pi + \theta(\ell^*)}, \tag{12}$$

where

$$\theta(l^*) = \arcsin\left(\frac{\sqrt{l^*(4 - l^*)}}{2}\right).$$

Proof. The proof is a direct consequence of Theorem 5.1 in Ref. 3. In fact, we obtain that the solution u_n belongs to $H^s(\Omega_n^{(\xi)})$ for every $s \leqslant 2$ such that

$$s < 1 + \frac{\pi}{\omega},$$

where ω is the amplitude of the largest of the internal angles in $\Omega_n^{(\xi)}$. By considering the amplitude of all the angles in $\Omega_n^{(\xi)}$, we obtain estimates (12). □

Remark 4.1. It is obvious that if $(\xi_1, \xi_2, \ldots, \xi_n) = (1, 1, \ldots, 1)$ or $(\xi_1, \xi_2, \ldots, \xi_n) = (2, 2, \ldots, 2)$, we obtain the results of Ref. 3.

By the previous regularity results, we can improve the statement (10) of Theorem 4.1 in the following way.

Theorem 4.3. *For any $f \in L^2(\Omega_n^{(\xi)})$, for every n in $\mathbb{N}$, the weak solution u_n of the prefractal problem (9) solves the following problem*

$$\begin{cases} -\Delta u_n = f & \textit{in } L^2(\Omega_n^{(\ell)}) \\ u_n = 0 & \textit{in } C^0(\Gamma_0) \\ \frac{\partial u_n}{\partial \nu} = 0 & \textit{in } L^2(\Gamma_1) \\ \frac{\partial u_n}{\partial \nu} + c_n u_n = d_n & \textit{in } L^2(K_n^{(\xi)}) \\ \frac{\partial u_n}{\partial \nu} = 0 & \textit{in } L^2(\Gamma_3). \end{cases} \tag{13}$$

From Theorem 5.2 in Ref. 3, we deduce in the following Theorem 4.4 that the regularity of the solutions can be expressed also in terms of weighted Sobolev spaces, where the weight is the distance from the vertices of reentrant corners (for the definition, see Refs. 8 and 9). These results play a key role in the numerical analysis (see Refs. 5, 20, 22).

Theorem 4.4. *Let u_n be the weak solution of the prefractal problem (9). Then, set $\ell^* = \min(\ell_1, \ell_2)$,*

$$u_n \in H^{2,\alpha}(\Omega_n^{(\xi)}), \text{ with } \quad \alpha > \frac{\theta(\ell^*)}{\pi + \theta(\ell^*)}, \tag{14}$$

where

$$\theta(l^*) = \arcsin\left(\frac{\sqrt{l^*(4 - l^*)}}{2}\right).$$

Remark 4.2. In Ref. 3 we have proved that, when the interface is the usual Koch curve ($\ell_1 = \ell_2 = 3$ and hence $\theta = \frac{\pi}{3}$), the solution of the related mixed Dirichlet-Robin problem belongs to the fractional Sobolev space H^s with $s < \frac{7}{4}$. Now, we can construct a class of scale irregular Koch curves $K^{(\xi)}$ which have Hausdorff dimensions equal to the Hausdorff dimension of the Koch curve, that is,

$$d^{(\xi)} = \frac{\ln 4}{\ln 3}.$$

It suffices to choose p_1, p_2, ℓ_1, ℓ_2 such that

$${\ell_1}^{p_1} {\ell_2}^{p_2} = 3,$$

where $2 < \ell_1 < 4$, $2 < \ell_2 < 4$, $\ell_1 \neq 3$, $\ell_2 \neq 3$, $0 < p_1 < 1$, $0 < p_2 < 1$, $p_1 + p_2 = 1$. Then, by Theorem 4.2, we obtain that the solution of the corresponding mixed Dirichlet-Robin problem belongs to the fractional Sobolev space H^s with $s < 1 + \frac{\pi}{\pi + \theta(\ell^*)} < \frac{7}{4}$, where $\ell^* = \min(\ell_1, \ell_2)$. Therefore, we conclude that the regularity of the solution of the corresponding problems depends on the minimum contraction factors ℓ_a of the family (2) uniquely. These results fit the numerical study of the transfer across fractal surfaces given in Ref. 6, where, within very good approximation, it is shown that only the fractal dimension and the lower and higher cutoffs influence the behavior of the system.

References

1. M.T. Barlow, B.M.Hambly, Transition density estimates for Brownian motion on scale irregular Sierpinski gasket, *Ann. Inst. H. Poincaré*, 33 (1997), pp. 531–556 .
2. Brezzi F., Gilardi G. Finite Elements Mathematics, in *Finite Element Handbook*, Eds. Kardestuncer H., D.H.Norrie, MacGraw-Hill Book Co., New York, 1987.
3. R. Capitanelli, Mixed Dirichlet-Robin problems in irregular domains, *Communications to SIMAI Congress,* 2 (2007) DOI:10.1685/CSC06035.

4. R. Capitanelli, Asymptotics for mixed Dirichlet-Robin problems in irregular domains, preprint.
5. E. Evans, A novel finite element meshing technique driven by fractal Koch curve, preprint.
6. M. Filoche, B. Sapoval, Transfer across random versus Deterministic Fractal Interfaces, *Phys. Rev. Lett.*, 84 (2000), pp. 5776–5779.
7. D. S. Grebenkov, M. Filoche, B. Sapoval, Mathematical Basis for a General Theory of Laplacian Transport towards Irregular Interfaces, *Phys. Rev. E* 73 021103, 2006.
8. P. Grisvard, *Elliptic problems in nonsmooth domains,* Pitman, Boston, 1985.
9. P. Grisvard, *Singularities in boundary value problems,* Paris, Masson, 1992.
10. A. Jonsson, H. Wallin, Function spaces on subsets of $\mathbb{R}^n$, *Math. Rep.*, 2 (1984), no. 1, xiv+221.
11. A. Jonsson, H. Wallin, The dual of Besov spaces on fractals, *Studia Math.*, 112 (1995), no. 3, pp. 285–300.
12. M.R. Lancia, A transmission problem with a fractal interface, *Z. Anal. und Ihre Anwend.*, 21 (2002), pp. 113–133.
13. M.R. Lancia, M.A. Vivaldi, Asymptotic convergence of transmission energy forms, *Adv. Math. Sc. Appl.*, 13 (2003), pp. 315–341.
14. J.L Lions, E. Magenes, *Non-Homogeneous Boundary Value Problems and Applications,* Springer-Verlag, Berlin,1972.
15. U. Mosco, Harnack inequalities on scale irregular Sierpinski gaskets. *Nonlinear problems in mathematical physics and related topics, II, Int. Math. Ser.*, 2 (2002), pp. 305–328.
16. U. Mosco, Gauged Sobolev inequalities. *Appl. Anal.*, 86 (2007), no. 3, pp. 367–402.
17. J. Necǎs, *Les méthodes directes en théorie des équationes elliptiques,* Paris, Masson, 1967.
18. B. Sapoval, General formulation of laplacian transfer across irregular surfaces, *Phys. Rev. Lett.*, 73 (1994), pp. 3314–3316.
19. H. Triebel, *Fractals and spectra. Related to Fourier analysis and function spaces.* Monographs in Mathematics, 91. Birkhäuser Verlag, Basel, 1997.
20. E. Vacca, *Galerkin approximation for highly conductive layers.* PhD thesis, Università degli Studi di Roma “La Sapienza”, 2005.
21. H. Wallin, The trace to the boundary of Sobolev spaces on a snowflake, *Manuscripta Math.*, 73 (1991), no. 2, pp. 117–125.
22. R.D. Wasik, *Numerical Solution of a transmission Problem with a prefractal interface.* PhD thesis, Worchester Polytechnic Institute, 2007.

A NON-COMMUTATIVE OPERATOR-HIERARCHY OF BURGERS EQUATIONS AND BÄCKLUND TRANSFORMATIONS

SANDRA CARILLO

Dipartimento di Metodi e Modelli Matematici per le Scienze Applicate
Sapienza University of Rome, Rome, Italy
E-mail: carillo@dmmm.uniroma1.it
www.dmmm.uniroma1.it/∼carillo

CORNELIA SCHIEBOLD

Department of Natural Sciences, Engineering, and Mathematics
Mid Sweden University, Sundsvall, Sweden
E-mail: Cornelia.Schiebold@miun.se

An operator equation on a Banach space, which represents the operator analog of Burgers equation, is here considered. The well known Cole-Hopf transformation, a particular case of the wider class of Bäcklund transformations, which connects the classical nonlinear Burgers equation to the linear heat equation, is extended to the case of operator valued equations. Then, since the operator Burgers equation admits a recursion operator, a whole hierarchy of Burgers operator equations is generated. Notably, each member of such a *Burgers operator* hierarchy is related, via Cole-Hopf transformation to the corresponding member of a *heat operator* hierarchy. Indeed, also the recursion operator admitted by the Burgers operator equation, is related, via Cole-Hopf transformation, to the (trivial) recursion operator admitted by the linear heat operator equation. Furthermore, the Burgers recursion operator is not Abelian, hence, the whole hierarchy does not enjoy commutativity properties.

Keywords: Bäcklund transformations, Burgers equation, operator valued equations in Banach spaces.

1. Introduction

The Cole-Hopf transformation is named after Cole [15] and Hopf [19] who solved an initial boundary value Burgers problem recognizing that Burgers equation is related, via such a Bäcklund Transformation, to the linear heat equation. The origin of Bäcklund Transformations goes back to the nineteens century when Bäcklund, in a series of papers [2,3,4,5,6], was investigating surface transformations to generalize Lie contact transformations. Since then, a wide variety of results has been obtained on applications of Bäcklund Transformations which not only allow to find solutions of initial boundary value problems for nonlinear evolution equations, but are also applied to establish structural properties enjoyed by a nonlinear system such as the Hamiltonian and/or bi-Hamiltonian structure and symmetry properties, in general. Many interesting results, as well as historical details and a wide bibliography on Bäcklund Transformation, are comprised in Rogers and Shadwick [25], and, subsequently, in Rogers and Ames [23]. Therein, furthermore, applications of Bäcklund Transformation to nonlinear initial boundary value problems, transformed into other related initial boundary value problems amenable of solution, are given.

On the other hand, non linear evolution equations, often termed *soliton equations* in the literature, can be studied and, on adopting the approach proposed by Aden and Carl [1], which consists in introduce operator valued equation in a suitable Banach space. Thus, hierarchies of noncommutative nonlinear operator valued equations [11], are considered. They represent a generalization to the operator level of corresponding hierarchies of noncommutative nonlinear evolution equations. Such hierarchies can, indeed, be obtained from the operator ones via a suitable projection. Bäcklund Transformations have been applied [11] to connect different hierarchies of operator valued hierarchies of noncommutative nonlinear evolution equations. Indeed, in the case when hierarchies of nonlinear evolution equations in $1+1$-dimensions are considered, a wide Bäcklund Chart depicts all the links relating the Korteweg deVries hierarchy of evolution equations, to the modified Korteweg deVries and Dym hierarchies [17]. In addition, algebraic as well as structural properties of the equations which appear in the Bäcklund Chart, can be revealed. According to [22], many results, and in particular the links established among hierarchies of nonlinear evolution equations, can be extended to the corresponding nonlinear evolution equations in $2+1$-dimensions, and also to hierarchies whose base member is of fifth order [24]. Such generalizations, however, are still under investigation as far as the extension to the operator hierarchies are concerned.

The material is organized as follows. Section 2, which opens reminding the definition of Bäcklund transformation, is concerned about the Cole-Hopf link between the heat equation with nonlinear partial differential equations. The subsequent Section 3 is devoted to a brief overview concerning operator approach to nonlinear evolution equations and, in addition, to some results obtained by the authors in the case of the Bäcklund Chart involving operator equations analog of Korteweg deVries and modified Korteweg deVries equations. Section 4 is devoted to the Cole-Hopf link connecting the operator Burgers equation to the linear operator heat equation. Some perspectives as well as work in progress are briefly mentioned in the closing Section 5.

2. Bäcklund Transformations: background notions

In this Section few notions are recalled which are needed in the subsequent parts. Accordingly, here throughout an evolution equation is denoted via $u_t = K(u)$, where $u(x,t) \in M$ where M presents a manifold modeled on a linear topological space so that the typical fiber $T_u M$ of M can be identified with M itself. Specifically, M is assumed to be the Schwartz space $\mathcal{S}$ of C^∞-functions rapidly vanishing at infinity [20]; $K : M \to TM$ denotes an appropriate C^∞-vector field on the manifold M. Then, according to [16],

Definition 2.1. Given two evolution equations in $1+1$-dimensions, say $u_t = K(u)$ and $s_t = G(s)$, then $B(u,s) = 0$ represents a Bäcklund transformation between them if , whenever, given two solutions, in turn, denoted as $u(x,t)$ and $s(x,t)$, if $B(u(x,t), s(x,t))|_{t=0} = 0$, it follows that

$$B(u(x,t), s(x,t)) \equiv 0, \quad \forall t > 0, \quad \forall x \in \mathbb{R}. \tag{1}$$

Hence, a Bäcklund transformation establishes a $1-1$ correspondence between solutions of the two evolution equations. Furthermore, when an evolution equations, say $u_t = K(u)$, admits a hereditary recursion operator [16], denoted as $\Phi(u)$, then, also the second one can be proved to admit such an operator. Indeed, the recursion operator which refers to second evolution equations can be obtained from the recursion operator $\Phi(u)$ of the other one, again, via the Bäcklund transformation[a]. Thus, let $\Psi(s)$ denote the recursion operator of the second equation, then the two equations can be written, in turn

$$u_t = \Phi(u) u_x \quad , \qquad s_t = \Psi(s) s_x \ , \tag{2}$$

and the corresponding hierarchies, which read

$$u_t = \Phi(u)^n u_x \quad , \qquad s_t = \Psi(s)^n s_x \ , \quad n \in \mathbb{N}, \tag{3}$$

[a]a detailed explanation concerning how recursion operators and also the Hamiltonian and bi-Hamiltonian structures are transformed on application of Bäcklund as well as reciprocal transformations is comprised in Refs. 16 and 17.

are connected via the same Bäcklund transformation which relates the base members [16]. Among the many important consequences implied by the link via Bäcklund transformation between the two hierarchies, such as the possibility to find solutions to initial boundary value problems, on one side, and to investigate structural properties of nonlinear evolution equations, on the other one, here the attention is focussed only on some aspects related to this second point of view. Consider, now, the linear heat equation

$$u_t = u_{xx} \quad , \tag{4}$$

where the thermal conductivity is set equal to 1; it can be written under the form (2) where [16]

$$\Phi(u) := D \quad , \quad D := \frac{\partial}{\partial x} \; . \tag{5}$$

Then, a hierarchy of linear heat equations can be constructed, via

$$u_t = \Phi(u)^n u_x \quad , \quad n \in \mathbb{N} \; . \tag{6}$$

The Cole-Hopf transformation, defined when $B(u, v) = 0$ reads [15,19]

$$u_x - uv = 0 \; ; \tag{7}$$

relates the linear heat equation to the nonlinear Burgers equation

$$v_t = v_{xx} + 2vv_x \quad . \tag{8}$$

The link allows to find new solutions to assigned problems. Notably, in general, when two initial boundary value problems are connected via a Bäcklund Transformation, their solutions turn out to be related via the link established. Hence, solutions to a problem can be constructed on transforming the corresponding ones of the other problem. This, is exactly the framework of Cole's [15] and Hopf's [19] work. Following the same lines, in [18], the solution of a problem of water infiltration in soils , modeled via a Burgers initial boundary value problem, is explicitly obtained in terms of complementary error functions.

On the other hand, as far as the structural viewpoint is concerned, the nonlinear Burgers equation (8) can also be written under the form (2), namely

$$v_t = \Psi(v) v_x \quad , \quad n \in \mathbb{N} \; , \tag{9}$$

where the recursion operator, obtained by Fokas and Fuchssteiner [16], reads

$$\Psi(v) := D + v + v_x D^{-1} \quad , \quad D^{-1} := \int_{-\infty}^{x} \tag{10}$$

wherein the *inverse derivative operator* is well defined since $u \in \mathcal{S}$, the Schwartz space of rapidly decreasing function [20]. Thus, the Burgers hierarchy follows

$$v_t = \Psi(v)^n v_x \quad , \quad n \in \mathbb{N} . \tag{11}$$

Notably, the Cole-Hopf transformation relates corresponding members in the two hierarchies of evolution equations.

3. Operator Soliton Equations

This section provides background on operator integrable equations with focus on a method linking their examination to operator theory. As a first illustration for an application to hierachies, an outline is given on forthcoming work of the authors [11] on KdV, mKdV, and potential KdV hierarchies. This is closely related in spirit to what is done in the sequel in the case of the Burgers and heat hierarchies, but technically too involved to be treated in detail in this note. The following diagram illustrates the general strategy.

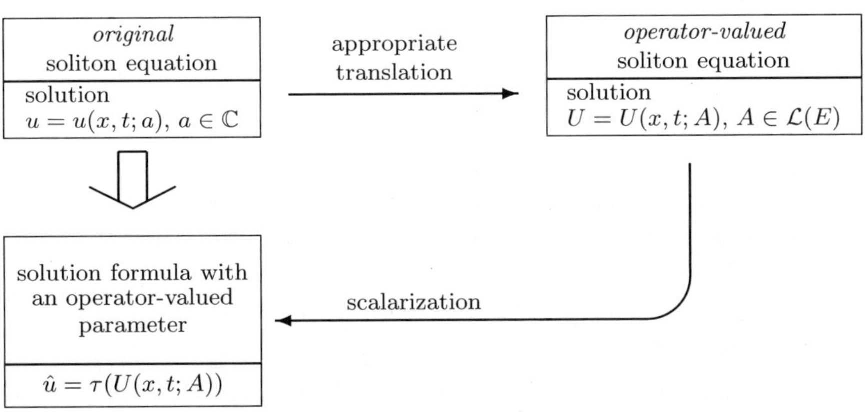

Typically, the solution u we start with depends on a scalar parameter $a \in \mathbb{C}$. The goal is now to construct solution formulae depending on an operator-valued parameter $A \in \mathcal{L}(E)$, E some Banach space, which can be viewed as a blow-up of a. As indicated in the diagram above, this can be achieved via a detour through the operator-level where operator-valued soliton equations and corresponding solutions are considered.

This strategy bases on pioneering work of Marchenko [21]. In [1] Aden and Carl introduce a crucial new idea to place the strategy into the frame of Banach operator ideals and to establish a link between soliton theory and the geometry of Banach spaces. This was done for the KdV equation. In the sequel, the method proved to be very flexible. As shown by Carl and the second author, it works for the most prominent soliton equations, including discrete ones as the Toda lattice, and the complete AKNS system [13,14,27]. However it turns out that there is no universal algorithm to produce the right translation to the operator level. In another way, the efficiency of the method is confirmed by the work of Aden, Blohm [7], who show that all solutions covered by the inverse scattering method can be realized in this frame. The way back to the scalar case will stay in the background in this note, but it should be mentioned that it provides powerful tools in the construction and qualitative study of complex solution families like multipole solutions and countable superpositions of solitons [1,26,27,28].

One of the most remarkable properties of integrable systems is the existence of a rich family of symmetries. For the classical equations in one space variable this is displayed by infinite hierarchies, countable families of integrable systems which may be viewed as infinitesimal generators of the symmetry group of the original system. As a rule, hierarchies are generated by iterative application of a recursion operator. If one tries to lift hierarchies to the noncommutative setting, one has first to find an noncommutative analogue to the recursion operator (additionally to noncommutative equations and candidates for solutions) and then to give an inductive argument to verify the solution property, the latter being the most involved part. In the sequel, this is outlined for the potential KdV hierarchy. Then it is indicated how to derive corresponding results for the KdV hierarchy and the mKdV hierarchy via Bäcklund transformations. The reader is referred to [11] for the details.

The recursion operator formulation of the noncommutative potential KdV hierarchy reads

$$V_{t_{2n-1}} = \Psi(V)^{n-1} V_x, \tag{12}$$

$n \geq 1$, where

$$\Psi(V) = D^2 - A_{V_x} - D^{-1} A_{V_x} D + D^{-1} C_{V_x} D^{-1} C_{V_x}, \tag{13}$$

and V is a function taking its values in the bounded linear operators on some Banach space. Here D stands for the derivative with respect to x, C_T stands for the commutator, A_T for the anti-commutator with respect to T,

$$C_T(S) = [T, S], \qquad A_T(S) = \{T, S\}.$$

Moreover V is supposed to decay sufficiently fast for $x \to \infty$, which allows to define D^{-1} by $D^{-1}V(x) = \int_{-\infty}^{x} V(\xi) d\xi$. In particular, the lowest members

$n = 1, 2, 3$ of the hierarchy read

$$\begin{aligned} V_{t_1} &= V_x, \\ V_{t_3} &= V_{xxx} - 3V_x^2, \\ V_{t_5} &= V_{xxxxx} - 5\{V_x, V_{xxx}\} - 5V_{xx}^2 + 10V_x^3. \end{aligned}$$

As usual one sets $t_1 = x$, $t_3 = t$. Note that (12) explicitly depends only on t_1 and t_{2n-1}.

Let E be a Banach space and A, B bounded linear operators on E, and consider the operator-valued function

$$V_N = -(I + L_N)^{-1}(AL_N + L_N A) \tag{14}$$

with

$$L_N = L_N(t_1, \dots, t_{2N-1}) = \exp\Big(\sum_{k=1}^{N} A^{2k-1} t_{2k-1}\Big) B \tag{15}$$

(and I the identity operator on E). It is known [1] that V_2 provides a solution of the noncommutative potential KdV equation, which can be understood as an operator-analogue of the 1-soliton. A direct proof of the corresponding fact that V_3 solves the noncommutative potential fifth order KdV [14] is already very involved. The fundamental result of [11], requiring a complicated inductive argument, is that the operator-valued function V_N given by (14) solves the system of noncommutative potential KdV equations obtained by setting $n = 1, \dots, N$ in (12).

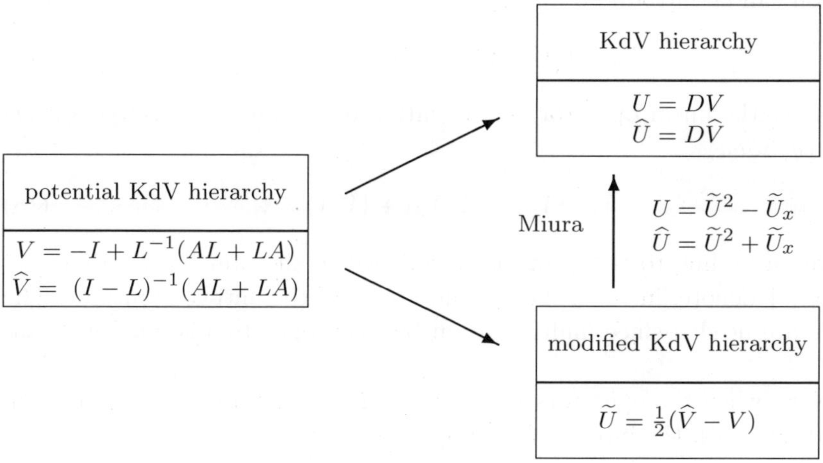

Fig. 1: Noncommutative solitons and their links.

It seems that the noncommutative potential KdV hierarchy is considerably more accessible than its relatives for KdV and mKdV. Hence it is advisable to use Bäcklund transformations and Miura links instead of trying to extend the above device to the more complex situations. The above diagram indicates the required transformations, see [11] for the details.

4. Burgers - Heat Operator Bäcklund Chart

This Section is devoted to briefly recall how to construct the operator equation analog of the Burgers equation, later denoted as *Burgers operator equation.* Then, a whole hierarchy of Burgers operator equations is generated on application of the recursion operator. The latter is also obtained via a generalization of the recursion operator admitted by the Burgers equation. Subsequently, the link between the Burgers operator equation and the operator analog of linear heat equation is shown. The Section closes showing that the link established remains valid for the whole hierarchies of operator equations; specifically, a hierarchy of linear heat operator equations is generated. Each member of such a hierarchy is related via Cole-Hopf transformation to the corresponding member of the Burgers operator hierarchy.

Consider the linear operator heat equation

$$U_t = U_{xx} \quad , \tag{16}$$

where U belongs to a suitable Banach space. In this case, letting U^{-1} represent the inverse operator of U, the following operator Cole-Hopf transformation can be introduced

$$U_x - UV = 0 \ , \quad \text{or} \quad V = U^{-1}U_x \quad . \tag{17}$$

It relates the linear operator heat equation to the *nonlinear Burgers operator equation*

$$V_t = V_{xx} + [V, V_x] + \{V, V_x\} \quad , \tag{18}$$

where, according to the notation introduced in the previous Section, $[\cdot,\cdot]$ and $\{\cdot,\cdot\}$ denote, in turn, the commutator and the anti-commutator. The latter are needed since multiplication between operators is not commutative.

Now, when the operators D as well as D^{-1} are those defined, in turn, in (5) and (10), the introduction of the operator

$$\Phi(U) := D \tag{19}$$

allows to construct the hierarchy of linear operator heat equations by

$$U_t = \Phi(U)^n U_x \quad , \quad n \in \mathbb{N} \; , \tag{20}$$

which, formally, is the analog of (6). On application of the transformation formulae [12], obtained on generalization of the corresponding ones in [16], the recursion operator follows from

$$\Psi(V)\,(D + [V, \cdot]) = (D + [V, \cdot])\,(D + V) \quad , \tag{21}$$

and, therefore, the whole hierarchy of Burgers operator equations is generated. Hence

$$V_t = \Psi(V)^n V_x \quad , \quad n \in \mathbb{N} \; , \tag{22}$$

which, again, shows the analogy with the case of hierarchies of nonlinear equations in 1+1-dimensions. However, it should be remarked that he analogy is formal since, the operator equation, base member of such a hierarchy does not enjoy commutativity property and, then, also the whole hierarchy is constituted by non Abelian operator equations. The details concerning the structural properties of the obtained hierarchy of operator equations is comprised in [12].

5. Conclusions and Perspectives

A systematic study concerning Bäcklund transformations and heat conduction in materials with and without memory is currently under investigation [10] and [12]. Indeed, some preliminary results are comprised in [8,9] where the linear Volterra type integro-differential heat equation is proved to be connected, via Bäcklund transformation, to a nonlinear equation, which can be termed *wave Burgers equation* since it represents a wave-type equation generalizing Burgers equation. The idea is to connect a linear integro-differential equation to a hyperbolic nonlinear differential equation, to provide a new justification of its well known hyperbolic behaviour. This viewpoint is currently under investigation [10] to reveal connections with other works as well as, possibly, explicit solutions to some boundary value problems of interest.

Acknowledgements

The partial support of the Italian National Mathematical Physics Research Group (G.N.F.M.-I.N.D.A.M.) is gratefully acknowledged.

References

1. H. Aden and B. Carl, *On realizations of solutions of the KdV equation by determinants on operator ideals*, J. Math. Phys.,**37** (4), 1833 - 1857 (1996).
2. A.V. Bäcklund, *Einiges über Curven- und Flächen-Transformationen*, Lunds Univ. Årsskr., **X**, 12 p., (1872-73).
3. A.V. Bäcklund, *Über Flächentransformationen*, Math. Ann., **IX**, 297 - 320, (1875).
4. A.V. Bäcklund, *Zur Theorie der partiellen Differentialgleichungen erster Ordnung*, Math. Ann., **XVII**, 285 - 328, (1880).
5. A.V. Bäcklund, *Zur Theorie der Flächentransformationen*, Math. Ann., **XIX**, 387 - 422, (1881).
6. A.V. Bäcklund, *Om ytor med konstant negativ krökning*, Lunds Univ. Årsskr., **XIX**, 48 p., (1883-84).
7. H. Blohm, *Solution of nonlinear equations by trace methods*, Nonlinearity, **13**, 1925 - 1964, (2000).
8. S. Carillo, *Bäcklund Transformations & Heat Conduction with Memory*, in *New Trends in Fluid and Solid Models*, M. Ciarletta et al. Ed.s, (2008).
9. S. Carillo, *Nonlinear Hyperbolic Equations and Linear Heat Conduction with Memory*, preprint (2009).
10. S. Carillo,, *Bäcklund transformations and Evolution Problems in Heat Conduction with Memory* , in progress, (2009).
11. S. Carillo and C. Schiebold, *Noncommutative Korteweg-de Vries and modified Korteweg-de Vries hierarchies via recursion methods*, J. Math. Phys., **50**, 073510 (2009).
12. S. Carillo and C. Schiebold, *Hierarchies of Burgers and Heat Operator Equations: a generalized Cole-Hopf Link*, (in progress), (2009).
13. B. Carl and C. Schiebold, *Nonlinear equations in soliton physics and operator ideals*, Nonlinearity, **12**, 333 - 364, (1999).
14. B. Carl and C. Schiebold, *Ein direkter Ansatz zur Untersuchung von Solitonengleichungen*, Jber. d. Dt. Math.-Verein., **102**, 102 - 148, (2000).
15. J. D. Cole, *On a quasilinear parabolic equation occuring in aerodynamics*, Quart. Appl. Math., **9**, 225 - 236, (1951).
16. A.S. Fokas and B. Fuchssteiner, *Bäcklund Transformations for Hereditary Symmetries*, Nonlinear Analysis TMA, **5**, 423 - 432, (1981).
17. B. Fuchssteiner and S. Carillo, *Soliton structure versus singularity analysis: Third order completely integrable nonlinear differential equations in 1+1 dimensions*, Physica A, **154**, 467 - 510, (1989).
18. B.-Y. Guo and S. Carillo, *Infiltration in soils with prescribed boundary concentration: a Burgers model*, Acta Math. Appl. Sinica, **6**, 365 - 369, (1990).
19. E. Hopf, *The partial differential equation* $u_t + uu_x = \mu u_{xx}$, Comm. Pure Appl. Math., **3**, 201 - 230, (1950).
20. L. Hörmander, *The Analysis of Linear Partial Differential Operators I (Distribution theory and Fourier Analysis)*, 2nd ed, Springer-Verlag, (1990).
21. V. A. Marchenko, *Nonlinear Equations and Operator Algebras*, Reidel, Dordrecht, (1988).

22. W. Oevel and S. Carillo, *Squared Eigenfunction Symmetries for Soliton Equations: Part I and Part II*, J. Math. Anal. and Appl., **217**, no.1, 161 - 178 and 179 - 199, (1998).
23. C. Rogers and W. F. Ames, *Nonlinear Boundary Value Problems in Science and Engineering*, Academic Press, Boston - San Diego - New York - Berkeley - London - Sydney - Tokyo - Toronto, (1989).
24. C. Rogers and S. Carillo, *On Reciprocal Properties of the Caudrey-Dodd-Gibbon and Kaup-Kupershmidt Hierarchies*, Physica Scripta, **36**, 865 - 869, (1987).
25. C. Rogers and W. F. Shadwick, *Bäcklund Transformations and their Applications*, Mathematics in Science and Engineering Vol. 161, Academic Press, New York - London - Paris - Sydney - Tokyo - Toronto, (1982).
26. C. Schiebold, *Solitons of the sine-Gordon equation coming in clusters*, Revista Matemática Complutense **15**, 265 - 325, (2002).
27. C. Schiebold, *Integrable Systems and Operator Equations*, Habilitation thesis, Jena, (2004) (http://apachepersonal.miun.se/~corsch/Habilitation.PDF).
28. C. Schiebold, *Explicit solution formulas for the matrix-KP*, Glasgow Math. J., **51**, Issue A, 147 - 155, (2009).

Probability of Detection for Penetrant Testing in Industrial Environment

Gennaro Caturano, Giovanni Cavaccini, Antonio Ciliberto, Vittoria Pianese,

Alenia Aeronautica S.p.A.
viale dell'Aeronautica s.n.c.
80038 Pomigliano d'Arco – Napoli, Italy
gcaturano@alenia.it gcavaccini@alenia.it aciliberto@alenia.it vpianese@alenia.it

and

Riccardo Fazio

Department of Mathematics, University of Messina
Salita Sperone 31, 98166 Messina, Italy
rfazio@dipmat.unime.it

Introduced at the end of 60's by NASA, Probability of Detection (PoD) is becoming more and more one of the main approach in order to assess, quantitatively, the general detection capabilities of a Non Destructive Inspection process. In spite of its importance, PoD can be elaborated in a variety of ways and can lead to some misinterpretations. Alenia Aeronautica assessed a specific approach for *liquid penetrant inspection* that is strictly connected to the estimation of the inspection sensitivity and it can be aimed at various targets, such as: inspection procedure validation, evaluation of personnel proficiency, comparative analysis of penetrant inspection processing materials, equipment and procedures, and evaluation of automated inspection systems. To this purpose, PoD is conceived as the probability, at a fixed confidence level, to detect a discontinuity belonging to a predefined class. Experimental PoD curves are obtained by processing metallic samples with defects generated and developed under controlled conditions.

Keywords: Probability of detection, liquid penetrant inspection.

1. Introduction

The aim of this study is to evaluate the performance of non-destructive inspection (NDI in short) called Penetrant Testing (PT) used in the production of airplane parts and in many industrial applications where the detection of open defects is of interest. Liquid penetrants are used to locate surface-accessible defects in solid parts.

The Probability of Detection (PoD) concept, is strictly related to the one of reliability used within NDI, can be defined as follows: "The probability of finding a crack of given dimension, under precise conditions, by prescribed inspection procedures". Even more generally, a suitable definition of PoD is: "The probability of finding an anomaly of given characteristics, under precise conditions, using a specific inspection procedure".

The PoD concept was introduced by NASA.[1] Indeed, by 1969, the USA spatial agency started to develop a sequence of researches aiming to evaluate the reliability of NDI controls. These studies were used mainly within the development of space shuttle project and ended up with a huge number of experimental data and PoD plots. Since then, the concept of PoD has been further defined and specialized to various cases, becoming an usual method for definition of NDI process characteristics and performance. In general, PoD applications include methodologies: to establish acceptance criteria, to validate inspection procedures, to assess the performance of the inspectors, to compare more inspection techniques, to select/qualify inspection processes, to quantify inspection process sensitivity.

2. Liquid Penetrant Inspection

One of the criteria for the design of a critical component (as aeronautic structures) is the inspection-ability: the component is designed in such a way that his compliance with structural and functional requirements is verifiable through available and qualified non-destructive methods. This approach is applied to the military aircraft (fail-safe approach), and to the civil aircraft (damage tolerance approach). The fail-safe approach provides that the component is free from not acceptable defects, it is designed to remain intact for the whole planned life. The component is inspected after manufacturing and in case of accident. A damage tolerance design requires that a component affected by an anomaly is capable of sustaining the damage, provided it does not exceed a critical size and is revealed after manufacturing or during programmed inspections.

Of course, NDI have limits of detection and application: the diagnosed defects are dependent on various factors (intrinsic limitations of the applied physical principles, instrumentation adopted, accessibility and characteristics of the component, inspection procedure, human factors, etc.).

The problem of the inspection-ability requires to answer on three questions:

(1) What types of defects are detectable?

(2) What is the smallest detectable flaw?
(3) What is the biggest flaw that may not be revealed?

The PoD is a way to answer those questions, especially the third.

3. General settings

There are essentially two methods for producing PoD curves (or the contrary, "acceptance curve", defined as 1-PoD):

- hit/miss (discrete data curve): an indication of anomaly can be detected, not detected, or doubtful;
- response curve (continuous data curve): the response to an indication of anomaly is dependent on an instrumental continuous parameter.

The method hit / miss[1] is applicable and sufficient for penetrant testing.

In any case, the methodology for obtaining PoD curves is based on the following steps:

(1) definition of parameters for defect classification;
(2) finding of samples and materials with representative defects;
(3) inspection of the samples with a defined technics and procedure;
(4) recording of results;
(5) analysis and best-fit of results.

In case of data hit/miss, different distributions can be considered for the best-fit. The distribution of Fisk[2]

$$PoD(d) = \frac{e^{\frac{\pi}{\sqrt{(3)}}\left(\frac{\ln(d)-m}{\sigma}\right)}}{1+e^{\frac{\pi}{\sqrt{(3)}}\left(\frac{\ln(d)-m}{\sigma}\right)}}, \qquad (1)$$

where d, m and s are the size of the defect, the median and standard deviation, respectively, is often used. The Fisk (or log-logistic) is a typical continuous distribution for not negative random variables, often used in economics. These approaches have much limits because they are based on continuous quantities (size of the defect), that don't take into account other conditions that typically characterize an inspection.

In a Penetrant Testing the detection of an indication depends not only by its size, but also by its visibility, by the exposed volume of penetrating and by lighting conditions. For this reason, often (as with Alenia) ad hoc approaches are defined.

4. Alenia Aeronautics Approach

Alenia Aeronautics approach is oriented to estimate the inspection process sensibility, by determining the PoD. To this end we provide the following definitions. Visibility level: the lowest value of the intensity of black light at which a discontinuity appears detectable by certified personnel. There are four levels of visibility L_v:

$$\mathbf{0} : L_v > 1\ \mathrm{W/m^2} \quad \mathbf{1} : 0.7\ \mathrm{W/m^2} < L_v \leq 1\ \mathrm{W/m^2}$$
$$\mathbf{2} : 0.4\ \mathrm{W/m^2} < L_v \leq 0.7\ \mathrm{W/m^2} \quad \mathbf{3} : Lv \leq 0.4\ \mathrm{W/m^2}\ . \tag{2}$$

Class of discontinuities: set of discontinuity whose indications have defined size and morphological characteristics and a given level of visibility.
PoD: for a defined inspection process, probability of detection of a discontinuity, with a defined level of confidence, depending on the class to which the discontinuity belongs. A PoD curve is given by PoD showing the dependency upon discontinuity classes.
DETEX (DETectability indeX): index of deceasing order of the detectability "difficulty". The different types of morphological defects are shown in figure 1. As far as the morphological types shown in figure 1 are concerned,

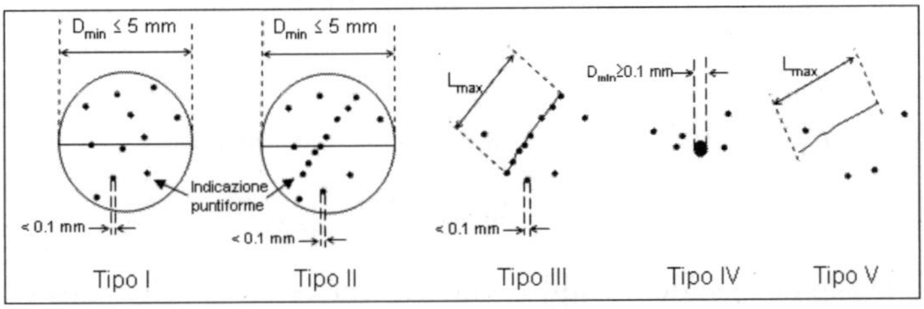

Fig. 1. Types of morphological defects.

the related DETEX indexes are defined as shown in table 1. Several discon-

Table 1. DETEX index.

	Type I non-lined spots	Type II lined and non-lined spots	Type III lined spots	Type IV round indications	Type V linear indications
DETEX	1	1	1	2	3

tinuity classes can be defined. In table 2 an exercise of classification based on test with a known fatigue discontinuity, inspected by certified experts,

Table 2. Visibility level (DETEX) classification.

Class	Morphological Type	Dimension	DETEX
X1	I (1)	all D_{min}	0 or 1
	IV (2)	$D_{min} \leq 0.2$ mm	0 or 1
	V (3)	all L_{max}	0
X2	I (1)	$D_{min} < 2.5$ mm	2
	II (1)	$D_{min} < 1.5$ mm	2
	II (1)	$D_{min} < 3.5$ mm	1
	III (1)	all L_{max}	1
	IV (2)	$D_{min} \leq 0.2$ mm	2
	V (3)	all L_{max}	1
X3	I (1)	$D_{min} > 2.3$ mm	2
	II (1)	1.5 mm $\leq D_{min} < 2.3$ mm	2
	III (1)	1.5 mm $\leq L_{max} \leq 2.3$ mm	2
	IV (2)	0.2 mm $< D_{min} \leq 0.8$ mm	2
	V (3)	$L_{max} = 2.3$ mm	2
X4	II (1)	2.3 mm $\leq D_{max} \leq 3.5$ mm	2
	V (3)	$L_{max} \geq 3.5$ mm	2
X5	III (1)	$L_{max} \geq 2.3$ mm	3
	V (3)	$L_{max} = 2.3$ mm	3

is shown. The index of each class increases when the difficulty of detection decreases. To assign a discontinuity to a class, an analytical formulation has been defined. The characteristic size of the discontinuity d (in mm), D_{min} or L_{max} in table 2, the inverse of the square root of the black light intensity I in W/m^2), and the DETEX index are used as classification parameters. In the following the formulation obtained by the best fit:

$$C(x) = A_0 \left[\alpha e^{\beta \log(x)}\right] , \tag{3}$$

where C is the class index, $A_0[\cdot]$ is the integer operator (it takes real numbers and returns the integer part of them), $\alpha = 0.44$, $\beta = 0.87$, and $x = \text{DETEX} + 4d + 8\sqrt{(I)}$ a linear combination of the classification parameters. As expected, class index increases when DETEX and discontinuity dimension increase and black light intensity to detect the discontinuity decreases. Starting from ≈ 10 W/m^2, the influence due to the intensity of black light becomes negligible.

4.1. *PoD and process sensibility*

The inspection process sensitivity is defined as the discontinuities detection capacity related to the inspected materials, the number of inspector personnel, inspective materials (Penetrant, Emulsifier, Developer, Solvents), their application and parameters process. The Alenia approach is devoted to estimate the inspection process sensitivity, by determining the PoD. The PoD should be defined for each type of inspected material in the process considered. Typically the types of material are listed below:

Material Type	STANDARD
AL (league of Alluminium)	7075T73 not-plated
TI (Titanium and its league)	6Al-4V
S (Steel)	series 300

4.2. *Standards manufacturing*

For each type of material at least 15 standards should be manufactured, with dimensions between 80 mm and 300 mm and thickness between 2 mm and 5 mm. For each type of material, with the exception of some (from 2 to 4) standards without discontinuity, a number of discontinuity (from 2 to 10) should be realized with a depth less than 1 mm, open to the surface (e.g., fatigue cracks). In the following a method to generate discontinuity (generation fatigue) is described:

- use raw with thickness of 4-6 mm;
- select a square of side of between 80 mm and 300 mm;
- select a point in the square to achieve the discontinuity;
- apply in this point a stress through a punch, with constant cycles of loading;
- monitor the zone and wait for a sufficient time (typically corresponding to 15000-40000 cycles);
- mill the part along the edge;
- remove at least 2 μm thick and wash with de-ionized water.

Standard classification procedure: according to Alenia technical specifications NTA94151/-1[3] and NTA94151/-2[4] and,[5] after standard processing, it is necessary:

a) to inspect the standard in the inspection cabin, with ambient illumination light less than 5 lx (lx is the SI unit of illuminance and luminous emittance), and at a distance of 100 mm from the standard;
b) to put the standard at a distance from the Wood's lamp so that an intensity less than 0.4 W/m^2 is measured on its surface;
c) to identify all the relevant visible information, indicating type and dimension, to record indications – a level of visibility equal to 3 will be given to them (DETEX);

d) to bring gradually the standard near Wood's lamp. Just an indication is visible, to indicate the type and shape of the corresponding sizes measuring the black light intensity (to fix relative visibility level: DETEX); to record indications.
e) to classify each discontinuity according to figure 1 and table 1.

4.3. *PoD and process sensibility determination*

(1) Inspection.
According to Alenia technical norms NTA94151/-1[3] and NTA94151/-2[4] and[5] the inspection procedure of standards must be approved by Level III.[6] Each inspector fulfill the standards inspection, filling a report where the revealed discontinuity are listed along with any uncertainty or dubious case; it is not allowed to reprocess locally the ambiguous cases. At the end of each inspection, it is possible to proceed with the PoD determination.
(2) Detecting frequencies determination.
For each class of discontinuities X_j and for each inspector i, we have the total numer $S(X_j, i)$ of the detected discontinuities of class X_j found by the ispector i-th, so that it is possible to compute the Detecting frequencies $F_R(X_j, i)$ according to the following formula

$$F_R(X_j, i) = \frac{S(X_j, i)}{N_j}\,, \tag{4}$$

where N_j is the total number of defects X_j for all classes $X_j, j = 1, 2, \ldots D$, $D = 5$ in our case see table 2.
(3) Cumulative frequency $F_R(X)$.
Moreover, we can compute the Cumulative frequency $F_R(X)$ according to the rule:

$$F_R(X_j) = \sum_{i=1}^{K} F_R(X_j, i)\,, \tag{5}$$

where K is the operators total number.
(4) For each class of discontinuity we get an estimate of the PoD

$$PoD(X_j) = \frac{F_R(X_j)}{K}\,, \tag{6}$$

as well as the standard deviation

$$\sigma(X_j) = \sqrt{\frac{1}{K-1}\sum_{i=1}^{K}\left[F_R(X_j, i) - PoD(X_j)\right]^2}\,. \tag{7}$$

(5) Determination of the confidence threshold 95% (under a Gaussian hypotesis) $T(X_j)$.
Furthermore, the confidence threshold 95% $T(X_j)$ can be fuond by the formula:

$$T(X_j) = PoD(X_j) - 2\frac{\sigma(X_j)}{\sqrt{K}} \ . \tag{8}$$

In literature, $T(X_j)$ as a function of X_j is known as the PoD's graphics of the related process. Note that in the last four formulas $j = 1, 2, \ldots, D$.

(6) Estimate of the process sensibility.
An estimate of the process sensibility is given by the value $S(X_j)$ given by

$$S = \sum_{j=1}^{D} c(j)T(X_j) \quad \text{with} \quad \sum_{j=1}^{D} c(j) = 4 \ , \tag{9}$$

where D is the number of classes and $c(j)$ is the class X_j specific weight that grow when the class index decreases. The standard normalization to the maximum value of 4, for $c(j)$ is used in analogy to the "classical" penetrants sensibility's classification (this is a part of the process sensibility), according to the table 3. As a rule of thumb, the values of

Table 3. Penetrants sensibility's classification

1/2	Very low
1	Low
2	Medium
3	High(this is used in aeronautics)
4	Ultrahigh

$c(j)$ are given by the following relations

$$c(j) = \frac{4}{\sum_{j=1}^{D}(D+1-j)}(D+1-j) \quad \text{for } j = 1, 2, \ldots, D \ . \tag{10}$$

Individual performance.
An estimate of the i-th inspector individual performance can be determined on the basis of the partial sensibility defined by

$$S_i = \sum_{j=1}^{D} c(j)F_R(X_j, i) \quad \text{for } i = 1, 2, \ldots, K \ . \tag{11}$$

To get the mentioned estimate we take the following steps:

By a permutation of indexes we rewrite the S_i in decreasing order, that is $S_1 \geq S_2 \geq \cdots \geq S_K$).
Then we compute the cumulative partial sensibilities $SC(h)$, $h = 1, 2, \ldots, K$ according to the formula:

$$SC(h) = \frac{1}{h} \sum_{k=1}^{h} S_k \quad \text{for} \quad h = 1, 2, \ldots, K \; . \tag{12}$$

Note that, owing to the ordering $S_i \geq S_{i+1}$, $i = 1, 2, \ldots, K-1$ we also have that $SC_h \geq SC_{h+1}$, $h = 1, 2, \ldots, K-1$.
Because of this ordering property, the difference between $SC(1)$ and $SC(K)$ is an indicator of the dependency of the process sensibility on the individual performance. The greater is the this difference, the higher is the process sensibility.

4.4. *A case study*

Here we report a case study where 27 samples (leagues of) Titanium were inspected by 5 NDI inspectors. The statistics reported in table 4 were computed by the formula (11). Now, by using the data from table 4 we can

Table 4. Sum of the defects detected by the each inspector (i) for each morphological class, see figure 1.

i	$S(X_1, i)$	$S(X_2, i)$	$S(X_3, i)$	$S(X_4, i)$	$S(X_5, i)$
1	17	14	13	9	6
2	20	15	12	11	7
3	17	14	12	11	7
4	19	17	13	11	7
5	17	14	12	11	7

compute the values $F_R(X_j, i)$ according to the formula (4) and reported in table 5. In our case the partial sensibility of the inspectors can be computed

Table 5. Detecting frequencies.

i	$F_R(X_1, i)$ with $N_1 = 27$	$F_R(X_2, i)$ with $N_2 = 21$	$F_R(X_3, i)$ with $N_3 = 13$	$F_R(X_4, i)$ with $N_4 = 11$	$F_R(X_5, i)$ with $N_5 = 7$
1	0.63	0.67	1.00	0.82	0.86
2	0.74	0.71	0.92	1.00	1.00
3	0.63	0.67	0.92	1.00	1.00
4	0.70	0.81	1.00	1.00	1.00
5	0.63	0.67	0.92	1.00	1.00

according to the formulas (10)-(11). In this way we get the following values: $S(1) = 3.015552$, $S(2) = 3.288021$, $S(3) = 3.089079$, $S(4) = 3.401764$, and $S(5) = 3.089079$. In our case study the permutation of indexes used, in order to put the $S(i)$, $i = 1, 2, \ldots, 5$ in decreasing order, is given by $(1, 2, 3, 4, 5) \to (4, 2, 3, 5, 1)$. An evaluation of the individual performance of each inspector ($K = 5$ in our case) is computed by the formula (12). Hence, the following data can be found: $SC(1) = 3.401764$, $SC(2) = 3.344892$, $SC(3) = 3.259621$, $SC(4) = 3.216985$, and $SC(5) = 3.176699$.

On account of the small variation between $SC(1)$ and $SC(5)$, we can conclude that the process sensibility does not depend upon the individual performance. In table 6 we report the Pod, standard deviation, and partial sensibilities for each class of defects, computed by the formulas (6), (7), and (8), respectively. By the data reported in table 6, the overall sensibility of

Table 6. PoD, standard deviation $\sigma(X_j)$, and confidence threshold at 95% $T(X_j)$.

	X_1	X_2	X_3	X_4	X_5
PoD	0.67	0.70	0.95	0.96	0.97
$\sigma(X_j)$	0.02	0.03	0.02	0.00	0.00
$T(X_j)$	0.62	0.64	0.92	0.96	0.97

the process, according to the formula (9) is given by $S = 3$ and it represents the High level more often used in aeronautics.

Acknowledgment.
The used technical data are "Alenia Aeronautics property information".

References

1. W. R. Rummel, P. Todd, R. A. Rathke and W. L. Castner, *Materials Evaluation* **32**, 205 (1974).
2. A. P. Berens and P. W. Hovey, *Evaluation of NDE reliability characterisation.* AFWAL-TR-81-4160, Vol. 1, Air Force Wright-Aeronautical Laboratories, Wright-Patterson Air Force Base, 1981.
3. G. Cavaccini, *Controlli di qualità per la verifica del processo di ispezione con liquidi penetranti*, tech. rep., Alenia Aeronautics (2007).
4. G. Cavaccini, *Liquidi penetranti fluorescenti – determinazione della PoD e della sensibilità del processo*, tech. rep., Alenia Aeronautics (2006).
5. G. Caturano, G. Cavaccini, A. Ciliberto and V. Pianese, Liquid penetrant testing: Industrial process SIMAI 2008.
6. Alenia Aeronautics, `www.aeronautica.alenia.it`, *Training, qualifications, and certification of personnel assigned to non destructive testing*, (2008).

Wave Propagation in Continuously-Layered Electromagnetic Media

Giacomo Caviglia[1], Angelo Morro[2]
[1] *Dipartimento di Matematica, Facoltà di Scienze, Università di Genova, Italy*
caviglia@dima.unige.it

[2] *DIBE, Facoltà di Ingegneria, Università di Genova, Italy*
angelo.morro@unige.it

The paper is devoted to the propagation of electromagnetic waves in continuously and planarly inhomogeneous media. Transversely-polarized waves are investigated. The wave solution is sought in the form of a generalized plane wave. The approach is based on a Volterra integral equation that arises from a direct integration of a first-order system of differential equations. In connection with the reflection-transmission process, the wave solution is obtained as a power-series of the angular frequency.

Keywords: Integral equation, Electromagnetic waves, Layered media.

1. Introduction

The paper is devoted to the propagation of electromagnetic waves in inhomogeneous media. Inhomogeneity is an interesting case of complexity and here it is meant as one-dimensional and planar in that the material properties vary only in one direction, say z (see [1]).

For simplicity we look at transversely-polarized waves, which means that the electric or the magnetic field are perpendicular to the z-axis. The wave solution is sought in the form of generalized plane waves and the pertinent equation is investigated in the Fourier-transform domain. Despite the inhomogeneity, no approximation (e.g. piecewise constancy) is introduced. The approach is based on a Volterra integral equation namely that arising from a direct integration of a first-order system of differential equations.

For definiteness the inhomogeneity is confined to a dielectric layer $z \in [0, L]$. The half-spaces $z < 0$ and $z > L$ are then homogeneous though with different material properties. The wave generated in the layer, by a known plane wave coming from $z < 0$ at any incidence angle, is established along with the solution in the whole space $z \in \mathbb{R}$. We then solve the reflection-transmission problem and, as an application, we find the reflection and transmission coefficients associated with the whole layer. As a check on the general results so obtained, three particular cases are considered, namely the uniform layer, the low-frequency limit, the thin-layer limit.

It is an advantage of the present approach that, owing to the Volterra type of the integral equation, the wave solution is obtained as a power-series, relative to the angular frequency, by only assuming the boundedness of the material properties. The Fredholm-type formulation often applied in the literature (see e.g. [2]) does not seem to be equally efficient to determine wave solutions in inhomogeneous media.

2. Waves in continuously-layered media

Let $\Omega \subseteq \mathbb{R}^3$ be the space region occupied by the body and $\mathbf{x} \in \Omega$ a position vector. A function f, on $\Omega \times \mathbb{R}$, is a plane wave propagating in the direction of the constant vector $\mathbf{q}$, with speed $1/|\mathbf{q}|$, if $f(\mathbf{x}, t) = \mathcal{F}(t - \mathbf{q} \cdot \mathbf{x})$ for some function $\mathcal{F}$, on $\mathbb{R}$. We look for generalized plane waves [3]

$$g(\mathbf{x}, t) = \mathcal{G}(z, t - \mathbf{p} \cdot \mathbf{x}),$$

where $\mathbf{p}$ is perpendicular to the z-axis, for any function $\mathcal{G}$ on $\mathbb{R}^2$. By an appropriate choice of the Cartesian axes we write

$$g(\mathbf{x}, t) = \mathcal{G}(z, t - \xi x), \qquad \xi \in \mathbb{R}.$$

By the Fourier transform $\tilde{g}$, with respect to t, we have

$$\tilde{g}(\mathbf{x}, \omega) = \int_{-\infty}^{\infty} g(\mathbf{x}, t) \exp(i\omega t) dt = \exp(i\omega \xi x) \tilde{\mathcal{G}}(z, \omega).$$

2.1. *Derivation of the scalar wave equations*

Let ϵ and μ be the z-dependent electric permittivity and magnetic permeability. The electric field $\mathbf{E}$, the electric displacement $\mathbf{D}$, the magnetic field $\mathbf{H}$ and the magnetic induction $\mathbf{B}$ are related by

$$\mathbf{D}(\mathbf{x}, t) = \epsilon(z) \mathbf{E}(\mathbf{x}, t), \qquad \mathbf{B}(\mathbf{x}, t) = \mu(z) \mathbf{H}(\mathbf{x}, t).$$

The waves are transversely polarized in that $\mathbf{E}$ or $\mathbf{H}$ are directed in the y-direction, $\mathbf{E} = E\mathbf{e}_y$ or $\mathbf{H} = H\mathbf{e}_y$. By Maxwell's equations

$$\nabla \cdot (\epsilon \mathbf{E}) = 0, \qquad \nabla \cdot (\mu \mathbf{H}) = 0 \tag{1}$$

the transversality conditions $\nabla\epsilon\cdot\mathbf{E} = 0$, $\nabla\mu\cdot\mathbf{H} = 0$ give $\nabla\cdot\mathbf{E} = 0$, $\nabla\cdot\mathbf{H} = 0$. By the remaining Maxwell's equations, in the form

$$\frac{1}{\mu}\nabla\times\mathbf{E} = -\partial_t\mathbf{H}, \qquad \frac{1}{\epsilon}\nabla\times\mathbf{H} = \partial_t\mathbf{E}, \tag{2}$$

evaluation of $\nabla\times$ and substitution yield

$$\mu\partial_z(\frac{1}{\mu}\partial_z E) + \partial_x^2 E = \mu\epsilon\partial_t^2 E, \qquad \epsilon\partial_z(\frac{1}{\epsilon}\partial_z H) + \partial_x^2 H = \mu\epsilon\partial_t^2 H.$$

Wave solutions of the form

$$E(\mathbf{x}, t) = \mathcal{E}(z, t - \xi x),$$

and analogously for H, are allowed provided $\tilde{\mathcal{E}}(z, \omega)$ satisfies

$$\mu\partial_z(\frac{1}{\mu}\partial_z\tilde{\mathcal{E}}) + \omega^2(\mu\epsilon - \xi^2)\tilde{\mathcal{E}} = 0. \tag{3}$$

Letting

$$\psi(z, \omega) = \mu^{-1/2}(z)\tilde{\mathcal{E}}(z, \omega)$$

we find that (3) takes the form

$$\psi'' + [\omega^2(\epsilon\mu - \xi^2) + \frac{1}{2}\frac{\mu''}{\mu} - \frac{3}{4}(\frac{\mu'}{\mu})^2]\psi = 0, \tag{4}$$

a prime denoting differentiation with respect to z. Equation (4) occurs also in acoustics, though there different inequivalent models are applied [4,5].

It is convenient to let

$$k^2 := \omega^2(\overline{\epsilon\mu} - \xi^2)$$

where $\overline{\epsilon\mu}$ is the value of $\epsilon\mu$ at a point $\overline{z}$ such that the quantity in brackets is positive. Also let

$$\alpha(z, \omega) := 1 - \frac{\epsilon\mu - \xi^2}{\overline{\epsilon\mu} - \xi^2} - \frac{\mu''/2\mu - 3(\mu')^2/4\mu^2}{\omega^2(\overline{\epsilon\mu} - \xi^2)}.$$

Hence the differential equation for ψ can be written as

$$\psi'' + k^2\psi = k^2\alpha(z, \omega)\psi.$$

If, instead, we let v be the pair

$$v = \begin{bmatrix} \tilde{\mathcal{E}} \\ \mu^{-1}\tilde{\mathcal{E}}' \end{bmatrix}$$

we find that the differential equation for $\tilde{\mathcal{E}}$ is equivalent to the first-order system

$$v' = M(z,\omega)v, \qquad M := \begin{bmatrix} 0 & \mu \\ -\omega^2(\epsilon - \xi^2/\mu) & 0 \end{bmatrix}.$$

The matrix $M \in \mathbb{R}^{2\times 2}$ is then parameterized by ω.

3. Integral-equation formulations

We now show that the equations for ψ and v can be given the form of integral equations. An obvious integration gives

$$v(z,\omega) = v(0,\omega) + \int_0^z M(s,\omega)v(s,\omega)ds, \tag{5}$$

where $v(0,\omega)$ stands for the limit as $z \to 0_+$. This is a Volterra integral equation and, for any $\omega \in \mathbb{R}$, $v(z,\omega) : [0,L] \to \mathbb{C}^2$. By the definition of v it follows that $v(0,\omega)$ is the pair of values $[\tilde{\mathcal{E}}(0,\omega), \tilde{\mathcal{E}}'(0,\omega)/\mu(0)]$. The conceptual advantage of using v is that it consists of two components with a direct physical meaning.

Another integral equation of the Volterra type is now obtained for the unknown function ψ. A two-point function $G(z,s)$ on $(0,L)\times(0,L)$ satisfying the distribution equation

$$\partial_z^2 G(z,s) + k^2 G(z,s) = \delta(z-s), \qquad z,s \in (0,L),$$

is a *fundamental solution* of the operator $\partial_z^2 + k^2$ (see [6]). By means of the fundamental solution

$$G(z,s) = \begin{cases} (1/k)\sin k(z-s), & z \in (0,s), \\ 0, & z \in (s,L). \end{cases}$$

we find that ψ satisfies the integral equation

$$\psi(z) = \phi(z) + \int_0^z A(z,s)\psi(s)ds, \tag{6}$$

where $\phi(z) = \phi_1 \exp(ikz) + \phi_2 \exp(-ikz)$, $A(z,s) = k\sin(k(z-s))\,\alpha(s)$. The parameters ϕ_1, ϕ_2 are related to the initial values $\psi(0), \psi'(0)$ by $\phi_1 = [\psi'(0)/ik + \psi(0)]/2$, $\phi_2 = [-\psi'(0)/ik + \psi(0)]/2$.

Existence and uniqueness of the solution ψ hold if $A(z,s)$ is a continuous function of (z,s) on $S = \{(z,s)|0 \le s \le z \le L\}$, and hence if α is continuous on $[0,L]$.

An integral equation of the form (6) is applied in [3] in connection with acoustics. Here, instead, we apply eq. (5) for v.

4. Existence and uniqueness of $\tilde{\mathcal{E}}$

We first show that if M is bounded, as $z \in [0, L]$, then the solution v is bounded too, as $z \in [0, L]$. Since only the off-diagonal entries of M are non-zero then

$$v_1(z) = v_1(0) + \int_0^z M_{12}(s)v_2(s)ds$$

and the like by interchanging the subscripts $1, 2$. Taking the absolute value we have

$$|v_1(z)| \leq |v_1(0)| + \int_0^z |M_{12}(s)|\,|v_2(s)|ds$$

and the like for v_1. If $m(z) = \max\{|M_{12}(z)|, |M_{21}(z)|\}$, $\|v(z)\| = |v_1(z)| + |v_2(z)|$ then

$$\|v(z)\| \leq \|v(0)\| + \int_0^z m(s)\|v(s)\|ds.$$

The Gronwall inequality provides

$$\|v(z)\| \leq \|v(0)\|\big[1 + \int_0^t m(s)\exp(\int_s^t m(u)du)ds\big].$$

As a consequence, since $z \in [0, L]$, the boundedness of m implies that of v on $[0, L]$. This is the proof of the following

Proposition 4.1 (Uniqueness). *If M is bounded on $[0, L]$ then the solution v, on $[0, L]$, with $v(0) = V$, is unique.*

Let $\mathcal{M}(z) : [0, L] \to \mathbb{R}^{2\times 2}$ be defined by

$$\mathcal{M}(z) = \mathbb{I} + \sum_{h=1}^{\infty} \int_0^z \int_0^{s_1} \cdots \int_0^{s_{h-1}} M(s_1)\cdots M(s_h)ds_1...ds_h, \tag{7}$$

where $\mathbb{I}$ is the identity matrix, in $\mathbb{R}^{2\times 2}$. We now show that the boundedness of M yields that of $\mathcal{M}$ and moreover that $\mathcal{M}(z)$ gives $v(z)$ in terms of the initial value $v(0)$.

Proposition 4.2 (Existence). *If M is bounded on $[0, L]$, and hence v is bounded too, then*

$$v(z) = \mathcal{M}(z)v(0). \tag{8}$$

Proof. Substitution of $v(s)$ in the integrand of (5) and subsequent iterations allow us to write

$$v(z) = \{\mathbb{I} + \sum_{h=1}^{q} \int_0^z \int_0^{s_1} \cdots \int_0^{s_{h-1}} M(s_1)\cdots M(s_h) ds_1 ... ds_h\} v(0)$$
$$+ \int_0^z \int_0^{s_1} \cdots \int_0^{s_q} M(s_1)\cdots M(s_{q+1}) v(s_{q+1}) ds_1 ... ds_{q+1}. \quad (9)$$

Because of the off-diagonal character of M, the products of matrices $M(s_1)M(s_2)\cdots M(s_h)$ take a diagonal form if h is even and a off-diagonal form if h is odd. Hence

$$M(s_1)...M(s_{h+1})v(s_{h+1}) = \begin{bmatrix} M_{12}(s_1)M_{21}(s_2)...M_{21}(s_{h+1})v_1(s_{h+1}) \\ M_{21}(s_1)M_{12}(s_2)...M_{12}(s_{h+1})v_2(s_{h+1}) \end{bmatrix},$$

or

$$M(s_1)...M(s_{h+1})v(s_{h+1}) = \begin{bmatrix} M_{12}(s_1)M_{21}(s_2)...M_{12}(s_{h+1})v_2(s_{h+1}) \\ M_{21}(s_1)M_{12}(s_2)...M_{21}(s_{h+1})v_1(s_{h+1}) \end{bmatrix},$$

according as h is odd or even. We then need an estimate of

$$I_h = \int_0^z \int_0^{s_1} \cdots \int_0^{s_h} M_{12}(s_1)M_{21}(s_2)\cdots M_{21}(s_{h+1})v_1(s_{h+1}) ds_1 ... ds_{h+1},$$

and of the other analogous terms. Since M and v are bounded, $m(z) \leq \gamma$, $\|v(z)\| \leq \nu$ as $z \in [0, L]$, for finite values of γ, ν. Hence we find that, for any $z \in [0, L]$,

$$|I_h| \leq \nu \frac{(\gamma L)^{h+1}}{(h+1)!}.$$

As a consequence, for any value of γ, ν and L it follows that $I_h \to 0$ as $h \to \infty$. The limit as $h \to \infty$ of (9) provides the result (8). □

In wave propagation problems the initial value $v(0)$ is unknown. The two components v_1, v_2 are complex-valued and parameterized by $\omega \in \mathbb{R}$. By (8) we have

$$v_1(z) = \mathcal{M}_{11}(z)v_1(0) + \mathcal{M}_{12}(z)v_2(0),$$

$$\tilde{\mathcal{E}}(z) = \mathcal{M}_{11}(z)\tilde{\mathcal{E}}(0) + \mathcal{M}_{12}(z)\tilde{\mathcal{E}}'(0)/\mu(0).$$

Letting

$$a := \tilde{\mathcal{E}}(0), \qquad b := \tilde{\mathcal{E}}'(0)/\mu(0), \quad (10)$$
$$g_1(z) := \mathcal{M}_{11}(z), \qquad g_2(z) := \mathcal{M}_{12}(z), \quad (11)$$

we can express $\tilde{\mathcal{E}}$ in the layer $z \in [0, L]$ as

$$\tilde{\mathcal{E}}(z) = a g_1(z) + b g_2(z), \qquad a, b \in \mathbb{C}. \tag{12}$$

Let $\eta = \xi^2/\mu - \epsilon$ so that

$$M = \begin{bmatrix} 0 & \mu \\ \omega^2 \eta & 0 \end{bmatrix}.$$

Hence by (7) we have

$$\mathcal{M}_{11}(z) = 1 + \sum_{h=1}^{\infty} \omega^{2h} \times$$
$$\int_0^z \int_0^{s_1} \dots \int_0^{s_{2h-1}} \mu(s_1)\eta(s_2)\mu(s_3)\eta(s_4)...\eta(s_{2h}) ds_1 ... ds_{2h},$$

$$\mathcal{M}_{12}(z) = \int_0^z \mu(s_1) ds_1 + \sum_{h=1}^{\infty} \omega^{2h} \times$$
$$\int_0^z \int_0^{s_1} \dots \int_0^{s_{2h}} \mu(s_1)\eta(s_2)\mu(s_3)...\mu(s_{2h+1}) ds_1 ... ds_{2h+1},$$

and the like for $\mathcal{M}_{21}$ and $\mathcal{M}_{22}$. In addition, (8) becomes

$$\begin{bmatrix} \tilde{\mathcal{E}}(z) \\ \mu^{-1}(z)\tilde{\mathcal{E}}'(z) \end{bmatrix} = \begin{bmatrix} \mathcal{M}_{11}(z) & \mathcal{M}_{12}(z) \\ \mathcal{M}_{21}(z) & \mathcal{M}_{22}(z) \end{bmatrix} \begin{bmatrix} a \\ b \end{bmatrix}. \tag{13}$$

Since $g_1(0) = 1, \quad g_1'(0) = 0, \quad g_2(0) = 0, \quad g_2'(0) = \mu(0)$ we let $f_1(z) = g_1(z)$, $f_2(z) = g_2(z)/\mu(0)$, so that eq. (12) becomes

$$\tilde{\mathcal{E}}(z) = a f_1(z) + \beta f_2(z), \tag{14}$$

where $a, \beta = b\mu(0) \in \mathbb{C}$ and $z \in [0, L]$ while

$$f_1(0) = 1, \quad f_1'(0) = 0, \qquad f_2(0) = 0, \quad f_2'(0) = 1. \tag{15}$$

5. Jump conditions

The Fourier transform of (1) and (2) provides

$$\nabla \cdot (\epsilon \tilde{\mathbf{E}}) = 0, \qquad \nabla \cdot (\mu \tilde{\mathbf{H}}) = 0,$$

$$\nabla \times \tilde{\mathbf{H}} = -i\omega\epsilon\tilde{\mathbf{E}}, \qquad \nabla \times \tilde{\mathbf{E}} = -i\omega\mu\tilde{\mathbf{H}}.$$

At an interface with unit normal $\mathbf{n}$, $\tilde{\mathbf{E}}, \tilde{\mathbf{H}}$ are subject to

$$\mathbf{n} \cdot [\![\epsilon \tilde{\mathbf{E}}]\!] = 0, \qquad \mathbf{n} \cdot [\![\mu \tilde{\mathbf{H}}]\!] = 0,$$

$$\mathbf{n} \times [\![\tilde{\mathbf{E}}]\!] = 0, \qquad \mathbf{n} \times [\![\tilde{\mathbf{H}}]\!] = 0.$$

For TE waves, with $\mathbf{n} = \mathbf{e}_z$, the condition $\mathbf{n} \cdot [\![\epsilon\tilde{\mathbf{E}}]\!] = 0$ is true whereas $\mathbf{n} \times [\![\epsilon\tilde{\mathbf{E}}]\!] = 0$ requires that $[\![\tilde{\mathcal{E}}]\!] = 0$. Also, by replacing $\tilde{\mathbf{H}}$ with $(i/\omega\mu)\nabla \times \tilde{\mathbf{E}}$ we have

$$\mathbf{n} \cdot [\![\nabla \times \tilde{\mathbf{E}}]\!] = 0, \qquad \mathbf{n} \times [\![\frac{1}{\mu}\nabla \times \tilde{\mathbf{E}}]\!] = 0.$$

As a consequence we find again the continuity of $\tilde{\mathcal{E}}$ and also $[\![(\partial_z\tilde{\mathcal{E}})/\mu]\!] = 0$.

6. Reflection-transmission of TE waves

Restrict attention to TE waves. The jump conditions amount to $[\![v]\!] = 0$.

We consider a continuously-layered dielectric such that

$$\epsilon(z) = \begin{cases} \epsilon_0, & z < 0, \\ \epsilon(z), & z \in (0, L), \\ \epsilon_1, & z > L. \end{cases} \qquad \mu(z) = \begin{cases} \mu_0, & z < 0, \\ \mu(z), & z \in (0, L), \\ \mu_1, & z > L. \end{cases}$$

Let the incident wave come from the half-space $z < 0$ and

$$\tilde{\mathcal{E}}_{\text{inc}} = \exp(ik_0 z).$$

The Fourier transform $\tilde{\mathcal{E}}$ then takes the form

$$\tilde{\mathcal{E}}(z) = \begin{cases} \exp(ik_0 z) + R\exp(-ik_0 z), & z < 0, \\ a f_1(z) + \beta f_2(z), & z \in (0, L), \\ T\exp(ik_1(z - L)), & z > L. \end{cases}$$

Here $a, \beta \in \mathbb{C}$ and $R, T \in \mathbb{C}$ are unknown parameters. Moreover, R and T represent the reflection and transmission coefficients.

Applying the jump conditions $[\![v]\!] = 0$ to the chosen form of $\tilde{\mathcal{E}}$ and using (15) provide

$$1 + R = a, \qquad i\frac{k_0}{\mu_0}(1 - R) = \beta,$$

$$a\hat{f}_1 + \beta\hat{f}_2 = T, \qquad \frac{a}{\mu(L)}\hat{f}_1' + \frac{\beta}{\mu(L)}\hat{f}_2' = i\frac{k_1}{\mu_1}T,$$

where $\hat{f}_1, \hat{f}_2, \hat{f}_1', \hat{f}_2'$ denote the values of f_1, f_2, f_1', f_2' at $z = L$. Upon substitutions of a, b and rearrangements we find that

$$R = -\frac{F_1 + i\lambda_0 k_0 F_2}{F_1 - i\lambda_0 k_0 F_2}, \qquad T = 2i\lambda_0 k_0 \frac{F_1\hat{f}_2 - F_2\hat{f}_1}{F_1 - i\lambda_0 k_0 F_2}, \tag{16}$$

where

$$\lambda_0 = \frac{\mu(0)}{\mu_0}, \qquad \lambda_1 = \frac{\mu(L)}{\mu_1}, \qquad F_j = \hat{f}_j + i\frac{1}{\lambda_1 k_1}\hat{f}_j', \quad j = 1, 2.$$

Moreover we find that

$$a = \tilde{\mathcal{E}}(0) = -\frac{2i\lambda_0 k_0 F_2}{F_1 - i\lambda_0 k_0 F_2}, \quad \mu(0)b = \tilde{\mathcal{E}}'(0) = \frac{2i\lambda_0 k_0 F_1}{F_1 - i\lambda_0 k_0 F_2}.$$

Equations (16) provide the reflection and transmission coefficients in terms of the μ and η (or ϵ) profiles and the ratios $\mu(0)/\mu_0$, $\mu_1/\mu(L)$ of μ at the interfaces. The functionals F_1, F_2 of μ, η on $(0, L)$ are evaluated through $\mathcal{M}_{11}(L)$ and $\mathcal{M}_{12}(L)$.

7. Limit values of R and T

7.1. *Uniform layer*

Let $\mu(z) = \mu, \epsilon(z) = \epsilon$ be constants while no requirements are placed on the ratios λ_0, λ_1. To begin with, let $\xi^2 < \epsilon\mu$ which means that $\eta = (\xi^2 - \epsilon\mu)/\mu < 0$.

Equation (3) simplifies to

$$\tilde{\mathcal{E}}'' + \omega^2(\mu\epsilon - \xi^2)\tilde{\mathcal{E}} = 0. \tag{17}$$

Since $\mu\epsilon$ is a constant, by (14) and (15) the solutions f_1, f_2 are given by

$$f_1(z) = \cos(\omega\sqrt{\mu\epsilon - \xi^2}\, z), \quad f_2(z) = \frac{1}{\omega\sqrt{\mu\epsilon - \xi^2}}\sin(\omega\sqrt{\mu\epsilon - \xi^2}\, z).$$

As a check, we show that $\mathcal{M}_{11}(z) = \cos(\omega\sqrt{\mu\epsilon - \xi^2}\, z)$, $\mathcal{M}_{12}(z) = (\mu/\omega\sqrt{\mu\epsilon - \xi^2})\sin(\omega\sqrt{\mu\epsilon - \xi^2}\, z)$. Letting $\eta = -|\eta|$ we find that

$$\mathcal{M}_{11}(z) = 1 + \sum_{h=1}^{\infty}(-1)^h\frac{1}{(2h)!}(\omega\sqrt{\mu|\eta|}z)^{2h} = \cos(\omega\sqrt{\mu|\eta|}\, z),$$

$$\mathcal{M}_{12}(z) = \mu\{z + \sum_{h=1}^{\infty}(-1)^h(\omega\sqrt{\mu|\eta|}z)^{2h}\frac{z^{2h+1}}{(2h+1)!}\} = \frac{\mu\sin(\omega\sqrt{\mu|\eta|}\, z)}{\omega\sqrt{\mu|\eta|}}.$$

By definition, $\hat{f}_1 = f_1(L)$, $\hat{f}_2 = f_2(L)$ and hence

$$F_1 = \cos(\omega\sqrt{\mu\epsilon - \xi^2}\, L) - i\frac{\lambda_1}{k_1}\omega\sqrt{\mu\epsilon - \xi^2}\sin(\omega\sqrt{\mu\epsilon - \xi^2}\, L),$$

$$F_2 = \frac{1}{\omega\sqrt{\mu\epsilon - \xi^2}}\sin(\omega\sqrt{\mu\epsilon - \xi^2}\, L) + i\frac{\lambda_1}{k_1}\cos(\omega\sqrt{\mu\epsilon - \xi^2}\, L).$$

We now let $\xi^2 > \epsilon\mu$, which means that the incidence angle is beyond the critical value, that is $\sin\theta_0 > \sqrt{\epsilon\mu/\epsilon_0\mu_0}$. The solution $\tilde{\mathcal{E}}$ to (17) then involves

$$f_1(z) = \frac{1}{2}[\exp(\omega\sqrt{\mu\eta}z) + \exp(-\omega\sqrt{\mu\eta}z)],$$

$$f_2(z) = \frac{1}{2\omega\sqrt{\mu\eta}}[\exp(\omega\sqrt{\mu\eta}z) - \exp(-\omega\sqrt{\mu\eta}z)].$$

Now, with η constant and positive we find that

$$\mathcal{M}_{11}(z) = \frac{1}{2}[\exp(\omega\sqrt{\mu|\eta|}\, z) + \exp(-\omega\sqrt{\mu|\eta|}\, z)],$$

whence we have $f_1(z) = \mathcal{M}_{11}(z)$. Likewise, $f_2(z) = \mathcal{M}_{12}(z)/\mu$.

7.2. *Low frequencies*

The knowledge of the entries of $\mathcal{M}$ as series of powers in the frequency ω, indeed in ω^2, allows us to obtain easily the predominant terms of f_1 and f_2, and hence of R and T, for small values of ω. It is essential that μ and η depend on the depth but are independent of ω. Hence we find that

$$\mathcal{M}_{11}(z) = 1 + I_z(\mu,\eta)\omega^2 + o(\omega^2),$$

$$\mathcal{M}_{12}(z) = \int_0^z \mu(s_1)ds_1 + I_z(\mu,\eta,\mu)\omega^2 + o(\omega^2),$$

where

$$I_z(\mu,\eta) = \int_0^z \int_0^{s_1} \mu(s_1)\eta(s_2)ds_1 ds_2,$$

$$I_z(\mu,\eta,\mu) = \int_0^z \int_0^{s_1} \int_0^{s_2} \mu(s_1)\eta(s_2)\mu(s_3)ds_1 ds_2 ds_3.$$

Hence we have

$$\hat{f}_1 = 1 + I_L(\mu,\eta)\omega^2 + o(\omega^2), \qquad \hat{f}_2 = \frac{1}{\mu(0)}[L\overline{\mu} + I_L(\mu,\eta,\mu)\omega^2] + o(\omega^2),$$

where $\overline{\mu}$ is the mean value of μ in $[0, L]$. Hence we find that

$$F_1 = 1 + i\frac{\lambda_1\omega^2}{k_1}\mu(L)L\overline{\eta} + o(\omega^2),$$

$$i\lambda_0 k_0 F_2 = -\frac{\mu_1 k_0}{\mu(0)k_1} + i\frac{L\overline{\mu}k_0}{\mu_0} + O(\omega^2).$$

Upon substitution we obtain R and T at small values of ω.

7.3. *Thin layers*

As a limit case we now evaluate R and T as the thickness L approaches zero. By the (exact) series solution we find that

$$\hat{f}_1 = 1 + O(L^2),\ \hat{f}_2 = O(L),\ F_1 = 1 + O(L),\ F_2 = i\frac{1}{\lambda_1 k_1}\frac{\mu(L)}{\mu(0)} + O(L).$$

Substitution in (16) provides

$$R = \frac{\lambda_0\mu(L)k_0 - \lambda_1\mu(0)k_1}{\lambda_0\mu(L)k_0 - \lambda_1\mu(0)k_1}, \qquad T = \frac{2\lambda_0 k_0\mu(L)}{\lambda_0 k_0\mu(L) + \lambda_1 k_1\mu(0)}, \tag{18}$$

to within $O(L)$ terms. A further simplification arises by letting μ be continuous at $z = 0, L$, whence $\lambda_0 = \mu(0)/\mu_0 = 1$, $\lambda_1 = \mu(L)/\mu_1 = 1$, but allowing for different values of $\mu(0)$ and $\mu(L)$. In such a case, eqs (18) simplify to the Fresnel form of the reflection and transmission coefficients

$$R = \frac{\mu(L)k_0 - \mu(0)k_1}{\mu(L)k_0 - \mu(0)k_1}, \qquad T = \frac{2\mu(L)k_0}{\mu(L)k_0 + \mu(0)k_1}.$$

8. Possible applications

The results (16) provide the reflection and transmission coefficients of the layer, parameterized by the frequency ω. As a future development, we may indicate the inverse problem of determining the profile of the constitutive parameters χ, μ from reflection (and transmission) data - see [7] and [1], ch. 9. We expect that the knowledge of appropriate derivatives of R with respect to ω provide the mean value and various moments of χ and μ. This view is motivated also by a previous analysis of the inverse problem for a stack of elastic, homogeneous layers [8].

Acknowledgements

The research leading to this paper has been supported by the Italian MIUR through the Research Project COFIN 2005 "Mathematical Models and Methods in Continuum Physics".

References

1. W. C. Chew, *Waves and Fields in Inhomogeneous Media*, IEEE Press, New York 1995.
2. A. B. Weglein, F. V. Araujo, P. M. Carvalho, R. H. Stolt, K. H. Matson, R. T. Coates, D. Corrigan, D. J. Foster, S. A. Shaw, and H. Zhang, Inverse scattering series and seismic exploration, *Inverse Problems* **19** (2003), pp. R27–R83.

3. G. Caviglia and A. Morro, Acoustic and elastic scattering by continuously stratified media, *Acta Mech.* DOI 10.1007/s00707-008-0100-0.
4. G. Caviglia and A. Morro, Sound-wave equation in the acoustic approximation, *Ist. Lombardo Sci. Lett.*, **140** (2006), pp. 233–242.
5. P. A. Martin, Acoustic scattering by inhomogeneous spheres, *J. Acoust. Soc. Am.* **111** (2002), pp. 2013–2018.
6. W. Cheney, *Analysis for Applied Mathematics*, Springer, New York, 2000, p. 273.
7. S. He, S. Ström, and V. H. Weston, *Time Domain Wave-Splitting and Inverse Problems.* Oxford University Press, Oxford, 1998.
8. G. Caviglia and A. Morro, Direct and inverse problems in elastic multilayers with reflection data, *J. Elasticity*, **83** (2006), pp. 75–94.

COMPETITIVE NUCLEATION IN METASTABLE SYSTEMS

E. N. M. CIRILLO

Dipartimento Me.Mo.Mat., Università degli Studi di Roma "La Sapienza",
via A. Scarpa 16, I–00161, Roma, Italy
E-mail: cirillo@dmmm.uniroma1.it

F. R. NARDI

Department of Mathematics and Computer Science, Eindhoven University of Technology,
P.O. Box 513, 5600 MB Eindhoven, The Netherlands
E-mail: F.R.Nardi@tue.nl

C. SPITONI

Department of Medical Statistics and Bioinformatics, Leiden University Medical Centre,
Postal Zone S–05–P, PO Box 9600, 2300 RC Leiden The Netherlands
E-mail: C.Spitoni@lumc.nl

Metastability is observed when a physical system is close to a first order phase transition. In this paper the metastable behavior of a two state reversible probabilistic cellular automaton with self–interaction is discussed. Depending on the self–interaction, competing metastable states arise and a behavior very similar to that of the three state Blume–Capel spin model is found.

Keywords: Metastability, phase transition, cellular automata.

1. Introduction

Metastable states are observed when a physical system is close to a first order phase transition. Well known examples are super-saturated vapor states and magnetic hystereses.[1] In the Figure 1 the isotherms of a ferromagnet are depicted on the left; T denotes the temperature, m the magnetization, i.e., the density of total magnetic moment, h the external magnetic field, $m^* > 0$ the spontaneous magnetization, and T_c the Curie temperature. At temperature higher than T_c the magnetization is zero for $h = 0$; it is said that the system is in the *paramagnetic* phase.[2] Below the critical temperature at $h = 0$ the system can exhibit the nonzero values m^* and $-m^*$; it is said that the system is in the *ferromagnetic* phase. For $h = 0$, when the temperature reaches the *critical* value T_c the system undergoes a *continuous* (*second order*) phase transition; the name is justified since the *order parameter* m varies continuously when T_c is crossed.

In the graph on the right in Figure 1 the behavior of the ferromagnet at T smaller than T_c is illustrated. When $h = 0$ the system jumps from the positive magnetization phase to the negative magnetization one, or vice-versa; the transition is called *first order* since the order parameter m, which is the first derivative of one of the thermodynamical potential, undergoes an abrupt variation.[2] Sometimes, provided the value $h = 0$ is crossed sweetly in the experiment, the system persists in the same phase and the *hysteresis* in the picture is observed. It is then said that the phase with negative (resp. positive) magnetization is *metastable* for $T < T_c$ and $h > 0$ (resp. $h < 0$) small.

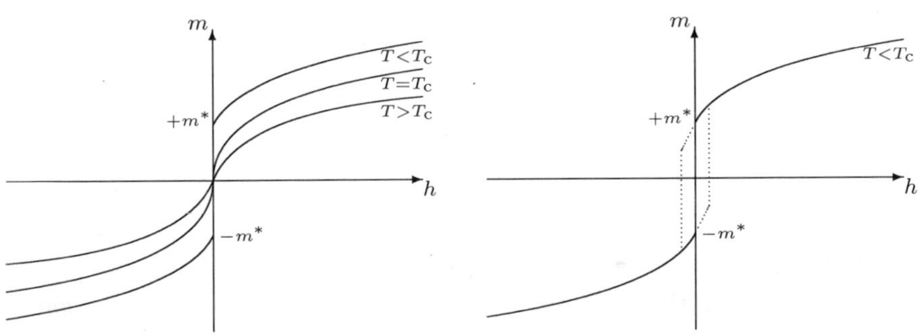

Fig. 1. Isotherms of a ferromagnet on the left; ferromagnetic hysteresis on the right. The temperature is denoted by T, m is the magnetization, h is the external magnetic field, m^* is the spontaneous magnetization, T_c is the Curie temperature.

The rigorous mathematical description of this phenomenon is relatively recent. Not completely rigorous approaches based on equilibrium states have been developed in different fashions. The purely dynamical point of view revealed more powerful and leaded to a pretty elegant definition and characterization of the metastable states; the most important results in this respect have been summed up in.[1]

In this paper we stick to the dynamical description and investigate competing metastable states. This problem shows up in connection with many physical processes, such as the crystallization of proteins[3] and in glasses, in which the presence of a huge number of minima of the energy landscape prevents the system from reaching the equilibrium.[4] The study of these systems is difficult, since the minima of the energy and the decay pathways between them change when the control parameters are varied. It is then of interest the study of models in which a complete control of the variations induced on the energy landscape by changes in the parameters is possible. In Section 2 we discuss the metastable behavior of the Blume–Capel model relying on results in [7]. In Section 3 the obtained result will be compared with the known metastable behavior of reversible Probabilistic Cellular automata with self–interaction.

2. The Blume–Capel model

The Blume–Capel model has been introduced in [5,6] in connection with the liquid Helium transition. In the context of metastability this model revealed very interesting for the three–fold nature of its ground states, see [7,8]. Consider the two–dimensional torus $\Lambda = \{0, \ldots, L-1\}^2$, with L even, endowed with the Euclidean metric; $x, y \in \Lambda$ are *nearest neighbors* iff their mutual distance is equal to 1. Associate a variable $\sigma(x) = 0, \pm 1$ with each site $x \in \Lambda$ and let $\Omega = \{-1, 0, +1\}^\Lambda$ be the *configuration space*. The *energy* associated to the configuration $\sigma \in \Omega$ is

$$H(\sigma) = \sum_{<x,y>} (\sigma(x) - \sigma(y))^2 - \lambda \sum_{x \in \Lambda} (\sigma(x))^2 - h \sum_{x \in \Lambda} \sigma(x) \tag{1}$$

where $< x, y >$ denotes a generic pair of nearest neighbors sites in the torus Λ, $\lambda \in \mathbb{R}$ is the *chemical potential*, $h \in \mathbb{R}$ is the *external magnetic field*, and $|h|, |\lambda| < 1$. The function H will be also called *Hamiltonian.* The equilibrium behavior of the system is described by the *Gibbs measure* $\mu(\sigma) := \exp\{-\beta H(\sigma)\}/Z$, where β is the inverse of the temperature and the normalization constant Z is called *partition function.*

It is possible to introduce the stochastic version of the model by defining a serial dynamics reversible w.r.t. the Hamiltonian (1). It will be a discrete time Glauber dynamics, that is a Markov chain with state space Ω and *transition matrix* $p : \Omega \times \Omega \to [0, 1]$ such that

$$p(\sigma, \eta) := \frac{1}{2|\Lambda|} e^{-\beta \max\{H(\eta) - H(\sigma), 0\}} \tag{2}$$

for $\sigma, \eta \in \Omega$ such σ and η are *nearest neighboring configurations*, i.e., σ is equal to η excepted for the value of the spin associated to a single site; $p(\sigma, \eta) := 0$ for $\sigma, \eta \in \Omega$ such that $\sigma \neq \eta$ and σ and η are not nearest neighboring, that is to say they differ for the values of the spins associated to at least two sites. To ensure the correct normalization of the transition matrix, we also set $p(\sigma,\sigma) = 1 - \sum_{\eta\neq\sigma} p(\sigma,\eta)$ for any $\sigma \in \Omega$.

This dynamics, called *Metropolis algorithm*, satisfies the two following important properties: (i) only transitions between nearest neighboring configurations are allowed; (ii) the dynamics is *reversible* w.r.t. the Hamiltonian (1), i.e.,

$$\mu(\sigma)p(\sigma,\eta) = \mu(\eta)p(\eta,\sigma) \tag{3}$$

for any $\sigma, \eta \in \Omega$. The equation (3) is called *detailed balance* condition.

The definition (2) of the dynamics implies that transitions decreasing the energy happen with finite probability," while transitions increasing the energy are performed with probability tending to zero for $\beta \to \infty$, that is when the temperature tends to zero. This means that when the temperature is small, the system takes a time exponentially large in β to leave a *local minimum* of the Hamiltonian, i.e., a configuration $\sigma \in \Omega$ such that $H(\eta) > H(\sigma)$ for any $\eta \in \Omega$ nearest neighbor of σ. We can then expect that, wherever started, the systems tends to reach the *ground state* of the energy, i.e., the minimum of H, in a *tunneling time* depending on the initial condition. Supposing that there exists initial data for which the tunneling time is exponentially large in β, it is rather natural to define the metastable state as the configuration to which corresponds the maximum tunneling time.

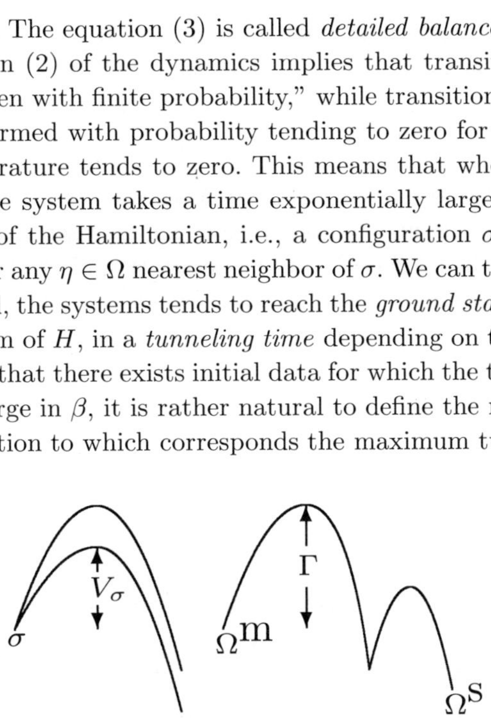

Fig. 2. Definition of metastable states.

More precisely, following [1] and referring to the Figure 2 for a description of the following definitions, given a sequence of configurations $\omega = \omega_1, \ldots, \omega_n$, with $n \geq 2$, we define the *energy height* along the path ω as $\Phi_\omega = \max_{i=1,\ldots,|\omega|} H(\omega_i)$. Given $A, A' \subset \Omega$, we let the *communication energy* $\Phi(A, A')$ between A and A' be the minimal energy height Φ_ω over the set of paths ω starting in A and ending in A'. For any $\sigma \in \Omega$, we let $\mathcal{I}_\sigma \subset \Omega$ be the set of configurations with energy strictly below $H(\sigma)$ and $V_\sigma = \Phi(\sigma, \mathcal{I}_\sigma) - H(\sigma)$ be the *stability level of* σ, that is the energy barrier that, starting from σ, must be overcome to reach the set of configurations with energy smaller than $H(\sigma)$; we set $V_\sigma = \infty$ if $\mathcal{I}_\sigma = \emptyset$. We denote by Ω^{s} the set of global minima of the energy (1), i.e., the collection of the ground states, and suppose that the *communication energy* $\Gamma = \max_{\sigma\in\Omega\setminus\Omega^{\mathrm{s}}} V_\sigma$ is strictly positive. Finally, we define the set of *metastable states* $\Omega^{\mathrm{m}} = \{\eta \in \Omega : V_\eta = \Gamma\}$. The set Ω^{m} deserves its name, since in a rather general framework it is possible to prove (see, e.g., [9, Theorem 4.9]) the following: pick $\sigma \in \Omega^{\mathrm{m}}$, consider the chain σ_n started at $\sigma_0 = \sigma$, then the *first hitting time* $\tau_{\Omega^{\mathrm{s}}} = \inf\{t > 0 : \sigma_t \in \Omega^{\mathrm{s}}\}$ to the ground states is a random variable with mean exponentially large in β, that is

$$\lim_{\beta\to\infty} \frac{1}{\beta} \log \mathbb{E}_\sigma[\tau_{\Omega^{\mathrm{s}}}] = \Gamma \tag{4}$$

with $\mathbb{E}_\sigma$ the average on the trajectories started at σ. In the considered regime, finite volume and temperature tending to zero, the description of metastability is then reduced to the computation of Ω^{s}, Γ, and Ω^{m}.

After this rather general discussion on the definition of metastable states we get back to the study of the Blume–Capel model and note that rigorous results have already been found in [7] in the region $h > \lambda > 0$. In this section we review those results on heuristic grounds and extend the discussion to the whole region $h > 0$ and $h > -\lambda$.

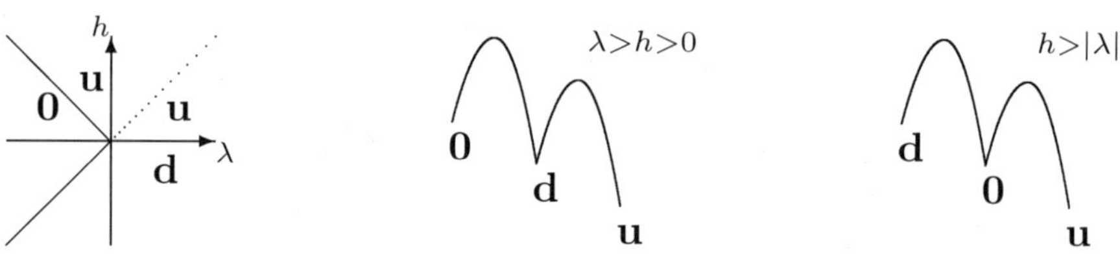

Fig. 3. Ground states of the Blume–Capel model.

First of all we describe the structure of the ground states of the Hamiltonian. Denote by **d**, **u** and **0** the configurations with all the spins in Λ equal respectively to -1, $+1$ and 0, and remark that $E(\mathbf{u}) = -L^2(\lambda + h)$, $E(\mathbf{d}) = -L^2(\lambda - h)$, and $E(\mathbf{0}) = 0$. It is not difficult to prove that for $\lambda = h = 0$ the ground state is three times degenerate and the configurations minimizing the Hamiltonian are **d**, **u** and **0**; for $h > 0$ and $h > -\lambda$, the ground state is **u**; for $h < 0$ and $h < \lambda$ the ground state is **d**; for $\lambda < 0$ and $\lambda < h < -\lambda$ the ground state is **0**; for $h = 0, \lambda > 0$ the ground state is two times degenerate and the configurations minimizing the Hamiltonian are **d** and **u**; for $h = \lambda < 0$ the ground state is two times degenerate and the configurations minimizing the Hamiltonian are **d** and **0**; for $h = -\lambda > 0$ the ground state is two times degenerate and the configurations minimizing the Hamiltonian are **u** and **0**. These results are summarized in the graph in the left in Figure 3. Note, also, that $E(\mathbf{0}) > E(\mathbf{d}) > E(\mathbf{u})$ for $0 < h < \lambda \leq 1$, $E(\mathbf{0}) = E(\mathbf{d}) > E(\mathbf{u})$ for $0 < h = \lambda \leq 1$, and $E(\mathbf{d}) > E(\mathbf{0}) > E(\mathbf{u})$ for $h > |\lambda|$, see the two graphs on the right in the Figure 3.

The obvious candidates to be metastable states are the configurations **d** or **0**; in particular the situation in the region $h > \lambda > 0$ looks really intriguing. In order to prove rigorously that one of them is the metastable state, one should compute Γ and prove that either $V_{\mathbf{d}}$ or $V_{\mathbf{0}}$ is equal to Γ. This is a difficult task, indeed all the paths ω connecting **d** and **0** to **u** should be taken into account and the related energy heights Φ_ω computed. This problem has been solved rigorously in [7] in the region $h > \lambda > 0$ under the technical restriction $2(h/\lambda)^2 + h/\lambda - 1 < 2J/\lambda$. There it has been proven that the metastable state is **d** and that, depending on the ration h/λ, during the tunneling from the metastable to the stable state the configuration **0** is visited or not visited.

As mentioned above we develop an heuristic argument to characterize the behavior of the system in the whole region $h > 0$ and $h > -\lambda$. To characterize the local minima of the Hamiltonian, it is necessary to compute the energy variation under the flip of a single spin. Then consider $\sigma \in \Omega$, $x \in \Lambda$, $a \in \{-1, 0, +1\}$, and denote by σ_x^a the configuration such that $\sigma_x^a(y) = \sigma(y)$ for all $y \neq x$ and $\sigma_x^a(x) = a$; note that $\sigma_x^a = \sigma$ iff $a = \sigma(x)$. By using (1) we easily get

$$H(\sigma_x^a) - H(\sigma) = -2(a - \sigma(x))S_\sigma(x) - (\lambda - 4)(a^2 - \sigma(x)^2) - h(a - \sigma(x)) \quad (5)$$

where $S_\sigma(x)$ is the sum of the four spins of σ associated to the nearest neighbors of the site x. Equation (5) can be used to compute the energy difference involved in all the possible spin flips; the results are summarized in the Table 1. Note that the three cases not listed in the table can be deduced by changing the sign accordingly, for instance if $\sigma(x) = -1$ and $a = +1$, we get $H(\sigma_x^a) - H(\sigma) = -4S_\sigma(x) - 2h$ whose sign is positive for $S_\sigma(x) \leq -1$ and negative for $S_\sigma(x) \geq 0$. It is also worth remarking that the results on the sign of the energy differences listed in the third column of the Table 1 strongly depend on the assumption $|\lambda|, |h| < 1$.

From the results in Table 1 it follows that for $h > \lambda$ the local configurations in which a minus can appear in a local minimum are those such that the sum of the neighboring spins is smaller than or equal to -3, see the two

Table 1. Spin flip energy costs. In the last column the sign of the energy difference is discussed.

$\sigma(x)$	a	$H(\sigma_x^a) - H(\sigma)$	sign
$+1$	-1	$4S_\sigma(x) + 2h$	> 0 if $S_\sigma(x) \geq 0$ < 0 if $S_\sigma(x) \leq -1$
$+1$	0	$2S_\sigma(x) - 4 + \lambda + h$	> 0 if $S_\sigma(x) \geq +2$ < 0 if $S_\sigma(x) \leq +1$
0	1	$2S_\sigma(x) + 4 - \lambda + h$	> 0 if $S_\sigma(x) \geq -1$ > 0 if $S_\sigma(x) \geq -2$ and $h > \lambda$ < 0 if $S_\sigma(x) \leq -2$ and $h < \lambda$ < 0 if $S_\sigma(x) \leq -3$

configurations on the left in Figure 4. For $h < \lambda$ the local configurations in which a minus can appear in a local minimum are those such that the sum of the neighboring spins is smaller than or equal to -2, see the four configurations in the Figure 4.

$$\begin{array}{ccc} & - & \\ - & - & - \\ & - & \end{array} \qquad \begin{array}{ccc} & 0 & \\ - & - & - \\ & - & \end{array} \qquad \begin{array}{ccc} & 0 & \\ 0 & - & - \\ & - & \end{array} \qquad \begin{array}{ccc} & + & \\ - & - & - \\ & - & \end{array}$$

Fig. 4. Minus spins allowed in a local minimum; for $h > \lambda$ only the two configurations on the left are allowed, while for $h < \lambda$ all the four depicted configurations are possible.

From the first two lines in Table 1 it follows that the sole local configurations in which a plus spin can appear in a local minimum are those such that the sum of the neighboring spins is greater than or equal to 2, see Figure 5.

$$\begin{array}{ccc} & + & \\ + & + & + \\ & + & \end{array} \qquad \begin{array}{ccc} & 0 & \\ + & + & + \\ & + & \end{array} \qquad \begin{array}{ccc} & - & \\ + & + & + \\ & + & \end{array} \qquad \begin{array}{ccc} & 0 & \\ 0 & + & + \\ & + & \end{array}$$

Fig. 5. Plus spins allowed in a local minimum.

We discuss in detail the case $h > \lambda$; the analogous results in the region $\lambda > h > 0$ will be summarized in the Figure 7. From the necessary condition for a minus in a local minimum, see the two graphs on the left in the Figure 4, we have that for a configuration to be a local minimum it is necessary that the zeroes form well separated rectangles possibly winding around the torus. To verify that this condition is sufficient for the configuration to be a local minimum we note that, in this case $h > \lambda$, the local configurations in which a zero can appear in a local minimum are those such that the sum of the neighboring spins is greater than or equal to -2 and smaller than or equal to $+1$. In the Figure 6 the possible local configuration for a zero with at least a neighboring plus are shown. This condition is surely met in a configuration in which the zeroes form separated rectangular clusters plunged in a sea of minuses with side lengths larger or equal to two. Moreover, see the Figure 4, in a local minimum direct interfaces between minuses and pluses are forbidden, then the pluses must necessarily be located in the bulk of the zero rectangular droplets. From the results in the Figure 6, see in particular the two graphs on the right, it follows that the pluses must a form well separated rectangular clusters, possibly winding around the torus, inside a rectangular zero cluster. Note that the plus cluster can be separated by the minus component even by a single layer of zeroes.

$-$	$-$	$-$	0	0	0
$-\ 0\ +$	$0\ 0\ +$	$-\ 0\ -$	$-\ 0\ -$	$0\ 0\ -$	$0\ 0\ 0$
$+$	$+$	$+$	$+$	$+$	$+$

Fig. 6. Zero spins with at least a neighboring plus allowed in a local minimum for $h > \lambda$.

In order to study the nucleation of the stable state starting from the possibly metastable states $\mathbf{0}$ and $\mathbf{d}$ the interesting local minima, in the case $h > \lambda$, are the *zero rectangular droplets* in the see of minuses, the *plus rectangular droplets* in the sea of zeroes, and the *frames* made of a plus rectangular droplet plunged in the sea of minuses and separated by the minus component by a single layer of pluses (the frame). The local minima can be used to construct the optimal paths connecting $\mathbf{d}$ and $\mathbf{0}$ to the ground state $\mathbf{u}$.

Consider, first, the paths from $\mathbf{d}$ to $\mathbf{0}$. Optimal paths can be reasonably constructed via a sequence of zero droplets. The difference of energy between two zero droplets with side lengths respectively given by $\ell, m \geq 2$ and $\ell, m+1$ is equal to $2-(h-\lambda)\ell$. It then follows that the energy of a such a droplet is increased by adding an ℓ–long slice iff $\ell < \lfloor 2/(h-\lambda) \rfloor + 1 = \ell_{\mathbf{d}}^{\mathbf{0}}$, where $\lfloor x \rfloor$ denotes the largest integer smaller than the real x. The length $\ell_{\mathbf{d}}^{\mathbf{0}}$ is called the *critical length.* It is reasonable that the energy barrier $V_{\mathbf{0}}$ is given by the difference of energy between the smallest supercritical zero droplet, i.e., the square zero droplet with side length $\ell_{\mathbf{d}}^{\mathbf{0}}$, and the configuration $\mathbf{0}$; by using (1) we get that such a difference of energy is equal to

$\Gamma_{\mathbf{d}}^{\mathbf{0}} = 4/(h-\lambda)$.

A path from **0** to **u** can be constructed with a sequence of plus droplets. By using (1) we get that the difference of energy between two plus droplets with side lengths respectively given by $\ell, m \geq 2$ and $\ell, m+1$ is equal to $2 - 2(h+\lambda)\ell$. It then follows that the energy of a plus droplet is increased by adding an ℓ–long slice iff $\ell < \lfloor 2/(h+\lambda) \rfloor + 1 = \ell_{\mathbf{0}}^{\mathbf{u}}$. The length $\ell_{\mathbf{0}}^{\mathbf{u}}$ is the critical length for the plus droplets; the difference of energy between the smallest supercritical plus droplet and **0** is equal to $\Gamma_{\mathbf{0}}^{\mathbf{u}} = 4/(h+\lambda)$.

A path from **d** to **u** can be constructed via a sequence of frames. It is not difficult to prove that the difference of energy between two frames with internal (rectangle of pluses) side lengths respectively given by $\ell, m \geq 2$ and $\ell, m+1$ is equal to $4 - 2(h-\lambda) - 2h\ell$, so that the critical length for those frames is given by $\ell_{\mathbf{d}}^{\mathbf{f}} = \lfloor (2-(h-\lambda))/h \rfloor + 1$ and the difference of energy between the smallest supercritical frame and **d** is equal to $\Gamma_{\mathbf{d}}^{\mathbf{f}} = 8 + 2(\ell_{\mathbf{d}}^{\mathbf{f}})^2 h - 4h\ell_{\mathbf{d}}^{\mathbf{f}}\varepsilon - 4(h-\lambda)$, where $\varepsilon = \ell_{\mathbf{d}}^{\mathbf{f}} - [(2-(h-\lambda))/h]$.

Remarked that for $h, \lambda \ll 1$ one has $\Gamma_{\mathbf{d}}^{\mathbf{f}} \sim 8/h$, by comparing the energy barriers computed above, it is possible to find the communication energy Γ and to deduce all the results summarized in Figure 7.

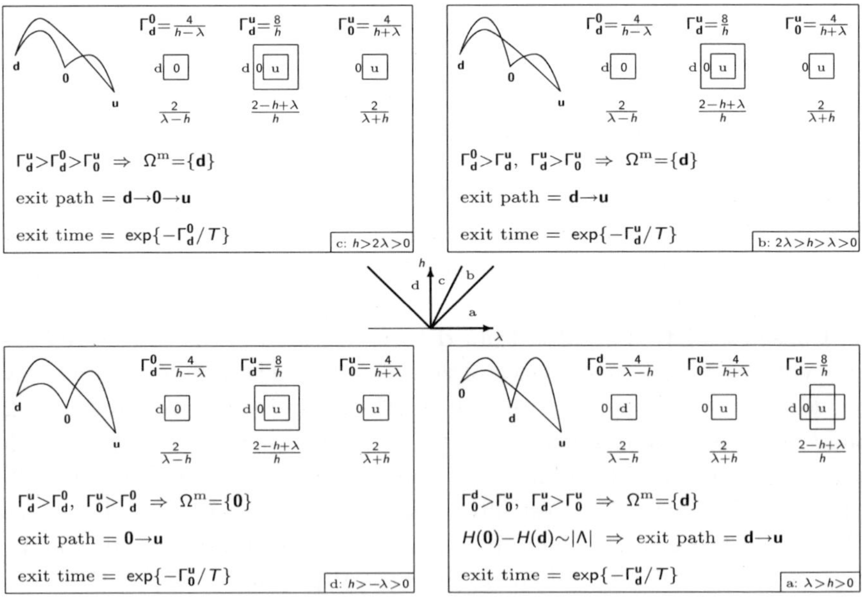

Fig. 7. Summary of results for the Blume–Capel model.

3. Probabilistic cellular automata with self–interaction

We have seen above how in the case of a three–state model as the Blume–Capel model competing metastable states shows up. In some sense this result is natural because the single site configuration space is three–state. In the framework of Probabilistic Cellular Automata it has been shown, see,[10–13] how competing metastable states arise in the context of a genuine two–state model.

Consider the two–dimensional torus $\Lambda = \{0, \ldots, L-1\}^2$, with L even, endowed with the Euclidean metric. Associate a variable $\sigma(x) = \pm 1$ with each site $x \in \Lambda$ and let $\Omega = \{-1, +1\}^\Lambda$ be the *configuration space.* Let $\beta > 0$ and $\kappa, h \in [0, 1]$. Consider the Markov chain σ_n, with $n = 0, 1, \ldots,$ on Ω with *transition matrix*

$$p(\sigma, \eta) = \prod_{x\in\Lambda} p_{x,\sigma}\left(\eta(x)\right) \quad \forall \sigma, \eta \in \Omega \tag{6}$$

where, for $x \in \Lambda$ and $\sigma \in \Omega$, $p_{x,\sigma}(\cdot)$ is the probability measure on $\{-1, +1\}$ defined as $p_{x,\sigma}(s) = 1/[1 + \exp\{-2\beta s(S_\sigma(x) + h)\}]$ with $s \in \{-1, +1\}$ and $S_\sigma(x) = \sum_{y\in\Lambda} K(x-y)\,\sigma(y)$ where $K(x-y)$ is 0 if $|x-y| \geq 2$, 1 if $|x-y| = 1$, and κ if $|x-y| = 0$. The probability $p_{x,\sigma}(s)$ for the spin $\sigma(x)$ to be equal to s depends only on the values of the spins of σ in the five site cross centered at x. The metastable behavior of model (6) has been studied in Ref.[11] for $\kappa = 0$ and in Ref.[10,12] for $\kappa = 1$.

The Markov chain (6) is a *probabilistic cellular automata* (PCA); the chain σ_n, with $n = 0, 1, \ldots,$ updates all the spins simultaneously and independently at any time. The chain is *reversible* with respect to the Gibbs measure $\mu(\sigma) = \exp\{-\beta H(\sigma)\}/Z$ with $Z = \sum_{\eta\in\Omega} \exp\{-\beta H(\eta)\}$ and

$$H(\sigma) = -h \sum_{x\in\Lambda} \sigma(x) - \frac{1}{\beta} \sum_{x\in\Lambda} \log\cosh\left[\beta\left(S_\sigma(x) + h\right)\right] \tag{7}$$

that is *detailed balance* $p(\sigma, \eta)\, e^{-\beta H(\sigma)} = p(\eta, \sigma)\, e^{-\beta H(\eta)}$ holds and, hence, μ is stationary; $1/\beta$ is called the *temperature* and h the *magnetic field.*

Although the dynamics is reversible w.r.t. the Gibbs measure associated to the Hamiltonian (7), the probability $p(\sigma, \eta)$ cannot be expressed in terms of $H(\sigma) - H(\eta)$, as usually happens for Glauber dynamics. Given $\sigma, \eta \in \Omega$, we define the *energy cost*

$$\Delta(\sigma, \eta) = -\lim_{\beta\to\infty} \frac{\log p(\sigma, \eta)}{\beta} = \sum_{\substack{x\in\Lambda:\\ \eta(x)[S_\sigma(x)+h]<0}} 2|S_\sigma(x) + h| \tag{8}$$

Note that $\Delta(\sigma,\eta) \geq 0$ and $\Delta(\sigma,\eta)$ is not necessarily equal to $\Delta(\eta,\sigma)$; it can be proven, see [12, Section 2.6], that

$$e^{-\beta\Delta(\sigma,\eta)-\beta\gamma(\beta)} \leq p(\sigma,\eta) \leq e^{-\beta\Delta(\sigma,\eta)+\beta\gamma(\beta)} \tag{9}$$

with $\gamma(\beta) \to 0$ in the zero temperature limit $\beta \to \infty$. Hence, Δ can be interpreted as the cost of the transition from σ to η and plays the role that, in the context of Glauber dynamics, is played by the difference of energy.

In this context the ground states are those configurations on which the Gibbs measure μ concentrates when $\beta \to \infty$; hence, they can be defined as the minima of the *energy*

$$E(\sigma) = \lim_{\beta\to\infty} H(\sigma) = -h\sum_{x\in\Lambda}\sigma(x) - \sum_{x\in\Lambda}|S_\sigma(x)+h| \tag{10}$$

For $X \subset \Omega$, we set $E(X) = \min_{\sigma\in X} E(\sigma)$. For $h > 0$ the configuration $\mathbf{u}$, with $\mathbf{u}(x) = +1$ for $x \in \Lambda$, is the unique ground state, indeed each site contributes to the energy with $-h-(4+\kappa+h)$. For $h = 0$, the ground states are the configurations such that all the sites contribute to the sum (10) with $4+\kappa$. Hence, for $\kappa \in (0,1]$, the sole ground states are the configurations $\mathbf{u}$ and $\mathbf{d}$, with $\mathbf{d}(x) = -1$ for $x \in \Lambda$. For $\kappa = 0$, the configurations $\mathbf{c}^{\mathrm{e}}, \mathbf{c}^{\mathrm{o}} \in \Omega$ such that $\mathbf{c}^{\mathrm{e}}(x) = (-1)^{x_1+x_2}$ and $\mathbf{c}^{\mathrm{o}}(x) = (-1)^{x_1+x_2+1}$ for $x = (x_1,x_2) \in \Lambda$ are ground states, as well. Notice that $\mathbf{c}^{\mathrm{e}}$ and $\mathbf{c}^{\mathrm{o}}$ are chessboard–like states with the pluses on the even and odd sub–lattices, respectively; we set $\mathbf{c} = \{\mathbf{c}^{\mathrm{e}}, \mathbf{c}^{\mathrm{o}}\}$. Since the side length L of the torus Λ is even, then $E(\mathbf{c}^{\mathrm{e}}) = E(\mathbf{c}^{\mathrm{o}}) = E(\mathbf{c})$. By studying those energies as a function of κ and h, recalling that periodic boundary conditions are considered, we get $E(\mathbf{u}) = -L^2(4+\kappa+2h)$, $E(\mathbf{d}) = -L^2(4+\kappa-2h)$, and $E(\mathbf{c}) = -L^2(4-\kappa)$; hence $E(\mathbf{c}) > E(\mathbf{d}) > E(\mathbf{u})$ for $0 < h < \kappa \leq 1$, $E(\mathbf{c}) = E(\mathbf{d}) > E(\mathbf{u})$ for $0 < h = \kappa \leq 1$, and $E(\mathbf{d}) > E(\mathbf{c}) > E(\mathbf{u})$ for $0 < \kappa < h \leq 1$.

In[13] the metastable behavior of this model has been studied with an heuristic argument very similar to the one developed in the Section 2 to discuss the metastable behavior of the Blume–Capel model. For the details we refer the interested reader to the quoted paper, we just mention here, that quite surprisingly results very similar to the ones obtained in the framework of the Blume–Capel model are found, provided the different parameters are interpreted according to the correspondences in Table 2.

Notice that the role of the zero state of the Blume–Capel model is played, in the context of the PCA, by the flip–flopping chessboard–like configurations. As (4) shows, the discussed results are valid in the limit $\beta \to \infty$. Their validity at finite temperature can be tested with Monte Carlo simulations, see the configurations in Figure 8 observed in a run

Table 2. Correspondence between the Blume–Capel model and the PCA.

Blume–Capel	**u**	**d**	**0**	h	λ
PCA	**u**	**d**	**c**	$h/2$	$\kappa/2$

of the dynamics of the PCA with the parameters specified in the caption and with starting configuration **d**. On the left it is shown that if the self–interaction is present the nucleation of the plus phase is achieved directly; the plot on the right shows that, if the self–interaction is zero, than the chessboard–like phase is visited before the plus phase is nucleated.

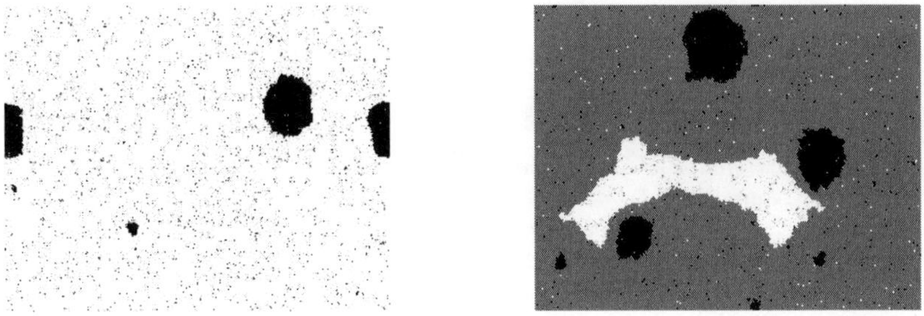

Fig. 8. On the left, typical configuration of the PCA with $\kappa = 1$; simulation performed on a 380×230 torus with $\beta = 0.7$. White and black points represent respectively minus and plus spins. On the right, the same with $\kappa = 0$; the chessboard region looks grey.

References

1. E. Olivieri and M. E. Vares, *Large deviations and metastability*, Cambridge University Press, UK, 2004.
2. K. Huang, *Statistical Mechanics*, Wiley, 1987.
3. P. R. ten Wolde and D. Frenkel, *Science* **277**, 1975 (1997).
4. G. Biroli and J. Kurchan, *Phys. Rev. E* **64**, 016101 (2001).
5. M. Blume, *Phys. Rev.* **141**, 517 (1966).
6. H. W. Capel, *Physica* **32**, 966 (1966).
7. E. N. M. Cirillo and E. Olivieri, *J. Stat. Phys.* **83**, 473 (1996).
8. T. Fiig, B. M. Gorman, P. A. Rikvold, and M. A. Novotny, *Phys. Rev. E* **50**, 1930 (1994).
9. F. Manzo, F. R. Nardi, E. Olivieri, and E. Scoppola, *J. Stat. Phys.* **115**, 591 (2004).
10. S. Bigelis, E. N. M. Cirillo, J. L. Lebowitz, and E. R. Speer, *Phys. Rev. E* **59**, 3935 (1999).
11. E. N. M. Cirillo and F. R. Nardi, *J. Stat. Phys.* **110**, 183 (2003).
12. E. N. M. Cirillo, F. R. Nardi, and C. Spitoni, *J. Stat. Phys.* **132**, 431 (2008).
13. E. N. M. Cirillo, F. R. Nardi, and C. Spitoni, *Phys. Rev. E* **78**, 040601 (2008).

MATHEMATICAL MODELS FOR BIOFILMS ON THE SURFACE OF MONUMENTS

F. CLARELLI

Istituto per le Applicazioni del Calcolo "M.Picone", CNR,
c/o Dip. di Matematica, Università di Roma "Tor Vergata"
Via della Ricerca Scientifica 1, I-00133, Roma, Italy
E-mail: clarelli@iac.rm.cnr.it

C. DI RUSSO

Dipartimento di Matematica, Università degli Studi Roma Tre,
Largo San Leonardo Murialdo 1, I-00146, Roma, Italy
E-mail: dirusso@mat.uniroma3.it

R. NATALINI

Istituto per le Applicazioni del Calcolo "M.Picone", CNR,
c/o Dip. di Matematica, Università di Roma "Tor Vergata"
Via della Ricerca Scientifica 1, I-00133, Roma, Italy
E-mail: r.natalini@iac.cnr.it
http://www.iac.rm.cnr.it/~natalini/

M. RIBOT

Laboratoire "J.A.Dieudonné", Universitè de Nice Sophia Antipolis,
Parc Valrose F-06108 Nice Cedex 02, France
E-mail: ribot@unice.fr
http://math.unice.fr/~ribot/

In this article, a system of nonlinear hyperbolic-elliptic partial differential equations is introduced to model the formation of biofilms. First, a short introduction to some basic concepts about biofilms is given. Then a detailed derivation of the model is presented, which is mainly based on the theory of mixtures, also in comparison with previous models. Adapted numerical schemes will be presented and numerical simulations will be discussed.

Keywords: Biofilms; Cyanobacteria; Theory of mixtures; Hyperbolic equations.

1. Introduction.

Although chemical degradation has been a main concern for conservation and restoration studies, there is now an increasing experimental evidence that biodegradation should also be taken into account since almost fifty percent of deterioration is due to biological factors. In this note, we are interested in particular on the formation and evolution of biofilms, with a special regard to their development on fountains walls, i.e.: on stone substrates and under a water layer. These biofilms cause many damages, such as unaesthetic biological patinas, decoesion and loss of substrate material from the surface of monuments or degradation of the internal structure.

A biofilm is a complex gel-like aggregation of microorganisms like bacteria, cyanobacteria, algae, protozoa and fungi, embedded in an extracellular matrix of polymeric substances (EPS). EPS develops resistance to antibiotics, to our immune system, to disinfectants or cleaning fluids. Even if a biofilm contains water, it is mainly a solid phase. Biofilms can develop on surfaces which are in permanent contact with water, i.e. solid/liquid interfaces, but the growth of microorganisms also occurs in different types of interfaces such as air/solid, liquid/liquid or air/liquid.

Let us summarize the main phases of biofilm formation as follows: first, some bacteria approach the surface and get attached. Then, during a phase of colonization, bacteria lose flagella and produce EPS. During the growth's phase, bacteria build the 3D biofilms and specialize themselves as fixed cells, motile cells or monolayer cells. In the end, a part of the biofilm may detach itself in order to colonize other parts of the surface.

Here we describe and analyze a new model of this process. The paper is organized as follows: in Section 2 we present the derivation of the model and in Section 3 we introduce an explicit-implicit numerical scheme to approximate the system and we report some numerical simulations.

2. Derivation of the model.

2.1. *Previous models.*

For a so huge topic, it is not surprising to find that there is a large literature of mathematical models of biofilms. It is possible to divide these models into two classes: the discrete one and the continuous one.

The models proposed by the Delft's team[4] are mainly multidimensional, multispecies and multisubstrates spatially discrete models, which are solved by individual-based approach or cellular automata (even if sometimes chemical components are taken continuous); thanks to these techniques it is possible to simulate biofilm's detachment. They are exhaustive from the biological point of view, but not fully satisfactory. On one side it is difficult to simulate large colonies of millions of individuals, while on the other side there is no way to prove stability results or to give a precise description of the asymptotic time behavior of the solutions.

Another class of models was proposed by Alpkvist and Klapper in [1], and it is based on a multidimensional and multispecies description, where biofilms are divided into biomass and liquid. This model is diffusive, and so the velocity of the biomass is infinite and this aspect makes the model unrealistic. Moreover, due to diffusivity, it is also difficult obtain sharp interfaces and finger like structures which characterize biofilms.

A recent very interesting model is the one proposed by Zang, Cogan, and Wang.[3] They consider two phases: the polymer network and the solvent, and analyze numerically the case of detachment under different initial conditions. This model does not consider the different biological components and it is unable to describe the specific evolution of bacteria.

A continuous model that includes more biological details is the one proposed by Anguine, King and Ward.[5] It concerns the biofilm produced by the *Pseudomonas aeruginosa*, a bacterium that causes serious infections. This is an one-dimensional and multispecies PDEs model, where four different phases are considered: live cells, dead cells, EPS, and liquid. The influence of nutrients is also taken into account, as well as quorum sensing (a sort of signaling of cells), and also some different medical treatments, like antibiotics and antiQS drugs. Actually, transport equations are introduced to model the four phases and advection-diffusion equations for nutrients, antibiotics and antiQS. A common velocity for bacteria, dead bacteria and EPS is assumed, while a different velocity is taken for the liquid. To close the system, the no-void condition is assumed besides a quite unrealistic relation between the liquid and EPS, that is: a local increment of EPS causes a local increment of liquid. Thanks to these assumptions, no equation for velocities is needed.

Considering the previous models, we propose here to consider a multidimensional and multispecies PDEs model. We have focused our attention on a specific type of microorganisms, namely the cyanobacteria. They grow in water films on the stone's surface and need for warmth, light and water to develop. They die during cold and dry season and deposit dead cells which lead to a rapid new growth at warm time, moreover their nutrients are CO2, N2 and salt minerals traces.

In our model we consider four phases: live cells (cyanobacteria), dead cells, EPS, and liquid. Moreover we introduce two macroscopic velocities: $\mathbf{v_S}$ for the solid phases and $\mathbf{v_L}$ for the liquid. We consider mass balance equations for the four phases and force balance equations. Using the theory of mixtures, proposed in [8] and used also by Preziosi in [6,7], and making some modeling assumptions, it's possible close the system, obtaining a system of nonlinear hyperbolic partial differential equations, coupled with an elliptic equation.

2.2. *The new hydrodynamical model.*

We consider four different phases: Live cells (B), Dead cells (D), EPS (E), and Liquid (L). We denote the concentration of biomass by $C_\phi = \rho_\phi \phi$, where ρ_ϕ is the mass density of the phase in $[g/cm^3]$ and $\phi = B, D, E, L$ is the volume fraction of the phases.

We assume that the biomasses are incompressible and Newtonian, then ρ_B, ρ_D, ρ_L and ρ_E are positive constants, and also that the phases have all the same constant density. Since the EPS encompasses the cells, we can assume that live cells, dead cells, and EPS have the same transport velocity $\mathbf{v_S}$. We denote instead by $\mathbf{v_L}$ the velocity of liquid, and by Γ_ϕ , with $(\phi = B, D, E, L)$, the mass exchange rate.
The equations expressing the mass balance are:

$$\partial_t B + \nabla \cdot (B\mathbf{v_S}) = \Gamma_B, \tag{1a}$$

$$\partial_t D + \nabla \cdot (D\mathbf{v_S}) = \Gamma_D, \tag{1b}$$

$$\partial_t E + \nabla \cdot (E\mathbf{v_S}) = \Gamma_E, \tag{1c}$$

$$\partial_t L + \nabla \cdot (L\mathbf{v_L}) = \Gamma_L. \tag{1d}$$

Assuming the volume constraint

$$B + D + E + L = 1 \tag{2}$$

and the total conservation of mass, we have

$$\Gamma_B + \Gamma_D + \Gamma_E + \Gamma_L = 0. \tag{3}$$

Denoting by T the temperature, by I the light intensity and by N the nutrient, we introduce the mass production terms:

$$\Gamma_B = B\left(Lk_B(I,T,N) - k_D(I,T,N)\right), \tag{4}$$

$$\Gamma_D = \alpha B k_D(I,T,N) - Dk_N(T), \tag{5}$$

$$\Gamma_E = BLk_E(I,T,N) - \varepsilon E, \tag{6}$$

where k_B and k_D are respectively a birth term and a death term for the active bacterial cells, α is the fraction of active cells that gives rise to dead cells (the remaining proportion becoming liquid), k_N is the natural decay of dead cells, k_E represents the production of EPS, and εE, with ε constant, is the natural decay of EPS.

In order to enforce condition (3), we take

$$\Gamma_L = B\left((1-\alpha)k_D(I,T,N) - L(k_B(I,T,N) + k_E(I,T,N))\right) + Dk_N(T) + \varepsilon E. \tag{7}$$

Adding the four equations of system (1) and using equations (2) and (3), yields:

$$\nabla \cdot ((1-L)\mathbf{v_S} + L\mathbf{v_L}) = 0 \tag{8}$$

which means that the divergence of the average hydrodynamic velocities is equal to zero.

Next, let us write the equations for the force balance. We denote by $\tilde{\mathbf{T}}_\phi$ the partial stress tensor relative to the component ϕ, and by $\tilde{\mathbf{m}}_\phi$ the respective interaction force. So we can write the equation of force balance for the component ϕ ($\phi = B, D, E, L$) as follows:

$$\partial_t(\phi\mathbf{v}_\phi) + \nabla \cdot (\phi\mathbf{v}_\phi \otimes \mathbf{v}_\phi) = \nabla \cdot \tilde{\mathbf{T}}_\phi + \tilde{\mathbf{m}}_\phi + \Gamma_\phi \mathbf{v}_\phi. \tag{9}$$

The total conservation of momentum yields:

$$\sum_\phi (\tilde{\mathbf{m}}_\phi + \Gamma_\phi \mathbf{v}_\phi) = 0. \tag{10}$$

From the theory of mixtures,[8] it is possible to decompose the interaction forces as $\tilde{\mathbf{m}}_\phi = P\nabla\phi + \mathbf{m}_\phi$, where P is the hydrostatic pressure, the same scalar for all the phases, and $\mathbf{m}_\phi$ is the force exerted by the phase ϕ on the other phases. It is also possible decompose the partial stress tensor as $\tilde{\mathbf{T}}_\phi = -\phi P I + \phi \mathbf{T}_\phi$, where $\mathbf{T}_\phi$ is the excess stress tensor. Hence, equation (9) can be rewritten as:

$$\partial_t(\phi\mathbf{v}_\phi) + \nabla \cdot (\phi\mathbf{v}_\phi \otimes \mathbf{v}_\phi) = \mathbf{m}_\phi - \phi\nabla P + \nabla \cdot (\phi\mathbf{T}_\phi) + \Gamma_\phi \mathbf{v}_\phi. \tag{11}$$

Taking the sum of equations (11) for $\phi = B, D, E$, and using (2), we obtain:

$$\partial_t((1-L)\mathbf{v_S}) + \nabla\cdot((1-L)\mathbf{v_S}\otimes\mathbf{v_S}) = -(1-L)\nabla P + \nabla\cdot(\sum_{\phi\neq L} \phi\mathbf{T}_\phi) - \mathbf{m}_L - \Gamma_L \mathbf{v_L}, \tag{12}$$

since, from (10), (2) and (3)

$$\sum_{\phi\neq L} \mathbf{m}_\phi = -\mathbf{m}_L + \Gamma_L(\mathbf{v_S} - \mathbf{v_L}).$$

For the liquid phase we have:

$$\partial_t(L\mathbf{v_L}) + \nabla \cdot (L\mathbf{v_L} \otimes \mathbf{v_L}) = -L\nabla P + \nabla \cdot (L\mathbf{T}_L) + \mathbf{m}_L + \Gamma_L \mathbf{v_L}. \quad (13)$$

Now we make some assumptions on the form of the excess stress tensors, namely

$$\sum_{\phi \neq L} \phi \mathbf{T}_\phi = \Sigma \mathbf{I} \quad \text{and} \quad \mathbf{T}_L = 0, \quad (14)$$

where Σ is a scalar function of the sum of volume ratios $B+D+E = 1-L$. We are assuming that the excess stress tensor is only present in the solid component, so in the liquid there is only the hydrostatic pressure; this means that if in the liquid there are no bacteria or EPS, then the liquid is at rest. We also assume that the interaction forces for the liquid follow the Darcy law:

$$\mathbf{m}_L = -M(\mathbf{v_L} - \mathbf{v_S}), \quad (15)$$

where M is an experimental constant.

Thanks to these assumptions, we can rewrite the equations for the velocities as:

$$\begin{aligned} \partial_t((1-L)\mathbf{v_S}) + \nabla \cdot ((1-L)\mathbf{v_S} \otimes \mathbf{v_S}) &= -(1-L)\nabla P + \nabla \Sigma \\ &\quad + M(\mathbf{v_L} - \mathbf{v_S}) - \Gamma_L \mathbf{v_L}, \\ \partial_t(L\mathbf{v_L}) + \nabla \cdot (L\mathbf{v_L} \otimes \mathbf{v_L}) &= -L\nabla P - M(\mathbf{v_L} - \mathbf{v_S}) + \Gamma_L \mathbf{v_L}. \end{aligned} \quad (16)$$

Now we have to describe the scalar equation satisfied by the pressure P. Summing the equations in (16) and taking their divergence, we have:

$$-\Delta P = \nabla \cdot (\nabla \cdot ((1-L)\mathbf{v_S} \otimes \mathbf{v_S} + L\mathbf{v_L} \otimes \mathbf{v_L})) - \Delta \Sigma. \quad (17)$$

So, equations (1), (16), and (17) give a closed system of equations

$$\begin{cases} \partial_t B + \nabla \cdot (B\mathbf{v_S}) = B\left(Lk_B(I,T,N) - k_D(I,T,N)\right), \\ \partial_t D + \nabla \cdot (D\mathbf{v_S}) = \alpha B k_D(I,T,N) - D k_N(T), \\ \partial_t E + \nabla \cdot (E\mathbf{v_S}) = BLk_E(I,T,N) - \epsilon E, \\ \partial_t L + \nabla \cdot (L\mathbf{v_L}) = B((1-\alpha)k_D(I,T,N) - Lk_B(I,T,N) \\ \qquad -Lk_E(I,T,N)) + Dk_N(T) + \epsilon E, \\ \partial_t((1-L)\mathbf{v_S}) + \nabla \cdot ((1-L)\mathbf{v_S} \otimes \mathbf{v_S}) + (1-L)\nabla P = \nabla \Sigma \\ \qquad +(M - \Gamma_L)\mathbf{v_L} - M\mathbf{v_S}, \\ \partial_t(L\mathbf{v_L}) + \nabla \cdot (L\mathbf{v_L} \otimes \mathbf{v_L}) + L\nabla P = -(M - \Gamma_L)\mathbf{v_L} - M\mathbf{v_S}, \\ -\Delta P = \nabla \cdot (\nabla \cdot ((1-L)\mathbf{v_S} \otimes \mathbf{v_S} + L\mathbf{v_L} \otimes \mathbf{v_L})) - \Delta \Sigma. \end{cases} \tag{18}$$

To complete the system, we have to find the form of the stress function Σ, the values of the coefficients of the source terms and of the Darcy constant M. A first approximation, useful for numerical tests, is a linear form of the stress function as

$$\Sigma = -\gamma(1-L). \tag{19}$$

We impose Neumann boundary conditions for the volume ratios:

$$\nabla B \cdot n|_{\partial\Omega} = \nabla E \cdot n|_{\partial\Omega} = \nabla D \cdot n|_{\partial\Omega} = 0,$$

and no-flux boundary conditions for the velocities:

$$\mathbf{v_S} \cdot n|_{\partial\Omega} = \mathbf{v_L} \cdot n|_{\partial\Omega} = 0.$$

2.3. *Influence of light, temperature and nutrients.*

The growth rate of cyanobacteria k_B is influenced by light, nutrient and temperature.

We indicate $k_B = k_{B0} g(I,T,N)$, where $g(I,T,N) = g_1(I) \cdot g_2(T) \cdot g_3(N)$ is a function of light, temperature and nutrient.

For simplicity we assume that k_{B0} is the optimal growth rate, so $0 \leq g(I,T,N) \leq 1$.

We assume instead, for simplicity, that the other rates are constants:

$$k_D = k_D(I,T,N) \approx k_{D0}$$
$$k_E = k_E(I,T,N) \approx k_{E0}$$
$$k_N = k_N(T) \approx k_{N0}$$

where k_{D0}, k_{E0} and k_{N0} are the optimal rates.

2.3.1. *Light dependence*

To describe the light influence on the biofilm growth, we indicate by I_0 the light intensity on the upper surface of water, and by $I(x,y,t)$ the intensity in the water. We assume that the light intensity is attenuated following the law of photon absorption in the matter. Thus, we assume that $I(x)$ is constant for every fixed y and t. Then, assuming $y \in [0,H]$ we have:

$$\frac{I(y,t)}{I_0(t)} = e^{-\mu(H-y)}, \qquad (20)$$

where the absorption coefficient μ depends on the matter and on the frequency of radiation. By experimental observations, it has been estimated that $\mu \approx 0.9\ m^{-1}$ if the water is turbid, and $\mu \approx 0.2\ m^{-1}$ if the water is clear.

We still need of a correction factor because, when the light reaches the biomasses, its value decreases exponentially, so we assume

$$\mu = \mu_0 \left(1 + h_\mu \left(B + E + D\right)\right), \qquad (21)$$

where μ_0 is the absorption coefficient when the water is clear, and h_μ is a coefficient in the biomasses.

Finally, we assume that the specific growth rate as function of irradiation $I(x,z,t)$ is given by

$$g_2(I) = 2w_0 \left(1 + \beta_2\right) \frac{\hat{I}}{\hat{I}^2 + 2\beta_2 \hat{I} + 1}, \qquad (22)$$

where $\hat{I} = I/I_0$, w_0 is the maximum specific growth rate and β_2 is a shape coefficient.

2.3.2. *Temperature dependence*

We assume that cyanobacteria have an optimal temperature where the growth rate is maximal, when the temperature is far from this optimal value the growth rate diminishes. Following,[9] we choose that the specific growth rate as function of temperature T, which is given by:

$$g_1(T) = 2h_1\,(1+\beta_1)\,\frac{\theta}{\theta^2 + 2\beta_1\theta + 1}, \tag{23}$$

where

$$\theta = \frac{T - T_{min}}{T_{opt} - T_{min}}. \tag{24}$$

Here h_1 corresponds to the maximum growth rate, β_1 is a shape parameter, and T_{min} is the minimal temperature for the model working.

2.3.3. *Nutrient dependence*

Nutrient is essential for the duplication of cyanobacteria and is consumed by them. We suppose that the nutrient diffuses according to a fickian diffusion, and we also neglect the transport term due to the low velocity of water $\mathbf{v_L}$ compared with the diffusion of a gas in water.

We indicate the nutrient mass density by C_N (g/cm^3), while by C_{ref} a reference value of nutrient, which can be the concentration of nutrient on the upper boundary between water and air.

Then, we introduce $N = C_N/C_{ref}$ which is an adimensional variable, so the nutrient equation is

$$\partial_t N = D_N \Delta N - q_N B N, \tag{25}$$

where D_N (cm^2/sec) is the diffusion coefficient of nutrient in the water and in the biomasses and q_N $(1/sec)$ is the consumption rate. The specific growth rate as function of nutrient N is given by

$$g_3(N) = N.$$

Thus, $k_B(I,T,N) = k_{B0}g_1(I)g_2(T)N$.

3. Numerical approximation and simulations.

Let us explain now how we solve numerically the system (18). For the first three equations, that is to say equations (1a), (1b), (1c), we use a relaxation scheme with flux limiters for the spatial discretization of the transport part,[2] and an explicit Euler scheme for the time discretization. Let us denote by Y the volume ratio of one of the solid phase, i.e. $Y = B, D$ or E. We also denote the time step by δt and the approximation of function Y at time $t_n = n\delta t$ by Y^n.

In the following, we shall write this scheme for the balance equation of solid phases in a condensed form as :

$$Y^{n+1} = Y^n - \delta t \left(\nabla \cdot (Y \mathbf{v_S})\right)^n + \delta t \, \Gamma_Y^n.$$

Then, thanks to the condition (2), it is possible to compute an approximation L^{n+1} of the liquid volume ratio at time t_{n+1} as:

$$L^{n+1} = 1 - (B^{n+1} + D^{n+1} + E^{n+1}).$$

We cannot proceed in the same way to solve numerically the equations of velocities. As a matter of fact being:

$$\begin{aligned}
&(1 - L^{n+1} \mathbf{v_S}^{n+1}) = G_1^n - \delta t \left((1-L)\nabla P\right)^n + \delta t (M - \Gamma_L^n) \mathbf{v_L}^n - M \mathbf{v_S}^n, \\
&L^{n+1} \mathbf{v_L}^{n+1} = G_2^n - \delta t (M - \Gamma_L^n) \mathbf{v_L}^n - M \mathbf{v_S}^n,
\end{aligned}$$

where

$$\begin{aligned}
G_1^n &= (1 - L^n) \mathbf{v_S}^n - \delta t \left(\nabla \cdot (1-L) \mathbf{v_S} \otimes \mathbf{v_S})\right)^n + \delta t \left(\nabla \Sigma\right)^n, \\
G_2^n &= L^n \mathbf{v_L}^n - \delta t \left(\nabla \cdot (L \mathbf{v_L} \otimes \mathbf{v_L}) + L \nabla P\right)^n,
\end{aligned}$$

we need to compute $\mathbf{v_S}^{n+1}$ and $\mathbf{v_L}^{n+1}$ to continue after time t_{n+1}. However, it is not possible to compute $\mathbf{v_S}^{n+1}$ if $L^{n+1} = 1$, nor $\mathbf{v_L}^{n+1}$ if $L^{n+1} = 0$.

Usually, in multiphases simulations, data are taken to avoid vanishing phases. However, in the present case, these phases are physically relevant. Fortunately, we can can overcome this problem by using an implicit approximation for the reaction terms of the equations of velocities, which can be written as:

$$\begin{aligned}
&(1 - L^{n+1}) \mathbf{v_S}^{n+1} = G_1^n - \delta t \left((1-L)\nabla P\right)^n + \delta t (M - \Gamma_L^{n+1}) \mathbf{v_L}^{n+1} - M \mathbf{v_S}^{n+1}. \\
&L^{n+1} \mathbf{v_L}^{n+1} = G_2^n - \delta t (M - \Gamma_L^{n+1}) \mathbf{v_L}^{n+1} - M \mathbf{v_S}^{n+1}.
\end{aligned}$$

In this way we obtain values of the velocities even in the particular cases when one of the phases, liquid or solid, vanishes. To compute the terms G_1^n and G_2^n we adopt the same relaxation scheme used to solve the equations for B, D, E.

In the end, we solve the elliptic equation for the pressure P thanks to classical finite differences.

We have solved the system in the 1D case with constant reaction rates for k_y i.e. optimal rates. In Figure 1 (resp. Figure 2), we display the volume ratio of B (resp. E) with respect to the height for different times. We notice the formation of a front which spreads in the whole domain for the propagation of bacteria, but also for the production of EPS.

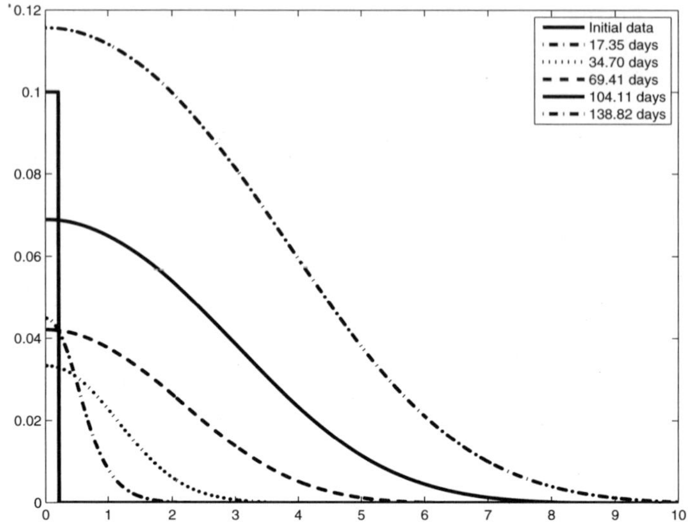

Fig. 1. Evolution with respect to time of the density of bacteria.

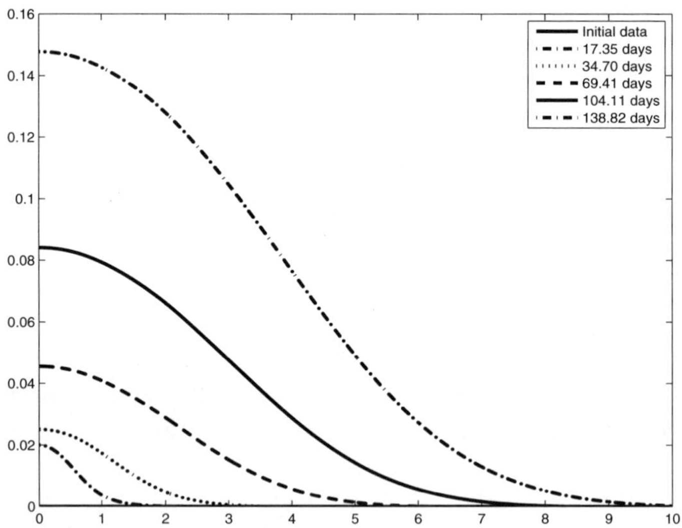

Fig. 2. Evolution with respect to time of the density of EPS.

Acknowledgements.

This work has been partially supported by the project "Mathematical problems for the biological damage of monuments" in the CNR-CNRS 2008-2009 agreement, and the INdAM-GNAMPA project 2008 "Hyperbolic models for chemotaxis".

References

1. E. Alpkvist and I. Klapper, *A multidimensional multispecies continuum model for heterogeneous biofilm development, Bull. Math. Biol.*, **69** (2007), pp. 765–789.
2. D. Aregba-Driollet and R. Natalini,*Discrete kinetic schemes for multidimensional systems of conservation laws, SIAM J. Numer. Anal.*, **37** (2000), pp. 1973–2004.
3. T. Zhang, N. Cogan, and Q. Wang, Phase-field models for biofilms II. 2-D numerical simulations of biofilm-flow interaction, *Commun. Comput. Phys.*, **4** (2008), pp. 72–101.
4. O. Wanner, H. J. Eberl, E. Morgenroth, D. Noguera, C. Picioreanu, B. E. Rittmann and M. C. M. Van Loosdrecht, *Mathematical Modeling of Biofilms*, IWA Scientific and Technical Report No.18, IWA Publishing (2006).
5. K. Anguige K, J. R. King and J. P. Ward, *A multiphase mathematical model of quorum sensing in a maturing Pseudomonas aeruginosa biofilm, Math. Biosci.*, **203** (2006), pp. 240–276.
6. L. Preziosi and A. Tosin, *Multiphase modelling of tumour growth and extracellular matrix interaction: mathematical tools and applications, J. Math. Biol.*, **58** (2009), pp. 625–656.
7. D. Ambrosi and L. Preziosi, *On the closure of mass balance models for tumour growth, Math. Models Methods Appl. Sci.*, **12** (2002), pp. 737–754.
8. K. R. Rajagopal and L. Tao, *Mechanics of mixtures.* Series on Advances in Mathematics for Applied Sciences, 35. World Scientific Publishing Co., River Edge, NJ, 1995.
9. J. M. Thebault and S. Rabouille, *Comparison between two mathematical formulations of the phytoplankton specific growth rate as a function of light and temperature, in two simulation models* (ASTER & YOYO), *Ecol. Model.*, **163** (2003), pp. 145–151.

A Mathematical Model for Consolidation of Building Stones

Fabrizio Clarelli[1], Roberto Natalini[3], Carlo Nitsch[4], Maria Laura Santarelli[2]
[1] *Dipartimento di Matematica Pura ed Applicata, Università degli studi de L'Aquila; Via Vetoio, I-67010 Coppito (L'Aquila) AQ - Italy*
clarelli@iac.rm.cnr.it

[2] *Cistec, Università degli studi di Roma "'La Sapienza"', Via Eudossiana, 18; I–00184 Rome, Italy*
marialaura.santarelli@uniroma1.it

[3] *Istituto per le Applicazioni del Calcolo "M. Picone", Consiglio Nazionale delle Ricerche, Viale del Policlinico 137, I–00161 Rome, Italy*
r.natalini@iac.cnr.it

[4] *Dipartimento di Matematica e Applicazioni "R. Caccioppoli" Università degli Studi Napoli "Federico II", Via Cintia, Monte S. Angelo I-80126 Napoli, Italy*
carlo.nitsch@dma.unina.it

We introduce a mathematical model to improve our understanding of consolidation processes and to take into account the fine scale evolution of reaction pathways. We focus on a silicone called TEOS (Tetraethyl Orthosilicate). The model is based on differential equations, inspired by the theory of porous media, which describes the process of consolidation in terms of filtration and solidification. Our main goal is the prediction of the ultimate depth of filtration of TEOS, according to the environmental and material data. The model has been calibrated on laboratory tests with a very good agreement with experimental data.

Keywords: Consolidation, ethyl silicate, differential models, porous media theory, numerical simulations.

1. Introduction

The consolidation of stone monuments (like marble, tuff etc) is one of the most common way to recover their structural integrity. The restoring of structural continuity depends on the material used to obtain the consolidation, and on its capacity to penetrate inside the structure to be recovered. The restoring of modern artefacts is usually done by suitable cements, but this is not possible on monumental rocks and building structures. Thus, from 1970s, it is quite popular the employment of polymers to consolidate monumental artefacts. At the beginning, these products were made by acrylic polymers or epoxy resins, but their high viscosity does not allow a deep impregnation of the stones, and they contribute only to obtain a consolidation near the external surface. Later, mixtures based on siliceous products have been developed, and, in particular, good results have been obtained using a mixture of solvents and ethyl silicate. The silicate polymerizes in situ entering in contact with water (humidity), then restoring the microscopic structural integrity. In siliceous rocks, it establishes an out-and-out chemical bond, while in calcareous rocks the consolidation happens only around the detached and powdered stone grains. Using this kind of products, it is necessary to evaluate the real impregnation of the rocks, to evaluate the effectiveness of the consolidation.[5]

In this paper we propose a first mathematical model for the evaluation of the penetration of consolidants (ethyl silicate). We present its derivation and some numerical tests will be discussed against experimental data.

2. A mathematical model for consolidation of building stones

Consolidation of monuments is achieved by in-depth impregnation of rock with resins. Among the adhesives commonly used, we are interested in a silicone called TEOS (Tetraethyl Orthosilicate often denoted simply by Ethyl Silicate) which recently has become very popular. In pristine state TEOS is made by monomers. They undergo to hydrolysis once they come into contact with moisture inside the stones, and subsequently they start a polymerization and solidification process.

Here we propose a first mathematical model which describes the process of consolidation in terms of filtration and solidification. The main goal would be the prediction of the ultimate depth of filtration of TEOS.

Our approach comes from the theory of fluid flows in porous media. For a detailed introduction on this subject we refer the reader to the comprehensive books[1] and.[2]

In order to build the model, we point out some preliminary observations.

- **O1** TEOS is usually dissolved in a solvent called White Spirit (WS). TEOS concentration can be up to 75% in volume.
- **O2** The pores of the stones are only partially occupied by the liquid phase. More precisely in our problem a liquid (white spirit solution) and a gas (air) flow simultaneously inside the stone. Under such a

condition the process is called *unsaturated flow*.

O3 Impregnation of the stones is often achieved using a brush and therefore the boundary pressure of the liquid on the surface of the stones is the atmospheric one.

O4 A consequence of the previous observation is that the solution is absorbed only by means of capillarity.

O5 The Water–TEOS reaction takes place inside the rocks and only at the air–liquid interface. Hydrolysis of monomers is strongly heterogeneous at small scale due to the presence of micro–drops of water, but homogeneous at a macro scale.

We make the following assumptions:

A1 The fluid density is constant throughout the process.

A2 We neglect, for simplicity, the variation of moisture content inside the rocks assuming that new water is continuously supplied. The reaction can vary the moisture content.

A3 Due to **O5**, we can assume that the concentration of hydrolyzed monomers of TEOS (which are the bricks of polymeric structure) is inhomogeneous. Polymerization does not take place everywhere but only in some localized region according to the water drops distribution. In such places monomers immediately react to form huge polymeric structures that are caught by the pores walls a soon as they are generated. Therefore we avoid all the complications coming from the study of polymeric flow in porous media, and we neglect the effects of the polymerization on the fluids properties, such as viscosity and density.

As usual in porous media theory, the physical quantities at a point x shall be interpreted as averaged within the representative element of volume (REV) centered at that point. We denote by n the porosity i.e.: the fraction of volume occupied by voids. While the fraction of volume occupied by the fluid within a REV is represented by θ_f, often called fluid content.

For the liquid we use the subscript l and for the gas the subscript g. The following relation easily holds

$$n = \theta_l + \theta_g. \tag{1}$$

Since, assuming **O1**, we deal with unsaturated flow, for a liquid of density

ρ_l the mass balance equation, taking into account the Darcy's Law, is (see[2])

$$\partial_t(\theta_l\rho_l) = -\nabla\left[\rho_l\frac{k(\theta_l/n)}{\mu_l}\left(\frac{n}{n_0}\right)^2\nabla\left(P_c - \rho_l g z\right)\right] - G_l. \tag{2}$$

Here $G_l \geq 0$ is a sink term representing the amount of liquid which in a unit of time becomes solid and n_0 is the initial porosity; P_c is the capillarity pressure i.e.: the pressure drop on the interface between liquid and gas

$$P_g - P_l = P_c.$$

As usual in literature,[1,2] we fix the capillary pressure to be a function of the fluid saturation (θ_l/n) alone

$$P_c \equiv P_c(\theta_l/n).$$

Let us denote by c_l the molar concentration of TEOS (i.e. mass of silicate per unit of volume of white spirit solution). Then, following [2], the mass balance equation for the TEOS within the liquid phase is given by

$$\partial_t(\theta_l c_l) = -\nabla\left[c_l\frac{k(\theta_l/n)}{\mu_l}\left(\frac{n}{n_0}\right)^2\nabla\left(P_c - \rho_l g z\right) - \theta_l D^T\nabla c_l\right] - G_l. \tag{3}$$

Here, D^T is the dispersion-diffusion coefficient; while k and μ_l (in equations 2 and 3) are respectively the permeability of the rock and the viscosity of the fluid. We assume that all of them are known functions. In particular, during the reaction, which changes liquid in solid, the porosity n diminishes; thus, the permeability diminishes as n^2.

Observing that the rate of growth of the volume of solid is exactly the opposite of the rate of increase of the volume of fluid, we get

$$\dot{n} = -G_l/\rho_l. \tag{4}$$

Finally, we assume that the rate of liquid absorbed by the solid matrix is proportional to the concentration of TEOS times that of water

$$G_l = K c_l c_a \theta_l (n - \theta_l). \tag{5}$$

where K is the mass exchanged between liquid and solid per unit of mass of the reagents and unit of time.and for simplicity the molar concentration c_a of moisture in the air is supposed to be constant.

Our mathematical model is the collection of (2)-(3)-(4)-(5), which constitute a closed system of equations, provided we know the dependence of P_c on the ratio $\dfrac{\theta_l}{n}$.

3. Fitting the model with the experiments

In this section we present our first attempt to fit the model with experimental data. Unfortunately a fine tuning of all the unknown parameters involved in the equations requires a certain amount of experimental data and we are still working on it. In this short paragraph we outline our method by providing some preliminary results, which give a first confirmation of our approach.

Fig. 1. Experimental setting for the diffusion test of Ethyl Silicate.

Experimental tests have been realized using a tuff specimen in a vertical way, to observe the ethyl silicate diffusion along the vertical axis. The yellow tuff of Lazio specimen has the size of 15 x 7.5 x 3.5 [cm], and it is put as shown in Figure 1. A continuously impregnated sponge feeds the specimen from one side, the four sides perpendicular to the flux direction have been sealed by epoxy resin to obtain a one-directional diffusion. In a first experiment, the specimen has been treated by a 70/30 volume of white spirit and TEOS; in a second experiment the tuff has been completely impregnated by white spirit, after that it has been treated by TEOS, as the previous case. Experiments are stopped when a drop appears on the other side of the specimen. In the case in which has been used ethyl silicate, it has been observed the first drop after 31 hours. It has been analyzed by thermo-gravimetry analyses (TGA). It permits to evaluate the weight loss of the specimen, during a warming up to 1000^oC. This way, it has been possible to estimate the percentage of solvent and ethyl silicate present in the specimen. In the case of the directly treated tuff, the ethyl silicate gone out of the specimen has the same percentage of solvent present initially.

In the case of saturated (white spirit) specimen, the first drop is appeared after 3 hour, and the solvent/silicate ratio was of 80/20. The second drop has gone out of the specimen after 24 hours with a ratio 70/30. These experimental tests are useful to estimate the parameter for the mathematical model. To perform the calibration we decided to consider the simple case of a one dimensional flow, a condition which is usually achieved in laboratory tests performed on prismatic specimens. We observe that for horizontal flow as well as for vertical flow in small specimen the effect of gravity can be neglected. As a result the model system becomes

$$\begin{cases} \partial_t \theta_l = \partial_x \left[\left(\dfrac{n}{n_0} \right)^2 \partial_x B(\theta_l/n) \right] - K u c_a \theta_l (n - \theta_l), \\[2ex] \partial_t(\theta_l u) = \partial_x \left[u \left(\dfrac{n}{n_0} \right)^2 \partial_x B(\theta_l/n) + \theta_l D^T \partial_x u \right] - K u c_a \theta_l (n - \theta_l), \\[2ex] \dot{n} = -K u c_a \theta_l (n - \theta_l). \end{cases} \tag{6}$$

Here $u = c_l/\rho_l$ is a "dimensionless concentration" and B is chosen such that $\nabla B = -\frac{k(\cdot)}{\mu_l} \nabla P_c(\cdot)$ and $B(0) = 0$.

In this model there are basically three unknowns: the function B and the constants D^T and K. Since K should be deduced from the chemical rate of reaction between TEOS and H_2O, we focus our attention on the first two unknowns. The idea is to find those experimental conditions which "isolate" the unknowns of the problem in order to fit them separately. To this aim let us consider two ideal experiments.

E.1 Diffusion of TEOS in a prismatic specimen of a porous rock which is completely imbibited by WS. The moisture content inside the rock is supposed to be so small that we are allowed to neglect TEOS hydrolysis;

E.2 Capillary rise of WS in a small prismatic specimen of a porous rock.

According to system (6) experiment E.1 is modelled just by the equation

$$\partial_t u = D^T \partial^2_{xx} u,$$

while experiment E.2 is modelled by the equation

$$\partial_t \theta_l = \partial^2_{xx} \left[B(\theta_l/n) \right]. \tag{7}$$

3.1. *Fitting the diffusion constant*

We performed an experiment on a specimen of tuff 15cm × 7.5cm × 3.5cm. After a pretreatment consisting in drying out the moisture followed by in depth impregnation with WS, a solution of 70/30 volume of WS and TEOS was placed on one of the sides 15cm × 7.5cm of the specimen. After 9600s the volume concetration of WS/TEOS on the opposite side was 83/17. We set $\Delta x = 3.5cm$, $\Delta t = 9600s$, $c = \frac{17}{30}$, and a straightforward computation yields a rough estimate of D^T (see [3]):

$$D^T \lesssim \frac{(\Delta x)^2}{2\Delta t N^{-1}\left(1 - \frac{c}{2}\right)} = 2 \times 10^{-5}\mathrm{m}^2\mathrm{s}^{-1}$$

Here N^{-1} is the inverse of the normal standard distribution

$$N(s) = \int_{-\infty}^{s} \frac{1}{\sqrt{2\pi}} e^{-\frac{r^2}{2}}\, dr$$

3.2. *Fitting the function B*

The determination of the function B is by far the most complicated problem we deal with. Clearly each time the dependence of the rock permeability and capillary pressure are assigned as function of the wetting fluid saturation then the function B is also known. However the determination of these two functions is by itself a difficult task and it is beyond our goal. We decided to face an inverse problem which consists in deducing informations on the function B directly from experimental data. Our ansatz was that

$$B'(s) = \max\{0, c(a - s)(s - b)\}, \tag{8}$$

were $0 \le a \le b \le 1$ and $c \ge 0$ are unknown constants. Let us try to justify this ansatz. We observe that B' is different from zero only in a range between a and b. From the analytical point of view we could infer that a solution of equation (7) has non zero flux only in this range of values. The physical interpretation is that the fluid flow is allowed only in the range of fluid saturation between a and b. The constant b, in fact, should correspond to that value of saturation such that the capillary pressure goes to zero. It correspond to the case where a residual content of air is entrapped in the unsaturated zone. The value $1-b$ is usually called *residual non wetting fluid saturation*. The value a, on the other hand, according to the experimental results, should correspond to the so called *irreducible wetting fluid saturation*, a threshold value of saturation below which the rock permeability drop to zero and a fluid flow would be impossible. The reason is that at low saturation values the wetting fluid (WS in our case) is in the form of isolated pendular rings and the pressure can not be transmitted.

At this point it looks like there are three unknown constant. However, for what we said before, there is a fraction of the pores which are unreachable for the wetting fluid. The bulk porosity can be replaced by the actual porosity i.e.: the fraction of volume occupied by voids that can be reached by the wetting fluid. Such a porosity can be easily measured by complete imbibition of a specimen sample of the porous rock in the wetting fluid. Equations (7) has the same form when replacing the bulk porosity with the actual porosity but now the constant b is equal to 1.

It remains to fit the constants a and c. A typical capillary rise experiment performed on a prismatic specimen consists in measuring, at certain scheduled time intervals (t_i with $1 \leq i \leq N$, N is the number of measures), the amount of fluid absorbed ($X(t_i)$). For any feasible values of a and c it is possible to perform a numerical experiment and compute the total amount of fluid absorbed ($Y(t_i)$) at the same time intervals (t_i). Then we can consider the error function

$$\mathcal{E}(a,c) = \sum_{i=1}^{N} \left| \frac{X(t_i) - Y(t_i)}{X(t_i)} \right|,$$

which sums up all the relative errors between experimental data and numerical simulation. The function $\mathcal{E}$ can be minimized over the parameters a and c to obtained the function B which provides the better agreement with the experimental data.
To this aim we used experiments performed on a tuff specimen of 15cm of length. We implemented a simulated annealing process to optimize the function $\mathcal{E}$ and the minimum was achieved for $a = 0.87$ and $c = 1.77 \cdot 10^{-4}$. The agreement between experimental result and numerical simulation is represented in Figure 2.

4. Numerical tests for the consolidation model

The initial conditions for this system (6) are $\theta_l|_{t=t_0} = 0$, $u|_{t=t_0} = 0$ and $n|_{t=t_0} = n_0$; t_0 and n_0 being respectively the initial time and the initial porosity of the stones. The system have been studied on the half line $x \geq 0$ and the boundary conditions are provided by $O3$ to be

$$\theta_1|_{x=0} = n$$

and

$$u\big|_{x=0} = \frac{c_{01}}{\rho_1},$$

where c_{01} is the initial concentration of TEOS in the white spirit solution. The remaining unknown constants involved in the equations have been chosen in a reasonable range, even if they cannot still be considered fully realistic.[3] Some special attention should be given to the function B. In the literature, several different forms of B have been proposed. After several tests performed on different kinds of building stones and different fluids (water, TEOS, white spirit), we decided to choose the equation (8). Using the parameters we obtained in Subsections 3.1 and 3.2, we performed some simulations.

Here, two simulations for a period of 30 minutes are presented. In the first one it has been used a solution of 30/70 volume of WS and TEOS, in Figure 3 it is shown its behaviour after 15 minutes, and in Figure 4 after 30 minutes. In the second case, a solution of 70/30 volume of WS and TEOS has been simulated, obtaining the results after 15 minutes in Figure 5, and after 30 minutes in Figure 6. Clearly in the first case we obtain a faster penetration, which has to be evaluated against economical factors. Also, we observed a better and faster penetration when the specimen was already impregnated by WS. However, more experimental and numerical results are needed to support this conclusion.

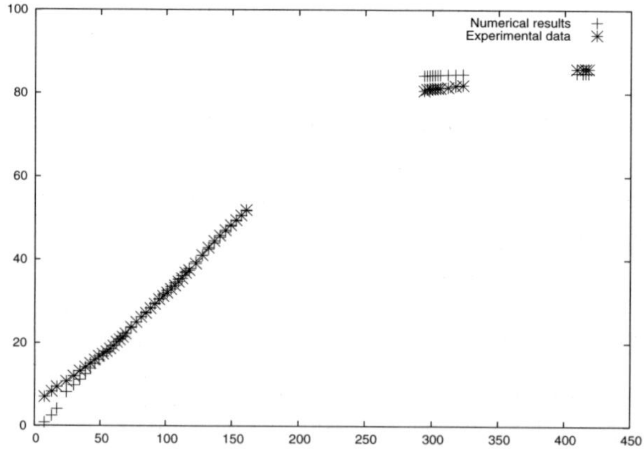

Fig. 2. Total amount of WS absorbed by the specimen [g] versus time $[(sec)^1/2]$. Experimental results against numerical simulations performed using the calibrated model (6).

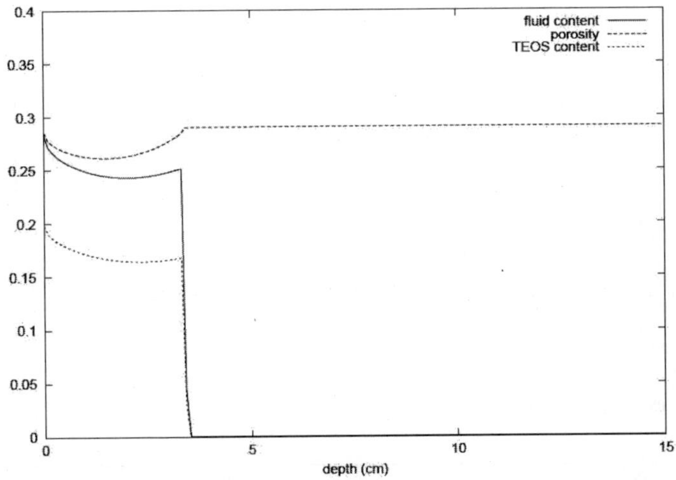

Fig. 3. Numerical simulation showing the evolution of the liquid content, the porosity and the TEOS content after 15 min, TEOS/WS ratio= 70/30.

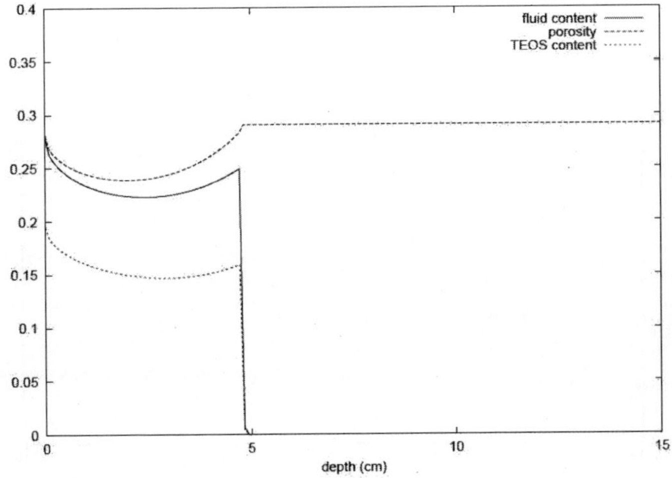

Fig. 4. Numerical simulation showing the evolution of the liquid content, the porosity and the TEOS content after 30 min, TEOS/WS ratio= 70/30.

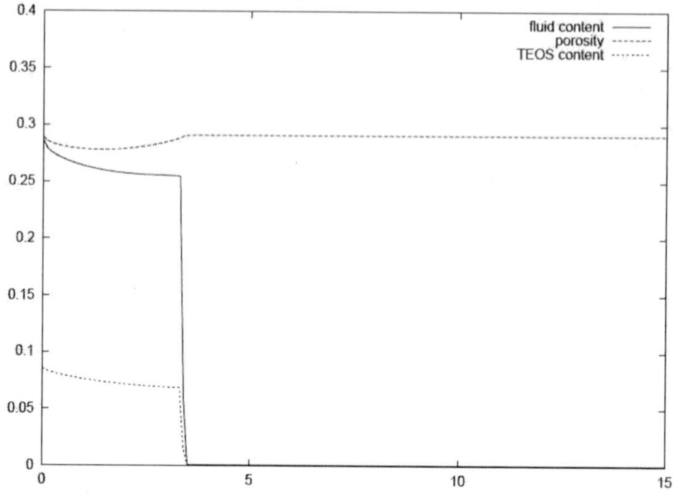

Fig. 5. Numerical simulation showing the evolution of the liquid content, the porosity and the TEOS content after 15 min, TEOS/WS ratio= 30/70.

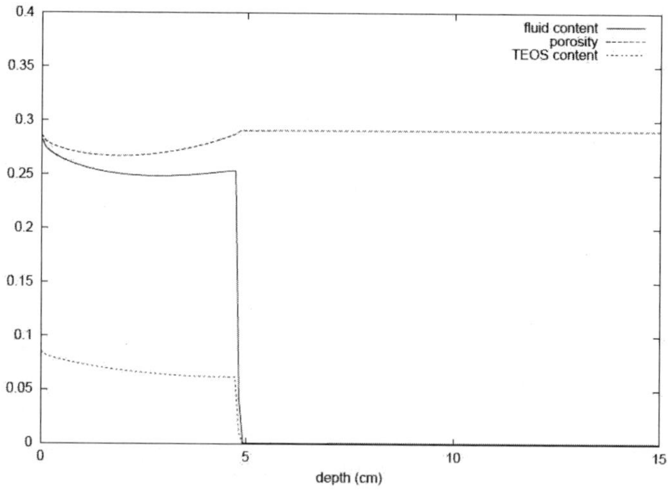

Fig. 6. Numerical simulation showing the evolution of the liquid content, the porosity and the TEOS content after 30 min, TEOS/WS ratio= 30/70.

5. Conclusions

We have introduced a differential model of the flow of WS and TEOS solution inside a rock, which describes the consolidation process. We have calibrated the model against experimental data. As expected the concentration of TEOS influences the behavior of the porosity and hence it certainly effects the ultimate penetration of the consolidant. More experiments, which will be described in,[3] are still needed to give a final assessment to the model, and in particular to establish if the actual consolidation is well predicted by the mathematical simulations, however, it is already clear that a there is a good agreement between the model and a first set of tests.

References

1. G.I. Barenblatt, V.M. Entonov, V.M. Ryzhik, *Theory of fluid flows through natural rocks*, Kluwer Academic Publ., Dordrecht, 1990.
2. J. Bear, Y. Bachmat, *Introduction to modelling of transport phenomena in porous media*, Kluwer Academic Publisher, 1991.
3. Clarelli, F., Giavarini. C., Natalini, R., Nitsch, C., Santarelli, M.L. - Experimental and numerical tuning of a mathematical model for consolidation, work in progressing paper, 2008.
4. Gauri, K.L., Bandyopadhyay, J.K. *Carbonate stone: chemichal behavior, durability and conservation.* John Wiley & Sons, New York (1999).
5. Italian Group International Institute for Conservation (IGIIC), *The silicates in Conservation Treatments.* Log Editrice, Genova, 2004.

Conservation Laws with Unilateral Constraints in Traffic Modeling

Rinaldo M. Colombo[1], Paola Goatin[2], Massimiliano D. Rosini[3]

[1] *Dipartimento di Matematica*
Via Branze 38, I-25123 Brescia, Italy
rinaldo@ing.unibs.it

[2] *I.S.I.T.V.*
B.P. 56, 83162 La Valette du Var Cedex, France
goatin@univ-tln.fr

[3] *IM PAN*
ul. Sniadeckich 8, 00956 Warszawa, Poland
mrosini@univaq.it

Macroscopic models for both vehicular and pedestrian traffic are based on conservation laws. The mathematical description of toll gates along roads or of the escape dynamics for crowds needs the introduction of unilateral constraints on the observable flow. This note presents a rigorous approach to these constraints, and numerical integrations of the resulting models are included to show their practical usability.

Keywords: Traffic Modeling, Conservation Laws, Nonclassical Shocks

1. Introduction

The classical Lighthill–Whitham[22] and Richards[23] traffic model is based on the following simple assumptions:

(1) The total number of vehicles is conserved.
(2) The speed at a point is a function of the density at that point.

Both assumptions appear perfectly suitable also for the macroscopic modeling of crowds. From the analytical point of view, in 1D, these assumptions lead to the conservation law

$$\partial_t \rho + \partial_x \left[\rho\, v(\rho)\right] = 0 \tag{1}$$

where the speed law $v = v(\rho)$ plays a role analogous to that played by the equation of state for Euler equations of gas dynamics.

Along a highway, a toll gate acts as a unilateral constraint on the flow. Indeed, call q the given maximal flow of vehicles that can pass through the gate sited at, say, $x = 0$. Then, the resulting traffic can be reasonably described through (1) supplemented with

$$\rho(t,0)\, v\left(\rho(t,0)\right) \leq q\,. \tag{2}$$

(All **BV** functions, such as solutions to conservation laws, are assumed to be continuous from the left, so that their value is well defined at any point.)

Similarly, consider a crowd escaping along a corridor through an exit. Let q denote the maximal through flow at the exit sited, say, at $x = 0$. Then, again, equation (1) supplemented with (2) provides a simple though effective description of the escape dynamics. For other related descriptions of pedestrian dynamics, see [14–19].

Remark, however, that there is a key qualitative difference in the real evolutions of the two preceding examples. On one hand, the efficiency of the toll gate is not affected by the amount of vehicles lining up. On the other hand, as is well known, the crowd heading towards the exit may cause over-compression phenomena, possibly leading to panic, that may well reduce the efficiency of the exit. Therefore, in the case of pedestrians, we let the evolution of (1) be governed by *nonclassical shocks.*

The next section is devoted to the analytical results on (1)–(2). Section 3 deals with the toll gate model while Section 4 is devoted to the case of pedestrians.

2. Analytical Results

We consider the Constrained Cauchy Problem[8] (CCP)

$$\begin{cases} \partial_t \rho + \partial_x f(\rho) = 0 \\ \rho(0,x) = \rho_o(x) \\ f\left(\rho(t,0)\right) \leq q(t) \end{cases} \tag{3}$$

under the standard assumptions

(CCP1) $f \in \mathbf{C^{0,1}}\left([0,R];\mathbb{R}\right)$, $f(0) = f(R) = 0$, there exists a $\bar\rho \in \,]0,R[$ such that $f'(\rho)(\bar\rho - \rho) > 0$ for a.e. $\rho \in [0,R]$.

(CCP2) $q \in \mathbf{BV}\left(\mathbb{R}^+;[0,f(\bar\rho)]\right)$.

Note that in the case $f(\rho) = \rho\, v(\rho)$, (3) reduces to the Cauchy Problem for (1)–(2).

To state the next well posedness result, it is useful to introduce the translation $\mathcal{T}_t$ through $(\mathcal{T}_t q)(\tau) = q(\tau + t)$. Below we define a map $\mathcal{S}\colon \mathbb{R}^+ \times \mathcal{D} \mapsto \mathcal{D}$, $\mathcal{D}$ being a suitable subset of $\mathbf{L^1}$ containing the initial data of (3). We then denote by $\bar{\mathcal{S}}$ the map $\bar{\mathcal{S}}\colon \mathbb{R}^+ \times \bar{\mathcal{D}} \mapsto \bar{\mathcal{D}}$ defined by $\bar{\mathcal{S}}_t(\rho, q) = (\mathcal{S}_t\rho, \mathcal{T}_t q)$ with $\bar{\mathcal{D}} = \mathcal{D} \times \mathbf{BV}$.

We use the nonlinear mapping[7,8,25]

$$\Psi(\rho) = \operatorname{sign}(\rho - \bar{\rho})\ \left(f(\bar{\rho}) - f(\rho)\right)\ .$$

Recall that by *Riemann Problem* we mean the following particular Cauchy Problem, see also [3, Chapter 5]:

$$\begin{cases} \partial_t \rho + \partial_x f(\rho) = 0 \\ \rho(0,x) = \begin{cases} \rho_l & \text{if } x < 0 \\ \rho_r & \text{if } x > 0\,. \end{cases} \end{cases} \tag{4}$$

Theorem 2.1. *[8, Theorem 3.4] Let (CCP1) and (CCP2) hold. Then, for every constraint* $q \in \mathbf{BV}\left(\mathbb{R}^+; [0, f(\bar{\rho})]\right)$ *there exists a map* $\mathcal{S}\colon \mathbb{R}^+ \times \mathcal{D} \mapsto \mathcal{D}$ *such that*

(S1) $\mathcal{D} \supseteq \left\{\rho \in \mathbf{L^1}\left(\mathbb{R}; [0,R]\right) : \Psi(\rho) \in \mathbf{BV}(\mathbb{R};\mathbb{R})\right\}$;
(S2) $\bar{\mathcal{S}}$ *is a Semigroup, i.e.* $\bar{\mathcal{S}}_0 = \mathrm{Id}$ *and* $\bar{\mathcal{S}}_{t_1} \circ \bar{\mathcal{S}}_{t_2} = \bar{\mathcal{S}}_{t_1+t_2}$;
(S3) $\mathcal{S}$ *is non expansive in* ρ*, i.e. for all* $\rho_1, \rho_2 \in \mathcal{D}$

$$\left\|\mathcal{S}_t\rho_1 - \mathcal{S}_t\rho_2\right\|_{\mathbf{L^1}} \leq \left\|\rho_1 - \rho_2\right\|_{\mathbf{L^1}}\,;$$

(S4) if ρ_o *and* q *are piecewise constant, then for* t *sufficiently small,* $\mathcal{S}_t\rho_o$ *coincides with the gluing of the solutions to standard Riemann problems centered at the points of jump of* ρ_o *and to (3) at* $x = 0$;
(S5) for all $\rho_o \in \mathcal{D}$*, the orbit* $t \mapsto \mathcal{S}_t\rho_o$ *yields a weak entropy solution to (3), according to [8, definitions 3.1 and 3.2].*

The proof uses the standard technique of wave front tracking, see [3, Chapters 6 and 7], and is deferred to [8]. The above statements *(S1)–(S4)* are clearly modeled on the definition of *Standard Riemann Semigroup*, see [3, Definition 9.1] and provide an analogue to it in the present constrained (and non autonomous) setting. The Lipschitz estimate *(S3)* is proved with suitable modifications of the classical techniques[20] or [3, Section 6.3].

It is now easy to tackle the Initial Boundary Value Problem

$$\begin{cases} \partial_t \rho + \partial_x f(\rho) = 0 & (t,x) \in \mathbb{R}^+ \times \mathbb{R}^- \\ \rho(0,x) = \rho_o(x) & x \in \mathbb{R}^- \\ f\left(\rho(t,0)\right) \leq q(t) & t \in \mathbb{R}^+\,. \end{cases} \tag{5}$$

Indeed, as in the case of the Riemann Problem, a solution to (5) is obtained restricting to $x < 0$ a solution to (3) with initial data, say, $\rho_o(x) = 0$ for $x > 0$. The extension to the general Initial Boundary Value Problem

$$\begin{cases} \partial_t \rho + \partial_x f(\rho) = 0 & (t,x) \in \mathbb{R}^+ \times [0, L] \\ \rho(0,x) = \rho_o(x) & x \in [0, L] \\ f\left(\rho(t,0)\right) = f_o(t) & t \in \mathbb{R}^+ \\ f\left(\rho(t,L)\right) \leq q(t) & t \in \mathbb{R}^+ \end{cases}$$

is immediate, see [9,13] and the references therein as a general reference on Initial Boundary Value Problems for scalar conservation laws.

The case of pedestrians[10,11,24], is based on a flow like that in Figure 1. Note that this particular shape of the fundamental diagram was first pos-

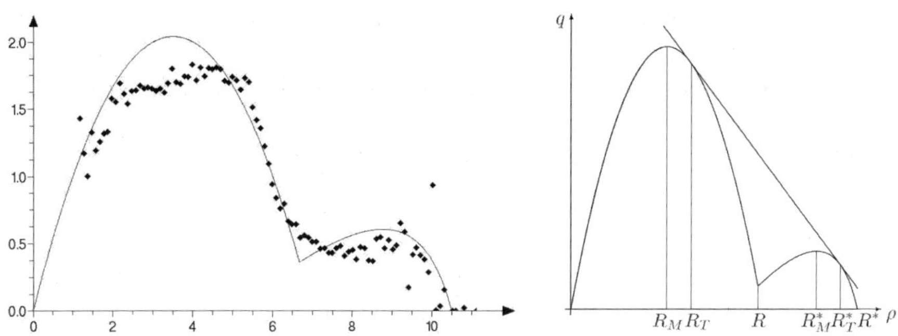

Fig. 1. Left, a flow satisfying the requirements of Theorem 2.2. Superimposed are experimental measurements[16]. Crowd density, ρ, is on the horizontal axis and flow, ρv, on the vertical one. Right, another possible flow and the notation.

tulated in [10] and then experimentally verified in [16]. Moreover, the evolution of the solutions to (1) is governed through the introduction of nonclassical shocks. That is, a Riemann Solver $\mathcal{R}$ is defined, which assigns to every pair ρ_l, ρ_r the self similar weak solution to the Riemann Problem (4) computed at time, say, $t = 1$.

As it is usual when dealing with nonclassical scalar conservation laws, see [21, Chapter II], we need first to introduce the auxiliary function ψ, see Figure 2, left. Let $\psi(R) = R$ and, for $\rho \neq R$, let $\psi(\rho)$ be such that the straight line through $\left(\rho, q(\rho)\right)$ and $\left(\psi(\rho), q \circ \psi(\rho)\right)$ is tangent to the graph of q at $\left(\psi(\rho), q \circ \psi(\rho)\right)$. Besides, for $\bar\rho \in [0, R_T[$ the line through $\left(\bar\rho, q(\bar\rho)\right)$ and $\left(\psi(\bar\rho), q \circ \psi(\rho)\right)$ has a further intersection with the graph of q, which we call $\left(\phi(\bar\rho), q \circ \phi(\bar\rho)\right)$. Introduce two thresholds s and Δs such that

$$s > 0\,,\ \Delta s > 0\,,\ s < R_M \text{ and } R > s + \Delta s \geq \phi(s) > R_T > R - \Delta s\,. \quad (6)$$

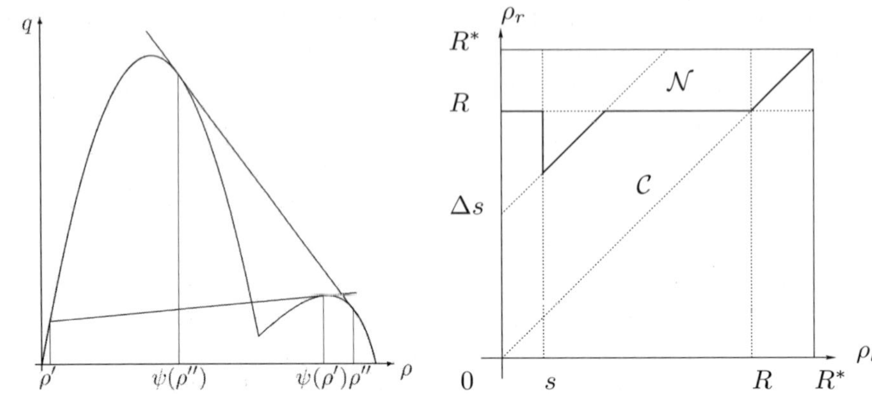

Fig. 2. Left, the function ψ: its geometrical meaning. Right, the Riemann Solver: in $\mathcal{C}$, the solution consists of classical waves only; in $\mathcal{N}$, also nonclassical shocks are present.

We are now ready to state the properties that allow to define $\mathcal{R}$[10,11].

(R.1) If $\rho^l, \rho^r \in [0, R]$, then $\mathcal{R}(\rho^l, \rho^r)$ selects the classical solution unless

$$\rho^l > s \quad \text{and} \quad \rho^r - \rho^l > \Delta s\,.$$

In this case, $\mathcal{R}(\rho^l, \rho^r)$ consists of a nonclassical shock between ρ^l and $\psi(\rho^l)$, followed by the classical solution between $\psi(\rho^l)$ and ρ^r.

(R.2) If $\rho^r < \rho^l$, then $\mathcal{R}(\rho^l, \rho^r)$ is the classical solution.

(R.3) If $R \leq \rho^l < \rho^r$ or $\rho^l < R < \rho^r$ and the segment between $(\rho^l, q(\rho^l))$ and $(\rho^r, q(\rho^r))$ does not intersect $q = q(\rho)$, then the solution is a shock between ρ^l and ρ^r.

(R.4) If $\rho^l < R < \rho^r$ and the segment between $(\rho^l, q(\rho^l))$ and $(\rho^r, q(\rho^r))$ intersects $q = q(\rho)$, then $\mathcal{R}(\rho^l, \rho^r)$ consists of a nonclassical shock between ρ^l and a panic state followed by a possibly null classical wave. More precisely,

$\rho^r \in \left]R, \psi(\rho^l)\right[$**:** $\mathcal{R}(\rho^l, \rho^r)$ consists of a nonclassical shock between ρ^l and $\psi(\rho^l)$, followed by a decreasing rarefaction between $\psi(\rho^l)$ and ρ^r;

$\rho^r \in \left[\psi(\rho^l), R^*\right[$**:** $\mathcal{R}(\rho^l, \rho^r)$ consists of a single nonclassical shock.

The Riemann Solver defined above is represented in Figure 2, right. According to [10, Theorem 2.1], there exists a unique Riemann Solver defined by (R.1)–(R.4) which is *consistent* and $\mathbf{L}^1_{\text{loc}}$ continuous in $\mathring{\mathcal{C}}$ and $\mathring{\mathcal{N}}$, see Figure 2, right. Recall that *consistency* is a necessary condition for the $\mathbf{L}^1$

stability of the Cauchy Problem for (1). Refer to [10, I and II] or [8, Definition 2.2] for its precise statement.

In the next theorem, we refer to [11,24] for the precise statement of the conditions on the flow function. Here, it is sufficient to recall that both diagrams in Figure 1 do satisfy to all of them.

Theorem 2.2. *[11, Theorem 3.4] Let f satisfy [10, **(Q.1)**–**(Q.9)**]. Let s, Δs satisfy (6) and assume that there exists a constant W satisfying*

$$W > 1 \quad \text{and} \quad \frac{\phi(0)}{\psi(0)} \leq \frac{W+1}{2W} \leq \frac{\Delta s}{\psi(s) - s}. \tag{7}$$

Then, for any initial datum $\bar{\rho} \in (\mathbf{L^1} \cap \mathbf{BV})\big(\mathbb{R}; [0, R^*]\big)$, the Cauchy Problem for (1) admits a nonclassical weak solution $\rho = \rho(t,x)$ generated by the nonclassical Riemann Solver $\mathcal{R}$ and defined for all $t \in \mathbb{R}^+$. Moreover:

$$\mathrm{TV}\big(\rho(t)\big) \leq W \cdot \mathrm{TV}(\bar{\rho}), \qquad \text{for all } t \in \mathbb{R}^+. \tag{8}$$

3. Passing through a Toll Gate

This section is devoted to some numerical integrations of (3). Our aim is only to show that this model features reasonable qualitative properties, hence we choose *normalized* parameters, leaving to future works the quantitative agreement with experimental data.

Let the real interval $[0, 2]$ describe a segment of a highway with a toll gate at its center $x = 1$, as in Figure 3 The evolution of traffic is described

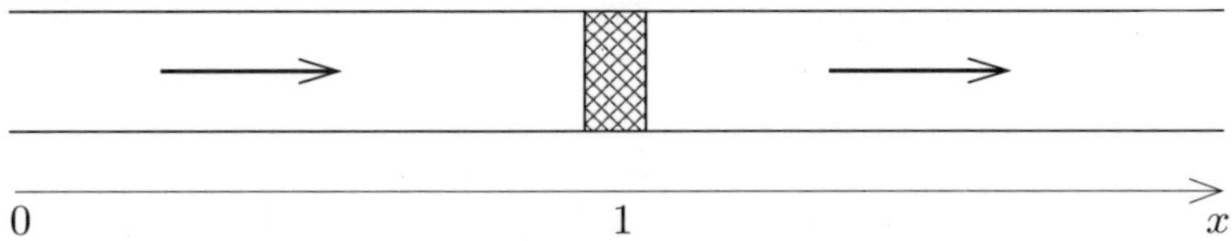

Fig. 3. A toll gate sited at $x = 1$.

by (3) with, for instance, $f(\rho) = \rho\,(1 - \rho)$. For simplicity, we assume that at time $t = 0$ vehicles are uniformly distributed between $x = 0.2$ and the site of the gate $x = 1$.

Assume that the initial density distribution is $\rho_o(x) = 0.3$ for $x \in [0.2, 1]$ and $\rho_o(x) = 0$ for $x \in [0, 2] \setminus [0.2, 1]$. The threshold of the through flow at the gate is 0.1. Then, the computed time necessary for all the vehicles to pass the toll gate is $t \approx 2.4$ and the evolution described by (3) is displayed

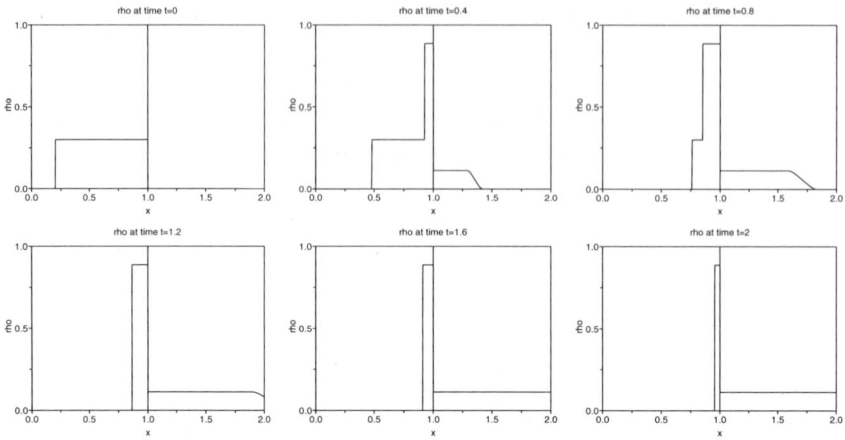

Fig. 4. Numerical integrations of (3) using Rusanov scheme, with $f(\rho) = \rho(1-\rho)$; $\rho_o(x) = 0.3$ for $x \in [0.2, 1]$, $\rho_o(x) = 0$ for $x \in \mathbb{R} \setminus [0.2, 1]$ and $q(t) = 0.1$. The constraint at $x = 1$ is treated as suggested in [1].

in Figure 4. As it has to be expected, the toll gate causes the rise of a queue to the left of the gate. This queue first increases an then decreases, finally disappearing when all vehicles passed the gate.

We now let the initial density ρ_o of vehicles and the efficiency of the gate q vary, while keeping the other parameters fixed. The time T that is necessary for all vehicles to pass the gate is then a function of ρ_o and q, that is $T = T(\rho_o, q)$. As it has to be expected, this function is monotone in both variables, see Figure 5. Note that as $q \to 0$, obviously, $T \to +\infty$. Hence, in

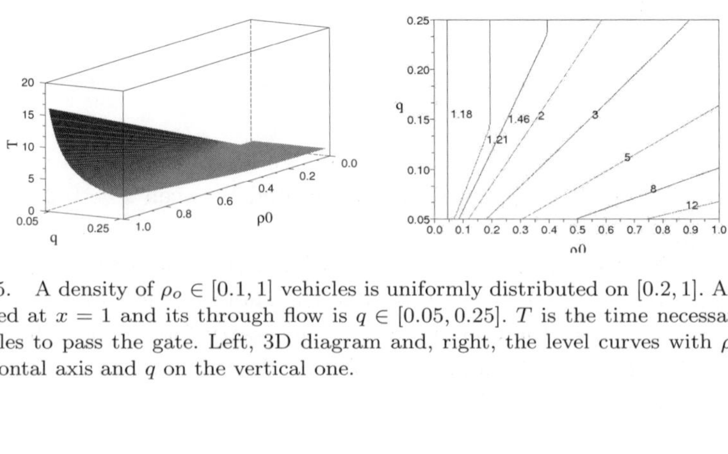

Fig. 5. A density of $\rho_o \in [0.1, 1]$ vehicles is uniformly distributed on $[0.2, 1]$. A toll gate is sited at $x = 1$ and its through flow is $q \in [0.05, 0.25]$. T is the time necessary for all vehicles to pass the gate. Left, 3D diagram and, right, the level curves with ρ_o on the horizontal axis and q on the vertical one.

Figure 5, T is computed only for $q \geq 0.04$. Note the vertical segments in the level curves of T in Figure 5, right. They realistically correspond to the gate being sufficiently efficient to avoid the rising of queues. On the contrary, as soon as the toll gate influences the traffic flow, T is well approximated by a function of the ratio ρ_o/q, as also dimensional considerations suggest.

4. Pedestrian Evacuation and Braess Paradox

Consider now a corridor with space coordinate, say, $x \in [0, L]$ with an exit at $x = D$, with $0 < D < L$. Then, the dynamics of the crowd exiting the corridor is described by (3), with the standard Riemann Solver substituted by that prescribed through (R.1)–(R.4), see [11, Section 4.1] or [24, Section 2]. In emergency situations, it is well known that the pressure of the people seeking to exit may dramatically reduce the door efficiency. To prevent this, often suitable obstacles, typically columns, are posed in front of the exit, at a suitable distance, to partially sustain the crowd pressure.

In fact, the presence of an obstacle may avoid the insurgence of panic among the people, therefore keeping the door efficiency at a higher level. Paradoxically, the insertion of this obstacle may reduce the evacuation time, although most individuals may have a slightly longer path to reach the exit. This remarkable behavior reminds of the Braess paradox[2] typical of networks and is captured by the present model[10,11,24].

Assume now that an obstacle is placed in the corridor above at, say, $x = d$, with $b < d < D$. A group of people is uniformly distributed on the

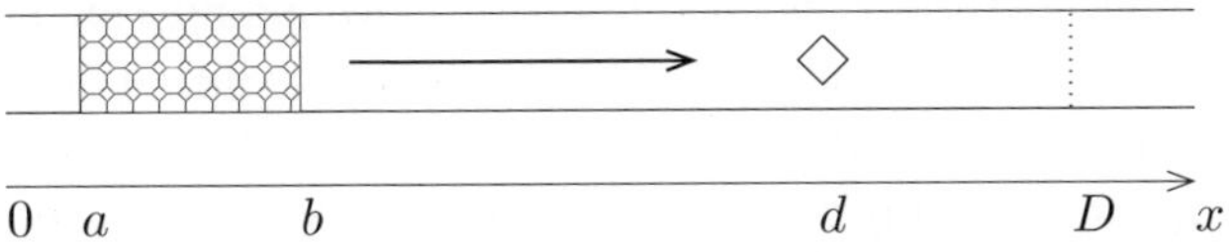

Fig. 6. A corridor with an obstacle before the exit.

segment $[a, b]$, with $0 < a < b < d$, see Figure 6. Following Section 2 above and [11, Section 4], one is thus lead to integrate

$$\begin{cases} \partial_t \rho + \partial_x f(\rho) = 0 & f\left(\rho(t, d-)\right) \leq q\left(\rho(t, d-)\right) \\ \rho(0, x) = \rho_o(x) & f\left(\rho(t, D-)\right) \leq Q\left(\rho(t, D-)\right) \end{cases} \tag{9}$$

with the evolution prescribed by the Riemann Solver defined by (R.1)–(R.4). The result is shown in Figure 7. As is well known, the efficiency of the exit may well be dramatically reduced when the crowd is panicking.

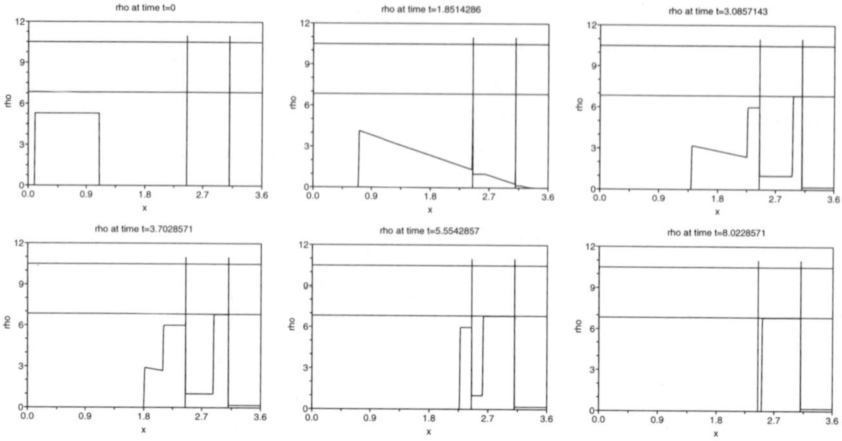

Fig. 7. Numerical integrations of (9) using the scheme in [4]. The vertical segments denote the positions of the obstacle and of the exit. The horizontal segments denote the maximal non-panic density R and the overall maximal density R^*. Here: $a = 0.1$, $b = 1.1$, $d = 2.45$, $D = 3.1$ and $L = 3.6$. The constraints at $x = d, D$ are treated as in [1].

Therefore, we assume that

$$\begin{aligned} q(\rho) &= \begin{cases} \hat{q} & \text{if} \quad \rho \in [0, R] \\ \check{q} & \text{if} \quad \rho \in \,]R, R^*] \end{cases} \qquad \text{with} \qquad \hat{q} > \check{q}, \\ Q(\rho) &= \begin{cases} \hat{Q} & \text{if} \quad \rho \in [0, R] \\ \check{Q} & \text{if} \quad \rho \in \,]R, R^*] \end{cases} \qquad \text{with} \qquad \hat{Q} > \check{Q}. \end{aligned} \tag{10}$$

The evacuation time T is particularly relevant and can be computed integrating (9)–(10) numerically following the procedure used in Section 3. Having a simple initial datum, i.e. uniformly distributed on a given segment, an analytical study is also possible. Indeed, the wave front tracking technique applied to (9)–(10) yields Figure 8, right. In Figure 8, left, the same problem (9)–(10) is considered, but the obstacle is removed, i.e. $q > \max\big(\rho\, v(\rho)\big)$. Remarkably, in this particular situation, the evacuation time with no obstacle is larger than the evacuation time with the obstacle. The detailed construction of these solutions can be found in [11, Section 4.2]. Note that the darker regions in Figure 8, left, represent where the crowd density attains panic values, i.e. $\rho \in \,]R, R^*]$. The presence of the obstacle avoids the density to reach these high values, thus allowing for a faster evacuation from the room. In particular, we get the diagram in Figure 9 for the evacuation time, T, as a function of the position of the obstacle, $d \in \,]b, D[$. Note that when the obstacle is too near to the exit, i.e. to

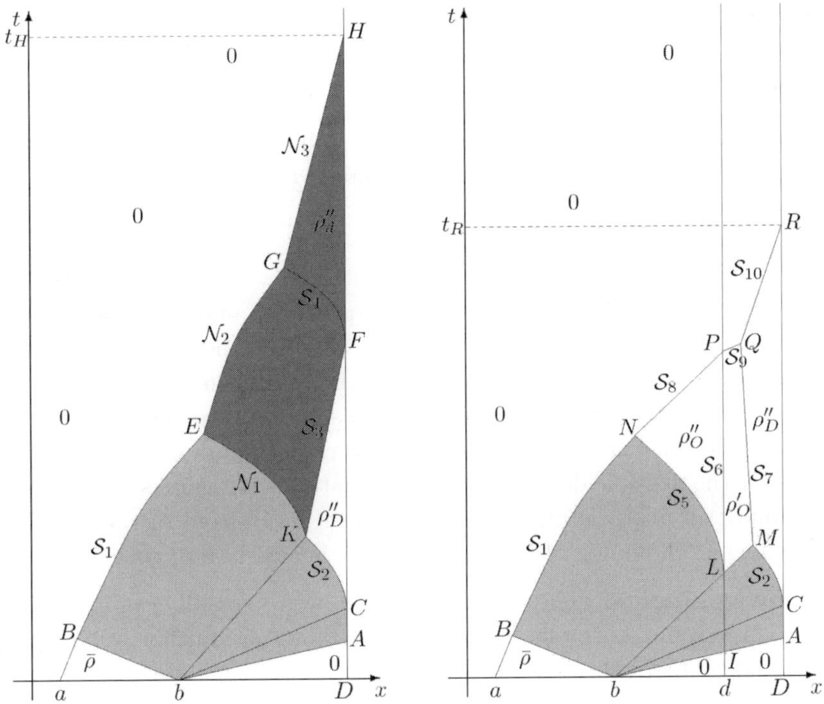

Fig. 8. Left, the structure of the solution to (9)–(10) with no obstacle; the evacuation time is t_H. Right, the case with the obstacle, the evacuation time is t_R.

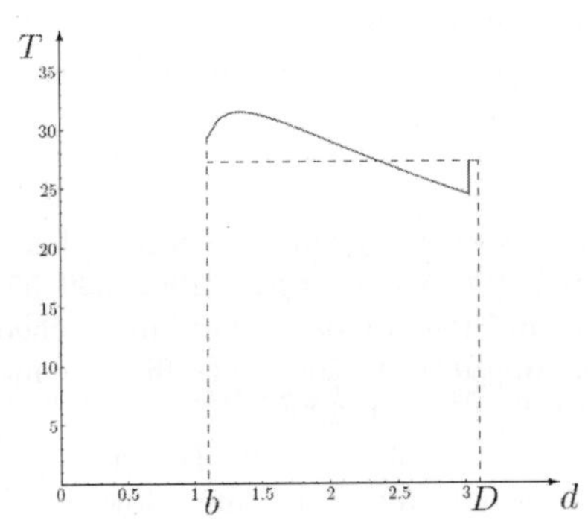

Fig. 9. The dotted line is the evacuation time with no obstacle. The solid line is the evacuation time, T, as a function of the position of the obstacle, d. When the obstacle is too near to the exit, its presence becomes negligible.

the right of the point K in Figure 8, left, its presence becomes negligible. Indeed, at K panic arises, while the efficiency of the obstacle is primarily dependent on avoiding it.

5. Conclusions

We presented some analytical results on unilateral constraints in scalar conservation laws. They can be used to model real situations, such as

(1) traffic flowing across a toll gate;
(2) crowd evacuating a corridor.

In both cases, numerical integrations are possible, allowing a detailed description of the phenomena. Furthermore, the times necessary for vehicles to cross the gate or for the people to exit the room can be computed.

Reasonable qualitative behaviors of the solutions are described. In particular, in the case of pedestrians exiting a room, the model presented accounts for the possible decrease in the evacuation time thanks to the careful insertion of an obstacle at a well chosen position in front of the exit. This phenomenon, an analog of Braess paradox[2], is clearly non generic.

Acknowledgments. The second author was partially supported by *I.N.R.I.A.* Sophia Antipolis - Méditerranée. The third author was partially supported by INdAM.

References

1. Boris Andreianov, Paola Goatin and Nicolas Seguin. Finite volume schemes for locally constrained conservation laws. In preparation.
2. Dietrich Braess. Über ein Paradoxon aus der Verkehrsplanung. *Unternehmensforschung*, 12:258–268, 1968.
3. Alberto Bressan, *Hyperbolic systems of conservation laws.* Oxford University Press, 2000.
4. Christophe Chalons. Numerical approximation of a macroscopic model of pedestrian flows. *SIAM J. Sci. Comput.*, 29(2):539–555 (electronic), 2007.
5. Christophe Chalons and Paola Goatin. Godunov scheme and sampling technique for computing phase transitions in traffic flow modeling. *Interfaces and Free Boundaries*, 10(2):195–219, 2008.
6. Christophe Chalons and Paola Goatin. Transport-equilibrium schemes for computing contact discontinuities in traffic flow modeling. *Commun. Math. Sci.*, 5(3):533–551, 2007.
7. Giuseppe M. Coclite and Nils H. Risebro. Conservation laws with time dependent discontinuous coefficients. *SIAM J. Math. Anal.*, 36(4):1293–1309 (electronic), 2005.
8. Rinaldo M. Colombo and Paola Goatin. A well posed conservation law with a variable unilateral constraint. *J. Differential Equations*, 234(2):654–675, 2007.

9. Rinaldo M. Colombo and Graziano Guerra On General Balance Laws with Boundary. Preprint, arXiv:0810.5246v1, 2008.
10. Rinaldo M. Colombo and Massimiliano D. Rosini. Pedestrian flows and non-classical shocks. *Math. Methods Appl. Sci.*, 28(13):1553–1567, 2005.
11. Rinaldo M. Colombo and Massimiliano D. Rosini. Existence of nonclassical solutions in a pedestrian flow model. *Nonlinear Analysis: Real World Applications*, 10, 2716-2728, 2009.
12. Constantine M. Dafermos, *Hyperbolic conservation laws in continuum physics.* Springer-Verlag, 2000.
13. François Dubois and Philippe Lefloch. Boundary conditions for nonlinear hyperbolic systems of conservation laws. *J. Differential Equations*, 71(1):93–122, 1988.
14. Dirk Helbing. A fluid-dynamic model for the movement of pedestrians. *Complex Systems*, 6(5):391–415, 1992.
15. Dirk Helbing, Illès Farkas and Tamàs Vicsek. Simulating dynamical features of escape panic. *Nature*, 407, September 28th 2000.
16. Dirk Helbing, Anders Johansson and Habib Zein Al-Abideen. Dynamics of crowd disasters: An empirical study. *Physical Review E*, 75(2), 2007.
17. Serge Hoogendoorn and Piet H. L. Bovy. Pedestrian route-choice and activity scheduling theory and mode. *Transp. Res. B*, 2002.
18. Serge Hoogendoorn and Piet H. L. Bovy. Simulation of pedestrian flows by optimal control and differential games. *Optimal Control Appl. Methods*, 24(3):153–172, 2003.
19. Roger L. Hughes. A continuum theory for the flow of pedestrians. *Transportation Research Part B*, 36:507–535, 2002.
20. Stanislav N. Kružhkov. First order quasilinear equations with several independent variables. *Mat. Sb. (N.S.)*, 81 (123):228–255, 1970.
21. Philippe G. Lefloch. *Hyperbolic systems of conservation laws.* Lectures in Mathematics ETH Zürich. Birkhäuser Verlag, Basel, 2002. The theory of classical and nonclassical shock waves.
22. Michael J. Lighthill and Gerald B. Whitham. On kinematic waves. II. A theory of traffic flow on long crowded roads. *Proc. Roy. Soc. London. Ser. A.*, 229:317–345, 1955.
23. Paul I. Richards. Shock waves on the highway. *Operations Res.*, 4:42–51, 1956.
24. Massimiliano D. Rosini. Nonclassical interactions portrait in a pedestrian flow model. *J. Differential Equations*, 246:408–427, 2009.
25. Blake Temple. Global solution of the Cauchy problem for a class of 2×2 nonstrictly hyperbolic conservation laws. *Adv. in Appl. Math.*, 3(3):335–375, 1982.

ON A MODEL FOR THE CODIFFUSION OF ISOTOPES

E. COMPARINI* and A. MANCINI**

Dipartimento di Matematica, Facoltà di Scienze M.F.N.
Università degli Studi di Firenze, Firenze, I-50134, Italy
**E-mail: elena.comparini@math.unifi.it, **E-mail: alberto.mancini@math.unifi.it*

C. PESCATORE

OECD/Nuclear Energy Agency I12 Blvd des Iles,
F-92130 Issy-les-Moulineaux, (France)
E-mail: claudio.pescatore@oecd.org

M. UGHI

Dipartimento di Matematica e Informatica,
Università di Trieste, Trieste, I-34127, Italy
E-mail: ughi@units.it

We consider a model for the distribution of radionuclides in the ground water around a deep repository for used nuclear fuel, based on the assumption that different isotopes of the same chemical element A contribute jointly to the chemical potential of A. In this hypothesis, the total flux J_i of a particular isotope A_i of an element A has two components. The corresponding problem consists in a parabolic system strongly coupled. In the physically relevant assumption that one of these components is negligible, the model reduces to a parabolic equation for the total concentration of the element A, possibly coupled with hyperbolic equations for the concentrations of the single isotopes.

Keywords: Isotopes; parabolic and hyperbolic differential equations.

1. Introduction

Most chemical elements in Nature consist of more than one isotopic component and have at least one radioactive isotope. Many of them (particularly, hydrogen, carbon, oxygen and sulphur) are very important in terms of their participation and abundance in rock-forming, ore-forming, water-rock interaction and life-processes. Radioactive isotopes are also widely utilized in a range of applications in medicine and technology, like e.g. tracer diffusion by using radioactively-labeled molecules.

In many studies about isotope composition, it is often assumed that the isotopes ratios are constant, at the so called "secular equilibrium", which has proved "to be a rule with many exceptions" (see Ref. 1). However a recent review of the analysis of isotopes in long-term experiments in bedrock and buffer materials (see Ref. 15) calls for increased attention to changes in the original isotope composition, an effect termed as "fractionation". This phenomenon may occur in the vicinity of breached radioactive waste canisters, for example deep repositories for used nuclear fuel, in a geological setting. Typically, the breached canisters would be in an environment (the buffer) that would still preclude water movement, however transport of those species occurs also in stagnant water due e.g. to diffusion, one of the processes that may lead to "fractionation". Moreover there are examples of discordance between predictions and experiments in the co-diffusion of isotopic molecules (see Refs. 9, 10).

Some studies Ref. 15 have pointed out the existence, next to fractures surfaces, of a narrow zone where the U^{234}/U^{238} activity ratio (roughly speaking the ratio of the concentrations of the two isotopes) are significantly different from secular equilibrium. Moreover it was noticed that, in many cases, while the profile of the total Uranium concentration can be readily rationalized in terms of classical diffusion (Fick's law), being reasonably smooth, the activity ratio U^{234}/U^{238} tends to be very irregular, showing an uncharacteristic succession of areas of varying enrichment or depletion of one of the two isotopes versus the other. These data are quite striking, since the two isotopes are the same chemical specie and even if a jugged activity profile were created at some time, diffusion should tend to mix the two radionucleides eliminating the areas of strong variation.

Recently Pescatore[20] has proposed a new model whose main point is that different isotopes of the same chemical element A should not be considered as distinct chemical species and should contribute jointly to the chemical potential of A, that is all isotopes of the same family do not diffuse separately from one another and hence it is the gradient in the total concentration of the isotopic family that drives the diffusion process.

Precisely this new model proposes that the total flux J_i of a particular isotope of an element A has two components, one representing the effects of the interaction of A_i with the solvent molecules B ($A_i - B$ interactions), the other the interaction with the kin isotopes ($A - A$ interactions). The former component obeys to classical Fick's law, while the latter depends also on the total concentrations of the A-molecules. Namely:

$$A - B\,\text{interaction} \longrightarrow \text{flux} = -\tilde{D}_i \nabla c_i, \tag{1}$$

$$A - A\,\text{interaction} \longrightarrow \text{flux} = -D_i \frac{c_i}{c} \nabla c, \tag{2}$$

where c_i is the concentration of A_i, c is the total concentration of A, $\tilde{D}_i$ is the diffusivity of A in the solvent B, D_i is a measure of the mobility of the A_i molecules due to the $A - A$ interactions within the solvent B.[3]

Intuitively, just because the different isotopes are indistinguishble, the total flux of A obeys Fick's law, as usual, while the portion of total flux costituted by the A_i isotopes is the local ratio of the concentration of A_i to the total, at least as a first approximation.

It has to be noted that the specialization of relation (2) to a single isotope solute reduces to a Fick's law expression, where the diffusion coefficient is the sum of $\tilde{D}_i + D_i$. In some sense equation (2) completes Fick's law but adds a sensitivity to the whole concentration of A and its gradient.

It would be reasonable, for solutes, that the coefficients $\tilde{D}_i$ are pratically the same for all isotopic molecules of the element A, as they have the same partial molar volume and the same electronic configuration. The same may be true also for the coefficients D_i, especially for the heavier chemical elements.

As relations (1)-(2) reduce to Fick's law in the case of a single isotope solute (where the diffusion coefficient is $\tilde{D}_i + D_i$), it is not clear at present whether the diffusion coefficient that is measured classically is dominated by the term $\tilde{D}_i$ or D_i. If the latter is the case, then there might be important non-linear effects, especially in the near field of radioactive waste repository, where the diffusion regime is created and elemental concentrations are highest. In any event, although it is not clear if the new term may or may not make the difference in any circumstances, this term is important in some problems, among which let us mention the case of self-diffusion, meaning by this term the diffusion of a tagged solute in an untagged but otherwise chemically identical solvent; in this case $\tilde{D}_i$ must be taken zero.

The issue of the relevance of the term (2) is thus, by itself, an important problem. Finally the reasoning provided for getting (2) seems to be sufficiently general to apply to any non-equilibrium situation.

There are in literature various problems in which the diffusion of one specie out of a family depends on the total gradient of the whole family. Namely some population dynamic problem, see Refs. 2, 12, 13, 17 and tumor growth problems, see Refs. 8, 1. The modeling of these problems leads to mathematical equations similar but still very distinct, mathematically, to those of Pescatore's couple-diffusion model.

Pescatore's model consists in a parabolic system strongly coupled, whose mathematical analysis was the object of Ref. 7.

In Ref. 5 the problem obtained assuming all diffusion coefficient D_i to be equal and $\tilde{D}_i = 0$ has been studied . This simplifying assumption allowed us to reduce the system to a parabolic equation for the total concentration c possibly coupled with hyperbolic linear equations for the ratioes $r_i = \frac{c_i}{c}$.

We have taken into account here also the effect of "loss of regularity" that one has in the solution for the single components c_i, due to the possible presence of zeros in the initial total concentration, even if from a physical point of view the only case of interest is the one with positive initial concentration. This "loss of regularity" consists in the onset of regions depleted of a component c_i or in strong oscillations also asymptotically (say a "asymptotic localization property"). A similar result is in agreement with the observation.

Numerical simulations have been performed in order to describe this effects: in particular (see Ref. 6) we have tested some explicit examples.

2. Statement of the problem

The model proposed in Ref. 20 for the diffusion of n species of isotopes of the same element in a medium is based on the assumption that the flux of the i component J_i is given by

$$J_i = -\left(\tilde{D}_i \nabla c_i + D_i \frac{c_i}{c} \nabla c\right), \quad i = 1, ..., n, \quad c = \sum_{i=1}^{n} c_i. \tag{3}$$

In the case of radioactive isotopes, we have to take into account the radioactive decay law, which for spacially homogeneous distributions is a linear ODE system

$$\frac{d\underline{C}}{dt} = \Lambda \underline{C}, \qquad \underline{C} \in \mathbb{R}^n, \tag{4}$$

with Λ a suitable $n \times n$ constant matrix.

In Ref. 7 we proved the existence and uniqueness of classical solution of the resulting system in the general case of positive diffusion coefficients, possibly all different, with Dirichlet boundary conditions:

$$\frac{\partial c_i}{\partial t} = -\mathrm{div} J_i + \sum_{j=1}^{n} \Lambda_{ij} c_j, \; i = 1, ..., n, \; \text{in}\, \Omega \times (0, T), \tag{5}$$

J_i given by (3), Ω bounded domain of $\mathbb{R}^N$ with regular boundary $\partial\Omega$.

The result was proved in the physically relevant assumption that
$K \geq c_i \geq k > 0, \quad i = 1, ..., n, \quad k, K$ constant.

In the paper Ref. 5 we have studied some qualitative properties of the solution in the physically relevant assumption that the diffusion coefficients $\tilde{D}_i$ are much smaller than the D_i, thus showing the appearance of a "hyperbolic" behaviour for the c_i, quite interesting for the applications.

For n species the system (5) with $\tilde{D}_i = 0$ reduces to

$$c_{it} = \left(D_i \frac{c_i}{c} c_x\right)_x + \Sigma_j \Lambda_{ij} c_j, \; i, j = 1, ..., n, \; c = \Sigma_j c_j, \; \Lambda_{ij} \text{ constants}, \quad (6)$$

If we assume that $D_1 = \max D_j$, $D_n = \min D_j$, then summing up the above equations and considering that $c_n = c - \Sigma_j c_j$ we get the system

$$\begin{cases} c_{it} = \left(D_i \dfrac{c_i}{c} c_x\right)_x + \Sigma_{j=1}^{n-1} \beta_{ij} c_j + \beta_{in} c, & i, j = 1, ..., n-1 \\ c_t = (a(c, c_1, ..., c_{n-1}) c_x)_x + \Sigma_{j=1}^{n-1} \beta_{nj} c_j + \beta_{nn} c, & \end{cases} \quad (7)$$

where $a = D_n + \Sigma_{j=1}^{n-1} (D_j - D_n) \dfrac{c_i}{c}$ and β_{ij} are constants depending on Λ_{ij}.

In the physical assumption that $0 \leq c_i \leq c$, $i = 1, ..., n$, we have that the total concentration c satisfies a uniformly parabolic quasilinear equation in divergence form, since $0 < D_n \leq a \leq D_1$, $\forall c_i, c \geq 0$. Hence $c(x, t)$ has a "parabolic" behaviour while, once c is given, the equations for the single species c_i are first order linear equations, so that we expect for c_i a "hyperbolic" behaviour. In particular c_i will have finite speed of propagation and will in general be non smooth for $t > 0$; this makes the model promising from the point of view of accordance with experimental data.

The problem without radioactive decay (i.e. $\Lambda_{ij} \equiv 0$ and hence $\beta_{ij} \equiv 0$), with $D_i = 1$ $\forall i$, gives quite a good portrait of the qualitative behaviour of the solutions in general: after scaling on t, the system reduces to

$$\begin{cases} c_{it} = \left(\dfrac{c_i}{c} c_x\right)_x, & i = 1, ..., n-1 \\ c_t = c_{xx}. & \end{cases} \quad (8)$$

Therefore the diffusion of the total concentration c is governed by the classical heat equation and it is uniquely defined by the initial and boundary data. The system for the c_i is uncoupled, and each equation is a linear first order equation for c_i. We can have a similar situation also for $\Lambda_{ij} \neq 0$. Let us mention two examples

Example 1. All the species decay with almost the same coefficient λ. The decay law is $-\lambda c_i$ for any i (i.e. $\Lambda_{ij} = -\lambda \delta_{ij}$). We get system (8), where we replaced c_i, c with $c_i e^{\lambda t}, c e^{\lambda t}$.

Example 2. A triplet c_1, c_2, c_3 with decay law respectively $-\lambda_1 c_1,\ \lambda_1 c_1 - \lambda_2 c_2,\ \lambda_2 c_2$. In this case we get

$$\begin{cases} c_{1t} = \left(\dfrac{c_1}{c} c_x\right)_x - \lambda_1 c_1, \\ c_{2t} = \left(\dfrac{c_2}{c} c_x\right)_x + \lambda_1 c_1 - \lambda_2 c_2, \\ c_t = c_{xx}. \end{cases} \tag{9}$$

Also in this case c is recovered from the data on the parabolic boundary and c_1, c_2 are solutions of a first order hyperbolic linear system.

Systems similar to the one studied here have been considered by various authors in quite a number of different applications (see Refs. 2, 12, 13, 17) ranging from populations dynamics to tumor growth.

What seems striking here is that some qualitative behaviour like segregation for all time, irregularity of solutions, finite speed of propagation appear also if the equation for c is uniformly parabolic (and in the simplest case the heat equation) while similar behaviour is easier to understand in the case of the previous papers for which the total concentration obeys the porous media equation, whose solutions are not classical and have finite speed of propagation.

A common feature of this class of models is that the ratio $r_i = \frac{c_i}{c}$ has an evolution law simpler than the one of c_i. In the case of isotopes, r_i is related to the activity ratio which e.g. for two species is $\frac{c_1}{c_2}$, but it is simpler to deal with mathematically, since it is bounded by 1 (of course assuming $c_i \geq 0$).

For the simplest problem (8) the equations for r_i are

$$r_{it} = r_{ix} \frac{c_x}{c}, \qquad i = 1, ..., n-1, \tag{10}$$

which means that r_i are constant on the characteristics (the same is true for the general equation with no source). For the problem with decay (7) with $D_i \equiv 1$, $a = 1$) after scaling we have

$$r_{it} = r_{ix} \frac{c_x}{c} + P_i(r_1, ..r_{n-1}), \qquad i = 1, ..., n-1, \tag{11}$$

where P_i is a polinomial of 2 degree.

Then each r_i evolves along the characteristics with the same law as the spatially homogeneous solutions, which is known a priori (since one knows the solution of the ODE system $\dot{\underline{C}} = \Lambda \underline{C}$, $\Lambda = (\Lambda_{ij})$, for any initial data $\underline{C}(0)$. Again let us remark that the same is true also for more general equations and it is essential in obtaining existence and uniqueness results.

Example 3 A couple (U^{238}, U^{234}) for which the decay law is respectively

$$\dot{c}_1 = -\lambda_1 c_1, \quad \dot{c}_2 = \lambda_1 c_1 - \lambda_2 c_2, \qquad \lambda_1 << \lambda_2. \tag{12}$$

We obtain (11) with $n = 2$, $P_1 = \lambda_2 r_1 (r_E - r_1)$, $r_E = \dfrac{\lambda_2 - \lambda_1}{\lambda_2}$.
This implies that along the characteristics:

$$\begin{aligned}
&r_1(0) = 0 \quad \rightarrow \quad r_1(t) \equiv 0, \quad t > 0 \\
&0 < r_1(0) \le 1 \quad \rightarrow \quad r_1(t) = \frac{r_E r_1(0)\exp(\lambda_2 - \lambda_1)t}{r_E - r_1(0) + r_1(0)\exp(\lambda_2 - \lambda_1)t}, \\
&r_1(t) \rightarrow r_E \quad \text{as} \quad t \rightarrow \infty,
\end{aligned} \tag{13}$$

in accordance with the physical fact that for the couple (U^{238}, U^{234}) has a "secular equilibrium" positive and attractive (i.e. normally the two isotopes are found in a precise positive ratio).

3. Previous results

In Ref. 5 we considered various intial-boundary value problems, namely Cauchy Problem, Neumann Problem (especially with flux null at the boundary i.e. isolated domain) and Dirichlet Problem, possibly allowing c_0 to reach zero. The initial-boundary data are either c_i, $i = 1, ..., n$, with $C > c_i \ge 0, \tilde{C} \ge c \ge 0$, or their fluxes $\frac{c_i}{c} c_x$ and from this we recover the data for either $c = \Sigma c_i$ or $c_x = \Sigma \frac{c_i}{c} c_x$. For the Neumann Problem the natural assumptions on the data are that the fluxes are such that $c \ge 0$ and bounded everywhere (in particular null flux is ok). Then from the strong maximum principle we have that c is bounded and strictly positive in $\Omega \times (0, T)$, Ω open. Moreover, due to the regularizing effect of uniformly parabolic equations, c is smooth in $\Omega \times (0, T)$. Let us remark that in the cited models for Porous Media equation (e.g. Ref. 2), c can be zero in $\Omega \times (0, T)$.

Therefore for any $(x_0, t_0) \in \Omega \times (0, T)$ fixed, we define the characteristic $X(t; x_0, t_0)$, starting in (x_0, t_0), whose equation is (see (10))

$$\frac{dX(t; x_0, t_0)}{dt} = -\frac{c_x}{c}\bigg|_{x = X(t; x_0, t_0)}, \quad X(t_0; x_0, t_0) = x_0, \tag{14}$$

with $r_i(X(t; x_0, t_0), t) = r_i(x_0, t_0)$, $\quad i = 1, ..., n - 1$.

Remark that, since $c(x_0, t_0) > 0$, $t = t_0$ is not a characteristic itself, and that, when the initial boundary data are regular and c is strictly positive, we can take $t_0 = 0$.

In a bounded domain $\Omega = (-L, L)$ in general there will be a set $\Omega_1(t_0) = (-L, l_1)$ such that the characteristics starting from (x_0, t_0) will reach the lateral boundary $x = -L$, a set $\Omega_2(t_0) = (l_1, l_2)$ such that they go to the initial set $\Omega \times \{t = 0\}$, and a set $\Omega_3(t_0) = (l_2, L)$ such that they go to $x = L$.

Let us remark that there are problems for which we know a priori that $\Omega_1 = \Omega_2 = \emptyset$, such are the homogeneous Neumann Problem (i.e. isolated domain for which the two lateral boundaries are itself two characteristics), the Neumann Problem with outgoing flux (i.e. $c_x(-L,t) \geq 0$, $c_x(L,t) \leq 0$), the homogeneous Dirichlet Problem (in fact $c = 0$ is a minimum for c hence from the strong maximum principle $c_x(-L,t) > 0,\ c_x(L,t) < 0$).

Of course for the Cauchy Problem in the entire space all the characteristics arrive to the initial set $t = 0$.

A main problem arises if the data for the total concentration c can be equal zero, especially from the mathematical point of view. Since $c = 0$ is a minimum, no characteristic can arrive to a point of the boundary $x = -L$ or $x = L$ where $c = 0$ from the interior of the domain. So we are left to study the cases in which the initial datum $c(x,0) = c_0(x)$ has zeros.

Let us recall some results obtained in Ref. 5.

1. Assume that two characteristics, denoted by $X_1(t), X_2(t)$ with $X_1(t_0) < X_2(t_0)$, starting from $t_0 > 0$, reach $t = 0$, then:

$$X_1(0) < X_2(0) \text{ and } c_0(x) \not\equiv 0, \qquad x \in (X_1(0), X_2(0)).$$

In other words if $c_0(x) \equiv 0$, $x \in I$, I interval of Ω, then there cannot be two distinct characteristics ending in I. This fact, which makes the present problem quite different from the one of Ref. 2, is due to the "infinite speed of propagation" of the total concentration (solution of a uniformly parabolic equation) i.e. to the fact that $c(x,t) > 0,\ \forall t > 0,\ x \in \Omega$.

2. Let us assume that

$$c_0(x) \equiv 0, \quad x \in I_0 = (a,b) \subset \Omega,$$

$$c_0(x) > 0, \quad x \in I = (a-\delta, b+\delta) - \overline{I}_0, \quad \delta > 0 \text{ such that } I \subset \Omega,$$

then there exists a curve $x = s(t)$ separating two regions C_- and C_+, respectively the sets of the characteristics from $(a-\delta, a)$, and from $(b, b+\delta)$.

3. Let us assume that $a \in \Omega = (-L, L)$, and that

$$c_0(x) > 0, \quad x \in [-L, a], \qquad c_0(x) \equiv 0, \quad x \in (a, L],$$

then $\Omega_2(t) \to (-L, a)$ as $t \to 0$.

4. Assume $c_0 \equiv 0$ in $\Omega = (-L, L)$, then Ω_2 is at most one curve, and

- if the boundary data are such that both Ω_1 and Ω_3 are not empty (e.g. incoming flux at both boundaries), then Ω_2 is precisely a line;
- if either Ω_1 or Ω_3 is empty (e.g. incoming flux from only one boundary) then Ω_2 is empty.

5. Concerning the behaviour of each component c_i in corrispondance with the data, since $r_i = \dfrac{c_i}{c}$ is constant along the characteristics, and hence the value of $c_i(x,t)$ inside the domain can be recovered from the initial and boundary data, we obtained that the oscillations of the initial and boundary data persist in the interior of the domain, which is consistent with the experimental results.

In particular we emphasized the possible existence of interior regions depleted of c_i. In few words we proved that the "holes" of c_i remain in time only if they do not coincide with the "holes" of c_0. This result differs from the results of Ref. 2 because of the "infinite speed of propagation" of c.

Indeed, if $c_i(x,0) \equiv 0$ in $I = (a,b) \subset \Omega$ and $c_0(x) \not\equiv 0$ in I, then, for any time t there exists an interval $\tilde{I}$ where $c_i(x,t) \equiv 0$ for $x \in \tilde{I}$.

On the contrary, suppose that $\Omega_1 = \Omega_3 = \emptyset$ and that supp $c_i(x,0) \equiv$ supp $c_0(x)$, then $c_i(x,t) > 0$ a.e. for any $t > 0$.

6. Concerning the behaviour of c_i when it has an initial compact support, we proved that if

$c_0(x) > 0$ in $\Omega \equiv \mathbb{R}$, $c_i(x,0) > 0$ in $I = (a,b) \subset \Omega$, $c_i(x,0) = 0$ in $\Omega - I$, then there exist two curves, $s_1(t) < s_2(t)$, with $s_1(0) = a$, $s_2(0) = b$, such that for any t

$$c_i(x,t) > 0 \text{ in } (s_1(t),\, s_2(t)), \quad c_i(x,t) = 0 \text{ in } \mathbb{R} \setminus (s_1(t),\, s_2(t)).$$

This result is proved in Theorem 3.1 of Ref. 5 in the case of Cauchy problem, but the same results hold for any problem such that the data guarantee that the characteristics start from $t = 0$, that is $\Omega \equiv \Omega_2$.

7. Concerning the asymptotic behaviour we will show in a forthcoming paper that, while the asymptotics of the total concentration c depends only on the total mass $M = \int_\Omega c_0(x)\,dx$ but not on the pointwise shape of $c_0(x)$, the asymptotic of c_i strongly depends on the pointwise value of $\dfrac{c_{i0}}{c_0}$. Hence there can be regions depleted of a component c_i, or strong oscillations also asymptotically (say a "asymptotic localization property"). A similar result is obtained in Ref. 2, but it seems interesting here with c solution of the heat equation and it is in agreement with the observation.

8. In the case with decay ($\Lambda \neq 0$) c will depend on c_i but the characteristics for each c_i are defined as before by (14), see (11). Then we can say similar things as before, provided we assume reasonably:

(H1) $\exists$ solution of $\dot{\underline{u}} = \Lambda \underline{u}$, $u \in R^n$ for any initial datum $\underline{u}_0$;

(H2) if $u_{i0} \geq 0$, then $u_i(t) \geq 0$, $i = 1, ..., n$ (positive property).

Therefore, if $0 \leq r_{i0} \leq 1$ then $0 \leq r_i(X(t),t) \leq 1$, and $c(x,t)$ satisfies the following equation (recall (7)) in $\overline{\Omega} \times [0,T]$

$$c_t = c_{xx} + c(\beta_{nn} + \Sigma_i^{n-1}\beta_{ni}r_i) = c_{xx} + b(x,t)c, \quad |b(x,t)| < B. \tag{15}$$

From the classical theory we get then explicit bounds for c.

As for the behaviour of the "holes" of c_i, we have similar results as in the case without decay iff the elements of Λ satisfy the following assumption: **(H3)** if $u_{i0} = 0$ and $u_{j0} \geq 0$, $u_{j0}(x) \neq 0 \quad j \neq i$, then $u_i(t) = 0$ for any t.

Remark 3.1.

In order to make the situation clearer, let us consider the Examples 2 and 3 of Section 2. There the assumption (H3) holds for $i = 1$, but not for $i \neq 1$. Hence, if we have initially an interval where $c_1 = 0$, $c_2 > 0$, we will have for any time a region where $c_1 = 0$.

Consider the Cauchy Problem for the equations of Example 3, and suppose that initially c_1 and c_2 are separated, i.e.

$$c_0(x) > 0, \quad c_{10}(x) = c_0(x)H(x), \text{ where } H(x) \text{ is the Heaviside function.}$$

Then there exists a free boundary $x = s(t)$, that is the characteristic starting at $x = 0, t = 0$, such that

$$r = \frac{c_1}{c} = \tilde{r}H(x - s(t)), \ \tilde{r}(t) = \frac{r_E \exp(\lambda_2 - \lambda_1)t}{r_E - 1 + \exp(\lambda_2 - \lambda_1)t}, \ r_E = \frac{\lambda_2 - \lambda_1}{\lambda_2}.$$

Once c is given (c, c_x continuous), we will have

$$c_1(x,t) = \tilde{r}(t)c(x,t)H(x - s(t)), \quad c_2(x,t) = c(x,t) - c_1(x,t),$$

with c_1 discontinuous across $x = s(t)$ and zero for $x < s(t)$.

Let us remark that if we drop the assumption of c_0 positive, e.g. we take $c_0(x) = 0$, $x < 0$ so that $c_{10}(x) = c_0(x)$, $x \in R$, then we will have instead

$$c_1(x,t) = \tilde{r}c(x,t), \quad c_2(x,t) = (1 - \tilde{r}(t))c(x,t), \quad x \in R, \ t \geq 0,$$

and $c(x,t)$ is the known solution of

$$c_t = c_{xx} - \lambda_2 c(1 - \tilde{r}(t)), \quad \text{in } R \times (0,t),$$

which can be calculated explicitely and is positive for $t > 0$.

Therefore c_1 will be strictly positive everywhere.

4. Regularity and weak solutions

Let us remark that in general we cannot expect classical solutions not only because this is a typical feature of hyperbolic first order equations (i.e. non smooth initial data give no smooth solutions, i.e. no parabolic effect) but also because for our problem it can happen that also if the data are C^∞ the solution has discontinuities. To be clear let us consider the following example for the simplest case (8):

Example 4 Consider the Cauchy problem for (8) with c_0 symmetric w.r.to $x = 0$, $c_0 \in C^\infty$, $c_0(x) \equiv 0$, $x \in I = [-L, L]$, $c_0(x) > 0$, $x \in \mathcal{R}\backslash I$, and

$$c_{i0}(x) = \begin{cases} \gamma_{1i} c_0(x), & x \leq -L \\ 0, & x \in I, \\ \gamma_{2i} c_0(x), & x \geq L, \end{cases}$$

for a given i, with $\gamma_{1i} \not\equiv \gamma_{2i}$ constants in $[0, 1]$. Hence $c_{i0}(x) \in C^\infty$.
For $t > 0$ we have the explicit solution

$$c_i(x,t) = \begin{cases} \gamma_{1i} c(x,t), & x < 0, \\ \gamma_{2i} c(x,t), & x > 0, \end{cases} \qquad \gamma[c_i^+ - c_i^-] = (\gamma_{2i} - \gamma_{1i}) c(0,t) \neq 0.$$

Note that c_i has a jump $\forall t > 0$ in $x = 0$ increasing in time.
We remark that a similar behaviour can be showed if c_0 is not symmetric and in the case with $\Lambda \neq 0$.

When we have a problem in a bounded domain then the question arises of understanding what means "assuming c_i on the lateral boundary", since c_i are solutions of first order equations. This can be done in a similar way as in Refs. 4, 21, 5.

Remark 4.1.

Let us remark that if the total concentration is strictly positive, we can apply to equation (8) the result of Ref. 4, thus showing that the solution constructed along the characteristics is the "viscosity solution" obtained as the limit of the complete physical model (1)-(2), with $\tilde{D}_i = \tilde{D} \neq 0$, $D_i = D = 1$ as $\tilde{D} \to 0$. The numerical simulations all confirm the convergence also if c_0 is allowed to become zero.

Therefore one can approximate the parabolic problem $\tilde{D} \neq 0$ with the hyperbolic one $\tilde{D} = 0$. In the limit, boundary layers will appear (see again Ref. 4), whose asymptotic dimension will be studied in a forthcoming paper.

References

1. D. Ambrosi, L. Preziosi, On the closure of mass balance models for tumor growth, *Math.Models Methods Appl. Sci.* **12** (2002) 737–754.
2. M. Bertsch, M.E. Gurtin, D. Hilhorst, On the interacting populations that disperse to avoid crowding: the case of equal dispersal velocities, *Nonlinear Anal.Th.Meth.Appl.* vol II, **4** (1987) 493–499.
3. H.F. Bremer, E.l. Cussler, Diffusion in the Ternary System d-Tartaric Acid, c-Tartaric Acid, Water at 25°C, *AIChE Journal* **16**, 9 (1980) 832–838.
4. C. Bardos, A.Y. Leroux, J.C. Nedelec, First order quasilinear equations with boundary conditions, *Comm. In PDE* **4**, 9 (1979) 1017–1034.

5. E. Comparini, R. Dal Passo, C. Pescatore, M. Ughi *On a model for the propagation of isotopic disequilibrium by diffusion Math.Models Methods Appl. Sci.* **19**, 8 (2009).
6. E. Comparini, A. Mancini, C. Pescatore, M. Ughi, Numerical results for the Codiffuson of Isotopes, Communications to SIMAI Congress (2009).
7. E. Comparini, C. Pescatore, M. Ughi On a quasilinear parabolic system modelling the diffusion of radioactive isotopes, *RIMUT* XXXIX (2007) 127–140.
8. M.A.J. Chaplain, L. Graziano, L. Preziosi, Mathematical modelling of the loss of tissue compression responsiveness and its role in solid tumor development, *Math.Medicine Biol.* **23** (2006) 197–229.
9. Rock matrix diffusion as a mechanism for radionuclide retardation: natural radioelement migration in relation to the microfractography and petrophysics of fractured crystalline rock, Report EUR 15977 EN, European Commission, Brussels, (1995).
10. Rock matrix diffusion as a mechanism for radionuclide retardation: natural radioelement migration in relation to the microfractography and petrophysics of fractured crystalline rock, Report EUR 17121 EN, European Commission, Brussels, (1997).
11. T. Gimmi, H.N. Waber, A. Gautschi, A. Rübel, Stable water isotopes in pore water of Jurassic argillaceous rocks as tracers for solute transport over large spatial and temporal scales, *WATER RESOURCES RESEARCH* **43** (2007).
12. G.E. Hernandez, Existence of solutions in a population dynamics problem, *Quarterly of Appl. Math.* vol. XLIII, 4 (1986) 509–521.
13. G.E. Hernandez, Localization of age-dependent anti-crowding populations, *Quarterly of Appl. Math.* vol. LIII 1 (1995) 35–52.
14. F.A. Howes, Multidimensional initial-boundary value problems with strong nonlinearities, *Archiv. Math.* **91**, 2 (1986) 153–168.
15. *KASAM Nuclear Waste state of the art reports 2004*, Swedish Government Official Reports SOU 2004-67 (2005).
16. A.G. Latham, Diffusion-sorption Modelling of Natural U in Weathered Granite Fractures: Potential Problems, *Proceedings of Migration '93* (1993) 701–710
17. R.C. MacCamy, A population model with nonlinear diffusion, *J.Diff.Eq.* **39** (1981) 52–72.
18. E.C. Percy, J.D. Prikryl, B.W. Leslie, Uranium transport through fractured silicic tuff and relative retention in areas with distinct fracture characteristics, *Applied Geochemistry* **10** (1995) 685–704.
19. A.J. Perunpanami, J.A. Sherrat, J. Norbury, Mathematical modelling of capsule formation and multinodularity in benign tumor growth, *Nonlinearity* **10** (1997) 1599–1614.
20. C. Pescatore, Discordance in understanding of isotope solute diffusion and elements for resolution, in Proc. OECD/NEA "Radionuclide retention in geological media", Oskarsham, Sweden, 7-9 May 2001, OECD Publishing, Paris (2002) 247-255.
21. A. Terracina, Comparison properties for scalar conservation laws with boundary conditions, *Nonlinear Anal. Th. Meth. Appl.* **28**, 4 (1997) 633–653.

Modified Collocation Techniques for Volterra Integral Equations

Dajana Conte[1], Raffaele D'Ambrosio[1], Maria Ferro [2], Beatrice Paternoster [1]

[1] *Dipartimento di Matematica e Informatica , Facoltà di Scienze MM.FF.NN.*
Università degli Studi di Salerno, Italy
dajconte@unisa.it, rdambrosio@unisa.it, beapat@unisa.it

[2] *Dipartimento di Matematica e Applicazioni "R. Caccioppoli"*
Università di Napoli "Federico II", Italy
maria.ferro@dma.unina.it

The aim of this paper is the analysis of some new modified collocation based numerical methods for solving Volterra Integral Equations (VIEs), which turn out to be at the heart of many modern applications of Mathematics to natural phenomena and are used more and more for the description of complex systems, in particular evolutionary problems with memory. The developed methods have strong stability properties and higher order of convergence than the classical one-step collocation methods, without any increase of the computational cost, which is an important request in order to approach real problems.

Keywords: Volterra Integral equations, two-step collocation methods, order conditions, A-stability.

1. Introduction

In this paper we analyze the construction of high order, highly stable new two-step collocation methods for Volterra Integral Equations (VIEs) of the form

$$y(t) = g(t) + \int_0^t K(t, \eta, y(\eta))d\eta, \quad t \in I, \tag{1}$$

with $I \subseteq \mathbb{R}_+$, $K \in C(D \times \mathbb{R})$, $D = \{(t, \eta) : 0 \leq \eta \leq t \leq T\}$, K satisfying the uniform Lipschitz condition with respect to the third variable and $g \in C(I)$. VIEs are models of evolutionary problems with memory arising in many applications. In fact, the spread of diseases, the growth of biologic populations, the brain dynamics, elasticity and plasticity, wave problems, heat conduction, fluid dynamics, scattering theory, sismology, biomechanics, game theory, control, queuing theory, design of electronic filters and many other problems from physics, chemistry, pharmacology, medicine, economics can be modelled through systems of VIEs, Ref. 1–6. The following books and survey papers contain sections with various applications of VIEs in the physical and biological sciences and also include extensive lists of references: Brunner, Ref. 7,8, Agarwal and O'Regan, Ref. 9, Corduneanu and Sandberg, Ref. 10, Zhao, Ref. 11. Due to the high variety of applications, it gets more and more important to develop efficient numerical methods in order to solve these problems and make some special requirements on these methods, such as high order and strong stability properties.

In the literature many authors (see Ref. 7,8 and references therein contained) have analyzed one-step collocation methods for VIEs. As it is well known, a collocation method is based on the idea of approximating the exact solution of a given integral equation with a suitable function belonging to a chosen finite dimensional space, usually a piecewise algebraic polynomial, which satisfies the integral equation exactly on a certain subset of the integration interval (called the set of collocation points). As done in Ref. 12 for Ordinary Differential Equations (ODEs), in the papers Ref. 13,14 we derived a general classe of m–stage r–step collocation methods for VIEs, with the aim of increasing the order of classical one-step collocation methods without any additional computational cost. The resulting high order methods had, however, bounded stability regions. For this reason, in Ref. 15, in analogy to the case of ODEs, Ref. 16, we introduced a modification in the technique, leading to two-step almost-collocation methods. Such methods have been obtained by relaxing some of the collocation conditions and by introducing some previous stage values, in order to further increase the order and to get A-stability.

In this paper we analyze a modified class of high order two-step collocation methods, providing A-stable methods of uniform order $p = 2m$ on the whole integration interval, where m is the number of collocation points, without relaxing any interpolation or collocation condition.

The paper is organized as follows. In Section 2 we describe the new two-step collocation methods and analyze the order. In Section 3 we carry out the linear stability analysis. In Section 4 we provide examples of one-stage and two-stage A-stable methods. Finally in Section 5 some concluding remarks are given and plans for future research are briefly outlined.

2. Construction of the methods and order conditions

We divide the interval I in N subintervals of fixed length h, obtaining the set of *grid points* $I_h = \{t_n : 0 = t_0 < t_1 < ... < t_N = T\}$ and we define the set of *collocation points* $X_h = \{t_{n,j} := t_n + c_j h : 0 \leq c_1 < c_2 < \cdots < c_m \leq 1, n = 0, 1, ..., N-1\}$. The equation (1) can be rewritten, by relating it to this mesh, as

$$y(t) = F_n(t) + \Phi_n(t), \quad t \in [t_n, t_{n+1}],$$

where $F_n(t) := g(t) + \int_0^{t_n} k(t, \tau, y(\tau))d\tau$ and $\Phi_n(t) := \int_{t_n}^{t} k(t, \tau, y(\tau))d\tau$ represent respectively the *lag term* and the *increment function*.

The collocation polynomial is considered of the form

$$P(t_n + sh) = \varphi(s)y_n + \sum_{j=1}^{m} \chi_j(s)Y_{n-1,j} + \sum_{j=1}^{m} \psi_j(s)Y_{n,j}, \tag{2}$$

with $s \in [0, 1]$, where

$$Y_{n-1,j} := P(t_{n-1,j}), \quad Y_{n,j} := P(t_{n,j}), \tag{3}$$

and the polynomials $\varphi(s)$, $\chi_j(s)$, $\psi_j(s)$ are determined by imposing the interpolation condition $P(t_n) = y_n$, and by satisfying (3). The collocation polynomial (2) differs from the polynomial introduced in Ref. 15, because we drop the previous time step y_{n-1}, mantaining only the previous stages $Y_{n-1,j}$, as it is usually done in two-step collocation and Runge-Kutta methods for ODEs, in order to get better stability properties and an efficient implementation, see Ref. 17–23.

By imposing the collocation conditions, i.e. that the collocation polynomial (2) exactly satisfies the VIE (1) at the collocation points $t_{n,i}$ and by computing $y_{n+1} = P(t_{n+1})$, the two-step collocation method takes the form

$$\begin{cases} Y_{n,i} = F_{n,i} + \Phi_{n,i} \\ y_{n+1} = \varphi(1)y_n + \sum_{j=1}^{m} \chi_j(1)Y_{n-1,j} + \sum_{j=1}^{m} \psi_j(1)Y_{n,j} \end{cases}, \tag{4}$$

The lag–term and increment–term approximations

$$F_{n,i} = g(t_{n,i}) + h \sum_{\nu=0}^{n-1} \sum_{l=1}^{\mu_1} b_l k(t_{n,i}, t_\nu + \xi_l h, P_\nu(t_\nu + \xi_l h)) \quad i = 1, ..., m \tag{5}$$

$$\Phi_{n,i} = h \sum_{l=1}^{\mu_0} w_{il} k(t_{n,i}, t_n + d_{il} h, P_n(t_n + d_{il} h)) \quad i = 1, ..., m \tag{6}$$

are obtained by using quadrature formulas of the form

$$(\xi_l, b_l)_{l=1}^{\mu_1}, \quad (d_{il}, w_{il})_{l=1}^{\mu_0}, \ i = 1, ..., m, \tag{7}$$

where the quadrature nodes ξ_l and d_{il} satisfy $0 \le \xi_1 < ... < \xi_{\mu_1} \le 1$ and $0 \le d_{i1} < ... < d_{i\mu_0} \le 1$, μ_0 and μ_1 are positive integers and w_{il}, b_l are suitable weights, as in Ref. 14.

The continuous order conditions and the convergence can be easily analyzed by looking at the method (4) as a subclass of the methods introduced in Ref. 15, with $\varphi_0(s) \equiv 0$, as shown in the following theorem.

Theorem 2.1. *Assume that the kernel $k(t,\eta,y)$ and the function $g(t)$ in (1) are sufficiently smooth. Then the method (4) has uniform order p for $s \in [0,1]$, if the following conditions are satisfied*

$$\begin{cases} 1 - \varphi(s) - \sum_{j=1}^{m} \chi_j(s) - \sum_{j=1}^{m} \psi_j(s) = 0, \\ s^k - \sum_{j=1}^{m} (c_j - 1)^k \chi_j(s) - \sum_{j=1}^{m} c_j^k \psi_j(s) = 0, \end{cases} \tag{8}$$

$s \in [0,1]$, $k = 1,2,...,p$. *Assume moreover that $c_i \neq c_j$, $c_i \neq c_j - 1$, $c_i \neq 0,1$. Then the system of continuous order conditions (8) is satisfied with $p = 2m$ if and only if the polynomials $\varphi(s)$, $\chi_j(s)$ and $\psi_j(s)$, $j = 1,2,...,m$ satisfy the interpolation conditions*

$$\varphi(0) = 1, \quad \chi_j(0) = 0, \quad \psi_j(0) = 0 \tag{9}$$

and the collocation conditions

$$\varphi(c_i) = 0, \quad \chi_j(c_i) = 0, \quad \psi_j(c_i) = \delta_{ij} \tag{10}$$

$$\varphi(c_i - 1) = 0, \quad \chi_j(c_i - 1) = \delta_{ij}, \quad \psi_j(c_i - 1) = 0, \tag{11}$$

i=1,2,...,m.

Proof. The order conditions (8) can be derived from Theorem 2.1 in Ref. 15. An argument similar to the proof of Theorem 2.2 in Ref. 15 leads to the a characterization (9)–(11) for the coefficients of the methods having order $p = 2m$. □

It can also be proved that the order of convergence is $2m$ if the conditions (9)–(11) are satisfied and the quadrature formulas (7) are of order at least $2m$.

3. Linear stability analysis

In this section we carry out the stability analysis of method (4) with respect to the basic test equation

$$y(t) = 1 + \lambda \int_0^t y(\eta)d\eta, \quad t \geq 0, \quad Re(\lambda) \leq 0, \tag{12}$$

usually employed in the literature for the stability analysis of numerical methods for VIEs (see Ref. 8,24 and their references). Let us consider the following vectors and matrices: $\mathbf{Y}_n = [Y_{n,1}, ..., Y_{n,m}]^T$, $\psi^T(1) = [\psi_1(1), ..., \psi_m(1)]$, $\chi^T(1) = [\chi_1(1), ..., \chi_m(1)]$, $\varphi(\xi) = [\varphi(\xi_1), ..., \varphi(\xi_{\mu_1})]^T$, $\mathbf{b} = [b_1, ..., b_m]^T$, $\beta = [\beta_1, ..., \beta_m]^T$, $\gamma = [\gamma_1, ..., \gamma_m]^T$, $\mathbf{v} = [v_1, ..., v_m]^T$, $\mathbf{\Omega} = [\Omega_{i,j}]_{i,j=1}^m$, $\mathbf{\Lambda} = [\Lambda_{i,j}]_{i,j=1}^m$, where

$$\beta_j = \sum_{l=1}^{\mu_1} b_l \chi_j(\xi_l), \quad \gamma_j = \sum_{l=1}^{\mu_1} b_l \psi_j(\xi_l),$$

$$v_i = \sum_{l=1}^{\mu_0} w_{il}\varphi(d_{il}), \quad \Omega_{i,j} = \sum_{l=1}^{\mu_0} w_{il}\chi_j(d_{il}) \quad \Lambda_{i,j} = \sum_{l=1}^{\mu_0} w_{il}\psi_j(d_{il})$$

and put $\mathbf{e} = [1, 1, .., 1]^T$. The following theorem provides the expression for the stability matrix of the method (4) with respect to the test equation (12).

Theorem 3.1. *The two-step collocation method (4), applied to the test equation (12), leads to the following matrix recurrence relation*

$$\begin{bmatrix} y_{n+1} \\ \mathbf{Y}_n \\ y_n \\ \mathbf{Y}_{n-1} \end{bmatrix} = \mathbf{R}(z) \begin{bmatrix} y_n \\ \mathbf{Y}_{n-1} \\ y_{n-1} \\ \mathbf{Y}_{n-2} \end{bmatrix},$$

$z = h\lambda$, *where the stability matrix* $\mathbf{R}(z)$ *is*

$$\mathbf{R}(z) = \mathbf{Q}^{-1}(z)\mathbf{M}(z), \tag{13}$$

with the matrices $\mathbf{Q}(z)$ *and* $\mathbf{M}(z)$ *defined by*

$$\mathbf{Q}(z) = \begin{bmatrix} 1 & -\psi^T(1) & 0 & 0 \\ 0 & I - z\mathbf{\Lambda} & 0 & 0 \\ 0 & 0 & 1 & 0 \\ 0 & 0 & 0 & \mathbf{I} \end{bmatrix},$$

$$\mathbf{M}(z) = \begin{bmatrix} \varphi(1) & \chi^T(1) & 0 & 0 \\ z\mathbf{v} & \mathbf{I} + z(u\gamma^T + \mathbf{\Omega} - \mathbf{\Lambda}) & z(\mathbf{e}\mathbf{b}^T\varphi(\xi) - \mathbf{v}) & z(\mathbf{e}\beta^T - \mathbf{\Omega}) \\ 1 & 0 & 0 & 0 \\ 0 & \mathbf{I} & 0 & 0 \end{bmatrix}.$$

Proof. By applying the method (4) to the test equation (12) we obtain

$$y_{n+1} = \varphi(1)y_n + \chi^T(1)\mathbf{Y}_{n-1} + \psi^T(1)\mathbf{Y}_n \tag{14}$$

where

$$\mathbf{Y}_n = \mathbf{F}_n + z(y_n\mathbf{v} + \mathbf{\Omega}\mathbf{Y}_{n-1} + \mathbf{\Lambda}\mathbf{Y}_n), \tag{15}$$

$$\mathbf{F}_n = \mathbf{e} + z\sum_{\nu=0}^{n-1}(\mathbf{b}^T\varphi(\xi)y_\nu + \beta^T\mathbf{Y}_{\nu-1} + \gamma^T\mathbf{Y}_\nu)\mathbf{e}. \tag{16}$$

From the expression (16) we derive

$$\mathbf{F}_n - \mathbf{F}_{n-1} = z(\mathbf{b}^T\varphi(\xi)y_{n-1} + \beta^T\mathbf{Y}_{n-2} + \gamma^T\mathbf{Y}_{n-1})\mathbf{e}. \tag{17}$$

The computation of the difference $\mathbf{Y}_n - \mathbf{Y}_{n-1}$ by substituting the expression (15) for both terms $\mathbf{Y}_n$ and $\mathbf{Y}_{n-1}$, and by using (17), leads to

$$\begin{aligned}(\mathbf{I}_m - z\mathbf{\Lambda})\,\mathbf{Y}_n &= z\mathbf{v}y_n + \left(\mathbf{I} + z(u\gamma^T + \mathbf{\Omega} - \mathbf{\Lambda})\right)\mathbf{Y}_{n-1} + \\ &+ z(\mathbf{e}\mathbf{b}^T\varphi(\xi) - \mathbf{v})y_{n-1} + z(\mathbf{e}\beta^T - \mathbf{\Omega})\mathbf{Y}_{n-2}.\end{aligned}$$

From the last equation and (14) the thesis immediately follows. □

We next consider the stability function of the method

$$p(\omega, z) = \det\Big(\lambda\mathbf{I} - \mathbf{R}(z)\Big), \tag{18}$$

where $\mathbf{I}$ is the identity matrix of order $2m+2$, and we investigate on the conditions to impose on the collocation abscissas $c_1, ..., c_m$ in order to get A-stable methods: this means that all the roots $\lambda_1, \ldots, \lambda_{2m+2}$ of (18) lie in the unit circle for all $z \in \mathbb{C}$ such that $\mathrm{Re}(z) \leq 0$. The investigation, carried out using the Schur criterionRef. 25 has shown the following results for $m = 1$ and $m = 2$.

Theorem 3.2. *Any one-stage collocation method of the type (4) is A-stable if and only if* $c > 1$.

Fig. 1 shows the A-stability region in the parameter space (c_1, c_2) for two stage collocation methods (4).

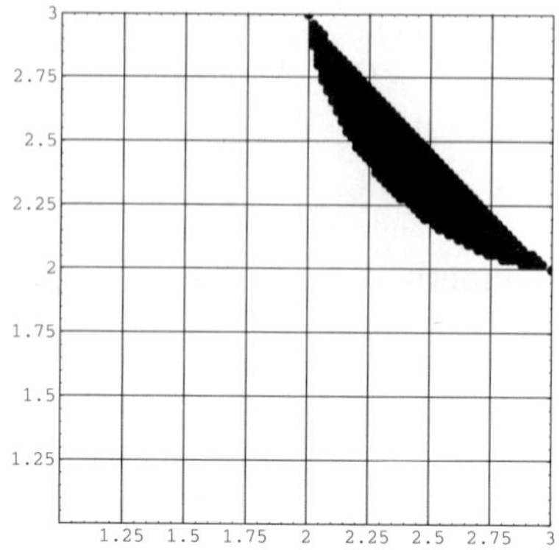

Fig. 1. Region of A-stability in the (c_1, c_2)-plane for two-step methods (4) with $m = 2$ and order 4.

4. Examples of methods

We first consider the case $m = 1$. According to Theorem 3.2, for any value of $c > 1$ we obtain A–stable methods of order 2. Solving the order conditions (8) for $m = 1$, $p = 2$, we obtain

$$\varphi(s) = \frac{s^2 + s(1-2c) + c(c-1)}{c(c-1)}, \quad \chi(s) = \frac{s(c-s)}{c-1}, \quad \psi(s) = \frac{s(1-c+s)}{c}.$$

The weights in (5) and (6) can be chosen by discretizing the lag-term by the trapezoidal rule and the increment term by the midpoint rule, i.e., $\mu_0 = 1$, $\mu_1 = 3$,

$$\xi = [0, c, 1]^T, \quad \mathbf{D} = c, \quad \mathbf{b} = \left[\frac{1}{2}, 0, \frac{1}{2}\right]^T, \quad \mathbf{W} = c.$$

This leads to a one parameter family of methods of order $p = 2$, depending on the collocation abscissa c.

We next consider the case $m = 2$. Solving the order conditions (8) for $m = 2$, $p = 4$, and choosing, according to Fig. 1, $c_1 = \frac{11}{5}$ and $c_2 = \frac{13}{5}$, we obtain

$$\varphi(s) = \frac{\left(126 - 115s + 25s^2\right)\left(66 - 85s + 25s^2\right)}{8316},$$

$$\chi_1(s) = -\frac{s\,(5s-11)\left(126-115s+25s^2\right)}{144}, \quad \chi_2(s) = \frac{s\,(5s-14)\left(66-85s+25s^2\right)}{54}$$

$$\psi_1(s) = -\frac{s(5s-6)\left(126-115s+25s^2\right)}{66}, \quad \psi_2(s) = \frac{s(5s-9)\left(66-85s+25s^2\right)}{336}$$

The weights in (5) and (6) can be chosen by considering $\mu_0 = 3$, $\mu_1 = 4$, and

$$\xi = \begin{bmatrix} 0 \\ c_1 \\ c_2 \\ 1 \end{bmatrix}, \quad \mathbf{D} = \begin{bmatrix} 0 & c_1 & c_2 \\ 0 & c_1 & c_2 \end{bmatrix},$$

$$\mathbf{b} = \begin{bmatrix} \frac{-1+2c_1+2c_2-6c_1c_2}{12c_1c_2} \\ \frac{1-2c_2}{12c_1(c_1-1)(c_1-c_2)} \\ \frac{2c_1-1}{12c_2(c_2-1)(c_2-c_1)} \\ \frac{-3+4c_1+4c_2-6c_1c_2}{12(c_1-1)(c_2-1)} \end{bmatrix}, \quad \mathbf{W} = \begin{bmatrix} -\frac{c_1^2-3c_1c_2}{6c_2} & \frac{c_1(2c_1-3c_2)}{6(c_1-c_2)} & \frac{c_1^3}{6c_2(c_1-c_2)} \\ -\frac{c_2^2-3c_1c_2}{6c_1} & -\frac{c_1^3}{6c_1(c_1-c_2)} & -\frac{c_2(2c_2-3c_1)}{6(c_1-c_2)} \end{bmatrix},$$

i.e., with $c_1 = \frac{11}{5}$ and $c_2 = \frac{13}{5}$,

$$\xi = \begin{bmatrix} 0 \\ \frac{11}{5} \\ \frac{13}{5} \\ 1 \end{bmatrix}, \quad \mathbf{D} = \begin{bmatrix} 0 & \frac{11}{5} & \frac{13}{5} \\ 0 & \frac{11}{5} & \frac{13}{5} \end{bmatrix}, \quad \mathbf{b} = \begin{bmatrix} \frac{233}{616} \\ -\frac{575}{2376} \\ \frac{425}{4536} \\ \frac{499}{648} \end{bmatrix}, \quad \mathbf{W} = \begin{bmatrix} \frac{431}{420} & \frac{22}{9} & -\frac{1331}{1260} \\ \frac{133}{165} & \frac{1372}{495} & -\frac{7}{9} \end{bmatrix}.$$

5. Concluding remarks

We have developed a class of modified two-step collocation methods (4) for the numerical solution of VIEs. These methods are of uniform order $p = 2m$ on the whole integration interval. We have discussed their stability properties, deriving A−stable methods. Examples of methods have also been provided.

The above methods seem to be promising for further investigations, because of their good properties of accuracy and stability. The uniform order and the continuous approximation to the solution make such methods particulary suitable for a variable stepsize implementation. The implementation issues related to these methods are subject of future work. They include the choice of appropriate starting procedures, estimation of local discretization error and stepsize changing strategies. Our aim is also to extend the results to other functional equations, such as Volterra integro-differential equations.

References

1. M. A. Abdou, and A. A. Badr, On a method for solving an integral equation in the displacement contact problem, *Appl. Math. Comput.*, **127**(1) (2002), pp. 65-78.

2. N. B. Franco, A Volterra integral equation arising from the propagation of nonlinear waves, *Rev. Mat. Estat.* **17** (1999), pp. 35-49.
3. H. W. Hetcote, and D. W. Tudor, Integral equation models for endemic infectious diseases, *J. Math. Biol.*, **9** (1980), pp. 37-47.
4. F. C. Hoppensteadt, Z. Jackiewicz, and B. Zubik-Kowal, Numerical solution of Volterra integral and integro-differential equations with rapidly vanishing convolution kernels, *BIT* **47**(2) (2007), pp. 325-350.
5. W. Huang, Y. Li, and W. Chen, Analysis of the dynamic response of a fluid-supported circular elastic plate impacted by a low-velocity projectile, Proceedings of the Institution of Mechanical Engineers. Part C, Journal of mechanicalengineering science, vol. 214, n. 5, pp. 719-727, 2000.
6. A. G. Nerukh, P. Sewell, and T. M. Benson, Volterra Integral Equations for Nonstationary Electromagnetic Processes in Time-Varying Dielectric Waveguides, *J. Lightwave Technol.*, **22**(5), 2004.
7. H. Brunner, *Collocation Methods for Volterra Integral and Related Functional Equations*, Cambridge University Press (2004).
8. H. Brunner, and P. J. van der Houwen, *The Numerical Solution of Volterra Equations*, CWI Monographs, Vol. 3, North-Holland, Amsterdam, (1986).
9. R. P. Agarwal, and D. O'Regan (eds.), *Integral and Integrodifferential Equations. Theory, Methods and Applications*, Ser. Math. Anal. Allp., 2, Amsterdam, Gordon and Breach (2000).
10. C. Corduneanu, and I. W. Sandberg (eds.), *Volterra Equations and Applications*, Stability Control, Theory, Methods Appl. 10, Amsterdam, Gordon and Breach (2000).
11. X. Q. Zhao, *Dynamical Systems in Population Biology*, CMS Books in Mathematics, New York, Springer–Verlag (2003).
12. R. D'Ambrosio, M. Ferro, and B. Paternoster, Collocation based two step Runge–Kutta methods for Ordinary Differential Equations, in ICCSA 2008, *Lecture Notes in Comput. Sci.*, Springer, New York, (Edited by O. Gervasi et al.), **5073**(2) (2008), pp. 736-751.
13. D. Conte, and B. Paternoster, A Family of Multistep Collocation Methods for Volterra Integral Equations, *AIP Conference Proceedings, Numerical Analysis and Applied Mathematics, Springer* T.E.Simos, G. Psihoyios, Ch. Tsitouras (Eds.), **936** (2007), pp. 128-131.
14. D. Conte, and B. Paternoster, Multistep Collocation Methods for Volterra Integral Equations, *App. Num.* (2009), in press.
15. D. Conte, Z. Jackiewicz, and B. Paternoster, Two-step almost collocation methods for Volterra integral equations, *Appl. Math. Comput.*, **204** (2008), pp. 839-853.
16. R. D'Ambrosio, M. Ferro, Z. Jackiewicz, and B. Paternoster, Two-step almost collocation methods for ordinary differential equations, *accepted for pubblication on Numer. Algorithms.*
17. N. H. Cong, A general family of pseudo two-step Runge–Kutta methods, *Southeast Asian Bull. Math.*, **25**(1) (2001), pp. 61-73.
18. N. H. Cong, H. Podhaisky, and R. Weiner, Numerical experiments with some explicit pseudo two-step RK methods on a shared memory computer, *Com-*

puters & Mathematics with Applications, **36** (1998), pp. 107-116.
19. Z. Jackiewicz, and J. H. Verner, Derivation and implementation of two-step Runge–Kutta pairs, *Jpn. J. Ind. Appl. Math.* **19** (2002), pp. 227-248.
20. H. Podhaisky, R. Weiner, and J. Wensch, High Order Explicit Two-Step Runge–Kutta Methods for Parallel Computers, Tech. Report 19, Universitat Halle, FB Mathematik/Informatik, 1999.
21. H. Podhaisky, and R. Weiner, A Class of Explicit Two-Step Runge–Kutta Methods with Enlarged Stability Regions for Parallel Computers, in Parallel Computation (P. Zinterhofer, M. Vajtersic, and Andreas Uhl, eds.), *Lecture Notes in Comput. Sci., Springer* , **1557** (1999), pp. 68-77.
22. J. H. Verner, Improved Starting methods for two-step Runge–Kutta methods of stage-order p-3, *Appl. Numer. Math.*, **10** (2006), pp. 388-396.
23. J. H. Verner, Starting methods for two-step Runge–Kutta methods of stage-order 3 and order 6, *J. Comput. Appl. Math.*, **185** (2006), pp. 292-307.
24. A. Bellen, Z. Jackiewicz, R. Vermiglio, and M. Zennaro, Stability analysis of Runge-Kutta methods for Volterra integral equations of the second kind, *IMA J. Numer. Anal.*, **10** (1990), pp. 103-118.
25. J. Schur, Über Potenzreihen die im Innern des Einheitskreises beschränkt sind, *J. Reine Angew. Math.*, **147** (1916), pp. 205-232.

Practical Construction of Two-Step Collocation Runge-Kutta Methods for Ordinary Differential Equations

Dajana Conte, Raffaele D'Ambrosio, Beatrice Paternoster,

Dipartimento di Matematica e Informatica
Università degli Studi di Salerno, Italy
E-mail: {dajconte,rdambrosio,beapat}@unisa.it

Maria Ferro

Dipartimento di Matematica e Applicazioni "R. Caccioppoli"
Università di Napoli "Federico II", Italy
E-mail: maria.ferro@dma.unina.it

It is the purpose of this paper to analyse a new class of two-step collocation methods for the numerical solution of ordinary differential equations (ODEs), possessing higher order of convergence than classical one step collocation methods and A-stability, without any increase of the computational cost.

These methods seem to be particularly promising for practical applications, e.g. efficient implementation, resolution of ODEs based problems belonging to models of evolutionary real life phenomena.

Keywords: Ordinary differential equations; Collocation methods; Two-step Runge–Kutta methods; A-stability.

1. Introduction

We are concerned with the numerical solution of initial value problems based on ordinary differential equations of the type

$$\begin{cases} y'(x) = f(x, y(x)), & x \in I = [x_0, X] \\ y(x_0) = y_0, \end{cases} \tag{1}$$

with $f : I \times \Omega \subset \mathbb{R}^d \to \mathbb{R}^d$ smooth enough to guarantee that the problem (1) is well-posed.

The methods we aim to derive belong to a special family of two-step

Runge–Kutta methods (TSRK), of the type

$$Y_j^{[n]} = y_n + h\sum_{i=1}^{m}[a_{ij}f(Y_i^{[n-1]}) + b_{ij}f(Y_i^{[n]})], \tag{2}$$

$$y_{n+1} = y_n + h\sum_{j=1}^{m}[v_j f(Y_j^{[n-1]}) + w_j f(Y_j^{[n]})], \tag{3}$$

with $j = 1, 2, ..., m$, depending on the stage derivatives at two consecutive step points. As pointed out in [12], methods of this type are interesting because it is possible to introduce more parameters to play with in order to obtain higher order and better stability properties than classical Runge–Kutta methods, without any increase in the computational cost: in fact, advancing from x_n to x_{n+1}, the stage derivatives $f(Y_j^{[n-1]})$, $j = 1, 2, ..., m$, have already been computed. Starting values are generally determined using a classical Runge–Kutta method of the same order.

In this paper we consider an extension of algebraic multistep collocation technique (see [9,10,15]), in order to obtain continuous, high effective order, highly stable methods of the type (2)-(3). It seems interesting and promising to look at TSRK methods of the type (2)-(3), because of what follows. In [5,6] we have extended this idea to a more general class of TSRK methods, introduced by Jackiewicz and Tracogna in [12], of the type

$$Y_j^{[n]} = u_j y_{n-1} + (1-u_j)y_n + h\sum_{i=1}^{m}[a_{ij}f(Y_i^{[n-1]}) + b_{ij}f(Y_i^{[n]})], \tag{4}$$

$$y_{n+1} = \theta y_{n-1} + (1-\theta)y_n + h\sum_{j=1}^{m}[v_j f(Y_j^{[n-1]}) + w_j f(Y_j^{[n]})], \tag{5}$$

deriving formulas of order $2m+1$, but having only bounded stability regions. In order to gain better stability properties, we have considered in [7] possible ways to relax the requirements imposed in [5,6], with a corresponding improvement in the stability (in fact we have obtained A-stable and L-stable formulas), but obtaining lower order of convergence. Therefore, in order to have a good balance between effective order and high stability, it is convenient to study methods of the type (2)-(3): Jackiewicz and Tracogna themselves derive in [12] examples of methods (4)-(5) with $\theta = 0$, $u_j = 0$, $j = 1, 2, ..., m$, emphasizing on the corresponding advantage in terms of stability and also simplifying the system of order conditions (see [12]).

Collocation methods of the type (2)-(3) are also very interesting in view of an efficient implementation (e.g. variable stepsize, variable order implementation, also in a parallel environment), as it has been showed, although outside collocation, in [2–4,13,20,22,23].

The paper is organized as follows: in Section 2 we discuss on the construction of two-step algebraic collocation methods of the type (2)-(3) and derive continuous order conditions for these methods in Section 3. Section 4 is focused on the linear stability analysis: in particular, we will discuss on the existence of A-stable methods within the class of collocation based methods (2)-(3). Examples of methods are provided in Section 5.

2. General two–step collocation methods

The classical collocation technique consists in the derivation of an algebraic polynomial $P(x_n + sh)$, satisfying suitable interpolation and collocation conditions: in particular, this polynomial exactly solves the equation (1) at some points. In order to derive collocation based methods of the type (2)-(3), we impose the following set of conditions:

$$\begin{aligned} P(x_n) &= y_n, && (6)\\ P'(x_{n-1}+c_ih) &= f(x_{n-1}+c_ih, P(x_{n-1}+c_ih)), \quad i=1,2,...,m, && (7)\\ P'(x_n+c_ih) &= f(x_n+c_ih, P(x_n+c_ih)), \quad i=1,2,...,m. && (8) \end{aligned}$$

Equation (6) is an interpolation condition, while conditions (7)-(8) form a set of $2m$ collocation conditions with respect to two consecutive step points. Introducing the dimensionless coordinate $t = \frac{x-x_n}{h}$, it is convenient to express the collocation polynomial in the following form

$$P(x_n+th) = \varphi(t)y_n + h\sum_{j=1}^{m}\left[\chi_j(t)P'(x_{n-1}+c_jh) + \psi_j(t)P'(x_n+c_jh)\right],$$

i.e. as linear combination of $2m+1$ basis polynomials

$$\{\varphi(t),\ \chi_j(t),\ \psi_j(t), \quad j=1,2,...,m\}.$$

In order to determine this $2m+1$ unknown basis functions, we apply the set of conditions (6)-(7)-(8), i.e.

$$\phi(0) = 1, \qquad \chi_i(0) = 0, \qquad \psi_i(0) = 0, \tag{9}$$

$$\phi'(c_i - 1) = 0, \ \chi_j'(c_i - 1) = \delta_{ij}, \ \psi_j'(c_i - 1) = 0, \tag{10}$$

$$\phi'(c_i) = 0, \qquad \chi_j'(c_i) = 0, \qquad \psi_j'(c_i) = \delta_{ij}, \tag{11}$$

for $i, j = 1, 2, \ldots, m$, where δ_{ij} is the usual Kronecker delta. Therefore, each basis polynomial is subject to $2m+1$ conditions, which uniquely define polynomials of degree at most $2m$. Therefore, the polynomial $P(x_n + th)$ is uniquely determined as linear combination of polynomials of degree at most $2m$ and, correspondingly, we compute $y_{n+1} = P(x_n + h)$. We omit for brevity the details of the resolution of the systems of equations arising from (9)-(10)-(11), which can be inferred, with simple modifications, from [5,6]. The expressions of the basis polynomials are available through e-mail from the authors.

We conclude this section with the following result of characterization of TSRK collocation methods.

Theorem 2.1. *The method defined by (9)-(10)-(11) is equivalent to a TSRK method (2)-(3), with*

$$\begin{aligned} v_j &= \chi_j(1), \quad w_j = \psi_j(1), \\ a_{ij} &= \chi_j(c_i), \ b_{ij} = \psi_j(c_i), \end{aligned}$$

for $i, \ j, = 1, \ ..., \ m$.

Proof. The proof follows along the line of Theorem 1 on p. 739 in [6]. □

3. Order conditions

In this section we derive continuous order conditions for collocation methods having the following form

$$\begin{cases} P(x_n + sh) = \varphi(s) y_n + \\ \qquad\qquad +h \sum_{j=1}^{m} \left[\chi_j(s) P'(x_{n-1} + c_j h) + \psi_j(s) P'(x_n + c_j h)\right], \\ y_{n+1} = P(x_{n+1}), \end{cases} \tag{12}$$

which are the continuous extensions of methods (2)-(3). We assume that $P(x_n + sh)$ is an uniform approximation to $y(x_n + sh)$, $s \in [0, 1]$, of order p.

As a consequence, the stage values $P(x_n + c_j h)$ have stage order $q = p$. We now introduce the local discretization error $\xi(x_n + sh)$, which is defined as the residuum obtained by replacing $P(x_n + sh)$ by $y(x_n + sh)$, $P(x_n + c_j h)$ by $y(x_n + c_j h)$, $j = 1, 2, \ldots, m$, y_{n-1} by $y(x_{n-1})$ and y_n by $y(x_n)$, where $y(x)$ is the exact solution. This leads to

$$\begin{aligned}\xi(x_n + sh) &= y(x_n + sh) - \varphi(s)y(x_n)\\ &\quad - h\textstyle\sum_{j=1}^{m}\Big(\chi_j(s)y'(x_n + (c_j - 1)h) + \psi_j(s)y'(x_n + c_j h)\Big),\end{aligned} \tag{13}$$

$s \in [0,1]$, $n = 1, 2, \ldots, N-1$. We have the following theorem.

Theorem 3.1. *Assume that the function $f(y)$ is sufficiently smooth. Then the method (12) has uniform order p if the following conditions are satisfied*

$$\begin{cases} \varphi_1(s) = 1, \\ \sum_{j=1}^{m}\left(\chi_j(s)\frac{(c_j-1)^{k-1}}{(k-1)!} + \psi_j(s)\frac{c_j^{k-1}}{(k-1)!}\right) = \frac{s^k}{k!}, \end{cases} \tag{14}$$

$s \in [0,1]$, $k = 1, 2, \ldots, p$.
Moreover, the local discretization error (13) takes the form

$$\xi(x_n + sh) = h^{p+1}C_p(s)y^{(p+1)}(x_n) + O(h^{p+2}), \tag{15}$$

as $h \to 0$, where the error function $C_p(s)$ is defined by

$$C_p(s) = \frac{s^{p+1}}{(p+1)!} - \sum_{j=1}^{m}\left(\chi_j(s)\frac{(c_j-1)^p}{p!} + \psi_j(s)\frac{c_j^p}{p!}\right). \tag{16}$$

Proof. We expand $y(x_n + sh)$, $y'(x_n + (c_j - 1)h)$ and $y'(n_n + c_j h)$ into Taylor series around the point x_n and next substitute them in 13. Collecting the terms appearing with the same powers of h, we obtain

$$\begin{aligned}\xi(x_n + sh) &= \Big(1 - \varphi_1(s)\Big)y(x_n) + \sum_{k=1}^{p+1}\frac{s^k}{k!}h^k y^{(k)}(t_n)\\ &\quad - \sum_{k=1}^{p+1}\sum_{j=1}^{m}\left(\chi_j(s)\frac{(c_j-1)^{k-1}}{(k-1)!} + \psi_j(s)\frac{c_j^{k-1}}{(k-1)!}\right)h^k y^{(k)}(t_n)\\ &\quad + O(h^{p+2}).\end{aligned}$$

Equating to zero the terms of order k, $k = 0, 1, \ldots, p$, we obtain order conditions (14). Comparing the terms of order $p+1$ we obtain (15) with error function $C_p(s)$ defined by (16). □

As a consequence of the above result, we can infer the following

Theorem 3.2. *Each method within the class (12) has order* $2m$*, for any value of* $s \in (0, 1]$.

Proof. The proof results as an extension of Theorems (2.3) and (2.4) in [7]. □

According to Theorem 3.2, each collocation method within the class (12) has uniform order $2m$, i.e. the order of convergence is $2m$ in any point of the integration interval. This aspect is very interesting in the applications, in particular for an accurate and efficient variable stepsize - variable order implementation of stiff systems, because having uniform order of convergence avoids the order reduction phenomenon (see [1]). Classical collocation based Runge–Kutta methods suffer of order reduction, because they have low stage order: for example, Gaussian Runge–Kutta methods have order $2m$ in the external stages, but only m in the internal ones.

4. Linear stability analysis

In order to investigate the linear stability properties of the methods, we recast the methods (2)-(3) as general linear methods (see [1,11]) of the form

$$\left[\begin{array}{c} Y^{[n]} \\ \hline y_{n+1} \\ hf(Y^{[n]}) \end{array}\right] = \left[\begin{array}{c|cc} A & e & B \\ \hline v^T & 1 & w^T \\ \mathbf{I} & \mathbf{0} & \mathbf{0} \end{array}\right] \left[\begin{array}{c} hf(Y^{[n]}) \\ \hline y_n \\ hf(Y^{[n-1]}) \end{array}\right], \tag{17}$$

where $\mathbf{I}$ is the identity matrix of dimension m and $\mathbf{0}$ is the zero matrix or vector of appropriate dimensions. The Butcher array of these methods is

$$\left[\begin{array}{c|c} \mathbf{A} & \mathbf{U} \\ \hline \mathbf{B} & \mathbf{V} \end{array}\right] = \left[\begin{array}{c|cc} A & e & B \\ \hline v^T & 1 & w^T \\ \mathbf{I} & \mathbf{0} & \mathbf{0} \end{array}\right]. \tag{18}$$

The corresponding stability (or amplification) matrix $M(z)$ takes the form (see [1,11])

$$M(z) = \mathbf{V} + z\mathbf{B}(I - z\mathbf{A})^{-1}\mathbf{U} \in \mathbb{R}^{(m+1)\times(m+1)} \in \mathbb{R}^{(m+1)\times(m+1)}. \tag{19}$$

We next consider the stability function of the method, it is

$$p(\omega, z) = \det\Big(\omega I - M(z)\Big), \tag{20}$$

where I is the identity matrix of order $m+1$. We investigate the conditions to impose on the collocation abscissas $c_1, ..., c_m$ in order to obtain A-stable methods, i.e. all the roots $\omega_1, \ldots, \omega_{m+1}$ of (20) lie in the unit circle for all $z \in \mathbb{C}$ such that $\mathrm{Re}(z) \leq 0$. The investigation, carried out using the Schur criterion (see [21]), has produced the following results for $m = 1$ and $m = 2$.

Theorem 4.1. *A-stability for $m = 1$ Any one-stage collocation based TSRK method of the type (12) is A-stable if and only if $c_1 > 1$.*

Figure 1 shows the A-stability region in the parameter space $(c1, c2)$ for collocation methods (12) with $m = 2$ and order 4.

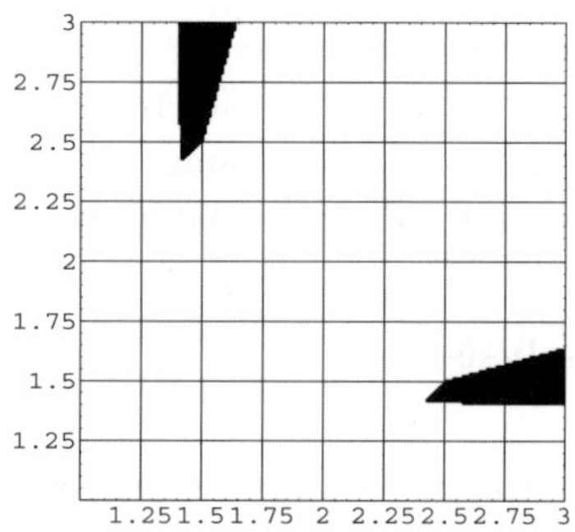

Fig. 1. Region of A-stability in the $(c1, c2)$-plane for two-step methods (12) with $m = 2$ and order 4.

5. Examples of methods

We now provide some examples of two step collocation methods (12), using the results carried out in this paper. We first consider the case $m = 1$. According to Theorem 4.1, we choose $c_1 = \frac{5}{4}$, obtaining an order 2 A-stable method of the type (12) with

$$\chi(s) = \frac{5s - 2s^2}{4}, \tag{21}$$

$$\psi(s) = \frac{s(-1 + 2s)}{4}, \tag{22}$$

whose Butcher array (18) is

$$\left[\begin{array}{c|c}\mathbf{A} & \mathbf{U} \\ \hline \mathbf{B} & \mathbf{V}\end{array}\right] = \left[\begin{array}{c|cc}\frac{25}{32} & 1 & \frac{15}{32} \\ \hline \frac{3}{4} & 1 & \frac{1}{4} \\ 1 & 0 & 0\end{array}\right]. \tag{23}$$

We next consider the case $m = 2$. According to Figure (1), we choose $c_1 = \frac{3}{2}$ and $c_2 = \frac{13}{5}$, obtaining an order 4 A-stable method of the type (12) with

$$\chi_1(s) = -\frac{s(-624 + 523s - 190s^2 + 25s^3)}{231}, \tag{24}$$

$$\chi_2(s) = \frac{-5s(-234 + 357s - 184s^2 + 30s^3)}{66}, \tag{25}$$

$$\psi_1(s) = \frac{s(-624 + 939s - 470s^2 + 75s^3)}{33}, \tag{26}$$

$$\psi_2(s) = \frac{5s(-48 + 79s - 48s^2 + 10s^3)}{462}, \tag{27}$$

whose Butcher array (18) is

$$\left[\begin{array}{c|c}\mathbf{A} & \mathbf{U} \\ \hline \mathbf{B} & \mathbf{V}\end{array}\right] = \left[\begin{array}{cc|ccc}\frac{1461}{1232} & \frac{225}{176} & 1 & -\frac{159}{176} & -\frac{75}{1232} \\ \frac{338}{275} & \frac{7267}{1650} & 1 & -\frac{2704}{825} & \frac{403}{1650} \\ \hline \frac{38}{33} & \frac{155}{66} & 1 & -\frac{80}{33} & -\frac{5}{66} \\ 1 & 0 & 0 & 0 & 0 \\ 0 & 1 & 0 & 0 & 0\end{array}\right]. \tag{28}$$

6. Numerical evidences

We now show some results arising from fixed stepsize numerical experiments, showing the effectiveness of the methods we have proposed.

We consider the following problem (see [14])

$$\begin{cases} y_1'(x) = -2y_1(x) + y_2(x) + 2\sin x \\ y_2'(x) = y_1(x) - 2y_2(x) + 2(\cos x - \sin x) \end{cases} \tag{29}$$

with $x \in [0, 10]$, with the initial condition $y(0) = [2, 3]^T$, whose exact solution is

$$\begin{cases} y_1(x) = 2e^{-x} + \sin x \\ y_2(x) = 2e^{-x} + \cos x. \end{cases} \tag{30}$$

We solve this problem using the two–step collocation method (28) with m=2 and uniform order 4. The results of the implementation are shown in table 1, where we have listed the number N of grid points we used, the global error $g(N)$ in the final point of the integration interval and the observed order of convergence p, computed from the formula

$$p = \frac{\log\left(ge(N)/ge(2N)\right)}{\log(2)}.$$

Table 1. Numerical results of the numerical solution of (29) with the method (28)

Method (28)		
N	ge	p
100	1.9705e-006	
200	1.0110e-007	4.2378
400	5.6576e-009	4.1594
800	3.3317e-010	4.0858
1600	1.9875e-011	4.0672

The table indicates the effectiveness of the derived methods and, moreover, confirms that the effective order of the method is 4. Further experiments, especially based on variable stepsize implementation of the derived methods, will be covered in papers in preparation.

7. Conclusions and future work

We have developed a new class of collocation based two-step Runge–Kutta methods (12) for the numerical solution of ordinary differential equations. These methods are of uniform order $p = 2m$ on the whole integration interval. We have discussed their stability properties, deriving A-stable methods with $m = 1, 2$ and order $p = 2, 4$ respectively. Examples of methods have also been provided and some preliminary numerical results have been shown.

Future work will address various implementation issues (e.g. the choice of appropriate starting procedures, stepsize and order changing strategy, solving nonlinear systems of equations by modified Newton methods and local error estimation) and also the extension of the collocation technique for methods of the type (2)-(3) using different basis of functions (e.g. trigonometrical, exponential or mixed basis), also in the context of second order ODEs (see [8,16,18,19]).

References

1. J. C. Butcher, *Numerical methods for Ordinary Differential Equations*, 2nd edition (John Wiley & Sons, Chichester, 2003).
2. N. H. Cong, A general family of pseudo two-step Runge–Kutta methods, *Southeast Asian Bull. Math.* **25**(1), pp. 61–73 (2001).
3. N. H. Cong, H. Podhaisky and R. Weiner, Numerical experiments with some explicit pseudo two-step RK methods on a shared memory computer, *Computers & Mathematics with Applications* **36**, pp. 107-116 (1998).
4. H. Podhaisky and R. Weiner, A Class of Explicit Two-Step Runge–Kutta Methods with Enlarged Stability Regions for Parallel Computers, in Parallel Computation (P. Zinterhofer, M. Vajtersic, and Andreas Uhl, eds.), *Lecture Notes in Comput. Sci.*, **1557**, pp. 68–77, Springer (1999).
5. R. D'Ambrosio, M. Ferro, B. Paternoster, A General Family of Two Step Collocation Methods for Ordinary Differential Equations, in Numerical Analysis and Applied Mathematics - *AIP Conf. Proc.* 936 (Edited by T. E. Simos et al.), pp. 45–48 (Springer, 2007).
6. R. D'Ambrosio, M. Ferro, B. Paternoster, Collocation based two step Runge–Kutta methods for Ordinary Differential Equations, in ICCSA 2008, *Lecture Notes in Comput. Sci.*, Part II 5073 (Edited by O. Gervasi et al.), pp. 736–751 (Springer, New York, 2008).
7. R. D'Ambrosio, M. Ferro, Z. Jackiewicz, B. Paternoster, Two-step almost collocation methods for ordinary differential equations, in press on *Numer. Algorithms*, doi: 10.1007/s11075-009-9280-5.
8. R. D'Ambrosio, M. Ferro, B. Paternoster, Two-Step Hybrid Collocation Methods for $y'' = f(x, y)$, *Appl. Math. Lett.* **22**, pp. 1076–1080 (2009).
9. A. Guillou & F.L. Soulé, La résolution numérique des problèmes differentiels aux conditions par des méthodes de collocation, *RAIRO Anal. Numér. Ser. Rouge* v. R-3, pp. 617-44 (1969).
10. E. Hairer and G. Wanner, *Solving Ordinary Differential Equations II – Stiff and Differential–Algebraic Problems*, Springer Series in Computational Mathematics **14** (Springer–Verlag, Berlin, 2002).
11. Z. Jackiewicz, *General linear methods*, in press on John Wiley & Sons.
12. Z. Jackiewicz and S. Tracogna, A general class of two–step Runge–Kutta methods for ordinary differential equations, *SIAM J. Numer. Anal.* **32**(5), pp. 1390–1427 (1995).
13. Z. Jackiewicz and J.H. Verner, Derivation and implementation of two-step Runge–Kutta pairs, *Jpn. J. Ind. Appl. Math.* **19**, pp. 227–248 (2002).
14. J. D. Lambert, Numerical methods for ordinary differential systems: The initial value problem (Wiley, Chichester, 1991).
15. I. Lie and S. P. Norsett, Superconvergence for Multistep Collocation, *Math. Comput.* **52**(185), pp. 65–79 (1989).

16. B. Paternoster, Runge-Kutta(-Nystrom) methods for ODEs with periodic solutions based on trigonometric polynomials, *Appl. Num. Math.* **28**(2-4), pp. 401-412 (1998).
17. B. Paternoster, Two step Runge-Kutta-Nystrom methods based on algebraic polynomials, *Rendiconti di Matematica e sue Applicazioni* Serie VII **23**, pp. 277-288 (2003).
18. B. Paternoster, Two step Runge-Kutta-Nystrom methods for oscillatory problems based on mixed polynomials, in Computational Science - ICCS 2003, *Lecture Notes in Comput. Sci.* 2658, Part II (P. M. A. Sloot, D. Abramson, A. V. Bogdanov, J. J. Dongarra, A. Y. Zomaya, Y. E. Gorbachev Eds.), pp. 131-138 (Springer, Berlin Heidelberg, 2003).
19. B. Paternoster, Two step Runge-Kutta-Nystrom methods for $y'' = f(x, y)$ and P-stability, in Computational Science - ICCS 2002, *Lecture Notes in Comput. Sci.* 2331, Part III (P. M. A. Sloot, C. J. K. Tan, J. J. Dongarra, A. G. Hoekstra Eds.), pp. 459-466 (Springer Verlag, Amsterdam, 2002).
20. H. Podhaisky, R. Weiner, and J. Wensch, High Order Explicit Two-Step Runge–Kutta Methods for Parallel Computers, Tech. Report 19, Universitat Halle, FB Mathematik/Informatik, 1999.
21. J. Schur, Über Potenzreihen die im Innern des Einheitskreises beschränkt sind, *J. Reine Angew. Math.* **147**, pp. 205–232 (1916).
22. J.H. Verner, Improved Starting methods for two-step Runge–Kutta methods of stage-order *p*-3, *Appl. Numer. Math.* **10**, pp. 388–396 (2006).
23. J.H. Verner, Starting methods for two-step Runge–Kutta methods of stage-order 3 and order 6, *J. Comput. Appl. Math.*, **185**, pp. 292–307 (2006).

A NEW COLLOCATION METHOD FOR A BVP

F. A. COSTABILE* and E. LONGO**

Dipartimento di Matematica,
Università degli Studi della Calabria, Italy
** E-mail: costabil@unical.it*
*** E-mail:longo@mat.unical.it*

A type collocation method for the global solution of the general non linear second order boundary value problem

$$\begin{cases} y''(x) = f(x, y(x), y'(x)) & x \in [a, b] \\ y(a) = y_a \\ y(b) = y_b \end{cases}$$

is proposed. An analysis of the global error and an algorithm for the numerical calculation are given. Finally, numerical examples and comparisons with other methods are presented.

Keywords: boundary value problem; collocation method; interpolation.

1. Introduction

In Ref. 3, after a discussion on the class of methods for numerical integration of the following non linear two-point boundary value problem:

$$\begin{cases} y''(x) = f(x, y(x), y'(x)) & x \in [a, b] \\ y(a) = y_a \\ y(b) = y_b \end{cases} \tag{1}$$

a spectral method is proposed which is a collocation method on the zeros of Chebyshev polynomials of the second kind. In particular, the following theorem is proven:

Theorem 1.1 ([3]). *Let $T_k(x)$ be the Chebyshev polynomial of the first kind of degree k and*

$$x_i = \cos \frac{\pi i}{n+1}, \quad i = 1, ..., n \tag{2}$$

the zeros of the Chebyshev polynomials of the second kind of degree n, by putting

$$\beta_{n,i}(x) = \frac{1}{n+1}\sin\frac{\pi i}{n+1}\left[\sum_{k=2}^{n}\frac{G_k(x)}{k}\sin\frac{k\pi i}{n+1} + (x^2-1)\sin\frac{\pi i}{n+1}\right], \tag{3}$$

where

$$G_k(x) = \frac{T_{k+1}(x)}{k+1} - \frac{T_{k-1}(x)}{k-1} + \begin{cases} \frac{2x}{k^2-1} & \text{even } k \\ \frac{2}{k^2-1} & \text{odd } k \end{cases},$$

then the polynomial of degree $n+1$ implicitly defined by

$$y_n(x) = \frac{y_b + y_a}{2} + \frac{y_b - y_a}{2}x + \sum_{i=1}^{n}\beta_{n,i}(x)f(x_i, y_n(x_i), y'_n(x_i)), \quad n > 1, \tag{4}$$

satisfies the relations

$$\begin{cases} y_n(-1) = y_a \\ y_n(1) = y_b \\ y''_n(x_i) = f(x_i, y_n(x_i), y'_n(x_i)) \quad i = 1, ..., n. \end{cases} \tag{5}$$

i.e., it is a collocation polynomial for the BVP on the set of nodes x_i.

Moreover, in Ref. 3 an efficient algorithm for numerical calculation and the estimation error are, also given.

In this note we will prove that the collocation polynomials for (1) in the form (4) are possible for any set of distinct nodes $x_i \in]-1,1[$. This is based on an interpolation problem which will be examined in Ref. 4, while now in section 2 we report the relevant results. The rest of the paper is organized as follows: in section 3 we provide the general method and an *a priori* estimation error; in Section 4 we propose a numerical algorithm for the calculation and, finally, present certain numerical examples to compare the method proposed with the Matlab build-in function bvp4c.

2. The general linear interpolation problem

In Ref. 4 there is the following main theorem:

Theorem 2.1 (4). *Let $\mathcal{P}_n$ be the space of polynomials of degree $\leq n$ and $x_1, ..., x_{n-1}$ $n-1$ distinct points in $]-1,1[$. Let $l_i(x)$ be the fundamental polynomials of Lagrange calculated on the $n-1$ nodes x_i and*

$$p_{n,i}(x) = \int_{-1}^{1} G(x,t)l_i(t)dt, \quad i = 1, ..., n-1 \tag{6}$$

where

$$G(x,t) = \begin{cases} \frac{(t+1)(x-1)}{2} & t \le x \\ \frac{(x+1)(t-1)}{2} & x \le t \end{cases} \tag{7}$$

Let $\omega_0, \omega_1, ..., \omega_{n-1}, \omega_n \in \mathbb{R}$, *then the polynomial*

$$y_n(x) = \frac{\omega_n + \omega_0}{2} + \frac{\omega_n - \omega_0}{2} x + \sum_{i=1}^{n-1} p_{n,i}(x)\omega_i \tag{8}$$

is the unique element of $\mathcal{P}_n$ *that satisfies the interpolation conditions*

$$\begin{cases} y_n''(x_i) = \omega_i, \quad i = 1, ..., n-1, \\ y_n(-1) = \omega_0, \\ y_n(1) = \omega_n. \end{cases} \tag{9}$$

An alternative representation of the polynomial basis is, also possible; in fact, the following theorem holds:

Theorem 2.2 (4). *With the notations already introduced we have*

$$p_{n,i}(x) = \frac{(-1)^i}{G} \begin{vmatrix} 1 & x^1 & x^2 & x^3 & \dots & x^n \\ 1 & -1 & 1 & -1 & \dots & (-1)^n \\ 1 & 1 & 1 & 1 & \dots & 1 \\ 0 & 0 & 2 & 6x_1 & \dots & n(n-1)(x_1)^{n-2} \\ \vdots & & & & & \vdots \\ 0 & 0 & 2 & 6x_{i-1} & \dots & n(n-1)(x_{i-1})^{n-2} \\ 0 & 0 & 2 & 6x_{i+1} & \dots & n(n-1)(x_{i+1})^{n-2} \\ \vdots & & & & & \vdots \\ 0 & 0 & 2 & 6x_{n-1} & \dots & n(n-1)(x_{n-1})^{n-2} \end{vmatrix}, \qquad i = 1, ..., n-1 \tag{10}$$

with

$$G = 2 * n!(n-1)! \prod_{i>j}^{n-1} (x_i - x_j), \tag{11}$$

3. The method

Let us consider the boundary value problem

$$\begin{cases} y''(x) = f(x, y(x), y'(x)) \;\; x \in [-1,1], \quad -\infty < y, y' < \infty \\ y(-1) = y_a \\ y(1) = y_b \end{cases} \tag{12}$$

We assume that $f(x, z_1, z_2)$ is a real function defined and continuous on the strip $S = [-1, 1] \times \mathbb{R}^2$ where it has continuous derivatives which satisfy, for some positive constant M,

$$\frac{\partial f}{\partial z_1} > 0, \qquad \left|\frac{\partial f}{\partial z_2}\right| \leq M.$$

Moreover, we assume that f satisfies a uniform Lipschit condition in z_1 e z_2, which means that there exist two nonnegative constants L and K such that, whenever (x, y_1, y_2) and $(x, \overline{y}_1, \overline{y}_2)$ are in the domain of f, the inequality

$$|f(x, y_1, y_2) - f(x, \overline{y}_1, \overline{y}_2)| \leq L\,|y_1 - \overline{y}_1| + K\,|y_2 - \overline{y}_2|\,.$$

holds. Under this hypotheses the BVP has a unique solution $y(x)$.

With the notation already introduced the following theorem holds:

Theorem 3.1. *For $n > 1$, the polynomial of degree n, implicitly defined by*

$$y_n(x) = \frac{y_b + y_a}{2} + \frac{y_b - y_a}{2}x + \sum_{i=1}^{n-1} p_{n,i}(x) f(x_i, y_n(x_i), y_n'(x_i)) \qquad (13)$$

satisfies the relations

$$\begin{cases} y_n''(x_i) = f(x_i, y_n(x_i), y_n'(x_i)), \quad i = 1, ..., n-1, \\ y_n(-1) = y_a, \\ y_n(1) = y_b, \end{cases} \qquad (14)$$

i.e., it is a collocation polynomial for the problem (12) on the set of nodes $x_1, ..., x_{n-1}$.

Proof. Follows from theorem 2.1 setting

$$\begin{cases} \omega_i = f(x_i, y_n(x_i), y_n'(x_i)),\ i = 1, ..., n-1, \\ \omega_0 = y_a, \\ \omega_n = y_b. \end{cases}$$

□

4. Global error

For the estimation of the global error

$$y^{(s)}(x) - y_n^{(s)}(x), \quad s = 0, 1, 2,$$

by putting

$$\Lambda_n = \max_{-1 \leq x \leq 1} \sum_{i=1}^{n-1} |l_i(x)| \qquad (15)$$

and

$$R = \max_{-1 \le x, \xi_x \le 1} \left| \frac{y^{(n+1)}(\xi_x)(x - x_1) \cdots (x - x_{n-1})}{(n-1)!} \right| \tag{16}$$

we have the following theorem:

Theorem 4.1 ([3]). *If* $y \in C^{n+3}([-1,1])$, L *and* K *are the Lipschitz constants of function* f *such that* $1 - 2\Lambda_n K > 0$ *and* $1 - \Lambda_n(2K + L) > 0$, *then*

$$\left\| y^{(s)}(x) - y_n^{(s)}(x) \right\|_\infty \le \frac{cR}{1 - \Lambda_n(2K + L)}, \qquad s = 0, 1, 2, \tag{17}$$

where

$$c = \begin{cases} 1 \; if \; s = 0, \; 2 \\ 2 \;\; if \;\; s = 1. \end{cases} \tag{18}$$

Remark 4.1 ([3]). *If* f *does not depend on* $y'(x)$, *then*

$$\left\| y^{(s)}(x) - y_n^{(s)}(x) \right\|_\infty \le \frac{cR}{1 - \Lambda_n L}. \tag{19}$$

5. The numerical algorithm and numerical comparison

In order to calculate the approximate solution of problem (1) by (13) at $x \in [-1,1]$, we need the values $y_i = y_n(x_i)$, $i = 1, ..., n-1$. To do this, we can solve the following system:

$$\begin{cases} y_i = \dfrac{y_b + y_a}{2} + \dfrac{y_b - y_a}{2} x_i + \sum\limits_{k=1}^{n-1} p_{n,k}(x_i) f(x_k, y_k, y'_k) & i = 1, ..., n-1 \\ \quad y'_i = \dfrac{y_b - y_a}{2} + \sum\limits_{k=1}^{n-1} p'_{n,k}(x_i) f(x_k, y_k, y'_k) & i = 1, ..., n-1 \end{cases} \tag{20}$$

Let us set

$$A = \begin{pmatrix} p_{n,1}(x_1) & \cdots & p_{n,n-1}(x_1) & 0 & \cdots & 0 \\ \vdots & & \vdots & \vdots & & \vdots \\ p_{n,1}(x_{n-1}) & \cdots & p_{n,n-1}(x_{n-1}) & 0 & \cdots & 0 \\ 0 & \cdots & 0 & p'_{n,1}(x_1) & \cdots & p'_{n,n-1}(x_1) \\ \vdots & & \vdots & \vdots & & \vdots \\ 0 & \cdots & 0 & p'_{n,1}(x_{n-1}) & \cdots & p'_{n,n-1}(x_{n-1}) \end{pmatrix},$$

$$f_k = f(x_k, y_k, y'_k), \qquad F(Y_n) = (f_1, ..., f_{n-1}, f_1, ..., f_{n-1})^T,$$
$$Y_n = (y_1, ..., y_{n-1}, y'_1, ..., y'_{n-1})^T, \qquad a_i = \frac{y_a - y_b}{2} x_i - \frac{y_a + y_b}{2},$$
$$b_i = \frac{y_a - y_b}{2}, \qquad C_n = (a_1, ..., a_{n-1}, b_1, ..., b_{n-1})^T,$$

the previous system can be written in the form

$$AF(Y_n) - Y_n = C_n,$$

or equivalently

$$Y_n = G(Y_n)$$

where

$$G(Z) = AF(Z) - C_n.$$

For the existence and uniqueness of the solution we provide the following theorem:

Theorem 5.1 ([3]). *If $\|A\| L < 1$, then the previous system has a unique solution that can be calculated with the iterative method*

$$\begin{aligned} &Y_n^{(0)} \ \textit{arbitrary}, \\ &Y_n^{(\nu+1)} = G(Y_n^{(\nu)}), \quad \nu = 1, ... \end{aligned}$$

Now we report certain numerical results for classical test problems and compare the error committed applying our method, with $n = 7$, on different sets of nodes, in particular between Chebyshev nodes of the second kind (already examined in Ref. 3) with equidistant, Chebyshev of the first kind and Legendre nodes. We also compare these results with the ones obtained by applying the Matlab ODE solver bvp4c.

Problem 5.1. *Let us consider the problem*

$$\begin{cases} y'' = y + 2e^x \\ y(-1) = 0 \\ y(1) = 2e \end{cases} \qquad \textit{with the solution } \ y(x) = (x+1)e^x.$$

The error in the approximation of the solution is plotted in Figure 1.

In the case of bvp4c (dashed line) 100 function evaluations are needed for this problem on a mesh of 13 points; our method applied on Chebyshev second kind nodes (dotted line) or on one of the other three set of nodes considered (solid line) requires only 36 function evaluations to obtain an error less than or equal to the requested tolerance $\epsilon = 10^{-5}$.

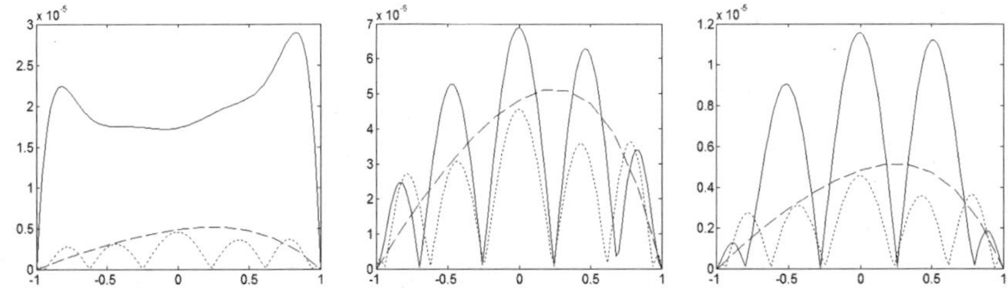

(a) Equidistant nodes (b) Legendre nodes (c) Cheb I kind nodes

Fig. 1: Error functions of problem 5.1.

Problem 5.2. *Let be*

$$\begin{cases} y'' = yy'^2 - 2(2x^4 - 1) \\ y(-1) = 1 \\ y(1) = 1 \end{cases} \qquad \textit{with the solution } \ y(x) = x^2.$$

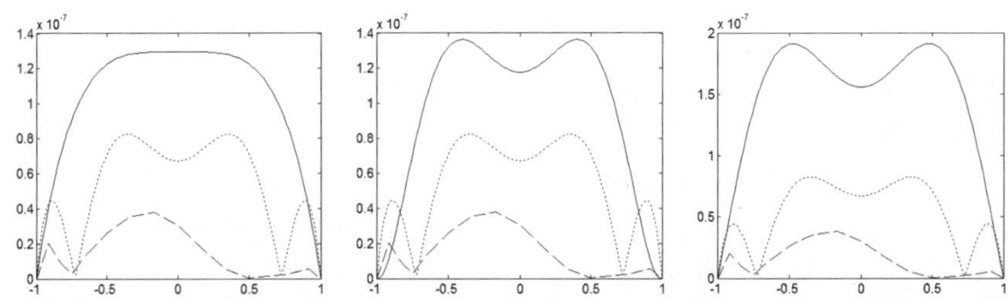

(a) Equidistant nodes (b) Legendre nodes (c) Cheb I kind nodes

Fig. 2: Error functions of problem 5.2.

For this problem our method applied on Chebyshev second kind nodes (dotted line) or on one of the other three set of nodes considered (solid line) requires only 102 function evaluations to obtain an error less than or equal to the requested tolerance $\epsilon = 10^{-7}$. *To obtain an error of the same order* (10^{-7}) *bvp4c (dashed line) requires 337 function evaluations on a mesh of 19 points.*

Problem 5.3. *Let us consider the problem*

$$\begin{cases} y'' = 2yy' \\ y(-1) = \frac{1}{3} \\ y(1) = 1 \end{cases} \quad \textit{with the solution } \ y(x) = \frac{1}{2-x}.$$

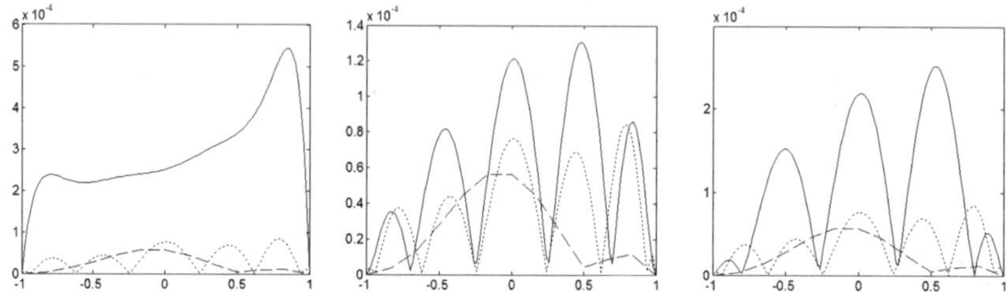

(a) Equidistant nodes (b) Legendre nodes (c) Cheb I kind nodes

Fig. 3: Error functions of problem 5.3.

In the case of bvp4c (dashed line) 201 function evaluations are needed for this problem on a mesh of 13 points; our method applied on Chebyshev second kind nodes (dotted line) or on one of the other three set of nodes considered (solid line) requires only 132 function evaluations to obtain an error less than or equal to the requested tolerance $\epsilon = 10^{-4}$.

Problem 5.4.

Let us consider the problem

$$\begin{cases} y'' = -(1+0.01y^2)y + 0.01\cos^3 x \\ \qquad y(-1) = \cos(-1) \\ \qquad y(1) = \cos(1) \end{cases} \qquad \textit{with the solution } \ y(x) = \cos(x).$$

In this case bvp4c (dashed line) requires 60 function evaluations on a mesh of 13 points; our method applied on Chebyshev second kind nodes (dotted line) or on one of the other three set of nodes considered (solid line) requires only 7 function evaluations to obtain an error less than or equal to the requested tolerance $\epsilon = 10^{-6}$.

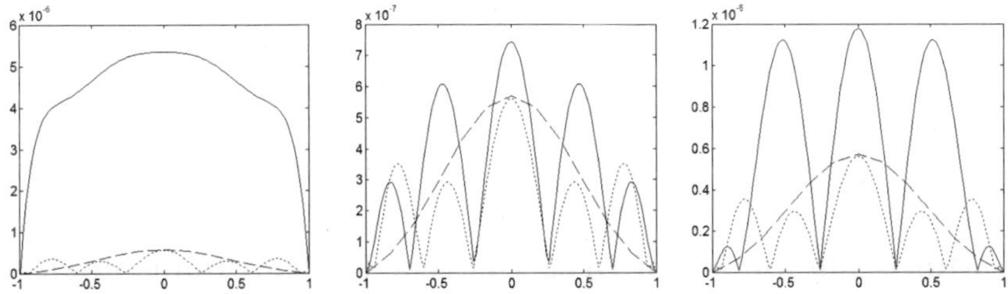

(a) Equidistant nodes (b) Legendre nodes (c) Cheb I kind nodes

Fig. 4: Error functions of problem 5.4.

References

1. U. M. Asher, R. M. M. Mattheij and R. D. Russell, *Numerical Solution of Boundary Value Problems for Ordinary Differential Equations*, (SIAM, Philadelphia, 1988).
2. J. P. Boyd, *Chebyshev and Fourier Spectral Methods*, (Dover Publications, Inc., New York, 2000).
3. F. Costabile and A. Napoli, A method for polynomial approximation of the solution of general second order BVPs, *Far East J. Appl. Math.*, **25**, 3 (2006), pp. 289–305.
4. F. Costabile and E. Longo, A Birkhoff type interpolation problem and applications, in preparation.
5. P. J. Davis, *Interpolation & Approximation*, (Dover Publication, Inc. New York, 1975).
6. L. Fox, *The Numerical Solution of Two-point Boundary Value Problems in Ordinary Differential Equations*, (Oxford University Press, 1957).
7. H. B. Keller, *Numerical Methods For Two-point Boundary-Value Problems*, (Dover Publications, New York, 1992).
8. J. Kierzenka and L. F. Shampine, A BVP solver based on residual control and the MATLAB PSE, *ACM TOMS*, **27**, 3 (2001), pp. 299–316.
9. L. F. Shampine, Solving ODEs and DDEs with residual control, *Appl. Num. Math.*, **52**, 1 (2005), pp. 113–127.

Multiscale Models of Drug Delivery by Thin Implantable Devices

C. D'Angelo and P. Zunino

MOX, Dipartimento di Matematica,
Politecnico di Milano, Italy
carlo.dangelo@polimi.it, paolo.zunino@polimi.it

In this paper, we address the numerical modelling of drug release from thin vascular devices, such as drug eluting stents (DES). We study blood flow and mass transport in blood vessels using a unified formulation of the coupled fluid dynamic problems in the arterial lumen and in the arterial wall, as well as of the respective transport problems for the drug concentration. Moreover, we consider suitable reduced 1D models for the release process from the device. This approach reduces the computational effort by avoiding to resolve the small space scales of a thin DES. Finally, we present numerical results for the fully coupled and three-dimensional problem on realistic geometries.

Keywords: drug eluting stents, multiscale models, heterogeneous media

1. Introduction

A stent is a medical device used during angioplasty to open occluded arteries. It is a wire metal mesh tube that is inserted permanently into the blocked blood vessel. The stent is initially crisped around a small inflatable balloon mounted on the tip of a catheter. The clinical procedure consists in moving the collapsed stent into the area of blockage; then, the balloon is inflated, causing the expansion of the device. Once expanded, the stent keeps the arterial lumen open, ensuring a physiological flow rate. Unfortunately, re-narrowing of the vessel after angioplasty is observed in some cases. This problem can be overcome using drug-eluting stents (DES). DES are normal metal stents coated with a pharmacologic agent (drug) that slows down the re-narrowing process. These devices have to respect severe requirements. Most of stents have a complex geometry consisting in thin interconnected struts (see figure 1, left). Moreover, DES must be charged with enough drug to be effective in healing the arterial wall and should release the drug in a possibly uniform manner. Hence, the design of such devices is a very complex task. Numerical simulation of drug delivery by DES may help engineers to choose between different solutions or to investigate new concepts.

To fully understand the relevant phenomena involved in the drug release and evaluate the performance of a given design, it is mandatory to solve the diffusion-advection equation for the drug concentration both in the blood stream around the DES and in the arterial wall. This in turn requires computing the corresponding advection fields, that are the blood velocity in the lumen and the plasma filtration velocity in the arterial wall. In this regard, we make use of a finite element scheme for the computation of incompressible flows in heterogeneous media, based on a stabilized mixed formulation, that automatically adapts to match different coupling conditions, such as the coupling between the viscous flow inside the arterial lumen and the Darcy's flow through the wall.

Starting from previous works about multiscale modelling of DES[1] and coupled 1D-3D diffusion-reaction problems,[2] we model a thin DES as a one-dimensional object at the interface between the lumen and the arterial wall. We neglect any fluid-dynamic effect caused by the presence of the device as an obstacle. Recent works[3,4] have shown that these effects may influence drug release; however, for thin devices and laminar regime, the perturbation of luminal flow is expected to be small. On the other side, this approach allows us to deal with complex DES geometries without having to resolve small length scales.

2. Mathematical models for blood flow and drug release

Let us consider the physical domains represented in figure 1. We denote by Ω_l the lumen, that is the (fluid) domain occupied by blood, and by Ω_w the arterial wall. Let $\partial\Omega_l = \Gamma_{\text{in}} \cup \Gamma_{\text{out}} \cup \Gamma_{\text{int}}$, where Γ_{in} is the inflow of the lumen, Γ_{out} is the outflow, and Γ_{int} is the interface with the arterial wall. Analogously, let $\partial\Omega_w = \Gamma_{\text{int}} \cup \Gamma_{\text{ext}} \cup \Gamma_{\text{cut}}$, where Γ_{ext} is the exterior part of the vessel, and Γ_{cut} is the "artificial" wall boundary. The exterior unit normal vectors on $\partial\Omega_l$, $\partial\Omega_l$, will be denoted respectively by $\mathbf{n}_l$, $\mathbf{n}_w$.

We assume that the stent has a thin cylindrical structure with (small) radius $\rho > 0$ and centerline Λ_s, lying on the interface Γ_{int} (see fig. 2). In our model the stent will be represented by the one-dimensional centerline Λ_s only. As a consequence, any fluid-dynamical effect of the device on the flow is neglected. This assumption is obviously critical, but can be reasonable for very thin DES. On the other hand, we will make use of suitable reduced models of drug release, avoiding to resolve the three-dimensional geometry of the device, and accounting only for its 1D structure.

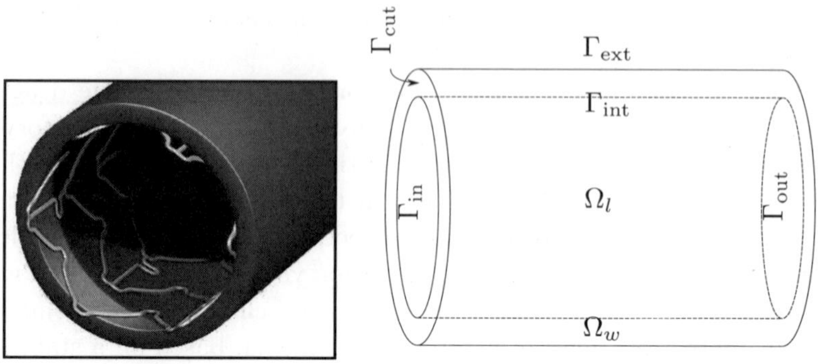

Figure 1. On the left: an example of (expanded) stent. Notice the small radius of the strut compared to the arterial radius. On the right: domains and boundaries considered by the model.

2.1. *Blood flow models*

Let us denote by $\mathbf{u}_i$ the blood velocity and by p_i the blood pressure (normalized with respect to the fluid density) in the arterial lumen ($i = l$) and in the arterial wall ($i = w$). Specifically, we will assume the steady Navier-Stokes equations for blood flow in the lumen,

$$\begin{cases} -\nu\Delta\mathbf{u}_l + (\mathbf{u}_l \cdot \nabla)\mathbf{u}_l + \nabla p_l = \mathbf{0} & \text{in } \Omega_l, \\ \nabla \cdot \mathbf{u}_l = 0 & \text{in } \Omega_l, \end{cases} \tag{1}$$

and the Darcy filtration model for the filtration of plasma in the wall,

$$\begin{cases} \eta\mathbf{u}_w + \nabla p_w = \mathbf{0} & \text{in } \Omega_w, \\ \nabla \cdot \mathbf{u}_w = 0 & \text{in } \Omega_w. \end{cases} \tag{2}$$

In equations (1) and (2), ν and η are respectively the fluid dynamic viscosity and the inverse permeability of the arterial wall. Regarding the boundary conditions, we assign the blood pressure jump across inflow and outflow of the lumen, and the external pressure on the exterior part of the wall, and impose vanishing normal velocity on the artificial boundaries. As interface conditions on Γ_{int}, we assume the continuity of the normal component of the velocity and the continuity of the normal component of the Cauchy stress tensor; moreover, a Beavers-Joseph-Saffman law[5] for the jump of the tangential components of the Cauchy stress tensor will be assumed. Precisely, we have

$$\begin{cases} p_l = p_{\rm in} & \text{on } \Gamma_{\rm in}, \\ p_l = p_{\rm out} & \text{on } \Gamma_{\rm out}, \\ p_w = p_{\rm ext} & \text{on } \Gamma_{\rm ext}, \\ \mathbf{u}_w \cdot \mathbf{n}_w = 0 & \text{on } \Gamma_{\rm cut}, \end{cases} \qquad \begin{cases} \mathbf{u}_l \cdot \mathbf{n}_l + \mathbf{u}_w \cdot \mathbf{n}_w = 0 & \text{on } \Gamma_{\rm int}, \\ \sigma_l \mathbf{n}_l + \sigma_w \mathbf{n}_w = \mathbf{0} & \text{on } \Gamma_{\rm int}, \\ \mathbf{t}^T(\sigma_l \mathbf{n}_l + \sigma_w \mathbf{n}_w) = \kappa \mathbf{t}^T \mathbf{u}_l & \text{on } \Gamma_{\rm int}, \end{cases} \tag{3}$$

where $\sigma_l = p_l \mathrm{I} - \nu \nabla \mathbf{u}_l$, $\sigma_w = p_w \mathrm{I}$ are the Cauchy stress tensors (we denote by I the identity tensor), $\kappa > 0$ is a friction coefficient, and $\mathbf{t}$ is a matrix whose columns form a basis for the tangential plane on $\Gamma_{\rm int}$. No-slip interface conditions are recovered for $\kappa \to \infty$, while no-friction conditions are obtained if $\kappa = 0$.

2.2. *Reduced 1D-3D models for drug release by thin DES*

Once the blood flow problem (1,2,3) is solved, the resulting blood velocity field can be used to solve the transport problem for the drug released by the stent. In Ref. 1, it has been shown that drug release from DES can be efficiently accounted for by means of special boundary conditions at the interface. Let Γ_s be the actual interface between the computational domain and the stent (see figure 2). The mass flux outgoing the stent is given by

$$\tilde{f}(t,c) = \varphi(t)(c - c_s) \quad \text{on } \Gamma_s, \tag{4}$$

where c_s is the initial drug concentration in the stent coating, that is the initial drug charge divided by the coating volume, and $\varphi(t)$ is a time-dependent coefficient depending on the thickness l_{coat} of the coating and the diffusivity D_{coat} of the drug in the coating, given by

$$\varphi(t) = 2\frac{D_{coat}}{l_{coat}} \sum_{n=0}^{\infty} e^{-(n+1/2)kt}, \quad k = \pi^2 \frac{D_{coat}}{l_{coat}^2}, \quad t > 0. \tag{5}$$

The term $\tilde{f}$ in (4) is a mass flux *per unit area*, being distributed over the surface Γ_s. In the spirit of the work presented in Ref. 2, this mass flux per unit area can be further reduced to a mass flux *per unit length*, distributed on the centerline of the stent, which reads

$$f(t,c) = \pi\rho\varphi(t)(\bar{c} - c_s) \quad \text{on } \Lambda_s,$$

where the "bar" operator gives the average value over half a circle $\gamma(s)$ having radius ρ and normal to Λ_s, that is, using cylindrical local coordinates (s, r, θ) near Λ_s,

$$\bar{c}(s) = \frac{1}{|\gamma(s)|} \int_{\gamma(s)} c \mathrm{d}\gamma = \frac{1}{\pi} \int_0^{\pi} c(s, \rho, \theta) \mathrm{d}\theta.$$

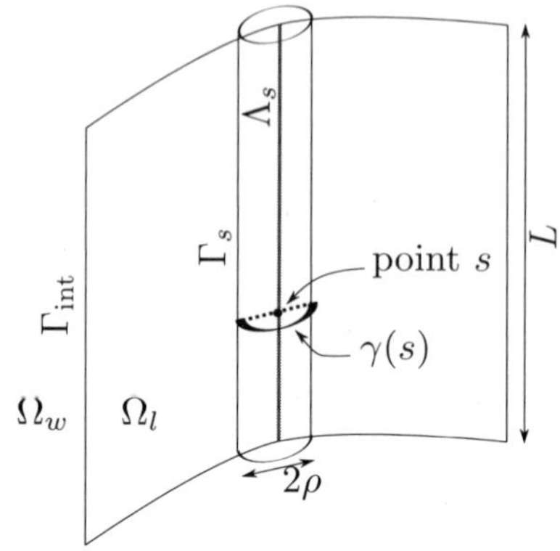

Figure 2. A small piece of stent of length L lying on the interface $\Gamma_{\rm int}$.

If we assume this one-dimensional representation of the mass flux, the corresponding transport problem reads

$$\begin{cases} \dfrac{\partial c_l}{\partial t} - D_l \Delta c_l + \mathbf{u}_l \cdot \nabla c_l + f(t, c_l)\delta_{\Lambda_s} = 0 & \text{in } \Omega_l, \\ \dfrac{\partial c_w}{\partial t} - D_w \Delta c_w + \mathbf{u}_w \cdot \nabla c_w + f(t, c_w)\delta_{\Lambda_s} = 0 & \text{in } \Omega_w, \end{cases} \tag{6}$$

where $D_{l,w}$ are the diffusivities of the drug respectively in the arterial lumen and in the arterial wall, and $f(c_l, c_w)\delta_{\Lambda_s}$ is the Dirac *measure* concentrated on Λ_s defined by

$$(f(t, c)\delta_{\Lambda_s}, \psi) = \int_{\Lambda_s} f(t, c)\psi \mathrm{d}s \tag{7}$$

for all continuous function ψ.

We assume $c_i = 0$ as initial conditions, and the following external boundary conditions and linear filtration law on the interface,

$$\begin{cases} c_l = c_{\rm in} & \text{on } \Gamma_{\rm in}, \\ -D_l \nabla c_l \cdot \mathbf{n}_l = 0 & \text{on } \Gamma_{\rm out}, \\ -D_w \nabla c_w \cdot \mathbf{n}_w = 0 & \text{on } \Gamma_{\rm ext} \cup \Gamma_{\rm cut}, \\ -D_l \nabla c_l \cdot \mathbf{n}_l = D_w \nabla c_w \cdot \mathbf{n}_w = P_{\rm int}(c_l - c_w) & \text{on } \Gamma_{\rm int}, \end{cases} \tag{8}$$

where $c_{\rm in}$ is the inflow concentration (typically zero), and $P_{\rm int}$ is the permeability of the *endothelium*, the membrane which lines the arterial wall. The resulting problem features *measure terms*; we refer to Ref. 2 for the analysis of the problem in the steady case using suitable weighted Sobolev spaces.

3. Numerical approximation

3.1. *Coupled fluid-dynamics*

In this section we set up the discrete variational formulation of the fluid model. We assume that Ω_i $(i \in l, w)$ are convex polygonal domains, equipped with a family of conforming triangulation $\mathcal{T}_{h,i}$ made of affine simplices K. We also denote with $\mathcal{F}_{h,i}$ the set of all interior faces F of $\mathcal{T}_{h,i}$.

As approximation spaces we consider the lowest order velocity-pressure finite element pair on each subdomain Ω_i, as follows,

$$V_{i,h} = \{v_h \in H^1(\Omega_i) \ : \ v_h|_K \in \mathbb{P}^1(K), \ \forall K \in \mathcal{T}_{h,i}\}, \quad \mathbf{V}_{h,i} = [V_{h,i}]^3,$$
$$Q_{i,h} = \{q_h \in L^2(\Omega_i) \ : \ q_h|_K \in \mathbb{P}^0(K), \ \forall K \in \mathcal{T}_{h,i}\}.$$

As global approximation spaces we thus consider

$$\mathbf{V}_h = \bigoplus_{i=1}^{N} \mathbf{V}_{i,h}, \quad Q_h = \bigoplus_{i=1}^{N} Q_{i,h},$$

and follow Ref. 6 where the coupled Stokes' and Darcy's equations are addressed. Nevertheless, such method can be easily extend to higher order $\mathbb{P}^k/\mathbb{P}^{k-1}_{\text{disc.}}$ spaces, with $k > 1$ and possibly depending on the subdomain. For these subjects we refer the reader to Ref. 7, where the general case of a multi-domain formulation is treated for any polynomial degree k.

We define the jump of the normal component of any finite element function $\mathbf{v}_h \in \mathbf{V}_h$ across any face F_{int} of the trace mesh at the interface Γ_{int}, and the jump of any $p_h \in Q_h$ across any internal face F in the usual way,

$$[\![\mathbf{v}_h \cdot \mathbf{n}]\!](\mathbf{x})_{|F_{\text{int}}} = \lim_{\delta \to 0} [\mathbf{v}_h(\mathbf{x} + \delta \mathbf{n}) - \mathbf{v}_h(\mathbf{x} - \delta \mathbf{n})] \cdot \mathbf{n},$$
$$[\![p_h]\!](\mathbf{x})_{|F} = \lim_{\delta \to 0} [p_h(\mathbf{x} + \delta \mathbf{n}) - p_h(\mathbf{x} - \delta \mathbf{n})].$$

The orientation of the normal $\mathbf{n}$ on the faces is arbitrary and does not influence the method. Finally, we denote $h_F = \text{diam}(F)$ the diameter of any face F. With little abuse of notation, we will also denote h_F a piecewise constant function defined on the boundaries of the subdomains, taking the value $\text{diam}(F)$ on each face F.

Now let us introduce the discrete variational formulation of the coupled problem by mixed finite elements with interior penalty (IP) stabilization, using Nitsche penalty techniques to treat the essential boundary/interface conditions. For $\mathbf{u}_h = (\mathbf{u}_{l,h}, \mathbf{u}_{w,h}) \in \mathbf{V}_h$, $\mathbf{v}_h = (\mathbf{v}_{l,h}, \mathbf{v}_{w,h}) \in \mathbf{V}_h$, $\mathbf{w}_{l,h} \in V_{l,h}$

and $p_h = (p_{l,h}, p_{w,h}) \in Q_h$, $q_h = (q_{l,h}, q_{w,h}) \in Q_h$ we define the following local forms,

$$a_l(\mathbf{u}_{l,h}, \mathbf{v}_{l,h}) = \int_{\Omega_l} \nu \nabla \mathbf{u}_{l,h} : \nabla \mathbf{v}_{l,h} + \int_{\Gamma_{\text{int}}} \kappa(\mathbf{t}_\Gamma^T \mathbf{u}_{l,h}) \cdot (\mathbf{t}_\Gamma^T \mathbf{v}_{l,h}), \quad (9)$$

$$a_w(\mathbf{u}_{w,h}, \mathbf{v}_{w,h}) = \int_{\Omega_w} \eta \mathbf{u}_{w,h} \cdot \mathbf{v}_{w,h} + \int_{\Gamma_{\text{cut}}} \gamma_u h_F^{-1} \mathbf{u}_h \cdot \mathbf{v}_h, \quad (10)$$

$$b_l(p_{l,h}, \mathbf{v}_{l,h}) = -\int_{\Omega_l} p_{l,h} \nabla \cdot \mathbf{v}_{l,h}, \quad (11)$$

$$b_w(p_{w,h}, \mathbf{v}_{w,h}) = -\int_{\Omega_w} p_{w,h} \nabla \cdot \mathbf{v}_{w,h}, + \int_{\Gamma_{\text{cut}}} p_{w,h} \mathbf{v}_{w,h} \cdot \mathbf{n}_w, \quad (12)$$

$$j_{i,p}(p_{i,h}, q_{i,h}) = \int_{\mathcal{F}_{i,h}} \gamma_p h_F [\![p_{i,h}]\!][\![q_{i,h}]\!], \quad i \in \{l, w\}, \quad (13)$$

where γ_u and γ_p are constant parameters that guarantee the stability of the method. For $\mathbf{w}_h = (\mathbf{w}_{l,h}, \mathbf{w}_{w,h}) \in \mathbf{V}_h$, consider the following global forms,

$$\begin{aligned} a(\mathbf{u}_h, \mathbf{v}_h; \mathbf{w}_h) &= \sum_i a_i(\mathbf{u}_{i,h}, \mathbf{v}_{i,h}) + \int_{\Omega_l} (\mathbf{w}_{l,h} \cdot \nabla \mathbf{u}_{l,h}) \cdot \mathbf{v}_{l,h} \quad (14) \\ &+ \int_{\Gamma_{\text{int}}} \gamma_u h_F^{-1} [\![\mathbf{u}_h \cdot \mathbf{n}]\!][\![\mathbf{v}_h \cdot \mathbf{n}]\!] \\ &+ \int_{\mathcal{F}_{l,h}} \gamma_{\mathbf{w}} \frac{h_F^2}{|\mathbf{w}_{l,h}|_F} [\![(\mathbf{w}_{l,h} \cdot \nabla)\mathbf{u}_{l,h}]\!][\![(\mathbf{w}_{l,h} \cdot \nabla)\mathbf{v}_{l,h}]\!], \\ b(p_h, \mathbf{v}_h) &= \sum_i b_i(p_{i,h}, \mathbf{v}_{i,h}) + \int_{\Gamma_{\text{int}}} p_{w,h} [\![\mathbf{v}_h \cdot \mathbf{n}]\!], \quad (15) \\ j_p(p_h, q_h) &= \sum_i j_{i,p}(p_{i,h}, q_{i,h}), \quad (16) \end{aligned}$$

where $\gamma_{\mathbf{w}}$ is the constant parameter associated with the IP stabilization of the advective terms.[8] Finally, consider the linear forms associated to boundary conditions (3),

$$F_u(\mathbf{v}_h) = \int_{\Gamma_{\text{in}}} p_{\text{in}} \mathbf{v}_{l,h} \cdot \mathbf{n}_l + \int_{\Gamma_{\text{out}}} p_{\text{out}} \mathbf{v}_{l,h} \cdot \mathbf{n}_w, + \int_{\Gamma_{\text{ext}}} p_{\text{ext}} \mathbf{v}_{w,h} \cdot \mathbf{n}_w, \quad (17)$$

$$F_p(q_h) = 0. \quad (18)$$

Given $\mathbf{w}_h$, consider the linear problem of *finding* $\mathbf{u}_h =: T(\mathbf{w}_h) \in \mathbf{V}_h$ *such that*

$$\begin{cases} a(\mathbf{u}_h, \mathbf{v}_h; \mathbf{w}_h) + b(p_h, \mathbf{v}_h) = F_u(\mathbf{v}_h), & \forall \mathbf{v}_h \in \mathbf{V}_h, \\ b(q_h, \mathbf{u}_h) - j_p(p_h, q_h) = F_p(q_h), & \forall q_h \in Q_h. \end{cases} \quad (19)$$

The discrete solution of the fluid model is the fixed point of the map T, which can be approximated by the standard Picard's iterations

$$\mathbf{u}_h^{(k+1)} = T(\mathbf{u}_h^{(k)}). \tag{20}$$

Each iteration requires solving the linear (unsymmetric) problem (19). The latter is a coupled problem on $\Omega_l \cup \Omega_w$, the coupling term being the interface integrals in equations (14) and (15). We employ an iterative splitting method that only requires solving on the subdomains separately. This approach is particularly useful in view of parallel applications and of extensions to problems with many subdomains. It has been studied in Ref. 7 for the symmetric case ($\mathbf{w} = \mathbf{0}$). The iterative method is obtained by testing (19) with test functions $(\mathbf{v}_h, q_h)$ having support in Ω_i only, $i \in \{l, w\}$. In each of the resulting decoupled problems, terms containing the interface values of $\mathbf{u}_{j,h}$ and $p_{j,h}$, $j \neq i$, are split and considered at the previous iteration.

In the sequel we consider $i \in \{l, w\}$, $\mathbf{u}_{i,h}, \mathbf{v}_{i,h}, \mathbf{w}_{i,h} \in V_{i,h}$ and $p_{i,h}, q_{i,h} \in Q_{i,h}$. Let us define the bilinear forms

$$\begin{aligned}
s_{i,p}(p_{i,h}, q_{i,h}) &= \int_{\Gamma_{\text{int}}} \sigma_p h_F^{-1} p_{i,h} q_{i,h}, \\
s_{i,u}(\mathbf{u}_{i,h}, \mathbf{v}_{i,h}) &= \int_{\Gamma_{\text{int}}} \sigma_u h_F^{-1} (\mathbf{u}_{i,h} \cdot \mathbf{n})(\mathbf{v}_{i,h} \cdot \mathbf{n}),
\end{aligned}$$

where $\sigma_p, \sigma_u > 0$ are relaxation constants for the iterative method. Consider the following forms,

$$\begin{aligned}
\tilde{a}_l(\mathbf{u}_{l,h}, \mathbf{v}_{l,h}; \mathbf{w}_h) &= a_l(\mathbf{u}_{l,h}, \mathbf{v}_{l,h}) + \int_{\Omega_l} (\mathbf{w}_{l,h} \cdot \nabla \mathbf{u}_{l,h}) \cdot \mathbf{v}_{l,h} \\
&\quad + \int_{\Gamma_{\text{int}}} \gamma_u h_F^{-1} (\mathbf{u}_{l,h} \cdot \mathbf{n})(\mathbf{v}_{l,h} \cdot \mathbf{n}) + s_{l,u}(\mathbf{u}_{l,h}, \mathbf{v}_{l,h}), \\
\tilde{b}_l(p_{l,h}, \mathbf{v}_{l,h}) &= b_l(p_{l,h}, \mathbf{v}_{l,h}), \\
\tilde{F}_{l,u}(\mathbf{v}_{l,h}; (\mathbf{u}_{l,}, \mathbf{u}_{w,h})) &= F_u((\mathbf{v}_{l,h}, \mathbf{0})) + \int_{\Gamma_{\text{int}}} \gamma_u h_F^{-1} (\mathbf{u}_{w,h} \cdot \mathbf{n})(\mathbf{v}_{l,h} \cdot \mathbf{n}) \\
&\quad - \int_{\Gamma_{\text{int}}} p_{w,l} \mathbf{v}_{l,h} \cdot \mathbf{n}_l + s_{l,u}(\mathbf{u}_{l,h}, \mathbf{v}_{l,h}), \\
\tilde{F}_{l,p}(q_{l,h}) &= F_p((q_{l,h}, 0)) - s_{l,p}(p_{l,h}, q_{l,h}),
\end{aligned}$$

$$\tilde{a}_w(\mathbf{u}_{w,h}, \mathbf{v}_{w,h}) = a_w(\mathbf{u}_{l,h}, \mathbf{v}_{l,h}) + \int_{\Gamma_{\text{int}}} \gamma_u h_F^{-1}(\mathbf{u}_{w,h} \cdot \mathbf{n})(\mathbf{v}_{w,h} \cdot \mathbf{n}) + s_{w,l}(\mathbf{u}_{w,h}, \mathbf{v}_{w,h}),$$

$$\tilde{b}_w(p_{w,h}, \mathbf{v}_{w,h}) = b_w(p_{w,h}, \mathbf{v}_{w,h}) - \int_{\Gamma_{\text{int}}} p_{w,h}\mathbf{v}_{w,h} \cdot \mathbf{n}_w,$$

$$\tilde{F}_{w,u}(\mathbf{v}_{w,h}; (\mathbf{u}_{l,}, \mathbf{u}_{w,h})) = F_u((\mathbf{0}, \mathbf{v}_{w,h})) + \int_{\Gamma_{\text{int}}} \gamma_u h_F^{-1}(\mathbf{u}_{l,h} \cdot \mathbf{n})(\mathbf{v}_{w,h} \cdot \mathbf{n}) + s_{w,u}(\mathbf{u}_{w,h}, \mathbf{v}_{w,h}),$$

$$\tilde{F}_{w,p}(q_{w,h}) = F_p((0, q_{w,h})) - s_{w,p}(p_{w,h}, q_{w,h}),$$

$$\tilde{j}_{i,p}(q_{i,h}) = j_{i,p}(q_{i,h}) + s_{i,p}(p_{i,h}, q_{i,h}).$$

Then, the iterative splitting scheme for the resolution of problem (19) reads: *given* $\mathbf{u}_{i,h}^{(k)} \in \mathbf{V}_{i,h}$, *find* $\mathbf{u}_{i,h}^{(k+1)} \in \mathbf{V}_{i,h}$ *such that*

$$\tilde{a}_l(\mathbf{u}_{l,h}^{(k+1)}, \mathbf{v}_{l,h}; \mathbf{w}_h) + \tilde{b}_l(p_{l,h}^{(k+1)}, \mathbf{v}_{l,h}) = \tilde{F}_{l,u}(\mathbf{v}_{l,h}; \mathbf{u}_h^{(k)}) \quad \forall \mathbf{v}_{l,h} \in \mathbf{V}_{l,h},$$

$$\tilde{b}_l(q_{l,h}, \mathbf{u}_{l,h}^{(k+1)}) - \tilde{j}_{l,p}(p_{l,h}^{(k+1)}, q_{l,h}) = \tilde{F}_p(q_{l,h}; p_{l,h}^{(k)}) \quad \forall q_{l,h} \in Q_{l,h};$$

$$\tilde{a}_w(\mathbf{u}_{w,h}^{(k+1)}, \mathbf{v}_{w,h}) + \tilde{b}_w(p_{w,h}^{(k+1)}, \mathbf{v}_{w,h}) = \tilde{F}_{w,u}(\mathbf{v}_{w,h}; \mathbf{u}_h^{(k)}) \quad \forall \mathbf{v}_{w,h} \in \mathbf{V}_{w,h},$$

$$\tilde{b}_w(q_{w,h}, \mathbf{u}_{w,h}^{(k+1)}) - \tilde{j}_{w,p}(p_{w,h}^{(k+1)}, q_{w,h}) = \tilde{F}_p(q_{w,h}; p_{w,h}^{(k)}) \quad \forall q_{w,h} \in Q_{w,h}.$$

Notice that the iterative method amounts to successive resolution of decoupled generalized saddle point problems. To solve these problems, we choose to use the generalized minimal residual method (GMRes) preconditioned by a threshold incomplete LU factorization (ILUT) on the full matrix $C_i = \begin{pmatrix} \tilde{A}_i & \tilde{B}_i^T \\ \tilde{B}_i & -\tilde{J}_i \end{pmatrix}$. Finally, we point out that it is also possible to include the Picard's fixed point iterations (20) in our iterative scheme by replacing $\mathbf{w}_h$ with $\mathbf{u}_h^{(k)}$.

3.2. *Coupled mass transport*

We consider a standard implicit Euler time advancing scheme together with $\mathbb{P}^1$ finite element space discretization of the time dependent mass transport problems, looking for solutions in the discrete spaces $V_{i,h}$, $i \in \{l, w\}$. We employ penalty techniques to enforce essential conditions, and introduce IP stabilization for the advection-dominated mass transport in the arterial lumen.[9] We resort to an iterative splitting method to reduce the solution of the coupled problem to a sequence of problems defined over each subdomain.[1] To this end, we consider the following forms,

$$
\begin{aligned}
d_w(t; c_{w,h}, \psi_{w,h}) = &\int_{\Omega_w} D_w \nabla c_{w,h} \cdot \nabla \psi_{w,h} + \int_{\Omega_w} \mathbf{u}_{w,h} \cdot \nabla c_{w,h} \psi_{w,h} \\
&+ \int_{\Gamma_{\text{int}}} P_{\text{int}} c_{w,h} \psi_{w,h} + \int_{\Lambda_s} \pi \rho \varphi(t) \bar{c}_{w,h} \psi_{w,h},
\end{aligned}
$$

$$
\begin{aligned}
d_l(t; c_{l,h}, \psi_{l,h}) = &\int_{\Omega_l} D_l \nabla c_{l,h} \cdot \nabla \psi_{l,h} + \int_{\Omega_l} \mathbf{u}_{l,h} \cdot \nabla c_{l,h} \psi_{l,h} \\
&- \int_{\Gamma_{\text{in}}} D_l \nabla c_{l,h} \cdot \mathbf{n}_l \psi_{l,h} + \int_{\Gamma_{\text{in}}} h_F^{-1} c_{l,h} \psi_{l,h} \\
&+ \int_{\Gamma_{\text{int}}} P_{\text{int}} c_{l,h} \psi_{l,h} + \int_{\Lambda_s} \pi \rho \varphi(t) \bar{c}_{l,h} \psi_{l,h} \\
&+ \int_{\mathcal{F}_{l,h}} \gamma_{\mathbf{u}} \frac{h_F^2}{|\mathbf{u}_{l,h}|_F} [\![(\mathbf{u}_{l,h} \cdot \nabla) c_{l,h}]\!] [\![(\mathbf{u}_{l,h} \cdot \nabla) \psi_{l,h}]\!].
\end{aligned}
$$

Let $\Delta t > 0$ be time step, $t_n = n\Delta t$ the n-th time step and $c_{i,h}^n \in V_{i,h}$ the numerical approximation of $c_i(t_n, \cdot)$, $i \in \{l, w\}$. The time advancing is defined by the following coupled problem: *for $i = l, w$ and respectively $j = w, l$, given $c_{i,h}^n \in V_{i,h}$, find $c_{i,h}^{n+1} \in V_{i,h}$ such that*

$$
\begin{aligned}
&\frac{1}{\Delta t} \int_{\Omega_i} c_{i,h}^{n+1} \psi_{i,h} + d_i(t_{n+1}; c_{i,h}^{n+1}, \psi_{i,h}) \\
&= \frac{1}{\Delta t} \int_{\Omega_i} c_{i,h}^n \psi_{i,h} + \int_{\Lambda_s} \pi \rho \varphi(t) c_s \psi_{i,h} + \int_{\Gamma_{\text{int}}} P_{\text{int}} c_{j,h}^{n+1} \psi_{i,h} \quad \forall \psi_{i,h} \in V_{i,h}.
\end{aligned} \tag{21}
$$

The iterations of the splitting method are defined as follows, *for $i = l, w$ and respectively $j = w, l$, given $c_{i,h}^{n,k}, c_{j,h}^{n,k} \in V_{i,h}$, find $c_{i,h}^{n,k+1} \in V_{i,h}$ such that*

$$
\begin{aligned}
&\frac{1}{\Delta t} \int_{\Omega_i} c_{i,h}^{n,k+1} \psi_{i,h} + d_i(t_{n+1}; c_{i,h}^{n,k+1}, \psi_{i,h}) \\
&= \frac{1}{\Delta t} \int_{\Omega_i} c_{i,h}^n \psi_{i,h} + \int_{\Lambda_s} \pi \rho \varphi(t) c_s \psi_{i,h} + \int_{\Gamma_{\text{int}}} P_{\text{int}} c_{j,h}^{n,k} \psi_{i,h} \quad \forall \psi_{i,h} \in V_{i,h}.
\end{aligned}
$$

The convergence of the iterative scheme can be easily analized in the framework proposed in Ref. 10 and applied in a similar case in Ref. 4.

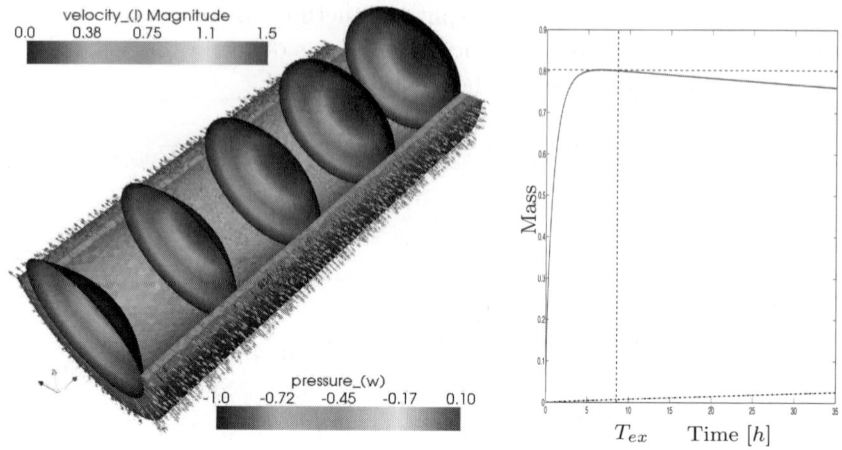

Figure 3. On the left: velocity profiles at various cross-sections in the arterial lumen, pressure values and velocity in the arterial wall. Notice the $\mathbb{P}^1/\mathbb{P}^0$ approximation spaces. Units are normalized respectively to 70 mmHg (pressure) and 240 mm/s (velocity). On the right: normalized drug release curves: shown are the total normalized drug mass in the arterial wall (solid curve), and the cumulative mass that has crossed the wall-lumen interface Γ_{int} (dot-dashed line). T_{ex} is the exit time at which the drug transported by the plasma reaches the external boundary of the arterial wall.

4. Numerical results

We present the study of drug release for a realistic 3D problem. We consider a simplified 3D geometry of a coronary artery. The lumen Ω_l is a cylinder having radius 1.6 mm and length 10 mm. The arterial wall Ω_w is represented by a hollow cylinder with thickness 0.55 mm. We consider standard data for blood[11] (viscosity 3 mm^2/s, density 1 mg / mm^3); the Darcy inverse permeability of the arterial wall is $\eta = 1 \cdot 10^{12}\text{s}^{-1}$.

The pressure drop $p_{\text{out}} - p_{\text{in}}$ is $1.34 \cdot 10^{-8}$ mmHg, corresponding to a peak velocity of about 300 mm/s. The external wall pressure p_{ext} is -70 mmHg; the computed wall filtration velocity is about $1.6 \cdot 10^{-5}$ mm/s.

Regarding the drug release process, we consider a DES loaded with heparin. The shape of the DES is shown in figure 4, and is based on common designs (for instance the Cypher models, Johnson & Johnson). The stent radius is $\rho = 0.15$ mm. The mesh size used for the simulation is $h \simeq 0.2$. Notice that a full three-dimensional analysis of the stent structure would have required a much smaller value for h, at least in proximity to the device.

According to experimental investigations,[12] we set $D_l = 1.5 \cdot 10^{-4}$ mm^2/s, $D_w = 7.7 \cdot 10^{-6}$ mm^2/s. The wall membrane permeability is

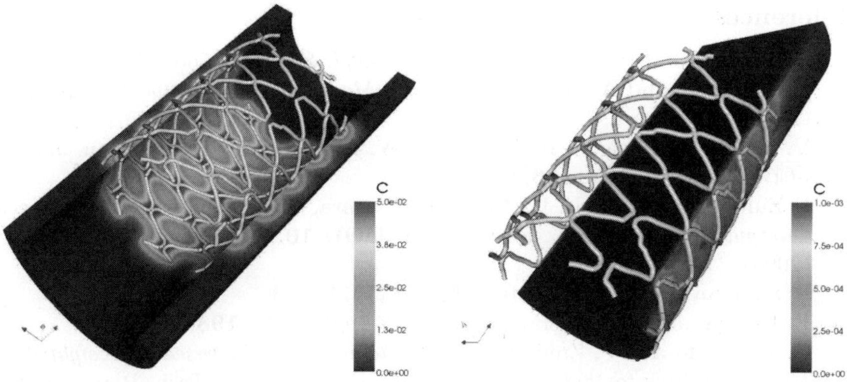

Figure 4. On the left: drug concentration in a section of the arterial wall after 1h. On the right: drug concentration in a section of the lumen at the beginning of the process (40 ms). The geometry of the 1D stent is also shown.

$P_{\text{int}} = 2 \cdot 10^{-7}$ mm/s. The diffusivity of the drug into the stent coating can assume very different values, and ranges from 10^{-8} to 10^{-12} mm^2/s, depending on the mechanical properties of the polymeric substrate. We set $D_s = 10^{-8}$ mm^2/s, which corresponds to a fast release profile. Figure 3 shows the velocity fields and the evolution of drug released into the wall (normalized with respect to the total charge of the stent). After the initial phase with considerable mass injection, a maximum is reached; then, the drug starts to be slowly washed away from the arterial wall. This is due to two mechanisms: some drug leaves the wall moving down the osmotic gradient induced by the concentration gap across the interface Γ_{int}, and, at the same time, plasma brings the drug away from the tissue. On the other hand, the concentration time course in the lumen is very fast: due to the dominant advection field, most of the drug leaving the stent is lost into the blood stream. Nevertheless, a thin layer near the interface is observed.

Acknowledgements

This work has been supported by the ERC Advanced Grant N.227058, *Mathematical Modelling and Simulation of the Cardiovascular System* (MATHCARD), and by the Italian Institute of Technology (IIT) with the project *NanoBiotechnology - Models and methods for degradable materials.*

References

1. C. Vergara and P. Zunino, *Multiscale Model. Simul.* **7**, 565 (2008).
2. C. D'Angelo and A. Quarteroni, *Math. Models Methods Appl. Sci.* **18**, 1481 (2008).
3. V. B. Kolachalama, A. R. Tzafriri, D. Y. Arifin and E. R. Edelman, *Journal of Controlled Release* **133**, 24 (2009).
4. P. Zunino, C. D'Angelo, L. Petrini, C. Vergara, C. Capelli and F. Migliavacca, *Comput. Methods Appl. Mech. Engrg.* **DOI: 10.1016/j.cma.2008.07.019** (2008).
5. P. Saffman, *Stud. Appl. Math.* **50**, 292 (1971).
6. E. Burman and P. Hansbo, *J. Comput. Appl. Math.* **198**, 35 (2007).
7. C. D'Angelo and P. Zunino, *A finite element method based on weighted interior penalties for heterogeneous incompressible flows*, Tech. Rep. 14/2008, MOX. Dept. of Mathematics, Politecnico di Milano (June 2008), Submitted.
8. E. Burman, M. A. Fernández and P. Hansbo, *SIAM J. Numer. Anal.* **44**, 1248 (2006).
9. E. Burman and P. Zunino, *SIAM J. Numer. Anal.* **44**, 1612 (2006).
10. A. Quarteroni, A. Veneziani and P. Zunino, *SIAM J. Sci. Comput.* **23**, 1959 (2002).
11. S. Tada and J. Tarbell, *Am. J. Physiol. Heart Circ. Physiol.* **287**, H905 (2004).
12. M. Lovich and E. Edelman, *Circ. Res.* **77**, 1143 (1995).

A Mathematical Model of Duchenne Muscular Dystrophy

Guido Dell'Acqua* and Filippo Castiglione

Institute for Computing Applications "M. Picone"
National Research Council (CNR)
c/o IASI, Viale Manzoni 130, 00185 - Rome, Italy
**E-mail: g.dellacqua@iac.cnr.it*

We present a mathematical model to investigate the role of the immune system in the Duchenne muscular dystrophy disease, based on the assumption that the immune system contributes to the tissue damage. Indeed, its interaction with the muscle tissue after an initial endogenous damage can be described as a predator-prey system showing typical oscillations. Moreover we investigate the dynamical properties of the system and we find that, for a biologically relevant parameters range, it shows two phase-transitions between qualitative different behaviors corresponding to complete recover or to a state where massive muscle regeneration and degeneration coexist.

Keywords: Biomathematics, mathematical modelling, immune system, nonlinear evolution problems.

1. Introduction

The Duchenne Muscular Dystrophy (DMD) is a genetic disorder that leads to the absence of the protein *dystrophin*. This protein is prominent in muscle cells as it binds with several other proteins in the muscle cell membrane to form what is called the *dystrophin glycoprotein complex*. The role of this complex is yet unknown, but many researchers believe that it provides mechanical stability for the muscle cell membrane or that it is involved in important intra-cellular signaling mechanisms. In a current view of the DMD pathogenesis it is believed that the immune response is to blame, at least in part, for the muscle damage. In favor of this hypothesis, microscopic analysis of dystrophic muscle tissue reveals that it is often infiltrated with a large numbers of immune cells including T lymphocytes (both helper and cytotoxic), macrophages and several others.[5]

We have constructed a mathematical model that follows this hypothesis, with the only difference that the cytotoxic activity of the immune system against the muscle tissue is initially ignited by a genetic defect worsened by, for example, mechanical stress. In this setting the pathology develops through a cascade of events that can be roughly summarized as follows: macrophages initially infiltrates the damaged muscle tissue and, while cleaning cellular debris, they call for CD4$^+$ T cells that in turn call for CD8$^+$ T cells. Also called cytotoxic cells, CD8$^+$ T lymphocytes have the ability to destroy targeted cells. In this case, they target also healthy muscle cells provoking more and more damage.

On the basis of this assumptions we have built a mathematical model that includes the concentration of macrophages (M), helper CD4$^+$ lymphocytes (H), of cytotoxic CD8$^+$ lymphocytes (C) and the percentage of normal (N), damaged (D) and regenerating (R) muscle cells.
The interaction between immune system and muscle cells follows a predator-prey scheme where the muscle cells are prey for the immune cytotoxic lymphocytes that, on the other hand, can be seen as the predators. This model has been previously validated against direct comparison with time course data from dystrophic muscles in "MDX" mice.[2] In the present work we look for qualitative dynamical properties of the system.

The paper is organized as follows: in section 2 we describe the system of ODEs and perform a non-dimensionalization of the model. In section 3 we study the linear stability of the "limiting system" near its stationary points. Our findings are corroborated in section 4, where we show the results of the numerical integration. Finally, section 5 summarizes and gives possible therapeutic hints.

2. Equations of the model

We model the mesoscopic interaction between the immune system (macrophages M, T helper H and T cytotoxic lymphocytes C) and the percentage of normal N, damaged D and regenerating R muscle cells. The model equations are

$$\frac{dM}{dt} = a_1 + b_1 MD - d_1 M \quad (1)$$
$$\frac{dH}{dt} = a_2 + b_2 MD - d_2 H \quad (2)$$
$$\frac{dC}{dt} = b_3 HD - d_3 C \quad (3)$$
$$\frac{dN}{dt} = a_4 R - b_4 CN - h\Delta N \quad (4)$$
$$\frac{dD}{dt} = b_4 CN - b_5 MD - d_4 D + h\Delta N \quad (5)$$

complemented by the conservation law

$$N + D + R = 100 \quad (6)$$

and by the following initial conditions

$$M(0) = M_0, \quad H(0) = H_0, \quad C(0) = 0, \quad N(0) = 100, \quad D(0) = 0, \quad R(0) = 0.$$

Equation (6) implies that the total number of cells is constant. Given that the experimental data for these quantities are expressed as percentages,[2] it is straightforward to fix this constant to 100. It follows also that $N(0) = 100$. Moreover, let us suppose that

$$a_1 = d_1 M(0) \quad (7)$$
$$a_2 = d_2 H(0). \quad (8)$$

Conditions (7)–(8) imply that if $h = 0$, i.e., there is no initial damage, then the initial state is a stationary stable state.

Notice that the system (1)–(6) is *non-autonomous* because of the presence in equation (5) of the function $\Delta = \Delta(t)$ that represents an impulse-like damage that initiates the immune response and then vanishes. The biological rationale for this choice is, for example, a mechanical stress of the muscle tissue on top of the genetic predisposition to the disease.

To model the process of accumulation of damage in the muscle tissue, that is composed by a large number of muscle cells, we resorted to a log-normal function. The reason for this choice is that log-normal distributions can be theoretically expected under the assumption of a "degradation process" resulting from failures.[3] Therefore we model the initial damage as the following function

$$\Delta(t, m, \sigma) = \frac{1}{\sigma t \sqrt{2\pi}} \exp(-\frac{1}{2\sigma^2} (\log t - m)^2) \quad (9)$$

with parameters m and σ.

Table 1. Parameters of the model in equations (1)–(6) estimated in.[2] Unit c is for cells and w for weeks.

Model parameters				
Entity	Parameter	Description	Value	Units
M	a_1	Turnover rate of macrophages	$d_1\, M(0)$	$c\ \ w^{-1}\ mm^{-3}$
	b_1	Infiltration rate of macrophages	0.950788	$c^{-1}\ w^{-1}\ mm^3$
	d_1	Death rate of macrophages	0.912485	w^{-1}
H	a_2	Turnover rate of CD4+ T cells	$d_2\, H(0)$	$c\ \ w^{-1}\ mm^{-3}$
	b_2	Damage-driven proliferation rate of CD4+ T cells	0.0483312	$c^{-1}\ w^{-1}\ mm^3$
	d_2	Death rate of CD4+ T-cells	0.999999	$c\ \ w^{-1}\ mm^{-3}$
C	b_3	Damage-driven proliferation rate of CD4+ T cells	0.08065	$c^{-1}\ w^{-1}\ mm^3$
	d_3	Death rate of CD8+ T cells	0.999999	w^{-1}
N	a_4	Generation rate of healthy fibers	0.17	w^{-1}
	b_4	Cytotoxicity degradation rate	0.0041864	$c^{-1}\ w^{-1}\ mm^3$
	h	Strength of impulse-like initial damage	0.19119	w^{-1}
	σ	Standard deviation of the initial damage	0.609114	
	m	Time of the peak of the initial damage	0.945333	
D	b_5	Cleaning rate by macrophages	0.000664372	$c^{-1}\ w^{-1}\ mm^3$
	d_4	Physiological cleaning rate	1.20001	$c^{-1}\ w^{-1}\ mm^3$

The system (1)–(6) can be described in few words as follows. Equations (1)–(3) represent the rate of change of the immune cell counts of macrophages, $CD4^+$ and $CD8^+$ T cells, respectively. For example, equation (1) says that at a given time t while d_1M macrophages age and die, they are replenished from their source at a rate a_1 cells per week. In presence of damage $D > 0$, the term b_1MD represents the local recruitment of cells by inflammatory mechanisms. Equation (2) and (3) are similar. Equations (4)–(5) represent the rate of change of the percentage of normal, damaged and regenerating muscle cells. Equation (5) reflects our assumption that the muscular tissue damage initially starts in a small percentage αN of the normal muscle fibers. Then, when $CD8^+$ T cells infiltrate the tissue they create more damage as a percentage of the normal cells b_4CN. Damaged fibers are cleared out both by macrophages at the rate b_5MD or by other degradation mechanisms at the rate d_4D. Equation (4) implies that healthy normal muscle cells N are replenished from the generating fibers R at the rate a_4R and, as already mentioned, are damaged at the rate $b_4CN + \alpha N$. Equation (6) says that a cell can be either normal, damaged or regenerating.

It is worth to note that the muscle damage raises in presence of $CD8^+$ T cells, which increase with the $CD4^+$ T cells (see equation (3)), which, in their turn, are boosted by a high level of macrophages (see equation (2)).

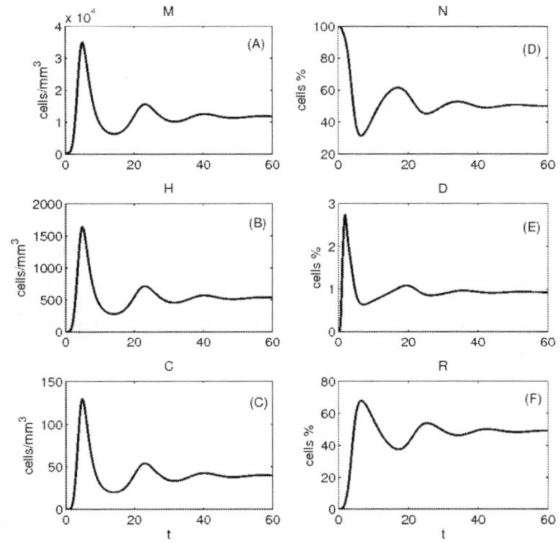

Fig. 1. Dynamics of the system (1)–(6) with values of the parameters from Table 1. Plots on the left show concentrations of: A) Macrophages; B) CD4+ T cells; C) CD8+ T cells. Plots on the right show percentages of: D) Normal cells; E) Damaged cells; F) Regenerating cells.

Therefore macrophages contribute to both physiological and pathological states since they clean cellular debris to favor cell regeneration and at the same time they foster an immune reaction that in its turn cause more damage to the tissue destroying healthy muscle fibers.

The system of equations has 14 free parameters (12 in equations (1)-(6) plus 2 in (9)) and 2 conditions. This leaves out 12 free parameters. The values of such parameters have been estimated using COPASI[1] by applying its built-in optimization methods, combining the time course data for several different dystrophic muscles of mice in the age range 14-84 days and identifying others time course data sets from the literature.[5–7] The resulting estimated parameters are shown in Table 1.

The initial conditions for M and H are taken from the literature as follows:

$$M_0 = 429 \; cells \; mm^{-3}, \qquad H_0 = 6.25 \; cells \; mm^{-3}$$

while $C_0 = 0$ means that in absence of damage no $CD8^+$ T cells infiltrate the muscle tissue.

To analyze the model, it is convenient to convert all equations to be

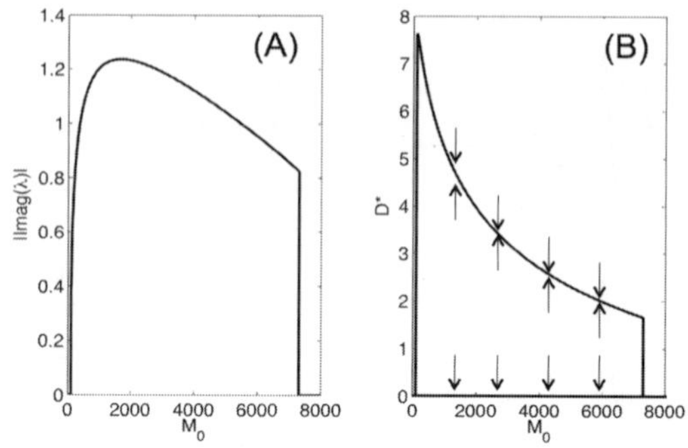

Fig. 2. Panel A: Absolute value of the imaginary part of the complex eigenvalue of (28) in function of M_0; Panel B: Corresponding plot of the stable stationary value of D; in both plots parameters are as in Table 1 except for $\beta_1 = 10.456$.

dimensionless. Let

$$t = t_c\tilde{t}, \quad M = M_c\widetilde{M} \quad H = H_c\widetilde{H}, \quad C = C_c\widetilde{C}, \quad N = N_c\widetilde{N}, \quad \Delta = \Delta_c\widetilde{\Delta} \tag{10}$$

and analogously for the initial conditions. Here $\widetilde{M}, \widetilde{H}, \widetilde{C}, \widetilde{N}, \widetilde{D}, \widetilde{R}$ and $\widetilde{\Delta}$ are dimensionless variables. Substituting (10) in (1)–(6) and incorporating conditions in (7)–(8) we get:

$$\begin{aligned}
\frac{d\widetilde{M}}{dt} &= d_1 t_c(\widetilde{M}_0 - \widetilde{M}) + b_1 D_c t_c \widetilde{M}\widetilde{D} \\
\frac{d\widetilde{H}}{dt} &= d_2 t_c(\widetilde{H}_0 - \widetilde{H}) + \frac{b_2 t_c M_c D_c}{H_c}\widetilde{M}\widetilde{H} \\
\frac{d\widetilde{C}}{dt} &= \frac{b_3 t_c H_c D_c}{C_c}\widetilde{H}\widetilde{D} - d_3 t_c\widetilde{C} \\
\frac{d\widetilde{N}}{dt} &= \frac{a_4 t_c R_c}{N_c}\widetilde{R} - b_4 t_c C_c\widetilde{C}\ Nt - h t_c \Delta_c \widetilde{\Delta}\widetilde{N} \\
\frac{d\widetilde{D}}{dt} &= \frac{b_4 t_c C_c N_c}{D_c}\widetilde{C}\widetilde{N} - b_5 t_c M_c \widetilde{M}\widetilde{D} - d_4 t_c \widetilde{D} + \frac{h t_c \Delta_c N_c}{D_c}\widetilde{\Delta}
\end{aligned}$$

and

$$N_c\widetilde{N} + D_c\widetilde{D} + R_c\widetilde{R} = 100\,.$$

Now we choose $t_c, M_c, H_c, C_c, N_c, D_c$ and R_c such that all variables are of the order of unity while reducing the number of free parameters. For what

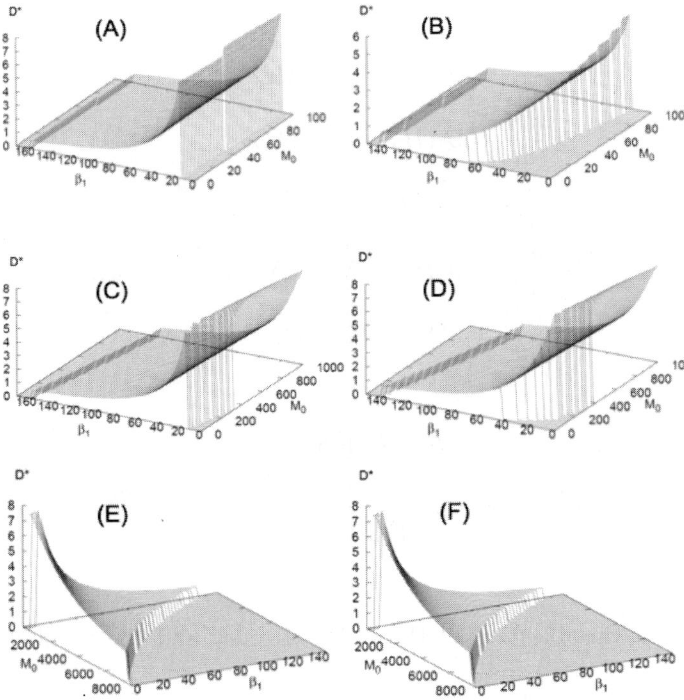

Fig. 3. Panels A, C and D: stationary values of D obtained by numerical integration of equations (1)–(6) in function of M_0 and β_1. Values of the parameters are as in Table 1. Panels B, E and F: corresponding stationary values obtained from the linear stability analysis of equations (18)–(22). Units: M_0 cells/mm^{-3}, D percent of muscle cells, $\beta_1 = \frac{10^2 b_1}{d_1} \sim 104.56$ non-dimensional.

concerns N_c, D_c and R_c, being percentages, we are left to set $N_c = D_c = R_c = 100$.
Moreover, we set:

$$t_c = \frac{1}{d_1}; \quad M_c = \frac{d_1}{b_5}; \quad H_c = \frac{10^2 b_2}{b_5}; \quad C_c = \frac{d_1}{b_4}; \quad \Delta_c = d_1;$$

Finally, having chosen a log-normal impulse Δ as in (9) we have

$$\Delta = \Delta(t, m, \sigma) = \frac{1}{t_c}\Delta(\tilde{t}, m - \log t_c, \sigma) = \frac{1}{t_c}\widetilde{\Delta}.$$

Therefore the dimensionless model can be written (dropping tildes):

$$\frac{dM}{dt} = M_0 - M + \beta_1 MD \tag{11}$$

$$\frac{dH}{dt} = \beta_2(H_0 - H) + MD \tag{12}$$

$$\frac{dC}{dt} = \beta_3 HD - \beta_4 C \tag{13}$$

$$\frac{dN}{dt} - \beta_5 R - CN - h\Delta N \tag{14}$$

$$\frac{dD}{dt} = CN - MD - \beta_6 D + h\Delta N \tag{15}$$

and

$$N + D + R = 1 \tag{16}$$

where

$$\beta_1 = \frac{10^2 b_1}{d_1}; \quad \beta_2 = \frac{d_2}{d_1}; \quad \beta_3 = \frac{10^4 b_2 b_3 b_4}{b_5 d_1^2}; \quad \beta_4 = \frac{d_3}{d_1}; \quad \beta_5 = \frac{a_4}{d_1}; \quad \beta_6 = \frac{d_4}{d_1}. \tag{17}$$

Note that all variables are of the order of unity and the number of free parameters is reduced to 7.

The dynamics of the model, with parameters values as in Table 1, is shown in Fig. 1.

3. Stability analysis of the "limiting system"

Let us consider the dynamical system (11)–(16) obtained by setting $h = 0$ in equations (14)–(15), i.e., in absence of the mechanical stress that starts the immune system reaction. This can be considered as the "limiting system" associated to (11)–(16), meaning that after a transient period for $t > \hat{t}$ the dynamics is "almost" fully governed by the system

$$\frac{dM}{dt} = M_0 - M + \beta_1 MD \tag{18}$$

$$\frac{dH}{dt} = \beta_2(H_0 - H) + MD \tag{19}$$

$$\frac{dC}{dt} = \beta_3 HD - \beta_4 C \tag{20}$$

$$\frac{dN}{dt} = \beta_5(1 - N - D) - CN \tag{21}$$

$$\frac{dD}{dt} = CN - MD - \beta_6 D \tag{22}$$

where we have incorporated the conservation law (16) in equation (21). Now we look for stationary solutions. Setting to zero the left hand sides of (18)–(22) it can be easily seen that depending on the parameters values the system admits two or four stationary solutions. Indeed, all variables can be expressed as functions of D and substituting in (22) we are left with an homogeneous polynomial of degree 4 that has, besides 0, one or three more roots.

To perform a linear stability analysis, let us linearize the limiting system (18)–(22) near a stationary state $(M^*, H^*, C^*, D^*, N^*)$ by setting

$$M = M^* + \varepsilon\tilde{m}\,, \quad H = M^* + \varepsilon\tilde{h}\,, \quad C = C^* + \varepsilon\tilde{c}\,,$$
$$N = N^* + \varepsilon\tilde{n}\,, \quad D = D^* + \varepsilon\tilde{d}\,,$$

with $\varepsilon \ll 1$. Substituting in (18)–(22) we get, at leading order,

$$\tilde{m}' = (\beta_1 D^* - 1)\tilde{m} + \beta_1 M^* \tilde{d} \tag{23}$$
$$\tilde{h}' = D^* \tilde{m} - \beta_2 \tilde{h} + M^* \tilde{d} \tag{24}$$
$$\tilde{c}' = \beta_3 D^* - \beta_4)\tilde{c} + \beta_3 H^* \tilde{d} \tag{25}$$
$$\tilde{n}' = -N^* \tilde{c} - (\beta_5 + C^*)\tilde{n} - \beta_5 \tilde{d} \tag{26}$$
$$\tilde{d}' = -D^* \tilde{m} - +N^* \tilde{c} + C^* \tilde{n} - (M^* + \beta_6)\tilde{d} \tag{27}$$

Then, setting

$$\tilde{m} = \hat{m}e^{\lambda t}\,, \quad \tilde{h} = \hat{h}e^{\lambda t}\,, \quad \tilde{c} = \hat{c}e^{\lambda t}\,,$$
$$\tilde{n} = \hat{n}e^{\lambda t}\,, \quad \tilde{d} = \hat{d}e^{\lambda t}\,,$$

and substituting in (23)–(27) we find

$$A \begin{pmatrix} \hat{m} \\ \hat{h} \\ \hat{c} \\ \hat{n} \\ \hat{d} \end{pmatrix} = \lambda \begin{pmatrix} \hat{m} \\ \hat{h} \\ \hat{c} \\ \hat{n} \\ \hat{d} \end{pmatrix}$$

where

$$A = \begin{pmatrix} \beta_1 D^* - 1 & 0 & 0 & 0 & \beta_1 M^* \\ D^* & -\beta_2 & 0 & 0 & M^* \\ 0 & \beta_3 D^* & -\beta_4 & 0 & \beta_3 H^* \\ 0 & 0 & -N^* & -C^* - \beta_5 & -\beta_5 \\ -D^* & 0 & N^* & C^* & -M^* - \beta_6 \end{pmatrix}. \tag{28}$$

The eigenvalues of A are the roots of its characteristic polynomial.

It is easy then to check that in the case of one root, it is either an unstable point or it is not admissible (i.e., value less than zero), while in the case there are three more roots we can have that only one of them is stable, or all three are unstable.

Summarizing, the limiting system can have regions of bistability. If $h = 0$ then the system remains in its initial state. If $h > 0$, the dynamics begins. When the resulting value of D rises above a certain threshold (e.g., one unstable stationary point) then the system is driven to the other stationary stable state.

If the stationary solution coincides with the initial state then all of the eigenvalues are real and negative whereas the situation changes when we study the other stable stationary solution.

Since it is widely accepted that the immune system promotes injury of dystrophin-deficient muscle,[7] we restricted our study to the 2-dimensional subspace of the parameters $M_0 \times \beta_1$, in view of finding possible therapeutic treatments. More in details, M_0 is the initial amount of M, i.e., the number of macrophages in a healthy muscle tissue, while $\beta_1 = 10^2 b_1/d_1$ (non-dimensional) represents the "sensitivity" of the immune system to the induced damage, normalized with respect to the death rate of these cells.

In Fig. (2) we show the imaginary part of the eigenvalues of the matrix associated to the linearization of the limiting system in equations (18)–(22) for different values of M_0 and for a fixed value of β_1. When M_0 gets closer to the first threshold value, the imaginary part of two eigenvalues becomes different from zero. This is where the system experiments a first phase transition. At the second threshold value, the imaginary part of these eigenvalues becomes zero, hence we have the other phase transition. We have checked that for larger values of β_1, the first part of this process is skipped, i.e., the system experiments only the second phase transition. For even larger values of β_1 the only stationary stable solution is the initial state. In Fig. (3), panels A, C and E, we show the stationary value of D as a function of the two parameters M_0 and β_1.

4. Comparison between numerical solution and stability analysis

Let us consider the non-autonomous system (1)–(6). We have used a standard variable-order solver of the MATLAB package, `ode15s`[8] to perform the numerical integration. The results agree with the linear stability analysis discussed above, as we show in Fig. (3), panels B, D and F, where we have depicted the stationary values of D as a function of the two parameters M_0

(dimensional) and β_1.

In Fig. (3), if we compare panels A, C and E with panels B, D and F, we see that the phase transitions in the latter cases are somehow smoother. This depends on the fact that the *strength* of the impulse is not always sufficient to drive the system to the basin of attraction of the other stable state. Therefore the linear stability analysis gives us a view on the possible stable stationary states of the system; the actual chance to reach those states depends on how we *stress* the system.

5. Conclusions

In this paper we have constructed a model of five ODEs, plus a conservation law, to investigate the role of the immune system in the Duchenne muscular dystrophy. The model has been validated against experimental data as described elsewhere.[2]

With a simple linear stability argument we have found that the model may present regions of bistability corresponding to either complete recover or to a state where massive cell degeneration and regeneration coexist. We have identified such region as a function of two parameters that are among those controlling the *strength* of the immune response. The numerical solution of the full system confirms our analysis.

In our model the pathology initiates with an impulse that may (or may not) drive the system from the healthy stationary state to the basin of attraction of the other stable stationary state.

The therapeutic insights of this analysis are interesting. It would suffice to drive the system (via a pharmacological or cytological treatment) into the mono-stability region of complete recover, or to a region where the basin of attraction of the other stable solution is too far to be reached with a physiological stress, so that the system falls back into the *healthy* region. To this end, it is worth to mention the work in[7] where the authors proved by *in vivo* experiments the active role of the macrophages in lysing the muscle membranes by depleting them and then assessing the occurrence of membrane lesions.

Acknowledgments

G. D. acknowledges support of the EC under the contract "P6-2005-NEST-PATH, No.043241 (ComplexDis)".

References

1. S. Hoops, S. Sahle, R. Gauges, C. Lee, J. Pahle, N. Simus, M. Singhal, L. Xu, P. Mendes, U. Kummer, COPASI–a COmplex PAthway SImulator, Bioinformatics **22** (2006), pp. 3067–3084.
2. A. Jarrah, N. Evans, R. Laubenbacher, A mathematical model of Duchenne Muscular Dystrophy in MDX mice, in preparation.
3. A. N. Kolmogorov, Uber das logarithmisch normale Verteilungsgesetz der Dimensionen der Teilchen bei Zerstuckelung Dockl. Akad. Nauk. SSSR **31** (1941), pp. 99–101.
4. J. D. Porter, W. Guo, A. P. Merriam, S. Khanna, G. Cheng, X. Zhou, F. H. Andrade, C. Richmonds and H. J. Kaminski, Persistent over-expression of specific CC class chemokines correlates with macrophage and T-cell recruitment in mdx skeletal muscle, Neurom. Dis., **13** 3 (2003), pp. 223–235.
5. M. J. Spencer, E. Montecino-Rodriguez, K. Dorshkind and J. G. Tidball, Helper (CD4+) and Cytotoxic (CD8+) T cells Promote the Pathology of Distrophin-Deficient Muscle, **98** 2 (2001), pp. 235–243.
6. M. J. Spencer, C. M. Walsh, K. A. Dorshkind, E. M. Rodriguez and J. G. Tidball, Myonuclear apoptosis in dystrophic mdx muscle occurs by perforin-mediated cytotoxicity, J Clin Invest **99** 11 (1997), pp. 2745–2751.
7. M. Wehling, M. J Spencer and J. G. Tidball, A nitric oxide synthase transgene ameliorates muscular distrophy in mdx mice J. Cell Biol., **15** 1 (2001), pp. 123–131.
8. The MathWorks, `www.mathworks.com`.

On Filippov and Utkin Sliding Solution of Discontinuous Systems

Luca Dieci

School of Mathematics,
Georgia Institute of Technology, Atlanta, GA 30332 U.S.A.
dieci@math.gatech.edu

Luciano Lopez

Dipartimento di Matematica
Università degli Studi di Bari, Via E. Orabona 5, 70125, Bari, Italy
lopezl@dm.uniba.it

Discontinuous dynamical systems with sliding modes are often used in Control Theory to model differential equations with discontinuous control. Filippov and Utkin (see[2,7]) have proposed two different approaches to define the solution of these dynamical systems. In case of linear systems, these two approaches are equivalent, but in case of nonlinear systems, the ways to extend the vector field on the sliding surface is generally different. In this note, we obtain a seemingly new approach, which lends support to Utkin's approach, but it is somewhat more general.

Keywords: Discontinuous dynamical systems, Filippov's and Utkin's diffe- rential systems, sliding modes, equivalent control.

1. Introduction and Motivation

In this brief paper, we consider a differential system with discontinuous right-hand-side, in the control setting. The model we consider is the classical one of Filippov or Utkin (see[2,6,7]) whereby one seeks a solution to a dynamical system in which the right-hand side (the vector field) varies discontinuously as solution trajectories reach a surface, called *discontinuity or switching surface.* Although several different behaviors of solutions of discontinuous systems are possible (e.g., impact systems or systems with tranverse, attractive, or repulsive, modes), here we are only interested in the case where the solution remains continuous, though not differentiable at each point, and slides on the discontinuity surface (*attractive sliding mode*) for some interval of time.

Systems with discontinuous right-hand sides appear pervasively in applications of various nature, but it is in Control Theory that this mathematical construction seems to have the most significat applications. For instance, in dynamical systems where a scalar control variable u switches between different values, the freedom afforded by a discontinuous control u can be used to obtain models achieving asymptotically stable sliding solution (see[4–8]).

Filippov and Utkin have proposed two different approaches to define the solution on the sliding surface, approaches which are equivalent in case of linear systems, but which are generally different in case of nonlinear systems. In this note we review the issue of sliding motion in control systems, and end up with an alternative to both Utkin and Filippov's constructions. In fairness, however, it remains to be investigated whether this theoretical alternative is of practical use.

1.1. *Filippov sliding modes*

Consider the discontinuous differential system:

$$x'(t) = f(x(t)) = \begin{cases} f_1(x(t)) \ , & h(x) < 0 \\ f_2(x(t)) \ , & h(x) > 0 \ , \end{cases} \qquad x(0) = x_0 \in \mathbb{R}^n \ . \tag{1}$$

Above, and henceforth, we suppose that the state space $\mathbb{R}^n$ is split into two subspaces Σ_1 and Σ_2 by a surface Σ such that $\mathbb{R}^n = \Sigma_1 \cup \Sigma \cup \Sigma_2$. Σ is defined by the scalar event function $h : \mathbb{R}^n \to \mathbb{R}$, so that the subspaces Σ_1 and Σ_2, and the Σ, are characterized as

$$\begin{aligned} \Sigma_1 &= \{x \in \mathbb{R}^n |\ h(x) < 0\} \, , \\ \Sigma_2 &= \{x \in \mathbb{R}^n |\ h(x) > 0\} \, , \\ \Sigma &= \{x \in \mathbb{R}^n |\ h(x) = 0\} \ . \end{aligned} \tag{2}$$

We will also assume throughout that the gradient $\nabla h(x) \neq 0$ at all $x \in \Sigma$, and write $\mathrm{n}(x) = \frac{\nabla h(x)}{\|\nabla h(x)\|}$ for the unit normal to Σ.

In (1), the right-hand side $f(x)$ can be assumed to be smooth on Σ_1 and Σ_2 separately, but it is usually discontinuous across Σ. More precisely, we will assume that f_1 is $C^k, k \geq 2$, on $\Sigma_1 \cup \Sigma$ and f_2 is $C^k, k \geq 2$, on $\Sigma_2 \cup \Sigma$.

As long as $x \notin \Sigma$, (1) is a standard differential equation. The interesting case is when $x \in \Sigma$. In this case, if the condition

$$[\mathrm{n}^{\,T}(x) f_1(x)] > 0 \quad \text{and} \quad [\mathrm{n}^{\,T}(x) f_2(x)] < 0, \tag{3}$$

is satisfied at $x \in \Sigma$, then one has so-called (attractive) sliding motion on Σ. Filippov defines this sliding motion as the solution of the differential system

$$x'(t) = \begin{cases} f_1(x(t)), & h(x) < 0, \\ f_F(x(t)), & h(x) = 0, \\ f_2(x(t)), & h(x) > 0, \end{cases} \tag{4}$$

where the vector field on the surface is taken to be in the convex hull $\text{co}(f_1, f_2)$ of f_1, f_2, that is:

$$f_F(x) = [\alpha f_2(x) + (1-\alpha) f_1(x)], \qquad x \in \Sigma ,$$

and $\alpha(x) \in [0,1]$ chosen so that $\text{n}^T(x) f_F(x) = 0$, that is

$$\alpha(x) = \frac{\text{n}^T(x) f_1(x)}{\text{n}^T(x)(f_1(x) - f_2(x))}, \quad x \in \Sigma . \tag{5}$$

We remark that Filippov's construction defines the vector field on the sliding surface as that vector in $\text{co}(f_1, f_2)$ which is tangent to Σ (see Figure 1). We stress that there is one unknown in Filippov's contruction, the scalar valued function α in (5).

1.2. *Utkin sliding modes*

To introduce Utkin's approach, we need to consider a system of the type (1), with additional control variables. To be precise, we consider the system

$$x' = f(x,u) \quad \text{where} \quad \begin{cases} u(x) = u_1(x) \text{ when } & h(x) < 0 \\ u(x) = u_2(x) \text{ when } & h(x) > 0 , \end{cases} \tag{6}$$

where u_1 and u_2 are C^k functions of x defined on $\Sigma_1 \cup \Sigma$ and $\Sigma_2 \cup \Sigma$, respectively, and taking values in $\mathbb{R}^p$. Notice that the system in (6) has the same functional form of f, and discontinuity in the vector field arises because of the discontinuous control. Again, the interest is in the case of attracting sliding motion, when $x \in \Sigma : [\text{n}^T(x) f(x, u_1(x))] > 0$ and $[\text{n}^T(x) f(x, u_2(x))] < 0$.

The problem can be trivially recast in the form of (1). In fact, letting $f_1(x) = f(x, u_1)$ and $f_2(x) = f(x, u_2)$, one obtains a system formally of the form (1). So doing, the Filippov extension in the case of sliding motion will be the same as before, that is:

$$x' = \begin{cases} f_1(x) = f(x, u_1(x)), & h(x) < 0, \\ f_F(x), & h(x) = 0, \\ f_2(x) = f(x, u_2(x)), & h(x) > 0, \end{cases} \tag{7}$$

where

$$f_F(x) = [\alpha f(x, u_2(x)) + (1-\alpha) f(x, u_1(x))] , \tag{8}$$

and α is such that $\mathrm{n}^T(x)f_F(x) = 0$, as in (5).

On the other hand, Utkin (see[7]) proposed a different way to deal with this problem, the *equivalent control approach*. For sliding motion (that is, for $x \in \Sigma$), he proposes to take the vector field obtained by choosing $f(x,u)$ so that $f(x,u)$ will lie in the tangent plane to Σ at x. Formally, Utkin's construction is as follows:

$$x'(t) = \begin{cases} f_1(x(t)) = f(x, u_1(x)), & h(x) < 0, \\ f_U(x(t)) = f(x, u_{\text{eq}}(x)), & h(x) = 0, \\ f_2(x(t)) = f(x, u_2(x)), & h(x) > 0, \end{cases} \tag{9}$$

where the equivalent control u_{eq} must be chosen so that $f(x, u_{\text{eq}}(x))$ lies in the tangent plane to Σ at x: $\mathrm{n}^T(x)f(x, u_{\text{eq}}(x)) = 0$. In general (i.e., when $u \in \mathbb{R}^p$, $p > 1$), this is an underdetermined system. Utkin's approach is to choose u_{eq} in the convex hull of u_1 and u_2:

$$u_{\text{eq}}(x) = \alpha u_2(x) + (1-\alpha)u_1(x) ,$$

and $\alpha \in [0,1]$ is chosen so that (this is a nonlinear system to be solved)

$$\mathrm{n}^T(x)f_U(x) \; = \; 0 . \tag{10}$$

So, there is one unknown in Utkin's construction, the scalar valued function α in (10) which must be found by solving a (single) nonlinear equation.

Utkin's approach is geometrically different from Filippov's approach. Among the vectors in $\{f(x,u)|\ u = \text{co}(u_1, u_2)\}$, Utkin takes the one tangent to Σ at x (see Figure 1). In general, Filippov's and Utkin's vector are different. They coincide directionally in case the convex hull is a straight line, and also in magnitude when f is linear with respect to the control u.

Example 1.1. In case f is linear with respect to u: $f(x,u) = g(x) + b(x)u$, then $f_F = f_U$. In fact, from (7), we get $f_F = \alpha f_2 + (1-\alpha)f_1 = g(x) + b(x)[\alpha u_2 + (1-\alpha)u_1] = f_U$. In the most important case when $p = 1$ (i.e., a scalar control), from (5) (or (10)) one ends up taking (assuming $\mathrm{n}^T(x)b(x) \neq 0$)

$$u_{\text{eq}}(x) = [\alpha u_2 + (1-\alpha)u_1] = -\frac{\mathrm{n}^T g}{\mathrm{n}^T b}(x) .$$

Example 1.2. To exemplify the difference between Filippov and Utkin approaches, consider the following problem in the plane[a]. We have

$$x' \; = \; f(x,u)$$

[a]This example was taken from a presentation of Prof. Heemels

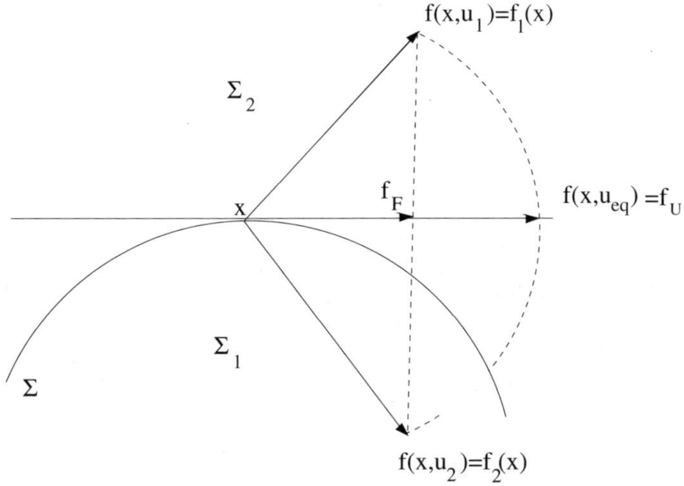

Fig. 1. Filippov's and Utkin's approach.

where

$$f(x) = \begin{pmatrix} -x_1 + x_2 - u \\ 2x_2(u^2 - u - 1) \end{pmatrix} \quad \text{and} \quad u(x) = \begin{cases} -1, \text{ when } & x_1 < 0 \\ \ \ 1, \text{ when } & x_1 > 0 \end{cases}.$$

The discontinuity surface $\Sigma := \{x : \ h(x) = 0\}$ is given by $x : \ x_1 = 0$, and a simple computation shows that there is attractive sliding motion on it if $-1 < x_2 < 1$. Again, simple computations show that the Filippov and Utkin vector fields on the sliding surface are (the equivalent control, here is $u_{\text{eq}}(x) = x_2$)

$$f_{\text{F}}(x) = \begin{pmatrix} 0 \\ 2x_2^2 \end{pmatrix}, \qquad f_{\text{U}}(x) = \begin{pmatrix} 0 \\ 2x_2(x_2^2 - x_2 - 1) \end{pmatrix}.$$

So, in the case of f_{F}, the origin is the only equilibrium, and is unstable, whereas in the case of f_{U} the origin is an asymptotically stable equilibrium and, on Σ, there is also the unstable equilibrium at $x_2 = (1 - \sqrt{5})/2$.

2. Reinterpretation

As we just remarked, in general (9) is not the same as (7), except in case the dependence on the control u is linear (see[2,6]). In the literature, this difference is observed to be essential and the approach based on (9) is often preferred in the Engineering field (see[1,3,5]).

Below, we derive an alternative way to look at sliding motion on Σ for systems such as (6).

To begin with, let us rewrite the original problem (6) in the following extended form

$$\begin{pmatrix} x' \\ u' \end{pmatrix} = \begin{pmatrix} f(x,u) \\ u_1' \end{pmatrix} = g_1(x,u) \text{ , when } h(x) < 0 \,,$$

and

$$\begin{pmatrix} x' \\ u' \end{pmatrix} = \begin{pmatrix} f(x,u) \\ u_2' \end{pmatrix} = g_2(x,u) \text{ , when } h(x) > 0 \,.$$

Notice that this rewriting takes into account the nature of the control variables as state (hence, time) dependent variables. According to our previous assumptions, we now have that $g_1 \in C^k$ for $x \in (\Sigma_1 \cup \Sigma)$ and $u \in \mathbb{R}^p$. Notice that by the chain rule we have

$$u_1' = \frac{du_1(x(t))}{dt} = (D_x u_1) f(x,u)|_{u=u_1} \,,$$

$$u_2' = \frac{du_2(x(t))}{dt} = (D_x u_2) f(x,u)|_{u=u_2} \,.$$

Also, observe that the surface of discontinuity $\Sigma = \{x \in \mathbb{R}^n \ : \ h(x) = 0\}$ has not changed, of course, but it has now become embedded in $\mathbb{R}^n \times \mathbb{R}^p$, and thus the normal to Σ is now more properly written as

$$\mathrm{v}(x,u) = \begin{pmatrix} \frac{\nabla h(x)}{\|\nabla h(x)\|} \\ 0 \end{pmatrix} .$$

Now, on the system we just rewrote, consider having attracting sliding motion on Σ. That is, a trajectory has reached the surface Σ and, for $x \in \Sigma$, one has

$$\mathrm{v}(x,u)^T g_1(x,u) > 0 \quad \text{and} \quad \mathrm{v}(x,u)^T g_2(x,u) < 0 \,,$$

that is

$$\mathrm{n}(x)^T f(x, u_1(x)) > 0 \quad \text{and} \quad \mathrm{n}(x)^T f(x, u_2(x)) < 0 \,.$$

The question is: What sliding motion does one have on Σ? Given the form of g_1 and g_2, the equation controlling evolution of the state variables x on Σ is simply, and certainly, going to be

$$x' = f(x,u) \,.$$

So, whichever extension, call it g_E, one proposes on Σ for $g_{1,2}$, it must have the form:

$$\begin{pmatrix} x' \\ u' \end{pmatrix} = g_E(x,u) = \begin{pmatrix} f(x,u) \\ u_E' \end{pmatrix} , \tag{11}$$

where the term $u'_{_E}$ needs not to be defined yet.

Now, since the extension in (11) must hold on the surface of discontinuity, then u must be taken so that the vector $g_{_E}(x,u)$ will be tangent to the discontinuity surface at (x,u). That is, we must have

$$\mathrm{v}^T(x,u)g_{_E}(x,u) = 0\ ,\ \ \mathrm{v}(x,u) = \begin{pmatrix} \frac{\nabla h(x)}{\|\nabla h(x)\|} \\ 0 \end{pmatrix}\ ,$$

which gives the condition

$$\Big(\frac{\nabla h(x)}{\|\nabla h(x)\|}\Big)^T f(x,u) = 0\ , \qquad \text{or} \qquad \mathrm{n}^T(x)f(x,u) \ = \ 0\,. \tag{12}$$

We stress that equation (12) is the key equation which must be satisfied: The control u must be chosen so that (12) holds. We notice that this equation expresses the same geometrical constraint as one has in Utkin's approach: In other words, working with the proposed extended vector fields validates the approach based on Utkin's equivalent control u_{eq} as the natural general approach.

We also notice that the requirement that (12) be satisfied, in general forces the control u to vary discontinuously when the trajectory arrives on Σ coming from $\Sigma_{1,2}$.

However, (12) is one equation in p unknowns (u), and thus generally one has $(p-1)$ free parameters in a solution to it. Utkin's idea is to require u to be a convex combination of the controls u_1 and u_2, which reduces to 1 the dimensionality of the nonlinear system, and generally yields (local) uniqueness of solutions. But other choices are also possible, at least in principle.

Example 2.1. To select a unique solution u to (12), one may want to require (one of) the following conditions.

(i) Min-norm solution. This arises from adding the constraint that $\|u(x)\|_2$ be minimized.

(ii) Min-variation solution. This is a requirement to minimize the variation (in norm) of $u(x)$ on Σ.

(iii) Require u to be in the convex hull of u_1 and u_2. This is Utkin's approach, giving $u'_{_E} = \alpha'(u_2 - u_1) + \alpha u'_2 + (1-\alpha)u'_1$.

(iv) Use Filippov's convexification idea on the derivative of u. Then, the evolution of the control variables on Σ is governed by the equation $u' = u'_{_E} := \alpha u'_2 + (1-\alpha)u'_1$, which represents a convex combination of derivatives (similarly to Filippov's construction), rather than function values. The two constraints $\mathrm{n}(x)^T f(x,u) = 0$, and $u' = \alpha u_2' + (1-\alpha)u_1'$, give a system of $p+1$ equations in $p+1$ unknowns (u and α).

References

1. G. Bartolini and T. Zolezzi, Control of nonlinear variable structure systems, *J. Math. Anal. Appl.*, **118** (1986), pp. 42–62.
2. A. F. Filippov, *Differential Equations with Discontinuous Right-Hand Sides*, Mathematics and Its Applications, Kluwer Academic, Dordrecht, 1988.
3. I. Haskare, U. Ozguner, and V. I. Utkin, On sliding modes observers via equivalent control approach, *Int. J. Control*, **71** (1998), pp. 1051–1067.
4. V. I. Utkin, Variable structure systems with sliding modes, *IEEE Transactions on Automatic Control*, **22** (1977), pp. 212–222.
5. V. I. Utkin, Application of equivalent control method to the systems with large feedback gain, *IEEE Transactions on Automatic Control*, **Ac-23** (1978), pp. 484–486.
6. V. I. Utkin, *Sliding Modes and Their Applications in Variable Structure Systems*, MIR Publisher, Moskow, Berlin, 1978.
7. V. I. Utkin, *Sliding Modes in Control and Optimization*, Springer, Berlin, 1992.
8. V. I. Utkin and S. Li, Window observers for linear systems, *Mathematial Problems in Enginnering*, **6** (2000), pp. 411–424.

On the Translation Groups and Non-iterative Transformation Methods

Riccardo Fazio and Salvatore Iacono

Department of Mathematics, University of Messina
Salita Sperone 31, 98166 Messina, Italy,
rfazio@dipmat.unime.it, iacono@dipmat.unime.it
First author home-page: http://mat520.unime.it/fazio/

As well-known, the invariance property with respect to a Lie group of transformation allows to solve numerically a boundary value problem, regardless its linearity, even in the case of intrinsic non-linearity, by means of non-iterative transformation methods. In literature this approach consists in transforming a boundary value problem into an initial value problem using the invariance with respect to a scaling or a spiral group. The main idea of this paper is to use the invariance with respect to a translational group, instead of the usual groups commonly encountered in literature. Then, two different applications are discussed: a free boundary problem for the optimal length estimation of tubular flow reactor where an isothermal chemical reaction occurs, and the motion of a parachutist expected to land after a prefixed time after the jump, respectively.

Keywords: Ordinary differential equation, boundary value problem, initial value methods, translation groups, tubular flow reactor, military parachuting.

1. Introduction

Several problems, that one come across over the applied mathematics, concern Boundary Value Problems (BVPs) or free BVPs. In most cases, it happens that these problems are non-linear and do not allow for an analytical solution. In such cases, the solution must be carried out by means of a numerical method. The approaches most frequently used for non-linear BVPs problems are of iterative type, such as shooting or relaxation methods (finite difference, finite element, or collocation). It is worth pointing out that as far as free BVPs are concerned, the free boundary is an additional unknown of the problem and, as a direct consequence, even if the governing equation is linear, according to Landau [1], the free BVP is indeed a non-linear problem.

This paper deals with the non-iterative solution of the mentioned BVPs, which are invariant with respect to a translation group, through the solution of a corresponding Initial Value Problem (IVP), also called auxiliary problem, associated to the original BVP. Non-iterative transformation methods (TMs) represents an elegant and interesting field of past and also recent research. Firstly, non-iterative TMs can be applied either to linear or non-linear problems so as to be more versatile than other methods like superposition [2, pp.135-145], chasing [3, pp. 30-51], and adjoint operator [3, pp. 52-69], that are applicable only to linear problems. Furthermore non-iterative TMs can be compared to the parameter differentiation method [3, pp. 233-288] or invariant embedding method [4] that are both applicable to non-linear problems as well, but the former is the only one funded on the geometric properties of the problem represented by its symmetry groups. For more details see Bluman and Cole [5], Dresner [6], or Bluman and Kumei [7].

From an historical viewpoint, the first to apply a non-iterative TM, to the Blasius problem and unconsciously about its symmetry properties, was Töpfer in [8]. Some time later Acrivos, Shah and Petersen [9] and Klamkin [10] reconsidered Töpfer's method in order to apply it to a more general problem and to a class of problems, respectively. By continuing on the conceptual framework set up by Klamkin, for a given problem Na [11,12] showed the relation between the invariance properties, with respect to a linear group of transformation (the scaling group), and the applicability of a non-iterative TM. Besides, in his work Na considered the invariance with respect to a nonlinear group of transformations, namely the spiral group. As long as eigenvalue problems are concerned, Belford [13] first, and Ames and Adams [14,15] later, defined non-iterative TMs for such a class of problems. Finally, a survey on the non-iterative TMs was written by Klamkin [16], whereas a survey book was written by Na [3] on the numerical solution of BVP, where he devoted the chapters from seven to nine to numerical TMs. However, non-iterative TMs are applicable only to particular classes of BVPs so that they have been considered as *ad hoc* methods, see Meyer [4, pp. 35-36], Na [3, p. 137] or Sachdev [17, p. 218]. On the other hand, extensions of non-iterative TMs, by requiring the invariance of one and of two or more physical parameters, when they are involved in the mathematical model, were respectively proposed by Na [18] and by Scott, Rinschler and Na [19], cf. also Na [3, Chapters 8 and 9]. Moreover, the introduction of a variable transformation linking two different invariant groups is a different way to extend the applicability of non-iterative TMs, as shown by Fazio

[20], and by Fazio and Iacono [21]. Finally, free BVPs governed by the most general second order differential equation, in normal form, can be solved iteratively by extending a scaling group via the introduction of a numerical parameter h so as to recover the original problem as h goes to one, see Fazio [22–25]. The application of this iterative TM to moving boundary problems governed by parabolic equations has been considered in [26].

This paper is organized as follows. The next section is devoted to deal with the group of translations in the independent variable and the definition of a related non-iterative TM algorithm. In section 3 we will focus on the group of translations in the dependent variable and the definition of a related non-iterative TM algorithm. Within each of these two sections an application, belonging to the characterized classes of BVPs, is reported in order to show the performance of the TMs. The numerical results were obtained by means of the `ODE45` integrator from the MATLAB ODE suite written by Samphine and Reichelt [27] and available with the latest releases of MATLAB.

Finally the last section is reserved for some concluding remarks

2. Translation of the independent variable

Let us consider the following class of free BVPs

$$
\begin{aligned}
&\frac{d^2u}{dx^2} = \Omega\left(u, \frac{du}{dx}\right) \\
&u(0) = A \\
&u(s) = B \quad , \quad \frac{du}{dx}(s) = C \quad x \in [0, s],
\end{aligned}
\tag{1}
$$

where A, B and C are arbitrary constants, and $s > 0$ is the unknown free boundary. The governing differential equation in (1) as well as the boundary conditions for the function u and all its derivatives are supposed to be invariant with respect to the translation Lie group

$$
x^* = x + \mu \quad ; \quad s^* = s + \mu \quad ; \quad u^* = u \; . \tag{2}
$$

This invariance allows us to define the following algorithm for the non-iterative numerical solution of (1):

- fix freely a value of s^*;

- backwards integration from s^* to x_A^* the following auxiliary IVP

$$\begin{aligned}
&\frac{d^2u^*}{dx^{*2}} = \Omega\left(u^*, \frac{du^*}{dx^*}\right) \\
&u^*(s^*) = B \\
&\frac{du^*}{dx^*}(s^*) = C \ ,
\end{aligned} \tag{3}$$

  where x_A^* is determined by means of an event locator so as to stop integrating as soon as the condition $u^*(x_A^*) = A$ is fulfilled;
- finally, through the invariance property, the similarity parameter

$$\mu = x_A^*$$

  is obtained and therefore the unknown free boundary

$$s = s^* - \mu \ ,$$

  is determined. As a consequence, the missing initial condition is given by

$$\frac{du}{dx}(0) = \frac{du^*}{dx^*}(x_A^*) \ .$$

2.1. *Length estimation for tubular flow reactors*

This is an engineering problem that consists in determining the optimal length of a tubular flow chemical reactor with axial missing and has been already treated by Fazio in [28], through an iterative TM based on scaling group, whereas in [22] it was made the comparison between the results obtained with a shooting method jointly with a non-iterative TM based on scaling group.

Roughly speaking, a tubular flow chemical reactor can be seen as a device where it is introduced a material A that along its passage inside the reactor undergoes a chemical reaction so that at the exit we get a product B plus a residual part of A. This problem can be formulated mathematically as

$$\begin{aligned}
&\frac{d^2u}{dx^2} = N_{Pe}\left(\frac{du}{dx} + Ru^n\right) \ , \\
&u(0) - \frac{1}{N_{Pe}}\frac{du}{dx}(0) = 1 \ , \\
&u(s) = \tau \quad \frac{du}{dx}(s) = 0 \ ,
\end{aligned} \tag{4}$$

where $u(x)$ is the ratio between the concentration of the reactant A at a distance x and the concentration of it at $x = 0$. N_{Pe}, R and τ are, the Peclet group, the reaction rate group and the residual fraction of reactant A at exit, respectively. Moreover, N_{Pe} and R are both greater than zero. Finally s is the length of the flow reactor we are trying to estimate and represents just the free boundary for the two points free BVP (4). Once

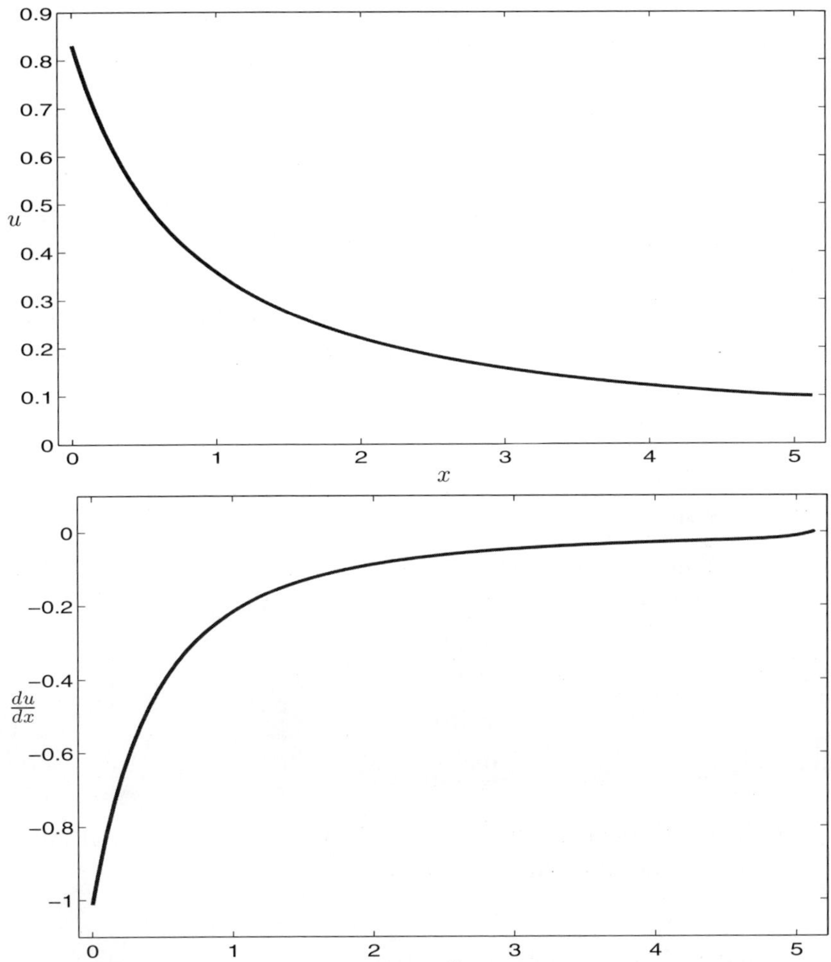

Fig. 1. Numerical solution for length estimation of a tubular flow reactor, $u(x)$ (top), $\frac{du}{dx}(x)$ (bottom).

applied the algorithm above outlined, through the MATLAB ODE suite integrator `ODE45`, with an absolute tolerance set up at 10^{-12}, jointly with its event locator, and having fixed $N_{Pe} = 6$, $R = 2$, $\tau = 0.1$ and $n = 2$, it was found that the value of the free boundary is $s = 5.119831$, whereas the initial conditions for the auxiliary problem result to be $u(0) = 0.831230$, $\frac{du}{dx}(0) = -1.012300$. These values are in good agreement with the ones found in [22] and [28] and the behavior of the solution can be seen in the figure 1.

3. Translation of the dependent variable

This time let us consider the class of BVPs

$$\begin{aligned} &\frac{d^2 w}{dx^2} = \Theta\left(x, \frac{dw}{dx}\right) \\ &\frac{dw}{dx}(0) = C \\ &w(b) = B \ , \end{aligned} \tag{5}$$

where B and C are arbitrary constants, and $b > 0$. In (5) the governing differential equation and the boundary condition at zero are invariant with regard to the translation Lie group

$$x^* = x \quad ; \quad w^* = w + \mu, \tag{6}$$

The non-iterative algorithm for the numerical solution of (5) is given by the following steps:

- fix freely a value of $w^*(0) = A$;
- integrate up to b the following auxiliary IVP

$$\begin{aligned} &\frac{d^2 w^*}{dx^{*2}} = \Theta\left(x^*, \frac{dw^*}{dx^*}\right) \\ &w^*(0) = A \\ &\frac{dw^*}{dx^*}(0) = C \ ; \end{aligned} \tag{7}$$

- finally, through the invariance property $w^*(b) = w(b) + \mu$, we can deduce the similarity parameter

$$\mu = w^*(b) - B \ ,$$

  and the missing initial condition for (7) to be equivalent to the given BVP

$$w(0) = w^*(0) - \mu \ .$$

3.1. *An application to a parachute model*

Among the differential equations invariant to the group (6), we found interesting to consider the one used for the well-known model describing the vertical motion of a parachutist through the atmosphere. In contrast with the usual approach to such a model, that is seen as an IVP, see for instance [29], we are interested in looking at it as a BVP where a zero initial velocity is prescribed and the parachutist is expected to touch the ground at a prefixed time.

Such a BVP model is expressed by

$$\begin{aligned} &\frac{d^2z}{dt^2} = \frac{K(t)}{m}\left(\frac{dz}{dt}\right)^2 - g \\ &\frac{dz}{dt}(0) = 0 \\ &z(b) = 0 \ , \end{aligned} \tag{8}$$

where z is the vertical coordinate taken positive going outward the ground that is also taken as the origin of the axes, t is the time variable, m is the total mass of parachutist's body plus his equipment, g is the acceleration of gravity, $K(t)$ is the damp coefficient of air, and b is a given final time when the parachutist is expected to land.

It is worth saying that such a problem, mathematically expressed in the BVP (8), has a practical importance for all those applications where the total time of fall must be kept under a given safety value. For instance, we can consider the military tactical throw of paratroopers behind the enemy lines where the shorter is the duration of the fall the lower is the probability to be targeted.

According to [29], because of a very high Reynolds numbers ($Re \approx 10^5 \div 10^6$), the air resistance against the motion of a parachutist through the atmosphere can be better modeled by means of a force proportional to the square of velocity, i.e.

$$F_{drag} = K(t)\left(\frac{dz}{dt}\right)^2 ,$$

instead of a force proportional to the velocity itself, which is more commonly used within an elementary modeling.

In order to define the parameter $K(t)$, we must take into account that the parachutist motion is essentially divided into a first phase relative to a free fall, a second one relative to the deployment of the parachute, and finally a third one relative to the dampened fall. Furthermore, instead of its usual piecewise definition, here we prefer to model it as the continuous function of time

$$K(t) = \frac{K_2 - K_1}{\pi} \arctan\left(P(t - t_c)\right) + \frac{K_1 + K_2}{2} .$$

The parameters involved can be defined as follows: according to the values reported in [29] for a parachute with cross sectional area of $43.8\,\mathrm{m}^2$, we can take $K_1 = 0.49\,\mathrm{Kg/m}$ and $K_2 = 29.16\,\mathrm{Kg/m}$ as the damp coefficient for the free fall and the dampened fall, respectively. With regard to the other parameters, by knowing that after 10 s the parachute activation is triggered and that the time spell for the complete deployment of the parachute is about $\Delta t_d = 3.2\,\mathrm{s}$, we chose to fix $t_c = 10 + \Delta t_d/2 = 11.6\,\mathrm{s}$. Finally, the value of P was determined by taking the average slope between the value of K_1 and K_2 over the time spell deployment Δt_d. For the reader convenience $K(t)$ is shown in the top frame of figure 2.

Having fixed $g = 9.81\,\mathrm{m/s}^2$, a final time $b = 40\,\mathrm{s}$ and $z^*(0) = 100\,\mathrm{m}$ as the initial missing condition for the auxiliary problem, it was carried out a numerical experiment with the model (8) for discrete increasing values of m ranging from 70 Kg to 120 Kg. The missing initial conditions found are reported in table 1. In figure 2, there are depicted the corresponding numerical solutions, for a few values of m. At a first glance, it is crystal

Table 1. Parachute model: numerical results for increasing m.

m	$z(0)$
70	425.1710
80	449.8336
90	472.4850
100	493.4470
110	512.9900
120	531.3088

clear that the higher is value of m the higher must be the drop altitude $z(0)$ in order to comply with the assigned boundary conditions.

4. Conclusion

It is worth pointing out that, according to what first conceived by S. Lie, the geometric features of any mathematical problem can be profitably used in finding its solution. This is also true whenever we solve BVPs by means

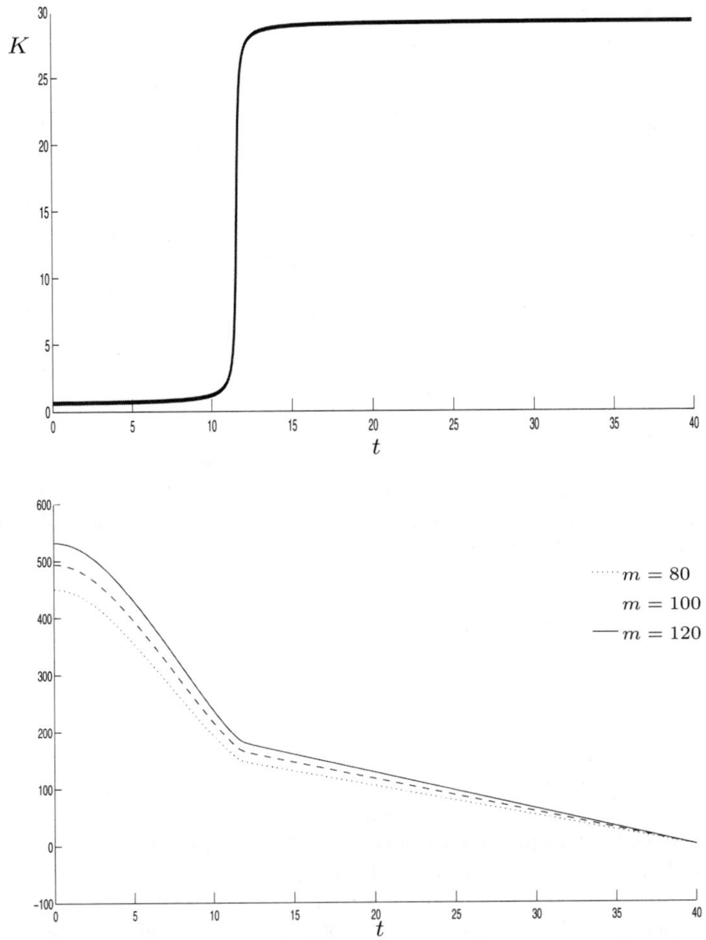

Fig. 2. On the top frame $K(t)$, on bottom one sample numerical results for the parachute model.

of non-iterative TMs. Indeed, the invariance with respect to a one parameter Lie group of translations of a second order governing equation, clearly allows for the application of this method, provided that also the values of the independent variable or of the dependent variable at the boundary conditions are invariant with respect to a translation of the independent or of the dependent variable, respectively.

Acknowledgments.
This work was supported by the University of Messina.

References

1. H. G. Landau, *Quart. Appl. Math.* **8**, 81 (1950).
2. U. M. Ascher, R. M. M. Mattheij and R. D. Russell, *Numerical Solution of Boundary Value Problems for Ordinary Differential Equations* (Prentice Hall, Englewood Cliffs, New Jersey, 1988).
3. T. Y. Na, *Computational Methods in Engineering Boundary Value Problems* (Academic Press, New York, 1979).
4. G. Meyer, *Initial Value Methods for Boundary Value Problems; Theory and Application of Invariant Imbedding* (Academic Press, New York, 1973).
5. G. W. Bluman and J. D. Cole, *Similarity Methods for Differential Equations* (Springer, Berlin, 1974).
6. L. Dresner, *Similarity Solutions of Non-linear Partial Differential Equations*, Research Notes in Math., Vol. 88 (Pitman, London, 1983).
7. G. W. Bluman and S. Kumei, *Symmetries and Differential Equations* (Springer, Berlin, 1989).
8. K. Töpfer, *Z. Math. Phys.* **60**, 397 (1912).
9. A. Acrivos, M. J. Shah and E. E. Petersen, *AIChE J.* **6**, 312 (1960).
10. M. S. Klamkin, *SIAM Rev.* **4**, 43 (1962).
11. T. Y. Na, *SIAM Rev.* **9**, 204 (1967).
12. T. Y. Na, *SIAM Rev.* **20**, 85 (1968).
13. G. G. Belford, *SIAM J. Numer. Anal.* **6**, 99 (1969).
14. W. F. Ames and E. Adams, *Nonlinear Anal.* **1**, 75 (1976).
15. W. F. Ames and E. Adams, *Int. J. Non-linear Mech.* **14**, 35 (1979).
16. M. S. Klamkin, *J. Math. Anal. Appl.* **32**, 308 (1970).
17. P. L. Sachdev, *Nonlinear Ordinary Differential Equations and their Applications* (Marcel Dekker, New York, 1991).
18. T. Y. Na, *J. Basic Engrg. Trans. ASME* **92**, 91 (1970).
19. T. C. Scott, G. L. Rinschler and T. Y. Na, *J. Basic Engrg. Trans. ASME* **94**, 250 (1972).
20. R. Fazio, *Int. J. Comput. Math.* **37**, 189 (1990).
21. R. Fazio and S. Iacono, Scaling and spiral equivalence from a numerical viewpoint, in *Proceedings World Congress on Engineering 2008*, eds. D. H. A. H. S.I. Ao, L. Gelman and A. Korsunsky (IAENG, KwunTong – Hong Kong, 2008).
22. R. Fazio, *Appl. Math. Comput.* **42**, 105 (1991).
23. R. Fazio, *Calcolo* **31**, 115 (1994).
24. R. Fazio, *SIAM J. Numer. Anal.* **33**, 1473 (1996).
25. R. Fazio, *Appl. Anal.* **66**, 89 (1997).
26. R. Fazio, *Int. J. Computer Math.* **78**, 213 (2001).
27. L. F. Shampine and M. W. Reichelt, *SIAM J. Sci. Comput.* **18**, 1 (1997).
28. R. Fazio, *J. Comput. Appl. Math.* **41**, 313 (1992).
29. D. B. Meade and A. A. Struthers, *Int. J. Engng. Ed.* **15**, 417 (1999).

Liquid Dynamics in a Horizontal Capillary: Extended Similarity Analysis

Riccardo Fazio*, Salvatore Iacono†, Alessandra Jannelli‡

Department of Mathematics, University of Messina
Salita Sperone 31, 98166 Messina, Italy
**rfazio@dipmat.unime.it*
†iacono@dipmat.unime.it
‡jannelli@dipmat.unime.it

Giovanni Cavaccini§, and Vittoria Pianese¶

Alenia Aeronautica S.p.A.
viale dell'Aeronautica s.n.c.
80038 Pomigliano d'Arco – Napoli, Italy
§gcavaccini@alenia.it
¶vpianese@alenia.it

The topic of this study is an extended similarity analysis for a one-dimensional model of liquid dynamics in a horizontal capillary. The bulk liquid is assumed to be initially at rest and is put into motion by capillarity, that is the only driving force acting on it. Besides the smaller is the capillary radius the steeper becomes the initial transitory of the meniscus location derivative, and as a consequence the numerical solution to a prescribed accuracy becomes harder to achieve.

Here, we show how an extended scaling invariance can be used to define a family of solutions from a computed one. The similarity transformation involves both geometric and physical feature of the model. As a result, density, surface tension, viscosity, and capillary radius are modified within the required invariance. Within our approach a target problem of practical interest can be solved numerically by solving a simpler transformed test case. The reference solution should be as accurate as possible, and therefore we suggest to use for it an adaptive numerical method. This study may be seen as a complement to the adaptive numerical solution of the considered initial value problems.

Keywords: Liquid penetrant testing, horizontal capillary, extended scaling invariance, adaptive numerical method.

1. Introduction

The present study was motivated by the non-destructive control named "liquid penetrant testing" used in the production of airplane parts and in many industrial applications where the detection of open defects is of interest. Liquid penetrants are used to locate surface-accessible defects in solid parts. The basic technique uses several stages. Among these stages, at least two involve capillary action, namely: application of a penetrant liquid and use of a developer, usually some kind of porous material (like an absorbing coating). These two stages can be modeled in the same way but at a completely different space scale. The visible dimension of a defect at the penetrant stage my be of the order of millimeters or microns; on the other hand, the porous material at the developer stage should have pores of the order of nanometers or Ångstrom. The present study may be seen as a possible way to use the results obtained within the penetrant stage to get results related to the developer stage.

2. Mathematical modeling

With reference to figure 1, we consider a liquid freely flowing within a horizontal cylindrical capillary of radius R. At the left end of the capillary we have a reservoir filled with the penetrant liquid. The 1D model governing

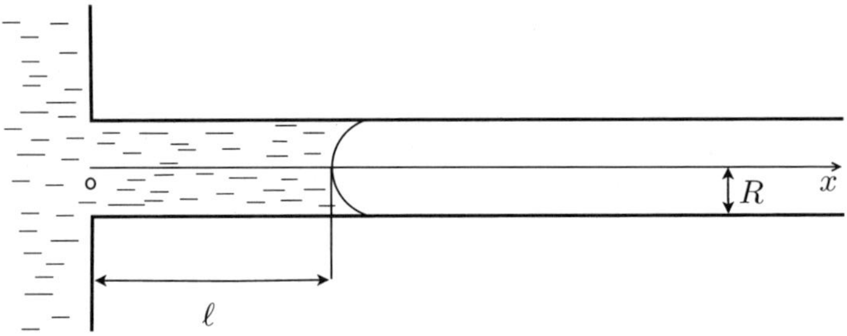

Fig. 1. Schematic drawing of a horizontal capillary section.

the dynamics of a liquid, through its liquid-gas interface, inside an open ended capillary is given by

$$\begin{aligned} \rho \frac{d}{dt}\left[(\ell + cR)\frac{d\ell}{dt}\right] &= 2\frac{\gamma\cos\vartheta}{R} - 8\frac{\mu\ell}{R^2}\frac{d\ell}{dt} \\ \ell(0) &= \frac{d\ell}{dt}(0) = 0 \ , \end{aligned} \tag{1}$$

where ℓ is the moving liquid-gas interface coordinate, $d\ell/dt$ can be interpreted as the average axial velocity U, t represents the time variable; ρ, γ, ϑ, and μ are the liquid density, surface tension, contact angle and viscosity, respectively, and $c = O(1)$ is the coefficient of apparent mass, introduced by Szekely et al. [1] to take into account the flow effects outside the capillary. Here we use $c = 7/6$. For a derivation of the governing equation we refer to Cavaccini et al. [2].

For the validity of this one-dimensional model, the Bond, capillary, and Weber numbers have to be small, that is $Bo = 4\rho g R^2/\gamma \ll 1$, $Ca = \mu U/\gamma \ll 1$, and $We = 2\rho R U^2/\gamma \ll 1$, where g is the acceleration due to gravity. Moreover, for the derivation of the model, we have to assume a quasi-steady Poiseuille velocity profile, and to simplify it we take the dynamic contact angle equal to the static one.

3. Extended scaling invariance

A classical similarity analysis would require the invariance of the model (1) with respect to a group involving only the dependent and independent variables, see for instance Dresner [3]. However, we are interested here to an extension of the classical analysis, obtained by requiring the invariance of physical parameters, proposed first by Na [4]. In particular, we are interested to consider the same model, but with different values of the capillary radius R, density ρ, surface tension γ, and viscosity parameter μ.

It is a simple matter to verify that the model (1) is invariant with respect to the following extended scaling group

$$t^* = \lambda^\delta t \;, \; \ell^* = \lambda \ell \;, \; R^* = \lambda R \;, \quad \rho^* = \lambda^{\alpha_1} \rho \;, \; \gamma^* = \lambda^{\alpha_2} \gamma \;, \; \mu^* = \lambda^{\alpha_3} \mu \;, \quad (2)$$

where λ is the group parameter, and α_1, α_2, α_3 and δ are group exponents to be determined. We note that the initial conditions are invariant with respect to the action of (2). Moreover, the invariance of the governing equation is granted, meaning that it is transformed by (2) into itself, on condition that

$$2 - 2\,\delta + \alpha_1 = \alpha_2 - 1 = \alpha_3 - \delta \;. \quad (3)$$

These are two linear equations for four unknowns. Henceforth, our model is left invariant by a scaling group depending on two arbitrary group exponents. It is worth noticing that if we limit our invariance analysis to space and time variables only (i.e. by fixing $\alpha_i = 0$, $i = 1, 2, 3$), then it cannot be attained any invariance of the model.

Our interest here is to define by the similarity invariance a family of solutions, each one identified by particular values of the capillary radius R, density ρ, surface tension γ and viscosity μ, from an approximated numerical one. In this way it can be possible to get an approximation also for the derivatives of the field variable ℓ. In fact, as a consequence of (2), we have

$$\frac{d^n \ell^*}{dt^{*n}}(t^*) = \lambda^{1-n\,\delta} \frac{d^n \ell}{dt^n}(t) \, , \qquad n = 1, 2, \ldots \, . \tag{4}$$

Cavaccini et al. have shown that in order to solve a capillary problem it would be advisable to apply an adaptive numerical method [2]. As a consequence, in [2] we used for the adaptive procedure the following monitor function

$$\eta(t_k) = \frac{\left| \frac{d\ell}{dt}(t_k + \Delta t_k) - \frac{d\ell}{dt}(t_k) \right|}{\Gamma_k} \tag{5}$$

where Δt_k is the current time-step and

$$\Gamma(t_k) = \begin{cases} \left| \frac{d\ell}{dt}(t_k) \right| & \text{if} \quad \frac{d\ell}{dt}(t_k) \neq 0 \\ \epsilon & \text{otherwise,} \quad \text{with } 0 < \epsilon \ll 1 \, . \end{cases} \tag{6}$$

The step size was modified in order to keep this monitor function between previously chosen tolerance bounds. More details on the adaptive strategy and the meaning of this type of monitor functions can be found in [5]. It is straightforward to verify that the monitor function (5) is invariant with respect to the scaling group (2) (with parameters verifying (3)) on condition that

$$\epsilon^* = \lambda^{1-\delta} \epsilon \, . \tag{7}$$

4. Numerical results

In this section we report on sample numerical results for viscous and for inviscid capillary flows. Besides the obtained results are compared to the celebrated Washburn asymptotic approximation [6]

$$\ell = \left(\frac{\gamma R \cos \vartheta}{2\mu} t \right)^{1/2} \, , \tag{8}$$

shown in all figures as a solid line. This approximation is valid only for $t >> t_\mu$, where $t_\mu = \rho R^2/\mu$ is a viscous time scale.

4.1. *The viscous case*

We consider first the case of a class of silicon oils (the so called PDMS series) according to the parameter values listed in table 1. It can be realized that

Table 1. Parameters for the series of PDMS silicone oils. For all cases we can consider these liquids as totally wetting, that is the related contact angle is equal to zero.

Liquid	ρ [Kg/m^3]	γ [mN/m]	μ [mPa $\cdot$ s]
PDMS5	918	19.7	5
PDMS10	935	20.1	10
PDMS20	950	20.3	20
PDMS50	960	20.8	50
PDMS500	971	21.5	500

all parameters within the PDMS series change with the liquid viscosity. Therefore, we apply the scaling group (2) with group exponents

$$\delta = 3/2 \ , \qquad \alpha_1 = \alpha_2 = 1 \ , \qquad \alpha_3 = 3/2 \ , \tag{9}$$

that satisfy the conditions (3).

Let us assume that our target is a capillary with $R^* = 0.5$ mm for the liquid parameters marked by PMDS5:

$$\rho^* = 918 \text{ Kg/m}^3 \ , \quad \gamma^* = 19.7 \text{ mN/m} \ , \quad \mu^* = 5 \text{ mPa} \cdot \text{s} \ , \quad \vartheta = 0^\circ \ . \tag{10}$$

The top frame of figure 2 displays the numerical results obtained by setting a reference value $\mu = 500$ mPa $\cdot$ s and computing all the other non-starred parameters from (10) according to the scaling group (2)-(9). This computation, using the adaptive procedure with criteria specified at the end of this section, required 5219 steps with 9 rejections for $t \in [0, 50]$; the used limits for the step size were $\Delta t_{min} \approx 2 \cdot 10^{-6}$ and $\Delta t_{max} = 0.1$. We note from the middle and bottom frame of figure 2 that the adaptive procedure concentrates most of the computational effort within the initial transient.

In the top frame of figure 3 we plot the target solution obtained by applying the scaling invariance to the computed numerical solution shown in figure 2.

It is interesting to note, that the same results were obtained starting with the reference value $\mu = 100$ mPa$\cdot$s and following the described scaling procedure. This means that the proposed approach does not depend upon the choice of the reference value of μ.

The reader can compare the top frame of figure 3 with its bottom frame, the bottom frame was obtained by plotting the numerical results of a second computation with the parameters in (10). In this case the computation, using the same adaptive criteria as in the first one, required 3942 steps

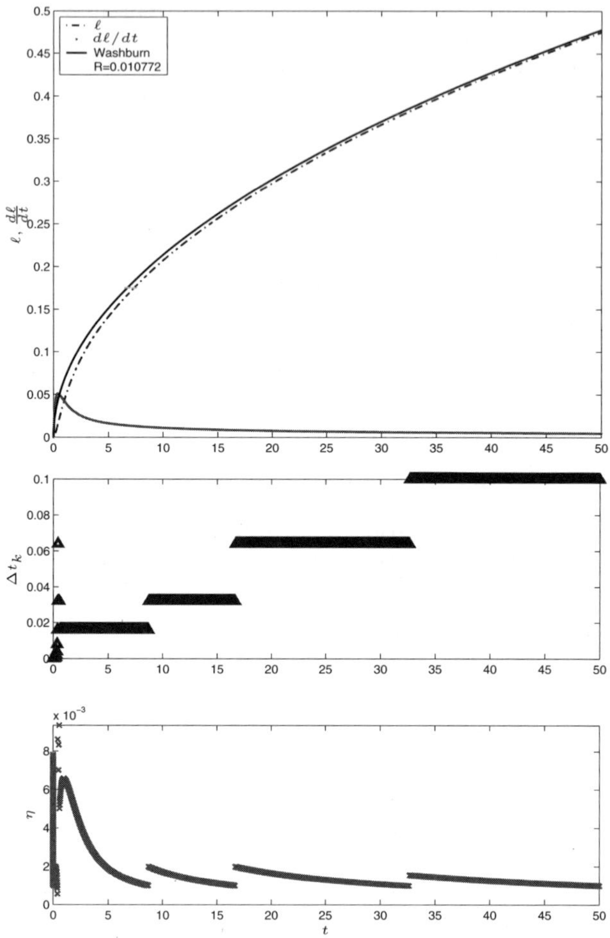

Fig. 2. Adaptive step-size results. Viscous case $\mu = 500$ mPa·s with $R = 0.010772$ mm. From top to bottom: $\ell(t)$ and its first derivative; adaptive step-size selection Δt_k; monitor function η.

with 8 rejections with $t \in [0, 0.5]$; the used limits for the step size were $\Delta t_{min} \approx 1 \cdot 10^{-7}$ and $\Delta t_{max} = 1.6 \cdot 10^{-3}$.

4.2. *The case of approximately invariant parameters*

We consider now the case of silicon oils according to the parameter values tabulated by Clanet and Quéré [7]. In what follows we refer to those values and, therefore, for the sake of completeness let us report them in table 2.

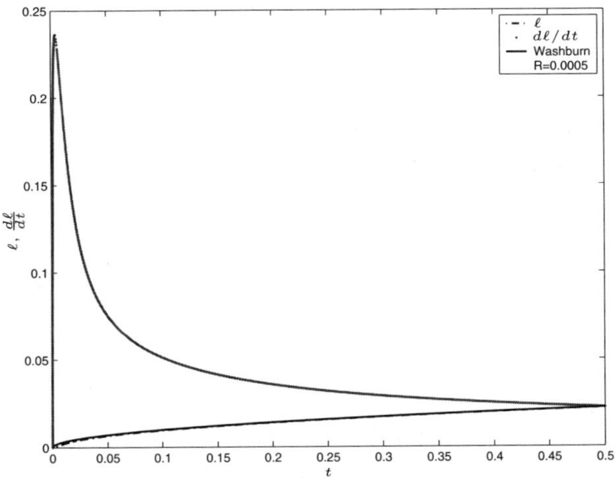

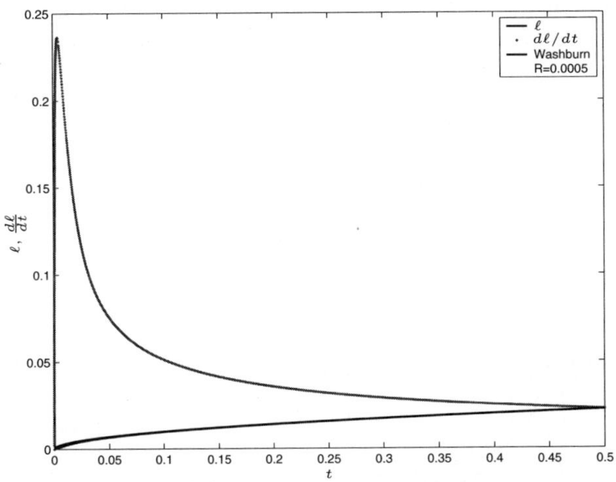

Fig. 3. Viscous case $\mu = 5$ mPa $\cdot$ s with $R = 0.5$ mm. Top frame: results found by invariance. Bottom frame: numerical results obtained with parameters in (10).

From this table we can easily realize that, for this particular series of silicon oils, the surface tension is invariant and the density may be considered as approximately constant for the oils ranging from V5 to V12500. Therefore, we apply now the scaling group (2) with group exponents

$$\delta = 3/2 \; , \qquad \alpha_1 = \alpha_2 = 0 \; , \qquad \alpha_3 = 1/2 \; , \tag{11}$$

that satisfy the conditions (3).

Figure 4 displays the numerical results obtained for the fluid parameters

Table 2. Parameters listed by Clanet and Quéré [7]. The V series correspond to silicone oils.

Liquid	ρ [Kg/m^3]	γ [mN/m]	μ [mPa $\cdot$ s]
Hexane	660	18	0.45
V1	815	21	1.0
V5	913	21	5.0
V10	930	21	10.0
V100	952	21	100.0
V1000	962	21	1000.0
V12500	965	21	12500.0
Water	1000	73	1.0

marked by V100:

$$\rho = 952\ \mathrm{Kg/m}^3\ , \quad \gamma = 21\mathrm{mN/m}\ , \quad \mu = 100\ \mathrm{mPa}\cdot\mathrm{s}\ , \quad \vartheta = 0^\circ\ . \qquad (12)$$

and $R = 0.01$ mm.

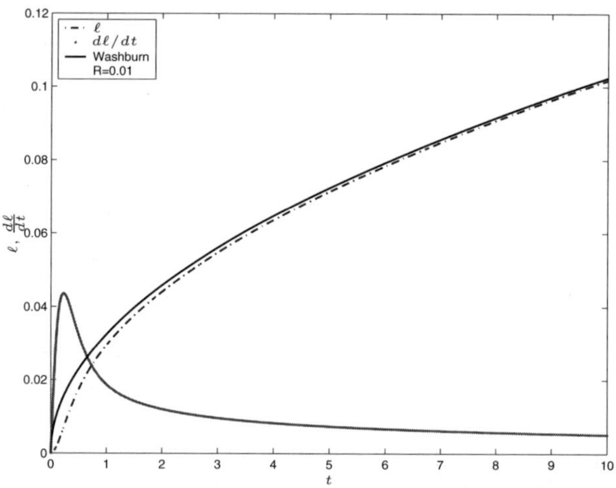

Fig. 4. Adaptive step-size results: $\ell(t)$ and its first derivative. Viscous case $\mu = 100$ mPa$\cdot$s with $R = 0.01$ mm.

The top frame of figure 5 shows one solution obtained by applying the scaling invariance to the numerical solution shown in figure 4.

The reader can compare the top frame of figure 5 with the bottom frame, the last was obtained by plotting the numerical results of a second computation with parameters

$$\rho = 940\ \mathrm{Kg/m}^3\ , \quad \gamma = 21\ \mathrm{mN/m}\ , \quad \mu = 50\ \mathrm{mPa}\cdot\mathrm{s}\ , \quad \vartheta = 0^\circ\ , \qquad (13)$$

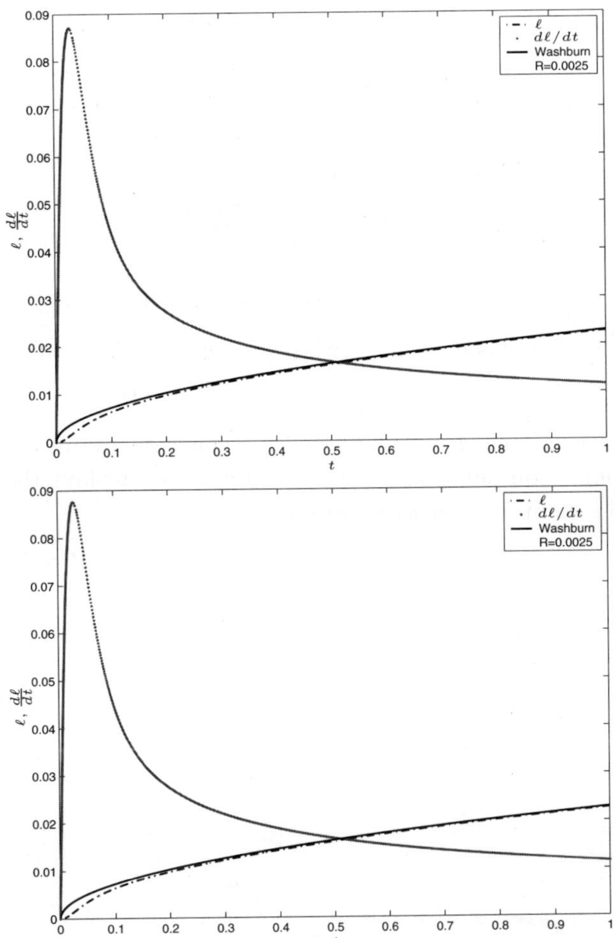

Fig. 5. Viscous case $\mu = 50$ mPa $\cdot$ s with $R = 0.0025$ mm. Top frame: results found by invariance. Bottom frame: numerical results.

and $R = 0.0025$ mm.

The differences in the two frames of figure 5 are due to the different value of the density used in the second numerical integration, namely 940Kg/m^3 instead of 952Kg/m^3. Despite this different value of the density, which is a factor in the inertial term, the agreement within the two frames is evident.

4.3. *The classical scaling invariance strategy*

Finally, we report other results obtained by ignoring low viscosity values of the considered liquids. This may be related to the classical way of generating a family of similarity solutions from a computed one.

The model should be modified as follows

$$\begin{aligned} \rho \frac{d}{dt}\left[(\ell + cR)\frac{d\ell}{dt}\right] &= 2\frac{\gamma \cos \vartheta}{R} \\ \ell(0) = \frac{d\ell}{dt}(0) &= 0 \,. \end{aligned} \tag{14}$$

This simplified model is invariant with respect to the classical scaling group

$$t^* = \lambda^{3/2} t \,, \quad \ell^* = \lambda \ell \,, \quad R^* = \lambda R \,. \tag{15}$$

As a consequence, for a given liquid, we can generate a family of similarity solutions from a computed reference one. Figure 6 displays the numerical results for the computation with parameters

$$\rho = 815 \text{ Kg/m}^3 \,, \quad \gamma = 21 \text{ mN/m} \,, \quad \mu = 0 \,, \quad \vartheta = 0^\circ \,, \tag{16}$$

and $R = 0.0025$ mm.

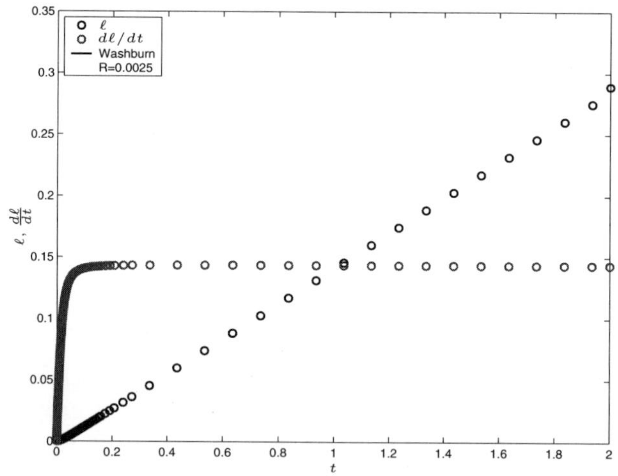

Fig. 6. Adaptive step-size results: $\ell(t)$ and its first derivative. Inviscid case $\mu = 0$ with $R = 0.0025$ mm.

The above results can be contrasted with those reported in Figure 7 for the parameters

$$\rho = 815 \text{ Kg/m}^3 \,, \quad \gamma = 21 \text{ mN/m} \,, \quad \mu = 1 \text{ mPa} \cdot \text{s} \,, \quad \vartheta = 0^\circ \tag{17}$$

and $R = 0.0025$ mm.

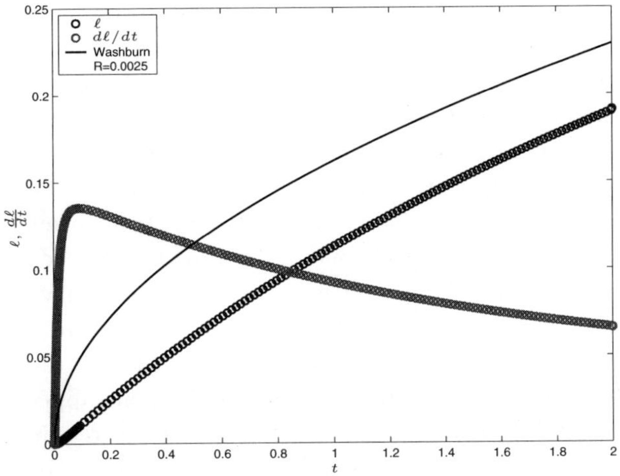

Fig. 7. Adaptive step-size results: $\ell(t)$ and its first derivative. Viscous case $\mu = 1$ mPa·s with $R = 0.0025$ mm.

We can easily realize that this strategy would be useless for the considered class of problems. Indeed, the generated family of similarity solution provides an upper bound for the real dynamics, but this is more expensive to get than the Washburn asymptotic solution which provides a more accurate and inexpensive upper bound.

4.4. *Numerical method and adaptive criteria*

The numerical method used for all the reported results was the classical fourth order Runge-Kutta's method implemented with the adaptive procedure developed by Jannelli and Fazio [5].

For the adaptive procedure we usually enforced the following conditions: $\Delta t_{\min} \leq \Delta t_k \leq \Delta t_{\max}$ with $\Delta t_{\min} = 10^{-6}$, $\Delta t_{\max} = 10^{-1}$, and $\eta_{\min} \leq \eta(t_k) \leq \eta_{\max}$ with $\eta_{\min} = 5 \cdot 10^{-2}$ and $\eta_{\max} = 2\,\eta_{\min}$. Only in the two target cases, namely for the silicone oils denoted by PDMS5 and V100 shown in the bottom frames of figures 3 and 5, we used a smaller value of $\Delta t_{\min}$. Moreover, the time step was modified in two cases: when $\eta(t_k) < \eta_{\min}$ we used $\Delta t_{k+1} = 2\,\Delta t_k$ as the next time step, or if $\eta(t_k) > \eta_{\max}$, then we repeated the same step using $\Delta t_k = \Delta t_k/2$.

5. Concluding remarks

In this paper we extend the original approach developed by Fazio in [8], and applied to the van der Pol model, to a class of problems. The topic of this study is an extended similarity analysis for a one-dimensional models of liquid dynamics a horizontal capillary. By requiring the invariance of all the physical parameters involved in the mathematical model we were able to show that the model itself is left invariant by a scaling group depending on two arbitrary group exponents. As a consequence we can choose a particular group that allows us to compute by the scaling transformation a target solution from a computed simpler one. Indeed, several choices are available.

We have to remark that in the case of a vertical capillary it is necessary to add to the right-hand side of the model (1) the term $-\rho g \ell$ where g represents the acceleration of gravity. Moreover, further terms should be added to the model in order to take into account the entrapped gas action in the case of a closed capillary. All these extended models require an extended scaling group that can be defined following the analysis outlined in the present study.

Acknowledgments.
This work was partially supported by the Italian MIUR and the Messina University.

References

1. J. Szekely, A. W. Neumann and Y. K. Chuang, *J. Colloid Interf. Sci.* **69**, 486 (1979).
2. G. Cavaccini, V. Pianese, S. Iacono, A. Jannelli and R. Fazio, Mathematical and numerical modeling of liquids dynamics in a horizontal capillary, in *Recent Progress in Computational Sciences and Engineering, ICCMSE 2006, Lecture Series on Computerand Computational Sciences*, eds. T. Simos and G. Maroulis (Koninklijke Brill NV, Leiden, The Netherlands, 2006).
3. L. Dresner, *Similarity Solutions of Non-linear Partial Differential Equations*, Research Notes in Math., Vol. 88 (Pitman, London, 1983).
4. T. Y. Na, *J. Basic Engrg. Trans. ASME* **92**, 91 (1970).
5. A. Jannelli and R. Fazio, *J. Comput. Appl. Math.* **191**, 246 (2006).
6. E. W. Washburn, *Phys. Rev.* **17**, 273 (1921).
7. C. Clanet and D. Quéré, *J. Fluid Mech.* **460**, 131 (2002).
8. R. Fazio, *Acta Appl. Math.* **104**, 107 (2008).

Ill and Well-Posed One-Dimensional Models of Liquid Dynamics in a Horizontal Capillary

Riccardo Fazio* and Alessandra Jannelli†

Department of Mathematics, University of Messina
Salita Sperone 31, 98166 Messina, Italy
**rfazio@dipmat.unime.it*
†jannelli@dipmat.unime.it

In this paper, we report a mathematical and numerical study of liquid dynamics models in a horizontal capillary. In particular, we prove that the classical model is ill-posed at initial time, and we present two different approaches in order to overcome this ill-posedness. By numerical viewpoint, we apply an adaptive strategy based on an one-step one-method approach, and we compare the obtained numerical approximations with suitable asymptotic solutions.

Keywords: Ill and well-posed models, horizontal capillary, asymptotic solutions, adaptive numerical method.

1. Introduction

The non-destructive control named "liquid penetrant testing" is routinely used by the aviation industry in the production of airplane parts as well as in the follow-up revisions during their operative life. In particular, liquid penetrants are used to locate surface-accessible defects in solid parts.

In this paper, we present some mathematical models that describe the dynamics of a liquid inside an open ended capillary. In particular, we prove that the classical model is ill-posed at initial time, and we present two different approaches in order to overcome this ill-posedness. A first model, by modifying the usual initial conditions, was obtained by Bosanquet [1], another one was derived by Szekely et al. [2], taking into account the flow effects outside the capillary. Finally, with reference to the academic test case presented in Cavaccini et al. [3], we compare the numerical approximations, obtained by an adaptive procedure based on an one-step one-method approach, with the solutions derived by an asymptotic study.

With reference to figure 1, we consider a liquid freely flowing within a horizontal cylindrical capillary of radius R. At the left end of the capillary

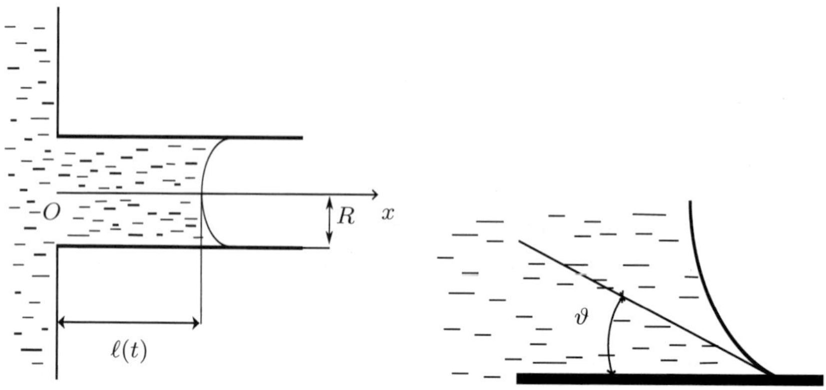

Fig. 1. Left frame: geometrical setup of a horizontal capillary section. Right frame: definition of the contact angle ϑ.

we have a reservoir filled with the penetrant liquid. The model governing the dynamics of a liquid inside an open ended capillary is given by

$$\rho \frac{d}{dt}\left[\ell \frac{d\ell}{dt}\right] = 2\frac{\gamma \cos\vartheta}{R} - 8\frac{\mu\ell}{R^2}\frac{d\ell}{dt} \tag{1}$$

where ℓ is the moving liquid-gas interface coordinate, $d\ell/dt$ can be interpreted as the average axial velocity. Moreover, ρ, γ, and μ are the liquid density, surface tension and viscosity, respectively. The contact angle between the liquid and the capillary wall is denoted by ϑ, with $0^\circ \leq \vartheta << 90^\circ$. The left-hand side term describes the inertial resistance. The effect of inertia are usually significant only in the early stages of penetration or when the radius R is large and/or μ is small. The first term of the right-hand side is the capillary driving force, the second one gives the viscous resistance of the liquid in the capillary. For the mathematical derivation of the governing equation we refer to Cavaccini et al. [3] and the references quoted therein.

2. Ill and Well-Posed One-Dimensional Models

Rewriting equation (1), and assuming that, at initial time, the liquid starts to flow inside the capillary, we have

$$\begin{cases} \ell \dfrac{d^2\ell}{dt^2} = 2\dfrac{\gamma\cos\vartheta}{\rho R} - 8\dfrac{\mu\ell}{\rho R^2}\dfrac{d\ell}{dt} - \left(\dfrac{d\ell}{dt}\right)^2, \\ \ell(0) = \dfrac{d\ell}{dt}(0) = 0 \, . \end{cases} \tag{2}$$

The right-hand side of the governing differential equation is a function of ℓ and its first derivative, that can be abbreviated by F:

$$F\left(\ell, \frac{d\ell}{dt}\right) = 2\frac{\gamma\cos\vartheta}{\rho R} - 8\frac{\mu\ell}{\rho R^2}\frac{d\ell}{dt} - \left(\frac{d\ell}{dt}\right)^2 , \tag{3}$$

so that we can rewrite the governing equation in (2) as

$$\ell\frac{d^2\ell}{dt^2} = F\left(\ell, \frac{d\ell}{dt}\right) \tag{4}$$

to be taken with zero initial conditions. Therefore, by evaluating F at the initial time we get

$$F(0,0) = 2\frac{\gamma\cos\vartheta}{\rho R} , \tag{5}$$

which is positive and ensures that the fluid flows inside the capillary. This implies that, at the same initial time, the left-hand side of the governing equation in (2) should be positive as well. However, because of the initial conditions the left-hand side of the governing equation in (2) must be zero which hints at a contradiction. However, this alone does not invalidate the derived modeling. In fact, the considered equation may possibly hold for all $t > 0$, and there may be a solution which assumes the required initial values. Here we show that this is not the case for the considered model.

In order to understand the ill-posedness of our model, we examine the simple model defined by

$$\begin{cases} \ell\dfrac{d^2\ell}{dt^2} = C \\ \ell(0) = \dfrac{d\ell}{dt}(0) = 0 , \end{cases} \tag{6}$$

where C is a positive constant. Now, if we assume that a solution $\ell(t)$ of (6) exists with $\ell(t) > 0$ for $t > 0$, then solving for the second derivative and multiplying both sides by the first derivative yields

$$\frac{d\ell}{dt}\frac{d^2\ell}{dt^2} = C\frac{\frac{d\ell}{dt}}{\ell} . \tag{7}$$

By considering solutions on the interval $[\delta, t]$ with $\delta > 0$, we integrate both sides of (7), with respect to t, to get

$$\frac{1}{2}\left[\frac{d\ell}{dt}(t)\right]^2 - \frac{1}{2}\left[\frac{d\ell}{dt}(\delta)\right]^2 = C\left[\ln\ell(t) - \ln\ell(\delta)\right] . \tag{8}$$

Fixing t and taking the limit as $\delta \to 0^+$, the left-hand side of equation (8) takes a positive finite value, but the right-hand side goes to infinity which is a contradiction. Therefore, there is no solution to the problem (6). This ill-posedness also applies to our problem with zero initial data.

Now, we define two different ways to revise the considered model in order to get a well-posed one.

As first revision, we can modify the given initial conditions. Assuming that, at the initial time, the liquid is inside the capillary, we obtain the model, so called Bosanquet one [1],

$$\begin{cases} \rho \dfrac{d}{dt}\left[\ell \dfrac{d\ell}{dt}\right] = 2\dfrac{\gamma \cos\vartheta}{R} - 8\dfrac{\mu \ell}{R^2}\dfrac{d\ell}{dt} \\ \ell(0) = R \,, \qquad \dfrac{d\ell}{dt}(0) = \left(2\dfrac{\gamma\cos\vartheta}{\rho R}\right)^{1/2} \,, \end{cases} \tag{9}$$

derived by rewriting the momentum balance for the moving column neglecting viscosity and external hydrodynamics. As shown below by numerical results, the Bosanquet velocity gives an accurate upper estimate of the initial velocity of liquid penetration into a horizontal capillary.

The second revision was already done by Szekely et al. [2] by taking into account the flow effects outside the capillary. These authors introduced a coefficient of apparent mass $c = O(1)$ and obtained the following model

$$\begin{cases} \rho \dfrac{d}{dt}\left[(\ell + cR) \dfrac{d\ell}{dt}\right] = 2\dfrac{\gamma \cos\vartheta}{R} - 8\dfrac{\mu \ell}{R^2}\dfrac{d\ell}{dt} \\ \ell(0) = \dfrac{d\ell}{dt}(0) = 0 \,. \end{cases} \tag{10}$$

Note that, starting with zeroth velocity, the liquid accelerates and attains a maximum velocity within a short interval. The most challenging model, by a numerical viewpoint, is the one proposed by Szekely et al. Therefore, we discuss numerical results comparing them with asymptotic ones obtained by solving an academic test case presented in [3]. Several numerical computations, related to real liquids, can be found, by the interested reader, in the paper by Cavaccini et al. [3]. Further numerical results were presented at the ICIAM congress held in Zurich, 16-20 July 2007, see Fazio et al. [4].

3. Academic test case

As an academic test case, we report on the numerical results for the model

$$\begin{cases} \dfrac{d}{dt}\left[(\ell + R) \dfrac{d\ell}{dt}\right] = 1 - \ell\dfrac{d\ell}{dt} \,, \\ \ell(0) = \dfrac{d\ell}{dt}(0) = 0 \,, \end{cases} \tag{11}$$

derived by equation (10) setting the following parameters

$$c = \rho = 2\frac{\gamma \cos \vartheta}{R} = 8\frac{\mu}{R^2} = 1 \ . \tag{12}$$

Now, we present asymptotic solutions of the model (11) that will be used, in the following, in order to validate the numerical approximations, of the models (9) and (10) with parameters defined in (12), obtained by an adaptive numerical approach.

3.1. *Washburn equation*

For very small radii, viscous forces are dominant, the inertial terms at the left-hand side of the (11) can be neglected, and one obtains the differential equation, named Washburn one [5]. As far as our academic test case is concerned, by using the relations (12), the Washburn equation specializes to

$$\ell \frac{d\ell}{dt} = 1 \ . \tag{13}$$

By integrating, taking into account the initial condition $\ell = 0$, at $t = 0$, we get the solution

$$\ell_W(t) = (2t)^{1/2} \ , \tag{14}$$

valid only for $t >> 8$. Note that, at initial time

$$\frac{d\ell_W}{dt}(0) = +\infty \ ,$$

and this is an evident paradox [6]. However, Washburn approximation is still considered as a valid approximation, although it fails to describe the initial transient, since it neglects the inertial effects which are relevant at the beginning of the process. In fact, this approximation has been confirmed by a lot of experimental data and also by molecular dynamic simulations, see for instance Martic et al. [7–9], and lattice-Boltzmann computations, see Chibbaro [10].

3.2. *Budd and Huang asymptotic analysis*

An asymptotic analysis has been developed by Budd and Huang [11], who used the method of matched asymptotic expansions to tackle the problem (11). Now, observe that the equation (11) has a first integral given by

$$(\ell + R)\frac{d\ell}{dt} = t - \frac{\ell^2}{2} \ . \tag{15}$$

In the inner region, where t is of order R, Budd and Huang rescaled the solution and then developed a series expansion of ℓ in powers of R, obtaining the asymptotic solution

$$\ell_{in}(t) = \sqrt{R^2 + t^2} - R \qquad \text{valid for } 0 < t << 1 \, . \tag{16}$$

In the outer region, they ignored the contributions involving R to leading order and integrated again. This led to the following asymptotic formula

$$\ell_{out}(t) = \sqrt{2}\sqrt{t - 1 - \exp(-t)} \qquad \text{valid for } R << t \, . \tag{17}$$

The two expansions match when $R << t$, where they both have the form

$$\ell(t) \approx t \, .$$

4. Validation of numerical results

In this section, we show the numerical approximations of the mathematical models, presented in section 2. These results, obtained by an adaptive approach developed by Jannelli and Fazio [12] and briefly described in the next section, are validated by comparing with the above asymptotic solutions. All computations were performed with MATLAB.

First, we consider the numerical results of the Bosanquet model (9), with the conditions (12) and $R = 0.01$. In this case, at the beginning of the

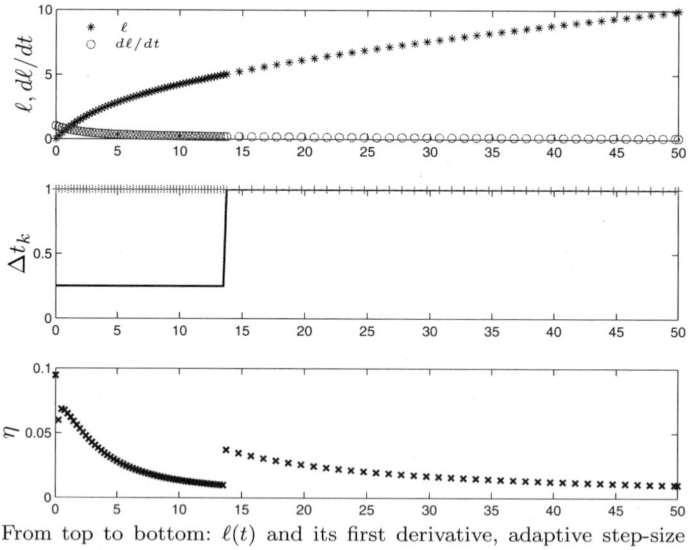

Fig. 2. From top to bottom: $\ell(t)$ and its first derivative, adaptive step-size selection Δt_k, and monitor function η.

process, there is not fast transitory of first derivative, the liquid is inside the capillary and it has a maximum velocity that, after decreases. Therefore, even a simple integration with constant step size would be suitable in this case. In the top frame of figure 2, we show the numerical solution ℓ and its first derivative. The step-size selection Δt_k and the monitor function $\eta(t_k)$ are reported in the middle and in the bottom frames, respectively. Note that, in this case our adaptive strategy used only 92 steps, with no rejections, to complete the integration in $[0, 50]$. The minimum value of Δt used was 0.25 and only in this case we magnified the time step by a factor of 4.

Now, we report the numerical results of the academic test case (11) with $R = 0.01$. In the top frame of figure 3, we show the numerical solution ℓ

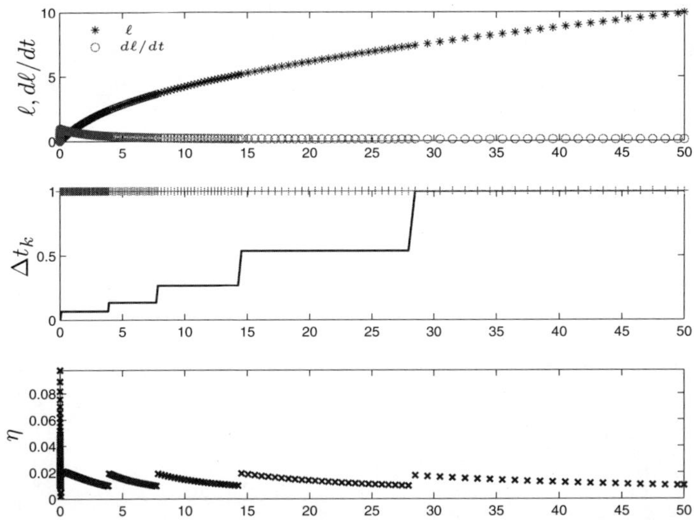

Fig. 3. Adaptive step-size results. From top to bottom: $\ell(t)$ and its first derivative, adaptive step-size selection Δt_k, and monitor function η.

and its first derivative. The step-size selection Δt_k and the monitor function $\eta(t_k)$ are reported in the middle and in the bottom frames. It is easy to note, how, the adaptive procedure modifies the time step in relation to the value of the monitor function. Initially, at the beginning of the process, the adaptive procedure reduces Δt_k corresponding to first derivative fast transitory. Then, when the solution becomes smooth, the procedure amplifies the step-size. Our adaptive strategy used 1603 successful steps, plus 10 rejections, to complete the integration in $[0, 50]$. For this test case, the minimum value of Δt used was about 10^{-12}.

In the figure 4, we report the numerical solutions obtained on the time interval $[0, 1]$. The solution of Bosanquet model (9) is shown in the top

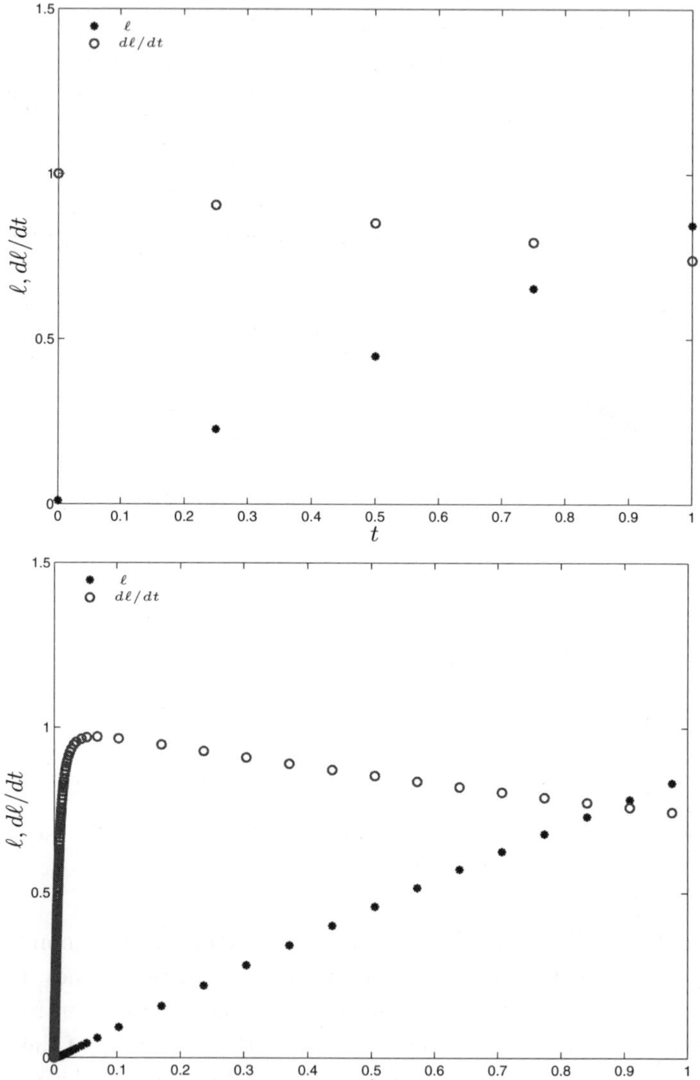

Fig. 4. Top frame: model (9) numerical solution: $\ell(t)$ and its first derivative. Bottom frame: model (10) adaptive step-size solution: $\ell(t)$ and initial transient of its first derivative.

frame, and the solution of model (10) developed by Szekely et al. is in the bottom frame. The physical parameters are given by (12) and $R = 0.01$. Note that, at the initial steps of the process, the model (10) describes the fast transient of first derivative of ℓ, developed by the increase of the velocity of the liquid entering the capillary. This large acceleration does not exist in the solution of the model (9), because, at initial time, the liquid is already inside the capillary. Note that, from the bottom frame of figure 4 the liquid entering the capillary is accelerated by capillary force ($\ell \approx t^2$ initially); but, soon thereafter the capillary force is compensated by the viscous drag so that a steady-state can be achieved ($\ell \approx t^{1/2}$ for t large enough). Moreover, we can notice how the Bosanquet velocity is an upper bound for the velocity of the Szekely et al. model.

In the top frame of figure 5, we report the asymptotic solution ℓ_{in} of (16), and its derivative, obtained for very short times. The numerical approximation, obtained by our adaptive procedure, is shown in the bottom frame of figure 5.

The asymptotic solution ℓ_{out}, given by (17), and its first derivative, with the numerical approximations are compared, for large times, in figure 6. The solution of the Washburn equation $\ell_W(t)$ is plotted for comparison. This figure shows a good agreement for large times between the two asymptotic solutions.

5. Numerical method and adaptive procedure

In this section we describe the considered adaptive procedure, developed by Jannelli and Fazio [12], used with the classical fourth order Runge-Kutta's method [13, p. 166]. For the reported test case we used the following monitor function

$$\eta(t_k) = \frac{\left| \dfrac{d\ell}{dt}(t_k + \Delta t_k) - \dfrac{d\ell}{dt}(t_k) \right|}{\Gamma_k} \tag{18}$$

where Δt_k is the current time-step and

$$\Gamma_k = \begin{cases} \left| \dfrac{d\ell}{dt}(t_k) \right| & \text{if} \quad \dfrac{d\ell}{dt}(t_k) \neq 0 \\ \epsilon & \text{otherwise} \; , \end{cases} \tag{19}$$

with $0 < \epsilon \ll 1$. The above monitor function has been chosen because, as far as the solution of the model (10) is concerned, we have found numerically that, for small values of R, initially the first derivative of $\ell(t)$ has a fast transient.

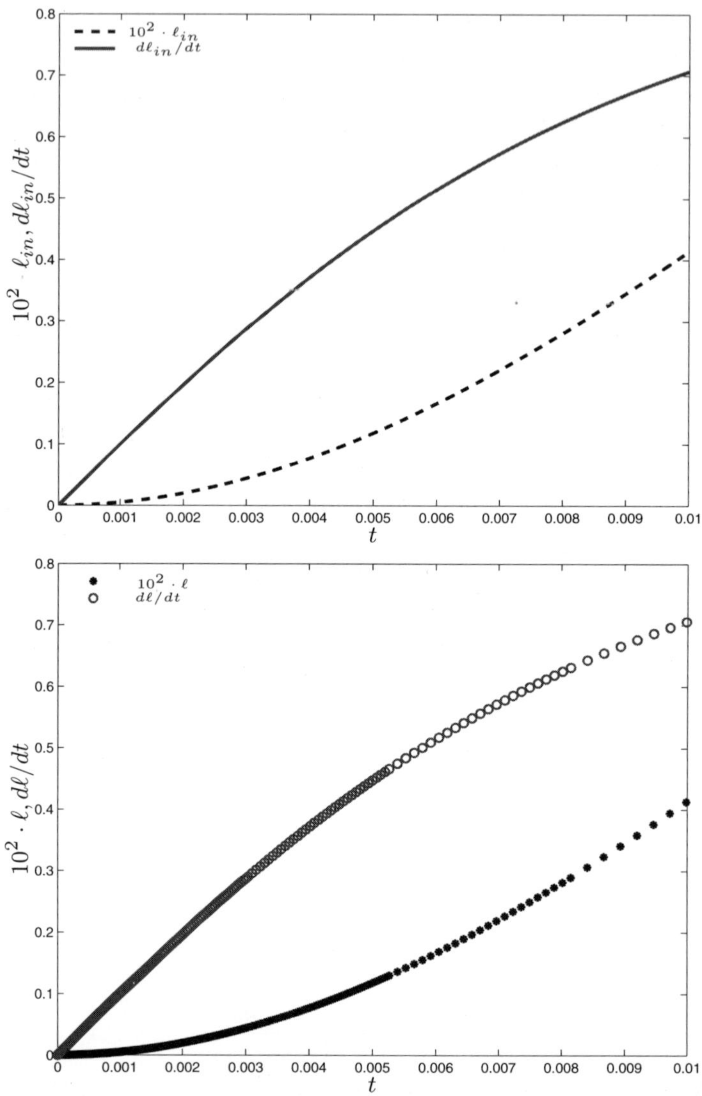

Fig. 5. Top frame: asymptotic solution: $10^2 \cdot \ell_{in}(t)$ and its derivative. Bottom frame: adaptive step-size results: $10^2 \cdot \ell(t)$ and its first derivative.

For the adaptive procedure we enforced the following conditions: $\Delta t_{\min} \leq \Delta t_k \leq \Delta t_{\max}$ with $\Delta t_{\min} = 10^{-15}$, $\Delta t_{\max} = 1$, and $\eta_{\min} \leq \eta(t_k) \leq \eta_{\max}$ with $\eta_{\min} = 10^{-2}$, $\eta_{\max} = 10\, \eta_{\min}$, and $\epsilon = 10^{-9}$. Moreover, the time step is modified in two cases: when $\eta(t_k) < \eta_{\min}$ we use $\Delta t_{k+1} = 2\, \Delta t_k$

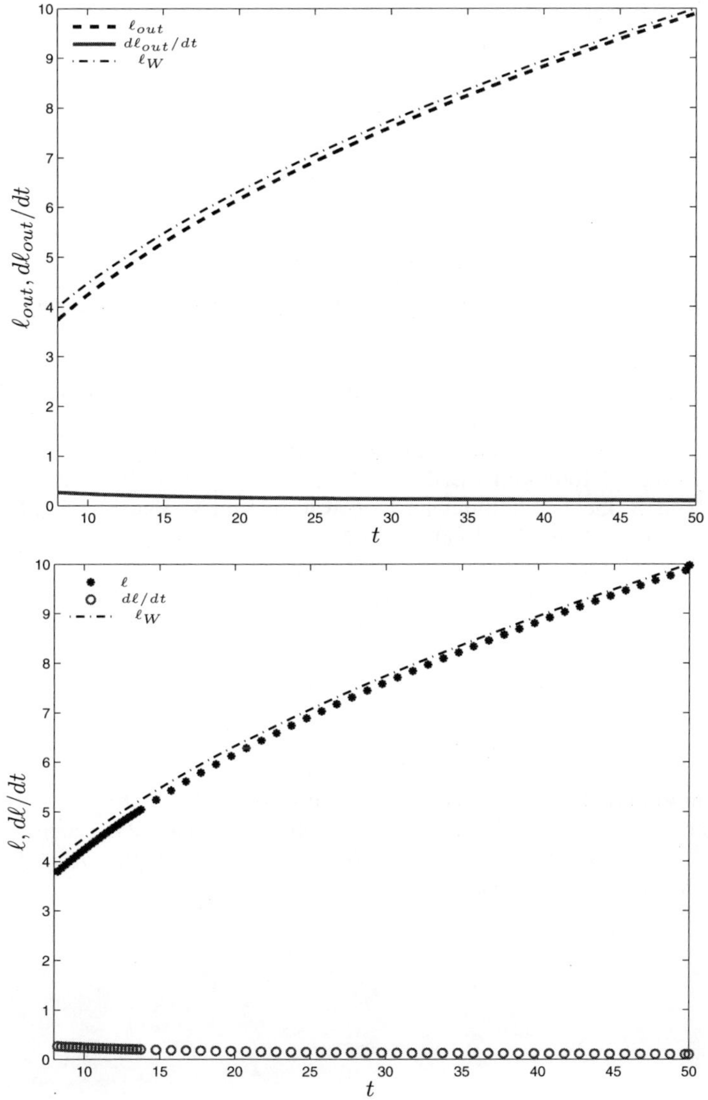

Fig. 6. Top frame: asymptotic solution: $\ell_{out}(t)$ and its derivative obtained for $R << t$. Bottom frame: adaptive step-size results: $\ell(t)$ and its first derivative. The solution of the Washburn equation $\ell_W(\mathrm{t})$ is shown for comparison.

as the next time step, whereas if $\eta(t_k) > \eta_{\max}$, then we repeat the same step using $\Delta t_k = \Delta t_k/2$. Full details on the adaptive strategy, as well as alternative monitor functions, can be found by the interested reader in [12].

Acknowledgments. The authors gratefully thank C. Budd (University of Bath, United Kingdom) and H. Huang (York University, Toronto, Canada), for fruitful discussions concerning the asymptotic solutions. This work was supported by the University of Messina and partially by the Italian MIUR.

References

1. C. H. Bosanquet, *Philos. Mag. Ser.* **6**, 525 (1923).
2. J. Szekely, A. W. Neumann and Y. K. Chuang, *J. Colloid Interf. Sci.* **69**, 486 (1979).
3. G. Cavaccini, V. Pianese, S. Iacono, A. Jannelli and R. Fazio, Mathematical and numerical modeling of liquids dynamics in a horizontal capillary, in *Recent Progress in Computational Sciences and Engineering, ICCMSE 2006, Lecture Series on Computerand Computational Sciences*, eds. T. Simos and G. Maroulis (Koninklijke Brill NV, Leiden, The Netherlands, 2006).
4. R. Fazio, S. Iacono, A. Jannelli, G. Cavaccini and V. Pianese, *PAMM (Proc. Appl. Math. Mech.)* **7**, 2150003 (2007).
5. E. W. Washburn, *Phys. Rev.* **17**, 273 (1921).
6. K. G. Kornev and A. V. Neimark, *J. Colloid Interf. Sci.* **235**, 101 (2001).
7. G. Martic, F. Gentner, D. Seveno, D. Coulon, J. D. Coninck and T. D. Blake, *Langmuir* **18**, 7971 (2002).
8. G. Martic, F. Gentner, D. Seveno, J. D. Coninck and T. Blake, *J. Colloid Interf. Sci.* **270**, 171 (2004).
9. G. Martic, T. D. Blake and J. D. Coninck, *Langmuir* **21**, 11201 (2005).
10. S. Chibbaro, *Eur. Phys. J. E* **27**, 99 (2008).
11. C. Budd and H. Huang. Private communication, (Bath, 2008).
12. A. Jannelli and R. Fazio, *J. Comput. Appl. Math.* **191**, 246 (2006).
13. J. C. Butcher, *Numerical Methods for Ordinary Differential Equations* (Whiley, Chichester, 2003).

DISSIPATIVE PROCESSES IN DEFECTIVE PIEZOELECTRIC CRYSTALS

D. Germanó, L. Restuccia*

Dipartimento di Matematica, Facoltà di Scienze MM.FF.NN.
Università degli Studi di Messina, Italy
**lrest@dipmat.unime.it*

In a previous paper a geometrization technique for thermodynamics of simple materials was applied to the case of piezoelectrics with defects of dislocations. Using a thermodinamical description, derived by one of us (L. R.) in the framework of extended irreversible thermodynamics with internal variables, a geometrical model for these media was derived. The dynamical system describing their behaviour, the entropy function and the entropy 1-form were derived. In this contribution, within the same thermodynamical model, we exploit the Clausius-Duhem inequality for these media and, using Maugin's technique, we work out the laws of state, the extra entropy flux, the residual dissipation inequality and the heat equation in a first and a second form.

Keywords: Non-equilibrium thermodynamics; Internal variables; Piezoelectrics; Defects of dislocations; Dissipative processes.

1. Introduction

The models for piezoelectrics with dislocations may have relevance in several fundamentals technological applications: in the construction of intermodulators, in the realization of thin films to produce very high frequency vibrations, in ultrasonics, in radar technology. Quartz crystals are used in radio transmitters. Rochelle salt, used in microphones, produces large voltages upon compression. Barium titanate, lead zirconate and lead titanate are ceramic materials which are used in ultrasonic transducers. In defective piezoelectric crystals the structure of the dislocation lines resembles a network of infinitesimally capillary tubes in a elastic solid (see Fig.1). Then, such defects acquired during a process of fabrication can self propagate because of changed conditions and favorable surrounding conditions, provoking a premature fracture. Thus, introducing a dislocation core tensor *à la* Maruszewski and its flux in the state space we can describe the geometry of the internal structure of these materials. In a previous paper [1], in a geometrized framework for the description of simple materials with internal variables [2-5], the specific example of defective piezoelectric crystals defective by dislocations (see [6-8]) was considered. The entropy function and the entropy 1-form were derived. In this contribution Clausius-Duhem inequality for these media is exploited and, using a Maugin technique [9] (see also Colemann-Noll procedure [10]), the laws of state, the extra entropy flux and the residual dissipation inequality are worked out. Also, following Maugin, the heat equation in a first and a second form is derived.

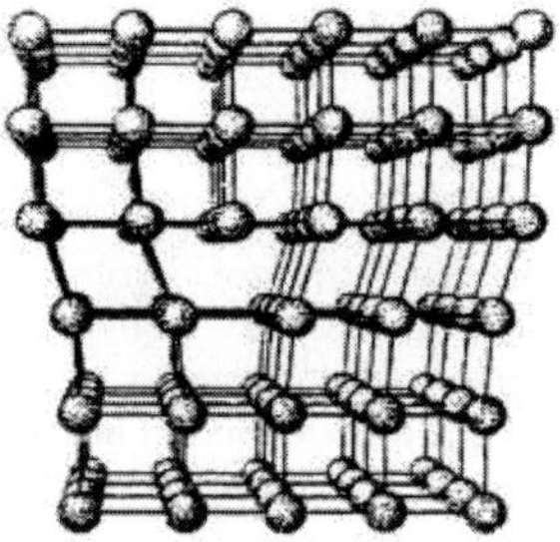

Fig. 1. An edge dislocation structure (after [15])

There are different approaches to describe the dislocations lines distributions [11]. In [12] a tensor quantity, called *dislocation density tensor*, is defined basing on the distortion field existing around dislocation lines and on the fact that dislocation cores form a kind of network of capillary tubes in the crystal. The definition and the introduction of this tensor is based on Kubik's idea concerning a very interesting geometrical model for a structure of a system of thin porous capillary channels [13]. Following [12], for any flux $\bar{\boldsymbol{\alpha}}$ of some physical field transported trough a network of lines, we postulate that

$$\bar{\alpha}(\mathbf{x})_i = r_{ij}(\mathbf{x}, \boldsymbol{\mu}) \overset{*}{\boldsymbol{\alpha}}_j (\mathbf{x}, \boldsymbol{\mu}), \tag{1}$$

where

$$\bar{\boldsymbol{\alpha}}(\mathbf{x}) = \frac{1}{\Omega} \int_{\Omega^{ch}} \boldsymbol{\alpha}(\boldsymbol{\xi}) d\Omega, \quad \boldsymbol{\xi} \in \Omega^{ch}, \tag{2}$$

and

$$\overset{*}{\boldsymbol{\alpha}} (\mathbf{x}, \boldsymbol{\mu}) = \frac{1}{\Gamma^{ch}} \int_{\Gamma} \boldsymbol{\alpha}(\boldsymbol{\xi}) d\Gamma, \quad \boldsymbol{\xi} \in \Gamma^{ch}. \tag{3}$$

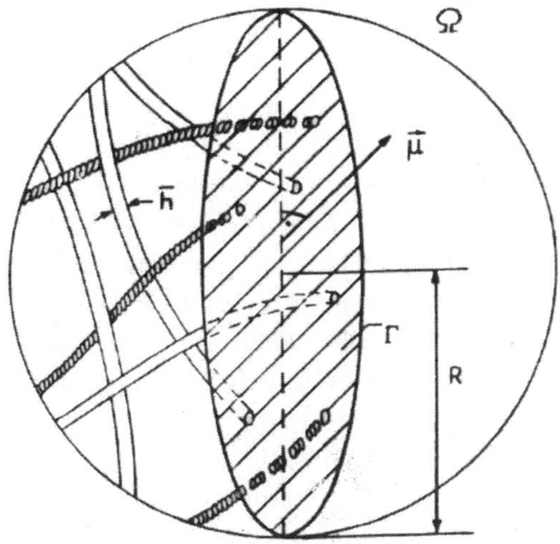

Fig. 2. Characteristics of a channel-core structure ($\bar{h} \ll R$) (after [12]).

$\bar{\boldsymbol{\alpha}}(\mathbf{x})$ is the bulk-volume average of the physical field flux under consideration, $\overset{*}{\boldsymbol{\alpha}}(\mathbf{x}, \boldsymbol{\mu})$ is the corresponding channel-area average of the same field flux . $\Omega = \Omega^s + \Omega^{ch}$ is a representative elementary sphere volume of a crystal with dislocation lines large enough to provide a representation of all the statistical properties of the channel space Ω^{ch}(see Fig.2). Ω^s is the solid space, Γ is the cross section of central sphere with normal vector $\boldsymbol{\mu}$ and Γ^{ch} is the channel-area of Γ. By definition the quantity $\boldsymbol{\alpha}(\xi)$ is zero in the solid space and on the solid surface Γ^s. Equation (1) gives a linear mapping between the bulk volume average of a physical field flux and the average of the same physical field flux on the channel area Γ^{ch}. In [12] a new tensor a_{ij}, called *dislocation core tensor*, was introduced that refers r_{ij} to the surface Γ as follows

$$a_{ij}(\mathbf{x}, \boldsymbol{\mu}) = \frac{r_{ij}(\mathbf{x}, \boldsymbol{\mu})}{\Gamma}. \tag{4}$$

The tensor a_{ij} expresses a structure of dislocation cores and its unit is m^{-2}. Moreover, the components of a_{ij} form a kind of continuous representation of the dislocation lines which cross the surface Γ. Investigations show that a_{ij} is also dependent on time.

2. A non conventional model for piezoelectric crystals with dislocations

In [6-8], in the framework of the extended rational thermodynamics with internal variables, using the standard Cartesian tensor notation in a rectangular coordinate system, a model was developed by one of us (L. R.) for piezoelectric media with dislocations, in which the following fields interact with each other: the thermal field described by the temperature θ and the heat flux q_i; the electromagnetic field described by the electromotive intensity $\mathcal{E}_i$ (the electric field referred to an element of matter at time t, i.e. the so called comoving frame) and the magnetic induction B_i for unit volume; the dislocation field described by the dislocation density tensor a_{ij} and the dislocation flux $\mathcal{V}_{ijk}$; the elastic field described by the total stress tensor T_{ij} (in general non symmetric) and the total strain tensor ε_{ij}. Thus, the independent variables are represented by the set

$$\boldsymbol{C} = \{\varepsilon_{ij}, \mathcal{E}_i, B_i, \theta, a_{ij}, \mathcal{V}_{ijk}, q_i, \theta_{,i}, a_{ij,k}\}\,. \tag{5}$$

This specific choice shows that the relaxation properties of the thermal field, the dislocation field are taken into account. All the processes occurring in the considered body are governed by three groups of laws:

Maxwell's equations:

$$\varepsilon_{ijk} E_{k,j} + \frac{\partial B_i}{\partial t} = 0, \quad D_{i,i} = 0, \tag{6}$$

$$\varepsilon_{ijk} H_{k,j} - \frac{\partial D_i}{\partial t} = 0, \quad B_{i,i} = 0, \tag{7}$$

where **E**, **B**, **D** and **H** denote the electric field, the magnetic induction, the electric displacement and the magnetic field per unit volume, respectively. Moreover,

$$H_i = \frac{1}{\mu_0} B_i, \qquad E_i = \frac{1}{\varepsilon_0} (D_i - P_i)\,. \tag{8}$$

ε_0, μ_0 denote the permittivity and permeability of vacuum, **P** is the electric polarization per unit volume. Moreover, we assume that the magnetic properties of the medium are neglected, so that the magnetization vector per unit volume has zero value: $\mathbf{M} = 0$. Furthermore, we have

the continuity equation:

$$\dot{\rho} + \rho v_{i,i} = 0,$$

where ρ denotes the mass density, v_i is the velocity of the body point and an overimposed dot denotes the material derivative;

the momentum balance:

$$\rho \dot{v}_i - T_{ji,j} - \in_{ijk} \overset{\Delta}{P}_j B_k - P_j \mathcal{E}_{i,j} - f_i = 0, \tag{9}$$

where

$$\overset{\Delta}{P}_j = \dot{P}_i + P_i v_{k,k} - P_k v_{i,k}, \quad \mathcal{E}_i = E_i + \in_{ijk} v_j B_k.$$

T_{ij} denotes the total stress tensor, f_i is the body force (that will be neglected in this context);

the moment of momentum balance:

$$\varepsilon_{ijk} T_{jk} + c_i = 0,$$

where c_i is a couple for unit volume;

the internal energy balance

$$\rho \dot{U} - T_{ji} v_{i,j} - \rho \mathcal{E}_i \dot{\mathcal{P}}_i + q_{i,i} - \rho r = 0. \tag{10}$$

Here U is the internal energy density and r is the heat source distribution. Furthermore,

$$P_i = \rho \mathcal{P}_i, \qquad v_{i,j} = L_{ij} = L_{(ij)} + L_{[ij]}, \tag{11}$$

where $L_{(ij)}$ and $L_{[ij]}$ are respectively the symmetric and antisymmetric part of the velocity gradient. Introducing the deformation gradient $\mathbf{F}$, also we have

$$\mathbf{L} = \nabla \mathbf{v} = \dot{\mathbf{F}} \mathbf{F}^{-1}, \quad L_{ij} = v_{i,j} = \dot{F_{ik}} \left(F_{kj}\right)^{-1}; \tag{12}$$

the *evolution equations* concerning *the rate properties of the dislocation core tensor, the dislocation flux and the heat flux* are given by, respectively,

$$\overset{*}{a}_{ij} + \mathcal{V}_{ijk,k} - A_{ij}(C) = 0, \quad \overset{*}{\mathcal{V}}_{ijk} - V_{ijk}(C) = 0, \quad \overset{*}{q}_i - Q_i(C) = 0,$$

where

$$\overset{*}{a}_{ij} = \dot{a_{ij}} - \Omega_{il} a_{lj} - \Omega_{jl} a_{il}, \quad \overset{*}{\mathcal{V}}_{ijk} = \dot{\mathcal{V}}_{ijk} - \Omega_{il} \mathcal{V}_{ljk} - \Omega_{jl} \mathcal{V}_{ilk} - \Omega_{kl} \mathcal{V}_{ijl},$$

$$\overset{*}{q}_i = \dot{q}_i - \Omega_{ij} q_j.$$

A_{ij}, V_{ijk} and Q_i are the source-like term dealing with the annihilation of dislocations of apposite signs, the source of the dislocation flux and the source of the heat flux, respectively. $\Omega_{ij} = \frac{1}{2}(v_{i,j} - v_{j,i})$ is the antisymmetric part of the velocity gradient. In the above equations an overimposed asterisk indicates the Zaremba-Jaumann time derivative (see [14] for the form of these equations).

All the admissible solutions of the proposed evolution equations should be restricted by the following *entropy inequality*:

$$\rho\dot{S} + \nabla \cdot \mathbf{J}^s - \frac{\rho r}{\theta} \geq 0,$$

where S denotes the entropy per unit mass and $\mathbf{J}^s$ is the entropy flux given by

$$\mathbf{J}^s = \frac{1}{\theta}\mathbf{q} + \mathbf{k}. \tag{13}$$

$\mathbf{k}$ is an additional term called *extra entropy flux density.*
In [6-8] constitutive functions $\mathbf{Z} = \tilde{\mathbf{Z}}(C)$, with

$$\mathbf{Z} = \left\{T_{ij}, P_i, c_i, U, A_{ij}, V_{ijk}, Q_i, S, J_i^S, \pi_{ij}, \Pi_i^Q, \Pi_i^A\right\} \tag{14}$$

were obtained for defective piezoelectrics in different cases. In equ. (14) $\pi_{ij}, \Pi_i^Q, \Pi_i^A$ are affinities that will be defined in the following section. The entropy inequality was analyzed by Liu's theorem and the state laws were derived. In particular, in [6] several characteristic groups of expressions were deduced expanding the free energy density $\Psi = U - \theta S - \frac{1}{\rho}\mathcal{E}_i P_i$ in Taylor's series with respect to a particular equilibrium state and confining the consideration to the quadratic terms. Considering very small deviations with respect to the equilibrium state (indicated with the subscript "$_0$"), it was assumed

$$\theta = \theta_0 + T, \qquad \left|\frac{T}{\theta_0}\right| \ll 1, \qquad \text{and} \quad M = M_0 + m, \quad \left|\frac{m}{M_0}\right| \ll 1, \tag{15}$$

where θ_0 denotes the room temperature and M stands for all considered interacting fields $M = \{\varepsilon_{ij}, \mathcal{E}_i, B_i, a_{ij}, \mathcal{V}_{ijk}, q_i, \theta_{,i}, a_{ij,k}\}$. T and m are very small deviations and M_0 defines a natural state of the body in which

$$M_0 = 0. \tag{16}$$

For the dislocation field the small deviation m was called α_{ij}. The following expressions were derived

$$P_i = \rho p_i = h_{ijl}\varepsilon_{jl} + \chi_{il}\mathcal{E}_l + \rho\lambda_i^\theta T - \alpha_{ijl}^{a\mathcal{E}}\alpha_{jl}, \tag{17}$$

$$\sigma_{ij} = c_{ijlm}\varepsilon_{lm} - h_{ijl}\mathcal{E}_l - \lambda_{ij}^\theta T + \alpha_{ijlm}^{a\epsilon}\alpha_{lm} - \mathcal{E}_s P_s \delta_{ij}, \tag{18}$$

$$S = \frac{\lambda_{ij}^\theta}{\rho}\varepsilon_{ij} + \lambda_i^\theta \mathcal{E}_i + \frac{c}{\theta_0}T - \frac{\alpha_{ij}^{a\theta}}{\rho}\alpha_{ij}, \tag{19}$$

$$\pi_{ij} = \alpha_{ijlm}^{a\epsilon}\varepsilon_{lm} + \alpha_{ijl}^{a\mathcal{E}}\mathcal{E}_l + \alpha_{ij}^{a\theta}T + \alpha_{ijlm}^{aa}\alpha_{lm}, \tag{20}$$

$$\Pi^A_{ijk} = \alpha^{\nu\nu}_{ijklmn}\mathcal{V}_{lmn} + \alpha^{\nu q}_{ijkl}q_l, \qquad \Pi^Q_i = \alpha^{\nu q}_{ijkl}\mathcal{V}_{jkl} + \alpha^{qq}_{ij}q_j, \tag{21}$$

$$\overset{*}{q}_i = \delta^1_{ij}\mathcal{E}_j + \delta^2_{ij}T_{,j} + \delta^3_{ij}q_j + \delta^4_{ijk}\alpha_{jk} + \delta^5_{ijk}\varepsilon_{jk} + \delta^6_{ijkl}\mathcal{V}_{jkl} + \delta^7_{ijkl}\alpha_{jk,l}, \tag{22}$$

$$\overset{*}{\alpha}_{ij} = \beta^1_{ijk}\mathcal{E}_k + \beta^2_{ijk}T_{,k} + \beta^3_{ijk}q_k + \beta^4_{ijkl}\varepsilon_{kl} + \beta^5_{ijkl}\alpha_{kl} + \beta^6_{ijklm}\mathcal{V}_{klm} + \beta^7_{ijklm}\alpha_{kl,m}, \tag{23}$$

$$\overset{*}{\mathcal{V}}_{ijk} = \gamma^1_{ijkl}\mathcal{E}_l + \gamma^2_{ijkl}T_{,l} + \gamma^3_{ijkl}q_l + \gamma^4_{ijklm}\varepsilon_{lm} + \gamma^5_{ijklm}\alpha_{lm} + \gamma^6_{ijklmn}\mathcal{V}_{lmn} + \gamma^7_{ijklmn}\alpha_{lm,n}. \tag{24}$$

In equs. (17)-(24) the constant phenomenological coefficients introduced satisfy the following symmetric relations

$$c_{ijlm} = c_{lmij} = c_{jilm} = c_{ijml} = c_{jiml} = c_{mlij} = c_{mlji} = c_{lmji}, \tag{25}$$

$$\lambda^\theta_{ij} = \lambda^\theta_{ji}, \quad h_{ijl} = h_{lij} = h_{jil} = h_{lji}, \quad \alpha^{a\mathcal{E}}_{ijl} = \alpha^{a\mathcal{E}}_{lij}, \tag{26}$$

$$\alpha^{a\epsilon}_{ijlm} = \alpha^{a\epsilon}_{lmji} = \alpha^{a\epsilon}_{lmij} = \alpha^{a\epsilon}_{jilm}, \quad \chi_{il} = \chi_{li}, \quad \alpha^{aa}_{ijlm} = \alpha^{aa}_{lmij}, \tag{27}$$

$$\alpha^{\nu\nu}_{ijklmn} = \alpha^{\nu\nu}_{lmnijk}, \quad \alpha^{\nu q}_{ijkl} = \alpha^{\nu q}_{lijk}, \qquad \alpha^{qq}_{ij} = \alpha^{qq}_{ji}. \tag{28}$$

The following notations are introduced: c denotes the specific heat, c_{ijlm} is the elastic tensor, λ^θ_{ij} are the thermoelastic constants, χ_{il} is the electrical susceptibility tensor, h_{ijl} are the piezoelectric constants and other quantities express interactions among the various effects involved in the system. In equ. (23) $\mathcal{V}_{ijk,k} = 0$ was assumed. Furthermore, the rate equation for the heat flux (22) generalizes Vernotte - Cattaneo relation and, when it is possible to identify the Zaremba - Jaumann derivative with the material derivative, denoting by τ_{ij} a relaxation time tensor associated to the heat flux, it takes the form

$$\tau_{ij}\dot{q}_j = -q_i - \chi^1_{ij}T_{,j} - \chi^2_{ij}\mathcal{E}_j + \chi^3_{ijk}\alpha_{jk} + \chi^4_{ijk}\varepsilon_{jk} + \chi^5_{ijkl}\mathcal{V}_{jkl} + \chi^6_{ijkl}\alpha_{jk,l}. \tag{29}$$

In the case that $\tau_{ij} = \tau\delta_{ij}$ equ. (29) becomes

$$\tau\dot{q}_i = -q_i - \chi^1_{ij}T_{,j} - \chi^2_{ij}\mathcal{E}_j + \chi^3_{ijk}\alpha_{jk} + \chi^4_{ijk}\varepsilon_{jk} + \chi^5_{ijkl}\mathcal{V}_{jkl} + \chi^6_{ijkl}\alpha_{jk,l}. \tag{30}$$

Inserting the constitutive equations into the balance equations we obtain the so-called "balances on the state space", that form a system of partial differential equations, whose order depends on the special choice of constitutive equations. Such system governs the evolution of the "wanted fields". By solving the field equations analytically and/or numerically we can describe the physical reality in several situations.

3. Clausius-Duhem inequality analysis and heat equation

In this section, taking into account the results obtained in [6-8], we exploit the Clausius-Duhem inequality for piezoelectric media with dislocations. Following Maugin [9] (see also Colemann-Noll procedure [10]), we consider the entropy inequality

$$\rho\dot{S} + J^s_{i,i} \geq 0, \tag{31}$$

which takes the following form

$$\rho\theta\dot{S} + \left(\theta J^s_i\right)_{,i} - J^s_i\theta_{,i} \geq 0, \tag{32}$$

where $\theta > 0$ and J^s is given by equ.(13). Then, we analyze the dissipation inequality obtaining the laws state and the first and the second form of heat equation. By derivation with respect to time of the free energy Ψ we obtain the following result:

$$\rho\theta\dot{S} = \rho\dot{U} - \rho S\dot{\theta} - \rho\dot{\Psi} + \frac{1}{\rho}\dot{\rho}\mathcal{E}_i P_i - \dot{\mathcal{E}}_i P_i - \mathcal{E}_i\dot{P}_i. \tag{33}$$

Substituting equ.(33) in the entropy inequality we have

$$\rho\dot{U} - \rho S\dot{\theta} - \rho\dot{\Psi} + \frac{1}{\rho}\dot{\rho}\mathcal{E}_i P_i - \dot{\mathcal{E}}_i P_i - \mathcal{E}_i\dot{P}_i + \left(\theta J^s_i\right)_{,i} - J^s_i\theta_{,i} \geq 0 \tag{34}$$

and using the internal energy balance equation (10) (where the heat source distribution r is neglected) we obtain

$$-\rho\left(\dot{\Psi} + \dot{\theta}S\right) + T_{ji}v_{i,j} - \dot{\mathcal{E}}_i P_i + \left(\theta k_i\right)_{,i} - \theta_{,i}J^s_i \geq 0. \tag{35}$$

Finally, taking into account equations $(11)_1$ and (13), we derive

$$\rho\dot{\mathcal{P}}_i = -\frac{1}{\rho}\dot{\rho}P_i + \dot{P}_i, \tag{36}$$

$$\left(\theta J^s_i\right)_{,i} = q_{i,i} + \left(\theta k_i\right)_{,i}, \tag{37}$$

$$\left(\theta k_i\right)_{,i} = \rho\left(\frac{\theta}{\rho}k_i\right)_{,i} + \left(\frac{\theta}{\rho}k_i\right)\rho_{,i}. \tag{38}$$

Moreover, using the following relation

$$\frac{\partial\Psi}{\partial a_{ij,k}}\dot{a}_{ij,k} = \left(\frac{\partial\Psi}{\partial a_{ij,k}}\dot{a}_{ij}\right)_{,k} - \left(\frac{\partial\Psi}{\partial a_{ij,k}}\right)_{,k}\dot{a}_{ij}, \tag{39}$$

choosing as state variables $(F_{ij}, \mathcal{E}_i, B_i, \theta, a_{ij}, \mathcal{V}_{ijk}, q_i, \theta_{,i}, a_{ij,k})$, and calculating the material derivative of the free energy Ψ from equ.(35) we obtain the following Clausius Duhem inequality:

$$-\rho\left(\frac{\partial\Psi}{\partial\theta}+S\right)\dot{\theta}+\dot{F}_{ik}\left[-\rho\frac{\partial\Psi}{\partial F_{ik}}+\tau_{ji}\left(F_{kj}\right)^{-1}\right]-\rho\left[\frac{\partial\Psi}{\partial a_{ij}}-\left(\frac{\partial\Psi}{\partial a_{ij,k}}\right)_{,k}\right]\dot{a}_{ij}$$

$$-\rho\frac{\partial\Psi}{\partial B_i}\dot{B}_i-\rho\frac{\partial\Psi}{\partial\mathcal{V}_{ijk}}\dot{\mathcal{V}}_{ijk}-\rho\frac{\partial\Psi}{\partial q_i}\dot{q}_i-\rho\frac{\partial\Psi}{\partial\theta_{,i}}\dot{\theta}_{,i}$$

$$-\dot{\mathcal{E}}_i\left(\rho\frac{\partial\Psi}{\partial\mathcal{E}_i}+P_i\right)+\rho\left(\frac{\theta}{\rho}k_k-\frac{\partial\Psi}{\partial a_{ij,k}}\dot{a}_{ij}\right)_{,k}+\left(\frac{\theta}{\rho}k_i\right)\rho_{,i}-\theta_{,i}J_i^s\geq 0. \quad (40)$$

As $T_{ji}F_{kj}^{-1}$, P_i, $\frac{\partial\Psi}{\partial B_i}$ and S are assumed not to depend on $\dot{F}_{ik}$, $\dot{\mathcal{E}}_i$, $\dot{B}_i$, $\dot{\theta}$ and $\dot{\theta}_{,i}$, while the remaining coefficients may in general depend on their respective factors, from inequality (40) we obtain the following laws of state

$$\rho\frac{\partial\Psi}{\partial F_{ik}}=T_{ji}F_{kj}^{-1}, \qquad \frac{\partial\Psi}{\partial B_i}=0, \qquad \rho\frac{\partial\Psi}{\partial\theta_{,i}}=0, \quad (41)$$

$$\rho\frac{\partial\Psi}{\partial\mathcal{E}_i}=P_i, \qquad \frac{\partial\Psi}{\partial\theta}=-S, \qquad \pi_{ij}=\rho\frac{\partial\Psi}{\partial a_{ij}}. \quad (42)$$

At this point, assuming

$$\left(\frac{\theta}{\rho}k_k-\frac{\partial\Psi}{\partial a_{ij,k}}\dot{a}_{ij}\right)_{,k}=0, \quad (43)$$

"it is astute to select" (see Maugin [9])

$$k_k=\frac{\rho}{\theta}\frac{\partial\Psi}{\partial a_{ij,k}}\dot{a}_{ij}, \quad (44)$$

so that Clausius-Duhem inequality reduces itself to the following *residual dissipation inequality*

$$-\rho\left[\frac{\partial\Psi}{\partial a_{ij}}-\left(\frac{\partial\Psi}{\partial a_{ij,k}}\right)_{,k}\right]\dot{a}_{ij}-\rho\frac{\partial\Psi}{\partial\mathcal{V}_{ijk}}\dot{\mathcal{V}}_{ijk}-\rho\frac{\partial\Psi}{\partial q_i}\dot{q}_i+\left(\frac{\theta}{\rho}k_i\right)\rho_{,i}-\theta_{,i}J_i^s\geq 0 \quad (45)$$

Equ. (43) comes from the following standard bilinear form which the residual dissipation inequality has in classical thermodynamics of irreversible processes, i.e.

$$\sum_\beta X_\beta Y_\beta\geq 0,$$

where $\mathbf{X}$ and $\mathbf{Y}$ are the fluxes and the associated forces, respectively. Therefore, the value of divergence term is assumed zero.

Very often the residual dissipation inequality is split in two parts ("resulting thus in stronger conditions")

$$\Phi_{intr} = \rho \mathcal{A}_{ij} \dot{a}_{ij} - \Pi^A_{ijk} \dot{\mathcal{V}}_{ijk} + \left(\frac{\theta}{\rho} k_i\right) \rho,_{\,i} - \Pi^Q_i \dot{q}_i \geq 0, \tag{46}$$

where

$$\mathcal{A}_{ij} = -\left[\frac{\partial \Psi}{\partial a_{ij}} - \left(\frac{\partial \Psi}{\partial a_{ij,k}}\right)_{,k}\right]$$

and

$$\mathbf{\Pi}^A = \rho \frac{\partial \Psi}{\partial \boldsymbol{\mathcal{V}}}, \qquad \mathbf{\Pi}^Q = \rho \frac{\partial \Psi}{\partial \mathbf{q}} \tag{47}$$

are affinities. Furthermore,

$$\Phi_{th} = -\mathbf{J}_s \cdot \nabla \theta \geq 0, \tag{48}$$

where, in some sense, we recognize the different qualitative nature of the two classes of dissipative processes. Φ_{intr} and Φ_{th} are the intrinsic and thermal dissipations, respectively [9]. Thus, we have derived the thermodynamical state laws, the entropy flux and the inequality governing dissipative processes.

Now, in order to obtain the heat equation, we observe that it is none other than a form of energy balance equation (10). Indeed, on using the free energy expression $\Psi = U - \theta S - \frac{1}{\rho} \mathcal{E}_i P_i$, its time derivative and the laws of state in the energy balance equation or, equivalently, "just comparing entropy inequality (32) and the residual inequality" (45), we deduce *the first general form of the heat equation*

$$\rho \theta \dot{S} + (\theta J^s_i)_{,i} = \Phi_{intr}, \tag{49}$$

where the intrinsic dissipation acts like a body source of heat.

Now, taking into account that the entropy S is a constitutive function $S = S\left(F_{ij}, \mathcal{E}_i, B_i, \theta, a_{ij}, \mathcal{V}_{ijk}, q_i, \theta_{,i}, a_{ij,k}\right)$, and the state laws $(41)_2$, $(41)_3$ and $(42)_2$, we have

$$\dot{S} = -\left[\frac{\partial^2 \Psi}{\partial \boldsymbol{F} \partial \theta} \dot{\boldsymbol{F}} + \frac{\partial^2 \Psi}{\partial \boldsymbol{\mathcal{E}} \partial \theta} \dot{\boldsymbol{\mathcal{E}}} + \frac{\partial^2 \Psi}{\partial \theta^2} \dot{\theta} + \frac{\partial^2 \Psi}{\partial \mathbf{q} \partial \theta} \dot{\mathbf{q}} + \frac{\partial^2 \Psi}{\partial \mathbf{a} \partial \theta} \dot{\mathbf{a}} \right.$$

$$\left. + \frac{\partial^2 \Psi}{\partial \nabla \mathbf{a} \partial \theta} \dot{\nabla a} + \frac{\partial^2 \Psi}{\partial \boldsymbol{\mathcal{V}} \partial \theta} \dot{\boldsymbol{\mathcal{V}}} \right]. \tag{50}$$

Finally, we set

$$C = -\rho\theta\frac{\partial^2\Psi}{\partial\theta^2}, \qquad \boldsymbol{\tau} = \frac{\partial^2\Psi}{\partial\mathbf{F}\partial\theta}, \qquad \mathbf{m} = \frac{\partial^2\Psi}{\partial\nabla\mathbf{a}\partial\theta} \tag{51}$$

$$\mathbf{l} = \frac{\partial^2\Psi}{\partial\mathbf{a}\partial\theta}, \qquad \mathbf{e} = \frac{\partial^2\Psi}{\partial\boldsymbol{\mathcal{E}}\partial\theta}, \qquad \mathbf{n} = \rho\frac{\partial^2\Psi}{\partial\mathbf{q}\partial\theta}, \qquad \boldsymbol{\nu} = \rho\frac{\partial^2\Psi}{\partial\boldsymbol{\mathcal{V}}\partial\theta}. \tag{52}$$

Using the definitions of affinities $\mathbf{\Pi}^A$ and $\mathbf{\Pi}^Q$ (see equ.(47)) and substituting equ.(50) in (49), we obtain "the second form of the heat equation" as follows

$$C\dot{\theta} + \nabla\cdot(\theta\mathbf{J}^{\mathrm{s}}) = \Phi_{te} + \Phi_{td} + \Phi_{tel} + \Phi_t + \Phi_\rho. \tag{53}$$

In (53) Φ_{te} is the dissipation given by the interaction between thermal and elastic phenomena, Φ_{td} is the dissipation due to interaction between thermal and dislocations phenomena, Φ_{tel} represents the dissipation due to the interaction between thermal and electric phenomena, Φ_d is the dissipation due to dislocation phenomena, Φ_t is the dissipation due to thermal phenomena, Φ_ρ is the dissipation due to the interaction between the entropy flux and mass phenomena. They are given by the following expressions

$$\Phi_{te} = \boldsymbol{\tau}\cdot\dot{\boldsymbol{F}}\theta, \quad \Phi_{td} = (\mathbf{l}\cdot\dot{\mathbf{a}} + \boldsymbol{\nu}\cdot\dot{\boldsymbol{\mathcal{V}}} + \mathbf{m}\cdot\dot{\nabla\mathbf{a}})\theta, \tag{54}$$

$$\Phi_{tel} = \theta\mathbf{e}\cdot\dot{\boldsymbol{\mathcal{E}}}, \quad \Phi_d = \rho\mathcal{A}\cdot\dot{\mathbf{a}} - \rho\frac{\partial\Psi}{\partial\boldsymbol{\mathcal{V}}}\cdot\dot{\boldsymbol{\mathcal{V}}}, \tag{55}$$

$$\Phi_t = \mathbf{n}\cdot\dot{\mathbf{q}}\theta - \rho\frac{\partial\Psi}{\partial\mathbf{q}}\cdot\dot{\mathbf{q}}, \quad \Phi_\rho = \left(\frac{\theta}{\rho}\mathbf{k}\right)\cdot\nabla\rho. \tag{56}$$

The non-negative of the specific heat C follows from the concavity of Ψ with respect to θ. In semiconductor crystals ρ is practically constant, so that the terms $\left(\frac{\theta}{\rho}\mathbf{k}\right)\cdot\nabla\rho$, $\frac{1}{\rho}\dot{\rho}\mathcal{E}_i P_i$ and $\frac{1}{\rho}\dot{\rho}P_i$ can be considered disregarding in the above obtained results. Substituting in this second form of the heat equation suitable constitutive equations we can describe real phenomena in different cases.

References

1. D. Germanó and L. Restuccia, *Thermodynamics of piezoelectric media with dislocations,* in *Applied and Industrial Mathematics in Italy II, Series on Advances in Mathematics for Applied Sciences,* eds. V. Cutello, G. Fotia, L. Puccio, Vol. 75 (Word Scientific Publishing, 2007), pp. 387-398.

2. M. Dolfin, M. Francaviglia and P. Rogolino, *J. Non-Equilib. Thermodyn.* **23**, 250 (1998).
3. M. Dolfin, M. Francaviglia and P. Rogolino, *Periodica Polytechnica Serie Mech. Eng.* **43**, 29 (1999).
4. W. Noll, *Arch. Rat. Mech. Anal.* **48**, 1 (1972).
5. B. D. Coleman, D. R. Owen, *Arch. Rat. Mech. Anal.* **54**, 1 (1974).
6. S. Giambò , B. Maruszewski, L. Restuccia, *Journal of Technical Physics* (Polish Academy of Sciences, Warszawa), **43** (2), 155 (2002).
7. L. Restuccia, B. T. Maruszewski, *Thermomechanics of piezoelectrics defective by dislocations*, in: *Applied and Industrial Mathematics in Italy, Series on Advances in Mathematics for Applied Sciences*, eds. M. Primicerio, R. Spigler and V.Valente, Vol. 69 (World Scientific 2005), pp.475-486.
8. L. Restuccia, B. T. Maruszewski, *Supplemento ai Rendiconti del Circolo Matematico di Palermo* **80**, 275 (2008).
9. G. A. Maugin, *J. Non-Equilib. Thermodyn.* **15**, 173 (1990).
10. B. D. Coleman and W. Noll, *Arch. Rational Mech. Anal.* **3**, 167 (1963).
11. E. Kroner, *Erg. Angew.* Math. **5**, 1 (1958).
12. B. Maruszewski, *Phys. stat. sol. (b)* **168**, 59 (1991).
13. J. Kubik, *Int. J. Engng. Sci.* **24**, 971 (1986).
14. C. Truesdell, R. A. Toupin, *The classical field theories*, in *Encyclopedia of Physcs*, ed. S. Flugge, Vol. III (1) (Springer Verlag, Berlin,1960), pp.226-793.
15. C. Kittel ,*Introduction to Solid State Physics*, 3rd edn. (John Wiley and Sons, New York, 1966).

A Dissipative System Arising in Strain-gradient Plasticity

LORENZO GIACOMELLI

Dipartimento MeMoMat, Università di Roma "La Sapienza"
Via A. Scarpa 16, I–00161 Rome, Italy
E-mail: giacomelli@dmmm.uniroma1.it

GIUSEPPE TOMASSETTI

Dipartimento di Ingegneria Civile, Università di Roma "Tor Vergata"
Via Politecnico 1, I–00133 Rome, Italy
E-mail: tomassetti@ing.uniroma2.it

We discuss a nonlocal and fully nonlinear system of partial differential equations which arises in a strain-gradient theory of plasticity proposed by Gurtin (J. Mech. Phys. Solids, 2004). The problem couples an elliptic equation to a parabolic system which exhibits two types of degeneracies: the first one is caused by the nonlinear structure, the second one by the dependence of the principal part on twice the curl of a planar vector field $\boldsymbol{\gamma}$. Furthermore, the elliptic equation depends on the divergence of $\boldsymbol{\gamma}$ — which is not controlled by twice its curl — and the boundary conditions suggested by Gurtin are of mixed type. We outline two reformulations of the system which are the key to obtain the existence and the uniqueness of solutions (proved elsewhere by M. Bertsch, R. Dal Passo and the authors): the first one enlightens a monotonicity property which is more "robust" than the dissipative structure inherited as an intrinsic feature of the mechanical model; the second one —based on a suitable, time-dependent representation of a divergence-free vector field which plays the role of the elastic stress — is essential to overcome the lack of control over the divergence of the vector field $\boldsymbol{\gamma}$.

Keywords: Degenerate parabolic system, fully nonlinear parabolic system, nonlocal parabolic system, strain-gradient plasticity.

1. The problem

A number of experimental evidences shows that the plastic behavior of metals undergoing plastic strain exhibits scale-effects, typically described by strain gradient theories. In this note we are concerned with the one proposed by Gurtin in Ref. 4, which deals with small strains, and in particular with the case of *anti-plane shear*: that is to say, the body is a right cylinder of axis x_3 and cross section $\Omega \subset \mathbb{R}^2$, all fields of interest do not depend on the coordinate x_3, and the displacement vector is parallel to the axis of the cylinder. The unknowns u, a scalar field, and $(\boldsymbol{\gamma}, \boldsymbol{\nu})$, a pair of planar vector fields, incorporate the relevant information on, respectively, the displacement vector and the plastic strain tensor. The flow rule, which in Ref. 4 follows from a balance of *microforces* combined with suitable constitutive assumptions, is given (in dimensionless form) by [a]

$$\begin{cases} |d|^{m-1}(\boldsymbol{\gamma}+\boldsymbol{\nu})_t = \varepsilon^2(-\,\mathbf{curl}^2\,\boldsymbol{\gamma} + \Delta\boldsymbol{\nu}) + 2(\nabla u - \boldsymbol{\gamma} - \boldsymbol{\nu}) \\ \chi\,|d|^{m-1}(\boldsymbol{\gamma}-\boldsymbol{\nu})_t = \varepsilon^2(-\,\mathbf{curl}^2\,\boldsymbol{\gamma} - \Delta\boldsymbol{\nu}), \end{cases} \tag{1}$$

where $\varepsilon > 0$, $\chi > 0$, and $m \in (0,1)$ are dimensionless parameters, and

$$d \;:=\; \sqrt{\tfrac{1}{2}|(\boldsymbol{\gamma}_t + \boldsymbol{\nu}_t)|^2 + \tfrac{\chi}{2}|(\boldsymbol{\gamma}_t - \boldsymbol{\nu}_t)|^2}\,. \tag{2}$$

The bulk system is closed by a balance of standard forces (where inertial and non-inertial distance forces are set to null), namely

$$\operatorname{div}\boldsymbol{\tau} = 0, \qquad \text{where}\;\; \boldsymbol{\tau} := \nabla u - \boldsymbol{\gamma} - \boldsymbol{\nu} \tag{3}$$

(this shorthand notation is maintained throughout this note). The vector field $\boldsymbol{\tau}$ plays the role of the stress and, in dimensionless form, may be identified with the elastic strain. The initial conditions are given by

$$\boldsymbol{\gamma} \;=\; \boldsymbol{\gamma}_0, \quad \boldsymbol{\nu} \;=\; \boldsymbol{\nu}_0 \quad \text{on} \;\; \partial_0 Q := \{0\} \times \Omega.$$

Finally, the boundary conditions proposed in Ref. 4 are (in general) of mixed type: to formulate them, we assume that

$$\begin{cases} \textit{$\Omega \subset \mathbb{R}^2$ is open, bounded, connected and simply connected} \\ \textit{domain of class $C^{0,1}$; $\partial\Omega$ is such that $\partial\Omega = \overline{\partial_H\Omega} \cup \overline{\partial_F\Omega}$ and} \\ \textit{$\partial_H\Omega \cap \partial_F\Omega = \emptyset$; if $\partial_H\Omega \neq \emptyset$ and $\partial_F\Omega \neq \emptyset$, both $\partial_H\Omega$ and} \\ \textit{$\partial_F\Omega$ consist of a finite number of smooth connected compo-} \\ \textit{nents, each of them having positive length and positive dis-} \\ \textit{tance from the others.} \end{cases} \tag{4}$$

The integer $\ell \in \mathbb{N}$ counts the number of the connected components $\partial_H\Omega^i$ of $\partial_H\Omega$:

$$\begin{aligned} &\ell = 0 \;\text{ and }\; \partial_H\Omega^0 = \emptyset && \text{if } \partial_H\Omega = \emptyset; \\ &\partial_H\Omega = \partial_H\Omega^1 \cup \cdots \cup \partial_H\Omega^\ell && \text{if } \partial_H\Omega \neq \emptyset. \end{aligned}$$

[a] The subscript t indicates partial differentiation with respect to t. The $\mathbf{curl}^2$ operator on the right-hand sides of system (1) is defined as follows: if $\boldsymbol{\gamma} = (\gamma_1, \gamma_2)$, then

$$\mathbf{curl}^2\,\boldsymbol{\gamma} := \mathbf{curl}\,\omega := \left(\frac{\partial\omega}{\partial x_2}, -\frac{\partial\omega}{\partial x_1}\right), \quad \text{where}\;\; \omega := \operatorname{curl}\boldsymbol{\gamma} := \frac{\partial\gamma_2}{\partial x_1} - \frac{\partial\gamma_1}{\partial x_2}.$$

Then, the boundary conditions read as follows[b]:

$$\begin{cases} \operatorname{curl}\gamma = 0, & \nabla\boldsymbol{\nu}\cdot\mathbf{n} = \mathbf{0}, \quad \boldsymbol{\tau}\cdot\mathbf{n} = 0 \quad \text{on } \partial_F\mathcal{Q} := (0,\infty)\times\partial_F\Omega, \\ \gamma_t\cdot\mathbf{t} = 0, & \boldsymbol{\nu}_t = \mathbf{0}, \qquad\quad u \overset{*}{=} u_H \quad \text{on } \partial_H\mathcal{Q} := (0,\infty)\times\partial_H\Omega \end{cases} \tag{5}$$

where u_H is a time-independent boundary datum.

Two features of Gurtin's theory are helpful in the analytical understanding. The first one is the fact that the free energy density depends not only on the elastic strain, but also on the so-called *Burgers tensor* (see Sec. 2.1 in Ref. 4) through a special combination of the derivatives of γ and $\boldsymbol{\nu}$. The (dimensionless) free energy here is:

$$E_\varepsilon[\gamma,\boldsymbol{\nu},\boldsymbol{\tau}] = \frac{\varepsilon^2}{2}\int_\Omega \{|\operatorname{curl}\gamma|^2 + |\nabla\boldsymbol{\nu}|^2\} + \frac{1}{2}\int_\Omega |\boldsymbol{\tau}|^2,$$

where ε is the ratio between a microscopic lengthscale and a characteristic dimension of the body. Its magnitude is of course essential in a quantitative study of the problem. However, it turns out to be irrelevant as far as well-posedness is concerned. Hence, to lighten all subsequent notation we hereafter let

$$\varepsilon = 1 \quad \text{and} \quad E[\gamma,\boldsymbol{\nu},\boldsymbol{\tau}] = E_1[\gamma,\boldsymbol{\nu},\boldsymbol{\tau}].$$

The second feature is the fact that, unlike standard (gradient) plasticity theories, the plastic strain tensor is not assumed to be symmetric: as a consequence, a contribution to dissipation comes from both its symmetric part and its skew-symmetric part (the *plastic rotation*), which in the present two-dimensional setting are represented by the vectors $\gamma+\boldsymbol{\nu}$ and $\gamma-\boldsymbol{\nu}$, in the order. In particular, the *plastic spin* $(\gamma-\boldsymbol{\nu})_t$ contributes to dissipation through the parameter χ which appears in the definition (2) of d: Indeed, it is not difficult to check formally that (1), (3) and (5) yield

$$\frac{d}{dt}E[\gamma(t),\boldsymbol{\nu}(t),\boldsymbol{\tau}(t)] = -\int_\Omega \left((-\,\mathbf{curl}^2\,\gamma+\boldsymbol{\tau})\cdot\gamma_t + (\Delta\boldsymbol{\nu}+\boldsymbol{\tau})\cdot\boldsymbol{\nu}_t\right).$$

A simple algebraic calculation (using (2) and (1)) shows that

$$|d|^{m+1} = (-\,\mathbf{curl}^2\,\gamma+\boldsymbol{\tau})\cdot\gamma_t + (\Delta\boldsymbol{\nu}+\boldsymbol{\tau})\cdot\boldsymbol{\nu}_t, \tag{6}$$

[b]Here $\mathbf{n}$ is the outward unit normal, $\mathbf{t} = \mathbf{e}_3\times\mathbf{n}$ is the unit tangent vector on $\partial\Omega$, and the symbol $\overset{*}{=}$ means that if a Dirichlet boundary condition is required on an empty set, then it is replaced by the normalizing zero-mean condition (which is necessary for uniqueness):

$$\left[f \overset{*}{=} g \text{ on } D \iff \int_\Omega f = 0\right] \quad \text{if } D = \emptyset.$$

and therefore

$$\frac{d}{dt}E[\gamma(t),\boldsymbol{\nu}(t),\boldsymbol{\tau}(t)] = -\int_\Omega |d|^{m+1} . \tag{7}$$

In this respect, let us mention that a related model, restricted to materials with zero plastic rotation, has been developed in Ref. 5 and studied in Ref. 7. The power-law d^{m+1} (which is standard in the modeling of metallic materials) is the simplest one which accounts for nonlinear effects, possibly close to the case of rate-independence: $0 < m \ll 1$.

2. The monotone structure

The system may be reformulated in such a way that a monotone structure is enlightened. Summing and subtracting the equations in (1), divided respectively by 2 and 2χ, we obtain

$$\begin{cases} |d|^{m-1}\gamma_t = -A\,\mathbf{curl}^2\,\gamma + B\,\Delta\boldsymbol{\nu} + \boldsymbol{\tau} \\ |d|^{m-1}\boldsymbol{\nu}_t = -B\,\mathbf{curl}^2\,\gamma + A\,\Delta\boldsymbol{\nu} + \boldsymbol{\tau}, \end{cases} \tag{8}$$

where $A = \frac{\chi+1}{2\chi}$ and $B = 1 - A$. Recalling (6) and (1), we may write[c]

$$\begin{aligned} |d|^{2m} = |d|^{m+1}|d|^{m-1} &= |d|^{m-1}\left(\left(-\,\mathbf{curl}^2\,\gamma + \boldsymbol{\tau}\right)\cdot\gamma_t + (\Delta\boldsymbol{\nu}+\boldsymbol{\tau})\cdot\boldsymbol{\nu}_t\right) \\ &= \begin{pmatrix} -\,\mathbf{curl}^2\,\gamma + \boldsymbol{\tau} \\ \Delta\boldsymbol{\nu}+\boldsymbol{\tau} \end{pmatrix}\cdot\mathbb{Q}\begin{pmatrix} -\,\mathbf{curl}^2\,\gamma + \boldsymbol{\tau} \\ \Delta\boldsymbol{\nu}+\boldsymbol{\tau} \end{pmatrix}, \end{aligned}$$

where

$$\mathbb{Q} = \begin{pmatrix} A & 0 & B & 0 \\ 0 & A & 0 & B \\ B & 0 & A & 0 \\ 0 & B & 0 & A \end{pmatrix} .$$

The matrix $\mathbb{Q}$ is symmetric and positive definite, hence it induces a scalar product and a norm on $\mathbb{R}^4$:

$$|\mathbb{v}|_{\mathbb{Q}} := (\mathbb{v}\cdot\mathbb{Q}\,\mathbb{v})^{1/2} \quad \mathbb{v}\in\mathbb{R}^4.$$

[c] Vectors are always written in columns, and

$$\text{given } \mathbf{v} = \begin{pmatrix} v_1 \\ v_2 \end{pmatrix} \text{ and } \mathbf{w} = \begin{pmatrix} w_1 \\ w_2 \end{pmatrix}, \text{ we write } \mathbb{v} = \begin{pmatrix} \mathbf{v} \\ \mathbf{w} \end{pmatrix} = \begin{pmatrix} v_1 \\ v_2 \\ w_1 \\ w_2 \end{pmatrix}.$$

Therefore, setting

$$p = 1 + \frac{1}{m} \quad \text{and} \quad \mathbb{A}(\mathbb{v}) := |\mathbb{v}|_{\mathbb{Q}}^{p-2}\, \mathbb{Q}\,\mathbb{v},$$

we see that (8), and thus the original system (1), can be rewritten in a compact form as

$$\begin{pmatrix} \gamma_t \\ \boldsymbol{\nu}_t \end{pmatrix} = \mathbb{A}\begin{pmatrix} -\,\mathbf{curl}^2\,\gamma + \boldsymbol{\tau} \\ \Delta\boldsymbol{\nu} + \boldsymbol{\tau} \end{pmatrix} \tag{9}$$

In this formulation the nonlinearity involving time-derivatives is transformed into a nonlinearity on the right-hand side. Also, (7) reads as

$$\frac{d}{dt}E[\gamma(t), \boldsymbol{\nu}(t), \boldsymbol{\tau}(t)] = -\int_{\Omega} \left|\begin{pmatrix} -\,\mathbf{curl}^2\,\gamma + \boldsymbol{\tau} \\ \Delta\boldsymbol{\nu} + \boldsymbol{\tau} \end{pmatrix}\right|_{\mathbb{Q}}^{p} . \tag{10}$$

In view of the inequality

$$(\mathbb{A}(\mathbb{v}_1) - \mathbb{A}(\mathbb{v}_2)) \cdot (\mathbb{v}_1 - \mathbb{v}_2) \geq C^{-1}\,|\mathbb{v}_1 - \mathbb{v}_2|_{\mathbb{Q}}^{p} \quad \text{for some } C > 1$$

(see e.g. Lemma I.4.4 in Ref. 3), (10) enlightens the monotone structure of the system.

3. The representation of $\boldsymbol{\tau}$

We will now describe a reformulation of (3) in which $\operatorname{div}\boldsymbol{\gamma}$ (which is not controlled by $\mathbf{curl}^2\,\gamma$) does not appear. Since $\boldsymbol{\tau}$ is required to be a divergence-free vector field, we use as Ansatz its Hodge decomposition,

$$\boldsymbol{\tau} = \mathbf{curl}\,V + \nabla p \;\text{ in }\; \Omega, \tag{11}$$

where p is harmonic in Ω. Taking the div, respectively the curl, on both sides of (11) and recalling (3), we see that

$$\Delta p = 0, \quad \Delta V = \operatorname{curl}(\gamma + \boldsymbol{\nu}) \quad \text{in } \Omega.$$

In addition, the boundary conditions (5) imply that

$$\nabla V \cdot \mathbf{t} + \nabla p \cdot \mathbf{n} = 0 \quad \text{on } \partial_F\Omega, \quad -\nabla V \cdot \mathbf{n} + \nabla p \cdot \mathbf{t} = f_H \quad \text{on } \partial_H\Omega$$

where

$$f_H := \frac{du_H}{ds} - (\gamma_0 + \boldsymbol{\nu}_0) \cdot \mathbf{t} \quad \text{on } \partial_H\Omega \tag{12}$$

(we maintain this shorthand throughout this note). We spend all degrees of freedom, except one, by letting V be the unique solution[d] of

$$\begin{cases} \Delta V = \mathrm{curl}(\gamma + \boldsymbol{\nu}) & \text{in } \Omega \\ V \stackrel{*}{=} 0 & \text{on } \partial_F \Omega \\ \nabla V \cdot \mathbf{n} = -f_H & \text{on } \partial_H \Omega \,. \end{cases} \tag{13}$$

The last one is used to normalize p so that $p \stackrel{*}{=} 0$ on $\partial_H \Omega^\ell$. Then $p = 0$ if $\ell \in \{0,1\}$, whereas if $\ell \geq 2$ p is a linear combination of the solutions ζ^i $(i = 1, \ldots, \ell - 1)$ of

$$\begin{cases} \Delta \zeta^i = 0 & \text{in } \Omega \\ \partial_n \zeta^i = 0 & \text{in } \Omega \\ \zeta^i = \delta^{ij} & \text{on } \partial_H \Omega^j,\ j = 1, \ldots, \ell \\ \nabla \zeta^i \cdot n = 0 & \text{on } \partial_F \Omega \end{cases} \tag{14}$$

(δ^{ij} denotes the Kronecker function). For notational convenience, we let

$$L = \begin{cases} 1 & \text{if } \ell \in \{0,1\} \\ \ell - 1 & \text{if } \ell \geq 2 \end{cases}$$

and define $\underline{\zeta} : \Omega \to \mathbb{R}^L$ by

$$\underline{\zeta} = \begin{cases} 0 & \text{if } \ \ell = \{0,1\} \\ (\zeta^1, \ldots \zeta^L) & \text{if } \ \ell \geq 2 \,, \end{cases}$$

so that we may write in compact form

$$\boldsymbol{\tau} = \mathbf{curl}\, V + \nabla(\underline{\alpha} \cdot \underline{\zeta}), \quad \underline{\alpha} \in \mathbb{R}^L \,. \tag{15}$$

Remark 3.1. This characterization of divergence-free planar vector fields has been recently proved by Auchmuty and Alexander[1] in the more general case of a non-simply connected domain. However, we will not use this result directly, since it requires a $C^{1,1}$ regularity of the domain which we do not need to impose.

In order to characterize $\underline{\alpha} = (\alpha^1, \ldots, \alpha^L)$ for $L \geq 1$, let $\overline{u}_H$ be any function whose trace on $\partial_H \Omega$ is u_H. Then for all $j = 1, \ldots, L$

$$\int_\Omega \nabla u \cdot \nabla \zeta^j = \int_{\partial_H \Omega} u_H \nabla \zeta^j \cdot \mathbf{n} = \int_\Omega \nabla \overline{u}_H \cdot \nabla \zeta^j,$$

[d]Note that if $\partial_F \Omega = \emptyset$, the compatibility condition $\int_{\partial\Omega} f_H = -\int_\Omega \mathrm{curl}(\boldsymbol{\gamma}+\boldsymbol{\nu})$ is satisfied.

and therefore, at a given time t,

$$\sum_{i=1}^{L} \alpha^i \int_\Omega \nabla\zeta^i \cdot \nabla\zeta^j \overset{(15)}{=} -\int_\Omega \mathbf{curl}\, V \cdot \nabla\zeta^j + \int_\Omega \boldsymbol{\tau} \cdot \nabla\zeta^j$$
$$\overset{(3),(13),(14)}{=} \int_\Omega (\nabla\bar{u}_H - \boldsymbol{\gamma} - \boldsymbol{\nu}) \cdot \nabla\zeta^j. \qquad (16)$$

Let $\underline{\underline{M}}$ be the $L \times L$ matrix with entries

$$M^{i,j} = \begin{cases} 1 & \text{if } \ell \in \{0,1\} \\ \int_\Omega \nabla\zeta^i \cdot \nabla\zeta^j & \text{if } \ell \geq 2 \end{cases} \qquad i,j = 1,\dots,L\,.$$

It is easily proved (see Lemma B.2 in Ref. 2) that $\underline{\underline{M}}$ is positive definite, so that (16) uniquely determines[e]

$$\underline{\alpha} = \underline{\underline{M}}^{-1} \int_\Omega ((\nabla\bar{u}_H - \boldsymbol{\gamma} - \boldsymbol{\nu}) \cdot \nabla)\underline{\zeta}\,. \qquad (17)$$

In conclusion, we have formally shown that the representation (15) of $\boldsymbol{\tau}$, with V and $\underline{\alpha}$ given respectively by (13) and (17), depends on $\boldsymbol{\gamma}$ and on its curl but not on its divergence.

4. The V-problem

Motivated by the previous discussion, we introduce a reformulation of the original problem, a "V-problem", in which the unknown is a triplet $(\boldsymbol{\gamma}, \boldsymbol{\nu}, V)$ — i.e. the unknown u is replaced by V — and the shear stress $\boldsymbol{\tau}$ is given by (15) rather than by (3). To emphasize that the representation (15) is now the *definition* of the shear stress, we replace $\boldsymbol{\tau}$ with $\boldsymbol{T}$ in the formulation of the V-problem, which is the following: in the bulk

$$\begin{cases} \begin{pmatrix} \boldsymbol{\gamma}_t \\ \boldsymbol{\nu}_t \end{pmatrix} = \mathbb{A}\begin{pmatrix} -\mathbf{curl}^2\, \boldsymbol{\gamma} + \boldsymbol{T} \\ \Delta\boldsymbol{\nu} + \boldsymbol{T} \end{pmatrix} & \text{in } \mathcal{Q} \\ \Delta V = \mathrm{curl}(\boldsymbol{\gamma} + \boldsymbol{\nu}) & \text{in } \mathcal{Q}, \end{cases} \qquad (18)$$

where

$$\begin{cases} \underline{\alpha} := \underline{\underline{M}}^{-1} \int_\Omega ((\nabla\bar{u}_H - \boldsymbol{\gamma} - \boldsymbol{\nu}) \cdot \nabla)\underline{\zeta} \\ \boldsymbol{T} := \mathbf{curl}\, V + \nabla(\underline{\alpha} \cdot \underline{\zeta}); \end{cases} \qquad (19)$$

[e]Here and after, as the notation suggests, for any $\mathbf{v} : \Omega \to \mathbb{R}^2$ we let $(\mathbf{v} \cdot \nabla)\underline{\zeta} = (\mathbf{v} \cdot \nabla\zeta^1, \dots, \mathbf{v} \cdot \nabla\zeta^L)$.

at the boundary

$$\begin{cases} \operatorname{curl}\boldsymbol{\gamma} = 0, & \nabla\boldsymbol{\nu}\cdot\mathbf{n} = 0, \quad V \stackrel{*}{=} 0 \qquad\qquad \text{on } \partial_F Q \\ \boldsymbol{\gamma}\cdot\mathbf{t} = \boldsymbol{\gamma}_0\cdot\mathbf{t}, & \boldsymbol{\nu} = \boldsymbol{\nu}_0, \quad \nabla V\cdot\mathbf{n} = -f_H \quad \text{on } \partial_H Q\,. \end{cases} \tag{20}$$

Note that because of the lack of control on $\operatorname{div}\boldsymbol{\gamma}$, $\boldsymbol{\gamma}$ will in general have just as much spatial regularity as its time derivative has, i.e. (looking at (10)) $\boldsymbol{\gamma}(t) \in W^{1,p'}(\Omega)$. For this reason, for the right-hand side of (19) to make sense we need an additional assumption on Ω:

$$\begin{cases} \textit{if } \ell \geq 2, \textit{ then } \Omega, \partial_F\Omega \textit{ and } \partial_H\Omega \textit{ are such that the} \\ \textit{weak solutions } \zeta^i \textit{ of (14) belong to } W^{1,p}(\Omega). \end{cases} \tag{21}$$

Remark 4.1. Assumption (21) is empty if $\ell \in \{0,1\}$. However, it is not always satisfied if $\ell \geq 2$: for a generic domain such that (4) holds, the best regularity one may expect is $\zeta^i \in H^{3/2-\varepsilon}(\Omega)$ for all $\varepsilon > 0$ (see e.g. Ref. 8), which means that (21) may not hold if $p \geq 4$. Generally speaking, the closer the angles at $\partial_F\Omega \cap \partial_H\Omega$ are to $\pi/2$, the better is the regularity: for instance, (21) holds true for any p if Ω is a rectangle with F-boundaries on top and bottom and H-boundaries on the sides, a quite natural case for the underlying mechanical model (see the discussion in Sec. 11.5 of Ref. 4).

5. Existence and uniqueness of solutions

In Ref. 2, existence and uniqueness of a solution to both the *V-problem* — (18)–(20) — and the *original problem* — (9), (3) and (5) — have been obtained. In order to present these results we need a few preliminaries[f]. Concerning the data, we assume that

$$\begin{cases} \boldsymbol{\gamma}_0 \in \mathbf{L}^{p'}(\Omega),\ \operatorname{curl}\boldsymbol{\gamma}_0 \in L^2(\Omega),\ \boldsymbol{\nu}_0 \in \mathbf{H}^1(\Omega), \\ \qquad \boldsymbol{\gamma}_0\cdot\mathbf{t} \in H^{\frac{1}{2}}(\partial_H\Omega),\ \overline{u}_H \in H^2(\Omega). \end{cases} \tag{22}$$

The following notion of "ammissible triplet" collects features which are common to both the V-formulation and the original formulation:

Definition 5.1 (Admissible triplet). *Let $T > 0$, let Ω satisfy (4), and assume (22). A triplet $(\boldsymbol{\gamma}, \boldsymbol{\nu}, \boldsymbol{\tau})$ is admissible in Ω_T if:*

[f]We let $p' = \frac{p}{p-1}$. For brevity, we omit the domains in "parabolic" function spaces whenever they coincide with $[0,T)$ and Ω: for example, $L^q(W^{1,p}) := L^q((0,T);W^{1,p}(\Omega))$. The subscripts 0_F (resp. $*_F$) denote spaces of functions u defined in Ω such that $u = 0$ on $\partial_F\Omega$ (resp. $u \stackrel{*}{=} 0$ on $\partial_F\Omega$); the meaning of 0_H and $*_H$ is analogous.

(a) $\gamma \in W^{1,p'}(\mathbf{L}^{p'})$, $\operatorname{curl}\gamma \in L^\infty(L^2)$, $\boldsymbol{\nu} \in W^{1,p'}(\mathbf{L}^{p'}) \cap L^\infty(\mathbf{H}^1)$, $\boldsymbol{\tau} \in L^\infty(\mathbf{L}^2)$;

(b) $-\mathbf{curl}^2\gamma + \boldsymbol{\tau} \in L^p(\mathbf{L}^p)$, $\Delta\boldsymbol{\nu} + \boldsymbol{\tau} \in L^p(\mathbf{L}^p)$;

(c) *for a.e.* $t \in (0,T)$, $\operatorname{curl}\gamma(t) = 0$ *on* $\partial_F\Omega$ *and* $\nabla\boldsymbol{\nu}(t)\cdot\mathbf{n} = 0$ *on* $\partial_F\Omega$ *in the sense of*

$$\int_\Omega \Delta\boldsymbol{\nu}(t)\cdot\boldsymbol{\varphi} = -\int_\Omega \nabla\boldsymbol{\nu}(t)\cdot\nabla\boldsymbol{\varphi}, \quad \forall\boldsymbol{\varphi}\in\mathbf{H}^1_{0_H}(\Omega);$$

(d) *for a.e.* $t \in (0,T)$, $\boldsymbol{\nu}(t) = \boldsymbol{\nu}_0$ *on* $\partial_H\Omega$ *and* $\gamma(t)\cdot\mathbf{t} = \gamma_0\cdot\mathbf{t}$ *on* $\partial_H\Omega$ *in the sense of*

$$\int_\Omega \varphi\,\operatorname{curl}\gamma(t) - \int_\Omega \mathbf{curl}\,\varphi\cdot\gamma(t) = \int_{\partial_H\Omega}\varphi\,\gamma_0\cdot\mathbf{t} \quad \forall\varphi\in W^{1,p}_{0_F}(\Omega).$$

Remark 5.1. It follows from (*a*) and (*b*) that $\operatorname{curl}\gamma(t) \in H^1$ for a.e. $t \in (0,T)$, hence its trace on $\partial_F\Omega$ is well defined.

We are now ready to state the two results:

Theorem 5.1 (Existence and uniqueness for the V-problem[2]). *Let* Ω *satisfy (4) and (21). For all* $(\gamma_0, \boldsymbol{\nu}_0, \overline{u}_H)$ *satisfying (22), there exists a unique triplet* $(\gamma, \boldsymbol{\nu}, V)$ *such that, for all* $T > 0$:

(a) $V \in L^\infty(H^1_{*_F})$, *and for a.e.* $t > 0$

$$\int_\Omega \nabla V\cdot\nabla\varphi = -\int_{\partial_H\Omega} f_H\varphi - \int_\Omega \varphi\operatorname{curl}(\gamma+\boldsymbol{\nu}) \quad \forall\varphi\in H^1_{0_F}(\Omega);$$

(b) *the triplet* $(\gamma, \boldsymbol{\nu}, \boldsymbol{T})$ *is admissible in* Ω_T, *with* $\boldsymbol{T}$ *as in (19);*

(c) *the triplet* $(\gamma, \boldsymbol{\nu}, \boldsymbol{T})$ *verifies*

$$\begin{pmatrix}\gamma\\ \boldsymbol{\nu}\end{pmatrix}_t = \mathbb{A}\begin{pmatrix}-\mathbf{curl}^2\gamma + \boldsymbol{T}\\ \Delta\boldsymbol{\nu} + \boldsymbol{T}\end{pmatrix} \quad \text{in } L^{p'}_{\mathrm{loc}}(\mathbb{L}^{p'});$$

(d) $\gamma(0) = \gamma_0$ *and* $\boldsymbol{\nu}(0) = \boldsymbol{\nu}_0$.

In addition $\operatorname{curl}\gamma \in C(L^2)$, $\boldsymbol{\nu} \in C(\mathbf{H}^1)$, $V \in C(H^1)$ *and* $\boldsymbol{T} \in C(\mathbf{L}^2)$.

Note that the traces of γ and $\boldsymbol{\nu}$ at $t = 0$ are well defined in view of Definition 5.1 (*a*).

Theorem 5.2 (Existence and uniqueness for the original problem[2]). *Let* Ω *satisfy (4) and (21). For all* $(\gamma_0, \boldsymbol{\nu}_0, \overline{u}_H)$ *satisfying (22) there exists a unique triplet* $(\gamma, \boldsymbol{\nu}, u)$ *such that for all* $T > 0$:

(a) $u \in L^\infty_{\text{loc}}(W^{1,p'})$ *and, for a.e.* $t > 0$, $u(t) \stackrel{*}{=} \overline{u}_H$ *on* $\partial_H\Omega$ *and*

$$\int_\Omega \boldsymbol{\tau}(t) \cdot \nabla\varphi = 0 \quad \forall \varphi \in H^1_{0_H}(\Omega)$$

with $\boldsymbol{\tau}$ *as in (3);*

(b) the triplet $(\boldsymbol{\gamma}, \boldsymbol{\nu}, \boldsymbol{\tau})$ *is admissible in* Ω_T*;*

(c) the triplet $(\boldsymbol{\gamma}, \boldsymbol{\nu}, \boldsymbol{\tau})$ *verifies*

$$\begin{pmatrix} \boldsymbol{\gamma} \\ \boldsymbol{\nu} \end{pmatrix}_t = \mathbb{A}\begin{pmatrix} -\,\mathbf{curl}^2\,\boldsymbol{\gamma} + \boldsymbol{\tau} \\ \Delta\boldsymbol{\nu} + \boldsymbol{\tau} \end{pmatrix} \quad \text{in } L^{p'}_{\text{loc}}(\mathbb{L}^{p'});$$

(d) $\boldsymbol{\gamma}(0) = \boldsymbol{\gamma}_0$ *and* $\boldsymbol{\nu}(0) = \boldsymbol{\nu}_0$.

In addition $\text{curl}\,\boldsymbol{\gamma} \in C(L^2)$, $\boldsymbol{\nu} \in C(\mathbf{H}^1)$, $u \in C(W^{1,p'})$ *and* $\boldsymbol{\tau} \in C_T(\mathbf{L}^2)$.

6. Outline of the proofs

In this last section, we briefly outline the main ideas used in Ref. 2 for the proof of the two results. The existence part of Theorem 5.1 exploits the fact that the right-hand side of (18) depends on $\boldsymbol{\gamma}$ only through its curl. Letting

$$\omega := \text{curl}\,\boldsymbol{\gamma}, \qquad \mathbb{v} := \begin{pmatrix} -\,\mathbf{curl}\,\omega + \boldsymbol{T} \\ \Delta\boldsymbol{\nu} + \boldsymbol{T} \end{pmatrix}, \qquad \mathbb{A}(\mathbb{v}) =: \begin{pmatrix} \mathbf{A}_\gamma(\mathbb{v}) \\ \mathbf{A}_\nu(\mathbb{v}) \end{pmatrix}$$

and taking the curl on both sides of the $\boldsymbol{\gamma}_t$ equation in (18), yields

$$\begin{cases} \omega_t = \text{curl}(\mathbf{A}_\gamma(\mathbb{v})) \\ \boldsymbol{\nu}_t = \mathbf{A}_\nu(\mathbb{v}) \end{cases}$$

In addition, the definition of $\underline{\alpha}$ in (19) is rewritten as

$$\begin{cases} \dfrac{d}{dt}\underline{\alpha} = -\underline{\underline{M}}^{-1} \displaystyle\int_\Omega ((\mathbf{A}_\gamma(\mathbb{v}) + \mathbf{A}_\nu(\mathbb{v})) \cdot \nabla)\underline{\zeta} \\ \underline{\alpha}(0) = \underline{\underline{M}}^{-1} \displaystyle\int_\Omega ((\nabla\overline{u}_H - \boldsymbol{\gamma}_0 - \boldsymbol{\nu}_0) \cdot \nabla)\underline{\zeta} \end{cases}$$

and the essential boundary condition on $\boldsymbol{\gamma}$ as

$$\mathbf{A}_\gamma(\mathbb{v}) \cdot \mathbf{t} = 0 \quad \text{on } \partial_H Q.$$

In this way, the resulting problem depends only on $\omega = \text{curl}\,\boldsymbol{\gamma}$: it is solved using a Galerkin scheme, coupled with Minty's argument to identify the nonlinear term. Both $\boldsymbol{\gamma}$ and its missing boundary condition are eventually recovered by integration with respect to time:

$$\boldsymbol{\gamma}(t) = \boldsymbol{\gamma}_0 + \int_0^t \mathbf{A}_\gamma(\mathbb{v}) \,.$$

As to uniqueness, the essential point is to show that weak solutions satisfy the energy balance (10) with *equality*. More precisely, the following holds:

Lemma 6.1.[2] *Under the assumptions of Theorem 5.1, let Ω satisfy (4) and (21), if a triplet $(\boldsymbol{\gamma}, \boldsymbol{\nu}, V)$ satisfies (a) and (b) in Theorem 5.1, then* $\operatorname{curl}\boldsymbol{\gamma}$, $\nabla\boldsymbol{\nu}$, ∇V *and* $\boldsymbol{T}$ *are continuous from* $[0,T]$ *to* L^2, *and for all* $0 \le t_1 < t_2$, *it holds:*

$$E[\boldsymbol{\gamma},\boldsymbol{\nu},\boldsymbol{T}]\Big|_{t_1}^{t_2} = -\int_{t_1}^{t_2}\int_\Omega \begin{pmatrix}\boldsymbol{\gamma}_t\\ \boldsymbol{\nu}_t\end{pmatrix}\cdot\begin{pmatrix}-\,\mathbf{curl}\operatorname{curl}\boldsymbol{\gamma}+\boldsymbol{T}\\ \Delta\boldsymbol{\nu}+\boldsymbol{T}\end{pmatrix}.$$

Given a solution of the V-problem, a solution of the original problem is recovered by integrating at each time

$$\nabla u(t) \;=\; \mathbf{curl}\,V(t)+\nabla(\underline{\alpha}(t)\cdot\underline{\zeta})+\boldsymbol{\gamma}(t)+\boldsymbol{\nu}(t). \tag{23}$$

The fact that such u exists is formally obvious, since the curl of the right-hand side vanishes, but in fact non-trivial for weak solutions: it follows from the extension of Poincaré's Lemma proved by Kesavan in Ref. 6. To identify the value of u at the hard boundary $\partial_H\Omega$ (if not empty, else we are already done), the idea is to reverse the reasoning which led to the definition of $\underline{\alpha}$ in Sec. 3. Let us sketch the formal argument: on $\partial_H\Omega$,

$$\begin{aligned}\nabla u(t)\cdot\mathbf{t} &\overset{(23)}{=} \big(\mathbf{curl}\,V(t)+\nabla(\underline{\alpha}(t)\cdot\underline{\zeta})+\boldsymbol{\gamma}(t)+\boldsymbol{\nu}(t)\big)\cdot\mathbf{t}\\ &\overset{(20),(14)}{=} f_H+(\boldsymbol{\gamma}_0+\boldsymbol{\nu}_0)\cdot\mathbf{t}\overset{(12)}{=}\frac{du_H}{ds}\,.\end{aligned}$$

which implies that $u(t)-u_H = \beta^i(t)$ on $\partial_H\Omega^i$ for all $t>0$. Of course, if u solves (23) then also $u+\beta(t)$ does: we use this degree of freedom to set $\beta^\ell(t)=0$, which already characterizes u if $\ell=1$. If $\ell>1$, let us set $\varphi(t) := u(t)-\overline{u}_H-\underline{\beta}(t)\cdot\underline{\zeta}$. Then $\varphi(t)$ has null trace on $\partial_H\Omega$, and (14) implies that $\int_\Omega \nabla\varphi(t)\cdot\nabla\zeta^i=0$, that is,

$$\begin{aligned}\sum_{j=1}^L M^{i,j}\,\beta^j(t) &= \sum_{j=1}^L \beta^j(t)\int_\Omega\nabla\zeta^j\cdot\nabla\zeta^i=\int_\Omega\nabla(u(t)-\overline{u}_H)\cdot\nabla\zeta^i\\ &\overset{(23)}{=}\int_\Omega(\mathbf{curl}\,V(t)+\nabla(\underline{\alpha}(t)\cdot\underline{\zeta})+\boldsymbol{\gamma}(t)+\boldsymbol{\nu}(t)-\nabla\overline{u}_H)\cdot\nabla\zeta^i\\ &=\sum_{j=1}^L M^{i,j}\alpha^j(t)-\int_\Omega(\nabla\overline{u}_H-\boldsymbol{\gamma}(t)-\boldsymbol{\nu}(t))\cdot\nabla\zeta^i\overset{(19)}{=}0,\end{aligned}$$

hence $\underline{\beta}\equiv 0$ (since $\underline{\underline{M}}$ is positive definite).

Finally, to prove the uniqueness part of Theorem 5.2, arguing as in Sec.3 a solution of the V-problem is constructed from a solution to the original problem, and then the uniqueness of the former is exploited.

Acknowledgements.

We are indebted to Roberta Dal Passo, who died prematurely in August 2007, and to Michiel Bertsch for their fundamental contribution to the results exposed in this note. We are also grateful to Morton Gurtin for having proposed the problem to us, and to Giles Auchmuty, Haïm Brezis, Morton Gurtin, Paolo Podio-Guidugli, and Giuseppe Savaré for useful discussions.

References

1. G. Auchmuty, J.C. Alexander, L^2 well-posedness of planar div-curl systems, *Arch. Ration. Mech. Anal.* **160** (2001), 91–134.
2. M. Bertsch, R. Dal Passo, L. Giacomelli, G. Tomassetti, A nonlocal and fully nonlinear degenerate parabolic system from strain-gradient plasticity, Preprint Dipartimento Me.Mo.Mat. 2/2008, submitted.
3. E. DiBenedetto, *Degenerate parabolic equations*, Springer–Verlag, New York, 1993.
4. M.E. Gurtin, A gradient theory of small-deformation isotropic plasticity that accounts for the Burgers vector and for dissipation due to plastic spin, *J. Mech. Phys. Solids*, **52** (2004), 2545–2568.
5. M.E. Gurtin, L. Anand, A theory of strain-gradient plasticity for isotropic, plastically irrotational materials. Part I: Small deformations, *J. Mech. Phys. Solids* **53** (2005), 1624–1649.
6. Kesavan, S., On Poincaré's and J. L. Lions' lemmas, *C. R. Math. Acad. Sci. Paris* **340** (2005), 27–30.
7. B.D. Reddy, F. Ebobisse, A. McBride, Well-posedness of a model of strain gradient plasticity for plastically irrotational materials, *Int. J. Plasticity* **24** (2008), 55–73.
8. G. Savaré, Regularity results for elliptic equations in Lipschitz domains, *J. Funct. Anal.* **152** (1998), 176–201.

Numerical Simulation of Capillary Flows Through Molecular Dynamics

Salvatore Iacono

Department of Mathematics, University of Messina
Salita Sperone 31, 98166 Messina, Italy,
iacono@dipmat.unime.it

Capillary pore imbibition represents a very challenging field of research because of its crucial role in several physical phenomenona. This paper deals with capillary dynamics in connection with non destructive test technique used to evidence eventual defects present in mechanical parts whose perfect integrity is vital for the reliability of all the systems (car, planes, etc.) which they are part of. For the description of such phenomenona, essentially two different approaches are possible. The first approach consists in considering the fluid motion as a continuum to be described in terms of mathematical models governed by ordinary or partial differential equation. Subsequently, according to the prescribed initial and boundary conditions to these models, the solution can be worked out numerically through one of the available numerical techniques such as finite difference, finite elements, finite volumes, meshless methods, etc. Conversely, the second approach considers the fluid motion as an atomistic ensemble of discrete particles (atoms or molecules), whose macroscopic features are strongly determined by the inner interactions occurring among them. In this context, the two most important conceptual available tools are the Lattice Boltzmann theory, mainly used by physicists and the Molecular Dynamics, MD for the sequel, that jointly with its numerical treatment represents the topic of the present paper. Capillary imbibition represents a specific aspect of the wider area of wetting solid by liquids and its description in terms of MD has revealed to be a very promising approach for the description of the capillary flows alternative to continuous differential models. At the beginning, MD was used to simulate essentially the behavior of biological materials, but over the years MD has proved to be a powerful tool to model and simulate nanotechnology structures. In this paper large scale MD is used in order to model the intrusion of water into a defect of a few given solid materials. Finally numerical results have been obtained for cylindrical defects of Titanium or Aluminum both internally smooth or rough.

Keywords: Capillary dynamics, molecular dynamics, hamiltonian systems

1. Continuous model

The simplest approach consists in considering the liquid as a continuous and incompressible bulk, penetrating an open-ended horizontal capillary. In this case, its motion can be described by a single lagrangian coordinate and the only two contrasting forces acting on it are the surface tension and the viscous drag. At the macroscopic level, what we are always interested in is the depth and the velocity of penetration as a function of time. Under these assumption, this system is very well described by the simple 1-D ordinary differential model

$$\rho\left[x\frac{d^2x}{dt^2}+\left(\frac{dx}{dt}\right)^2\right]=\frac{2}{R}\gamma\cos(\theta)-\frac{8}{R^2}\eta x\frac{dx}{dt} \tag{1}$$

where x is the coordinate of the liquid-air interface, γ is the surface tension relative to the liquid-solid interface, η is the viscosity of the liquid, ϑ is the contact angle, R is the radius of the capillary. For further details on this model, where the contact angle is considered static, see [1,2]. Furthermore, another generalization of this model exists where the contact angle is considered dynamic as well. For instance see [3,4], for further details on this model. However this model represents a generalization of the Lucas-Washburn equation known since the beginning of the last century, whose solution is the asymptotic solution of (1). Indeed the Lucas-Washburn equation can be obtained, by neglecting the inertial contribution on the left hand side of (1), so as to read

$$\frac{2}{R}\gamma\cos(\theta)=\frac{8}{R^2}\eta x\frac{dx}{dt}. \tag{2}$$

This equation can be solved analytically and, under the assumption $x(0)=0$, its solution is given by

$$x(t)=\sqrt{\frac{\gamma R\cos(\theta)t}{2\eta}} \quad . \tag{3}$$

Washburn asymptotic solution (3) is valid only for $t >> \rho R^2/\mu$. This limitation is also confirmed by the non physical behavior of such a solution at the origin, where its first derivative becomes infinite. However, there exists two very interesting and promising microscopic alternative approaches to the continuous modeling. The atomistic description of the phenomenon can be carried out through either MD or Lattice-Boltzmann approach. If on the one hand, capillary flow has been studied through the latter approach, used mainly by physicist, see [5,6], on the other hand the present paper is just devoted to inquire capillary flow through MD. In section 2, the most essential MD concepts are outlined. Section 3 is focused on the numerical treatment of MD integration. In section 4, a capillary flow MD simulation is described, according to a predesigned set of parameters. Furthermore section 5 contains the root mean square deviation (RMSD) results of the simulation for a few capillary flows and finally, in section 6, some conclusions are reported.

2. Molecular Dynamics model

MD consists in considering the interaction among particles, that can be atoms or molecules, constituting a given physical system whose time and space evolution is responsible for its macroscopic behavior. The motion of these particles is intrinsically chaotic and is described by a huge number of degrees of freedom, but the final goal of any MD simulation is just the deduction of 'averaged' macroscopic features, such as temperature, pressure, interface displacement, surface-tension, etc. Because of the very high complexity of calculation needed to determine the trajectory of any particle constituting the whole system, the use of MD is feasible only for at most tens of thousands of atoms, that corresponds to very small volumes not higher than $10 \div 20\ nm^3$, and for hundreds of thousands of time-steps, that correspond to total integration times of about tens of ns. For higher numbers of particles and integration times MD is viable only through massive parallel computing. Another fundamental aspect of MD simulation is the thermodynamical conditions under which the simulation itself is carried out. In literature it is usual to distinguish among NPT (fixed number of particles N, pressure P and temperature T), NVE (fixed particles N, volume V and energy E), NVT (fixed particles N, volume V and temperature T).

Classical or Newtonian MD consists essentially in considering the system under inspection as a dynamical system where the degrees of freedom (or state variables) are represented by position and velocity of any particle belonging to it. The particles trajectories are found by using as governing equation the second law of dynamics jointly with the external forces acting on the system and the due initial conditions. In more formal terms, for a system of N particles, classical MD is carried out through the solution of the mathematical model

$$m_i \frac{d^2\vec{r_i}}{dt^2} = \vec{F_i}(\vec{r_1}, \cdots, \vec{r_N}) \tag{4}$$

$$\vec{r_i}(0) = \vec{r_i} \qquad \frac{d\vec{r_i}}{dt}(0) = \frac{d\vec{r_i}}{dt}$$

for $i = 1, 2..N$, where m_i is the mass of the ith particle, F_i is the resulting of forces acting on it.

The forces usually considered in classical MD are divided into bond type or non-bond type. The first ones concern the direct connection among some of the particles forming the system, whereas the second ones concern the distance interactions like van der Waals attraction and hard-core repulsion, or the electrostatic interaction, both of them at least within a prefixed range as it will be cleared in the sequel. In particular, bonded forces are usually modeled as a spring-like connection, so that their potential is always of quadratical type, whereas distance forces are modeled as a function proportional to the reciprocal of the squared distance, so that their potential is a decreasing function of distance. As matter of fact all of them are conservative forces. Force fields usually considered for MD are derived from quantum mechanics calculations, or spectroscopic measurement of small molecules that are similar to parts of macromolecules, or by fitting to measured constitutive properties such as diffusion coefficients and dielectric constants. The most common force fields are CHARMM, AMBER, GROMOS, and OPLS.

In literature there exist several kinds of two body interaction potential, the most known are: Square Well, Yukawa, Buckingham or Lennard-Jones (L-J). All of them are usually defined through two parameters, the radius of the particles (considered rigid and impenetrable) σ and the depth of the potential well itself ϵ. The generic L-J potential can be defined as

$$V(r) = \epsilon \left[\left(\frac{\sigma}{r}\right)^n - \left(\frac{\sigma}{r}\right)^m \right].$$

By choosing $n = 12$ and $m = 6$ the potential above reduces to the L-J potential of wide use in applications L-J (12-6). It can be observed that the first term corresponds to a repulsive force predominant in a short range, whereas the second one is an attractive force predominant in the long range. The theoretical range of this potential extends up to infinity, but for practical computational reason it is defined the concept of "cut-off" range, meaning that for each particle only particles located below $r_{cut-off}$ are involved in computation of interactions. For each force, a potential function $U_i(q)$ can be defined so that $\vec{F_i}(q) = -\nabla U_i$, equation (4) is rewritable as an

Hamiltonian system whose Hamiltonian function is just the total initial energy $E = K + U$ that is conserved along with motion, where K is the kinetic energy and U the potential energy. The two most outstanding properties arising from energy conservation are symplectiness of flow and time reversibility (or time symmetry).

Symplectiness consists in the volume conservation property according to the the Liouville's theorem, for further details see [7]. Furthermore it is a simple matter to show that the composition of symplectic maps results to be still a symplectic map. In particular, Hamiltonian systems own the remarkable property of generating a flow that is symplectic. This means that any Hamiltonian flow can be seen as a symplectic map from the phase space to itself. Time reversibility for conservative system is easily proved by inverting the direction of time so as to get still the same Hamiltonian and the same Hamiltonian equations. Finally, Hamiltonian systems own also the property of conservation of linear momentum and/or angular momentum, whether they are invariant with respect to linear translation and/or to rotation respectively.

3. Numerical methods

As it can be easily predicted, the main problem arising in numerical integration of Hamiltonian system is the demand for numerical schemes which are able to preserve their geometrical properties also in the resulting discrete numerical solution. These methods are called symplectic. From a numerical viewpoint, whenever we apply a symplectic integration to an Hamiltonian system it is experienced that the corresponding numerical Hamiltonian remains always bounded for arbitrarily long integration time, whereas it diverges for all the other non-symplectic integrators. For further details see [8] or [9].

In the context of numerical MD integrators, the most used numerical scheme coupling a good accuracy of second order with symplectiness is the celebrated Verlet method. For a further and deeper explanation of its geometrical features see also the paper by Hairer, Lubich and Wanner [12]. It can be easily derived through a central Taylor expansion of q, obtained by adding its forward and backward expansion both centered in t_n

$$\begin{cases} q^{n+1} = q^n + \Delta t \dfrac{dq^n}{dt} + \dfrac{\Delta t^2}{2}\dfrac{d^2q^n}{dt^2} + \dfrac{\Delta t^3}{6}\dfrac{d^3q^n}{dt^3} + \cdots \\ q^{n-1} = q^n - \Delta t \dfrac{dq^n}{dt} + \dfrac{\Delta t^2}{2}\dfrac{d^2q^n}{dt^2} - \dfrac{\Delta t^3}{6}\dfrac{d^3q^n}{dt^3} + \cdots \end{cases}$$

As a result, the scheme for the position q is

$$q^{n+1} = 2q^n - q^{n-1} + \Delta t^2 M^{-1} F(q^n),$$

whereas for the velocity, it can be deduced by subtracting the two above expansions, so as to get

$$v^n = \frac{q^{n+1} - q^{n-1}}{2\Delta t}.$$

This method can be also met with different names, like leap-frog method in the context of partial differential equation, or Stormer method in the context of astronomy. It is worth noticing that this method is a two-step one so that, in order to trigger it, we need to provide also the numerical scheme with the position and velocity at times $t = 0$ and $t = \Delta t$.

4. Capillary flows MD simulation

Firstly, all of the different simulations were carried out through the two softwares called NAMD for the numerical integration and VMD for the visualization. Both of them were developed by the Theoretical Biophysics Group at Illinois' Beckman Institute for Advanced Science and Technology at the University of Illinois [10]. Each simulation was of NVT type and consisted in considering the imbibition of a cylindrical pore, made up of a metallic-like material, put in contact with a water reservoir. In particular, cylinders of Aluminum-like and Titanium-like with different radii (R=10, R=30, R=50)(dimension in Angstrom Å) were considered with their internal wall smooth or rough. Besides, in each simulation the cylinder was positioned so that its axis is a segment of the positive z axis with one end into the origin and it was generated by fixing the radius R and the number N of atoms for each layer. In particular, in each layer these N atoms were located according to a uniform equipartition of the angle 2π with a phase shift of π/N between two any consecutive layers. Rough cylinders were created in the same way as smooth ones, but for each layer the radius was assumed to be $R - 0.1R\alpha$, where $\alpha \in [0, 1]$ is an aleatory variable with uniform distribution. In order to simplify the computational effort, all of the solid atoms were kept fixed in their original location and also water molecule bonds were considered as fixed. As a matter of fact, this implies that only non-bonded forces act on the system and they are responsible for dynamics of capillary flows. These forces are also called distance forces or in alternative van der Walls forces and in NAMD they are defined through the L-J(12-6) potential expressed by

$$V_{L-J}(r_{i,j}) = \epsilon_{i,j}\left[\left(\frac{Rmin_{i,j}}{r_{i,j}}\right)^{12} - 2\left(\frac{Rmin_{i,j}}{r_{i,j}}\right)^{6}\right],$$

where $r_{i,j}$ is the distance between particles i and j, ϵ_i is related to the depth of potential well and $Rmin_i$ is the minimum interaction distance equal to the effective diameter of the particle i (as the particles are considered rigid and impenetrable). In order to take into account both the contribution of the liquid and solid particles, it is assumed $\epsilon_{i,j} = \frac{\epsilon_i+\epsilon_j}{2}$ and $Rmin_{i,j} = \frac{1}{2}(Rmin_i + Rmin_j)$.

For this MD simulation, the physical features of each type of particle involved had to be fixed and they are reported in table 1. Then it had to be fixed the geometrical structure of the whole system, as well as the parameters related to the bonded forces for liquid material and non bonded interaction (distance or van der Walls interaction) both for solid and liquid. Before starting the simulation, the initial conditions for the dynamical system (4) had to be defined, consisting in position and velocity of the particles of the liquid and all of the integration parameters and options had to be set up and are reported in table 2. Finally it is worth specifying that for cylinders with radius R=10, we used 2600 solid atoms and 659 water molecules, for radius R=30, we used 7750 solid atoms and 3739 water molecules, whereas for R=50, we used 9030 solid atoms and 9835 water molecules. Figure 1 shows a picture of longitudinal section of initial and final configuration for an Al-like cylinder with R=10Å, smooth and rough. In both of them, according to the usually observed macroscopic capillary imbibition, it can be seen the rise of the liquid inside the capillary and the typical convex shape of the interface meniscus.

Table 1. Physical parameters for the solid materials considered.

	Ti-like	Al-like
Rmin/2=σ	1.4	1.25
ϵ	-0.01	-0.01
mass	47.867	26.981

5. Results

From a quantitative viewpoint, it was calculated the Root Mean Square Deviation (RMSD) of water molecules from their initial configuration. In formal mathematical terms, RMSD is defined as

Table 2. Peculiar simulation parameters used.

integration parameter	value	explanation
time-step	$2fs$	
equilibration time	$0.5ps$	
temperature	$310K$	
switching	on	artificial correction to make potential null at $r_{cut-off}$
switchdist	10	distance from which to apply the switching correction
$r_{cut-off}$	12	
pairlistdist	14	distance between pairs for inclusion in pair lists
Langevin	on	kinetic energy of the system kept constant by adding damping and random forces
Langevintemp	$310K$	

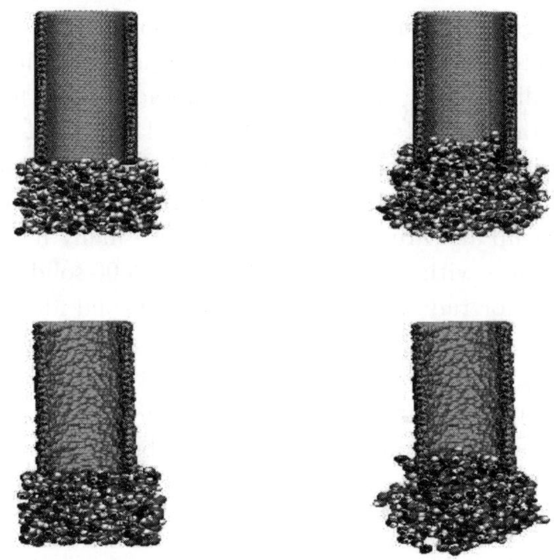

Fig. 1. Simulation graphical results for: a smooth capillary with R=10Å (top), the same capillary but rough (bottom). On the left the initial configuration, on right the final configuration after $20ps$.

$$RMSD_\alpha = \sqrt{\frac{\sum_{j=1}^{N_t} \sum_{\alpha=1}^{N_\alpha} (\vec{r_\alpha}(t_j) - \langle \vec{r_\alpha} \rangle)^2}{N_\alpha}},$$

where N_α is the number of atoms whose positions are compared, N_t is the number of time steps over which atomic positions are being compared, $\vec{r}_\alpha(t_j)$ is the position of atom α at time t_j, whereas the average value of atom α is defined as: $\langle \vec{r}_\alpha \rangle = \frac{1}{N_t} \sum_{j=1}^{N_t} \vec{r}_\alpha(t_j)$. For internally smooth Al-like and Ti-like cylinders, the results are reported in table 3. From these data it can be deduced that at the beginning of the simulation there is a steep rise of penetration. This is due essentially to the search of the system for the configuration with the minimum of energy, according to the natural behavior of any physical system. Furthermore, it is evident that the real initial rate of penetration is very high, so that the assumption of null initial velocity, for the problem (1), makes it ill-conditioned and, paradoxally, the Wahsburn solution (3) is closer to such a behavior. From a graphical viewpoint, it is enough to look at figures from 2 to 5 to have the confirmation of the said capillary rise that is as fast as small is the capillary radius. Finally from the comparison between a smooth and rough capillary in figure 6 it can be observed that at the beginning the imbibition is slightly faster for the smooth cylinder whereas then the situation is the other way round.

Table 3. RMSD values for the first *ns* of simulation for internally smooth cylinders

	Al-like RMSD			Ti-like RMSD		
time	R=10	R=30	R=50	R=10	R=30	R=50
0,2	0,262	0,245	0,219	0,262	0,246	0,219
0,4	0,651	0,619	0,596	0,651	0,61	0,603
0,6	1,169	1,071	1,043	1,107	1,062	1,056
0,8	1,462	1,347	1,307	1,427	1,338	1,314
1	1,641	1,538	1,479	1,675	1,541	1,493

Table 4. RMSD values for the first *ns* of simulation for internally rough cylinders

	Al-like RMSD			Ti-like RMSD		
time	R=10	R=30	R=50	R=10	R=30	R=50
0,2	0,25	0,242	0,221	0,217	0,224	0,215
0,4	0,671	0,611	0,598	0,543	0,614	0,595
0,6	1,17	1,075	1,046	1,032	1,088	1,042
0,8	1,52	1,355	1,303	1,346	1,363	1,308
1	1,743	1,543	1,474	1,553	1,534	1,493

6. Conclusions

MD simulation confirms its usefulness in featuring some aspects of capillary dynamics. It reveals to be a very helpful alternative to differential models.

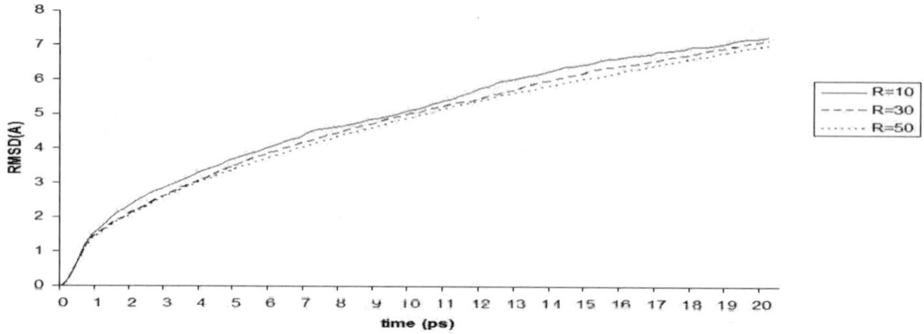

Fig. 2. RMSD curves vs time for a smooth cylinder of Aluminum-like material.

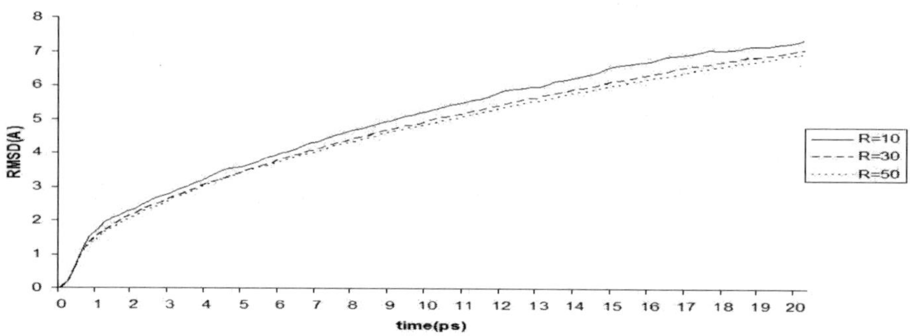

Fig. 3. RMSD curves vs time for a rough cylinder of Aluminum-like material.

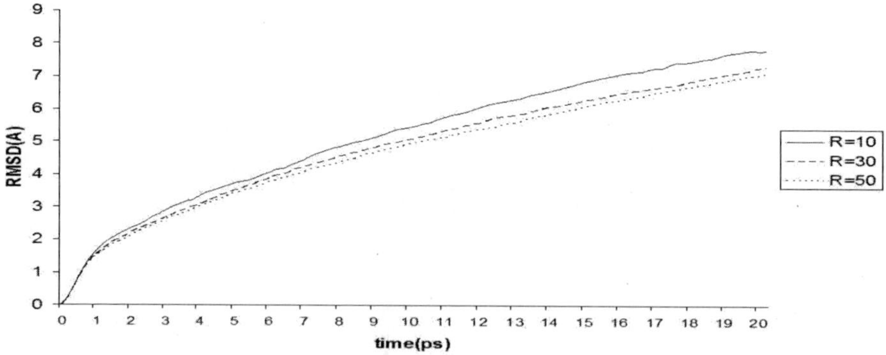

Fig. 4. RMSD curves vs time for a smooth cylinder of Titanium-like material.

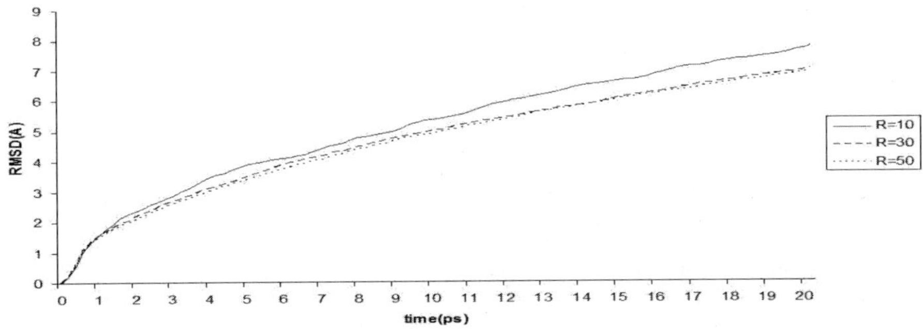

Fig. 5. RMSD curves vs time for a rough cylinder of Titanium-like material.

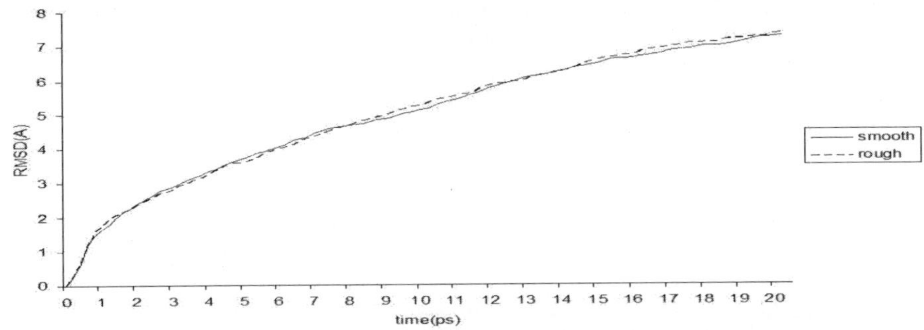

Fig. 6. RMSD curves vs time for a comparison between a smooth and a rough cylinder with $R = 10$ of Aluminum-like material.

Very interesting applications are the ones that can be found in some papers by Martic et al. [3,4,11], where the authors exploit MD to explore the dynamics of capillary flows, taking into account also a time variable contact angle. The most outstanding result obtained in the above simulations is the confirmation of the macroscopic behavior of getting a capillary rise as fast as small is the capillary radius. From a physical viewpoint this is explainable by considering that the distance force is proportional to the square of the distance so that a reduction of a factor 10 in the distance results in an increase of the force (and hence of acceleration) of a factor 100.

References

1. *G. Cavaccini, V. Pianese, S. Iacono, A. Jannelli, R. Fazio*, Mathematical and numerical modeling of liquids dynamics in a horizontal capillary. Lecture Series on Computer and Computational Sciences, Recent Progress in Computational Sciences and Engineering, vol 7A, 66-70, ISSN: 9004155422, Leiden:Koninlijke Brill NV (Netherlands).
2. *R. Fazio, S. Iacono, A. Jannelli, G. Cavaccini, V. Pianese*, A two immiscible liquids penetration model for surface-driven capillary flows. PAMM, 7,2150003-2150004 (2007)/DOI 10.1002/pamm.200700151
3. *G. Martic, F. Gentner, D. Seveno, D. Coulon, J. De Connick*, A molecular Dynamics Simulation of Capillary Imbibition. Langmuir 2002,18,7971-7976.
4. *G. Martic, T.D. Blake, J. De Connick*, Dynamics of Imbibition into a Pore with a Heterogeneus Surface. Langmuir 2005, 21 11201-11207
5. *S. Chibbaro*, Capillary filling with pseudo-potential binary Lattice-Boltzmann model. European Physical Journal E, 27, 99106 (2008). DOI 10.1140/epje/i2008-10369-4
6. *F. Diotallevi, L. Biferale, S. Chibbaro, A. Puglisi, S.Succi*, Front pinning in capillary filling of chemically coated channels. Physical Review E, 78, 036305 (2008)
7. *V. I. Arnold*, Mathematical Methods of Classical Mechanics. Springer-Verlag, 2nd edition (1989)
8. *C. J. Budd and M. Piggott*, Geometric integration and its applications. Foundations of Computational Mathematics XI, ed. F. Cucker. Elsevier, (2003), 35 139
9. *E. Hairer, C. Lubich, G. Wanner*, Geometric numerical integration. Springer Series in Computational Mathematics. 2005
10. *J.C. Phillips, R. Braun, W. Wang, J. Gumbart, E. Tajkhorshid, E. Villa, C. Chipot, R. D. Skeel, L. Kale, and K. Schulten*, Scalable molecular dynamics with NAMD. Journal of Computational Chemistry, 2005, 26,1781-1802.
11. *G. Martic, F. Gentner, D. Seveno, J. De Connick, T.D. Blake*, The possibility of Different Time Scales in the Dynamics of Pore Imbibition. Journal of Colloid and Interface Science 2004, 270 171-179
12. *E. Hairer, C. Lubich, G. Wanner*, Geometric numerical integration illustrated by the Stormer/Verlet method. Acta numerica, pp 1-51, 2003.

MATERIAL SYMMETRY AND INVARIANTS FOR A 2D FIBER-REINFORCED NETWORK WITH BENDING STIFFNESS

G. INDELICATO

Dipartimento di Matematica
Università di Torino,
Via Carlo Alberto 10, 10123 Torino, Italy
E-mail: giuliana.indelicato@unito.it

In this work we review some results, proved in Indelicato and Albano,[13] on the material symmetry of a woven fabric, in order to characterize the invariance properties of its deformation energy. The woven fabric is described as a fiber-reinforced surface: we consider four basic weave patterns, depending on the angle between the yarns/fibers and their material properties. To each of these pattern is associated a material symmetry group under which the elastic energy is invariant. We compute the polynomial invariants under the action of the material symmetry group of the network and we discuss a representation for deformation energies depending on the shear between the yarns and their curvature. Our approach yields invariant polynomials for higher grade elastic shells in general, but the symmetry restrictions on the energy select special invariants that are related to the geometry of the weave and the yarns.

Keywords: 2D fibered networks, fiber-reinforced materials, woven fabrics, material symmetry, anisotropic invariants.

1. Introduction

Material symmetry is a powerful piece of information that allows to correlate the structure of a material to its mechanical response. This is particularly true for fabrics and 2-dimensional fiber-reinforced networks, which are materials made of two families of fibers (the yarns) and whose response is strongly dependent on the geometry of the weave pattern. Fiber-reinforced networks have a wide-range of applications, ranging from textiles to surgical materials and aerospace engineering (cf. e.g., Aimène,[1] Christensen[5]): for instance, an effective way of manufacturing shells with a required shape is to build a network of semi-rigid fibers and subsequently embed them into a polymeric matrix.

In this paper we review some of the the work published in Indelicato and Albano[13] (all proofs are omitted here): specifically, we discuss material symmetry for surfaces made by networks of fibers, in order to characterize the deformation energy in dependence of the geometry of the network and the material properties of the fibers. We are interested in networks of inextensible fibers for which the elastic energy depends on the shear and bending of its components (cf. Wang and Pipkin[19] for the typical models, and Indelicato[12]).

In the simplest case (alternating intersections between the fibers), four weave patterns are possible: the square structure (orthogonal equivalent fibers), the rectangular structure (orthogonal inequivalent fibers), the rhombic structure (non-orthogonal equivalent fibers), and the parallelogram structure (non-orthogonal inequivalent fibers). We generalize to this case a technique, based on the so-called Rychlewski's theorem (cf. for instance, Boehler[3,4]), that allows to explicitly compute the basic polynomial invariants of the action of the symmetry groups and, by consequence, the general form of an invariant energy density (cf. Spencer,[18] Holzapfel[11]). Rychlewski's theorem is extensively used to compute the invariants depending on the Cauchy-Green tensor for fiber-reinforced materials (cf. for instance Hartmann,[10] Aksel and Itskov,[2] Schröder and Neff[17] , the applications of Aimène[1] to textiles and of Gasser, Ogden and Holzapfel[7] to arterial walls).

The classical form of Rychlewski's theorem, however, does not account for symmetry operations that interchange the fibers, and cannot distinguish, for instance, the square and the rectangular structure. The result has been extended to cover these cases (cf. Zhang and Rychlewski[21]), but we prove here an easier generalization that allows to compute the polynomial invariants that depend on up to the first derivatives of the deformation, and discuss the general form of an invariant energy depending on the shear between the fibers and their curvature. Our approach yields invariants for high grade elastic shells in general: textiles are described in such a framework once special constitutive choices are made in terms of these invariants.

The invariants have been computed using the computer package 'Singular', available online at http://www.singular.uni-kl.de (cf. Greuel, Pfister and Schoenemann[8]).

2. Material symmetries for fiber-reinforced materials: Rychlewski's theorem

The classical approach to material symmetry for fiber-reinforced materials is based on the so–called Rychlewski theorem. Consider, to fix ideas, a fiber-reinforced composite made of an isotropic matrix and N families of reinforcing fibers. We know that, for instance, the presence of a single family of reinforcing fibers in an isotropic matrix gives the material transversely isotropic properties: when the matrix is hyperelastic, transverse isotropy is just the invariance of the deformation energy W under rotations that preserve the fibers' direction.

Rychlewski's theorem allows to compute the anisotropic invariants of a fiber-reinforced material in terms of suitable isotropic invariants of an extended energy function. The classical form of Rychlewski's theorem, however, does not distinguish between symmetry operations that interchange the fibers. The result has been extended to cover these cases also (cf. Zhang and Rychlewski[21]), but we prove here an alternative generalization that allows to compute the polynomial invariants that depend on up to the first derivatives of the deformation, and discuss the general form of an invariant energy depending on the shear between the fibers, and their curvature.

In this section we work in arbitrary dimension n. For $\boldsymbol{A}_1, \ldots, \boldsymbol{A}_N \in \mathbb{R}^n$ unit vectors parallel to the fibers, define the structure tensors by $\boldsymbol{M}_i = \boldsymbol{A}_i \otimes \boldsymbol{A}_i$, and let G be the group of orthogonal transformations which leave the structure tensor invariant (the restricted symmetry group of the structure)

$$G = \left\{ \boldsymbol{R} \in O(n) \;/\; \boldsymbol{R}\boldsymbol{M}_i\boldsymbol{R}^\top = \boldsymbol{M}_i,\ i = 1, \ldots, N \right\}.$$

Also, denote by M the orbit of $(\boldsymbol{M}_1, ..., \boldsymbol{M}_N)$ under $O(n)$ (the group of orthogonal transformations over $\mathbb{R}^n$)

$$\begin{aligned} M &= \left\{ \left(\boldsymbol{Q}\boldsymbol{M}_1\boldsymbol{Q}^\top, ..., \boldsymbol{Q}\boldsymbol{M}_N\boldsymbol{Q}^\top \right) \right\}_{\boldsymbol{Q}\in O(n)} \\ &= \left\{ (\boldsymbol{Q}\boldsymbol{A}_1 \otimes \boldsymbol{Q}\boldsymbol{A}_1, ..., \boldsymbol{Q}\boldsymbol{A}_N \otimes \boldsymbol{Q}\boldsymbol{A}_N) \right\}_{\boldsymbol{Q}\in O(n)}. \end{aligned}$$

Let $\boldsymbol{C} \in \mathrm{Sym}(n)$, with $\mathrm{Sym}(n)$ the group of symmetric transformations of $\mathbb{R}^n$, be a symmetric tensor, and consider an elastic energy density of the form $W = W(\boldsymbol{C})$. We say that W is invariant under G if $W(\boldsymbol{C}) = W(\boldsymbol{R}\boldsymbol{C}\boldsymbol{R}^\top)$ for all $\boldsymbol{R} \in G$ and $\boldsymbol{C} \in \mathrm{Sym}(n)$.

Theorem 2.1 (Rychlewski). *(i) If $W = W(\boldsymbol{C})$ is invariant under G, then the extended energy function $\widehat{W}$, defined on $Sym(n) \times M$ by*

$$\widehat{W}(\boldsymbol{C}, \boldsymbol{P}_i) := W(\boldsymbol{Q}\boldsymbol{C}\boldsymbol{Q}^\top), \tag{1}$$

where $\boldsymbol{Q} \in O(n)$ is such that $\boldsymbol{Q}\boldsymbol{P}_i\boldsymbol{Q}^\top = \boldsymbol{M}_i$ for every $i = 1, ..., N$, is an isotropic function of its arguments (i.e. is invariant under $O(n)$).

(ii) Given an isotropic function $\widehat{W}$ *on Sym(n)* $\times M$*, the function* W *on Sym(n) defined by*

$$W(\boldsymbol{C}) := \widehat{W}(\boldsymbol{C}, \boldsymbol{M}_i), \tag{2}$$

is invariant under G.

The above result shows that an energy function invariant under G can be expressed as a function of the isotropic invariants of $(\boldsymbol{C}, \boldsymbol{M}_i)$. However the restricted symmetry group G of a structure is in general too small to describe its symmetry properties, since it does not allow for symmetry operations that interchange the structure tensors. Hence, we define the full symmetry group of a given structure by:

$$\begin{aligned} H =& \Big\{ \boldsymbol{R} \in O(n) \ / \ \boldsymbol{R}\boldsymbol{A}_i = (-1)^{\lambda(i)} \boldsymbol{A}_{\sigma(i)}, \ i = 1, \ldots, N, \\ & \qquad \text{for some } \sigma \in S_N, \lambda(i) \in \{0, 1\} \Big\} \\ =& \Big\{ \boldsymbol{R} \in O(n) \ / \ \boldsymbol{R}\boldsymbol{M}_i\boldsymbol{R}^\top = \boldsymbol{M}_{\sigma(i)}, \ i = 1, \ldots, N, \ \text{for some } \sigma \in S_N \Big\}, \end{aligned} \tag{3}$$

where σ is a permutation of the indices $i = 1, \ldots, N$. We denote by $H' \subset S_N$ the set of permutations defined by (3).

The following result generalizes Thm. 2.1.

Theorem 2.2. *(i) If* $W = W(\boldsymbol{C})$ *is invariant under* H*, then the extended energy function* $\widehat{W}$*, defined on Sym(n)* $\times M$ *by*

$$\widehat{W}(\boldsymbol{C}, \boldsymbol{P}_i) := W(\boldsymbol{Q}\boldsymbol{C}\boldsymbol{Q}^\top), \tag{4}$$

where $\boldsymbol{Q} \in O(n)$ *is such that* $\boldsymbol{Q}\boldsymbol{P}_i\boldsymbol{Q}^\top = \boldsymbol{M}_i$ *for every* $i = 1, ..., N$*, is an isotropic function of its arguments, and is invariant under* H'*, i.e.,*

$$\widehat{W}(\boldsymbol{C}, \boldsymbol{P}_{\sigma(i)}) = \widehat{W}(\boldsymbol{C}, \boldsymbol{P}_i), \qquad \text{for every } \sigma \in H'. \tag{5}$$

(ii) Given an isotropic function $\widehat{W}$ *on Sym(n)* $\times M$*, invariant under* H' *(in the sense of (5)), the function* W *on Sym(n) defined by*

$$W(\boldsymbol{C}) := \widehat{W}(\boldsymbol{C}, \boldsymbol{M}_i), \tag{6}$$

is invariant under H.

The above result shows that an energy function invariant under the full symmetry group H can be expressed as a function of the isotropic invariants of $(\boldsymbol{C}, \boldsymbol{M}_i)$ that is, in addition, invariant under permutations in H'.

An application of the above theorem leads to the following somewhat trivial result, formulated in terms of deformation energies depending on the deformation gradient. Consider an energy function of the form $W = W(\boldsymbol{F})$, with $\boldsymbol{F} \in \mathrm{GL}(\mathbb{R}^n)$, and the action of the symmetry group defined by $\boldsymbol{F} \mapsto \boldsymbol{F}\boldsymbol{R}$ for $\boldsymbol{R} \in H$.

Theorem 2.3. *Assume that $n = N$ and $(\boldsymbol{A}_i)$ a basis of $\mathbb{R}^n$, and let $W = W(\boldsymbol{F})$, $\boldsymbol{F} \in GL(\mathbb{R}^n)$, be an objective function (i.e., such that $W(\boldsymbol{F}) = W(\boldsymbol{Q}\boldsymbol{F})$ for every $\boldsymbol{Q} \in SO(n)$) invariant under H. Then W can be written in the form*

$$W = \widetilde{W}(\boldsymbol{a}_i \cdot \boldsymbol{a}_j) \tag{7}$$

where $\boldsymbol{a}_i = \boldsymbol{F}\boldsymbol{A}_i$ and $\widetilde{W}$ is invariant under the action of H on its arguments defined by

$$\boldsymbol{a}_i \mapsto (-1)^{\lambda(i)} \boldsymbol{a}_{\sigma(i)}. \tag{8}$$

It is straightforward to extend Thms. 2.2 and 2.3 to functions depending on higher-order tensors, and hence on higher derivatives of the deformation, after properly defining the action of the symmetry group H.

3. Four basic structures for surfaces formed by two families of yarns

In what follows we consider a surface S formed by two families of yarns. The yarns are here described by one-dimensional curves, meant to model the behavior of rods under the costumary simplifying assumptions that the cross sections remain orthogonal to the central line, do not change shape under deformation and are transversely isotropic relative to the yarn central axis.

We assume that the yarns are continuously distributed and those of each family have the same direction in the reference configuration. Let $S_0 \subset \mathbb{R}^2$ be a plane domain, and $\boldsymbol{r} : S_0 \to S \subset \mathbb{R}^3$ a parametrization of S. Denote by $\boldsymbol{A}_1$ and $\boldsymbol{A}_2$ the unit vectors associated to the yarn directions in the reference configuration S_0, and by $\boldsymbol{A}^1$ and $\boldsymbol{A}^2$ their dual (such that $\boldsymbol{A}_\alpha \cdot \boldsymbol{A}^\beta = \delta_\alpha^\beta$). The basic structures that characterize the symmetry properties of a network of two families of yarns are four, and are described below.

- In case the two families of yarns are mutually orthogonal in the reference configuration and have the same material properties (Fig. 1a), we refer to the structure as the **square structure**. The full symmetry group H_{sq}

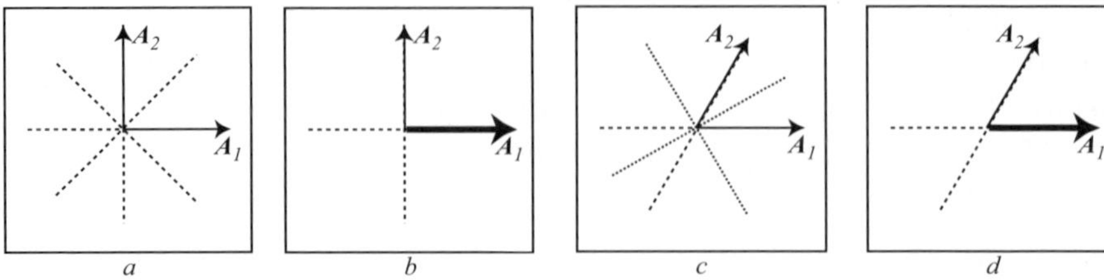

Fig. 1. The four basic structures of a weave pattern. a: square, b: rectangular, c: rhombic, d: parallelogram

describing the anisotropy of this material is generated by the orthogonal transformations of $\mathbb{R}^2$ whose matrix representation in the orthonormal basis $(\boldsymbol{A}_1, \boldsymbol{A}_2)$ is

$$\boldsymbol{R}_{\pi/2} = \begin{pmatrix} 0 & -1 \\ 1 & 0 \end{pmatrix}, \qquad \boldsymbol{R}_1 = \begin{pmatrix} -1 & 0 \\ 0 & 1 \end{pmatrix};$$

the set of permutations associated to the full symmetry group of the square structure is $H'_{sq} = S_2$.

- In case the two families of yarns are mutually orthogonal and have different material properties, and $\boldsymbol{A}_1$ and $\boldsymbol{A}_2$ are not interchangeable, we refer to the resulting structure as the **rectangular structure** (Fig. 1b). The full symmetry group describing the anisotropy of this material is generated by

$$\boldsymbol{R}_1 = \begin{pmatrix} -1 & 0 \\ 0 & 1 \end{pmatrix}, \qquad \boldsymbol{R}_2 = \begin{pmatrix} 1 & 0 \\ 0 & -1 \end{pmatrix},$$

where the matrix representation is taken in the orthonormal basis $(\boldsymbol{A}_1, \boldsymbol{A}_2)$. In this case, the yarns are not interchangeable, so that $H'_{rt} = \{(1)(2)\}$ reduces to the identity permutation.

- In case the two families of yarns, with the same material properties, are not orthogonal, we refer to the resulting structure as the **rhombic structure** (Fig. 1c). The symmetry group describing the anisotropy of this material is generated by

$$\boldsymbol{R}_3 = \begin{pmatrix} \cos\delta & \sin\delta \\ \sin\delta & -\cos\delta \end{pmatrix}, \qquad \boldsymbol{R}_4 = \begin{pmatrix} -\cos\delta & -\sin\delta \\ -\sin\delta & \cos\delta \end{pmatrix}$$

with δ the angle between the yarns, and the matrix representation is taken in the orthonormal basis $(\boldsymbol{A}_1, \boldsymbol{j})$, with $\boldsymbol{j}$ orthogonal to $\boldsymbol{A}_1$. Notice that $\boldsymbol{R}_3$ and $\boldsymbol{R}_4$ are the reflections about the lines bisecting the angle between $\boldsymbol{A}_1$ and $\boldsymbol{A}_2$, and the axis orthogonal to it.

In this case the yarns are interchangeable and from definition (3) it follows that the set of permutations associated to the full symmetry group of the rhombic structure is $H'_{rb} = S_2$.

- Finally, if the two families of yarns are not orthogonal and have different material properties, we have the **parallelogram structure** of Fig. 1d. The symmetry group describing the anisotropy of this material is

$$H_{pr} = \left\{ \boldsymbol{I} = \begin{pmatrix} 1 & 0 \\ 0 & 1 \end{pmatrix}, -\boldsymbol{I} = \begin{pmatrix} -1 & 0 \\ 0 & -1 \end{pmatrix} \right\}$$

and H'_{pr} reduces to the identity.

4. Application to a fibered surface with bending stiffness

We consider as before a surface made of two families of continuously distributed yarns. Consider a cartesian coordinate system (X^α), $\alpha = 1, 2$ in $\mathbb{R}^2$ associated to the basis $(\boldsymbol{A}_\alpha)$. In terms of these coordinates, we can write the parametrization as $\boldsymbol{r} = \boldsymbol{r}(X^\alpha)$, and denote by

$$\boldsymbol{a}_\alpha := \boldsymbol{r}_{,\alpha} = \partial \boldsymbol{r} / \partial X^\alpha \qquad \alpha = 1, 2,$$

the vectors associated to the yarn directions in the current configuration. The deformation gradient has therefore the representation

$$\boldsymbol{F} = \boldsymbol{a}_1 \otimes \boldsymbol{A}^1 + \boldsymbol{a}_2 \otimes \boldsymbol{A}^2 \quad \Leftrightarrow \quad \boldsymbol{F}\boldsymbol{A}_\alpha = \boldsymbol{a}_\alpha, \quad \alpha = 1, 2, \tag{9}$$

and the second gradient of the deformation is

$$\nabla \boldsymbol{F} = \boldsymbol{a}_{\alpha,\beta} \otimes \boldsymbol{A}^\alpha \otimes \boldsymbol{A}^\beta.$$

Using the above representations, we can easily extend the action of the material symmetry group to higher gradients of deformation: H acts on $\boldsymbol{F}$ and $\nabla \boldsymbol{F}$ as follows

$$\begin{cases} \boldsymbol{F} \mapsto \boldsymbol{F}\boldsymbol{R}^\top = \boldsymbol{a}_\alpha \otimes \boldsymbol{R}\boldsymbol{A}^\alpha \\ \nabla \boldsymbol{F} \mapsto \boldsymbol{a}_{\alpha,\beta} \otimes \boldsymbol{R}\boldsymbol{A}^\alpha \otimes \boldsymbol{R}\boldsymbol{A}^\beta, \end{cases} \qquad \boldsymbol{R} \in H, \tag{10}$$

while the group $SO(3)$, related to observer changes, acts according to the rule

$$\begin{cases} \boldsymbol{F} \mapsto \boldsymbol{Q}\boldsymbol{F} = \boldsymbol{Q}\boldsymbol{a}_\alpha \otimes \boldsymbol{A}^\alpha \\ \nabla \boldsymbol{F} \mapsto \boldsymbol{Q}\boldsymbol{a}_{\alpha,\beta} \otimes \boldsymbol{A}^\alpha \otimes \boldsymbol{A}^\beta, \end{cases} \qquad \boldsymbol{Q} \in SO(3). \tag{11}$$

We assume from now on that the yarns are inextensible, so that X^1 and X^2 are arc parameters for the curves of each family, and $\boldsymbol{a}_1$ and $\boldsymbol{a}_2$ are unit vectors. We denote by $(\boldsymbol{a}_1, \boldsymbol{n}_1, \boldsymbol{b}_1)$ and $(\boldsymbol{a}_2, \boldsymbol{n}_2, \boldsymbol{b}_2)$ the Frenet basis associated to the first and second family of yarns respectively. The Frenet equations hold

$$\begin{cases} \boldsymbol{a}_{\alpha,\alpha} = \kappa_\alpha \boldsymbol{n}_\alpha, \\ \boldsymbol{n}_{\alpha,\alpha} = -\kappa_\alpha \boldsymbol{a}_\alpha + \tau_\alpha \boldsymbol{b}_\alpha, \\ \boldsymbol{b}_{\alpha,\alpha} = -\tau_\alpha \boldsymbol{n}_\alpha, \end{cases} \qquad \alpha = 1, 2, \tag{12}$$

with κ_α and τ_α, $\alpha = 1, 2$ the curvature and torsion of the deformed yarns. Also, we denote by γ, defined by

$$\sin \gamma = \boldsymbol{a}_1 \cdot \boldsymbol{a}_2 \tag{13}$$

the angle of shear.

An immediate extension of Thm. 2.3 leads to the result that a function of the form

$$W = W(\boldsymbol{F}, \nabla \boldsymbol{F}) \tag{14}$$

that is objective, i.e., invariant under $SO(3)$ with action (11), and invariant under H with action (10), can be written in the form

$$W = \widetilde{W}(\boldsymbol{a}_\alpha \cdot \boldsymbol{a}_\beta, \boldsymbol{a}_\alpha \cdot \boldsymbol{a}_{\beta,\gamma}, \boldsymbol{a}_{\alpha,\beta} \cdot \boldsymbol{a}_{\gamma,\delta}) \tag{15}$$

which is invariant under the action of H on its arguments defined by

$$\begin{aligned} \boldsymbol{a}_\alpha &\mapsto (-1)^{\lambda(\alpha)} \boldsymbol{a}_{\sigma(\alpha)}, \\ \boldsymbol{a}_{\alpha,\beta} &\mapsto (-1)^{(\lambda(\alpha)+\lambda(\beta))} \boldsymbol{a}_{\sigma(\alpha),\sigma(\beta)}. \end{aligned} \tag{16}$$

We list below the invariants for the rectangular structure only. The lists of invariants and the action of the symmetry group for the other three structures can be found in Indelicato and Albano.[13] To compute the invariants, we have used the shareware computer package 'Singular' (cf. Greuel, Pfister and Schoenemann[8]).

$$
\begin{aligned}
L_1 &= (\boldsymbol{a}_1 \cdot \boldsymbol{a}_2)^2 && = \sin^2\gamma \\
L_2 &= (\boldsymbol{a}_1 \cdot \boldsymbol{a}_{2,2})^2 && = \kappa_2^2\,(\boldsymbol{a}_1 \cdot \boldsymbol{n}_2)^2 \\
L_3 &= (\boldsymbol{a}_{1,1} \cdot \boldsymbol{a}_2)^2 && = \kappa_1^2\,(\boldsymbol{n}_1 \cdot \boldsymbol{a}_2)^2 \\
L_4 &= \boldsymbol{a}_{1,1} \cdot \boldsymbol{a}_{1,1} && = \kappa_1^2 \\
L_5 &= \boldsymbol{a}_{2,2} \cdot \boldsymbol{a}_{2,2} && = \kappa_2^2 \\
L_6 &= \boldsymbol{a}_{1,1} \cdot \boldsymbol{a}_{2,2} && = \kappa_1\kappa_2\,(\boldsymbol{n}_1 \cdot \boldsymbol{n}_2) \\
L_7 &= \boldsymbol{a}_{1,2} \cdot \boldsymbol{a}_{1,2} && \\
L_8 &= (\boldsymbol{a}_{1,1} \cdot \boldsymbol{a}_{1,2})^2 && = \kappa_1^2\,(\boldsymbol{n}_1 \cdot \boldsymbol{a}_{1,2})^2 \\
L_9 &= (\boldsymbol{a}_{2,2} \cdot \boldsymbol{a}_{2,1})^2 && = \kappa_2^2\,(\boldsymbol{n}_2 \cdot \boldsymbol{a}_{2,1})^2 \\
L_{10} &= (\boldsymbol{a}_{2,2} \cdot \boldsymbol{a}_{2,1})\,(\boldsymbol{a}_{1,1} \cdot \boldsymbol{a}_{1,2}) && = \kappa_1\kappa_2\,(\boldsymbol{n}_2 \cdot \boldsymbol{a}_{1,2})\,(\boldsymbol{n}_1 \cdot \boldsymbol{a}_{1,2}) \\
L_{11} &= (\boldsymbol{a}_1 \cdot \boldsymbol{a}_2)\,(\boldsymbol{a}_{2,2} \cdot \boldsymbol{a}_{2,1}) && = \sin\gamma\,(\kappa_2\boldsymbol{n}_2 \cdot \boldsymbol{a}_{2,1}) \\
L_{12} &= (\boldsymbol{a}_1 \cdot \boldsymbol{a}_2)\,(\boldsymbol{a}_{1,1} \cdot \boldsymbol{a}_{2,1}) && = \sin\gamma\,(\kappa_1\boldsymbol{n}_1 \cdot \boldsymbol{a}_{2,1}) \\
L_{13} &= (\boldsymbol{a}_1 \cdot \boldsymbol{a}_{2,2})\,(\boldsymbol{a}_2 \cdot \boldsymbol{a}_{1,1})\,(\boldsymbol{a}_{2,2} \cdot \boldsymbol{a}_{2,1}) && = \kappa_1\kappa_2^2(\boldsymbol{a}_1 \cdot \boldsymbol{n}_2)(\boldsymbol{a}_2 \cdot \boldsymbol{n}_1)\,(\boldsymbol{n}_2 \cdot \boldsymbol{a}_{2,1}) \\
L_{14} &= (\boldsymbol{a}_1 \cdot \boldsymbol{a}_{2,2})\,(\boldsymbol{a}_2 \cdot \boldsymbol{a}_{1,1})\,(\boldsymbol{a}_{1,1} \cdot \boldsymbol{a}_{2,1}) && = \kappa_1^2\kappa_2(\boldsymbol{a}_1 \cdot \boldsymbol{n}_2)(\boldsymbol{a}_2 \cdot \boldsymbol{n}_1)(\boldsymbol{n}_1 \cdot \boldsymbol{a}_{1,2}) \\
L_{15} &= (\boldsymbol{a}_1 \cdot \boldsymbol{a}_2)\,(\boldsymbol{a}_1 \cdot \boldsymbol{a}_{2,2})\,(\boldsymbol{a}_2 \cdot \boldsymbol{a}_{1,1}) && = \kappa_1\kappa_2(\sin\gamma)(\boldsymbol{a}_1 \cdot \boldsymbol{n}_2)(\boldsymbol{a}_2 \cdot \boldsymbol{n}_1).
\end{aligned}
\tag{17}
$$

5. Discussion

We have derived in the preceding section the integrity bases for the action of the material symmetry group of a network of inextensible yarns[a]. This result allows to characterize all polynomial energies of such networks that depend on the curvature of the fibers. Specifically, notice that the listed invariants depend on the shear angle γ, the curvature of the yarns, the angle between the unit normal vectors of intersecting yarns, and the derivative $\boldsymbol{a}_{1,2} = \boldsymbol{a}_{2,1}$.

In constitutive theories for fabrics it seems reasonable to assume that the energy is independent of $\boldsymbol{n}_1 \cdot \boldsymbol{n}_2$ and $\boldsymbol{a}_{1,2}$, and only depends on the curvature of the yarns, referred to as yarns in the literature on textiles (cf. Luo,[14] Wang and Sun,[20] Hearle, Potluri and Thammandra,[15] Potluri and Thammandra[16]), even though experimental data suggest that inter-yarn forces have a very significant effect on the fabric bending rigidity (cf. Grosberg,[9] De Jong and Postle[6]).

Indeed, $\boldsymbol{n}_1 \cdot \boldsymbol{n}_2$ coincides with $\boldsymbol{a}_1 \cdot \boldsymbol{a}_2$ for plane deformations, in which case it is just a measure of shear, but otherwise it measures the angle between

[a]An integrity basis for a group action is a set of polynomials invariant under the group action, such that any other invariant polynomial is expressible as a polynomial in the elements of the integrity basis.

the osculating planes of yarns of different families, whose influence on the deformation of the fabric might be reasonably neglected. Also, it can be shown that $\boldsymbol{a}_{1,2}$ vanishes for plane deformations, and may be interpreted as a measure of the spatial spread of the first family of yarns for non-planar deformations.

Repeating the analysis of the previous section for the square, rhombic and parallelogram structures we may conclude that the basic polynomial invariants involving only the curvatures and shear are

(-) for the square symmetry,

$$\sin^2\gamma, \quad \kappa_1^2\kappa_2^2, \quad \kappa_1^2+\kappa_2^2. \tag{18}$$

(-) for the rectangular symmetry,

$$\sin^2\gamma, \quad \kappa_1^2, \quad \kappa_2^2. \tag{19}$$

(-) for the rhombic symmetry,

$$\sin\gamma, \quad \kappa_1^2\kappa_2^2, \quad \kappa_1^2+\kappa_2^2. \tag{20}$$

(-) for the parallelogram symmetry,

$$\sin\gamma, \quad \kappa_1^2, \quad \kappa_2^2. \tag{21}$$

Hence, for rhombic and parallelogram symmetries a shear of angle γ does not have the same effect as the shear of angle $-\gamma$, contrary to what happens for the square and rectangular structures.

We remark that the above results are an essential consequence of the extended version 2.2 of the classical Rychlewski theorem. Indeed, the restricted symmetry group of the square and rectangular structures is the same, and the same holds for the rhombic and parallelogram structures. Hence, these structures cannot be distinguished from each other by the classical Rychlewski theorem.

Acknowledgements

This work was supported by the grant 'Modelli Matematici per la Scienza dei Materiali' (PRIN 2005) of the Italian MIUR.

References

1. Y. Aimène, *Approche hyperèlastique pour la simulation des renforts fibreux en grandes transformations*, (Thése de doctorat, Institut National des Sciences Appliques de Lyon, 2007).
2. N. Aksel and M. Itskov, A class of orthotropic and transversely isotropic hyperelastic constitutive models based on a polyconvex strain energy function, *I. J. Sol. Str.* **41**, pp. 3833–3848 (2004).
3. J. P. Boehler, Lois de comportement anisotrope del milieux continus, *Journal de Mécanique* **17**, pp. 153–170 (1978).
4. J. P. Boehler, A simple derivation of representation for non-polynomial constitutive equations in some cases of anisotropy, *Z. Angew. Math. Mech.* **59**, pp. 157–167 (1979).
5. R. M. Christensen, *Mechanics of composite materials*, (Wiley, New York, 1979).
6. S. de Jong and R. Postle, An energy analyis of woven-fabric mechanics by means of optimal control theory. Part II: pure bending properties, *J. Text. Inst.* **68**, pp. 362–369 (1977).
7. T. C. Gasser, R. W. Ogden and G. A. Holzapfel, Hyperelastic modelling of arterial layers with distributed collagen fibre orientations, *J. Royal Soc. Lond. Interface* **3**, pp. 15–35 (2006).
8. G. M. Greuel, G. Pfister and H. Schoenemann, *Singular 3.0.3, a computer algebra system for polynomial computations*, Center for Computer Algebra, University of Kaiserslautern. (2007) Available at: http://www.singular.uni-kl.de.
9. P. Grosberg, The mechanical properties of woven fabric. Part II: the bending of woven fabrics, *Text. Res. J.* **36**, pp. 205–211 (1966).
10. S. Hartmann and P. Neff, Polyconvexity of generalized polynomial-type hyperelastic strain energy functions for near-incompressibility, *I. J. Sol. Str.* **40**, pp. 2768–2791 (2003).
11. G. A. Holzapfel, *Nonlinear solid mechanics, A continuum approach for engineering*, (John Wiley & Sons Ltd., Chichester, 2000).
12. G. Indelicato, The influence of the twist of individual fibers in 2D fibered networks, *I. J. Sol. Str.* **46**, pp. 912–922 (2009).
13. G. Indelicato and A. Albano, Symmetry Properties of the Elastic Energy of a Woven Fabric with Bending and Twisting Resistance *J. Elast.* **94**(1), pp. 33–54 (2009).
14. C. Luo, Determination of Loads in an Inextensible Network According to Geometry of Its Wrinkles, *J. Appl. Mech.* **71**, pp. 298–300 (2004).
15. J. W. S. Hearle, P. Potluri and V. S. Thammandra, Modelling fabric mechanics, *J. Text. Inst.* **92**(3), pp. 53–69 (2001).
16. P. Potluri and V. S. Thammandra, Influence of uniaxial and biaxial tension on meso-scale geometry and strain fields in a woven composite, *Composite Structures* **77**(3), pp. 405–418 (2007).
17. J. Schröder and P. Neff, Invariant formulation of hyperelastic transverse isotropy based on polyconvex free energy functions, *I. J. Sol. Str.* **40**, pp. 401–445 (2003).

18. A. J. M. Spencer, Theory of invariants, *Continuum Phys.* **1**, pp. 239–353 (1971).
19. W. B. Wang and A. C. Pipkin, Inextensible networks with bending stiffness, *Quart. J. Mech. Appl. Math.* **39**(3), pp. 343–359 (1986).
20. Y. Wang and X. Sun, Digital-element simulation of textile processes, *J. Comp. Sci. Tech.* **61**, pp. 311–319 (2001).
21. J.M. Zhang and J. Rychlewski, Structural tensors for anisotropic solids, *Arch. Mech.* **42** pp. 267–277 (1990).

KINETIC TREATMENT OF CHARGE CARRIER AND PHONON TRANSPORT IN GRAPHENE

P. LICHTENBERGER and F. SCHÜERRER

Institute of Theoretical and Computational Physics, Graz University of Technology, Petersgasse 16, 8010 Graz, Austria
E-mail: lichtenb@itp.tugraz.at, ferdinand.schuerrer@tugraz.at
www.itp.tugraz.at/AG/TP/

S. PISCANEC and A. C. FERRARI

Cambridge University, Engineering Department, Trumpington Street, Cambridge CB2 1PZ, United Kingdom
E-mail: sp344@cam.ac.uk, acf26@hermes.cam.ac.uk

We present a kinetic transport model describing the transport of charge carriers and optical phonons in graphene. The Boltzmann transport equations are solved numerically by a set of deterministic methods. Our simulations provide a deep insight into the coupled dynamics of electrons and phonons including features unique to graphene. We demonstrate the importance of such effects when turning to the high field transport regime.

Keywords: Boltzmann equation, graphene, kinetic transport, hot phonons, deterministic methods

1. Introduction

In the ongoing search for materials and technologies for post-silicon nano-electronics carbon based electronics holds a prominent position. Carbon nanotubes (CNTs) have been regarded as the mainstay of future carbon based nano-electronics for a long time. This position has been challenged by graphene in the recent years. Graphene, a single hexagonal layer of carbon atoms, was long seen as a useful theoretical construct for the description of graphite and later on for CNTs, but thought not to exist in a free form. This view had to be dramatically changed after the first successful preparation of single free graphene sheets in 2004 by Geim's group.[1]

Since its introduction, graphene has demonstrated remarkable electric properties and has been experimentally investigated by virtually all methods available. Its two-dimensional, planar structure facilitates the integration of graphene into classical semiconductor technology and numerous experimental devices have been fabricated, including suspended and gated graphene.

For the charge carrier transport, the main difference to classical semiconductor materials is the radically different band structure of graphene. The concept of an effective mass must be replaced by a model of massless Dirac-fermions with an effective speed of light. Additionally, the valence and the conduction band meet at the charge neutrality level called the Dirac points. This leads to the descriptions of graphene as zero gap semiconductor or semimetal.

Right from the start, measured mobilities seemed to be very promising and with the constant improvement in sample preparation, values higher than 2×10^5 cm^2/Vs have been recently reported at room temperature.[2,3] This value significantly exceeds that of conventional semiconductors and other low-dimensional systems.

In very high quality samples, phonons become the major source of resistivity at room temperature. An accurate understanding of the fundamental transport processes is essential for possible future applications of graphene in electronic devices. In single walled CNTs, a strong impact of non-equilibrium phonon occupation numbers on the charge carrier transport was found.[4] This effect was modeled theoretically and numerically in detail by kinetic transport equations for dynamic charge carrier and optical phonon distribution functions.[5] Due to this experience we were motivated to develop a kinetic model for investigating the dynamics of electrons and phonons in graphene.

In this paper, we introduce the relevant kinetic equations governing the charge carrier transport. We demonstrate how to obtain an efficient and accurate numerical scheme to simulate graphene devices. Our numerical model is of general form, allowing for the study of a wide range of transport setups. Here, we will use it to present some fundamental investigation of the high-field transport in graphene. It turns out that special properties of graphene lead to very interesting unexpected effects.

2. Kinetic Transport Model

Corresponding to the lattice of carbon atoms in real space, the reciprocal lattice of graphene is also hexagonal. The band structure including the highest valence and the lowest conduction band is shown in the left panel of Fig. 1. Around the Dirac points at the six corners of the first Brillouin

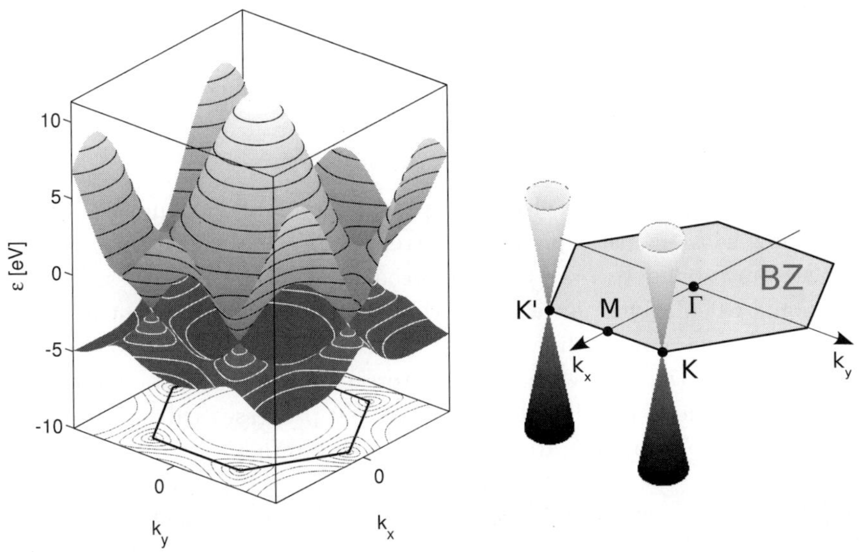

Fig. 1. Left: Full dispersion relation for the highest valence and the lowest conduction band of graphene. Right: Geometry of the Brillouin zone and position of the double-cones approximating the band structure around the Dirac points K and K′.

zone, the band structure can be linearly approximated by two cone-shaped surfaces. The corresponding dispersion relation reads

$$\varepsilon_\alpha(\boldsymbol{k}) = \pm\hbar v_F |\boldsymbol{k}| \tag{1}$$

with the Fermi velocity v_F and quasi momentum vectors $\boldsymbol{k}$ measured relative to the Dirac point. This is the graphene equivalent to the effective mass approximation in classical semiconductors.[6] Due to symmetry and periodicity of the reciprocal lattice, the six valleys at the corners of the Brillouin zone reduce to two double cones at the points K and K′, which can be treated as one effective valley.

In the context of kinetic theory, the state of the electronic system is determined by distribution functions $f_\alpha = f_\alpha(\boldsymbol{k}, \boldsymbol{x}, t)$. The index α distinguishes the conduction ($\alpha = 1$) from the valence band ($\alpha = 2$). The time evolution of these functions is governed by the Boltzmann transport equation (BTE)

$$\partial_t f_\alpha + \boldsymbol{v}_\alpha\, \partial_{\boldsymbol{x}} f_\alpha - \frac{e_0}{\hbar}\boldsymbol{E}\, \partial_{\boldsymbol{k}} f_\alpha = \mathcal{C}_\alpha \tag{2}$$

with the group velocity $\boldsymbol{v}_\alpha(\boldsymbol{k}) = \partial_{\boldsymbol{k}}\varepsilon_\alpha/\hbar = \pm v_F \boldsymbol{k}/|\boldsymbol{k}|$, the electric field strength $\boldsymbol{E}$ and the elementary charge e_0. The vector $\boldsymbol{x}$ denotes the position and $\boldsymbol{k}$ stands for the wave vector. The collision operator

$$\mathcal{C}_\alpha = \mathcal{C}_\alpha^{ac.} + \sum_{\beta=1}^{2}\sum_{\eta=1}^{3} \mathcal{C}_{\alpha\beta}^{\eta} \tag{3}$$

accounts for scattering of electrons with different phonon modes. Elastic scattering with acoustic phonons is treated by $\mathcal{C}_\alpha^{ac.}$ and $\mathcal{C}_{\alpha\beta}^{\eta}$ represents inelastic intraband ($\alpha = \beta$) and interband ($\alpha \neq \beta$) processes. Density functional calculations[7] have shown that there are three relevant modes of optical phonons: longitudinal optical (LO, $\eta = 1$) and transversal optical (TO) phonons ($\eta = 2$) with wave vectors close to the Γ-point and zone boundary phonons responsible for intervalley processes called K-phonons ($\eta = 3$). The collision operator modeling such interactions reads

$$\begin{aligned}\mathcal{C}_{\alpha\beta}^{\eta}(\boldsymbol{k}) = \int_{\mathcal{B}} d\boldsymbol{k}' \Big\{ & w_{\alpha\beta,\eta}^{+}(\boldsymbol{k},\boldsymbol{k}',\boldsymbol{q}^{+})\big[(g_\eta^{+}+1)f_\beta'(1-f_\alpha) - g_\eta^{+}(1-f_\beta')f_\alpha\big] \\ & + w_{\nu\mu,\eta}^{-}(\boldsymbol{k},\boldsymbol{k}',\boldsymbol{q}^{-})\big[g_\eta^{-} f_\beta'(1-f_\alpha) - (g_\beta^{-}+1)(1-f_\beta')f_\alpha\big]\Big\}.\end{aligned} \tag{4}$$

We use the abbreviations $f_\alpha = f_\alpha(\boldsymbol{k},\boldsymbol{x},t)$, $f_\beta' = f_\beta(\boldsymbol{k}',\boldsymbol{x},t)$ and for the optical phonon distribution functions $g_\eta^{\pm} = g_\eta(\boldsymbol{q}^{\pm},\boldsymbol{x},t)$ with $\boldsymbol{q}^{\pm} = \pm(\boldsymbol{k}'-\boldsymbol{k})$. The scattering rates $w_{\alpha\beta,\eta}^{\pm}$ for optical phonons are given by

$$w_{\alpha\beta,\eta}^{\pm}(\boldsymbol{k},\boldsymbol{k}',\boldsymbol{q}) = \frac{|D_{\alpha\beta}^{\eta}(\boldsymbol{k},\boldsymbol{k}')|^2}{8\pi\rho\omega_\eta}\,\delta[\varepsilon_\alpha(\boldsymbol{k}) - \varepsilon_\beta(\boldsymbol{k}') \pm \hbar\omega_\eta] \tag{5}$$

with phonon wave vector $\boldsymbol{q}$, the optical phonon energy $\hbar\omega_\eta$, the area density ρ and the Dirac delta function δ stemming from Fermi's golden rule. The electron–phonon couplings $D_{\alpha\beta}^{\eta}$ for the Γ–LO, Γ–TO, and K-modes read[8]

$$|D_{\alpha\beta}^{\mathrm{LO/TO}}(\boldsymbol{k},\boldsymbol{k}')|^2 = D_\Gamma^2\left[1 \pm \cos(\theta+\theta')\right], \tag{6}$$

$$|D_{\alpha\beta}^{\mathrm{K}}(\boldsymbol{k},\boldsymbol{k}')|^2 = D_{\boldsymbol{K}}^2\left(1 \pm \cos\theta''\right), \tag{7}$$

with $\theta = \angle(\boldsymbol{k},\boldsymbol{k}'-\boldsymbol{k})$, $\theta' = \angle(\boldsymbol{k}',\boldsymbol{k}'-\boldsymbol{k})$ and $\theta'' = \angle(\boldsymbol{k},\boldsymbol{k}')$. In the case of LO- and K-phonons, the plus sign in Eqs. (6) and (7) refers to interband processes and for TO-phonons to intraband processes and vice versa for the minus sign. The scattering with equilibrium acoustic phonons is modeled quasi elastically by

$$\mathcal{C}_\alpha^{ac.}(\boldsymbol{k}) = \int_{\mathcal{B}} d\boldsymbol{k}'\, w_\alpha^{ac.}(\boldsymbol{k},\boldsymbol{k}')(f_\alpha - f_\alpha') \tag{8}$$

with the scattering rate

$$w_{\alpha}^{ac.}(\boldsymbol{k},\boldsymbol{k}') = \frac{D_{ac.}^2 k_B T}{4\pi\rho\, c_{ac.}} \left(1+\cos\theta''\right)\, \delta\big[\varepsilon_\alpha(\boldsymbol{k}) - \varepsilon_\alpha(\boldsymbol{k}')\big] \tag{9}$$

including the acoustic deformation potential $D_{ac.}$, the sound velocity $c_{ac.}$, Boltzmann's constant k_B and the environmental temperature T.

The dynamics of the optical phonons is determined by the evolution equation

$$\partial_t g_\eta = \mathcal{D}_\eta. \tag{10}$$

The collision operator

$$\mathcal{D}_\eta = \mathcal{D}_\eta^{pp} + \sum_{\alpha=1}^{2}\sum_{\beta=1}^{2} \mathcal{D}_\eta^{\alpha\beta} \tag{11}$$

describes the decay of optical phonons into acoustic ones by $\mathcal{D}_\eta^{pp}$ and the interaction with electrons by $\mathcal{D}_\eta^{\alpha\beta}$. The latter is given by

$$\mathcal{D}_\eta^{\alpha\beta}(\boldsymbol{q}) = 4\int_{\mathcal{B}} d\boldsymbol{k}\; w_{\alpha\beta,\eta}^{+}(\boldsymbol{k},\boldsymbol{k}^{+},\boldsymbol{q})\big[(g_\eta+1) f_\beta^{+}(1-f_\alpha) - g_\eta(1-f_\beta^{+}) f_\alpha\big] \tag{12}$$

with $g_\eta = g_\eta(\boldsymbol{q},\boldsymbol{x},t)$, $f_\alpha = f_\alpha(\boldsymbol{k},\boldsymbol{x},t)$, $f_\beta^{+} = f_\beta(\boldsymbol{k}^{+},\boldsymbol{x},t)$ and $\boldsymbol{k}^{+} = \boldsymbol{k}+\boldsymbol{q}$. It is the direct counterpart to (4). The factor 4 accounts for spin and valley degeneracy. The decay of optical phonons due to phonon-phonon processes is modeled by the relaxation time approximation

$$\mathcal{D}_\eta^{pp}(\boldsymbol{q}) = -\frac{1}{\tau_\eta}\big[g_\eta(\boldsymbol{q},\boldsymbol{x},t) - g_\eta^{0}(T)\big] \tag{13}$$

with the optical phonon lifetimes τ_η and the Bose-Einstein distribution $g_\eta^0(T) = [\exp(\hbar\omega_\eta/k_B T) - 1]^{-1}$ for the same environmental temperature T as in Eq. (8).

3. Numerical Approach

To solve the transport Eqs. (2) and (10) numerically our method is based on a complete discretization of the phase space. It was our goal to carry over essential features of the BTEs to discrete model equations, namely the conservative properties of charge carrier density, momentum and energy as well as the equilibrium behavior. To facilitate this endeavor we move to a polar, energy based coordinate system

$$\boldsymbol{k}(\varepsilon,\varphi) = \frac{\varepsilon}{\hbar v_F}(\cos\varphi, \sin\varphi) \tag{14}$$

for the momenta of electrons and ordinary polar coordinates for the optical phonon momenta. The choice of energy based coordinates is motivated by the constant energy model for optical phonons and leads to a straight forward analytical evaluation of the Dirac delta function in the collision operator.

Our approach is centered around a fixed uniform discretization of the phase-space variables ε, q, φ and x. The discretization length Δ_ε of the energy variable is chosen in such a way that $\hbar\omega_\eta \approx \sigma_\eta \Delta_\varepsilon$ with $\sigma_\eta \in \mathbb{N}^+$. Based on this assumption, the energy grid is defined by $\varepsilon_n = (n - 1/2 - N_\varepsilon)\Delta_\varepsilon$ for $n = 0, \ldots, 2N_\varepsilon$. Hence, the energy domain is $[-N_\varepsilon\Delta_\varepsilon, N_\varepsilon\Delta_\varepsilon]$. The modulus of the optical phonon wave vectors $q = |\boldsymbol{q}|$ is discretized by $q_n = (n - 1/2)\Delta_q$ for $n = 1, \ldots, N_q$ and the angular variable φ by $\varphi_n = (n - 1/2)2\pi/N_\varphi$ for $n = 1, \ldots, N_\varphi$. For convenience, we use the same angular discretization for electrons and optical phonons and only one common set of discrete values q_n for all optical phonon modes.

The polar coordinates (14), together with the step size Δ_ε adjusted to the optical phonon energies, are convenient for the numerical integration of the collision integral. We calculate the remaining integral in (4), after having taken into account the Dirac delta function, by a midpoint rule over discrete angular values. The distribution function for charge carriers needs to be evaluated only at grid points, but this can not be assured for phonon distribution functions $g_\eta(\boldsymbol{q})$, which necessitates an interpolation of surrounding values on the grid to ensure the conservation of energy and quasi momentum.

The change of coordinates equally affects the left-hand side of the transport Eq. (2) and leads to derivatives with respect to ε, φ and x. Due to the type of a hyperbolic conservation law, we construct a conservative approximation of derivatives by applying a fifth-order weighted essentially non-oscillatory (WENO) finite-difference scheme.[9–11]

These discrete approximations of the advective terms on the left hand side and the collision terms on the right hand side of the BTEs (2) and (10) lead to a set of coupled first-order ordinary differential equations. This system describes the time evolution of the discretized distribution function for charge carriers and optical phonons. The integration of this system is performed by an explicit third-order Runge-Kutte scheme with total variation diminishing (TVD) properties.[12]

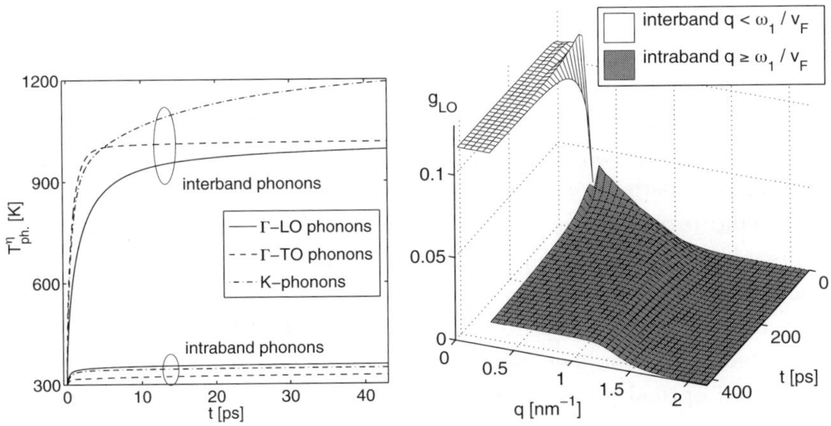

Fig. 2. Left: Temperatures $T^{\eta}_{ph.}$ of the optical phonon modes separated for the inter- and intraband subtypes. Right: Time evolution of the isotropic distribution function $g_{\rm LO} = g_1(q,t)$ for Γ LO-phonons.

4. Results

The following study of the electron and phonon dynamics in graphene is based on our model equations. The simulations focus on the special properties of graphene and how they influence the charge carrier transport.

In our simulation we assume a Fermi velocity of $v_F = 1000$ km/s and an area density $\rho = 7.63 \times 10^{-11}$ kg/cm^2. The optical phonons are modeled according to Ref. 8 and the acoustic ones as presented in Ref. 13. The electron–phonon coupling constants are $|D^{\Gamma}|^2 = 45.60$ (eV/Å)^2 and $|D^{\rm K}|^2 = 92.05$ (eV/Å)^2, respectively. The associated optical phonon energies are $\hbar\omega_{\rm LO/TO} = 196$ meV and $\hbar\omega_{\rm K} = 161$ meV. These values are rounded to 200 meV and 160 meV to be compatible with the energy discretization step size of $\Delta_\varepsilon = 20$ meV. We assume an optical phonon life times $\tau_\eta = 3.5$ ps for all phonon modes.[14] The acoustic deformation potential $D_{ac.} = 16$ eV together with the sound velocity $c_{ac.} = 20$ km/s.

Numerically we use, if not explicitly stated otherwise, an energy discretization of $\Delta_\varepsilon = 20$ meV and an angular resolution of $\Delta_\varphi = \pi/8$ ($N_\varphi = 16$). We use the same angular grid to discretize the optical phonon wave vectors. The discrete set of moduli is automatically obtained from the conditions $\Delta_q \leq 2\Delta_\varepsilon/\hbar v_f$ and $N_q\Delta_q \geq 2N_\varepsilon\Delta_\varepsilon/\hbar v_f$, ensuring that all occurring wave vectors can be handled with sufficient resolution.

First, we consider the collision operators (4) and (12) which are responsible for the interaction between electrons and optical phonons, forming the focus of our interest. Examining the possible collision geometries between linear dispersive electrons and constant energy phonons reveals an important fact: Optical phonons with $q < \omega_\eta / v_F$ can only participate in interband processes ($\alpha \neq \beta$) and phonons with $q \geq \omega_\eta / v_F$ are restricted to intraband processes ($\alpha = \beta$). This means that every phonon mode decomposes into two subtypes, consequently, called interband and intraband phonons.

We use an isotropic setup with hot electrons ($T_{el.} = 1000$ K), with $\varepsilon_F = 0$ meV and room temperature phonons ($T_{ph.} = 300$ K). We exclude the decay of optical phonons. Hence, the total energy of electrons and optical phonons is constant. Of course, when allowing for optical phonon decay we would end up at a common predetermined lattice temperature. We can expect that during the relaxation towards an equilibrium, the optical phonon system gains energy at the expense of the energy of the electron system. The former can clearly be seen in the left panel of Fig. 2. The temperatures are obtained from averaged occupation numbers $\langle g \rangle$ via

$$T^{\eta}_{ph.}(\langle g \rangle) = \frac{\hbar \omega_\eta}{k_B} \frac{1}{\log(\langle g \rangle^{-1} - 1)}. \tag{15}$$

The most striking feature is that the temperature of inter- and intraband phonons do not converge to the same value as expected. This clearly indicates how strongly these two subtypes of phonons are decoupled. They are not able to reach a common Bose–Einstein distribution, although they are indirectly coupled via the electron system. This inability to fully relax is due to an additional constrain. Any change of the number and therefore the energy and temperature of interband phonons must be linked to a transfer of electrons between the bands. Cooling the electron system via phonon interaction means transferring electrons to lower energy states by intraband and interband processes. The number of interband phonons is determined by the number of these processes. We further illustrate the decoupling of optical phonons within one mode in the right panel of Fig. 2 by plotting the time evolution of the distribution function for the LO-phonons. Decoupling of the two subtypes of phonons is evident. The interband as well as the intraband phonons for lower wave numbers reach a constant distribution forming a Bose–Einstein distribution. In the long time limit the intraband phonons would reach a constant distribution over the whole domain, but phonons with high wave numbers need to be scattered by electrons with high energies. Thus, the relaxation rate drops exponentially with higher

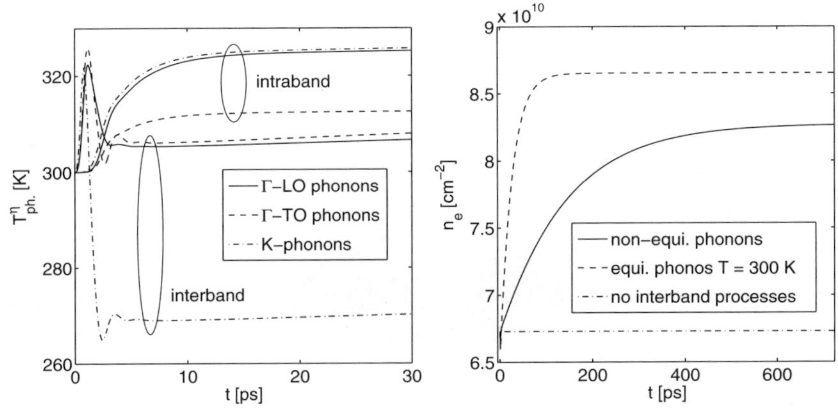

Fig. 3. Left: Time evolution of the optical phonon temperatures $T^{\eta}_{ph.}$. Right: Evolution of electron densities n_e towards a steady state depending on the modeling of the interband processes.

wave numbers like the electron occupation decreases with higher energies.

The isotropic, field-free setup clearly demonstrates the special behavior of graphene regarding the interaction of the electron and the phonon system. Nevertheless, it is a rather artificial setup, especially when excluding the decay of optical phonons. To study a more real world situation, we apply an electric field and include the decay of optical phonons. Again, we assume intrinsic graphene at a system temperature of 300 K. The field applied is 1 kV/cm, which can be clearly considered as a high field regime.

In the steady state, the energy transferred by the electric field to the electrons must leave the system via optical phonons, as this is the only energy drain in our setup. The time evolution of the optical phonon temperatures for the different modes and subtypes are shown in the left panel of Fig. 3. After a general initial reaction towards higher temperatures, the interband phonons start losing energy and cool down. In the case of the interband K-phonons this is rather dramatic, because they reach a steady state temperature below the environmental temperature of the acoustic phonon system that acts as a heat bath in our setup. Though unexpected, this behavior can be explained in the context of our previous findings. The energy fed to the electron system leads to the occupation of higher energy states, directly through drift and indirectly by heated intraband phonons. At the same time the collision operators try to bring the system to a new equilibrium while accommodating the additional energy provided by the

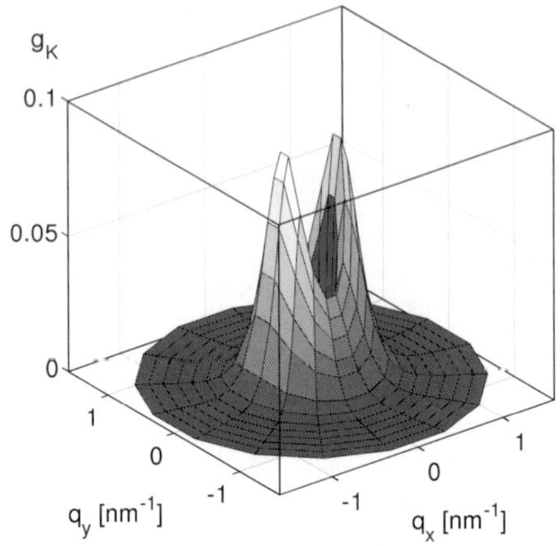

Fig. 4. Steady state distribution function $g_{\mathrm{K}} = g_3(q_x, q_y, t \to \infty)$ of the K-phonons. The electric field acts along the x-axis.

electric field. This includes interband processes moving electrons into the conduction band. As previously stated for the opposite case, this is directly linked to the occupation numbers of the interband phonons. The number of electrons is, of course, a crucial parameter for the electronic properties of graphene. Hence, a proper modeling of interband processes is essential. We further illustrate this in the right panle of Fig. 3. Assuming fixed equilibrium phonon distribution functions at 300 K leads to a significant overestimation of steady state carrier and current densities. In the given example, we obtain a current density of 0.14 μA/nm instead of 0.12 μA/nm and a mobility of 1.0×10^5 cm^2/Vs instead of 9.1×10^4 cm^2/Vs. When excluding interband effects, these values would be 0.10 μA/nm and 9.3×10^4 cm^2/Vs, respectively. In the right panel of Fig. 3 it can be seen that the interband processes act on a rather long time scale compared to the initial reaction shown in the left panel.

Turning to the intraband phonons, we observe the expected increase in temperature to a significantly higher, but not extreme level. In Fig. 4 the steady state distributions for the intraband part of the K-phonons is depicted. The strong deviation ($\max_{\boldsymbol{q}} g = 8.9 \times 10^{-2}$) from the equilibrium ($g_3^0(300K) = 2.5 \times 10^{-3}$) clearly shows the occurrence of non-equilibrium phonon distributions, that are not well described by averaged quantities. Though this maximum corresponds to a temperature of 720 K, it is still in the regime of $g \ll (g+1)$. Therefore, non-equilibrium effects stemming from intraband phonons are not very pronounced in this case. However, these effects become important for higher field strength or when using doped graphene with higher carrier concentrations. The shape of the phonon distribution already reveals its complex structure that must be handled by the presented method.

5. Conclusion

We present a coupled system of Boltzmann transport equations that model the non-equilibrium dynamics of electrons and optical phonons in graphene. The particularity of the electronic band structure of graphene gives raise to a decoupling of optical phonons. Our numerical studies clearly show that this effect must be treated carefully, when attempting to accurately simulate the charge carrier transport in graphene. We used a simplified, isotropic structure to demonstrate that our approach copes with this decoupling, and finally, we turned to a high field case, which is representative for a possible experimental setup, to confirm the practical applicability of our method.

Acknowledgments

This work has been performed within the NAWI Graz GASS project *Advanced Characterization of Nanostructured Materials*.

References

1. K. S. Novoselov, A. K. Geim, S. V. Morozov, D. Jiang, Y. Zhang, S. V. Dubonos, I. V. Grigorieva and A. A. Firsov, *Science* **306**, 666 (2004).
2. S. V. Morozov, K. S. Novoselov, M. I. Katsnelson, F. Schedin, D. C. Elias, J. A. Jaszczak and A. K. Geim, *Phys. Rev. Lett.* **100**, p. 016602 (2008).
3. K. I. Bolotin, K. J. Sikes, Z. Jiang, M. Klima, G. Fudenberg, J. Hone, P. Kim and H. L. Stormer, *Solid State Commun.* **146**, 351 (2008).
4. C. Auer, F. Schürrer and C. Ertler, *Phys. Rev. B* **74**, p. 165409 (2006).
5. C. Auer, F. Schürrer and C. Ertler, Deterministic solution of Boltzmann equations governing the dynamics of electrons and phonons in carbon nanotubes, in *Applied and Industrial Mathematics in Italy II*, eds. V. Cutello, G. Fotia and L. Puccio (World Scientific, Singapore, 2007).
6. A. H. Castro Neto, F. Guinea, N. M. Peres, K. S. Novoselov and A. K. Geim, *Rev. Mod. Phys.* **81**, 109 (2009).
7. S. Piscanec, M. Lazzeri, J. Robertson, A. C. Ferrari and F. Mauri, *Phys. Rev. B* **75**, p. 035427 (2007).

8. M. Lazzeri, S. Piscanec, F. Mauri, A. C. Ferrari and J. Robertson, *Phys. Rev. Lett.* **95**, p. 236802 (2005).
9. G. Jiang and C.-W. Shu, *J. Comput. Phys.* **126**, 202 (1996).
10. J. A. Carrillo, I. M. Gamba, A. Majorana and C.-W. Shu, *J. Comput. Phys.* **184**, 498 (2003).
11. M. Galler, A. Majorana and F. Schürrer, A multigroup-WENO solver for the non-stationary Boltzmann-Poisson system for semiconductor devices, in *Scientific Computing in Electrical Engineering*, eds. A. M. Anile, G. Ali and G. Mascali (Springer Verlag, Berlin, 2006).
12. S. Gottlieb and C.-W. Shu, *Math. Comp.* **67**, 73 (1998).
13. E. H. Hwang and S. D. Sarma, *Phys. Rev. B* **77**, p. 115449 (2008).
14. N. Bonini, M. Lazzeri, N. Marzari and F. Mauri, *Phys. Rev. Lett.* **99**, p. 176802 (2007).

Mathematical Models and Numerical Simulation of Controlled Drug Release

Sara Minisini*, Luca Formaggia†

Dipartimento di Matematica, "F. Brioschi"
MOX - Modeling and Scientific Computing
Politecnico di Milano, Italy
**sara.minisini@polimi.it, †luca.formaggia@polimi.it*

Drug-eluting stents (DES) are cardiovascular biomedical devices able to release a drug after implantation inside an artery. The medical drug is used to heal the vascular tissues injured after implantation and to prevent restenosis. DES are very complex medical devices whose development involves the design of complex geometrical shapes, the study of advanced biocompatible materials that store the drug, and complex pharmacokinetic processes involved in the release process. This study focus on a specific aspect of the drug-eluting stents. In particular, the phenomena involved in the drug-release from the polymeric stent coating, where the drug is initially cast, are analyzed. The aim of the work concerns with the identification and analysis of how the physical properties of the stent coating impact on the release rate of the drug.

Keywords: Drug diffusion and dissolution, matrix erosion, controlled drug release.

1. Introduction

Nowadays arteriosclerosis is one of the most common cardiovascular diseases. It causes a reduction of blood flow to the downstream tissue by narrowing or occluding large and medium sized arteries. A technique to restore the blood flow and reopen the artery is based on the implantation of a stent. The stent a metallic wire structure is driven and expanded into the stenosed site of the artery. One of the major limitations of this technique is the in-stent restenosis, where the treated vessel becomes blocked again few months after the implantation of the device. Hence, DES have been developed where a polymeric matrix is added to the metallic struts and loaded with a drug which is released after implantation. An optimal therapeutic effect of the released drug depends on the release rate into the arterial wall. While the shape of the stent is mainly determined by its mechanical properties, the release of the drug can be controlled by using different drugs and different type of polymers, among which the biodegradable ones are of particular interest.

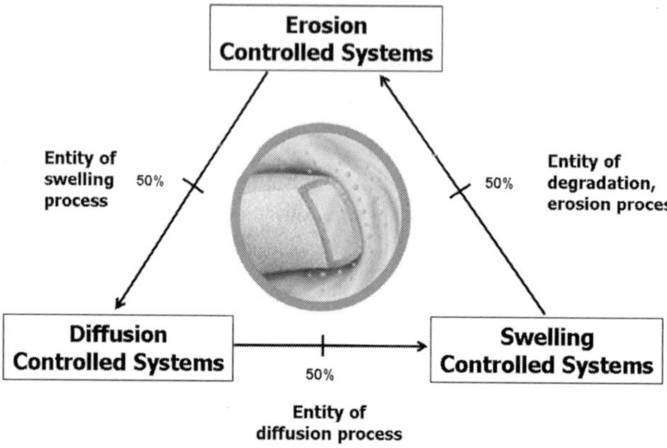

Fig. 1. Schematic representation of the main mechanisms of drug release from polymers, (classification proposed in [13]).

To define the main drug release mechanisms which take place inside the coating layer of the device, we focus on the structure. The drug contained in the material can be surrounded by a rate-controlling biodegradable membrane or uniformly distributed throughout the polymer matrix giving monolithic devices ([5]) such as microspheres, cylinders, films or layers. In the first case, drug release is simple to predict. When considering the drug release from monolithic devices the diffusion and dissolution, swelling, matrix erosion or combinations of these phenomena, as summarized in Fig.1, have to be taken into account. The swelling phase consist in a rearrangement of polymer chains due to the penetration of the solvent from the surrounding environment into the polymer. In our work we will focus on the diffusion/dissolution and erosion controlled systems. If drug release is diffusion controlled, delivery is described for monolithic devices by either Fick's second law (when drug loading is lower than drug solubility in the polymer) or by the Higuchi equation (when drug loading is higher than drug solubility in the polymer,[6]).
When erosion is faster than drug diffusion, drug release is controlled by erosion. In this case a part from the classical physical mass transport processes (drug dissolution and diffusion), chemical reaction due to degrada-

tion occurs. There are two main patterns of erosion. Erosion can take place throughout the whole matrix, this case is referred as homogeneous or bulk erosion, or can be restricted to the device surface. The latter process is named heterogeneous or surface erosion.
Bulk erosion is very complex because water uptake is much faster than polymer degradation. The entire device is rapidly hydrated and degradation, can take place throughout the complete matrix. This results in simultaneous drug diffusion and matrix erosion. However, the device retains its original shape and mass, until up to approximately 90 % of the polymeric chains are degraded. The release rate is difficult to predict, especially if the matrix disintegrates before drug is released completely. If surface erosion occurs, water penetration is slow compared to the polymer degradation rate. Degradation takes place only at the matrix boundaries and not in the core of the polymer.[11] Thus, except for the boundaries, the chemical integrity of the device is preserved and a consistent degradation rate of the polymer is the result. Therefore, surface erosion is considered as the preferable erosion mechanism in medical devices, because it is highly reproducible, the degradation rate can be manipulated by just changing the surface area and water labile drugs are protected within the device.[12] Thus, the aim of our mathematical model is to consider that process.

2. Mathematical model of dissolution

The dissolution is the physical process for which a solid element dissolves in the surrounding solution. In the following we go back over the derivation of the model that we will use to describe the drug release in our system. The basic step of the dissolution is the reaction that takes place at the liquid-solid interface and depends on three main factors, that are:[9] the flow rate of the dissolution medium inside the material, the reaction rate of the physical change from the solid to the dissolved state and the diffusivity of the dissolved drug inside the dissolution medium. In the mathematical model used the rate of the dissolution is limited by the rate of diffusion, and by the saturation solubility. The model is based on a reformulation of the Noyes-Whitney findings that at the beginning of the nineteenth century found as the governing force of the dissolution is the difference between the concentration solubility (c_s) and the concentration of the dissolved specie in the matrix. The diffusion dissolution phenomena, in its general form, can be described with the continuity equation:

$$\frac{\partial c}{\partial t} + \nabla \dot{j} = R \tag{1}$$

where $c(\mathbf{x}, t)$ is the concentration of the drug able to move inside the matrix, $j = (\mathbf{x}, t)$ is the drug flux and $R(\mathrm{x}, t)$ is a drug source to take into account the dissolution. We assume that the diffusion process dominates the mobility of the drug inside the matrix. Thus, j obeys Fick's law and can be expressed as:

$$\mathrm{j} = -D\nabla c \tag{2}$$

where, D is the diffusion coefficient. The source density, $R(\mathrm{x}, t)$ takes into account for the dissolution process that increases the total amount of mobile drug inside the matrix and is modeled using a reformulation of the empirical original Noyes-Whitney equation,[9] that reads:

$$\frac{dW}{dt} = \frac{D\tilde{A}(c_s - c)}{L} \tag{3}$$

The equation describes the rate of dissolution, dW/dt as a function of the properties of the medium: $\tilde{A}$ and L are respectively the contact surface area of the solid and the diffusion layer thickness. In the mathematical model we use the reaction term, R , averaged on a small volume, is obtained by a rearrangement of equation (3)and assumes the form:

$$R = k_d A(c_s - c) \tag{4}$$

where A is the average surface area of undissolved drug per unit volume, k_d is the dissolution rate constant introduced to account for the original variables D, L and c_s is the solubility of the drug. The drug source can be identified with the decrease of the solid drug concentration (or undissolved drug), $s(\mathrm{x}, t)$, within the matrix. Under that assumption equation (4) in terms of the solid/undissolved drug can be modeled as:

$$-\frac{\partial s}{\partial t} = k_d A(c_s - c) \tag{5}$$

where A is the average surface area of the undissolved drug. Using the continuity equation we define the variation on time of the undissolved drug component as:

$$\frac{\partial c}{\partial t} - D\nabla^2 c = -\frac{\partial s}{\partial t} \tag{6}$$

It is possible to relate the surface contact area, A, with the concentration $s(\mathrm{x}, t)$, supposing that the solid drug crystals have a uniform granulomery, so that the granes have approximately the same size and spherical shape. From geometrical considerations a relation between the surface area of the undissolved drug and the concentration can be established[2,7] . Further, scaling the equation by the initial undissolved concentration, the final system of equations is obtained as proposed in:[4]

$$\begin{aligned}
&\partial_t c + \mathbb{L}_c c = k_d\, s^{2/3}(c_s - c), && \text{in } (0,T] \times \Omega_c,\\
&\partial_t s = -k_d\, s^{2/3}(c_s - c), && \text{in } (0,T] \times \Omega_c,\\
&c = 0, && \text{on } \Gamma_D,\\
&\nabla c \cdot \mathbf{n} = 0, && \text{on } \partial\Omega \backslash \Gamma_D,\\
&c(0,\mathbf{x}) = c_0, \quad s(0,\mathbf{x}) = s_0. &&
\end{aligned} \tag{7}$$

$\mathbf{n}$ being the outward normal at the boundaries of the domain. The domain Ω_c represents the polymeric material that stores the drug, while $\mathbb{L}_c c = -D_c \Delta c$. The initial concentration of the solid and dissolved drug are respectively c_0, s_0 which we assume constant. The set of boundary conditions, prescribes no flux of the dissolved drug through the boundaries except on Γ_D where an homogeneous Dirichlet condition on c is imposed. The latter condition, also called sink condition states that the polymer matrix is immersed in a dilute and very large solution compared to the dimensions of the domain. D_c is the diffusion coefficient while k_d is the dissolution rate constant (1/sec). It has to be pointed out that all the parameters in the model, such as the diffusivity, the solubility have to be interpreted as effective parameters. That is a consequence of the porous nature of the polymeric matrix that can change its porosity $\phi(t)$ during the release process. Thus, $c_s(t) = \phi(t) c_s^*$ and $D(t) = \phi(t) D^*$ where c_s^* and D^* are the constant coefficients of the drug in the free medium. In an ideal configuration the volume occupied by the polymer matrix after the swelling phase is composed by the solid drug V_s, the polymer chains that we assume insoluble V_{ins} and void space that we expect to be filled by the solvent V_l. Thus, the total volume occupied by the matrix is:

$$V_{tot} = V_s(t) + V_{ins} + V_l(t).$$

Assuming no erosion (no mass loss) of the matrix, V_{tot} is constant in time as the volume of the insoluble part. Based on volumetric consideration, and assuming that a reduction in the drug mass corresponds to a proportional decrease in its volume[14] we can define a relation to describe the change of the porosity inside the matrix due to the dissolution of the solid drug as: $\phi(t) = \phi_0 + \dfrac{s_0 - s}{\rho_s}$, where ϕ_0 is the initial known porosity of the matrix and ρ_s is the density of the solid drug. Finally the effective solubility and diffusivity can then be defines as:

$$c_s(t) = \left(\phi_0 + \frac{s_0 - s}{\rho_s}\right) c_s^*, \qquad D(t) = \phi(t) D^*. \tag{8}$$

3. Mathematical model of dissolution and surface erosion

As highlighted in the introduction in this work we focus on the phenomenon of the surface erosion. During the process the material reduces its dimensions keeping the average molecular weight constant. This is the consequence of the fact that erosion is confined at the external boundaries of the material even though the fluid may penetrate inside the polymer. Inside the matrix, the two drug phases coexist, and so at each point drug concentrations are assigned for both liquid and solid phase. In an ideal surface erosion, the external solution should not be able to penetrate into the interior of the material, however in reality some of the solution will be soaked inside the material at the beginning of the release process. To simplify the problem we assume that the amount of the water initially absorbed is low and do not alter the surface erosion process significantly. This assumption is indeed in accordance with experimental trials.[10] Moreover, we assume that matrix boundaries in contact with the solution erode following a linear law with a constant erosion rate as experimental results show for a large class of polymers.[8]

The erosion model that we propose is equivalent to (7), but is defined on a moving domain $\Omega(t)$:

$$\begin{aligned}
&\partial_t c + \mathbb{L}_c c = k_d\, s^{2/3}(c_s - c), && \text{in } \Omega_c(t),\\
&\partial_t s = -k_d\, s^{2/3}(c_s - c), && \text{in } \Omega_c(t),\\
&c = 0, && \text{on } \Gamma_D,\\
&\nabla c \cdot \mathbf{n} = 0, && \text{on } \partial\Omega_c(t)\backslash\Gamma_D,\\
&c(0,\mathbf{x}) = c_0, \quad s(0,\mathbf{x}) = s_0. &&
\end{aligned} \tag{9}$$

The domain is chosen to have the form $\Omega_c(t) = L(t)\, H$, where H is the height of the matrix and is assumed to be constant, while $L(t) = L_c - v\, t$ with L_c initial length of the matrix.

4. Numerical method

The system of equations (9) is rewritten through the ALE formulation[3] in the reference domain $\Omega_0 = (0,1) \times (0, H/L_c)$. The velocity v is estimated by experiments.
Thus, the following transformation of coordinates holds,

$$\mathbf{x}(t) = \begin{bmatrix} L_c - vt & 0 \\ 0 & L_c \end{bmatrix} \mathbf{X}, \tag{10}$$

where $\mathbf{x}(t) \in \Omega(t)$ and $\mathbf{X} \in \Omega_0$.
The velocity of the moving domain is,

$$\mathbf{w} = \begin{bmatrix} -vX \\ 0 \end{bmatrix}, \tag{11}$$

and the Jacobian matrices of the transformation $\mathbf{J}_{\mathcal{M}_t^{-1}}$ and $\mathbf{J}_{\mathcal{M}_t^{-1}}$ assume the form :

$$\mathbf{J}_{\mathcal{M}_t^{-1}} = \begin{bmatrix} \dfrac{1}{(L_c - vt)} & 0 \\ 0 & \dfrac{1}{L_c} \end{bmatrix}, \quad \mathbf{J}_{\mathcal{M}_t} = \begin{bmatrix} (L_c - vt) & 0 \\ 0 & L_c \end{bmatrix}, \tag{12}$$

with determinants, $J_{\mathcal{M}_t^{-1}} = 1/(L_c(L_c - vt))$, and $J_{\mathcal{M}_t} = L_c(L_c - vt)$.
For a better control of the positivity of the solution at the numerical level we have introduced the so called Maculay operator $\langle x \rangle$ defined as:

$$\langle x \rangle = \frac{x + |x|}{2} \tag{13}$$

and we use it for the non linear reaction term.
The system of equations (9) becomes for, $t > 0$:

$$\begin{aligned}
&\frac{\partial c}{\partial t} J_{\mathcal{M}_t} + \frac{D}{J_{\mathcal{M}_t}} \left(L_c^2 \frac{\partial^2 c}{\partial X^2} + (L_c - vt)^2 \frac{\partial^2 c}{\partial Y^2} \right) \\
&\qquad + (v\, L_c\, X) \frac{\partial c}{\partial X} - k_d \, \langle s^{2/3}(c_s - c) \rangle (J_{\mathcal{M}_t}) = 0 && \text{in } \Omega_0, \\
&\frac{\partial s}{\partial t} J_{\mathcal{M}_t} + (v\, L_c\, X) \frac{\partial s}{\partial X} + k_d \, \langle s^{2/3}(c_s - c) \rangle \, (J_{\mathcal{M}_t}) = 0 && \text{in } \Omega_0, \\
&c = 0, && \text{on } \Gamma_D, \\
&\nabla c \cdot \mathbf{n} = 0, && \text{on } \partial\Omega_0 \backslash \Gamma_D, \\
&\nabla s \cdot \mathbf{n} = 0 && \text{on } \partial\Omega_{0^-},
\end{aligned}$$

and

$$c = c_0, \quad s = s_0 \quad \text{at } t = 0, \text{in } \Omega_0.$$

The third condition of implies that there is no flux of the solid drug leaving the domain, but only of the dissolved phase. We have set $\partial\Omega_{0^-} = \{\mathbf{x} \in \partial\Omega : \mathbf{w} \cdot \mathbf{n} > 0\}$. The first step in order to discretize the system of equations (4) is to consider its variational formulation. Then, the discretization of the variational problem is straightforwardly derived by the application of the finite element method. Precisely, we consider linear finite elements defined on a triangular unstructured mesh. The discretization of the time is achieved by finite difference schemes of implicit type, to be precise, by the implicit Euler method. The time step discretization is uniform for the interval of simulation. The non linear reactive term is solved by means of a fixed point iterative scheme. A regularization of the reaction term $f = s^{2/3}(c_s - c)$ when $s \to 0$ is introduced. The problem presented in (7) can be seen as a particular case of this description when $v = 0$.

5. Results and Discussion

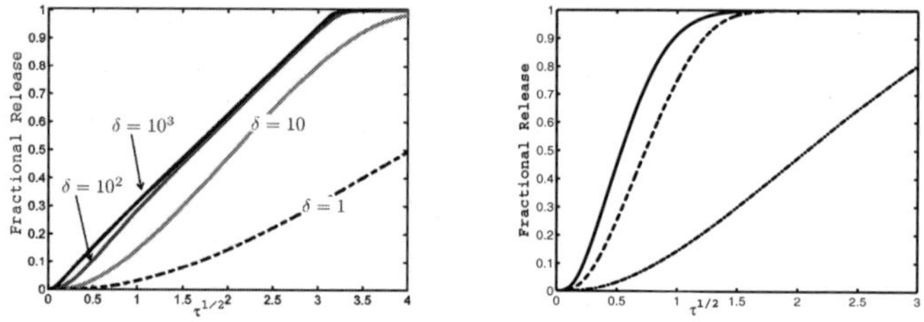

(a) (b)

Fig. 2. In (a) release profiles against adimensionalized time τ for $\delta = 1, 10, 100, 10^3$ from right to left. In (b) release profiles against adimensionalized square root of τ for different values of the solubility ratio c_s. From left to right $c_s = 1.5, 0.5, 0.05$ while $\delta = 10$.

In this section, we present some numerical experiments carried out to study the sensitivity of dissolution and erosion on the release curve. In particular we investigate the dependence of the release characteristics on the solubility and on the diffusion, dissolution and erosion coefficients.
We assume that the polymeric matrix is a slab of length $L_c = 20\mu m$ and height $H = 7\mu m$. The matrix is made by an homogeneous polymer with uniformly dispersed solid drug.

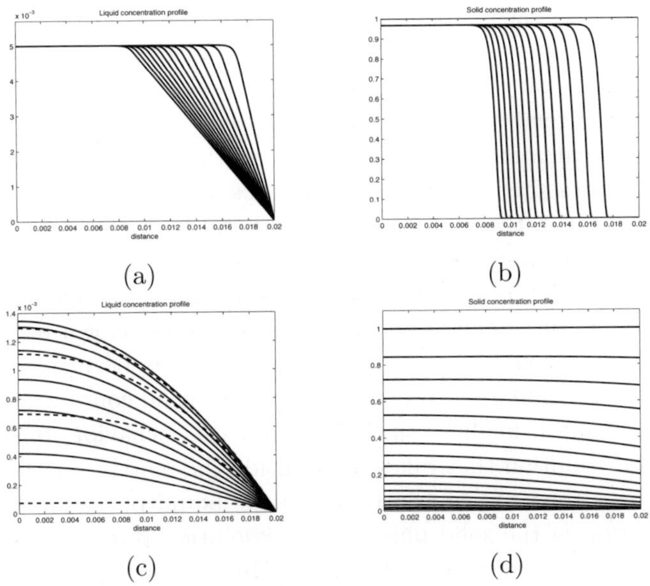

Fig. 3. Sensitivity analysis for the dissolution model (7). The profiles are those of the solid and liquid concentration s along the slab at $H/2$ for two different values of δ and $c_s = 0.005$. In (a) the liquid concentration c and in (b) the solid one s for $\delta = 10^4$. In (c) the liquid concentration for $\delta = 1$. The concentration c rapidly increases from $c_0 = 0$ to its maximum value. The dashed profiles highlight this behavior. Afterward, c decreases to zero, as the profiles with the solid line show. In (d) the profiles of the solid concentrations, that decreases uniformly in the matrix from its initial value.

We start showing the numerical results obtained in the case of diffusion and dissolution, i.e. the system of equations (7).
At first, we evaluate the effect of a finite dissolution rate on the release profile. The fraction of released drug as function of time is defined as:

$$Q(t) = 1 - \frac{1}{|\Omega_c|} \int_{\Omega_c} (c + s) \, d\Omega_c \tag{14}$$

where $|\Omega_c|$ is the area of the matrix.
We introduce the adimensional parameter $\delta = k_d \, L_c^2 / D_c$, and present the trend of the fraction of released drug for four different values of δ. The results presented are computed with the same value of solubility that we set to $c_s = 0.05$. In Fig.2(a) we show the trend of Q against the square root of the adimensionalized time $\tau = D_c \, t / L_c^2$. We plot the fractional release against $\tau^{1/2}$ to point out the initial delay in the release drug profile, due to the time taken from the dissolution process to build up. The entity of this delay depends on the dissolution kinetics. Thus, for small values of δ that

correspond to a slow dissolution process the delay is evident, while when the dissolution rates is higher, the effect is restricted to a relatively short time at the beginning of the process.
To explore the relationship between the drug release and the dissolution rate, the concentration profile along the length of the slab are considered. The profiles presented are in correspondence of $H/2$.
As shown in Fig.3, the profiles of the solid and liquid concentration inside the matrix are completely different. This behavior depends on the dominant effect between dissolution and diffusion, when the values of $c_s < 1$. When $\delta >> 1$, the matrix divided in two zones, as shown in Fig.3.b. We have a depleted zone, where all the solid drug has been dissolved and s approaches zero, and a second region where the solid concentration has the values of its initial loading, s_0. We explain this behavior considering the pattern of dissolution that takes place inside the matrix. At the beginning of the process the drug concentration in the liquid phase increases due to the dissolution of the solid phase. The dissolution process continues until c reaches the value of the solubility, c_s. Due to the rapid dissolution the amount of solid drug is reduced or even depleted, while the effect of the diffusion is to decrease the drug in the liquid phase. In the region near the boundary, where the liquid drug flows out from the matrix regulated by the sink condition ($c = 0$), $c_s - c$, namely the dissolution driving force, is greater that in any other points of the matrix. In fact the dissolution only happens when the liquid concentration is lower than the saturation concentration. Thus, the solid concentration is reduced essentially near the boundary. The consequence is that the profile of the solid concentration s is a progressive wave that moves from the boundary to the inside of the slab.
When δ is small, the solid phase decreases almost uniformly throughout the matrix, as shown in Fig.3(d). The liquid concentration c initially increases from 0, as shown in Fig.3(c) with the dashed line. The replenish due to the dissolved solid drug is very slow, while the effect of the diffusion is to uniformly distribute the liquid concentration inside the matrix. Thus, c never reaches its upper limit, c_s. As consequence we have very close dissolution driving forces at each position along the matrix that explains the uniform decrease of the solid phase.
The effect of a different value of the solubility c_s on the release rate is show in Fig.2(b). Increasing the solubility, the release rate is increased and that is in agreement with the fact that the solubility is the limiting step in the dissolution model. The kinetics of the dissolution is defined by k_d, but also by the value of the solubility that stops the dissolution process when

c approaches c_s. Moreover the initial delay in the release profile disappears gradually with increasing c_s . That is confirmed by the fact that as the initial drug loading does not exceed the solubility,($c_s > 1$) and the dissolution is rapid enough, the problem reduce to a standard diffusion problem,[1]
Finally, we present some numerical tests for a time dependent porosity and effective coefficients defined in 8. The initial porosity is $\phi_0 = 0.1$ and its final values after all the solid drug has been dissolved is $\phi(t) = 0.4$. These values are computed from the data in.[8] At the beginning $\delta = 1$ and $c_s = 0.05\,\phi_0$. The effect of the increase of the porosity is to speed up the release process. In Fig.4(a), we compare the drug release profile in the case of variable porosity, (dashed line), with the case of a constant porosity, (solid line). The increase of the release rate is due to the increase of the saturation solubility, and the driving force of the dissolution $(c_s - c)$.
We present a sensitivity analysis for the surface erosion model, (9). In Fig.4.b the release profiles are plotted against the square root of τ. The solid blue curves correspond to different values of the adimensional parameter, $B = vL/D$, where v is the erosion velocity measured experimentally, while δ is set to 1. The solid lines from left to right have a decreasing value of the B, and the dashed line corresponds to dissolution only, $B = 0$. As we can expect the effect of the erosion is to enhance the release rate, because both phases are released together. For high values of the B, the release tend to be linear in τ while decreasing the value of the velocity, the drug is mainly released by dissolution and diffusion and the effect of the erosion is less evident.

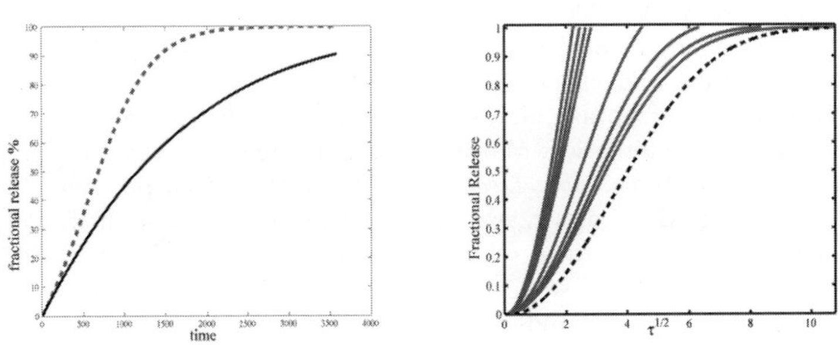

(a) (b)

Fig. 4. In (a) the effect of the porosity on the release profile against time in minutes. In (b) release profiles with a varying erosion velocity. From left to right with the solid line $v\,L/D$ is 0.1, 1, 5, 10, 20, 40, 70, 120 with $\delta = 1$. With the dashed line the release rate, with $\delta = 1$ and no erosion.

Acknowledgements

This work has been carried out at MOX-Modeling and Scientific Computing, Dipartimento di Matematica, Politecnico di Milano. The studies on matrix erosion are part of an ongoing research project financed by the Istituto Italiano di Tecnologia (IIT). The authors also acknowledge the italian MIUR for partial support by a grant PRIN07.

References

1. J. Crank, *The mathematics of diffusion*, Paperback edtion, Oxford University press, 1979.
2. L.J. Edward, The dissolution of aspirin in aqueous media, *Trans. Faraday*, **47** (1951), pp. 1191-1210.
3. L. Formaggia, F. Nobile, A Stability Analysis for the Arbitrary Lagrangian Eulerian Formulation with Finite Elements, *East WJ Numer. Math.*, **7** (1999), pp. 105-132.
4. G. Frenning, Theoretical investigation of drug release from palan matrix systems: effects of a finite dissolution rate, *J. Control. Rel.*, **92** (2003), pp. 331-339.
5. J. Heller, Biodegradable Polymers in Controlled Drug Delivery,*Crit. Rev. Ther. Drug Carrier Syst*, (1984), pp. 39-90.
6. T. Higuchi, Rate of release of medicaments from ointment bases containing drugs suspensions, *Int J. Pharm* , **50** (1961), pp. 874-875.
7. A.W. Hixon, J.H. Crowell, Dependence of reaction velocity upon surface and agitation, *Ind. Eng.Chem.*, **23** (1931), pp. 923-931.
8. R. Langer, J. Folkman, Sustained release of macromolecules from polymers, *Midland Macromolecular Monog.*, **5** (1978), pp. 175-196.
9. P. Macheras, A. Iliadis, *Modeling in biopharmaceutics, pharmacokinetics, and pharmacodynamics*, vol. 30 of Interdis- ciplinary Applied Mathematics. Springer, New York, 2006.
10. E. Park, M. Maniar, J. Shah, Water uptake into polyanhydride devices: Kinetics of uptake and effects of model compunds incorporated, and device geometry on water uptake, *J. Control. Rel*, **40** (1996), pp. 55-65.
11. J. Siepmann, A. Göpferich, Mathematical modeling of bioerodible polymeric drug delivery systems, *Adv. Drug Deliv. Rev.*, **48** (2001), pp. 229-247.
12. K.E. Uhrich, S.M. Cannizzaro, R.S. Langer, K.M. Shakesheff, Polymeric systems for controlled drug release,*Chemical Rev. (Washington, D.C.)*, **99** (1999), pp. 3181-3198.
13. G. Winzenburg, C. Schmidt, S. Fuchs, T. Kissel, Biodegradable Polymers and Their Potential Use in Parenteral Veterinary Drug Delivery Systems, *Adv. Drug Deliv. Rev.*, **56** (2004), pp. 1453-1466.
14. M. Zhang, Z. Yang, W. Chow, C. Wang, Simualtion of the drug release from biodegradable polymeric microspheres with bulk and surface erosion, *Int. J. Pharm*, **92(10)** (2003), pp. 2040-2056.

Hydrodynamic Equations of Anisotropic, Polarized, Turbulent Superfluids

Maria Stella Mongiovì[1], Michele Sciacca[1] and David Jou[2]

[1] *Dipartimento di Metodi e Modelli Matematici Università di Palermo, c/o Facoltà di Ingegneria, Viale delle Scienze, 90128 Palermo, Italy*
mongiovi@unipa.it
msciacca@unipa.it

[2] *Departament de Física, Universitat Autònoma de Barcelona, 08193 Bellaterra, Catalonia, Spain*
david.jou@uab.cat

Quantized vortices in superfluids have been mainly studied in two typical situations: rotating superfluids and counterflow experiments, the latter meaning the presence of a heat flux without matter flow. In these situations the vortices are modelled as an array of parallel rectilinear vortices or as an isotropic tangle, respectively. However, in partially polarized vortex tangles, the mutual friction force and the nonvanishing tension of the vortex lines are not sufficiently known. Here microscopic and macroscopic expressions for the friction force and vortex tension are given and an application of the model to the case of rotating counterflow turbulence is made. The model is completed by an evolution equation for the vortex line density per unit volume, which takes into account the anisotropy of the tangle.

Keywords: Turbulent superfluids, vortex tangles.

1. Introduction

The behavior of superfluid helium is very different from that of ordinary fluids. It has an extremely low viscosity and an extremely high thermal conductivity (three million times larger than that of helium I). Furthermore the second sound, a wave motion in which temperature and entropy oscillate and density and pressure remain essentially constant, can propagate in it.

In most literature, the motion of a superfluid is modelled using Landau's two-fluid model, which regards the fluid to be made of two completely mixed components: the normal fluid and the superfluid, with densities ρ_n and ρ_s respectively, and velocities $\mathbf{v}_n$ and $\mathbf{v}_s$ respectively, with total mass density ρ

and barycentric velocity $\mathbf{v}$ defined by $\rho = \rho_s + \rho_n$ and $\rho\mathbf{v} = \rho_s\mathbf{v}_s + \rho_n\mathbf{v}_n$. The first component is related to thermally excited states (phonons and rotons) that form a classical Navier-Stokes viscous fluid. The second component is related to the quantum coherent ground state and it is an ideal fluid, which does not experience dissipation neither carries entropy.

If the superfluid is put in rotation with angular velocity Ω higher than Ω_c, an ordered array of parallel quantized vortex lines of equal circulation κ

$$\kappa = \oint \mathbf{v}_s \cdot dl \tag{1}$$

is created. The vortex array is described by introducing the *line density* L, defined as the average vortex line length per unit volume, in this case equivalent to the areal density of vortex lines, which is proportional to the angular velocity,[1–4] namely

$$L = L_R = \frac{2\Omega}{\kappa}. \tag{2}$$

It is well known too that a disordered tangle of quantized vortex lines is created in the so-called counterflow superfluid turbulence,[1–4] characterized by no matter flow but only heat transport, exceeding a critical heat flux q_c. In this case the line density L depends on the square of the averaged counterflow velocity vector, defined as

$$\mathbf{V}_{ns} = [\mathbf{v}_{ns}]_{av} = \frac{1}{\Lambda}\int \mathbf{v}_{ns} d\Lambda, \tag{3}$$

($\mathbf{v}_{ns} = \mathbf{v}_n - \mathbf{v}_s$ being the microscopic counterflow velocity) related to heat flux $\mathbf{q}$ by $\mathbf{q} = \rho_s T s \mathbf{V}_{ns}$ (T being the temperature of the helium and s the entropy of the normal component). When the turbulence is fully developed, line density L becomes proportional to the square of the averaged counterflow velocity:[1–4]

$$L = L_H = \frac{\gamma_H^2}{\kappa^2}\mathbf{V}_{ns}^2, \tag{4}$$

the dimensionless coefficient γ_H being dependent on the temperature.

Due to the smallness of the quantum of circulation, even a relatively weak rotation or a small counterflow velocity produces a large density of vortex filaments. It is therefore possible to average over the presence of many individual vortex lines. In this macroscopic model a fluid particle is a small but mesoscopic region threaded by a high density of vortex lines. In (3) and in the following, capital letters denote local macroscopic velocities averaged over a small mesoscopic volume Λ.

A set of equations frequently used, is the Hall-Vinen-Bekarevich-Khalatnikov (HVBK) model.[5,6] These are the equations of the two fluid model, modified to take into account the presence of vortices. For our purposes it is sufficient to consider here the incompressible HVBK equations, which in an inertial frame, neglecting inhomogeneous terms are:

$$\rho_n \frac{d\mathbf{V}_n}{dt} = \mathbf{F}_{ns}, \tag{5}$$

$$\rho_s \frac{d\mathbf{V}_s}{dt} = -\mathbf{F}_{ns} + \rho_s \mathbf{T}. \tag{6}$$

In these equations $\mathbf{F}_{ns}$ is the mutual friction force exerted by the superfluid component on the normal component — which bears an opposite sign in the other evolution equation — and $\rho_s \mathbf{T}$, the vortex tension force, related to the average curvature of vortices.

In the regular vortex array produced by the rotation, the mutual friction force assumes the Hall-Vinen expression:[5]

$$\mathbf{F}_{ns} = 2\rho_s \alpha \hat{\mathbf{\Omega}} \times [\mathbf{\Omega} \times (\mathbf{V}_n - \mathbf{V}_s)] + 2\rho_s \alpha' \mathbf{\Omega} \times (\mathbf{V}_n - \mathbf{V}_s) \tag{7}$$

and vortex tension $\mathbf{T}$ vanishes, because the vortices are straight lines.

In counterflow superfluid turbulence, the vortex tangle can be assumed to be approximately isotropic and the mutual friction force is expressed by:

$$\mathbf{F}_{ns} = -\frac{2}{3} \rho_s \kappa \alpha L \mathbf{V}_{ns}. \tag{8}$$

Vortex tension $\mathbf{T}$ vanishes also in this case, owing of the supposed isotropy of the tangle.

However, in other situations, as rotating counterflow or non stationary Couette and Poiseuille flows,[7–10] one expects a partially polarized vortex tangle, as a compromise between the orienting effect of a rotation or of a macroscopic velocity gradient, and the randomizing effect of the relative velocity of normal and superfluid components. In these cases, the mutual friction force $\mathbf{F}_{ns}$ between these components, as well as the nonvanishing tension of the vortex lines $\rho_s \mathbf{T}$ are not sufficiently known. The aim of this paper is the exploration of $\mathbf{F}_{ns}$ and $\mathbf{T}$ for partially polarized tangles with nonvanishing average curvature of vortex lines. This study is completed by a generalization of the Vinen's equation, as an evolution equation for the vortex line density L.

2. Mutual friction force and vortex tension in anisotropic tangles

In this Section a statistical determination of the friction force and vortex tension is made, with the aim to check the limit of validity of the HVBK model in anisotropic situations; a more general expression for $\mathbf{F}_{ns}$ and $\mathbf{T}$ is given and an application of the model to the interesting case of rotating counterflow turbulence is made.

2.1. *Microscopic description of* $\mathbf{F}_{ns}$ *and* $\mathbf{T}$

To derive the expressions of $\mathbf{F}_{ns}$ and of $\mathbf{T}$ used in the HVBK model, and to determine their limit of validity, we consider the microscopic description of a vortex tangle in the vortex filament model.[11–13]

In this model the vortex lines are described by a vectorial function $\mathbf{s}(\xi, t)$, where ξ is the arc-length. The primes indicate differentiation with respect to ξ in such a way that $\mathbf{s}' = \partial\mathbf{s}/\partial\xi$ is the unit vector tangent to the vortex line, and $\mathbf{s}'' = \partial^2\mathbf{s}/\partial\xi^2$ the curvature vector.

The normal component reacts to a moving vortex by producing the *microscopic* mutual friction force $-\mathbf{f}_{MF}$, which can be written as

$$\mathbf{f}_{MF} = \alpha\rho_s\kappa\mathbf{s}' \times [\mathbf{s}' \times (\mathbf{v}_{ns} - \mathbf{v}_\mathbf{i})] + \alpha'\rho_s\kappa\mathbf{s}' \times (\mathbf{v}_{ns} - \mathbf{v}_\mathbf{i}), \tag{9}$$

where $\mathbf{v}_\mathbf{i}$ is the *self-induced velocity*, a flow due to all the other vortices. In the *local induction approximation*, the self-induced velocity $\mathbf{v}_\mathbf{i}$ is approximated by:

$$\mathbf{v}_\mathbf{i} \simeq \mathbf{v}_\mathbf{i}^{(loc)} = \tilde{\beta}\,[\mathbf{s}' \times \mathbf{s}'']_{s=s_0}\,. \tag{10}$$

The self-induced velocity is zero if the vortices are straight lines. In equation (9), $\tilde{\beta}$ is the vortex tension parameter,[1] defined as $\tilde{\beta} = \epsilon_V/\kappa\rho_s$, where

$$\epsilon_V = \frac{\rho_s\kappa^2}{4\pi}\ln\left(\frac{L^{-1/2}}{a_0}\right) \tag{11}$$

is the energy per unit length of vortex line. Coefficients $\alpha = B(\rho_n/2\rho)$ and $\alpha' = B'(\rho_n/2\rho)$ depend on the temperature and describe the interaction between the normal fluid and the vortices. B and B' are the well-known Hall-Vinen coefficients.[5]

To obtain the *macroscopic* mutual friction force $\mathbf{F}_{ns}$ *per unit volume*, which superfluid and normal components mutually exert, we first determine the average mutual friction force per unit length

$$< \mathbf{f}_{MF} >= \frac{\int \mathbf{f}_{MF} d\xi}{\int d\xi} = \frac{1}{\Lambda L}\int \mathbf{f}_{MF} d\xi, \tag{12}$$

where the integral is made over all the vortices contained in the volume Λ. To obtain $\mathbf{F}_{ns}$, we must multiply this average by L, which is the length of vortex lines per unit of volume; we obtain

$$\mathbf{F}_{ns} = \alpha\rho_s\kappa L < \mathbf{s}' \times [\mathbf{s}' \times (\mathbf{v}_{ns} - \mathbf{v_i})] > + \alpha'\rho_s\kappa L < \mathbf{s}' \times (\mathbf{v}_{ns} - \mathbf{v_i}) > . \quad (13)$$

The vortex tension force $\mathbf{T}$ arises from the microscopic form of the evolution equation of $\mathbf{v}_s$, which, neglecting the mutual friction force and inhomogeneous terms, is written:[1]

$$\frac{\partial \mathbf{v}_s}{\partial t} = \mathbf{v}_L \times \vec{\omega}_{micr}, \quad (14)$$

where $\mathbf{v}_L = \mathbf{v}_{sl} = \mathbf{v}_s + \mathbf{v_i}$ is the velocity of the vortex line.[1] Here $\mathbf{v}_s$ denotes the microscopic velocity of the superfluid around the vortex and $\vec{\omega}_{micr} = \nabla \times \mathbf{v}_s$ the vorticity of the single vortex line. When the average on the mesoscopic volume Λ is taken, one has

$$\frac{\partial \mathbf{V}_s}{\partial t} = [\mathbf{v}_s \times (\nabla \times \mathbf{v}_s)]_{av} + [\mathbf{v_i} \times (\nabla \times \mathbf{v}_s)]_{av}. \quad (15)$$

This allows to identify the tension $\mathbf{T}$ as

$$\mathbf{T} = [\mathbf{v_i} \times (\nabla \times \mathbf{v}_s)]_{av} = \kappa L < \mathbf{v_i} \times \mathbf{s}' > . \quad (16)$$

This relation may be rewritten in several equivalent forms, by taking into account some vectorial identities. For instance, taking into account that $\mathbf{s}'$ is a unit vector, we have

$$\mathbf{T} = \kappa L\tilde{\beta} < (\mathbf{s}' \times \mathbf{s}'') \times \mathbf{s}' >= \kappa L\tilde{\beta} < \mathbf{s}'' > . \quad (17)$$

By using $\mathbf{s}'' = (\mathbf{s}' \cdot \nabla)\mathbf{s}' = -\mathbf{s}' \times (\nabla \times \mathbf{s}')$, we can also write:

$$\mathbf{T} = -\kappa L\tilde{\beta} < \mathbf{s}' \times (\mathbf{s}' \times \mathbf{s}'') >= -\kappa L\tilde{\beta} < \mathbf{s}' \times (\nabla \times \mathbf{s}') > . \quad (18)$$

2.2. *Mutual friction and vortex tension in the HVBK model*

The mutual friction force in the HVBK model is expressed by:

$$\mathbf{F}_{ns} = \rho_s\alpha\hat{\omega} \times [\vec{\omega} \times (\mathbf{V}_{ns} - \tilde{\beta}\nabla \times \hat{\omega})] + \rho_s\alpha'\vec{\omega} \times (\mathbf{V}_{ns} - \tilde{\beta}\nabla \times \hat{\omega}), \quad (19)$$

where $\vec{\omega} = \nabla \times \mathbf{V}_s$ is the local averaged superfluid vorticity, and $\hat{\omega} = \vec{\omega}/|\omega|$ the unit vector along $\vec{\omega}$.

The vortex tension force $\rho_s\mathbf{T}$ is expressed by

$$\mathbf{T} = (\tilde{\beta}\nabla \times \hat{\omega}) \times \vec{\omega} = \tilde{\beta}(\vec{\omega} \cdot \nabla)\hat{\omega}. \quad (20)$$

2.3. *Limit of validity of HVBK's equations*

To determine the limit of validity of HVBK's equations in anisotropic situations, we introduce the tangle polarity as[14–16]

$$\mathbf{p} =< \mathbf{s}' >= \frac{1}{\Lambda L}\int \mathbf{s}' d\xi = \frac{\vec{\omega}}{\kappa L} = \frac{\nabla \times \mathbf{V}_s}{\kappa L}. \tag{21}$$

In a totally polarized tangle all the tangents $\mathbf{s}'$ are parallel to each other and $|\mathbf{p}| = 1$, whereas in an isotropic tangle $\mathbf{p} = \mathbf{0}$.

We consider first the two averaged quantities $< \mathbf{s}' \times (\mathbf{s}' \times \mathbf{v}_{ns}) >$ and $< \mathbf{s}' \times \mathbf{v}_{ns} >$. By indicating the fluctuations of $\mathbf{p}$ with $\delta \mathbf{s}'$, and neglecting the fluctuations of the counterflow velocity $\mathbf{V}_{ns}$, we obtain

$$< \mathbf{s}' \times (\mathbf{s}' \times \mathbf{v}_{ns}) >= \mathbf{p} \times (\mathbf{p} \times \mathbf{V}_{ns}) + < \delta\mathbf{s}' \times (\delta\mathbf{s}' \times \mathbf{V}_{ns}) >, \tag{22}$$

$$< \mathbf{s}' \times \mathbf{v}_{ns} >= \mathbf{p} \times \mathbf{V}_{ns}. \tag{23}$$

Because the first terms in the right-hand side of equations (22) and (23) can be written as

$$\mathbf{p} \times (\mathbf{p} \times \mathbf{V}_{ns}) = \frac{1}{\kappa L}\hat{\omega} \times [\vec{\omega} \times \mathbf{V}_{ns}] =< \mathbf{s}' \times (\mathbf{s}' \times \mathbf{v}_{ns}) >^{(HVBK)}, \tag{24}$$

$$\mathbf{p} \times \mathbf{V}_{ns} = \frac{1}{\kappa L}\vec{\omega} \times \mathbf{V}_{ns} =< \mathbf{s}' \times \mathbf{v}_{ns} >^{(HVBK)}, \tag{25}$$

we deduce that in the HVBK equation the quadratic term in the fluctuations has been neglected. In general, this approximation is too crude and unrealistic. For instance, in the limiting situations of an isotropic tangle, $\mathbf{p} = 0$, but $< \mathbf{s}' \times (\mathbf{s}' \times \mathbf{v}_{ns}) >= (2/3)\mathbf{V}_{ns}$.

We now study the quantities $< \mathbf{s}' \times (\mathbf{s}' \times \mathbf{v_i}) >$ and $< \mathbf{s}' \times \mathbf{v_i} >$. Using the local induction approximation, we have

$$< \mathbf{s}' \times (\mathbf{s}' \times \mathbf{v_i}) >= - < \mathbf{v_i} >= -\tilde{\beta} < (\mathbf{U} - \mathbf{s}'\mathbf{s}') \cdot (\nabla \times \mathbf{s}') >, \tag{26}$$

$$< \mathbf{s}' \times \mathbf{v_i} >= -\tilde{\beta} < \mathbf{s}'' >, \tag{27}$$

where $\mathbf{U}$ is the unit matrix and $\mathbf{s}'\mathbf{s}' = \mathbf{s}' \otimes \mathbf{s}'$ the diadic product.

From relation (26), we deduce that in the HVBK model the quantity $\mathbf{s}' \otimes \mathbf{s}' \cdot (\nabla \times \mathbf{s}') =< [\mathbf{s}' \cdot (\nabla \times \mathbf{s}')]\mathbf{s}' >$ is neglected.

The most critical hypothesis underlying the HVBK equations is the relation

$$\kappa L \simeq |\nabla \times \mathbf{V}_s| = |\vec{\omega}|, \tag{28}$$

which is used to evaluate line density L. In fact, if in a mesoscopic region Λ there are several vortex lines oriented in a random way, line density L in this point will be very different from $L \simeq |\nabla \times \mathbf{V}_s|/\kappa$, which corresponds to an extreme polarization of the tangle (in the small region Λ). For example, if a vortex loop is present, the average circulation relative to this loop is zero, but this is not true for line density L.

Summarizing, in the HVBK equations all the second-order moments of the fluctuations in $\mathbf{s}'$ are neglected and it is assumed that $\kappa L \simeq |\nabla \times \mathbf{V}_s|$ (i.e. $|\mathbf{p}| \simeq 1$). Hence, the HVBK equations correctly describe the local interaction between the normal component and the vortex tangle only in mesoscopic regions with an array of vortex lines all of them pointing in the same direction.

2.4. *Generalized expressions for* $\mathbf{F}_{ns}$ *and* $\mathbf{T}$

The expressions for $\mathbf{F}_{ns}$ and $\mathbf{T}$ which overcome the restrictions which we have mentioned, are:

$$\mathbf{F}_{ns} = -\rho_s\kappa L\alpha < \mathbf{U}-\mathbf{s}'\mathbf{s}' > \cdot[\mathbf{V}_{ns}-\tilde{\beta}(\nabla\times\mathbf{p})]+\rho_s\kappa L\alpha'[\mathbf{p}\times\mathbf{V}_{ns}-\tilde{\beta}\mathbf{p}\times(\nabla\times\mathbf{p})], \tag{29}$$

$$\rho_s\mathbf{T} = -\rho_s\kappa L\tilde{\beta}\mathbf{p}\times(\nabla\times\mathbf{p}). \tag{30}$$

These expressions may be written in the equivalent form:

$$\mathbf{F}_{ns} = \alpha\rho_s\kappa L\left[\frac{2}{3}\mathbf{\Pi}\cdot\mathbf{V}_{ns} + \tilde{\beta}c_1 L^{1/2}\left(\mathbf{I} + \frac{\alpha'}{\alpha}\mathbf{J}\right)\right], \tag{31}$$

$$\mathbf{T} = \kappa L^{3/2}c_1\mathbf{J}, \tag{32}$$

where we have introduced the tensor $\mathbf{\Pi} = \mathbf{\Pi}^s + \mathbf{\Pi}^a$,[14]

$$\mathbf{\Pi}^s \equiv \frac{3}{2} < \mathbf{U} - \mathbf{s}'\mathbf{s}' >, \qquad \mathbf{\Pi}^a \equiv \frac{3}{2}\frac{\alpha'}{\alpha} < \mathbf{W}\cdot\mathbf{s}' >, \tag{33}$$

(with $\mathbf{W}\cdot\mathbf{s}'\cdot\mathbf{V}_{ns} = -\mathbf{s}'\times\mathbf{V}_{ns}$), and the vectors $\mathbf{I}$ and $\mathbf{J}$[15]

$$\mathbf{I} \equiv \frac{\int \mathbf{s}'\times\mathbf{s}''d\xi}{\int|\mathbf{s}''|d\xi} \simeq \frac{1}{c_1 L^{1/2}} < \mathbf{U} - \mathbf{s}'\mathbf{s}' > \cdot\nabla\times\mathbf{p}, \tag{34}$$

$$\mathbf{J} \equiv \frac{\int \mathbf{s}''d\xi}{\int|\mathbf{s}''|d\xi} \simeq -\frac{1}{c_1 L^{1/2}}\mathbf{p}\times(\nabla\times\mathbf{p}), \tag{35}$$

with $c_1 \equiv \frac{1}{\Lambda L^{1/2}}\int|\mathbf{s}''|d\xi$. Tensor $\mathbf{\Pi}$ describes the geometrical properties of the tangle related to the orientational distribution of the vortex lines. Vectors $\mathbf{I}$ and $\mathbf{J}$ contain information on the curvature of the lines.

2.5. *Application to counterflow in rotation*

The combination of counterflow and rotation is especially interesting in the context of the present paper, because it provides intermediate situations between an isotropic tangle and a totally anisotropic array of parallel vortices. Indeed, rotation tends to align vortex lines parallel to the rotation axis, whereas counterflow velocity tends to produce a disordered tangle. In these situations one has partially polarized tangles, requiring the full detailed analysis presented here.[17,18]

We assume that the total ensemble of vortex lines is a superposition of both contributions

$$\mathbf{\Pi}^s = (1-b)\mathbf{\Pi}^s_H + b\mathbf{\Pi}^s_R, \qquad \mathbf{\Pi}^a = c\mathbf{\Pi}^a_R, \tag{36}$$

where b and c are parameters between 0 and 1, that describe the relative weight of the vortex array parallel to $\mathbf{\Omega}$ and the disordered tangle of counterflow: when $b = c = 0$ we recover an isotropic tangle, and when $b = c = 1$ we recover the ordered array. In general situations these coefficients depend on $\mathbf{\Omega}$ and $\mathbf{V}_{ns}$.

By assuming that the vortex tangle is isotropic in the plane orthogonal to the first axis, we obtain

$$\mathbf{\Pi}^s = \begin{pmatrix} 1-b & 0 & 0 \\ 0 & 1+\frac{b}{2} & 0 \\ 0 & 0 & 1+\frac{b}{2} \end{pmatrix}, \qquad \mathbf{\Pi}^a = \frac{3}{2}\frac{\alpha'}{\alpha}\begin{pmatrix} 0 & 0 & 0 \\ 0 & 0 & c \\ 0 & -c & 0 \end{pmatrix} \tag{37}$$

Coefficients b and c can be interpreted in terms of moments of the tangent vector $\mathbf{s}'$

$$< {s'_x}^2 >= \frac{1+2b}{3}, \qquad < {s'_y}^2 >=< {s'_z}^2 >= \frac{1-b}{3}, \qquad < s'_x >= c. \tag{38}$$

A microscopic evaluation of these coefficients was made in Ref. 14, based on a paramagnetic analogy, which reflects the competition between the orienting effects of $\mathbf{\Omega}$ and the randomizing effects of $\mathbf{V}_{ns}$, respectively analogous to the orienting effects of a magnetic field $\mathbf{H}$ on magnetic dipoles μ and the randomizing effects of thermal excitations.

Using the Langevin model of paramagnetism, we found:

$$< s'_x >= \coth x - \frac{1}{x}, \qquad < {s'_x}^2 >= 1 + \frac{2}{x}\left[\frac{1}{x} - \coth x\right], \tag{39}$$

where x is found to be $x = 11\frac{L_R}{L_H}$, where L_R would be the line density if we only had rotation and L_H would be the line density if we only had counterflow.

3. Evolution equation for line density L in anisotropic tangles

To complete our model, we must add an evolution equation for the line density L to equations for $\mathbf{V}_n$ and $\mathbf{V}_s$.

An evolution equation for L, in homogeneous counterflow superfluid turbulence, was formulated by Vinen.[19]

$$\frac{dL}{dt} = \alpha_v V_{ns} L^{3/2} - \beta_v \kappa L^2, \tag{40}$$

where α_v and β_v are dimensionless parameters. A microscopic derivation of this equation was given by Schwarz, who however neglected terms in the average curvature vector $\mathbf{s}''$.

In a previous work,[15] the following extension of Vinen's equation was written

$$\frac{dL}{dt} = \alpha c_1 \mathbf{V}_{ns} \cdot \mathbf{I} L^{3/2} + \alpha' c_1 \mathbf{V}_{ns} \cdot \mathbf{J} L^{3/2} - \alpha \tilde{\beta} c_2(\mathbf{p}) L^2, \tag{41}$$

where $c_2(\mathbf{p}) = \frac{1}{\Lambda L^2} \int |\mathbf{s}''|^2 d\xi$. In another work,[16] for the coefficient $c_2(\mathbf{p})$ was proposed the expression: $c_2(\mathbf{p}) = c_{20}\left[1 - \sqrt{|\mathbf{p}|}\right]\left[1 - B\sqrt{|\mathbf{p}|}\right]$, where B is a dimensionless coefficient lower than 1 and c_{20} is the value of $c_2(\mathbf{p})$ for $\mathbf{p} = 0$.

By using the expressions (34)-(35) for $\mathbf{I}$ and $\mathbf{J}$, we obtain the following evolution equation for L, in homogeneous situations:

$$\frac{dL}{dt} = \frac{2}{3} \alpha L \mathbf{V}_{ns} \cdot \mathbf{\Pi} \cdot (\nabla \times \mathbf{p}) - \alpha \tilde{\beta} c_2(\mathbf{p}) L^2. \tag{42}$$

3.1. *Onsager-Casimir reciprocity relation*

Another modification is necessary to insure the thermodynamic consistency of the equations for L and for $\mathbf{V}_s$, in accordance with the formalism of linear irreversible thermodynamics.[20–22]

We study the consequences of the Onsager-Casimir reciprocity relations on the evolution equations for $\mathbf{V}_s$ and L proposed, which we rewrite in the form:

$$\rho_s \frac{d\mathbf{V}_s}{dt} = \alpha \rho_s \kappa L \frac{2}{3} \mathbf{\Pi} \cdot \left[\mathbf{V}_{ns} - \tilde{\beta}(\nabla \times \mathbf{p})\right] - \rho_s \kappa L \tilde{\beta} \mathbf{p} \times (\nabla \times \mathbf{p}), \tag{43}$$

$$\frac{dL}{dt} = \frac{2}{3} \alpha L \mathbf{V}_{ns} \cdot \mathbf{\Pi} \cdot (\nabla \times \mathbf{p}) - \alpha \tilde{\beta} c_2(\mathbf{p}) L^2. \tag{44}$$

According to the formalism of nonequilibrium thermodynamics, in the right-hand side of the evolution equation for L an additional contribution must be included to make this equation consistent with the evolution equation for $\mathbf{V}_s$. This new term is due to the presence of the vortex tension in the equation for $\mathbf{V}_s$, and has the form:

$$-L\mathbf{V}_{ns}\cdot\mathbf{p}\times(\nabla\times\mathbf{p}) = -\frac{1}{\epsilon_V}\rho_s\mathbf{T}\cdot\mathbf{V}_{ns} = -L^{3/2}c_1\mathbf{J}\cdot\mathbf{V}_{ns} \tag{45}$$

After adding this term, one may write $d\mathbf{V}_s/dt$ and dL/dt in terms of their conjugate thermodynamic forces $-\rho_s\mathbf{V}_{ns}$ and ϵ_V. Indeed, we can write equations (43)-(44) in matrix form in the following system

$$\begin{pmatrix} \frac{d\mathbf{V}_s}{dt} \\ \frac{dL}{dt} \end{pmatrix} = L\begin{pmatrix} -\alpha\kappa\frac{2}{3}\mathbf{\Pi} & CT \\ CT & -\frac{\alpha\tilde{\beta}}{\epsilon_V}Lc_2(\mathbf{p}) \end{pmatrix}\begin{pmatrix} -\rho_s\mathbf{V}_{ns} \\ \epsilon_V \end{pmatrix} \tag{46}$$

where CT is the coupling term:

$$CT = -\frac{\alpha}{\rho_s}\frac{2}{3}(\nabla\times\mathbf{p})\cdot\mathbf{\Pi} + \frac{1}{\rho_s}\mathbf{p}\times(\nabla\times\mathbf{p}). \tag{47}$$

The new evolution equation for L, which takes into account the anisotropy and polarization of the tangle, is:

$$\frac{dL}{dt} = \frac{2}{3}\alpha L\mathbf{V}_{ns}\cdot\mathbf{\Pi}\cdot(\nabla\times\mathbf{p}) - L\mathbf{V}_{ns}\cdot\mathbf{p}\times(\nabla\times\mathbf{p}) - \alpha\tilde{\beta}c_2(\mathbf{p})L^2. \tag{48}$$

Note that, by introducing the tensor

$$\mathbf{\Pi}' = <\mathbf{U}-\mathbf{s}'\mathbf{s}'> + \frac{1+\alpha'}{\alpha} <\mathbf{W}\cdot\mathbf{s}'>, \tag{49}$$

system (46) can be written:

$$\begin{pmatrix} \frac{d\mathbf{v}_s}{dt} \\ \frac{dL}{dt} \end{pmatrix} = L\begin{pmatrix} -\alpha\kappa\frac{2}{3}\mathbf{\Pi} & \frac{\alpha}{\rho_s}\frac{2}{3}(\nabla\times\mathbf{p})\cdot\mathbf{\Pi}' \\ -\frac{\alpha}{\rho_s}\frac{2}{3}(\nabla\times\mathbf{p})\cdot\mathbf{\Pi}' & -\frac{\alpha\tilde{\beta}}{\epsilon_V}Lc_2(\mathbf{p}) \end{pmatrix}\begin{pmatrix} -\rho_s\mathbf{V} \\ \epsilon_V \end{pmatrix}. \tag{50}$$

4. Conclusions

Summarizing, in this work we propose to substitute in the HVBK equations the expression of the mutual friction force $\mathbf{F}_{ns}$ (equation (19)) with the equation (29), to take into account of the second-order moment of $\mathbf{s}'$, allowing in this way that not all the vortex lines in the small volume element under consideration have the same direction. Analogously, the expression (20) for the tension $\mathbf{T}$ has been replaced by equation (30). The coefficients b and c appearing in (37) can be related to the counterflow velocity and to the angular velocity by the relations (38), with $< s'_x >$ and $< {s'_x}^2 >$ expressed by (39). Another important modification consists in adding to the evolution equations for $\mathbf{V}_n$ and $\mathbf{V}_s$ an evolution equation for L including the effects of polarization and anisotropy. In fact, in a general situation, it is too crude and unrealistic in (28) to substitute κL with the modulus of the curl of $\mathbf{V}_s$.

Acknowledgments

The authors acknowledge the support of the University of Palermo (Progetto CoRI 2007, Azione D, cap. B.U. 9.3.0002.0001.0001) and the collaboration agreement between the "Università di Palermo" and the "Universitàt Autònoma de Barcelona". DJ acknowledges the financial support from the Dirección General de Investigación of the Spanish Ministry of Education under grant Fis2006-12296-c02-01 and of the Direcció General de Recerca of the Generalitat of Catalonia, under grant 2005 SGR-00087. MSM and MS acknowledge the financial support of the University of Palermo, under grant 2006 ORPA0642ZR. MS acknowledges the "Assegno di ricerca: Studio della turbolenza superfluida e della sua evoluzione" of the University of Palermo.

References

1. R.J. Donnelly, *Quantized vortices in helium II*, Cambridge University Press, Cambridge, UK, 1991.
2. S.K. Nemirovskii and W. Fiszdon, Chaotic quantized vortices and hydrodynamic processes in superfluid helium, *Rev. Mod. Phys.*, **67** (1995), pp. 37–84.
3. C.F. Barenghi, R.J. Donnelly and W.F. Vinen, *Quantized Vortex Dynamics and Superfluid Turbulence*, Springer, Berlin, 2001.
4. W.F. Vinen and J.J. Niemela, Quantum Turbulence, *J. Low Temp. Phys.*, **128** (2002) pp. 167–231.
5. H.E. Hall and W.F. Vinen, The rotation of liquid helium II, II. The theory of mutual friction in uniformly rotating helium II, *Proc. Roy. Soc.*, **A238** (1956), pp. 204–214.
6. I.L. Bekarevich and I.M. Khalatnikov, Phenomenological derivation of the equation of vortex motion in He II, *Sov.Phys. JETP*, **13** (1961), pp. 643–646.
7. C.J. Swanson and R.J. Donnelly, Instability of Taylor-Couette fluid in Helium II, *Phys. Rev. Lett.*, **67** (1991), pp. 1578-1581.
8. C.F. Barenghi, Vortices and the Couette fluid of helium II, *Phys. Rev. B* **45** (1992), pp. 2290–2293.
9. K.L. Henderson and C.F. Barenghi, End effects in rotating helium II, *Physica B* **184-288** (2000), pp. 67–68.
10. K.L. Henderson and C.F. Barenghi, Superfluid Couette flow in an enclosed annulus, *Theor. Comput. Fluid Dynam.* **18** (2004), pp. 183–196.
11. K.W. Schwarz, Generating superfluid turbulence from simple dynamical rules *Phys. Rev. Lett.* **49** (1982), pp. 283–285.

12. K.W. Schwarz, Three-dimensional vortex dynamics in superfluid ^{4}He, I. Line-line and line boundary interactions *Phys. Rev. B*, **31** (1985), pp. 5782–5804.
13. K.W. Schwarz, Three-dimensional vortex dynamics in superfluid ^{4}He *Phys. Rev. B*, **38** (1988), pp. 2398–2417.
14. D. Jou and M.S. Mongiovì, Description and evolution of anisotropy in superfluid vortex tangles with counterflow and rotation, *Phys. Rev. B*, **74** (2006), 054509 (11 pages).
15. M.S. Mongiovì D. Jou and M. Sciacca, Energy and Temperature of Superfluid Turbulent Vortex Tangles, *Phys. Rev. B* **75** (2007), 214514 (10 pages).
16. D. Jou, M. Sciacca and M.S. Mongiovì, Vortex dynamics in rotating counterflow and plane Couette and Poiseuille turbulence in superfluid Helium, *Physica B* **403** (2008), pp. 2215–2224.
17. D. Jou and M.S. Mongiovì, Phenomenological description of counterflow superfluid turbulence in rotating containers, *Phys. Rev. B* **69** (2004), 094513 (7 pages).
18. M. Tsubota, C.F. Barenghi, T. Araki and A. Mitani, Instability of vortex array and transitions to turbulence in rotating helium II *Phys. Rev. B* **69** (2004), 134515 (12 pages).
19. W.F. Vinen, Mutual friction in a heat current in liquid helium II. III. Theory of the mutual friction, *Proc. Roy. Soc. London* **A240** (1957), pp. 493–515.
20. S. R. de Groot and P. Mazur, *Nonequilibrium thermodynamics*, North Holland, Amsterdam, 1962.
21. G. Lebon, D. Jou and J. Casas-Vzquez, *Understanding non-equilibrium thermodynamics*, Springer, Berlin, 2008.
22. D. Jou and M.S. Mongiovì, Non-Equilibrium Thermodynamics in Counterflow and Rotating Situations *Phys. Rev. B*, **72** (2005), 144517 (8 pages).

An Atomic-scale Finite Element Method for Single-Walled Carbon Nanotubes

M. Morandi Cecchi* and V. Rispoli, M. Venturin

Dip. di Matematica Pura ed Applicata,
Università degli Studi di Padova,
Via Trieste 63, 35121 Padova, Italy
**E-mail: mcecchi@math.unipd.it*
www.math.unipd.it

In this paper the application of the atomic-scale finite element method is presented for the simulation of the mechanical behaviour of carbon nanotubes. The obtained numerical results for the Young's moduli, the Poisson's ratios, the shear moduli and the ground energy configurations show agreement with the available experimental data.

Keywords: Atomic scale FEM, Carbon nanotubes, Nanomechanical properties.

1. Introduction.

Carbon nanotubes (CNTs) are at the boundary between small structures and large molecules. Reflecting their condition, single–scale modeling approaches can be generally classified into two categories: atomistic and continuous modeling.

In this work an atomistic computational method is developed for CNTs computation, following the approach proposed in Ref. 1. Moreover, the developed method is based on the Atomic–Scale Finite Element Method (AFEM) recently proposed in Ref. 2 as an efficient alternative to molecular mechanics with comparable accuracy for the nonlinear minimization of the system energy. This is due to the use of both first and second derivatives of system energy which improves the convergence of the method allowing the reduction of the total computational time.

In particular, the AFEM provides an order N framework, N being the number of atoms, for efficient and accurate computation of mechanical properties of CNTs.[2] Another advantage of the AFEM is the same formal structure of the continuum Finite Element Method (FEM) permitting the combination of AFEM with FEM avoiding artificial interfaces. For example, this method is applied in multiscale simulation for CNTs computation where the FEM is used for the macro–scale problem while the AFEM is used for the micro–scale problem.[3] The idea is to combine the atomistic method which captures the nanoscale physical laws with the continuum FEM which collects the behavior of atoms reducing the degrees of freedom of the system.

This paper is organized as follows.

The first section presents the carbon nanotubes from the graphene sheets and its mathematical modeling for the numerical simulation of these structures. In the following section, the formal modeling framework is presented, including the geometrical aspects and the choice of the specific molecular potential. Then the numerical method based on the atomic–scale finite element method is presented together with the numerical results. These include the Young's moduli, the Poisson's ratio, the shear moduli and the ground energy configurations as a function of molecule's diameter and chirality. Moreover the obtained results are cross–compared with the available external data.

2. Carbon nanotubes and chirality

Graphene is the two–dimensional counterpart of graphite. A sheet of graphene consists of a hexagonal lattice of carbon atoms, where the atoms occupy the vertices of the hexagons. A CNT is a sheet of graphene rolled up into a tube of a few nanometers in diameter and it can be many microns long. The basic structure is given by Single–Walled Carbon Nanotubes (SWCNTs), that are the fundamental component of the more complex forms Multi–Walled Carbon NanoTubes (MWCNTs), which are formed by several nested SWCNTs.

SWCNTs can be mathematically classified according to their he li ci ty and radius. The classification is based on an ordered pair of integers $(n_1, n_2) \in \mathbb{N}^2$.

Let a graphene sheet at its ground energy configuration be on the plane $\mathbb{R}^2$. Let the origin coincide with one atom of the lattice, say a_0, and suppose that one of the three adjacent atoms bonded to a_0 lies on the x axis.

Two additional vectors can be specified in order to simplify the classification of SWCNTs: $\mathbf{a}_1 = \left(3/2, \sqrt{3}/2\right)$ and $\mathbf{a}_2 = \left(3/2, -\sqrt{3}/2\right)$ and they can be used to define the chirality vector $\mathbf{C}_h = n_1\mathbf{a}_1 + n_2\mathbf{a}_2$, as shown by Fig. 1. This vector connects two "crystallographically equivalent" atoms whose positions will coincide when rolled–up into a tube. Each pair of integers $(n_1, n_2) \in \mathbb{N}^2$ defines a way of rolling up a graphene sheet to obtain a SWCNT but the additional constraint $n_2 \leq n_1$ must be specified to eliminate the multiple definitions. The chiral angle θ and the diameter d are defined as:

$$\theta = \arctan\left(\frac{\sqrt{3}n_2}{(n_2 + 2n_1)}\right), \qquad d = \frac{\sqrt{3}}{\pi}a_{(C-C)}\sqrt{n_2^2 + n_2 n_1 + n_1^2},$$

where $a_{(C-C)}$ is the distance between two adjacent carbon atoms on the sheet at rest which is assumed to maintain the value $a_{(C-C)} = 1.42\,\text{Å}$.[4]

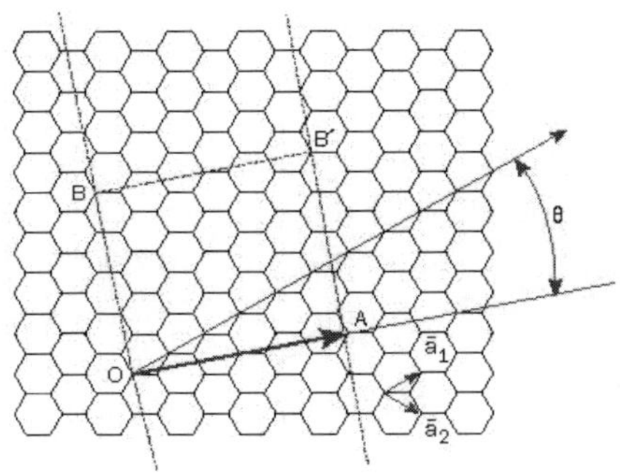

Fig. 1. Lattice of carbon atoms and the chirality vector $\mathbf{C}_h = \overline{OA}$.

According to their chirality, SWCNTs are named "zigzag", when identified by the couple $(n, 0)$, "armchair" by the couple (n, n), and "general chiral" otherwise. Moreover, chirality must be considered in computations of the mechanical behavior, since it influences the mechanical properties of the tubules.[5]

3. Numerical modeling

In this section the formal modeling framework is presented. This includes the geometrical aspects, the specific molecular potential and the atomic–scale finite element method for carbon nanotubes.

3.1. *Lattice modeling*

Reflecting the repetitive display of the lattice, a numbering scheme can be defined to model CNTs with reference to the unstressed configuration of the sheet, obtaining a structure that describes the entire sheet as the tiling repetition of an elementary Y–shaped cell.[1]

A sheet of graphene is defined as $\mathcal{G} = (\mathcal{A}, \mathcal{B}, \mathcal{C})$ where $\mathcal{A}$ is the set of all the atoms of the sheet, $\mathcal{B}$ is the set of all the binary bonds between pairs of adjacent atoms and $\mathcal{C}$ is the set of all the ordered couples of adjacent bonds.

Every atom $a \in \mathcal{A}$ is considered as a material point with two attributes: a label, lab_a, and its position at time t, $\mathrm{pos}_a(t)$.

Because of the hexagonal structure of the graphene sheet $\mathcal{G}$ another useful coordinate system that is not orthonormal can be specified for the physical space: $\mathbf{j}_1 = \sqrt{3}a_{(C-C)}(1,0,0)$, $\mathbf{j}_2 = \sqrt{3}a_{(C-C)}\left(1/2, \sqrt{3}/2, 0\right)$, $\mathbf{j}_3 = a_{(C-C)}(0,0,1)$. A rhomboidal grid on the graphene sheet is defined as $\mathcal{A}_1 \subset \mathcal{A}$:

$$\mathcal{A}_1 = \{a \in \mathcal{A}|\, \mathrm{pos}_a(t) = \alpha_1\mathbf{j}_1 + \alpha_2\mathbf{j}_2,\ \alpha_1, \alpha_2 \in \mathbb{Z}\}\,.$$

Every carbon atom $a \in \mathcal{A}_1$ will be uniquely identified by the integer couple $(\alpha_1, \alpha_2) \in \mathbb{Z}^2$ and will be referred to as $\mathrm{lab}_a = \mathcal{A}_1(\alpha_1, \alpha_2) = (\alpha_1, \alpha_2, 1)$. Let $\mathcal{A}_2 = \mathcal{A} \setminus \mathcal{A}_1$ be another rhomboidal grid with the same integer constraints. More precisely, $\mathcal{A}_2$ is the set whose atoms have their positions that are the shifted version of the atoms in $\mathcal{A}_1$. This shift is induced by the shifting vector $\mathbf{s} = a_{(C-C)}\left(\sqrt{3}/2, 1/2, 0\right)$. Hence,

$$\mathcal{A}_2 = \{a \in \mathcal{A}|\, \mathrm{pos}_a(t) = \alpha_1\mathbf{j}_1 + \alpha_2\mathbf{j}_2 + \mathbf{s},\ \alpha_1, \alpha_2 \in \mathbb{Z}\}\,.$$

The label for the atom $a \in \mathcal{A}_2$ is given by $\mathrm{lab}_a = \mathcal{A}_2(\alpha_1, \alpha_2) = (\alpha_1, \alpha_2, 2)$.

The $Y-$shaped cell is the elementary tile that covers the entire graphene sheet.[6] Every cell is uniquely identified by the ordered pair of integers $(\alpha_1, \alpha_2) \in \mathbb{Z}^2$ and consists of two atoms and three bonds. Every bond can be specified by an ordered pair of neighbouring atoms and, viceversa, one can refer to the first atom of the bond $b_u(\alpha_1, \alpha_2)$ as $b_u^1(\alpha_1, \alpha_2)$ for $u = 1, 2, 3$ and similarly for the second atom. The length vector $\mathbf{B}_u(t)$ can be associated with every bond $b_u(\alpha_1, \alpha_2)$ at time t as $\mathbf{B}_u(t) = \mathrm{pos}_{a_1}(t) - \mathrm{pos}_{a_2}(t)$ with length $l_u(t) = ||\mathbf{B}_u(t)||_2$.

In addition to atoms and bonds, the presented numbering scheme is extended to include couples of bonds that share a common atom. Six couples of bonds are associated to every $Y-$shaped elementary cell: let a generic cell be $Y(\alpha_1, \alpha_2)$, then these couples are named $c_i(\alpha_1, \alpha_2)$ with $i = 1, 2, \ldots, 6$.

Also in this case, one can refer to the first bond of the couple $c_u(\alpha_1, \alpha_2)$ as $c_u^1(\alpha_1, \alpha_2)$ for $i = 1, 2, \ldots, 6$ and similarly for the second bond. Let $c = (b_1, b_2) \in \mathcal{C}$ be a couple of bounds that share a common atom, then the angle between the bonds is given by $\rho_c(t) = (\mathbf{B}_1^T(t)\mathbf{B}_2(t))/(l_1(t)l_2(t))$, where $\mathbf{B}_i$ and l_i are, respectively, the vector and the length associated to the bond b_i for $i = 1, 2$. This numbering scheme applies also to SWCNTs.[7]

3.2. *Empiric potential*

The behavior and the energetics of molecules are fundamentally quantum mechanical. However, it is possible to employ the use of classical mechanics with a force–field approximation: the energy system is expressed only as a function of the nuclear positions, according with the Born–Oppenheimer approximation. Being an atomistic method, every atom that constitutes the system is modeled as a single material particle and an appropriate potential is introduced to predict the energy associated with the given conformation of the molecule.

Given a conformation vector $\boldsymbol{\Psi}(t) = [\mathrm{pos}_{a_1}, \ldots, \mathrm{pos}_{a_n}]^T$, $a_i \in \mathcal{A}$ of a SWCNT $\mathcal{S}$, the potential function V can be written as

$$V(\boldsymbol{\Psi}(t)) = V_c(\boldsymbol{\Psi}(t)) + V_n(\boldsymbol{\Psi}(t)),$$

where V_c is the partial term accounting for covalent bonds and V_n for noncovalent interactions. Moreover, the covalent term can be expressed as follows $V_c = V_b + V_a + V_d$, where V_b is the bond stretching potential, V_a is the angle bending potential and V_d is the dihedral interactions potential.

The noncovalent term accounts for the non–bonded electrostatic and van der Waals interactions and can be neglected:[8] only those accounting for bond stretching and angle variation are significant for the system potential. Therefore, the resulting potential energy is approximated as

$$V(\boldsymbol{\Psi}(t)) \simeq V_b(\boldsymbol{\Psi}(t)) + V_a(\boldsymbol{\Psi}(t)) .$$

Under the assumption of small local deformations, it is adequate to employ the simple harmonic approximation for angles and bonds.[9] In particular, the stretch potential of a single bond b can be expressed as $V_b(\boldsymbol{\Psi}(t)) = k_l/2 \left(l(t) - a_{(C-C)}\right)^2$ and the bending potential for the couple $c = (b_1, b_2)$ is given by $V_c(\boldsymbol{\Psi}(t)) = k_p/2 \left(\arccos p_c(t) - 2\pi/3\right)^2$. Hence, the potential of the molecule with the conformation $\boldsymbol{\Psi}(t)$ can be expressed as the summation of the term V_b and V_c for every bond and every couple of adjacent bonds at time t as $V = \sum_{b \in \mathcal{B}} V_b + \sum_{c \in \mathcal{C}} V_c$. The values of the harmonic parameters are $k_l = 652$ N/m and $k_p = 8.76 \cdot 10^{-19}$ Nm rad^{-2}.[6]

For every atom a, there is a force that can be described as $f_a(t)$. Thus, the external loads acting on the entire molecule are accounted by the sum

$$L(\boldsymbol{\Psi}(t)) = \sum_{a \in \mathcal{A}} \left(\mathrm{pos}_{a[\boldsymbol{\Psi}]}(t)^T f_a(t)\right) .$$

Finally, the formula for the potential energy and the loading term is:

$$I(\boldsymbol{\Psi}(t)) = V(\boldsymbol{\Psi}(t)) + L(\boldsymbol{\Psi}(t)) .$$

From this formulation it is possible to directly compute the gradient of the scalar potential. For the details see Ref. 1.

3.3. *The atomic–scale finite element method*

A state of ground energy corresponds to the equilibrium configuration of a solid. In standard FEM, a continuous solid is partitioned into a finite number of elements, each one with its set of discrete nodes. The energy minimization of the solid is obtained by the appropriate determination of the molecular conformation. Likewise, in molecular mechanics the calculation of the atom positions is based on a similar energy minimization.

Let a molecular system have N atoms, then the energy stored in the atomic bonds is denoted by $V_{\text{tot}}(\boldsymbol{\Psi})$ where $\boldsymbol{\Psi} = [\text{pos}_1, \ldots, \text{pos}_n]^T$ is a representation of the conformation vector. The total energy of the system is

$$E_{\text{tot}} = V_{\text{tot}}(\boldsymbol{\Psi}) - \sum_{i=1}^{N} \bar{\mathbf{F}} \cdot \boldsymbol{\Psi}_i\,,$$

with the external force $\bar{\mathbf{F}}$ applied to the i–th atom and the state of minimal energy is given by

$$\frac{\partial E_{\text{tot}}}{\partial \boldsymbol{\Psi}} = 0\,.$$

Let $\boldsymbol{\Psi}^{(0)}$ be an initial guess of equilibrium state and $\mathbf{u} = \boldsymbol{\Psi} - \boldsymbol{\Psi}^{(0)}$ the displacement, then the Taylor expansion of E_{tot} around $\boldsymbol{\Psi}^{(0)}$ is

$$E_{\text{tot}} \approx E_{\text{tot}}(\boldsymbol{\Psi}^{(0)}) + \left.\frac{\partial E_{\text{tot}}}{\partial \boldsymbol{\Psi}}\right|_{\boldsymbol{\Psi}^{(0)}} \mathbf{u} + \frac{1}{2}\mathbf{u}^T \left.\frac{\partial^2 E_{\text{tot}}}{\partial \boldsymbol{\Psi}^2}\right|_{\boldsymbol{\Psi}^{(0)}} \mathbf{u}.$$

The combination of the above two formulae yields the equation

$$\mathbf{K}(\boldsymbol{\Psi})\mathbf{u} = \mathbf{P}(\boldsymbol{\Psi})\,,$$

where $\mathbf{K}$ is the stiffness matrix and $\mathbf{P}$ is the non–equilibrium force vector, defined as

$$\mathbf{K} = \left.\frac{\partial^2 E_{\text{tot}}}{\partial \boldsymbol{\Psi}^2}\right|_{\boldsymbol{\Psi}^{(0)}} = \left.\frac{\partial^2 V_{\text{tot}}}{\partial \boldsymbol{\Psi}^2}\right|_{\boldsymbol{\Psi}^{(0)}}, \qquad \mathbf{P} = -\left.\frac{\partial E_{\text{tot}}}{\partial \boldsymbol{\Psi}}\right|_{\boldsymbol{\Psi}^{(0)}} = \bar{\mathbf{F}} - \left.\frac{\partial V_{\text{tot}}}{\partial \boldsymbol{\Psi}}\right|_{\boldsymbol{\Psi}^{(0)}}$$

where $\bar{\mathbf{F}}$ is the force vector.

When no bifurcations are present, the resulting nonlinear system can be solved with iterative methods and it is solved iteratively until $\mathbf{P}$ reaches zero. For atomistic interactions with pair potentials, $\mathbf{K}$ and $\mathbf{P}$ can be obtained from the continuous model as a representation of nonlinear spring elements.

In the application of the AFEM to carbon nanotubes, each carbon atom interacts with both first and second nearest neighbour atoms, since those are the most relevant interactions that must be considered.[10] These iterations are the result of the bond angle dependence in the interatomic potential. As a result, ten carbon atoms are considered in each element.

Since the potential has two components, namely the bond and the angle parts, also the stiffness matrix can be decomposed as follows $\mathbf{K} = \mathbf{K}_b + \mathbf{K}_a$ where the $\mathbf{K}_b$ accounts for $\mathbf{V}_b$ and $\mathbf{K}_a$ for $\mathbf{V}_a$. The simple analytical form of the potential permits the easy computation of the components of $\mathbf{K}$.

For a given element centered at the i–th atom, only the relation with other nine atom positions is significant. The position of every atom in $\mathbb{R}^3$ is identified by exactly three parameters and, hence, only thirty non–zero elements will appear in every row and every column of the stiffness matrix, since the topological distribution follows the honeycomb pattern. The same applies to the element stiffness matrix and the element non–equilibrium force vector, given by

$$\mathbf{K}^e = \begin{bmatrix} \left[\frac{\partial^2 V_{\mathrm{tot}}}{\partial \mathbf{\Psi}_1 \partial \mathbf{\Psi}_1}\right]_{3\times 3} & \left[\frac{\partial^2 V_{\mathrm{tot}}}{\partial \mathbf{\Psi}_1 \partial \mathbf{\Psi}_i}\right]_{3\times 27} \\ \left[\frac{\partial^2 V_{\mathrm{tot}}}{\partial \mathbf{\Psi}_i \partial \mathbf{\Psi}_1}\right]_{27\times 3} & [\mathbf{0}]_{27\times 27} \end{bmatrix}, \quad \mathbf{P}^e = \begin{bmatrix} \left[\bar{\mathbf{F}} - \frac{\partial V_{\mathrm{tot}}}{\partial \mathbf{\Psi}_1}\right]_{3\times 1} \\ [\mathbf{0}]_{27\times 1} \end{bmatrix}.$$

The nonlinear resolution of the system $\mathbf{K}(\mathbf{\Psi})\mathbf{u} = \mathbf{P}(\mathbf{\Psi})$ was obtained by the iteration of the four steps that follow: (i) explicitly compute $\mathbf{P}(\mathbf{\Psi}^{(i)})$; (ii) explicitly compute $\mathbf{K}(\mathbf{\Psi}^{(i)})$; (iii) solve $\mathbf{u}^{(i+1)}$ from $\mathbf{Ku} = \mathbf{P}$; (iv) update $\mathbf{\Psi}^{(i+1)}$, where $\mathbf{\Psi}^{(i+1)} = \mathbf{u}^{(i+1)} - \mathbf{\Psi}^{(i)}$, with $i = 0, 1, \ldots, \eta$ until $||\mathbf{P}(\mathbf{\Psi}^{(\eta)})|| < \varepsilon$, where $\varepsilon = 10^{-6}$ is a predefined tolerance.[2]

4. Numerical results

The obtained results permit the analysis of the dependence of length, diameter and elasticity of CNTs on curvature and helicity, compared and validated with the available experimental and computational data.

4.1. *Diameters and Lengths at the Energy Ground*

The results of the simulation of the length at rest of several SWCNTs are plotted as a function of the diameter in Figure 2 for all the armchair tubules $(n, 0)$ with $n = 3, \ldots, 11$ and for all zigzag SWCNTs (n, n) with $n = 5, \ldots, 13$. In these plots, the length discrepancy is defined as $\delta l = l_g - l_0$, where l_0 is the length on a graphene sheet approximation and l_g is the length at the energy ground for the tube. The length discrepancy is reduced with the increase of the diameter and follows the approximation $\delta l(d) \propto 1/d^2$ for all the tubules, almost vanishing for diameters smaller than 1.2 nm.

Another effect of the curvature is the discrepancy $\delta r = d_t - d_g$ between the estimation of the diameter with the norm of the chirality vector $d_t = 1/\pi||\mathbf{C}_h||$ and the real diameter at the energy ground d_g. In Figure 2 it is plotted the radial discrepancy for armchair and zigzag nanotubes, showning the smaller is the radius, the larger d_g becomes with respect to d_t.

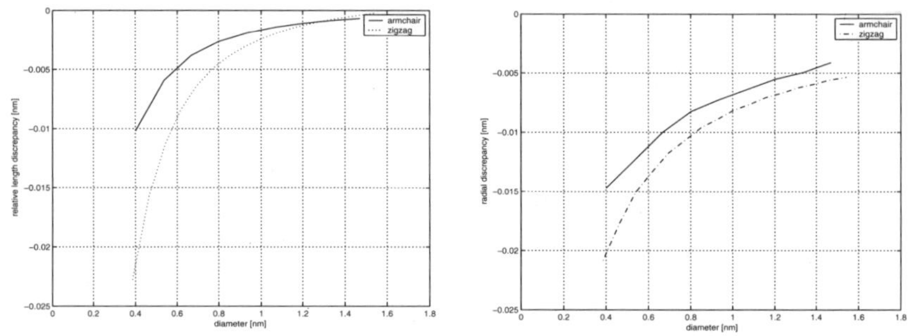

Fig. 2. Comparison of the lenght (left) and radial (right) discrepancies for nanotubes.

In brief, zigzag nanotubes present a double intensity of the length discrepancy when compared with armchair tubes. Similarly, the former class presents a radial discrepancy whose intensity is approximatively 4/3 of the value obtained for the latter. This results are comparable with existing investigations.[11]

4.2. *Young's Modulus*

The first step to describe a novel structural material is to present its Young's modulus $\bar{E}$ which, for a thin rod of isotropic and homogeneous material of cross–sectional area A_0 and of length l_g, $\bar{E}$, is given by $\bar{E} = (l_g F)/(A_0 \Delta l)$, where Δl is the elongation after the load and F is the force applied to the cross–sectional area.

These results are plotted as a function of the diameter for all the armchair tubules $(n, 0)$ with $n = 3, \ldots, 17$ and for all zigzag SWCNTs (n, n) with $n = 5, \ldots, 19$ and $A_0 = \pi(d_g/2)^2$. In the simulations, models with comparable length have been used, with an applied force on the top surface of -0.05 nN per atom in the axial direction of the tubule.

The results for the Young's modulus, obtained under the thin shell assumption, are presented in Fig. 3. From the plotted data, the Young's modulus increases with the diameter and, at the infinity tends to the same value of the Young's modulus for graphene sheet.

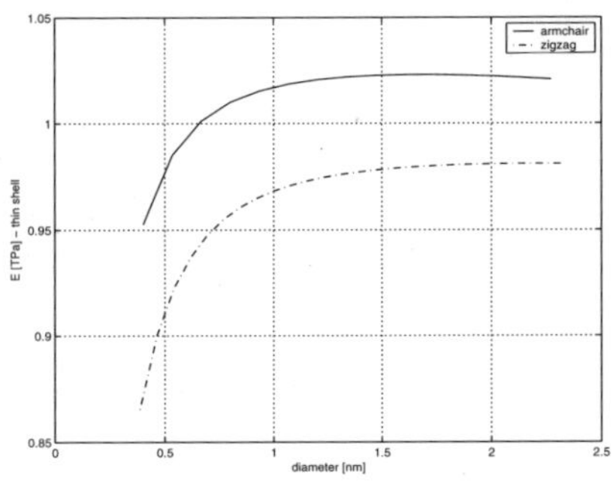

Fig. 3. Young's moduli for armchair and zigzag nanotubes as a function of their diameters, under the thin shell assumption.

In general, the experimentally and computationally predicted Young's moduli for SWCNTs range from 0.5 to 5.5 TPa .[5,12] The large scattering is not only due to the gap between computational and experimental approaches. From the computational point of view, several factors that may strongly influence the results can be summarized as follows: different methods, different molecular parameters, different reference models, different computational conditions. From the experimental point of view, the factors that may diffuse the results can be summarized as: different synthesis of the sample, different measurement techniques, presence of defects and their type, challenging identification of the sample.

Regardless of the reference model, the simulations suggest that armchair nanotubes are slightly stiffer than zigzag tubes, but the relative difference is upper bounded by 5% and, thus, can be considered negligible, as noted by several authors[13] with a wide range of methods. In brief, the Young's modulus approximatively does not dependent on chirality.

4.3. *Poisson's Ratio and Shear modulus*

The Poisson's ratio ν is a measure of the tendency of a material, when stretched in one direction, to get thinner in the remaining directions. More precisely, it is the ratio of the relative contraction strain normal to the applied load and the relative extension strain. In the case of CNTs one has

$\nu = -(d_g \Delta l)/(l_g \Delta d)$ where Δl is the elongation after the load and Δd is the difference in diameter after the load. In the numerical simulations, an axial force of -0.05 nN per atom on one surface is used and the minimization does not suffer bifurcations.

Figure 4 plots the Poisson's ratio for armchair and zigzag nanotubes.

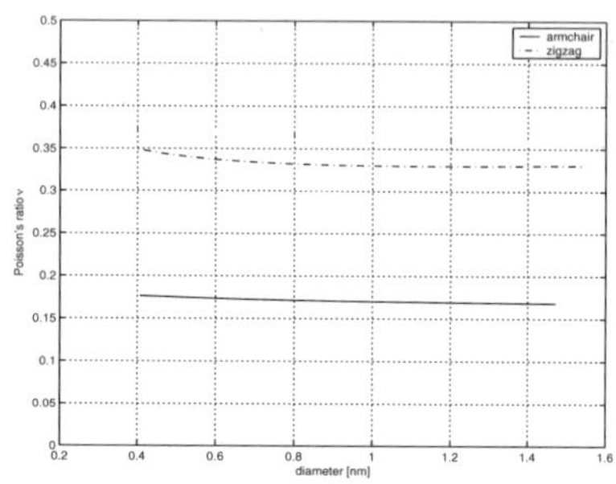

Fig. 4. Poisson's ratios for armchair and zigzag nanotubes.

The obtained Poisson's ratios are approximatively not dependent on the radius of the tubules. In fact, narrow tubes have slightly larger Poisson ratios that, however, are upper bounded by 5%. Viceversa, the Poisson's ratio seems to depend in a significant way on the chirality. Indeed, for armchair tubes it's value is $\nu_a = 0.18$, which is almost half of the value $\nu_z = 0.33$ for zigzag tubes. Consequently, it is always necessary to specify the chirality of the tube when the data is analyzed. The values for the Poisson's ratio range from 0.15[14] to 0.34.[15]

The shear modulus G, also called the modulus of rigidity, describes the tendency of a material to shear, that is to deform its shape at constant volume when acted upon by opposing forces. It is defined as shear stress over shear strain. For an isotropic elastic material the Young's modulus $\bar{E}$, the Poisson's ratio ν and the shear modulus G are related as follows $G = \bar{E}/(2 + 2\nu)$. Hence, under the assumption of an isotropic elastic material, Fig. 5 plots the results for the obtained shear moduli for armchair and zigzag nanotubes. Due to the explained difficulties with experimental techniques, there are still a small number of reports on the measured values of shear modulus of SWCNTs. They slightly depend both on chirality and on diameter, as observed in Ref. 16. Theoretical predictions[17] obtained the average result of 0.5 TPa, which is comparable to the values presented in this work. The lattice–dynamics model permitted the derivation of an analytical expression for the shear modulus, indicating that the shear modulus is about equal to that of graphene for large radii and smaller for narrow tubes.[16] Moreover, it made possible the observation of a weak dependence of the shear modulus on the chirality of the tubes for small radii.

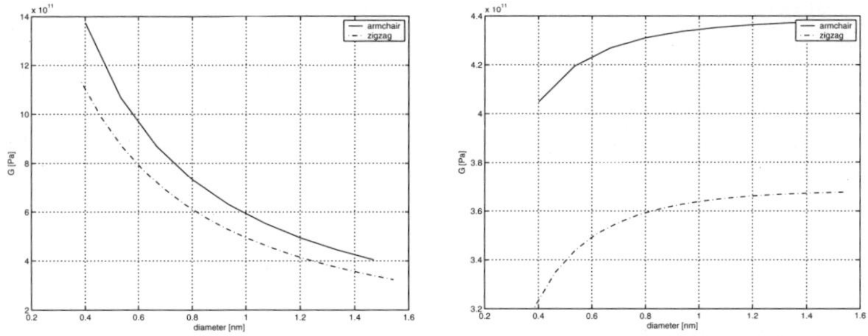

Fig. 5. (Left) Shear moduli for armchair and zigzag nanotubes; (Right) Shear moduli for the thin shell models.

5. Concluding remarks

In this article, the application of the atomic–scale finite element method with the use of the harmonic approximation for bond stretch and angle bending potentials is presented for the simulation of the mechanical behaviour of carbon nanotubes.

Several simulations have been performed for the numerical results for the Young's moduli, the Poisson's ratios, the shear moduli and the ground energy configurations, that include the evaluation of the lengths and of the diameters at rest applied to a broad range of carbon nanostructures. The obtained parameters and data show agreement with complementary results that come from the experimental data obtained by other authors.

References

1. M. Morandi Cecchi, A. Busetto, *Computational mechanical modeling of the behavior of carbon nanotubes*, in *ISTASC'07: Proc. 7th Int. Conf. WSEAS on Systems Theory and Scient. Comp.*, (2007), pp. 58–65.
2. B. Liu, Y. Huang, H. Jiang, S. Qu, K. Hwang, *The atomic–scale finite element method*, in *Comp. Meth. Appl. Mech. Eng.*, **193** (2004), pp. 1849–1864.
3. D. Qian, G. Wagner, W. Liu, *A multiscale projection method for the analysis of carbon nanotubes*, in *Comp. Meth. Appl. Mech. Eng.*, **193** (2004), pp. 1603–1632.

4. C. Sun, H. Bai, B. Tay, S. Li, E. Jiang, *Dimension, strength, chemical and thermal stability of a single C-C bond in carbon nanotubes*, in *J. Phys. Chem. B*, **31** (2003), pp. 7544–7546.
5. D. Sanchez–Portal, E. Artacho, J. Soler, A. Rubio, P. Ordejon, *Ab–initio structural, elastic and vibrational properties of Carbon Nanotubes*, in *Phy. Rev. B*, **59** (1999), pp. 12678-12688.
6. D. Caillerie, A. Mourad, A.Raoult, *Discrete homogenization in graphene sheet modeling*, in *J. Elast.*, **84** (2006), pp. 33–68.
7. R. Saito, G. Dresselhaus, M. Dresselhaus, *Physical Properties of Carbon Nanotubes*, Imp. College Press, (1998).
8. T. Chang, H. Gao, *Size–dependent elastic properties of a single–walled carbon nanotube via a molecular mechanics model*, in *J. Mech. and Phys. Solids*, **51** (2003), pp. 1059–1074.
9. N. Allinger, *Conformational analysis. 130. MM2. a hydrocarbon force field utilizing V1 and V2 torsional terms*, in *J. Am. Chem. Soc.*, **99** (1977), pp. 8127–8134.
10. D. Brenner, O. Shenderova, J. Harrison, B. Ni, S. Sinnott, *A second-generation reactive empirical bond order (REBO) potential energy expression for hydrocarbons*, in *J. Phys: Condensed Matter*, **14** (2002), pp. 783–802.
11. X. Zhou, H. Chen, J. Zhou, O.-Y. Zhong-Can, *The structure relaxation of carbon nanotube*, in *Physica B*, **304** (2001), pp. 86–90.
12. E. Hernández, C. Goze, P. Bernier, A. Rubio, *Elastic properties of C and $B_x C_y N_z$ composite nanotubes*, in *Phys. Rev. Lett.*, **80** (1998), pp. 4502–4505.
13. M. Arroyo, T. Belytschko, *Finite element methods for the non-linear mechanics of crystalline sheets and nanotubes*, in *Int. J. Num. Meth. Eng.*, **59** (2004), pp. 419–456.
14. K. N. Kudin, G. E. Scuseria, B. I. Yakobson, *c_2f, BN, and C nanoshell elasticity from ab–initio computations*, in *Phys. Rev. B*, **64** (2001), pp. 235406.
15. Z.-C. Tu, Z.-C. Ou-Yang, *Single–walled and multiwalled carbon nanotubes viewed as elastic tubes with the effective young's moduli dependent on layer number*, in *Phys. Rev. B*, **65** (2002), pp. 233407.
16. J. Kürti, G. Kresse, H. Kuzmany, *First–principles calculations of the radial breathing mode of single–wall carbon nanotubes*, in *Phys. Rev. B*, **58** (1998), pp. R8869–R8872.
17. J. Lu, *Elastic properties of carbon nanotubes and nanoropes*, in *Phys. Rev. Lett.* **79** (1997), pp. 1297–1300.

SYSTEMATIC VARIABLE LENGTH CHECK BINARY UNORDERED/AUED CODES

LAURA PEZZA

Dipartimento di Metodi e Modelli Matematici, Università di Roma "La Sapienza", Via Antonio Scarpa, 16. 00161 Roma. Italy.
E-mail: `pezza@dmmm.uniroma1.it`

LUCA G. TALLINI

Dipartimento di Scienze della Comunicazione, Università di Teramo, Coste Sant'Agostino. 64100 Teramo. Italy.
E-mail: `ltallini@unite.it`

BELLA BOSE

School of Electrical Engineering & Computer Science, Oregon State University, Corvallis, OR 97331. USA.
E-mail: `bose@eecs.orst.edu`

In an unordered code no codeword is contained into any other codeword. Unordered codes are All Unidirectional Error Detecting (AUED) codes. It is well known that among all systematic codes with k information bits, Berger codes are optimal unordered codes with $r = \lceil \log_2(k+1) \rceil$ check bits. This paper introduces a new class of systematic unordered codes with variable length check symbols whose average redundancy is $r \approx (1/2)\log_2(\pi e k/2) = (1/2)\log_2 k + 1.04709\ldots$. Such new codes are shown to be optimal in the class of systematic AUED codes with fixed length information symbols and variable length check symbols.

Keywords: unordered codes, All Unidirectional Error Detecting (AUED) codes, systematic codes, Berger codes, unidirectional errors, asymmetric errors.

1. Introduction

Let $\mathbb{Z}_2 = \{0, 1\}$ be the binary alphabet. Given any binary word $X \in \mathbb{Z}_2^n$ of length n, let $supp(X)$ indicate the support of X, that is, the subset of the index set, say, $[1, n] = \{1, 2, \ldots, n\}$, where X is different from 0. For example, if $n = 6$ and $X = 010110 \in \mathbb{Z}_2^6$ then $supp(X) = \{2, 4, 5\} \subseteq [1, 6]$, and the Hamming weight of X is $w(X) = w_H(X) = |supp(X)| = 3$. It is well known that the function *supp* defines an isomorphism between the Boolean algebra on $\mathbb{Z}_2^n$ defined by the usual bitwise logic operations (AND, OR, NOT, etc.) and the Boolean algebra on the family of subsets of $[1,n]$ (the power set of $[1,n]$) defined by the usual set operations (intersection, union, complement, etc., respectively). Usually, the inverse of the function *supp* is called the characteristic function of sets. Here, we will identify binary words $X \in \mathbb{Z}_2^n$ with their support $X \equiv supp(X) \subseteq [1,n]$. Given any two word $X, Y \in \mathbb{Z}_2^n$, we say that X is contained in Y and write $X \subseteq Y$ if, and only if, $X = X \cap Y$, where $\cap$ indicates set intersection. If X is not contained in Y and Y is not contained in X we say that X is unordered with Y (or X and Y are unordered). If X and Y are not unordered we say that they are ordered. For example, $001101 \equiv \{3,4,6\} \subseteq \{1,3,4,5,6\} \equiv 101111$ (and 001101 is ordered with 101111); whereas, $001101 \equiv \{3,4,6\}$ is unordered with $100101 \equiv \{1,4,6\}$. Note that the relation $\subseteq$ defines a partial ordering in the family of subsets of $[1,n]$ ($\equiv$ sets of binary words of length n). In this paper, systematic codes are primarily considered. So, let $\overline{I}$ indicate the complement of any subset $I \subseteq [1,n]$. A code $\mathcal{C}$ of length n over $\mathbb{Z}_2$ is called systematic with k information digits and $r = n - k$ check digits if there exists an index subset $I \subseteq [1,n]$ of cardinality $|I| = k$ such that for all (information) words/subsets $X \subseteq I$ there exists one codeword $E(X) \in \mathcal{C}$ (the encoding of X) such that $X = E(X) \cap I$. The word/subset $C(X) = E(X) \cap \overline{I}$ is called the check symbol of X in the encoding defined by $E(X)$. Note that $X \subseteq I$, with $|I| = k$, and $C(X) \subseteq \overline{I}$, with $|\overline{I}| = r$. Systematic codes are desirable because the information is readily available upon receiving a codeword which was sent through a communication channel.

In an unordered code no codeword is contained into any other codeword.[1,2] For example,

$$\mathcal{C} = \left\{ \begin{array}{l} 00\,10, \\ 01\,01, \\ 10\,01, \\ 11\,00 \end{array} \right\} \tag{1}$$

is an unordered code. Such codes can detect all the transmission errors in an asymmetric and/or unidirectional channel.[3–5] In the asymmetric channel only 1 to 0 (written as $1 \to 0$) type of errors can occur. In the unidirectional channel, within a transmitted word, either the errors are all of type $1 \to 0$ or they are all of type $0 \to 1$. For example, let $X = 0110$, $Y = 01\underline{0}0$, $Z = \underline{1}110$ and $T = \underline{1}0\underline{0}1 \in \mathbb{Z}_2^4$. If X is the transmitted word on the asymmetric channel then X and Y can be received but not Z or T. Whereas, if X is the transmitted word on the unidirectional channel then X, Y or Z can be received but not T. Hence, for both channel models if $X \in \mathbb{Z}_2^n$ is transmitted and $Y \in \mathbb{Z}_2^n$ is received, then $Y \subseteq X$ or $X \subseteq Y$ (i. e., X is ordered with Y). Asymmetric and/or unidirectional errors are typical in optical fibers, optical disks and VLSI systems.[2]

It is well known that among all systematic codes with k information bits, Berger codes are optimal unordered block codes with $r = r(k) = n - k = \lceil \log_2(k+1) \rceil$ check bits.[1] For all $w \in \mathbb{N}$, let $[w]_2$ indicate the binary digit representation of the number w. In a Berger code, a k-bit information word $X \subseteq I = [1, k]$ is encoded by appending the check symbol $C(X) = [w(I - X)]_2 \subseteq \overline{I} = [k+1, n]$ to X; that is,

$$\mathcal{E}_{Berger}(X) = E(X) = XC(X) = X[w(I - X)]_2,$$

where "$-$" indicates set difference. For example, the code in (1), is a Berger code with $k = 2$ information bits where $I = [1, 2]$. Note that the redundancy is $r = n - k = \lceil \log_2(k+1) \rceil = \lceil \log_2(3) \rceil = 2$ check bits, hence, the length is $n = 4$ and $\overline{I} = [3, 4]$. Section 2 of this paper, introduces a new class of systematic unordered codes where the length of the check symbol associated with an information word $X \in \mathbb{Z}_2^k$ may vary, and depend on X (through its weight). Such systematic variable length check symbol unordered codes have an average redundancy equal to $r = r_{prop}(k) = \mathbb{E}[r(X)]$, such that for all $k \in \mathbb{N}$,

$$0 \lesssim r_{prop}(k) - \frac{1}{2} \log_2 \frac{e\pi k}{2} \lesssim 0.08607\ldots \tag{2}$$

where $a(k) \lesssim b(k)$ indicates that $a(k) \leq b(k)$ up to infinitesimal quantities, and $(1/2)\log_2(e\pi k/2) = (1/2)\log_2 k + 1.04709\ldots$. Hence, this code design gives an improvement with respect to the optimal systematic block Berger codes of roughly $(1/2)\log_2 k + \Theta(1)$, as the rightmost column of Table 1 shows. It is also shown that such codes are optimal in the class of all systematic AUED codes with fixed length information symbols and variable length check symbols.

2. Code design

Given $k \in \mathbb{N}$, the code design idea is to append as the check symbol of any information word $X \in \mathbb{Z}_2^k$ a compressed encoding of the number of 0's of X $(= w(I - X))$. This can be simply accomplished by using a suitable variable length encoding $\mathcal{H} : [0, k] \to \mathcal{C}_{\mathcal{H}} \subseteq \mathbb{Z}_2^+$ which P1) makes sure that the resulting variable length code

$$\mathcal{C} = \{ X\mathcal{H}(w(I - X)) : X \in \mathbb{Z}_2^k \} \tag{3}$$

Table 1. Comparisons of the redundancies of the proposed code design with the lower bounds in Theorem 2.4 and the redundancies of the Berger code design.

k	r_{Berger}	r_{prop}	$r_{prop} - \frac{1}{2}\log_2 \frac{e\pi k}{2}$	$r_{Berger} - r_{prop}$
1	1	1.000000000	−0.047095585	0.000000000
2	2	1.750000000	0.202904415	0.250000000
3	2	2.000000000	0.160423164	0.000000000
4	3	2.312500000	0.265404415	0.687500000
5	3	2.375000000	0.166940367	0.625000000
6	3	2.562500000	0.222923164	0.437500000
7	3	2.578125000	0.127351954	0.421875000
8	4	2.722656250	0.175560665	1.277343750
9	4	2.726562500	0.094504414	1.273437500
10	4	2.851562500	0.143502867	1.148437500
11	4	2.852539063	0.075727668	1.147460938
12	4	2.965820313	0.126243477	1.034179688
13	4	2.966064453	0.068749009	1.033935547
14	4	3.068420410	0.117647364	0.931579590
15	4	3.070983887	0.070443004	0.929016113
16	5	3.138336182	0.091240596	1.861663818
18	5	3.207092285	0.075034199	1.792907715
20	5	3.274541855	0.066482222	1.725458145
22	5	3.340617180	0.063805785	1.659382820
24	5	3.400358915	0.060782080	1.599641085
26	5	3.449067324	0.051751880	1.550932676
28	5	3.497637291	0.046864244	1.502362709
30	5	3.545969152	0.045428269	1.454030848
32	6	3.593991189	0.046895604	2.406008811
40	6	3.751186305	0.043126673	2.248813695
48	6	3.886957021	0.047380186	2.113042979
56	6	4.005617662	0.054844616	1.994382338
64	7	4.100836770	0.053741185	2.899163230
128	8	4.573971210	0.026875625	3.426028790
256	9	5.091964423	0.044868838	3.908035577
512	10	5.567392870	0.020297285	4.432607130
1024	11	6.089518254	0.042422669	4.910481746
2048	12	6.566269852	0.019174266	5.433730148
4096	13	7.088597720	0.041502135	5.911402280
8192	14	7.565977237	0.018881651	6.434022763

is All Unidirectional Error Detecting (AUED), and P2) compresses

$$w(I - X) \in [0, k].$$

First, we need to make more precise the concept of being AUED for a variable length code. Some further notations are needed. Given $n \in \mathbb{N}$, an index subset $I = \{i_1, i_2, \ldots, i_l\} \subseteq [1, n]$ and $X = x_1 x_2 \ldots x_n \in \mathbb{Z}_2^n$, let X_I indicate the restriction of X to I; that is, the word $X_I = x_{i_1} x_{i_2} \ldots x_{i_l} \in \mathbb{Z}_2^l$ of length $l = |I|$. For example, if $n = 4$, $I = \{1, 2, 4\} \subseteq [1, 4]$ and $X = \underline{01}1\underline{0} \in \mathbb{Z}_2^4$ then $X_I = 010 \in \mathbb{Z}_2^3$. Further, let $l(X)$ indicate the length of $X \in \mathbb{Z}_2^+$. We give the following definition which strictly generalize the AUED property for block codes.

Definition 2.1. A variable length uniquely decodable code (see Ref. 6, page 80) is (instantaneous) AUED if, and only if, upon receiving the output of (the last bit of) any codeword from the channel, the receiver can either 1) output the received word if no error occurred during the transmission, or 2) detect errors otherwise (i. e., if some unidirectional errors occurred during the transmission).

For example, the variable length uniquely decodable code (in the rest of the paper the codewords are read/parsed from left to right)

$$\mathcal{C} = \left\{ \begin{array}{l} 010, \\ 0011, \\ 1001, \\ 1010 \end{array} \right\},$$

is AUED because no erroneous version of any codeword is a prefix of a sequence of codewords. This implies that any codeword received in error cannot be parsed by the decoder (because it is not a prefix of a sequence of codewords) and will cause a parsing error when or before the last bit of the codeword has been read. Viceversa, any codeword received not in error will be parsed by the decoder. In the first case, the receiver will detect a unidirectional error, in the second case the receiver will output the successfully parsed codeword. The following theorem holds.

Theorem 2.1. *Let $\mathcal{C}$ be a variable length uniquely decodable code. If $\mathcal{C}$ is a AUED code then $\mathcal{C}$ is a prefix code (i.e., no codeword is the prefix of any other codeword).*

Proof. If $\mathcal{C}$ is not a prefix code then there exists two codewords $X, Y \in \mathcal{C}$ such that X is a proper prefix of Y. Assume X is sent and received correctly. Once the the last bit of X is received, there can be the following two cases: C1) X was sent and received correctly (in this case the receiver should output X), and C2) Y was sent and the proper prefix X of Y was received correctly (in this case the receiver should wait and receive the remaining bits of Y). Since both cases are possible the code $\mathcal{C}$ is not AUED according to Definition 2.1. □

Theorem 2.1 states that we need to look for suitable codes in the class of the prefix (or, instantaneous) codes; namely, codes which can be represented by the set of leaves of labeled binary trees (see Ref. 6, page 81). The combinatorial property for a variable length prefix code to be AUED is that no codeword is ordered with (namely, is contained or contains) a prefix of any other codeword, as the following theorem states.

Theorem 2.2. *A variable length prefix code $\mathcal{C}$ is AUED if, and only if, for all $X, Y \in \mathcal{C}$ such that $X \neq Y$, the following relations hold:*

$$\begin{array}{l} X_{[1,\min\{l(X),l(Y)\}]} \not\subseteq Y_{[1,\min\{l(X),l(Y)\}]}, \text{ and} \\ Y_{[1,\min\{l(X),l(Y)\}]} \not\subseteq X_{[1,\min\{l(X),l(Y)\}]} \end{array} \iff$$

$$\begin{array}{l} X_{[1,\min\{l(X),l(Y)\}]} - Y_{[1,\min\{l(X),l(Y)\}]} \neq \emptyset, \text{ and} \\ Y_{[1,\min\{l(X),l(Y)\}]} - X_{[1,\min\{l(X),l(Y)\}]} \neq \emptyset \end{array} \quad (4)$$

Proof. Assume the prefix code $\mathcal{C}$ is not AUED. Then there exists $X \in \mathcal{C}$ (a sent codeword) and $X' \in \mathbb{Z}_2^{l(X)}$ (the received word), with $X' \neq X$, such that X' is obtained from X due to unidirectional errors and X' is a prefix of a sequence of codewords. Hence, $X' \subseteq$(or $\supseteq$) X and X' is a prefix of a sequence of codewords, say $Y_1 Y_2 \dots Y_s \in \mathcal{C}^+$, $s \in \mathbb{N}$. So, either $X' \subseteq$(or $\supseteq$) $X \in \mathcal{C}$ is a prefix of $Y_1 \in \mathcal{C}$ or $Y_1 \in \mathcal{C}$ is a prefix of $X' \subseteq$(or $\supseteq$) $X \in \mathcal{C}$. In both cases there exists two distinct codewords such that one codeword is ordered with a prefix of the other codeword. Viceversa, assume that there exists $X, Y \in \mathcal{C}$, with $X \neq Y$, such that $X_{[1,\min\{l(X),l(Y)\}]} \subseteq Y_{[1,\min\{l(X),l(Y)\}]}$ or $Y_{[1,\min\{l(X),l(Y)\}]} \subseteq X_{[1,\min\{l(X),l(Y)\}]}$. Without loss of generality assume that $l(X) = \min\{l(X), l(Y)\} \leq l(Y)$ so that $X = X_{[1,\min\{l(X),l(Y)\}]} \in \mathcal{C}$ is ordered with a prefix, say $X' \neq X$, of $Y \in \mathcal{C}$. If X is sent and X' is received then as soon as the receiver obtains X' it cannot decide between the following two cases: C1) X was sent and $X' \neq X$ was received (in this case the receiver should detect errors), and C2) Y was sent and the prefix X' of Y was received correctly (in this case the receiver should wait and receive the remaining bits of Y). So, the prefix code $\mathcal{C}$ is not AUED. □

Given the combinatorial characterization provided by Theorem 2.2 for the variable length AUED codes that **every codeword is unordered with the prefix of any other codeword**, we call variable length AUED codes also **variable length unordered codes**. For example, from Theorem 2.2, the code

$$\mathcal{C} = \left\{ \begin{array}{ll} 0010, & 100011, \\ 00011, & 110100, \\ 010010, & 1001011, \\ 010101, & 1100010 \end{array} \right\},$$

is a variable length AUED/unordered (prefix) code. Note that as a consequence of Theorem 2.1 and Definition 2.1, decoding is simply done as the usual prefix code decoding, but as soon as the decoder has a parse error on the received sequence, it detects an error on the word which is currently parsing (because it is not a codeword).

We are now ready to focus our attention to the code design. As we mentioned above, the structure of our code design is like the one given in (3) and the problem is to find which variable length encoding $\mathcal{H} : [0, k] \rightarrow \mathcal{C}_{\mathcal{H}} \subseteq \mathbb{Z}_2^+$ makes the code $\mathcal{C}$ to be AUED; that is, satisfy the combinatorial property given in Theorem 2.2. Recall that any prefix binary code $\mathcal{C}_{\mathcal{H}}$ can be represented by a binary tree where every left edge is labeled by 0 and every right edge is labeled by 1. In this representation, every codeword in $\mathcal{C}_{\mathcal{H}}$ is represented by the unique path form the root of the tree to a leave. One solution to our problem is given by Theorem 2.3 below. The theorem states that any encoding to a variable length prefix code $\mathcal{C}_{\mathcal{H}}$ which preserves the integer ordering by mapping it into the left to right ordering of the leaves of the tree representing $\mathcal{C}_{\mathcal{H}}$ is fine. Let $d : \mathbb{Z}_2^+ \rightarrow [0, 1] \subseteq \mathbb{R}$ be the function which associates every binary sequence $C = c_1 c_2 \ldots c_l$ with the number $d(C) = c_1(1/2)^1 + c_2(1/2)^2 + \ldots + c_l(1/2)^l \in [0, 1]$ whose binary expansion is C. For example, $d(1101) = (1/2) + (1/4) + (1/16) = 13/16$.

Theorem 2.3. *Given the number of information bits $k \in \mathbb{N}$, let $\mathcal{H}$: $[0, k] \rightarrow \mathcal{C}_{\mathcal{H}} \subseteq \mathbb{Z}_2^+$ be any encoding to a variable length prefix code $\mathcal{C}_{\mathcal{H}}$ which preserves the integer ordering; that is, such that,*

$$\text{for all } w_1, w_2 \in [0, k], \qquad w_1 < w_2 \implies d(\mathcal{H}(w_1)) < d(\mathcal{H}(w_2)). \quad (5)$$

Then the variable length prefix code in (3) is AUED.

Proof. Let $XC(X), YC(Y) \in \mathcal{C}$ be two distinct codewords and let $l = \min\{l(C(X)), l(C(Y))\}$. If X is unordered with Y then, $XC(X)$ is unordered with a prefix of $YC(Y)$ or viceversa. If instead $X \subseteq Y$ and $X \neq Y$, then $Y - X \neq \emptyset$ and so $[YC(Y)]_{[1,k+l]} - [XC(X)]_{[1,k+l]} \neq \emptyset$. Further, if $X \subseteq Y$ and $X \neq Y$, then $w(X) < w(Y)$ and $w(I - X) = k - w(X) > k - w(Y) = w(I - Y)$, and so $d(C(X)) = d(\mathcal{H}(w(I - X)) > d(\mathcal{H}(w(I - Y)) = d(C(Y))$. But, $d(C(X)) > d(C(Y))$ implies $[C(X)]_{[k+1,k+l]} - [C(Y)]_{[k+1,k+l]} \neq \emptyset$, and so $[XC(X)]_{[1,k+l]} - [YC(Y)]_{[1,k+l]} \neq \emptyset$. In any case, condition (4) of Theorem 2.2 holds, and so the variable length code $\mathcal{C}$ is AUED. □

For example, let $k = 8$, consider the variable length prefix code $\mathcal{C}_{\mathcal{H}_1} = \{0000, 0001, 001, 01, 100, 1010, 1011, 110, 111\}$, and let $\mathcal{H}_1 : [0, 8] \rightarrow \mathcal{C}_{\mathcal{H}_1}$ be defined as

$$\begin{array}{lll} \mathcal{H}_1(0) = 0000, & \mathcal{H}_1(3) = 01, & \mathcal{H}_1(6) = 1011, \\ \mathcal{H}_1(1) = 0001, & \mathcal{H}_1(4) = 100, & \mathcal{H}_1(7) = 110, \\ \mathcal{H}_1(2) = 001, & \mathcal{H}_1(5) = 1010, & \mathcal{H}_1(8) = 111. \end{array}$$

Since condition (5) is satisfied for $\mathcal{H} = \mathcal{H}_1$, the variable length code $\mathcal{C} = \mathcal{C}_1$ as defined in Theorem 2.3 with $k = 8$ information bits and $\mathcal{H} = \mathcal{H}_1$ is AUED and the following is a subset of codewords.

$$\left\{\begin{array}{lll} 00000000\,111, & 11100000\,1010, & 11111100\,001, \\ 10000000\,110, & 11110000\,100, & 11111110\,0001, \\ 11000000\,1011, & 11111000\,01, & 11111111\,0000 \end{array}\right\} \subseteq \mathcal{C}_1.$$

From (3), note that all the information words with the same weight have as check symbol the same codeword from $\mathcal{C}_{\mathcal{H}_1}$.

In general, decoding is done as in the classical block length case by checking if the received word is a codeword ($\iff$ no erorrs occurred during the transmission) or not ($\iff$ some erorrs occurred during the transmission). For the code $\mathcal{C}$ as defined in Theorem 2.3 above, decoding is done as follows. Assume the k bit information symbol is sent before the check symbol for every codeword. Upon receiving a word, the receiver computes the check symbol from the received k bit information part and checks if this is equal to the received check symbol (namely, the prefix of the remaining received sequence of bits). If the two check symbols are equal it can output the received word as the sent codeword; if they are different it detects an error (and, for example, in an ARQ (Automatic Repeat Request) protocol communication system[4,5] it asks the sender for a resend of the same codeword). For example, consider the above code $\mathcal{C}_1$ with $k = 8$ information bits. Assume the source needs to send the information sequence

$$X_1 = 11101101, X_2 = 11001011, X_3 = 10100010, \ldots$$

to the destination. The sender encodes the above sequence into

$$S = X_1\,C(X_1)\;X_2\,C(X_2)\;X_3\,C(X_3)\;\ldots =$$

$$11101101\,001\;11001011\,01\;10100010\,1010\;\ldots$$

and sends it to the destination; where,

$$\begin{array}{l} C(X_1) = \mathcal{H}_1(w(I - X_1)) = \mathcal{H}_1(2) = 001, \\ C(X_2) = \mathcal{H}_1(w(I - X_2)) = \mathcal{H}_1(3) = 01, \\ C(X_3) = \mathcal{H}_1(w(I - X_3)) = \mathcal{H}_1(5) = 1010, \;\ldots. \end{array}$$

Assume

$$S' = Y_1C_1Y_2C_2Y_3C_3\ldots = 1110110100111\underline{0}0010\underline{0}10\underline{0}101000101010\ldots$$

is received by the destination. Once the receiver has obtained the first $k = 8$ bits $Y_1 = 11101101$, it computes $C(Y_1) = \mathcal{H}_1(w(I - Y_1)) = \mathcal{H}_1(2) = 001$. Since $C(Y_1) = 001$ is the prefix of the remaining sequence

$$C_1Y_2C_2Y_3C_3\ldots = \boxed{001}1\underline{0}0010\underline{0}10\underline{0}101000101010\ldots,$$

it parses the codeword $Y_1C_1 = 11101101\,001$, outputs the information symbol $Y_1 = 11101101 = X_1$ and consider the remaining sequence

$$Y_2C_2Y_3C_3\ldots = 1\underline{0}0010\underline{0}10\underline{0}101000101010\ldots.$$

Once the receiver has obtained the first $k = 8$ bits $Y_2 = 1\underline{0}0010\underline{0}1$ of the above sequence, it computes $C(Y_2) = \mathcal{H}_1(w(I - Y_2)) = \mathcal{H}_1(5) = 1010$. Since $C(Y_2) = \mathbf{1}010$ is not a prefix of the remaining sequence

$$C_2Y_3C_3\ldots = \boxed{\mathbf{0}\underline{0}10}1000101010\ldots,$$

the receiver detects an error and calls the procedure to handle unidirectional errors. Note that the receiver is able to detect the errors even before the codeword $X_2\,C(X_2) = 11001011\,01$ has been received completely (as $Y_2C_2 = 1\underline{0}0010\underline{0}10\underline{0}$). In particular, the receiver can detect the errors as soon as it has received the 9-th bit of $X_2\,C(X_2)$ because the 9-th bit of $Y_2C_2Y_3C_3\ldots = 1\underline{0}0010\underline{0}1\mathbf{0}\underline{0}101000101010\ldots$ is a 0 whereas the 9-th bit of $Y_2C(Y_2) = 1\underline{0}0010\underline{0}1\mathbf{1}010$ is a 1.

Now we turn our attention to the redundancy of the above code design. Note that, unlike the block length codes, the number of check bits r varies and depends on the (weight of) the information symbol of a codewords. So to analyze the redundancy, one should consider and evaluate the average number of check bits used by the code design. For example, let us consider the redundancy performance of the variable length code $\mathcal{C}_1$ described above. Assume that all the $k = 8$ information bits are independent and equally distributed with the probability of being 0 equal to the probability of being 1 equal to 1/2. In this case, the average redundancy is equal to, $r_1 = r = \mathbb{E}[r(X)] = \left[\binom{8}{0}4 + \binom{8}{1}4 + \binom{8}{2}3 + \binom{8}{3}2 + \binom{8}{4}3 + \binom{8}{5}4 + \binom{8}{6}4 + \binom{8}{7}3 + \binom{8}{8}3\right]/2^8 = 3.14453\ldots$. Note that this redundancy is considerably better than the classical Berger code redundancy which is equal to $r_{Berger} = 4$ check bits for the given $k = 8$ information bits. Actually, it is possible to lower the redundancy even further by considering, for example, this other variable length prefix code: $\mathcal{C}_{\mathcal{H}_2} = \{00000, 00001, 0001, 001, 01, 10, 110, 1110, 1111\}$, and letting $\mathcal{H}_2 : [0, 8] \to \mathcal{C}_{\mathcal{H}_2}$ be defined as

$$\begin{array}{lll} \mathcal{H}(0) = 00000, & \mathcal{H}(3) = 001, & \mathcal{H}(6) = 110, \\ \mathcal{H}(1) = 00001, & \mathcal{H}(4) = 01, & \mathcal{H}(7) = 1110, \\ \mathcal{H}(2) = 0001, & \mathcal{H}(5) = 10, & \mathcal{H}(8) = 1111. \end{array}$$

Also in this case condition (5) is satisfied for $\mathcal{H} = \mathcal{H}_2$, and the code $\mathcal{C} = \mathcal{C}_2$ as defined in Theorem 2.3 with $k = 8$ information bits is AUED. The average redundancy of the code $\mathcal{C}_2$, is equal to $r_2 = 2.72265\ldots < r_1 = 3.14453\ldots < r_{Berger} = 4$. The following theorem gives a lower bound on the minimal average redundancy which is possible to achieve.

Theorem 2.4. *Given the number of information bits $k \in \mathbb{N}$, let the components of an information symbol $X \in \mathbb{Z}_2^k$ be independent random variables equally distributed with the probability of being* 0 *equal to the probability of being* 1 *equal to* 1/2*. The weight of X, $W_k = w(X) \in [0,k]$ remains defined as a binomial random variable with $p(w) = Pr(W_k = w) = \binom{k}{w}/2^k$, for all $w \in [0,k]$. Then the minimal average redundancy, $r_{min} = r_{min}(k) \in \mathbb{R}$, which is possible to obtain with any systematic variable length check unordered code satisfies the following relation*

$$r_{min} \geq H[W_k] \simeq \frac{1}{2}\log_2 \frac{e\pi k}{2} = \frac{1}{2}\log_2 k + 1.04709\ldots,$$

where $H[W_k] \in \mathbb{R}$ indicates the entropy of the random variable $W_k = w(X)$.

Proof. Consider the following set of $k+1$ information words

$$\mathcal{I} = \left\{ \begin{array}{l} 0000\ldots0000, \\ 1000\ldots0000, \\ 1100\ldots0000, \\ \qquad\vdots \\ 1111\ldots1100, \\ 1111\ldots1110, \\ 1111\ldots1111 \end{array} \right\}$$

and note that for all $X, Y \in \mathcal{I} \subseteq \mathbb{Z}_2^k$, X is ordered with Y. Hence, the check symbols associated with the information words in $\mathcal{I}$ must be all distinct. So there must be at least $|\mathcal{I}| = k+1$ distinct check symbols. Further, if two information words have the same weight then they are unordered and so they can have the same check symbol. Hence, a least redundant code design can have $k+1$ distinct check symbols each of which associated with all the information words of a certain weight; namely, an optimal code design is structured like the code in (3). This and Shannon's source coding Theorem (see Ref. 6, page 86), imply

$$r_{min} \geq H[W_k] = -\sum_{w=0}^{k} p(w)\log_2 p(w) = -\sum_{w=0}^{k} \frac{\binom{k}{w}}{2^k}\log_2 \frac{\binom{k}{w}}{2^k}.$$

Now, since the binomial distribution is asymptotically normal, for sufficiently large $k \in \mathbb{N}$, the entropy of the random variable $w(X)$ with mean $k/2$ and variance $k/4$ can be approximated with the entropy of a normal random variable with the same parameters. In particular, from Theorem 1 in Ref. 7, page 170, it can be shown that the following approximation holds:

$$H[W_k] \simeq -\sum_{w=0}^{k} \frac{\binom{k}{w}}{2^k}\log_2 \left\{ \frac{e^{-\frac{1}{2}\left[(k/2-w)/\sqrt{k/4}\right]^2}}{\sqrt{2\pi k/4}} \right\} = \frac{1}{2}\log_2 \frac{e\pi k}{2},$$

and the theorem is proved. □

At this point, to obtain the optimal code design whose average redundancy is as close as possible to the lower bound given in Theorem 2.4, one needs to find a variable length prefix encoding of $w(X)$ whose average length is minimal and closest to $H[w(X)]$; but, which preserves the integer ordering by satisfying condition (5). In other words, an alphabet $\mathcal{A} = [0, k]$ is given with $k+1$ ordered elements such that a probability $p(w) = Pr(w(X) = w) \in \mathbb{R}$ is defined for each $w \in [0, k]$. The problem is to find a labeled binary tree where the left edges are labeled with 0 and the right edges are labeled with 1, containing $k+1$ leaves each of which is associated with each element of $[0, k]$. The average path length from the root to the leaves of the sought tree should be minimal and such that if $w_1 < w_2$ then the leave associated with w_1 lies to the left of the leave associated with w_2, for all $w_1, w_2 \in [0, k]$. Note that without the restriction of preserving the ordering such problem can be simply solved by the Huffman Algorithm (see Ref. 6, pages 92–101). Variable length codes which preserve the integer ordering are well known in the literature and are called alphabetical codes.[8] The general problem of finding an optimal (= minimal average length) alphabetical code for an information source (our case) can be solved in polynomial time, for example, with the $O(n \log n)$ computation time algorithm by Hu and Tucker.[8] However, our goal is to solve the problem for the probability distribution of $W_k = w(X)$ which we know to be the binomial distribution. In the full paper, we give a simpler algorithm to find an optimal alphabetical encoding for any unimodal distributions such as the binomial distribution (recall that a distribution $\{p(w) : w \in [0, k]\}$ is unimodal if, and only if, $p(0) \leq p(1) \leq \ldots \leq p(h) \geq p(h+1) \geq p(h+2) \geq \ldots \geq p(k)$). The algorithm idea is to appropriately split W_k into the two sources $W' = \{W_k | W_k \leq \lfloor k/2 \rfloor\}$ and $W'' = \{W_k | W_k > \lfloor k/2 \rfloor\}$, to apply Huffman algorithm to W' and W'', and to combine the two Huffman encoding by adding as prefix to any codeword, a 0 if $W_k \leq \lfloor k/2 \rfloor$ and, a 1 if $W_k > \lfloor k/2 \rfloor$. Let $l_{opt}(W_k)$ be the average length of any optimal alphabetical encoding of W_k. We have

Theorem 2.5. *Let* $\sigma = 1 + \log_2[(\log_2 e)/e] = 0.08607\ldots$ *In the same hypothesis of Theorem 2.4, the following relations holds.*

$$r_{prop}(k) = l_{opt}(W_k) \leq H[W_k] + \sigma + \frac{1}{2^{k-1}} \binom{k}{\lfloor k/2 \rfloor}.$$

Proof. Let $l_{Huff}(W)$ be the average length of an Huffman encoding of an information source W. Since the distribution of W_k is unimodal, Theorem 2 of Ref. 8 implies $l_{opt}(W_k) \leq l_{Huff}(W_k) + p(\lfloor k/2 \rfloor)$. Since $p(\lfloor k/2 \rfloor)$ is the highest probability of the distribution of W_k, Theorem 2 of Ref. 9 implies $l_{Huff}(W_k) \leq H[W_k] + \sigma + p(\lfloor k/2 \rfloor)$. Hence, combining the last two inequalities, $l_{opt}(W_k) \leq H[W_k] + \sigma + 2p(\lfloor k/2 \rfloor)$. □

Note that $r_{min}(k) \leq r_{prop}(k)$ and so the relations in Theorem 2.4 and 2.5 imply (2). The third column of Table 1 gives the values of $r_{prop}(k)$ for some values of k. The fourth column of Table 1 gives the difference between $r_{prop}(k)$ and the asymptotic lower bound in Theorem 2.4. Indeed, it can be noticed that the upper bound in (2) seems to be a little bit loose and that $r_{prop}(k) \simeq (1/2)\log_2(e\pi k/2) = (1/2)\log_2 k + 1.04709\ldots$, as k increases.

Acknowledgment

This work was supported by the Italian MIUR under Grants PRA (ex 60%) and by the US NSF under Grants CCF-0701452 and CCF-0728810.

References

1. J. M. Berger, "A Note on Error Detection Codes for Asymmetric Channels", *Information and Control*, vol. 4, pp. 68–73, March 1961.
2. M. Blaum, *Codes for Detecting and Correcting Unidirectional Errors*, IEEE Computer Society Press, Los Alamitos, CA, 1993.
3. L. G. Tallini, "Bounds on the Capacity of the Unidirectional Channels", *IEEE Transactions on Computers*, vol. 54, pp. 232–235, Feb. 2005.
4. L. G. Tallini, S. Elmougy and B. Bose, "Analysis of Plain and Diversity Combining Hybrid ARQ Protocols over the $m(\geq 2)$-ary Asymmetric Channel", *IEEE Transactions on Information Theory*, vol. 52, pp. 5550–5558, Dec. 2006.
5. L. G. Tallini, S. Al-Bassam and B. Bose, "Feedback Codes Achieving the Capacity of the Z-Channel", *IEEE Transactions on Information Theory*, vol. 54, pp. 1357–1362, March 2008.
6. T. M. Covers and J. A. Thomas, *Elements of Information Theory*, New York: John Wiley and Sons Inc., 1991.
7. W. Feller, *An Introduction to Probability Theory and Its Applications* - Volume I - Second Edition, John Wiley & Sons, 1965.
8. N. Nakatsu, "Bounds on the Redundancy of Binary Alphabetical Codes", *IEEE Transactions on Information Theory*, vol. 37, pp. 1225–1229, July 1991.
9. R. G. Gallager, "Variations on a Theme by Huffman", *IEEE Transactions on Information Theory*, vol. IT-24, pp. 668–674, Nov. 1978.

A LATTICE BOLTZMANN MODEL ON UNSTRUCTURED GRIDS WITH APPLICATION IN HEMODYNAMICS

G. PONTRELLI* and S. SUCCI

Istituto per le Applicazioni del Calcolo - CNR
Via dei Taurini 19, 00185 Roma, Italy
**E-mail: g.pontrelli, s.succi@iac.cnr.it*
www.iac.rm.cnr.it/~pontrell

S. UBERTINI

Department of Technology, University "Parthenope"
Centro Direzionale, Isola C4, 80143 Napoli, Italy
E-mail: stefano.ubertini@uniparthenope.it

A finite-volume discretization over an unstructured grid is used to integrate the differential form of the Lattice Boltzmann equation with a variable viscosity using a cell-vertex finite-volume technique. The use of such technique for blood flow simulations combines the advantages of the LBM for Non-Newtonian fluids with the enhanced geometrical flexibility, allowing to suit the complex shapes typical of blood vessels with a limited number of nodes. Being the stress tensor expressed in terms of the distribution function, an explicit and accurate form for the shear stress is available at the wall surface. The methodology is applied to a selected number of test flow problems.

Keywords: Lattice Boltzmann equation, unstructured grid, finite volume, Non-Newtonian fluids, blood flow, stenosis.

1. Introduction

In the last years, the lattice Boltzmann method (LBM) has become an established numerical approach in computational fluid dynamics. Many models and extensions have been formulated that cover a wide range of complex flows.[1] This method possesses some advantages over conventional CFD methods, such as the simplicity of the stream-and-collide dynamics, that makes that makes LB very efficient computationally, its amenability to parallel computing, its ease in handling complex flows and the physical implementation of irregular boundaries. However, the essential restriction of the standard LBE to the lattice uniformity, which makes it macroscopically similar to a uniform Cartesian-grid solver, represents a severe limitation for many practical engineering problems. Therefore, recently, much research has been directed to the goal of enhancing the geometrical flexibility of the LB method.[2,3] Considering that for many practical problems an irregular grid or a meshless structure is always preferable due to the fact that curved boundaries can be described more accurately, and that computational resources can be used more efficiently, our recent effort is to extend the LBM order of accuracy and flexibility so that its spatial resolution requirements for various flow situations may be reduced and may be adapted to more general meshes.[4]

A particularly interesting development is represented by finite-volume formulations on fully unstructured grids.[5] The Unstructured Lattice Boltzmann schemes (ULBE for short) integrates the differential form of the Lattice Boltzmann equation (LBE) using a cell-vertex finite-volume technique in which the unknown fields are placed at the nodes of the mesh and evolve based on the fluxes crossing the surfaces of the corresponding control volumes.[6]

Up to now, ULBE implementations were limited to Newtonian fluids. However, in some flow applications, such as those arising in hemodynamics, the fluid cannot be treated as Newtonian, and a more realistic rheological model should be used. At low shear rates, the apparent viscosity of blood is high, whereas an increase in shear rate leads to a decrease in the apparent viscosity. This shear-thinning property results from the reversible formation of red blood cell aggregates where increased shear rates cause disaggregation of rouleaux. Non-newtonian models with shear-thinning viscosity are commonly used to solve blood flow problems.[7–9] The use of ULBE for blood flow simulations combines the advantages of the LBM for Non-Newtonian fluids to the enhanced flexibility, as the use of unstructured grid allows to accommodate the complex geometries (bifurcation, branching, and curvatures) typical of blood vessels, with a limited number of nodes. Moreover, as in ULBE the body surface is defined by grid points (while in standard LBM the body surface does not generally lie on lattice sites), both pressure and viscous forces are locally available at the walls and explicitly expressed in terms of the distribution function, thus making the wall shear stress calculation straightforward. This is particularly valuable in hemodynamics, where an accurate knowledge of the wall shear stress is highly valued.[10] The formation of atherosclerotic plaque to certain regions of the vascular tree is believed to be caused by alteration in the wall shear stress. Therefore it is important to provide a quantitative description of it in region of interest.

As a starting point, because of the complexity of the constitutive equation, only a steady state case is considered in this paper. A constant flow rate is imposed and the wall deformability is disregarded. The mathematical model is the shear-thinning Carreau model able to describe blood flows for the typical range of shear rates.[7,11]

Firstly, a couple of basic flow configurations are considered: an excellent agreement with the numerical solution in a long straight pipe in the Newtonian case is obtained and the typical flow fields in constricted tubes are well reproduced.

2. Formulation of ULBE method

Let us consider the classical single-time relaxation Lattice Boltzmann equation in differential form:

$$\partial_t f_i + \vec{v}_i \cdot \vec{\partial}_x f_i = -(f_i - f_i^e)/\tau \tag{1}$$

where $f_i(\vec{x}, t) \equiv f(\vec{x}, \vec{v} = \vec{c}_i, t)$, $i = 1, b$, is the probability of finding a particle at lattice site $\vec{x}$ at time t, moving along the lattice direction defined by the discrete speed $\vec{c}_i$. The second term of left-hand side of this equation denotes the molecular free-streaming, whereas the right-hand side represents molecular collisions via a single-time relaxation towards local equilibrium f_i^e on a typical timescale τ.[1] This local equilibrium is a (local) Maxwellian expanded to second order in the fluid speed:

$$f_i^e = \rho w_i [1 + \beta u_i + \frac{\beta^2}{2}(u_i^2 - u^2)] \tag{2}$$

where $\beta = 1/c_s^2$, c_s being the lattice sound speed, and w_i are the associated weight coefficients. In the above formulation, fluid density and velocity are $\rho = \sum_i f_i$ and $\vec{u} = \sum_i \vec{c}_i f_i / \rho$ respectively. The relaxation time τ fixes the fluid kinematic viscosity as $\nu = c_s^2(\tau - \Delta t/2)$. In order to recover fluid dynamic behaviour, the set of discrete speeds must be chosen such that mass, momentum and energy conservation are fulfilled.[1] In the present work, we shall refer to the two-dimensional nine-speed model D2Q9 with one zero-speed particle of weight $w_0 = 4/9$, four speed-1 particles of weight $w_1 = 1/9$ and four speed-$\sqrt{2}$ particles of weight $w_2 = 1/36$.

The ULBE approach consists in discretizing eqn. (1) by introducing a tessellation based on triangular elements. To each node P of the discrete grid, we associate a set of $b = 9$ discrete populations $f_i(P, t), i = 1, b$, representing the unknowns of the problem. The set of K triangles $T_k(P)$, $k = 1, .., K$, which share P as a common vertex, defines the finite element associated with node P (fig. 1). Each of these K triangles is defined by the three vertices $T_k(P) \equiv [P, P_k, P_{k+1}]$, K being the connectivity number of the unstructured mesh. Each triangle T_k associates with a control volume $\Omega_k(P)$, defined by the union of the two sub-triangles $\Omega_k^- = [P, E_k, C_k]$ and $\Omega_k^+ = [P, C_k, E_{k+1}]$, where C_k is the center of the triangle T_k and E_k,E_{k+1} are the midpoints of the edges PP_k,PP_{k+1}, respectively.[5]

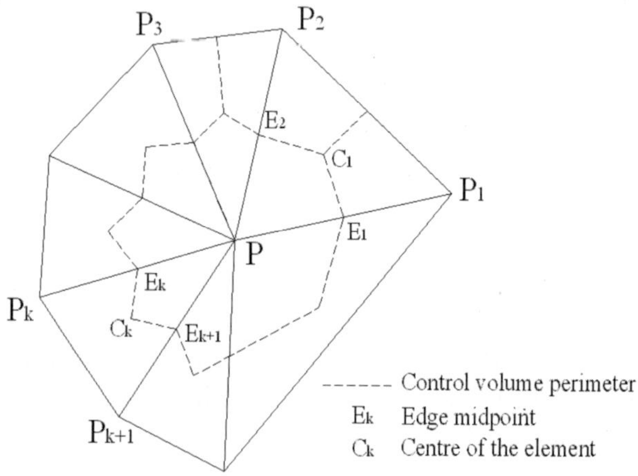

Fig. 1. Geometrical layout of the cell-vertex finite volume discretisation around a grid point P.

Application of the Gauss theorem to each finite volume $\Omega_k^{\mp}$, as combined with a first-order time marching, yields the following finite difference equation:[2]

$$f_i(P, t+dt) = f_i(P, t) + dt \sum_{k=0}^{K} S_{ik} f_i(P_k, t) - \frac{dt}{\tau} \sum_{k=0}^{K} C_{ik} [f_i(P_k, t) - f_i^e(P_k, t)] \tag{3}$$

where the index k runs over the spatial neighborhood of node P (corresponding to $k = 0$). In the above, we have defined:

$$f_i(P, t) = \frac{1}{V_P} \int_{\Omega(P)} f_i(\vec{x}, t) d\vec{x} \tag{4}$$

where V_P is the volume of the control volume $\Omega(P)$.

The detailed expressions of the streaming and collision matrices S_{ik} and C_{ik} are easily obtained once the interpolation rules are specified. Regardless of the specific values of these matrices, the following sum rules:

$$\sum_{k=0}^{K} S_{ik} = 0, \qquad \sum_{k=0}^{K} C_{ik} = 1, \ \forall i$$

hold. These sum rules play an important role in the theoretical analysis of the scheme.[5] It has been shown that the ULBE scheme recovers hydrodynamics with the same viscosity of the continuum, $\nu = c_s^2\tau$, and that numerical viscosity effects are within second order of accuracy in space. It is worth noting that with a explicit Euler scheme the maximum timestep allowed by the numerical stability condition is

$$\Delta t < 2\tau \tag{5}$$

This condition, combined with the expression of the fluid viscosity $\nu = c_s^2\tau$, indicates that small viscosities can be attained by making the time-step correspondingly small.[3]

Boundary conditions for ULBE need to cope with the fact that the corresponding control volumes do not close up leaving two external edges exposed on the boundary. To date, the best strategy to deal with this problem is provided by the so-called *covolume method*. In it the edge fluxes are evaluated explicitly by using interpolation at the boundary edges and thus taking part in the matrix S_{ik} definition.

The use of multiple regular buffers at inlet/outlet sections is found to be beneficial to the stability of ULBE computations for open flows. For further details on boundary conditions refer to Ref.[3]

3. ULBE for Non-Newtonian flows

It is well accepted that blood can no longer be treated as a single-phase Newtonian viscous fluid when flowing through small vessels. Its viscosity is a decreasing function of the shear rate (shear-thinning property). Accordingly, the ULBE model should be extended to incorporate the viscosity law $\mu(\dot{\gamma})$. To this aim, the constant time relaxation τ in eqns. (1) and (3) is replaced by a self-consistent, shear-dependent $\tau(\dot{\gamma})$, being $\dot{\gamma} = \dot{\gamma}[f_i]$ the shear rate, a function of the density distribution function. The current value $\dot{\gamma}$ is obtained by the following considerations. In LBM, the strain and stress tensors $\mathbf{\Gamma}$ and $\mathbf{\Pi}$ can be written explicitly in terms of the particle distribution functions, respectively as:[8,12]

$$\mathbf{\Gamma}_{\alpha\beta} = -\frac{1}{2\rho\tau c_s^2}\mathbf{\Pi}_{\alpha\beta} \tag{6}$$

where

$$\mathbf{\Pi}_{\alpha\beta} = \sum_i (f_i - f_i^e)\mathbf{c}_{i\alpha}\mathbf{c}_{i\beta} \tag{7}$$

and α, β run over spatial dimensions. Let us now introduce a measure of the magnitude of the previous tensors (norm):

$$\dot{\gamma} \equiv 2|\mathbf{\Gamma}| = 2\sqrt{\sum_{\alpha,\beta} \Gamma_{\alpha\beta}\Gamma_{\alpha\beta}}, \qquad \sigma \equiv |\mathbf{\Pi}| = \sqrt{\sum_{\alpha,\beta} \Pi_{\alpha\beta}\Pi_{\alpha\beta}} \tag{8}$$

The relationship strain-stress (6) can be normed as:

$$\sigma = \rho\tau(\dot{\gamma})c_s^2\dot{\gamma} = \mu(\dot{\gamma})\dot{\gamma} \tag{9}$$

In principle $\dot{\gamma}$ must be obtained by solving the above nonlinear eqn. by iteration at each lattice site. However, due to the slow variation of $\mu(\dot{\gamma})\dot{\gamma}$ on a time scale Δt, current practice shows that one can adjust τ along the time integration as follows:

$$\tau(t + \Delta t) = \frac{\mu\left[\dot{\gamma}(t), t\right]}{\rho(t)c_s^2} \tag{10}$$

Note that eqn. (9) yields the standard scalar relation $\sigma = \mu\dot{\gamma}$ in the Newtonian case.

4. Flow through a stenosed vessel

Several attempts have been made to study the effect of a stenosis on the blood flow characteristics. The augmented shear stress is believed to be responsible for deposition and adhesion of lipids and formation of atherosclerotic plaque. Most researches have studied the characteristics of blood flowing in a mild stenosed artery by considering it as a Newtonian fluid.[13] Here, to simulate a physiological case, we consider a shear-thinning viscosity that follows the Carreau model:

$$\mu(\dot{\gamma}) = \mu_\infty + (\mu_0 - \mu_\infty)\left(1 + (\lambda\dot{\gamma})^2\right)^{\frac{n-1}{2}} \tag{11}$$

with the following rheological parameters:[11,14]

$$\mu_0 = 0.56P \qquad \mu_\infty = 0.0345P \qquad \lambda = 3.313s \qquad n = 0.3568 \tag{12}$$

that give rise one order magnitude drop of viscosity for shear rates in the range $0.1 \div 100s^{-1}$. To emphasize the NN shear-thinning effects, let us consider physical variables pertaining to smaller vessels (arterioles):

$$U_{max} = 2cm/s \qquad H = 5 \cdot 10^{-3}cm \qquad \rho = 1g/cm^3$$

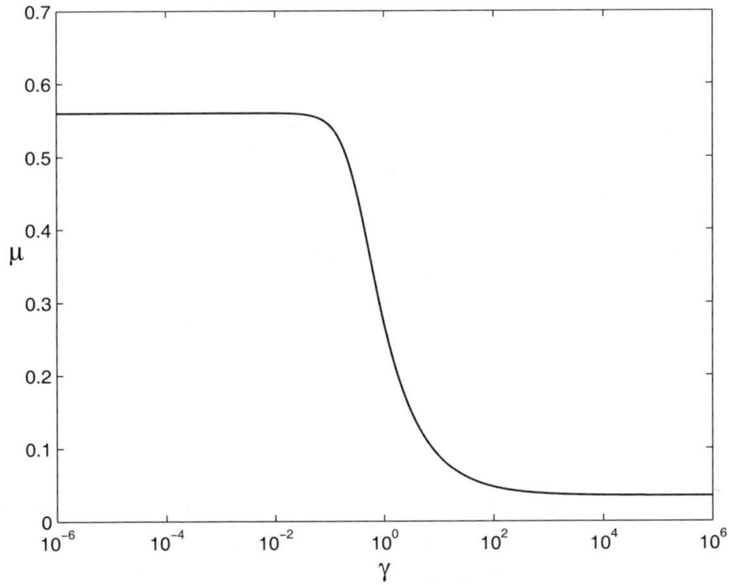

Fig. 2. Shear-thinning viscosity as a decreasing function of the shear rate in the Carreau model (11)–(12).

that combine to a rather small Reynolds number:

$$Re_0 = \frac{\rho U_{max} H}{\mu_0} \approx 0.017 \tag{13}$$

Firstly, to validate the method, a flow in a straight tube is modelled. To guarantee a mesoscopic flow regime comparable with the above physical values, a careful scaling between macroscopic and mesoscopic (overbarred) variables is carried out. Accordingly, let us consider a channel of semi-width $\bar{H}$ where a fluid of density $\bar{\rho}$ and viscosity $\bar{\mu}$ is flowing under a constant forcing $\bar{G}$. As LB variables, we set:

$$\bar{H} = 1 \qquad \bar{\rho} = 1 \qquad \bar{G} = 10^{-5}$$

and $\bar{\mu}_0$ is chosen in order that the correspondent Reynolds number matches Re_0 in eqn. (13). A channel $[-5, 5] \times [-1, 1]$ has been covered by 4460 triangular cells, constituting a uniform grid with 2351 equally distributed nodes. The time step has been fixed initially as $\Delta t = 0.01$, and possibly reduced (halved) in order to guarantee that condition (5) is verified. In the following, the over-bar denoting LB variables is omitted.

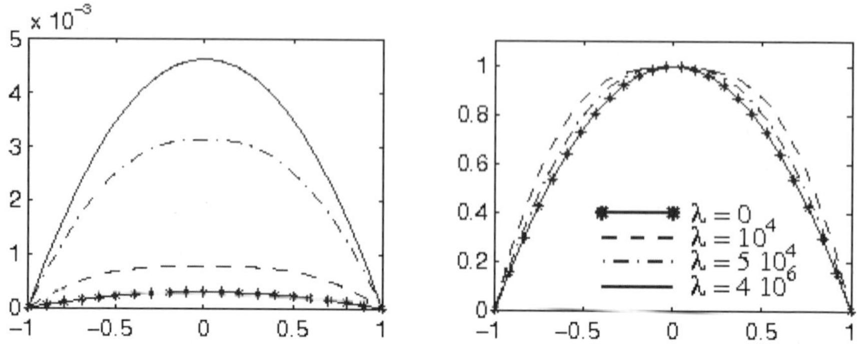

Fig. 3. Velocity profiles (normalized on the right) relative to the flow in a straight tube, with the Carreau model at three different $\lambda > 0$. Comparison with Newtonian solution ($\lambda = 0$, starred line) is made ($G = 10^{-5}$, LB units).

In fig. 3 the velocity profiles corresponding to four time constants λ's are displayed. The magnitude of the velocity raises dramatically with λ, passing from a Newtonian regime with higher viscosity μ_0 to another Newtonian regime with a lower μ_∞. In the transition shear-thinning region, the parabolic shape appears flattened in the center of the channel. It turns out that λ plays a critical role, since its value affects the viscosity decay as well as the maximum shear rate.

To simulate the flow through a stenosis, let us now consider a channel with the shape of a long rectangle except in a small region centered at $x = 0$ with a smooth symmetric contraction as described by the following bell-shaped curve:

$$H(x) = (1 - \delta e^{-\phi x^2})H_0 \tag{14}$$

where $H(x)$ is the height, $0 \le \delta < 1$ is a measure of the degree of contraction, ϕ of its length (fig. 4). The value of ϕ should be taken quite small to guarantee a slowly varying boundary profile. As a particular case, the rectangular channel is recovered for $\delta = 0$. The geometrical parameters have been fixed as:

$$H_0 = 5 \cdot 10^{-3} cm \qquad \phi = 0.8 \qquad \delta = 0.3$$

corresponding to a degree of contraction of about 50%.[15] The pressure gradient, the rheological parameters and LB settings are as in the straight channel flow.

A sensitivity analysis on the grid size has been carried out and the results confirm the ULBE capability to cluster the degrees of freedom in

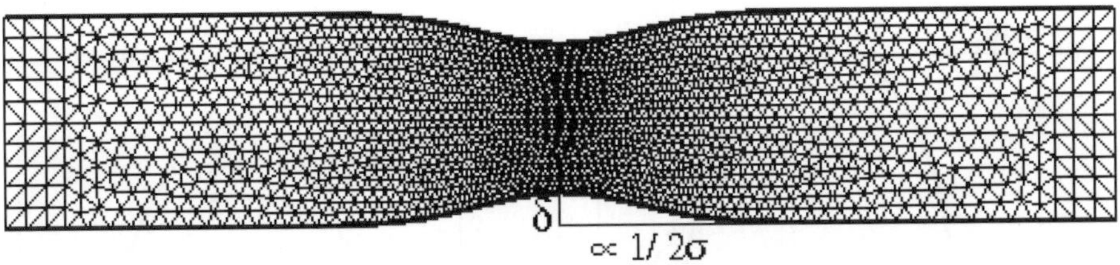

Fig. 4. Axisymmetric stenosed rectangular channel covered by a refined unstructured grid (1737 grid nodes and 3284 elements).

the critical regions of the flow, without suffering any loss of accuracy.[16] In the following simulations a nonuniform mesh consisting in 1237 refined nodes in the contraction has been chosen (fig. 4). In the present NN model the limiting values for LB viscosities are $\mu_0 = 16.7 \cdot 10^{-3}$ $\mu_\infty = 1.03 \cdot 10^{-3}$, but the local viscosity in all simulations remains above the lower bound $\mu_{min} = 2.011 \cdot 10^{-3}$.

5. Results and discussion

Numerical simulation of arterial hemodynamics offers a non invasive tool for obtaining detailed and realistic measurements for physiological and pathological flow conditions. The velocity profiles through a contraction are of great physiological interest, since they provide a simple and effective description of the flow field. While flow through high grade stenosis produces a separation region downstream of the throat, at mild contraction and/or at low Reynolds number neither recirculation nor flow reversal are present and the flow results symmetric upstream and downstream the stenosis.[17,18] The magnitude of velocity is much larger in NN case with a flattened profile (fig. 5). Most of the shear-thinning behaviour stems from the formation of rouleaux in the low shear flowing typical of the microcirculation. Thus, at medium-high shear rate the Non-Newtonian deviation should be minimal, except in a small area downstream the contraction, where the shear rate reduces. The velocity distribution is almost parabolic, both for expanding and contracting cross sections.

In presence of a narrowing, the flow exhibits a resistance and hence an enhanced shear stress (i.e. the wall vorticity) and a pressure drop. These are indicators of flow disturbances and are quantities of physiological relevance. Since there is no reliable method of determining the wall shear stress

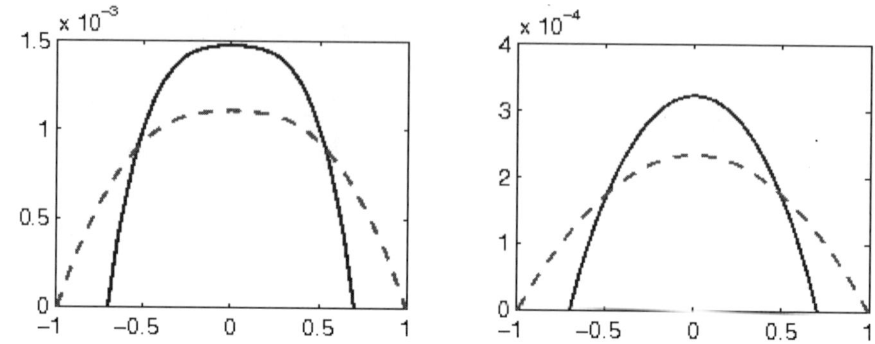

Fig. 5. Velocity profiles in the center of the throat (continuous line) and downstream a stenosis (dashed line) in the case of the Carreau model ($\lambda = 10^4$, left) and Newtonian fluid ($\lambda = 0$, right) ($G = 10^{-5}$, LB units).

experimentally near the regions of possible reversal flow, the numerical simulations provide a valuable tool, because they offer a sufficiently accurate approximation of the relevant variables.[13,15]

A pressure drop is observed as the occlusion is approached, while the wall shear stress (WSS) increases smoothly in correspondence of the contraction and has a peak value placed symmetrically at the center of the throat (fig. 6). Downstream, it goes back to the previous value. In the straight portion the WSS is higher in NN case, and the opposite in the contraction, because of the lowered viscosity. All these results are in qualitative agreement with those of other models existing in literature.[17] However, the flow model presented here has some limitations when compared to a real stenosis in the vascular system. Deviation from the physiological flow include the use of steady as opposed to pulsatile flow and the axysimmetric geometry. Neglecting wall distensibility is deemed not a serious restriction in the study, since our attention is confined to arterioles where the effects of elasticity are rather small.

Despite their preliminary nature, the numerical results are believed to be important in interpreting the measured velocity profiles and acquiring more information on the flow field in stenotic regions. The present study represents an initial attempt to provide results comparable with in vivo measurements. Moreover, the presented tests clearly demonstrate the capability of the ULBE method to gain accuracy and to reduce the computational overhead by clustering the lattice nodes in the critical regions of the flow or near the wall where an accurate evaluation of WSS is required.

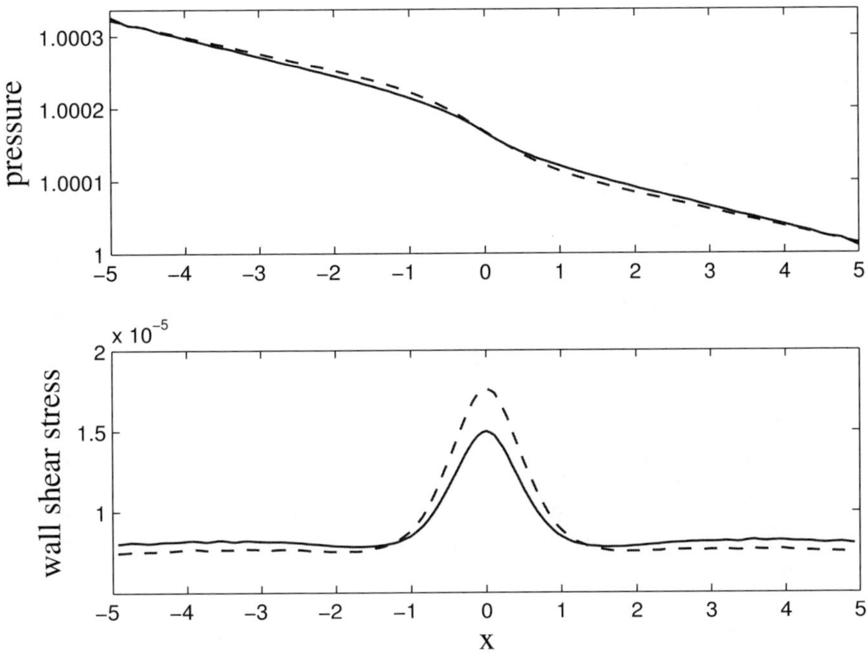

Fig. 6. Pressure drop along the centerline (above) and wall shear stress (below) in a contracted channel. Difference between N ($\lambda = 0$, dashed line) and NN ($\lambda = 10^4$, continuous line) cases ($G = 10^{-5}$, LB units).

Besides enhancing accuracy of the hemodynamic simulations, this property may also prove very beneficial for future multiscale applications, coupling macroscopic hemodynamics with the nano-particle transport within the blood stream. Work is currently directed to extend the ULBE models to three-dimensional flows in complex geometries.

Acknowledgements

This work has been partially supported by the bilateral CNR (Italy) - FCT (Portugal) grant: *Multiscale analysis and numerical simulation of mathematical models in hemodynamics and hemorheology*, 2007-2008. The italian project *CNR-Bioinformatics* is also greatly acknowledged.

References

1. R. Benzi, S. Succi and M. Vergassola, The lattice Boltzmann equation: theory and applications, *Phys. Rep.*, **222**, 145 (1992).

2. S. Ubertini, G. Bella and S. Succi, Lattice Boltzmann method on unstructured grids: further developments, *Phys. Rev. E*, **68**, 016701 (2003).
3. S. Ubertini and S. Succi, Recent advances of Lattice Boltzmann Techniques on Unstructured Grids, *Progress Comput. Fluid Dynamics*, **5 (1/2)**, 84–95 (2005).
4. J.M. Buick, J.A. Cosgrove, S.J. Tonge, A.J. Mulholland, B.A. Steves and M.W. Collins, The Lattice Boltzmann Equation for modelling arterial flows: reviews and applications, *Int. Medicine*, **11** 1–3 (2003).
5. G. Peng, H. Xi and C. Duncan, Lattice Boltzmann method on irregular meshes, *Phys. Rev. E*, **58**, R4124 (1998).
6. S. Ubertini and S. Succi, A generalised lattice Boltzmann equation on unstructured grids, *Commun. Comput. Phys.*, **3**, 342–356 (2008).
7. A.M. Artoli and A. Sequeira, Mesoscopic simulations of unsteady shear-thinning flows, *Lect. Notes Comp. Sci.*, **3992**, 78–85 (2006).
8. B. Chopard, R. Ouared and D.A. Rufenacht, A lattice Boltzmann simulation of clotting in stented aneursysms and comparison with velocity or shear rate reductions, *Math. and Comp. in Simulation*, **72**, 108–112 (2006).
9. A.G. Hoekstra, J. van't Hoff, A.M. Artoli and P.M.A. Sloot, Unsteady flow in a 2D elastic tube with the LBGK method, *Future Gen. Comp. Syst.*, **20**, 917–924 (2004).
10. G. Pontrelli, C. König, M.W. Collins, Q. Long, S. Succi, *Modelling wall shear stress in small arteries using LBM and FVM*, 2nd Micro and Nano Flows Conference, West London, UK, (2009).
11. J. Boyd, J.M. Buick and S. Green, Analysis of the Casson and Carreau-Yasuda non-Newtonian blood models in steady and oscillatory flow using the lattice Boltzmann method, *Phys. Fluids*, **19**, 093103 (2007).
12. O. Malaspinas, G. Courbebaisse and M. Deville, Simulation of generalized newtonian fluids with the Lattice Boltzmann method, *Int. J. Mod. Phys. C*, **18**, 1939–49 (2007).
13. U. Solzbach and A. Zeiher, Effect of stenosic geometry on flow behaviour across stenotic models, *J. Biomech.*, **25**, 543-550 (1987).
14. A.M. Artoli, J. Janela and A. Sequeira, The role of Womersley number in shear-thinning fluids, *WSEAS Trans. Fluid Mech.*, **1**, 133, 2006.
15. G. Pontrelli, Blood flow through an axisymmetric stenosis, *Proc. Instn. Mech. Engrs., Part H, J. Eng. in Medicine*, **215 (1)**, 1–10 (2001).
16. O. Filippova and D. Hanel, Grid refinement for lattice-BGK models, *J. Comp. Phys.*, **147**, 219 (1998).
17. H. Jung, J.W. Choi, and C.G. Park, Asymmetric flows of non-Newtonian fluids in symmetric stenosed artery, *Korea-Australia Rheol. J.*, 101–108 (2004).
18. K.W. Lee and X.Y. Xu Modelling of flow and wall behaviour in a mildy stenosed tube, *Med. Eng. Phys.*, **24**, 575–586 (2002).

THERMOMECHANICS OF POROUS SOLIDS FILLED BY FLUID FLOW

L. RESTUCCIA

Dipartimento di Matematica, Facoltà di Scienze MM.FF.NN.
Università degli Studi di Messina, Italy
lrest@dipmat.unime.it

In a previous paper, in the framework of the extended irreversible thermodynamics, a non conventional model for the description of a fluid flow through porous solids was given. A second order structural permeability tensor *à* la Kubik and its flux were introduced in the thermodynamical state vector as internal variables. Liu's theorem was used to analyze the entropy inequality and the laws of state and the extra entropy flux were derived. In this paper, within the same thermodynamical model, the constitutive relations are derived to close the balance equations governing the behaviour of the media under consideration. Smith's theorem is used to construct the constitutive theory by isotropic polynomial representations of proper functions obeying the principles of objectivity and material indifference. Interaction effects among a thermal field, a fluid flow and a structural permeability field have taken into consideration.

Keywords: Non-equilibrium Thermodynamics; Internal variables; Porous structures; Constitutive theory.

1. Introduction

The models of porous solids may have relevance in important advances, that have been made during the last decades to describe phenomena accompanying flows of mass in porous structures. They find applications in several fundamental sectors: geology, biology, medical sciences, technology of materials and like. It is well known that the permeability is the the basic property of porous media that allows fluid flow through interconnected pores. There exist several descriptions concerning fluid flow through porous materials (see for instance [1]-[6]). In most papers the volume porosity and the coefficient of permeability, which occur in Darcy's law, are used as macroscopic pore structure parmeters. In the case of anisotropic pore structure, permeability coefficients are considered as components of a symmetric second order tensor in a generalized Darcy's law, which represents the conductivity of the pore structure regarding the permeation by a fluid flow [1]. In some cases a surface ratio has also been considered as a pore structure parameter. However, the analysis of an outgoing fluid flux crossing a surface of a porous structure requires the use of the surface average velocity that in general may not coincide with the volume average velocity. In [6] a macroscopic characterization of a pore structure is given by Kubik accounting for its anisotropic character. The use of volume and area averaging procedures for fluid flow through porous media shows that, under the appropriate assumptions, an anisotropic pore structure can be described by two parameters: the volume porosity and the second order structural permeability tensor. In the framework of the extended irreversible thermodynamics with internal variables ([8]-[12]), a non conventional thermodynamical model for a fluid flow through porous solids was developed by the author in [7], introducing in the thermodynamical state space the second order structural permeability tensor *à* la Kubik and its flux as internal variables. In fact, these porous channels, sometime can self propagate because of changed and favorable surrounding conditions. For instance, in metallurgy during a process of fabrication, such defect propagation, can provoke a premature fracture. In this way the balance equations for the mass, the momentum, the moment of momentum and the internal energy by the constitutive equations and in particular by the evolution equation for the structural permeability tensor. In [7] the entropy inequality was analyzed by the application of Liu' s theorem [13] and the the state laws were investigated. Furthermore, in [14] and [15], in a geometrized framework ([6]-[18]) a geometric model for thermodynamics of these media was constructed and the dynamical system for a *simple material element* , the expression of the entropy function and the relevant entropy 1-form were derived. In this contribution, taking into account the model developed in [7], using Smith's theorem [19] the constitutive theory is built on isotropic polynomial representations of proper functions which must obey the principle of objectivity. Interaction effects among a thermal field, a fluid flow and a structural permeability field, coming from a network of porous channels in an elastic body, have taken into consideration Inserting constitutive equations into the balance equations a system field equations of differential equations may be obtained which allow to solve analitically and/or numerically physical real problems in particular situations.

2. A non conventional model for porous solids filled by fluid flow

Now, we recall the thermodynamical model developed [7] for describing a pore structure filled by a fluid flow. The structure of the porous channels resembles a network of infinitesimally capillary tubes in a elastic solid. Their existence should not be omitted in the analysis of kinetic processes as diffusion of mass or charges, transport of heat, energy, etc. The fluid and the elastic solid are considered as a two components mixture. To describe a pore structure in [6] Kubik considers a representative elementary sphere volume Ω of a porous skeleton filled with fluid (see Fig. 1), large enough to provide a representation of all the statistical properties of the pore space Ω^p. $\Omega = \Omega^s + \Omega^p$, where Ω^s is the solid space. Since all pores are considered to be interconnected the *effective volume porosity* is completely defined as $f_v = \frac{\Omega^p}{\Omega}$. The analysis is restricted to media which are homogeneous with respect to volume porosity f_v , i.e. f_v remains constant in the medium. In Fig. 1 to

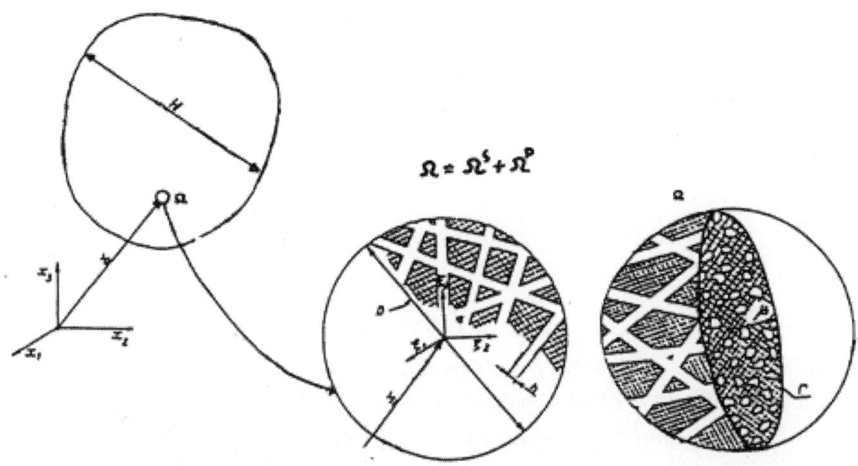

Fig. 1. The volume and surface averaging scheme. Characteristics of the given pore structure (after [6])

avoid confusion all microscopic quantities are described with respect to a ξ_i coordinate system, while macroscopic quantities are described with respect to a x_i coordinate system. Let $\boldsymbol{\alpha}(\xi)$ be any scalar, spatial vector or second order tensor a quantity describing some microscopic property of the fluid flowing through the pore space Ω^p. We assume that such quantity is zero in the solid space Ω^s. The volume averaging procedures give

$$\hat{\boldsymbol{\alpha}}(\mathbf{x}) = \frac{1}{f_v \Omega} \int_{\Omega_p} \boldsymbol{\alpha}(\boldsymbol{\xi}) d\Omega, \tag{1}$$

$$\bar{\boldsymbol{\alpha}}(\mathbf{x}) = \frac{1}{\Omega} \int_{\Omega_p} \boldsymbol{\alpha}(\boldsymbol{\xi}) d\Omega, \tag{2}$$

where $\hat{\boldsymbol{\alpha}}$ and $\bar{\boldsymbol{\alpha}}$ are average quantities on pore-volume and bulk-volume, respectively. Similarly, we define the average quantity of $\boldsymbol{\alpha}(\boldsymbol{\xi})$ on pore area as follows

$$\overset{*}{\boldsymbol{\alpha}}(\mathbf{x}, \boldsymbol{\mu}) = \frac{1}{\Gamma^p} \int_{\Gamma} \boldsymbol{\alpha}(\boldsymbol{\xi}) d\Gamma, \tag{3}$$

where Γ is the central sphere section and Γ^p represents the pore area of Γ. The orientation of Γ in Ω is given by the normal vector $\boldsymbol{\mu}$. Furthermore, $\Gamma = \Gamma^s + \Gamma^p$, where Γ^s is the solid area. By definition the quantity $\boldsymbol{\alpha}(\xi)$is zero in the solid surface Γ^s. In such medium Kubik defines the so called *structural permeability tensor*, describing a pore structure, as the following linear mapping

$$\bar{v}_i(\mathbf{x}) = \mathcal{R}_{ij}(\mathbf{x}, \boldsymbol{\mu}) \overset{*}{v}_j (\mathbf{x}, \boldsymbol{\mu}), \tag{4}$$

where

$$\bar{\mathbf{v}}(\mathbf{x}) = \frac{1}{\Omega} \int_{\Omega^p} \mathbf{v}(\boldsymbol{\xi}) d\Omega, \quad \boldsymbol{\xi} \in \Omega^p, \quad \overset{*}{\mathbf{v}}(\mathbf{x}, \boldsymbol{\mu}) = \frac{1}{\Gamma^p} \int_{\Gamma} \mathbf{v}(\boldsymbol{\xi}) d\Gamma, \quad \boldsymbol{\xi} \in \Omega^p \tag{5}$$

are the average fluid velocity on the bulk-volume and the the average fluid velocity on the pore-area, respectively. In [6] Kubik establishes the geometrical interpretation of $\boldsymbol{\mathcal{R}}$ considering a fluid flow having the average velocity on a bulk-volume $\bar{\mathbf{v}}(\mathbf{x})$ as the superposition of three one-dimensional fluid flows (along three mutually perpendicular channels) having average velocities on the areas of these channels $\overset{*}{v}_i (\mathbf{x}, \boldsymbol{\mu})$. In fact, equ.(4) gives a linear mapping between the average fluid velocity $\bar{\mathbf{v}}(\mathbf{x})$ on the bulk volume and the area average fluid velocities $\overset{*}{v}_i (\mathbf{x}, \boldsymbol{\mu})$. Only part of the fluid can flow unimpeded while the rest is trapped in the porous structure. Now, following Maruszewski in [20], using the previous definitions, for any flux α_i of some quantity transported through a cobweb of lines one postulates that

$$\bar{\alpha}(\mathbf{x})_i = \mathcal{R}_{ij}(\mathbf{x}, \boldsymbol{\mu}) \overset{*}{\alpha}_j (\mathbf{x}, \boldsymbol{\mu}), \tag{6}$$

where

$$\mathcal{R}_{ij}(\mathbf{x}, \boldsymbol{\mu}) = \Gamma r_{ij}(\mathbf{x}, \boldsymbol{\mu}).$$

In the above equations r_{ij} is a new tensor called *structural permeability core tensor* that refers $\mathcal{R}_{ij}$ to the surface Γ. Its unit is m^{-2}.

3. Balance and evolution equations

Now, we recall the thermodynamical model developed in [7] for describing reciprocal interactions among a thermal field, a fluid flow and a structural permeability field coming from a network of pores in an elastic body. The assumption that an anisotropic pore structure is continuously distributed within the medium was done. Since during deformation the porous structure evolves in time, the structural permeability field is described both by the state tensor r_{ij} (not necessarily symmetric) and by the flux of this tensor $\mathcal{V}_{ijk}$. Similarly the fluid flow is described by two variables: the concentration of the fluid c and the flux of this fluid j_i. The mass of density ρ_1 is the mass of the fluid transported through the elastic porous body of density ρ_2. The mass of the fluid and the elastic solid form a two-components mixture of density

$$\rho = \rho_1 + \rho_2. \tag{7}$$

We define the concentration c as follows

$$c = \frac{\rho_1}{\rho}. \tag{8}$$

For the mixture of continua as a whole and also for each constituent, separately, the continuity equations are satisfied

$$\dot{\rho} + \rho v_{i,i} = 0, \tag{9}$$

$$\frac{\partial \rho_1}{\partial t} + (\rho_1 v_{1i})_{,i} = 0, \tag{10}$$

$$\frac{\partial \rho_2}{\partial t} + (\rho_2 v_{2i})_{,i} = 0, \tag{11}$$

where a superimposed dot denotes the material derivative, v_{1i} and v_{2i} are the velocities of the fluid particles and the particles of the elastic body, respectively, so that the barycentric velocity is given by

$$\rho v_i = \rho_1 v_{1i} + \rho_2 v_{2i}. \tag{12}$$

Then the fluid flux has the following form

$$j_i = \rho_1(v_{1i} - v_i). \tag{13}$$

In (10) and (11) the sources of mass of the constituents are neglected because we assume that there are not chemical reactions between the constituents or coagulations. The mechanical properties of the considered system are described by the total stress tensor σ_{ij} (in general non symmetric), related to the whole body considered as a mixture, and the small-strain tensor ε_{ij}, which describes the deformation of the elastic solid, having the form

$$\varepsilon_{ij} = \frac{1}{2}(u_{i,j} + u_{j,i}). \tag{14}$$

The gradient of the velocity of the body may take the form

$$v_{i,j} = w_{ij} + \frac{\partial \varepsilon_{ij}}{\partial t}, \tag{15}$$

where w_{ij} is the antisymmetric part of $v_{i,j}$

$$w_{ij} = \frac{1}{2}(v_{i,j} - v_{j,i}). \tag{16}$$

The proposed model is based on the extended irreversible thermodynamics with internal variables. The thermal field is governed by the temperature T and the heat flux q_i.

The vector space is chosen as follows

$$C = \{\varepsilon_{ij}, c, T, r_{ij}, j_i, q_i, \mathcal{V}_{ijk}, c_{,i}, T_{,i}, r_{ij,k}\}, \tag{17}$$

where we have taken into consideration the gradients $c_{,i}, T_{,i}$ and $r_{ij,k}$ describing nonlocal effects. We ignore the viscoelastic effects, so that σ_{ij} is not in the set C. All the processes occurring in the considered body are governed by two groups of laws.
The first group concerns the classical balance equations.
The *balance of mass* resulting from (7)-(13)

$$\rho\dot{c} + j_{i,i} = 0. \tag{18}$$

The *momentum balance*

$$\rho\dot{v}_i - \sigma_{ij,j} - f_i = 0, \tag{19}$$

where f_i denotes a body force.

The *angular of momentum balance*

$$\varepsilon_{ijk}\sigma_{jk} + g_i = 0, \tag{20}$$

where g_i is a couple per unit volume.
The *internal energy balance*

$$\rho\dot{e} - \sigma_{ji}v_{i,j} + q_{i,i} - \rho r = 0, \tag{21}$$

where r is the heat source distribution.

The second groups of laws deals with the rate properties of the internal variable and the fluxes of mass, heat and field coming from a network of pores in an elastic body.

The evolution equations for the internal variable r_{ij} and for the fluxes q_i and j_i are given by

$$\overset{*}{r}_{ij} + \mathcal{V}_{ijk,k} - R_{ij}(C) = 0, \tag{22}$$

$$\overset{*}{j}_i - J_i(C) = 0, \tag{23}$$

$$\overset{*}{q}_i - Q_i(C) = 0, \tag{24}$$

$$\overset{*}{\mathcal{V}}_{ijk} - V_{ijk}(C) = 0. \tag{25}$$

In (22)-(25) the superimposed asterisk indicates the Zaremba-Jaumann derivative [21].

Generally for a tensor function $\psi_{k...q}$, one has

$$\overset{*}{\psi}_{k...m} = \dot{\psi}_{k...m} - w_{kq}\psi_{q...m} - w_{mq}\psi_{k...q}, \tag{26}$$

where in this model

$$w_{ij} = v_{i,j} - \frac{\partial \varepsilon_{ij}}{\partial t}. \tag{27}$$

4. Entropy inequality analysis

All processes considered here should be admissible from the thermodynamical point of view and thus they should not contradict the second law of thermodynamics. Thus, all the admissible solutions of the proposed evolution equations should be restricted by the following *entropy inequality*:

$$\rho\dot{S} + J^S_{k,k} - \frac{\rho r}{T} \geq 0, \tag{28}$$

where S denotes the entropy per unit mass and $\mathbf{J}^S$ is the entropy flux associated with the fields of the set C given by

$$\mathbf{J}^S = \frac{1}{T}\mathbf{q} + \mathbf{k}, \tag{29}$$

where **k** an additional term called *extra entropy flux density*. Constitutive functions $W = \tilde{W}(C)$, with

$$W = \{T_{ij}, \mu^c, \nu_{ij}, g_i, e, R_{ij}, J_i, Q_i, V_{ijk}, S, J_i^S), \tag{30}$$

for porous structure filled by a fluid flow have to be derived in order to close the system of balance and evolution equations. μ^c denotes the chemical potential of the diffusion mass field and $\mathcal{V}_{ij}$ is the similar constitutive quantity related to the field coming from a network of pores in an elastic body. In [7] the entropy inequality was analized by Liu's theorem [13], where all balance and evolution equations are considered as mathematical constraints for the general validity of the entropy inequality. From the mathematical point of view Liu's theorem applied to the above set of laws can prove its hyperbolicity. This means that set describes propagation of signals of all physical fields with finite velocities. In [7] several characteristic groups of expressions were deduced. In particular the *laws of state* and *the affinities* Π_i^J, Π_i^Q, Π_{ijk}^R were obtained in the following form

$$\sigma_{ij} = \rho\frac{\partial\psi}{\partial\varepsilon_{ij}}, \qquad \mu^c = \frac{\partial\psi}{\partial c}, \qquad \frac{\partial\psi}{\partial c_{,i}} = 0, \tag{31}$$

$$S = -\frac{\partial\psi}{\partial T}, \qquad \nu_{ij} = \frac{\partial\psi}{\partial r_{ij}}, \qquad \frac{\partial\psi}{\partial T_{,i}} = 0, \qquad \frac{\partial\psi}{\partial r_{ij,k}} = 0, \tag{32}$$

$$\Pi_i^J \equiv \rho\frac{\partial\psi}{\partial j_i}, \quad \Pi_i^Q \equiv \rho\frac{\partial\psi}{\partial q_i}, \quad \Pi_{ijk}^R \equiv \rho\frac{\partial\psi}{\partial\mathcal{V}_{ijk}}. \tag{33}$$

Moreover, the entropy flux has the form

$$J_k^S = \frac{1}{T}(q_k - \mu^c j_k + \nu_{ij}\mathcal{V}_{ijk}), \tag{34}$$

σ_{ij} is symmetric (and then the couple g_i vanishes) and the free energy is the following function

$$\psi = \psi(\varepsilon_{ij}, c, T, r_{ij}, j_i, q_i, \mathcal{V}_{ijk}). \tag{35}$$

5. Constitutive theory

The constitutive laws can be deduced using Smith's theorem [19] with the help of the isotropic polynomial representations of proper tensor, vector and scalar constitutive functions of tensor, vector and scalar arguments, which should satisfy the objectivity principle (see [21]- [24]). Inserting constitutive equations into the balance equations we obtain the so-called "balances on the state space", that form a system of partial differential equations, whose order depends on the special choice of constitutive equations. That system governs the evolution of the "wanted fields" [8]. To obtain field equations which allow to consider and solve analytically and/or numerically particular problems, the theory can be linearized, obtaining a mathematical model to describe the physical reality in many situations. In general, the arrangement of the porous channels is anisotropic one. However, there exist situations in which it is possible to assume that the thin porous tubes are randomly distributed in the solid. In this first approximation the quantities responsible for the pore structure field can be presented in the form

$$r_{ij} = r\delta_{ij}, \qquad \nu_{ij} = \nu\delta_{ij}, \qquad R_{ij} = R\delta_{ij}, \tag{36}$$

$$\mathcal{V}_{ijk} = \mathcal{V}_k\delta_{ij}, \qquad \Pi^R_{ijk} = \Pi_k\delta_{ij}, \qquad V_{ijk} = V_k\delta_{ij}. \tag{37}$$

In this way the evolution equations concerning that field are given by

$$\overset{*}{r} + \mathcal{V}_{k,k} = R(C), \qquad \overset{*}{\mathcal{V}_k} = V_k(C). \tag{38}$$

Three groups of constitutive functions (dependent variable) result from the analysis made in the paper [7] applying Smith's theorem [19]. The first one deals with the laws of state for the stress tensor σ_{ij}, for the chemical potential μ^c, the chemical pore structure potential ν and the entropy S.

They are functions of the following reduced set of indipendent variables

$$C_1 = \{\varepsilon_{ij}, c, T, r\}.$$

The general representation for them, i.e.

$$\sigma_{ij} = \sigma_{ij}(C_1), \qquad \mu^c = \mu^c(C_1),$$

$$\nu = \nu(C_1), \qquad S = S(C_1),$$

are as follows for the *isotropic tensor* σ_{ij}

$$\sigma_{ij} = \alpha^1_\sigma\delta_{ij} + \alpha^2_\sigma\varepsilon_{ij} + \alpha^3_\sigma\varepsilon_{ik}\varepsilon_{kj}, \tag{39}$$

and for the *scalar functions* μ^c, ν, S

$$\mu^c = \alpha^1_c c + \alpha^2_c T + \alpha^3_c r + \alpha^4_c\varepsilon_{kk} + \alpha^5_c\varepsilon_{ij}\varepsilon_{ij} + \alpha^6_c\varepsilon_{ij}\varepsilon_{jk}\varepsilon_{ki}, \tag{40}$$

$$\nu = \alpha^1_\nu c + \alpha^2_\nu T + \alpha^3_\nu r + \alpha^4_\nu\varepsilon_{kk} + \alpha^5_\nu\varepsilon_{ij}\varepsilon_{ij} + \alpha^6_\nu\varepsilon_{ij}\varepsilon_{jk}\varepsilon_{ki}, \tag{41}$$

$$S = \alpha^1_S c + \alpha^2_S T + \alpha^3_S r + \alpha^4_S\varepsilon_{kk} + \alpha^5_S\varepsilon_{ij}\varepsilon_{ij} + +\alpha^6_S\varepsilon_{ij}\varepsilon_{jk}\varepsilon_{ki}, \tag{42}$$

where the coefficients α_σ^m (with m = 1,2,3), α_c^n, α_ν^n and α_S^n (with n = 1,...,6) can be functions of the set of invariants built on the set C_1, i.e.

$$T, c, r, \varepsilon_{kk}, \varepsilon_{ij}\varepsilon_{ij}, \varepsilon_{ij}\varepsilon_{jk}\varepsilon_{ki}. \tag{43}$$

The second group concerns the affinities Π_i^J, Π_i^Q and Π_i^R as constitutive functions of the reduced set of independent variables C_2

$$C_2 = \{j_i, q_i, \mathcal{V}_i\}. \tag{44}$$

The general representations for such constitutive functions

$$\Pi_i^J = \Pi_i^J(C_2), \qquad \Pi_i^Q = \Pi_i^Q(C_2), \qquad \Pi_i^R = \Pi_i^R(C_2) \tag{45}$$

are as follows

$$\Pi_i^J = \beta_J^1 j_i + \beta_J^2 q_i + \beta_J^3 \mathcal{V}_i,\ \Pi_i^Q = \beta_Q^1 j_i + \beta_Q^2 q_i + \beta_Q^3 \mathcal{V}_i,\ \Pi_i^R = \beta_R^1 j_i + \beta_R^2 q_i + \beta_R^3 \mathcal{V}_i, \tag{46}$$

where the coefficients β_I^m, β_Q^m and β_R^m $(m = 1,\ 2,\ 3)$ can be functions of the irreducible set of invariants built on the set C_2, i.e.

$$j_i j_i,\ \ q_i q_i,\ \ \mathcal{V}_i \mathcal{V}_i,\ \ j_i q_i,\ \ j_i \mathcal{V}_i,\ \ q_i \mathcal{V}_i. \tag{47}$$

The last group of constitutive functions includes the evolution equations of the pore density and the fluxes of mass, heat and field coming from a network of pores

$$\overset{*}{r} + \mathcal{V}_{k,k} = R(C),\quad \overset{*}{j}_i = J_i(C),\quad \overset{*}{q}_i = Q_i(C),\quad \overset{*}{\mathcal{V}}_i = V(C). \tag{48}$$

In the linear approximation their general representation is the following

$$\begin{aligned}\overset{*}{r} = {} & -\mathcal{V}_{k,k} + \gamma_R^1 c + \gamma_R^2 T + \gamma_R^3 r + \gamma_R^4 \varepsilon_{kk} \\ & + \gamma_{Ri}^5 j_i + \gamma_{Ri}^6 q_i + \gamma_{Ri}^7 \mathcal{V}_i + \gamma_{Ri}^8 c_{,i} + \gamma_{Ri}^9 T_{,i} + \delta_{Ri}^{10} r_{,i},\end{aligned} \tag{49}$$

$$\overset{*}{j}_i = \gamma_J^1 j_i + \gamma_J^2 q_i + \gamma_J^3 \mathcal{V}_i + \gamma_J^4 c_{,i} + \gamma_J^5 T_{,i} + \gamma_J^6 r_{,i}, \tag{50}$$

$$\overset{*}{q}_i = \gamma_Q^1 j_i + \gamma_Q^2 q_i + \gamma_Q^3 \mathcal{V}_i + \gamma_Q^4 c_{,i} + \gamma_Q^5 T_{,i} + \gamma_Q^6 r_{,i}, \tag{51}$$

$$\overset{*}{\mathcal{V}}_i = \gamma_V^1 j_i + \gamma_V^2 q_i + \gamma_V^3 \mathcal{V}_i + \gamma_V^4 c_{,i} + \gamma_V^5 T_{,i} + \gamma_V^6 r_{,i}, \tag{52}$$

where the coefficients γ_R^α (with $\alpha = 1, 2, 3$), γ_R^4, γ_{Ri}^β (with $\beta = 5, 6, ..., 10$) γ_J^δ, γ_Q^δ, γ_V^δ, (with $\delta = 1, 2, ..., 6$), can be functions of the set of invariants built on the set C. The obtained relations together with the sets of invariants form the constitutive laws that we have been looking for. The last equation is the generalized Vernotte-Cattaneo relation. All the derived laws in their general form deal with many complex phenomena occurring in the porous body filled by a fluid flow.

References

1. A. E. Scheidegger, *The Physics of Flow through Porous Media*, 2nd edn. (Univ. of Toronto, 1960).
2. J. Bear, D. Zaslavsky and S. Irmay , *Physical Principles of Water Percolation and Seepage*, ed. S. Irmay (UNESCO, 1968).
3. F. A. L. Dullien, *Porous Media, Fluid Transport and Pore Structure* (Academic Press, New York, 1979).
4. J. C. Slattery, *Momentum, Energy and Mass Transfer in Continuo* (McGraw Hill, New York, 1972).
5. S. Whitaker, *Indus. Enging. Chem.* **61**, 14 (1969).
6. J. Kubik, *Int. J. Engng. Sci.* **24** (6), 971 (1986).
7. L. Restuccia, *Supplemento ai Rendiconti del Circolo Matematico di Palemo* **77** (II), 565 (2006).
8. W. Muschik and L. Restuccia, *Communications to SIMAI Congress*, ISSN: 1827-9015 (2006), DOI: 10.1685/CSC06120.
9. W. Muschik, C. Papenfuss, and H. Ehrentraut, *Concepts of continuum thermodynamics* (Kielce, Poland, 1996).
10. W. Muschik, *Aspects of non-equilibrium thermodynamics* (World Scientific, Singapore, 1990).
11. W. Muschik, Fundamentals of non-equilibrium thermodynamics, in *Non-equilibrium thermodynamics with applications to solids*, ed.W. Muschik, CISM Courses and Lectures, Vol. 336 (Springer Verlag, Wien - New York, 1993).
12. W. Muschik, C. Papenfuss, and H. Ehrentraut, *J. Non-Newtonian fluid Mech.* **96**, 255 (2001).
13. I-Shih Liu, *Arch. Rat. Mech. Anal.* **46**, 131 (1972).
14. M. E. Malaspina and L. Restuccia, *Communications to SIMAI Congress*, ISSN: 1827-9015 (2006), DOI: 10.1685/CSC06105.
15. M. E. Malaspina and L. Restuccia, *Communications to SIMAI Congress*, ISSN: 1827-9015, Vol 3 (2009), DOI: 10.1685/CSC09XXX.
16. M. Dolfin, M. Francaviglia and P. Rogolino, *J. Non-Equilib. Thermodyn.* **23**, 250 (1998) .
17. M. Dolfin, M. Francaviglia and P. Rogolino, *Periodica Polytechnica Series Mech. Eng.* **43**, 29 (1999).
18. W. Noll, *Arch. Rat. Mech. Anal.* **48**, 1 (1972).
19. G. F. Smith, *Int. J. Engng. Sci.* **9**, 899 (1971).
20. B. Maruszewski, *Phys.stat.sol.(b)* **168**, 59 (1991).
21. C. Truesdell, W. Noll, *The non-linear field theories of mechanics*, Handbuch der Physik, **III/3** (Springer, Berlin, Heidelberg, New York, 1965)
22. W. Muschik and L. Restuccia, *Technische Mechanik* **22**, 152 (2002).
23. H. Herrmann, W. Muschik, G. Rückner, and L. Restuccia, *Constitutive mappings and the non-objective part of material frame indifference*, in *Trends in Continuum Physics'04* (TRECOP'04) eds. B. T. Maruszewski, W. Muschik and A. Radowicz (WNPP, Poznan, Poland, 2004), pp.128-136
24. W. Muschik and L. Restuccia, *Archive of Applied Mechanics*, DOI: 10.1007/s00419-007-0193-2(2008), 180.

TOWARD ANALYTICAL CONTOUR DYNAMICS

G. RICCARDI and D. DURANTE
Dept. of Aerospace and Mechanical Engineering, *II* University of Naples
via Roma, 29 - 81031 Aversa (Ce), Italy
E-mails: giorgio.riccardi, danilo.durante@unina2.it

The classical contour dynamics approach to the numerical simulation of the two-dimensional motion of a uniform vortex in an inviscid fluid is here reread in the analytical framework of the Schwarz function of the vortex boundary. The key point of the present analysis lies in a nonlinear integrodifferential problem in that function, which follows from a new integral relation between Schwarz function and self-induced velocity of the vortex.

As a preliminar approach to the solution of such a problem, a procedure leading to small time solutions is presented and its results are compared with the numerical ones. A more satisfactory approach is also proposed: it consists in reducing the integrodifferential problem to an integral equation, by applying the Laplace transform in time. The resulting equation contains a singular linear dominant part and a regular nonlinear one, as well as the initial data. The use of successive approximations in the analytical solution of such a nonlinear integral equation is proposed. The first order approximation of the solution is also compared with numerical simulations.

Keywords: two-dimensional vortex dynamics, contour dynamics, Schwarz function, complex analysis

1. Introduction

The dynamics of two dimensional plane vortices having uniform vorticity in an inviscid fluid has been broadly investigated in Literature, *numerically* as well as through suitable *simplified models*. The numerical simulations are almost entirely based on the Contour Dynamics (CD) algorithm.[1,2] The motion of the vortex boundary is approximated through the numerical integration of the dynamics of a finite number of nodes, taking care about the node redistribution in presence of high stretching or curvature. On the other hand, the theoretical analysis of the vortex motion is approached by adopting strongly simplifying assumptions, in order to reduce the degrees of freedom of the system to a finite and (possibly) small number. Progenitors of such models are the point vortex[3] and the elliptical uniform vortex[4] (which rotates without changing its shape). In particular, one of the major achievements of the use of elliptical vortices lies in the explanation of the merging mechanism[5] and its rereading in terms of rotating strain.[6]

The present paper collocates in the middle between these two approaches. It uses the CD formulation to define the vortex motion, but without resorting to numerical discretization, interpolation and time integration. Moreover, simplifying assumptions are made, in order to reduce the degrees of freedom of the problem, in a sense that will be explained below. At the same time, the shape of the vortex is not constrained into a prescribed family of curves. The analysis is performed in term of the *Schwarz function* $\boldsymbol{\Phi}$ (complex quantities are indicated with bold symbols) of the vortex boundary ∂P, that is the function[7] which holds $\overline{\boldsymbol{x}}$ (overbar means complex conjugate) on any point $\boldsymbol{x} \in \partial P$, being defined through analytical continuation outside that curve. The key point of the present approach lies in using a new integral relation[8] between $\boldsymbol{\Phi}$ and the complex conjugate of the velocity $\overline{\boldsymbol{u}}$:

$$\overline{\boldsymbol{u}}(\boldsymbol{x}, \overline{\boldsymbol{x}}) = \frac{\omega}{2\boldsymbol{i}} \left[\chi_P(\boldsymbol{x})\, \overline{\boldsymbol{x}} + \frac{1}{2\pi \boldsymbol{i}} \int_{\partial P} d\boldsymbol{y}\, \frac{\boldsymbol{\Phi}(\boldsymbol{y})}{\boldsymbol{x} - \boldsymbol{y}} \right] , \tag{1}$$

ω being the vorticity level inside the vortex. The integral is understood as a Cauchy one (it will be indicated with a bar on the integral) when $\boldsymbol{x}$ lies on the curve ∂P. Moreover, χ_P is the characteristic function of the domain P: it holds 1 inside P, 0 outside P and just 1/2 on the boundary. The use of the formula (1) enables us to evaluate the self-induced velocity simply in terms of residues, overcoming in this way the difficulties related to the splitting of $\boldsymbol{\Phi}$ in the sum of two functions, one analytical inside and one outside the vortex, as proposed by Saffman.[9]

An outline of the paper is the following one. In Section 2 the equation of the dynamics of $\boldsymbol{\Phi}$ is written and an integrodifferential problem for $\boldsymbol{\Phi}$ is formulated, by using the velocity (1). In Section 3 the vortex dynamics is analyzed for times much smaller than the eddy turn-over ones, by writing the Schwarz function in a time series. A new integral formulation of the problem is deduced in Section 4, by applying the Laplace transform in time to a suitable rewriting of the above integrodifferential problem. The dynamics of a non-trivial sample family of vortices (having their Schwarz functions with two simple poles, see Appendix A) is investigated through both approaches in Section 5. Comparisons with CD simulations are also discussed. Finally, conclusions and the plan of the future work are offered in Section 6.

2. Schwarz function dynamics

In the present section an equation for the dynamics of the Schwarz function is written, by starting from the Lagrangian representation of the motion of the vortex boundary. A point $\boldsymbol{x} \in \partial P(t)$ of the vortex boundary at the current time is taken as a function of the position $\boldsymbol{\xi}$ at the reference time $(t = 0)$ and of the time itself: $\boldsymbol{x} = \boldsymbol{x}(\boldsymbol{\xi}, t)$. By conjugating this position and deriving in time:

$$\partial_t \overline{\boldsymbol{x}}(\boldsymbol{\xi}, t) = \overline{\boldsymbol{u}}[\boldsymbol{x}(\boldsymbol{\xi}, t), t] \; . \tag{2}$$

The conjugate of the position $\boldsymbol{x}$ on the vortex boundary is a function of the position itself $\boldsymbol{x}$ through the Schwarz function: $\overline{\boldsymbol{x}}(\boldsymbol{\xi}, t) = \boldsymbol{\Phi}[\boldsymbol{x}(\boldsymbol{\xi}, t), t] = \boldsymbol{\Phi}^{(\mathcal{L})}(\boldsymbol{\xi}, t)$ (the apex "$(\mathcal{L})$" indicates the Lagrangian form). By using this form of the Schwarz function in equation (2), the dynamics of $\boldsymbol{\Phi}^{(\mathcal{L})}$ follows:

$$\partial_t \boldsymbol{\Phi}^{(\mathcal{L})}(\boldsymbol{\xi}, t) = \overline{\boldsymbol{u}}[\boldsymbol{x}(\boldsymbol{\xi}, t), t] \; ,$$

for any $\boldsymbol{\xi} \in \partial P(0)$. Once the conjugate of the velocity on the vortex boundary is written through equation (1.1), the dynamics of $\boldsymbol{\phi}^{(\mathcal{L})}$ leads to the nonlinear integrodifferential problem:

$$\begin{cases} \partial_\tau \boldsymbol{\Phi}^{(\mathcal{L})}(\boldsymbol{\xi}, \tau) = -\boldsymbol{i} \left[\boldsymbol{\Phi}^{(\mathcal{L})}(\boldsymbol{\xi}, \tau) + \dfrac{1}{\pi \boldsymbol{i}} \displaystyle\int_{\partial P(0)} d\boldsymbol{\eta} \; \partial_{\boldsymbol{\eta}} \boldsymbol{x}(\boldsymbol{\eta}, \tau) \; \dfrac{\boldsymbol{\Phi}^{(\mathcal{L})}(\boldsymbol{\eta}, \tau)}{\boldsymbol{x}(\boldsymbol{\xi}, \tau) - \boldsymbol{x}(\boldsymbol{\eta}, \tau)} \right] \\ \boldsymbol{\Phi}^{(\mathcal{L})}(\boldsymbol{\xi}, 0) = \boldsymbol{\Phi}_0(\boldsymbol{\xi}) \quad \text{(given)}, \end{cases} \tag{3}$$

$\boldsymbol{x}$ on the vortex boundary being $\overline{\boldsymbol{\Phi}}^{(\mathcal{L})}$.

In the problem (3), time is nondimensionalized as $t\,\omega/4 = \tau$, for the following reasons. By introducing the diagnostic ellipse,[10] that is an elliptical uniform vortex having the same vorticity and second order moments as P (see also Section 6), the bulk angular velocity $\Theta' = d\Theta/dt$ of the vortex can be assumed as the angular velocity[4] of the ellipse:

$$\Theta' \simeq \frac{ab}{(a+b)^2} \, \omega \; , \tag{4}$$

a and b being the lengths of its semiaxes. It happens that Θ' closely approaches $\omega/4$, at least for not too elongated vortices. As a consequence, by nondimensionalizing the time with $\omega/4$, one obtains the simple result that the eddy turn-over time of the vortex is about 2π.

3. Small time dynamics

In the present section the Lagrangian Schwarz function $\mathbf{\Phi}^{(\mathcal{L})}$ is assumed as the series in time:

$$\mathbf{\Phi}^{(\mathcal{L})}(\boldsymbol{\xi},\tau) = \mathbf{\Phi}_0(\boldsymbol{\xi}) + \mathbf{\Phi}_1(\boldsymbol{\xi})\,\tau + \mathbf{\Phi}_2(\boldsymbol{\xi})\,\tau^2 + \dots \tag{5}$$

the conjugate giving also the Lagrangian position $\boldsymbol{x}(\boldsymbol{\xi},\tau)$. By inserting the series (5) as well as the corresponding one for $\boldsymbol{x}$ into the integrodifferential equation (1.3), the following hierarchy of equations are obtained (the tilde indicates dependence on $\boldsymbol{\eta}$, while the apex derivative in $\boldsymbol{\eta}$):

$$\begin{aligned}
\boldsymbol{i}\mathbf{\Phi}_1 &= \mathbf{\Phi}_0 + \frac{1}{\pi \boldsymbol{i}}\fint_{\partial P(0)} \frac{d\boldsymbol{\eta}}{\boldsymbol{\xi}-\boldsymbol{\eta}}\,\tilde{\mathbf{\Phi}}_0 \\
2\boldsymbol{i}\mathbf{\Phi}_2 &= \mathbf{\Phi}_1 + \frac{1}{\pi \boldsymbol{i}}\fint_{\partial P(0)} \frac{d\boldsymbol{\eta}}{\boldsymbol{\xi}-\boldsymbol{\eta}}\left[\tilde{\mathbf{\Phi}}_1 + \tilde{\mathbf{\Phi}}_0'(\boldsymbol{x}_1 - \tilde{\boldsymbol{x}}_1)\right] \\
&\vdots \qquad\qquad\qquad \vdots
\end{aligned} \tag{6}$$

In the present paper, solutions in the hierarchy (6) are evaluated up to second order. It can be shown that area, first and second order moments of the vortex (see Appendix B) are keept constant in time up to their second order terms by these solutions. Sample of calculations of the solutions (6) will be shown in Section 5, where they are compared with the numerical CD simulations, as well as with the results of the integral equation written in the next section.

4. An integral approach to vortex dynamics

The integrodifferential problem (2.3) is modified in the following one:

$$\begin{cases}
\partial_\tau \mathbf{\Phi}^{(\mathcal{L})}(\boldsymbol{\xi},\tau) = -\boldsymbol{i}\Bigg[\ \mathbf{\Phi}^{(\mathcal{L})}(\boldsymbol{\xi},\tau) - \dfrac{1}{\pi \boldsymbol{i}}\fint_{\partial P(0)} d\boldsymbol{\eta}\,\dfrac{\mathbf{\Phi}^{(\mathcal{L})}(\boldsymbol{\eta},\tau)}{\boldsymbol{\eta}-\boldsymbol{\xi}} \\
\qquad\qquad\qquad -\dfrac{1}{\pi \boldsymbol{i}}\displaystyle\int_{\partial P(0)} d\boldsymbol{\eta}\; \boldsymbol{g}(\boldsymbol{\eta},\boldsymbol{\xi};\tau)\;\mathbf{\Phi}^{(\mathcal{L})}(\boldsymbol{\eta},\tau)\Bigg] \\
\mathbf{\Phi}^{(\mathcal{L})}(\boldsymbol{\xi},0) = \mathbf{\Phi}_0(\boldsymbol{\xi}) \quad \text{(given)},
\end{cases} \tag{7}$$

where the geometry of the vortex at the current time is accounted for the function $\boldsymbol{g} : \partial P(0) \times \partial P(0) \times [0,+\infty)$ defined in the following way:

$$\boldsymbol{g}(\boldsymbol{\eta},\boldsymbol{\xi};\tau) = \frac{\partial}{\partial \boldsymbol{\eta}} \log \frac{\boldsymbol{x}(\boldsymbol{\eta};\tau) - \boldsymbol{x}(\boldsymbol{\xi};\tau)}{\boldsymbol{\eta}-\boldsymbol{\xi}}\,.$$

It is worth noticing that $\boldsymbol{g}(\boldsymbol{\eta},\boldsymbol{\xi};0) \equiv \boldsymbol{g}(\boldsymbol{\xi},\boldsymbol{\xi};\tau) \equiv 0$. The integrodifferential problem (7) exhibits the remarkable advantage of having a Cauchy dominant part which is linear in the Schwarz function $\boldsymbol{\Phi}^{(\mathcal{L})}$.

In the problem (7), the Laplace transform in time is applied to both sides of the equation. The use of Laplace transform is due to the lack of convergence to zero of the Schwarz function $\boldsymbol{\Phi}^{(\mathcal{L})}(\boldsymbol{\xi};\tau)$ at large values of τ. By indicating with hats the Laplace transformed functions, the integrodifferential problem (7) becomes the following integral one:

$$\underbrace{(\boldsymbol{i}\boldsymbol{\sigma}-1)\hat{\boldsymbol{\Phi}}^{(\mathcal{L})}(\boldsymbol{\xi};\boldsymbol{\sigma}) + \frac{1}{\pi \boldsymbol{i}} \fint_{\partial P(0)} d\boldsymbol{\eta}\, \frac{\hat{\boldsymbol{\Phi}}^{(\mathcal{L})}(\boldsymbol{\eta};\boldsymbol{\sigma})}{\boldsymbol{\eta}-\boldsymbol{\xi}}}_{\mathbf{SD}[\hat{\boldsymbol{\Phi}}^{(\mathcal{L})}]:\ \textit{singular dominant part (linear)}} + \underbrace{\frac{1}{\pi \boldsymbol{i}} \int_{\partial P(0)} d\boldsymbol{\eta}\, \widehat{\boldsymbol{g}\boldsymbol{\Phi}}^{(\mathcal{L})}(\boldsymbol{\eta},\boldsymbol{\xi};\boldsymbol{\sigma})}_{\mathbf{R}[\hat{\boldsymbol{\Phi}}^{(\mathcal{L})}]:\ \textit{regular part (non linear)}} = \underbrace{\boldsymbol{i}\boldsymbol{\Phi}_0(\boldsymbol{\xi})}_{\mathbf{ID}:\ \textit{data}} . \tag{8}$$

As stated above, the integral equation (8) is formed by a linear dominant part and a nonlinear regular one. Moreover, it also accounts for the initial data. In order to analytically integrate the above equation, the following iterative procedure is proposed: it is firstly solved by neglecting the nonlinear regular term $\mathbf{R}[\hat{\boldsymbol{\Phi}}^{(\mathcal{L})}]$:

$$\mathbf{SD}[\hat{\boldsymbol{\Phi}}^{(\mathcal{L})}_{(0)}] = \mathbf{ID}\ , \tag{9}$$

which leads to the first (linear) approximation $\boldsymbol{\Phi}^{(\mathcal{L})}_{(0)}$ of the solution. Then the hierarchy of singular integral problems is posed:

$$\mathbf{SD}[\hat{\boldsymbol{\Phi}}^{(\mathcal{L})}_{(k)}] = \mathbf{ID} - \mathbf{R}[\hat{\boldsymbol{\Phi}}^{(\mathcal{L})}_{(k-1)}]\ , \tag{10}$$

for $k = 1, 2, \ldots$, which leads to successive approximations of the solution of the original equation (8). In the next section the term $\boldsymbol{\Phi}^{(\mathcal{L})}_{(1)}$ is numerically evaluated and compared with the results of CD simulations, as well as with the second order series approach discussed in Section 3.

5. Application to the sample vortices

The issue of the integration of the integrodifferential problem (2.3) has been approched in two different ways: by using the time series (3.5) for the Lagrangian Schwarz function which leads to the hierarchy of solutions (3.6) or through a Laplace transform in time, which reduces the integrodifferential problem to the singular integral equation (4.8) (see Ref. 11 for a comprehensive analysis). A natural approach to the solution of that equation

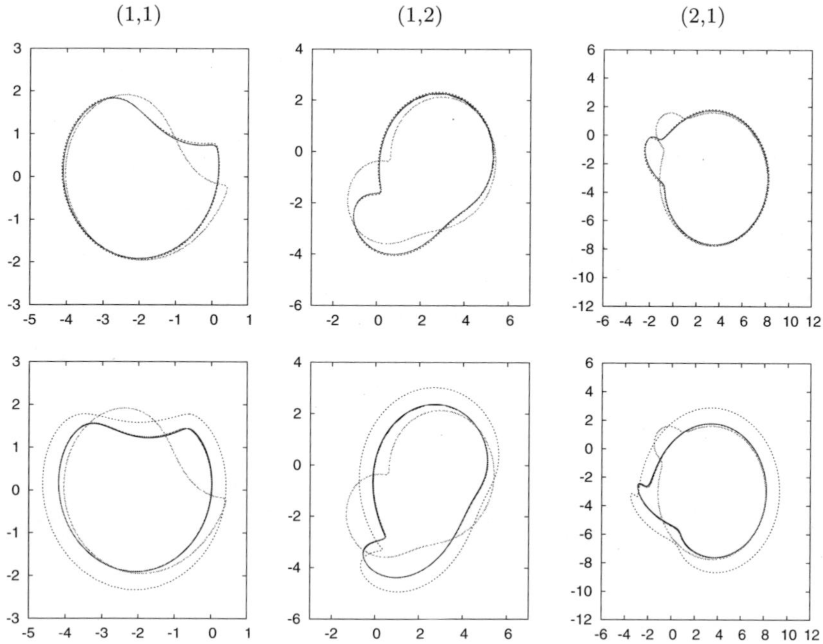

Fig. 1. Comparisons between the results of numerical simulation (black solid lines), small time (blue dashed) and integral (red solid) approaches for vortices of kind (1,1) (first column), (1,2) (second) and (2,1) (third). The time $\tau = 0$ is also shown (green dashed line). The time τ holds 0.3 in the first row and 0.6 in the second one.

appears to be the analytical iterative method (4.9, 4.10). Notice also that the calculation of the Laplace transform in time is not practically needed, the antitransform of each $\hat{\mathbf{\Phi}}^{(\mathcal{L})}_{(k)}$ being finally computed.

In the present paper, the series approach is truncated at the second order, while only the first order term is considered in the iterative method (4.9, 4.10), due to strong analytical difficulties. It is evaluated numerically, its analytical calculation being subject of the current research. The two approaches have been adopted in investigating the dynamics of the sample vortices, whose kinematics has been briefly accounted for in Appendix A. In Figs. 1, 2 comparisons with contour dynamics simulations are performed. The first order solution of the iterative method (4.9, 4.10) (drawn with a red solid line) exhibits an excellent agreement with the direct simulation (black line), at least at the moderate times used in the figures (remember that $\tau = 0.3$ is roughly $1/20$ of the eddy turn-over times of these vortices). On

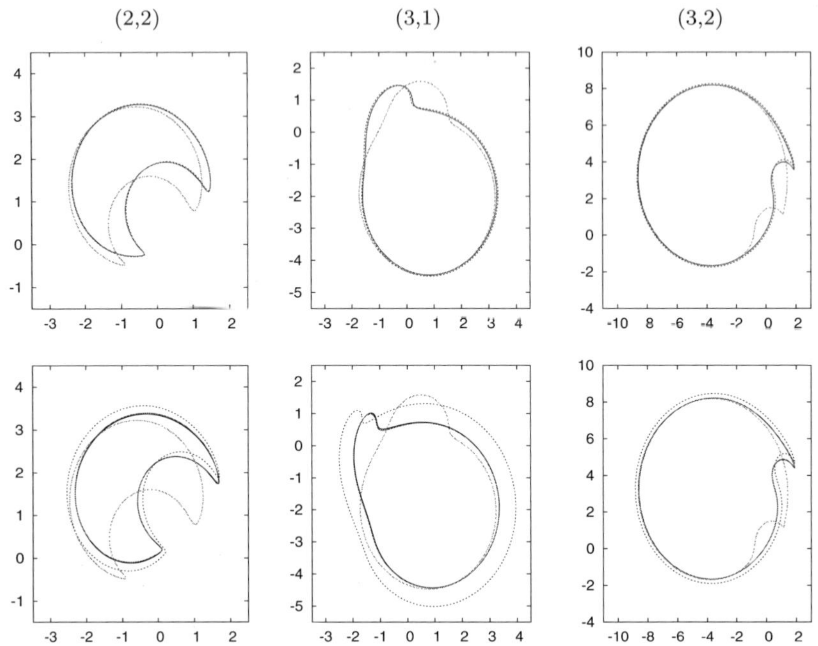

Fig. 2. As in Fig. 1, but for vortices of kind (2,2) (first column), (3,1) (second) and (3,2) (third). The time τ is fixed to 0.3 in the first row, while in the second one it holds 0.6 for the vortices (2,2) and (3,1) and 0.4 for the one (3,2).

the contrary, the second order series solution (blue dashed line) is affected by larger errors, due to the lack of conservation of the first integrals of the motion. In fact, the first and second order terms in the time series (B.B.1, B.B.2, B.B.3) vanish, but higher order contributions are still different from zero.

Numerical simulations of the motion of the vortex $(1,1)$ show that the high curvature region becomes flat and a filament develops in the arc which is approximatively $\pi/2$ ahead. This behaviour is well described by the integral approach, while higher order terms are needed for an accurate series solution. In the motion of the vortices $(1,2)$, $(2,1)$, $(3,1)$ and $(3,2)$, the arcs which lead to the formation of filaments are defined by the initial shapes: both approaches capture this dynamics. More complex appears to be the dynamics of the vortex $(2,2)$, where a large amount of circulation moves toward the lower edge from the upper one, in which a filament will be produced. Even in this case the integral equation behaves better than the series, the second order solution of which appears to be rather ahead to the numerical one.

6. Conclusions and future work

The comparison with the direct numerical simulations of the previous section shows that the integral equation promises to be an efficient analytical approach to the vortex dynamics in planar flows of inviscid fluids. About this issue, two points must be investigated in the future. First of all, the analytical evaluation of such solutions appears to be rather difficult, at least for non-trivial initial vortex shapes. Indeed, huge algebraic difficulties are produced by the analytical calculation of the integrals along the initial vortex boundary. At the present time, they are evaluated in a transformed plane, even if the calculation in the plane of the initial vortex shape is also under investigation. Moreover, a large amount of work is still needed in order to understand the time behaviour of the Schwarz function, even at the first order. In particular, the time evolution of its initial discontinuities (the circular arc between the points $\boldsymbol{x}_{io}^{+}$ and $\boldsymbol{x}_{io}^{-}$) is the main goal of such an analysis.

The use of the series (3.5) to solve the integrodifferential problem (2.3) leads to satisfactory solutions for very small times, only. This approach will be improved, also. The enhancement is under investigation at the present time and is based on the writing of the series (3.5) in a suitable rotating reference frame. To this aim, the diagnostic ellipse[10] is built as the uniform elliptical vortex which has the same vorticity and second order moments:

$$\begin{aligned}
I_{xx}(\tau) &= \int_{P(\tau)} dA(\boldsymbol{x})\ (x - x_{cv})^2 &&= -\frac{\mathrm{Im}[\boldsymbol{\chi}(\tau)]}{8} + \frac{I}{2} - |P|\, x_{cv}^2 \\
I_{xy}(\tau) &= \int_{P(\tau)} dA(\boldsymbol{x})\ (x - x_{cv})(y - y_{cv}) &&= -\frac{\mathrm{Re}[\boldsymbol{\chi}(\tau)]}{16} - |P|\, x_{cv} y_{cv} \qquad (11)\\
I_{yy}(\tau) &= \int_{P(\tau)} dA(\boldsymbol{x})\ (y - y_{cv})^2 &&= +\frac{\mathrm{Im}[\boldsymbol{\chi}(\tau)]}{8} + \frac{I}{2} - |P|\, y_{cv}^2
\end{aligned}$$

as the true vortex $P(t)$. In equations (11), $\boldsymbol{x}_{cv} = x_{cv} + \boldsymbol{i} y_{cv}$ is the position of the center of vorticity and I is the second order moment moment of the vortex in the absolute reference frame (for the evaluation of I, see Appendix B). Moreover, the function $\boldsymbol{\chi}(\tau)$ can be written in terms of the Schwarz one as follows:

$$\boldsymbol{\chi}(\tau) = \int_{\partial P(\tau)} d\boldsymbol{x} \left[\boldsymbol{x}^2 \boldsymbol{\Phi}(\boldsymbol{x},\tau) - \boldsymbol{\Phi}^3(\boldsymbol{x},\tau) \right] .$$

By expanding in Taylor series about the time $\tau = 0$ the above function, the series for the second order moments (11) and then for the ellipse axes are obtained, from which the series for the angular velocity (2.4): $\Theta'(\tau) = \Theta'_0 + \Theta''_0\tau + \Theta'''_0\tau^2/2 + \ldots$ follows. In this way, the time series for the position in the rotating frame is written as:

$$\begin{aligned}\boldsymbol{x}(\boldsymbol{\xi},\tau) &= \boldsymbol{x}_{cv} + (\boldsymbol{\xi} - \boldsymbol{x}_{cv})\exp[\boldsymbol{i}\Theta(\tau)] + \boldsymbol{x}_1(\boldsymbol{\xi})\tau + \frac{1}{2}\,\boldsymbol{x}_2(\boldsymbol{\xi})\tau^2 + \ldots \\ &= \quad \boldsymbol{\xi} + \left[\boldsymbol{i}\Theta'_0(\boldsymbol{\xi} - \boldsymbol{x}_{cv}) + \boldsymbol{x}_1(\boldsymbol{\xi})\right]\tau \\ &\quad + \frac{1}{2}\left[(\boldsymbol{i}\Theta''_0 - \Theta'^2_0)(\boldsymbol{\xi} - \boldsymbol{x}_{cv}) + \boldsymbol{x}_2(\boldsymbol{\xi})\right]\tau^2 + \ldots\end{aligned} \tag{12}$$

assuming $\Theta(0) = 0$. The use of the series (12) enables us to define the perturbation terms $\boldsymbol{x}_k$ in the rotating frame: they are only due to the vortex shape deformations, without any contribution of the solid body rotation.

As a final remark about the present approaches, it is worth stressing that a detailed knowledge of the initial self-induced velocity field is required, in order to investigate the vortex dynamics. Indeed, the Lagrangian mapping is used everywhere, in particular in evaluating integrals on the initial vortex boundary $\partial P(0)$.

Acknowledgements

The authors would like to express their gratitude to Enrico De Bernardis and Alessandro Iafrati, for many helpful discussions and for their kindliness.

Appendix A. Kinematics of vortices having Schwarz functions with two simple poles

In a previous paper,[8] the kinematics of the family of uniform vortices the Schwarz functions of which have two poles has been investigated. It will be briefly summarized here, for the Reader convenience.

For such vortices, the Schwarz function on the vortex boundary is assumed of the form:

$$\boldsymbol{\Phi}(\boldsymbol{z}) = \overline{\boldsymbol{x}}(\boldsymbol{z}) = \frac{\boldsymbol{a}_1}{\boldsymbol{z} - \boldsymbol{z}_1} + \frac{\boldsymbol{a}_2}{\boldsymbol{z} - \boldsymbol{z}_2}\,, \tag{A.1}$$

$\boldsymbol{z}$ running on the unit circle $\mathcal{C}$. In the definition (A.1), poles ($\boldsymbol{z}_1$, $\boldsymbol{z}_2$) and residues ($\boldsymbol{a}_1$, $\boldsymbol{a}_2$) are given in such a way that the vortex boundary $\partial P(0)$

results a simple curve. The vortex shape is obtained by conjugating the above definition as:

$$\boldsymbol{x}(\boldsymbol{z}) = -(\overline{\boldsymbol{a}}_1\boldsymbol{w}_1 + \overline{\boldsymbol{a}}_2\boldsymbol{w}_2) - \frac{\overline{\boldsymbol{a}}_1\boldsymbol{w}_1^2}{\boldsymbol{z} - \boldsymbol{w}_1} - \frac{\overline{\boldsymbol{a}}_2\boldsymbol{w}_2^2}{\boldsymbol{z} - \boldsymbol{w}_2}\,, \tag{A.2}$$

$\boldsymbol{w}_k$ being the image of $\boldsymbol{z}_k$ ($k = 1, 2$) through the unit circle, *i.e.* $\boldsymbol{w}_k = 1/\overline{\boldsymbol{z}}_k$. The analytical continuation of the above map is considered, as well as its inverse one $\boldsymbol{z} = \boldsymbol{z}(\boldsymbol{x})$.

A fundamental property of the function (A.2) is that the equation in $\boldsymbol{\zeta}$: $\boldsymbol{x}(\boldsymbol{\zeta}) = \boldsymbol{x}(\boldsymbol{z})$ possesses two solutions. Besides the trivial one ($\boldsymbol{\zeta} = \boldsymbol{z}$), another solution is given by the following linear fractional map:

$$\boldsymbol{\zeta}^\star(\boldsymbol{z}) = \frac{\boldsymbol{\beta}\boldsymbol{z} - \boldsymbol{\gamma}}{\boldsymbol{\alpha}\boldsymbol{z} - \boldsymbol{\beta}}\,, \tag{A.3}$$

$\boldsymbol{\alpha}$, $\boldsymbol{\beta}$ and $\boldsymbol{\gamma}$ being the quantities $\boldsymbol{\alpha} = \overline{\boldsymbol{a}}_1\boldsymbol{w}_1^2 + \overline{\boldsymbol{a}}_2\boldsymbol{w}_2^2$, $\boldsymbol{\beta} = \boldsymbol{w}_1\boldsymbol{w}_2(\overline{\boldsymbol{a}}_1\boldsymbol{w}_1 + \overline{\boldsymbol{a}}_2\boldsymbol{w}_2)$ and $\boldsymbol{\gamma} = \boldsymbol{w}_1^2\boldsymbol{w}_2^2(\overline{\boldsymbol{a}}_1 + \overline{\boldsymbol{a}}_2)$, respectively.

The map (A.2) results to be defined on a suitable (bounded or unbounded) neighbourhood of $\mathcal{C}$, inside or outside a certain circle $\mathcal{C}_i$, which is invariant for the map (A.3), *i.e.* $\boldsymbol{\zeta}^\star(\mathcal{C}_i) = \mathcal{C}_i$. Obviously, the inverse map $\boldsymbol{z}(\boldsymbol{x})$ is defined almost everywhere on the physical plane, being discontinuous only across an arc of circle, the endpoints of which are named as $\boldsymbol{x}_{io}^+$ and $\boldsymbol{x}_{io}^-$. This arc is just the image of $\mathcal{C}_i$ through the map (A.2).

A classification of the present vortices is possible, by looking at the position of the image $\boldsymbol{\zeta}^\star(\mathcal{C}) =: \mathcal{C}^\star$ with respect to the unit circle itself (first property) and to the behaviour of the inverse map $\boldsymbol{z}(\boldsymbol{x})$ (second property). A vortex is said to be of the first kind with respect to the first property if $\mathcal{C}^\star \subset \mathcal{C}$, of the second kind if $\mathcal{C} \subset \mathcal{C}^\star$ and of the third kind if $\mathcal{C}^\star$ is external to $\mathcal{C}$ and it does not include the unit circle. At the same time, the second property specifies where the inside of $\mathcal{C}$ goes through the map (A.2): the vortex is said of the first or of the second kind if it goes inside or outside P, respectively.

Once the analysis of the maps (A.2, A.3) has been carried out, the self-induced velocity field is straightforwardly calculated by using the relation (1.1), through a standard application of the residue theorem.

Appendix B. First integrals of motion

In the present appendix some of the first integrals of the motion are written in terms of the time series (3.5). The area of the vortex has the following series expansion in time (the apex indicates derivative in $\boldsymbol{z}$):

$$\begin{aligned} |P| &= \frac{1}{2\boldsymbol{i}} \int_{\partial P(t)} d\boldsymbol{x}\ \boldsymbol{\Phi} \\ &\propto \int_C dz\ \boldsymbol{\xi}' \boldsymbol{\Phi}_0 + \tau \int_C dz \left(\boldsymbol{x}_1' \boldsymbol{\Phi}_0 + \boldsymbol{\xi}' \boldsymbol{\Phi}_1 \right) \\ &\quad + \tau^2 \int_C dz \left(\boldsymbol{x}_2' \boldsymbol{\Phi}_0 + \boldsymbol{x}_1' \boldsymbol{\Phi}_1 + \boldsymbol{\xi}' \boldsymbol{\Phi}_2 \right) + \dots \end{aligned} \tag{B.1}$$

the series of the first order moment is written as:

$$\begin{aligned} \boldsymbol{M} &= \frac{1}{2\boldsymbol{i}} \int_{\partial P(t)} d\boldsymbol{x}\ \boldsymbol{x}\boldsymbol{\Phi} \\ &\propto \int_C dz\ \boldsymbol{\xi}' \boldsymbol{\xi} \boldsymbol{\Phi}_0 + \tau \int_C dz \left[\boldsymbol{\xi}\boldsymbol{\xi}' \boldsymbol{\Phi}_1 + \left(\boldsymbol{\xi} \boldsymbol{x}_1' + \boldsymbol{\xi}' \boldsymbol{x}_1 \right) \boldsymbol{\Phi}_0 \right] \\ &\quad + \tau^2 \int_C dz \left[\ \boldsymbol{\xi} \boldsymbol{\Phi}_2 + \left(\boldsymbol{\xi} \boldsymbol{x}_1' + \boldsymbol{\xi}' \boldsymbol{x}_1 \right) \boldsymbol{\Phi}_1 \right. \\ &\qquad \left. + \left(\boldsymbol{\xi} \boldsymbol{x}_2' + \boldsymbol{x}_1 \boldsymbol{x}_1' + \boldsymbol{\xi}' \boldsymbol{x}_2 \right) \boldsymbol{\Phi}_0 \right] + \dots \end{aligned} \tag{B.2}$$

and, finally, the second order moment is written in series of τ as:

$$\begin{aligned} I &= \frac{1}{4\boldsymbol{i}} \int_{\partial P(t)} d\boldsymbol{x}\ \boldsymbol{x}\boldsymbol{\Phi}^2 \\ &\propto \int_C dz\ \boldsymbol{\xi}' \boldsymbol{\xi} \boldsymbol{\Phi}_0^2 + \tau \int_C dz \left[2\boldsymbol{\xi}\boldsymbol{\xi}' \boldsymbol{\Phi}_1 + \left(\boldsymbol{\xi} \boldsymbol{x}_1' + \boldsymbol{\xi}' \boldsymbol{x}_1 \right) \boldsymbol{\Phi}_0 \right] \boldsymbol{\Phi}_0 \\ &\quad + \tau^2 \int_C dz \left[\ \boldsymbol{\xi}\boldsymbol{\xi}' (\boldsymbol{\Phi}_1^2 + 2\boldsymbol{\Phi}_0 \boldsymbol{\Phi}_2) + 2 \left(\boldsymbol{\xi} \boldsymbol{x}_1' + \boldsymbol{\xi}' \boldsymbol{x}_1 \right) \boldsymbol{\Phi}_0 \boldsymbol{\Phi}_1 \right. \\ &\qquad \left. + \left(\boldsymbol{\xi} \boldsymbol{x}_2' + \boldsymbol{x}_1 \boldsymbol{x}_1' + \boldsymbol{\xi}' \boldsymbol{x}_2 \right) \boldsymbol{\Phi}_0^2 \right] + \dots \end{aligned} \tag{B.3}$$

In correspondence with the series solutions shown in Section 3, the coefficients of all the positive powers of τ in equations (B.1, B.2, B.3) vanish.

The issue of rewriting the excess energy $\mathcal{E}$ in terms of the Schwarz function is under investigation at the present time, by starting from the following form of $\mathcal{E}$:

$$\mathcal{E} = \frac{1}{16\pi} \int_{\partial P(t)} ds(\boldsymbol{x}) \int_{\partial P(t)} ds(\boldsymbol{y})\ \boldsymbol{\tau}(\boldsymbol{x}) \cdot \boldsymbol{\tau}(\boldsymbol{y})\ |\boldsymbol{x} - \boldsymbol{y}|^2 \log |\boldsymbol{x} - \boldsymbol{y}| \ ,$$

$\boldsymbol{\tau}$ being the unit tangent vector on $\partial P(t)$ and the dot indicating a scalar product.

References

1. N.J. Zabusky, M.H. Hughes and K.V. Roberts, Contour Dynamics for the Euler equations in two dimensions, *J. Comp. Physics* **48**, 96 – 106 (1979).
2. D.G. Dritschel, Contour Dynamics and Contour Surgery: numerical algorithms for extended high-resolution modeling of vortex dynamics in two-dimensional, inviscid, incompressible flows, *Comp. Phys. Repts.* **10**, 77 – 146 (1989).
3. H. Aref and D.L. Vainchtein, Point vortices exhibit asymmetric equilibria, *Nature* **392**, 769 – 770 (1998).
4. H. Lamb, *Hydrodynamics*, Dover Publication 1930.
5. M.V. Melander, N.J. Zabusky and J.C. McWilliams, Symmetric vortex merger in two dimensions: causes and conditions, *J. Fluid Mechanics* **195**, 303 – 340 (1988).
6. G. Riccardi and R. Piva, Motion of an elliptical vortex under rotating strain: conditions for asymmetric merging, *Fluid Dynamics Research* **23**, 63 – 88 (1998).
7. P.J. Davis, *The Schwarz function and its applications*, Carus Mathematical Monographs, The Mathematical Association of America 1974.
8. G. Riccardi and D. Durante, Velocity induced by a plane uniform vortex having the Schwarz function of its boundary with two simple poles, accepted on J. of Applied Mathematics (Hindawi Pub.) (2008).
9. P.G. Saffman, *Vortex Dynamics*, Cambridge University Press 1992.
10. M.V. Melander, J.C. McWilliams and N.J. Zabusky, Axisymmetrization and vorticity-gradient intensification of an isolated two-dimensional vortex through filamentation, *J. of Fluid Mechanics* **178**, 137 – 159 (1987).
11. N.I. Muskhelishvili, *Singular integral equations*, Dover Publications 2008.

AN INTRODUCTION TO REDUCED BASIS METHOD FOR PARAMETRIZED PDEs

GIANLUIGI ROZZA*

Ecole Polytechnique Fédérale de Lausanne
Chair of Modelling and Scientific Computing
Station 8-MA, CH-1015 Lausanne, Switzerland
E-mail: gianluigi.rozza@epfl.ch
http://iacs.epfl.ch/~rozza

We provide an introduction on reduced basis (RB) method for the solution of parametrized partial differential equations (PDEs). We introduce all the main ingredients to describe the methodology and the algorithms used to build the approximation spaces and the error bounds. We consider a model problem describing a steady potential flow around parametrized bodies and we provide some illustrative results.

Keywords: Reduced basis method; parametrized PDEs; Galerkin method; error bounds; potential flows.

1. Introduction

In several optimization contexts arising, for example, in the aerospace industry, the problem of the resolution of PDEs in parametrized configurations is growing. The necessity to avoid to rebuild the geometry for each simulation and to have real-time computations in the many-query context provides a strong motivation for the development of the reduced basis method (Refs. 1–6) as a tool for the solution of parametrized problem built upon finite element (FE) method (Ref. 7).

In Sec. 2 we define a model problem, then in Sec. 3 we provide some basic results for the introduction of the methodology. Sec. 4 and Sec. 5 are devoted for the introduction of lower bounds for the coercivity constant and for a posteriori error bounds, respectively. In Sec. 6 we present numerical results considering steady potential flows around parametrized bodies, then

*Numerical simulations in collaboration with Gwenol Grandperrin (EPFL, Mathematics Section)

some conclusions follow.

2. Problem definition

We introduce an abstract model problem. We consider $\mathcal{D} \subset \mathbb{R}^p$ as the range of variation of p parameters and $\Omega \subset \mathbb{R}^d$ as a domain (p and d are integers). The functional space X^e is such that $(H_0^1(\Omega)) \subset X^e \subset (H^1(\Omega))$, with $H^1(\Omega)$ the Sobolev space defined as $H^1(\Omega) = \{f \in L^2(\Omega) \mid D^\alpha f \in L^2(\Omega), \alpha \leq 1\}$ and $D^\alpha f$ derivation of f in the distributions sense, $L^2(\Omega) = \{f : \Omega \rightarrow | \int_\Omega f(x)^2 dx < \infty\}$; $H_0^1(\Omega) = \{f \in H^1(\Omega) \mid f = 0 \text{ on } \partial\Omega\}$ and $H_D^1 = \{f \in H^1 \mid f = 0 \text{ on } \Gamma_D, D \subset \partial\Omega\}$.

We introduce $\forall \mu \in \mathcal{D}$ a bilinear and coercive form $a(.,.;\mu)$ and two linear and continuos functionals $f(.;\mu)$ and $l(.;\mu)$, then we consider the following "exact" problem

$$\begin{cases} \text{For } \mu \in \mathcal{D}, \text{ solve} \\ \qquad s^e(\mu) = l(u^e(\mu);\mu) \text{ s.t. } u^e(\mu) \in X^e \text{ satisfies} \\ a(u^e, v; \mu) = f(v;\mu) \qquad \forall v \in X^e. \end{cases} \tag{1}$$

We define a scalar product and a norm related with the energy of the system as: $\langle w, v\rangle_\mu \equiv a(w, v; \mu)$, and $|||w|||_\mu \equiv \langle w, w\rangle_\mu^{\frac{1}{2}}$ $\forall w, v \in X^e$, respectively. We also introduce a second scalar product and its norm defined on X^e ($\tau > 0$) for a selected $\overline{\mu}$ such that (Ref. 6) $(w, v)_X \equiv \langle w, v\rangle_{\overline{\mu}} + \tau(w, v)_{L^2(\Omega)}$, and $||w||_X \equiv (w, w)_X^{\frac{1}{2}}$ $\forall w, v \in X^e$, respectively. We introduce the crucial hypothesis that the bilinear form a can be expressed as

$$a(w, v; \mu) = \sum_{q=1}^{Q} \Theta^q(\mu) a^q(w, v) \tag{2}$$

such that for $q = 1, \ldots, Q$: $\Theta^q : \mathcal{C} \rightarrow \mathbb{R}$ is depending on μ and $a^q : X^e \times X^e \rightarrow \mathbb{R}$ is μ-independent.[2] This hypothesis on a allows us to significantly improve the computational efficiency in the evaluation of $a(w, v; \mu)$: the component $a_q(w, v)$ can be computed once and then stored to form (2). We are interested in geometrical parametrizations such that Ω will be a reference (and fixed) domain and it can be seen as the pre-image of $\Omega_0(\mu)$ (depending on the parameters) denoted original domain through the transformation $\mathcal{T} : \Omega \rightarrow \Omega_0$. We can define $a_0(w_0, v_0; \mu)$ as $a(w, v; \mu) = a_0(\mathcal{T}(w), \mathcal{T}(v); \mu)$.

We introduce a numerical discretization in our model problem given by finite element method such that the space $X^{\mathcal{N}} \subset X^e$ ($dim(X^{\mathcal{N}}) = \mathcal{N}$) and the problem is reformulated as

$$\begin{cases} \text{For } \mu \in \mathcal{D}, \text{ solve} \\ \qquad s^{\mathcal{N}}(\mu) = l(u^{\mathcal{N}}(\mu)) \text{ and } u^{\mathcal{N}}(\mu) \in X^{\mathcal{N}} \text{ satisfies} \\ a(u^{\mathcal{N}}, v; \mu) = f(v; \mu) \quad \forall v \in X^{\mathcal{N}}. \end{cases} \tag{3}$$

We recall the definition of the coercivity constant for the discretized problem as $\alpha^{\mathcal{N}}(\mu) \equiv \inf_{w \in X^{\mathcal{N}}} \frac{a(w,w;\mu)}{||w||_X^2}$, $\forall \mu \in \mathcal{D}$.

3. Reduced basis method

We introduce a principal set of parameters $\Xi = \{\mu^1, \ldots, \mu^{N_{max}}\} \subset \mathcal{D}$ and then for $1 \le N \le N_{max}$ we define the subsets $S_N = \{\mu^1, \ldots, \mu^N\}$ to which we associate the Lagrange reduced basis space (see Refs. 2,5,6) defined as $W_N^{\mathcal{N}} = vect\{u^{\mathcal{N}}(\mu^n), 1 \le n \le N\}$. It is clear that the nested (or hierarchical) condition is valid for $S_1 \subset S_2 \subset \ldots \subset S_{N_{max}}$ and for $W_1^{\mathcal{N}} \subset W_2^{\mathcal{N}} \subset \ldots \subset W_{N_{max}}^{\mathcal{N}} \subset X^{\mathcal{N}}$. The finite element solutions $u^{\mathcal{N}}(\mu^n)$ for $1 \le n \le N_{max}$ and for some properly selected values of the parameter μ^n are referred to as snapshots. By a Galerkin projection we can solve the reduced basis problem defined as

$$\begin{cases} \text{For a new } \mu \in \mathcal{D}, \text{ evaluate} \\ \qquad s_N^{\mathcal{N}}(\mu) = l(u_N^{\mathcal{N}}(\mu)) \text{ s.t. } u_N^{\mathcal{N}}(\mu) \in W_N^{\mathcal{N}} \subset X^{\mathcal{N}} \text{ satisfies} \\ a(u_N^{\mathcal{N}}, v; \mu) = f(v) \qquad \forall v \in W_N^{\mathcal{N}}. \end{cases} \tag{4}$$

The goal is to obtain a cheap evaluation of $s_N^{\mathcal{N}}(\mu)$ for many values of μ. Lemma 3.1 shows that $s_N^{\mathcal{N}}$ is converging to $s^{\mathcal{N}}$ quadratically with respect to the convergence[6] of $u_N^{\mathcal{N}}$ to $u^{\mathcal{N}}$.

Lemma 3.1. *Best approximation of RB method and quadratic convergence of the output for $l = f$ (compliance, see Ref. 2) is given by:*

(1) $|||u^{\mathcal{N}}(\mu) - u_N^{\mathcal{N}}(\mu)|||_\mu \le \inf_{w \in X_N^{\mathcal{N}}} |||u^{\mathcal{N}}(\mu) - w|||_\mu$;
(2) $s^{\mathcal{N}}(\mu) - s_N^{\mathcal{N}}(\mu) = |||u^{\mathcal{N}}(\mu) - u_N^{\mathcal{N}}(\mu)|||_\mu^2$.

When $l \neq f$ the "square" effect in the convergence is given by the solution of a dual problem.[6] To have a system from (4) which is not ill-conditioned we can use a Gram-Schmidt orthonomalization procedure (see Ref. 2) for the snapshots $u^{\mathcal{N}}(\mu^n)$, $1 \le n \le N_{max}$, with respect to the scalar product $(\cdot, \cdot)_X$ to obtain $\zeta_n^{\mathcal{N}}$, $1 \le n \le N_{max}$ as basis functions, so that $W_N^{\mathcal{N}} = \{\zeta_n^{\mathcal{N}}\}_{n=1,\ldots,N}$ for $1 \le N \le N_{max}$.

We can rewrite $u_N^{\mathcal{N}}$ as

$$u_N^{\mathcal{N}}(\mu) = \sum_{m=1}^{N} u_{Nm}^{\mathcal{N}}(\mu)\zeta_m^{\mathcal{N}}. \tag{5}$$

By posing $v = \zeta_n^{\mathcal{N}}$, $1 \leq n \leq N$ in (4) we get

$$\sum_{m=1}^{N} a(\zeta_m^{\mathcal{N}}, \zeta_n^{\mathcal{N}}; \mu) u_{Nm}^{\mathcal{N}}(\mu) = f(\zeta_n^{\mathcal{N}}) \qquad 1 \leq n \leq N, \tag{6}$$

and by (2) we may rewrite (6) as

$$\sum_{m=1}^{N} \left(\sum_{q=1}^{Q} \Theta^q(\mu) a^q(\zeta_m^{\mathcal{N}}, \zeta_n^{\mathcal{N}}) \right) u_{Nm}^{\mathcal{N}}(\mu) = f(\zeta_n^{\mathcal{N}}) \qquad 1 \leq n \leq N. \tag{7}$$

Here $\zeta^{\mathcal{N}}$ are independent by μ and so the quantities $f(\zeta_n^{\mathcal{N}})$ $(l(\zeta_n^{\mathcal{N}}))$, $1 \leq n \leq N_{max}$ and $a^q(\zeta_m^{\mathcal{N}}, \zeta_n^{\mathcal{N}})$, $1 \leq n \leq N_{max}, 1 \leq q \leq Q$ can be pre-computed and stored to decouple the offline computational part (parameter independent) from the online one (parameter dependent). The output can be computed as $s_N^{\mathcal{N}} = \sum_{m=1}^{N} u_{Nm}^{\mathcal{N}}(\mu) l(\zeta_m^{\mathcal{N}})$.

3.1. *Greedy algorithm for reduced basis space construction*

Let Ξ a subset of $\mathcal{D}$, used as a surrogate of $\mathcal{D}$ to test the reduced basis approximation. This subset has to be sufficiently rich and it can be built by using Monte-Carlo sampling (with uniform or log density, see Ref. 2).

In order to build the space $W_N^{\mathcal{N}}$, given Ξ and N_{max} we start by considering $S_1 = \{\mu^1\}$, $W_1^{\mathcal{N}} = span\{u^{\mathcal{N}}(\mu^1)\}$. Then, for $N = 2, \ldots, N_{max}$, we look for

$$\mu^N = \arg\max_{\mu \in \Xi} \Delta_{N-1}(\mu)$$

where $\Delta_N = \Delta_N^{en}$ or Δ_N^s are the error bounds for the energy or the output $s_N^{\mathcal{N}}$, respectively. These quantities will be introduced in Sec. 5. Instead of fixing N_{max} it is possible to set a tolerance ϵ as stopping criterium for a new μ^N when $\max_{\mu \in \Xi} \Delta_{N-1}(\mu) \leq \epsilon$. We define, in a recursive way, $S_N = S_{N-1} \cup \mu^N$ and $W_N^{\mathcal{N}} = W_{N-1}^{\mathcal{N}} \cup span\{u^{\mathcal{N}}(\mu^N)\}$. In this way only few FE solutions have to be computed (just the selected snapshots).

4. Lower bounds for the coercivity constant

We are interested in getting a fast and reliable method to compute a lower bound for the corcivity constant which is going to play a role in the error bounds. We use the so-called *successive constraint method* (SCM), briefly recalled here, see Refs. 6,8. This method uses online a linear programming algorithm with a number of operations independent by $\mathcal{N}$.

Let us introduce an *objective function* $\mathcal{J}^{obj} : \mathcal{D} \times \mathbb{R}^Q \rightarrow \mathbb{R}$, s.t. $(\mu, y) \mapsto \sum_{q=1}^{Q} \Theta^q(\mu) y_q$ where $y = (y_1, \ldots, y_Q)$. The coercivity constant can be expressed as $\alpha^{\mathcal{N}}(\mu) = \inf_{y \in \mathcal{Y}} \mathcal{J}^{obj}(\mu; y)$ where $\mathcal{Y} = \left\{ y \in \mathbb{R}^Q \mid \exists w_y \in X^{\mathcal{N}} \text{ such that } y_q = \frac{a^q(w_y, w_y)}{||w_y||_X^2}, 1 \leq q \leq Q \right\}$. We introduce also a *box* of constraints

$$\mathcal{B} = \prod_{q=1}^{Q} \left[\inf_{w \in X^{\mathcal{N}}} \frac{a^q(w, w)}{||w||_X^2}, \sup_{w \in X^{\mathcal{N}}} \frac{a^q(w, w)}{||w||_X^2} \right]. \tag{8}$$

We properly select a set of parameters in $\mathcal{D}$ denoted with $C_J = \{\mu_{SCM}^1 \in \mathcal{D}, \ldots, \mu_{SCM}^J \in \mathcal{D}\}$, and we indicate with $C_J^{M,\mu}$ the set of $M \geq 1$ elements of C_J nearest $\mu \in \mathcal{D}$; if $M > J$ then $C_J^{M,\mu} = C_J$. There are some techniques to build C_J as reported in Ref. 8. Given C_J, $M \in \mathbb{N} = \{1, 2, \ldots\}$ and $\mu \in \mathcal{D}$, we define $\mathcal{Y}_{LB}(\mu; C_J, M)$ as

$$\mathcal{Y}_{LB}(\mu; C_J, M) = \left\{ y \in \mathbb{R}^Q \mid y \in \mathcal{B}, \mathcal{J}^{obj}(\mu', y) \geq \alpha^{\mathcal{N}}(\mu'), \forall \mu' \in C_J^{M,\mu} \right\},$$

such that $\mathcal{Y} \subset \mathcal{Y}_{LB}(\mu; C_J, M) \quad \forall \mu \in \mathcal{D}$ (being $\mathcal{Y}_{LB}$ richer than $\mathcal{Y}$) and we define the lower bound of the coercity constant as $\alpha_{LB}^{\mathcal{N}}(\mu; C_J, M) = \min_{y \in \mathcal{Y}_{LB}(\mu; C_J, M)} \mathcal{J}^{obj}(\mu, y)$ such that $\alpha_{LB}^{\mathcal{N}}(\mu) \leq \alpha^{\mathcal{N}}(\mu)$, $\forall \mu \in \mathcal{D}$. The problem of computing a lower bound for the coercivity constant $\alpha_{LB}^{\mathcal{N}}$ is a linear programming minimization problem with Q variables $y_1, \ldots, y_Q$ and $2Q + M$ constraints. Each y_i is subject to two constraints from $\mathcal{B}$ (8) and then there are the M conditions $\mathcal{J}^{obj}(\mu', y) \geq \alpha^{\mathcal{N}}(\mu'), \forall \mu' \in C_J^{M,\mu}$ and $|C_j^{M,\mu}| \leq M$. Given $\mathcal{B}$ and $\{\alpha^{\mathcal{N}}(\mu') \mid \mu' \in C_J\}$, the evaluation of $\mu \rightarrow \alpha_{LB}^{\mathcal{N}}(\mu)$ is independent by $\mathcal{N}$.

5. A posteriori error estimation

We introduce here the a posteriori error bounds as described in Refs. 2,6,9. We are interested in a method which should be realible and efficient. We reconsider the finite element problem (3) and the Galerkin projection to get the problem (4). We define $e(\mu) \equiv u^{\mathcal{N}}(\mu) - u_N^{\mathcal{N}}(\mu) \in X^{\mathcal{N}}$. Thanks to the linearity of $a(.,.)$ we have

$$\begin{aligned} a(e(\mu), v; \mu) &= a(u^{\mathcal{N}}(\mu), v; \mu) - a(u_N^{\mathcal{N}}(\mu), v; \mu) \\ &= f(v; \mu) - a(u_N^{\mathcal{N}}(\mu), v; \mu) \qquad \forall v \in X^{\mathcal{N}}. \end{aligned}$$

We denote

$$r(v;\mu) = f(v;\mu) - a(u_N^{\mathcal{N}}(\mu), v;\mu), \tag{9}$$

to get the equation of the residual

$$a(e(\mu), v;\mu) = r(v;\mu). \tag{10}$$

Thanks to Riesz representation theorem we can write $r(v;\mu)$ as

$$r(v;\mu) = (\hat{e}(\mu), v)_X \qquad \forall v \in X^{\mathcal{N}}, \tag{11}$$

so that from (10) we get $a(e(\mu), v;\mu) = (\hat{e}(\mu), v)_X$. At the same time we have $||r(.;\mu)||_{(X^{\mathcal{N}})'} \equiv \frac{r(v;\mu)}{||v||_X} = ||\hat{e}(\mu)||_X$. By the coercivity lower bound $\alpha_{LB}^{\mathcal{N}}(\mu)$ introduced in Sec. 4 we define the following error bound for the energy norm

$$\Delta_N^{en}(\mu) = \frac{||\hat{e}(\mu)||_X}{\sqrt{\alpha_{LB}^{\mathcal{N}}(\mu)}} \tag{12}$$

and for the output

$$\Delta_N^{s}(\mu) = \frac{||\hat{e}(\mu)||_X^2}{\alpha_{LB}^{\mathcal{N}}(\mu)}. \tag{13}$$

Concerning Δ_N^s, if $l = f$ we have also that $|s^{\mathcal{N}}(\mu) - s_N^{\mathcal{N}}(\mu)| \leq \Delta_N^s = (\Delta_N^{en})^2$ for $N = 1, \ldots, N_{max}$ and for all $\mu \in \mathcal{D}$. See Ref. 6 for demonstrations.

5.1. *Offline-online computational procedures*

The residual (9) can be computed as

$$\begin{aligned} r(v;\mu) &\equiv f(v) - a(u_N^{\mathcal{N}}(\mu), v;\mu) \\ &= f(v) - \sum_{n=1}^{N} u_N^{\mathcal{N}}(\mu) \sum_{q=1}^{Q} \Theta^q(\mu) a^q(\zeta_n^{\mathcal{N}}, v) \text{ by (5)-(2)} \end{aligned}$$

$$\begin{aligned} \text{by (11) we get } (\hat{e}(\mu), v)_X &\equiv r(v;\mu) \\ &= f(v) - \sum_{n=1}^{N} \sum_{q=1}^{Q} \Theta^q(\mu) u_N^{\mathcal{N}}(\mu) a^q(\zeta_n^{\mathcal{N}}, v). \end{aligned}$$

We may rewrite $\hat{e}(\mu)$ as $\hat{e}(\mu) = \mathcal{C} + \sum_{q=1}^{Q} \sum_{n=1}^{N} \Theta^q(\mu) u_{N_n}^{\mathcal{N}}(\mu) a^q(\zeta_n^{\mathcal{N}}, v)$ where $\mathcal{C}$ and $\mathcal{L}_n^q$, $1 \leq n \leq N$, $1 \leq q \leq Q$ are given by the following problems

$$\begin{cases} (\mathcal{C}, v)_X = f(v) \quad \forall v \in X^{\mathcal{N}} \\ (\mathcal{L}_n^q, v)_X = -a^q(\zeta_n^{\mathcal{N}}, v) \quad \forall v \in X^{\mathcal{N}}. \end{cases} \tag{14}$$

The quantity $||\hat{e}(\mu)||_X^2$ is then given by

$$||\hat{e}(\mu)||_X^2 = (\mathcal{C}, \mathcal{C})_X + 2\sum_{q=1}^{Q}\sum_{n=1}^{N} \Theta^q(\mu) u_{N_n}^{\mathcal{N}}(\mu)(\mathcal{C}, \mathcal{L}_n^q)_X \tag{15}$$

$$+\sum_{q=1}^{Q}\sum_{n=1}^{N} \Theta^q(\mu) u_{N_n}^{\mathcal{N}}(\mu) \left\{ \sum_{q'=1}^{Q}\sum_{n'=1}^{N} \Theta^q(\mu) u_{N_n}^{\mathcal{N}}(\mu)(\mathcal{L}_n^q, \mathcal{L}_{n'}^{q'})_X \right\}.$$

In the offline part, we compute quantities μ-independent like $\mathcal{C}$, $\mathcal{L}_n^q$, $1 \leq n \leq N_{max}$, $1 \leq q \leq Q$ and we store $(\mathcal{C}, \mathcal{C})_X, (\mathcal{C}, \mathcal{L}_n^q)_X, (\mathcal{L}_n^q, \mathcal{L}_{n'}^{q'})_X \quad 1 \leq n, n' \leq N_{max}, 1 \leq q, q' \leq Q$. In the offline part, the computational complexity depends on N_{max}, Q and $\mathcal{N}$. In the online part we compute the μ-dependent quantities $\Theta^q(\mu)$, $1 \leq q \leq Q$, $u_{Nn}^{\mathcal{N}}(\mu)$, $1 \leq n \leq N$. To evaluate (15) we perform $O(Q^2N^2)$ operations, independently by $\mathcal{N}$.

6. Numerical results

We present here some numerical results as example of application of reduced basis method to potential flows around parametrized bodies (see Ref. 10), representing for example the bulb of a yacht (with or without the keel) or the nacelles of an aircraft or bodies placed under the wings and/or the fuselage. We provide the general parametrization, state equation, convergence results and some representative solutions. Our emphasys is on computational performance.

6.1. *Bulb and keel*

In Fig. 1 we report two very preliminary configurations used as a parametrized bodies for our first tests. The first configuration, describing just the bulb without the keel, has three parameters μ_1, μ_2, μ_3 for the non-symmetric ellipse, the second configuration, with the inclusion of the keel in the bulb configuration has five parameters including a parametrization on the keel height and width.

In Fig. 2 we report the two domains used for this first tests. We used a simple inviscid and irrotational fluid model described by a potential flow (Ref. 10) and represented by a Laplace problem, whose strong formulation

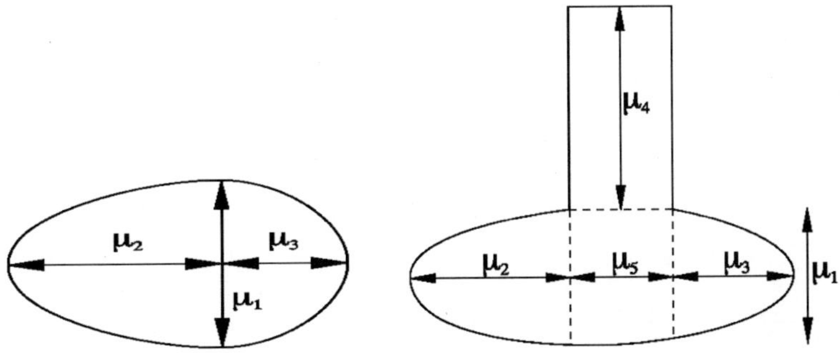

Fig. 1. Parametrized bodies: bulb (left) and keel with bulb (right).

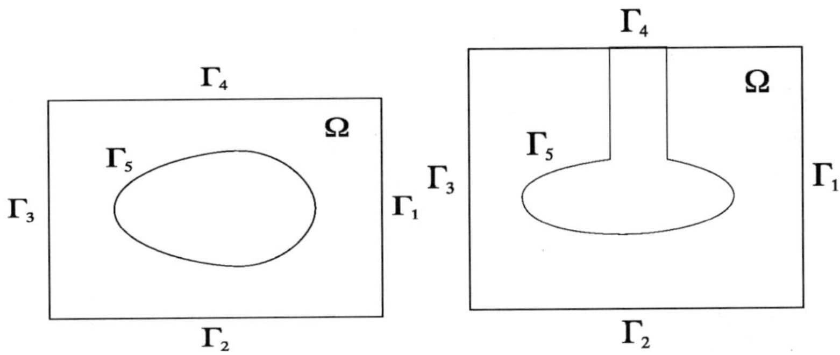

Fig. 2. Domains for the bulb (left) and for the keel with the bulb (right).

is

$$\begin{cases} \text{Find } u^{\mathcal{N}}(\mu) \in X^{\mathcal{N}} \text{ s.t.} & \\ \Delta u^{\mathcal{N}} = 0 & \text{on } \Omega \\ \quad u^{\mathcal{N}} = 1 & \text{on } \Gamma_1 \\ \partial_n u^{\mathcal{N}} = 1 & \text{on } \Gamma_3 \\ \partial_n u^{\mathcal{N}} = 0 & \text{on } \Gamma_2 \cup \Gamma_4 \cup \Gamma_5, \end{cases}$$

where ∂_n indicates the normal derivative. On the outflow Γ_1 we impose a non-homogeneous Dirichlet condition representing the potential level, on Γ_3 we impose a non-homogeneous Neumann condition representing the imposition of a (unity) velocity in the normal direction, on Γ_2, Γ_4, Γ_5 we impose a homogeneous Neumann condition representing a condition of zero velocity in the normal direction of the body and/or walls (non-penetration). The previous problem is then transfomed in the form introduced in (1). We report in Fig. 3 some convergence results plotting the quantity $max(\Delta_N^{en})$ (12) over a very large sample $\Xi \subset \mathcal{D}$ as a function of N during the greedy algorithm of Sec. 3.1 for the two test configurations. For the coercivity lower bounds of Sec. 4 we used an approximation with $J = 169$ and $J = 187$, respectively for the bulb and for the bulb with the keel. The parameters range is given by $\mathcal{D} = [1,4] \times [1,4] \times [1,4]$ for the first example (bulb) and by $\mathcal{D} = [1,7] \times [1,7] \times [1,7] \times [1,8] \times [1,5]$ for the second one (keel and bulb). Computational results were performed using the library `rbMIT`.[11]

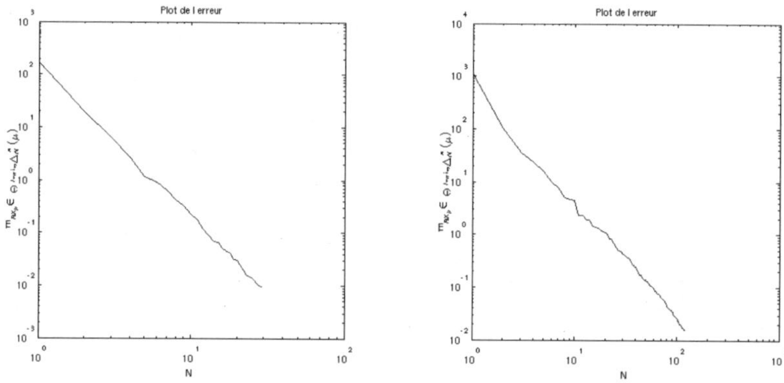

Fig. 3. Very fast convergence of the greedy algorithm over a large sample of parameters: maximum Δ_N^{en} (12) as a function of N for the bulb configuration (left) and for the keel and the bulb (right).

We report in Fig. 4 representative solutions of the bulb problem for a reference value of $\mu = [3,4,3]$. We report the potential solution with over the velocity field (on the left) and the pressure field (on the right) computed by the Bernoulli Theorem.[10]

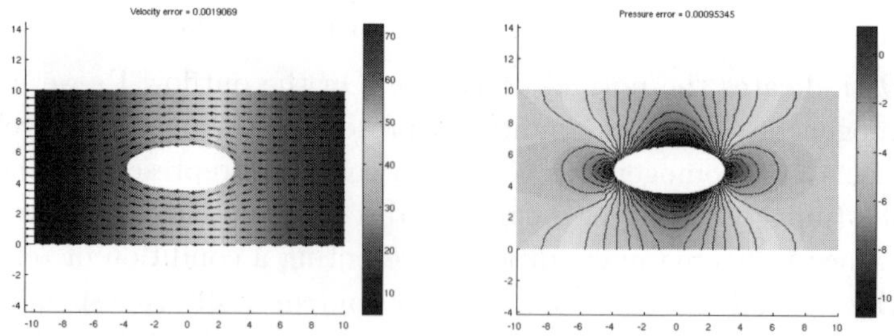

Fig. 4. A representative solution of velocity and potential (left) and pressure (right) for the bulb test.

Thanks to reduced basis method we can get real-time evaluations and visualizations of parametrized problems by testing a large number of different configurations, corresponding to many μ's belonging to $\mathcal{D}$. In Table 1 we report offline and online computational times for the two first tests. We can see that the average online computational times for the solution of the problem for a certain $\mu \in \mathcal{D}$ (with error bounds) is less than 1% compared with the offline computational times needed to set the geometry, build the mesh and compute FE solutions (and error bound ingredients preparation).

Table 1. Computational times for the two preliminary tests and comparison.

test	bulb	bulb and keel
phase	computational times (s)	computational times (s)
offline	553.8	763.4
online	5.2	7.4
ratio	0.94%	0.97%

6.2. *A more complex example: several parametrized bodies*

We consider here a more complex example where we study a parametrized configuration made up by three bodies with different shape, size and position. This test can be seen as a preliminary study for a trimaran configuration, a multihulled boat or for an aircraft with many nacelles.

In Fig. 5 we report a scheme with 8 parameters considered: parameters $\mu_1 - \mu_6$ are describing the bodies' geometry, which are non-symmetric ellipses (upper and lower bodies have the same parametrization $\mu_4 - \mu_6$). Parameters μ_7 and μ_8 are responsible of the position of the bodies in the domain and their mutual distance (horizontal and vertical distance). The range of variation of the parameters is given by $\mathcal{D} = [3,8] \times [3,20] \times [3,20] \times [3,8] \times [3,20] \times [3,6] \times [3,8] \times [-8,8]$. In the same figure we illustrate the domain Ω.

The state problem is the same as the one described in Sec. 6.1 where we considered steady potential flow; in this case we have just two more boundaries to consider: Γ_6 and Γ_7 with homogeneous Neumann condition (zero velocity). We report in Fig. 6 the mesh and triangulation of the considered configuration (using `rbMIT`[11]) and some convergence results plotting the quantity $max(\Delta_N^{en})$ (12) over a very large sample $\Xi \subset \mathcal{D}$ as a function of N, during the greedy algorithm of Sec. 3.1. For the coercivity lower bounds of Sec. 4 we used an approximation with $J = 200$.

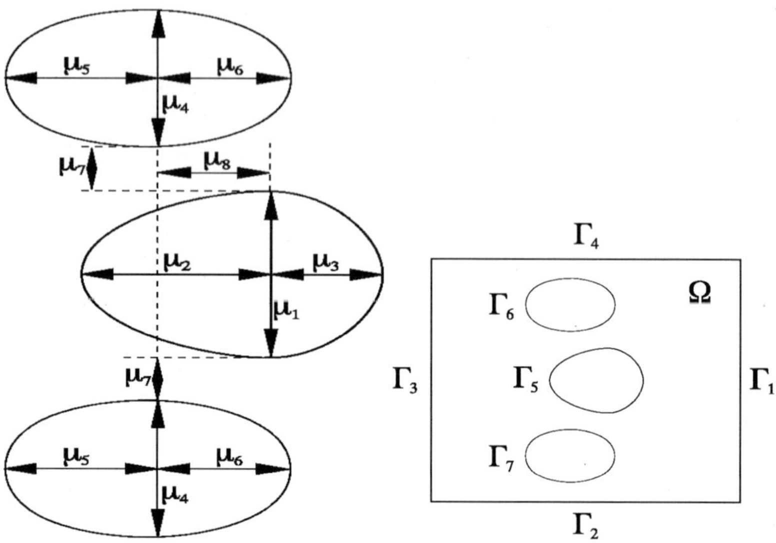

Fig. 5. Parametrized geometrical configuration with three bodies (left) and schematic domain (right).

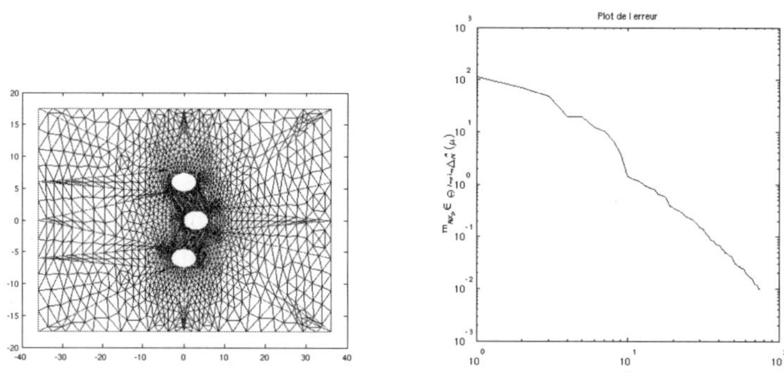

Fig. 6. Mesh and triangulation of the considered configuration (left); convergence of the greedy algorithm over a large sample of parameters: maximum Δ_N^{en} (12) as a function of N for the same configuration (right).

We report in Table 2 offline and online computational times for the last test. We can see that the average online times for the solution of the problem for a certain $\mu \in \mathcal{D}$ is less than 0.5% compared with the offline computational times needed to set the geometry, build the mesh and compute FE solutions (and error bound ingredients). The method is well suited, efficient and reliable for the solution of PDEs in parametrized geometries in the many query context.

Table 2. Computational times for the third test.

phase	computational times (s)
offline	4853.7
online	16.6
ratio	0.34%

7. Conclusion

We have described the basic elements of the reduced basis method and introduced simple problems dealing with a steady potential flow around parametrized bodies. The offline-online computational decomposition strategy is crucial to achieve computational economies of at least two orders of magnitude in the many query context and in a repetitive computational environment. A posteriori error bounds and greedy algorithm convergence prove the reliability of the methodology.

References

1. K. Ito and S. Ravindran, *Journal of Computational Physics* **143**, 403 (1998).
2. A. Patera and G. Rozza, *Reduced Basis Approximation and A Posteriori Error Estimation for Parametrized Partial Differential Equations,* (MIT Pappalardo Graduate Monographs Series in Mechanical Engineering, available at `http://augustine.mit.edu`, ©Massachusetts Institute of Technology 2006-08).
3. J. S. Peterson, *SIAM Journal of Scientific and Statistical Computing* **4**, 777 (1989).
4. T. A. Porsching, *SIAM Journal of Numerical Analysis* **24**, 1277 (1987).
5. C. Prud'homme, D. Rovas, K. Veroy, L. Machiels, Y. Maday, A. Patera and G. Turinici, *Journal of Fluids Engineering* **124**, 70 (2002).
6. G. Rozza, D. Huynh and A. Patera, *Archives Computational Methods in Engineering* **15**, 229 (2008).
7. A. Quarteroni and A. Valli, *Numerical Approximation of Partial Differential Equations* (Springer-Verlag, 1997).
8. D. Huynh, G. Rozza, S. Sen and A. Patera, *Comptes Rendus de l'Académie des Sciences Paris* **Sér I 345**, 473 (2007).
9. W. Rheinboldt, *Nonlinear Analysis, Theory, Methods and Applications* **21**, 849 (1993).
10. R. L. Panton, *Incompressible Flow,* 3rd edn. (John Wiley & Sons, Inc., 2005).
11. D. Huynh, N. Nguyen, G. Rozza and A. Patera, *rbMIT Software and Documentation.* ©Massachusetts Institute of Technology, (2007-08). Available at `http://augustine.mit.edu`.

Support Function Representation for Curvature Dependent Surface Sampling

Maria Lucia Sampoli
Department of Mathematics and Computer Science, University of Siena, Pian dei Mantellini 44, 53100 Siena, Italy
E-mail: sampoli@unisi.it

Bert Jüttler
[2] *Institute of Applied Geometry, Johannes Kepler University, Altenberger Str. 69, 4040 Linz, Austria*
E-mail: Bert.Juettler@jku.at

In many applications it is required to have a curvature-dependent surface sampling, based on a local shape analysis. In this work we show how this can be achieved by using the support function (SF) representation of a surface. This representation, a classical tool in Convex Geometry, has been recently considered in CAD problems for computing surface offsets and for analyzing curvatures. Starting from the observation that triangular Bézier spline surfaces have quite simple support functions, we approximate any given free-form surface by a quadratic triangular Bézier spline surface. Then the corresponding approximate SF representation can be efficiently exploited to produce a curvature dependent sampling of the approximated surface.

Keywords: Support function, triangular Bézier surfaces, quadratic patches, data sampling.

1. Introduction

One of the main tasks of Computer Aided Geometric Design is to represent curves and surfaces, satisfying some interpolations or approximation conditions, in a way which allows an easy manipulation for further applications (see for instance Ref. 1). The most important performed operations are usually offsetting, convolution computations, feature lines computations and extraction of information on curvatures and other geometric quantities.

Among all representations, NURBS (Non Uniform Rational B-Spline) are widely used and therefore a big deal of research has been done, for example, in order to detect subsets of NURBS closed under offsetting or the more general convolution operation, or to characterize specific tools for defining the geometric features of the curves and surfaces.

In this context quite recently it was noted that a simple approach to deal with offsets and convolutions is to use the support function (SF) representation of surfaces. The support function representation is a classical tool in the field of Convex Geometry;[2,3] it consists in describing a surface by the distance of its tangent planes to the origin of the coordinate system, and such a distance is seen as a function on the unit sphere. The surface can be then recovered from its support function by computing the envelope of the tangent planes.

The application of support function representation to problems from Computer Aided Design was noted for the first time by Sabin,[4] but only recently has been investigated effectively.[5–7] In these papers the shapes (curves and surfaces) which can be described by particular types of support functions -polynomial, rational, or piecewise linear- are considered and their geometric properties are discussed. In particular it is shown that the class of curves and surfaces with (piecewise) polynomial support functions is closed under convolutions, offsetting, rotations and translations. Indeed these operations correspond to simple algebraic operations of the corresponding support functions. Moreover we can see that the SF representation leads also to particularly simple expressions for quantities and mappings governing the differential geometry of the surfaces.

In this paper the approximation of a free-form surface with a quadratic triangular Bézier spline is considered. Such an approximation is done, following Ref. 8, by considering a C^1 quadratic spline quasi-interpolant. This implies the approximation of the corresponding support function. We see that each quadratic Bézier triangular patch has support function given by a rational function defined over a spherical triangle whose boundaries are conic sections. Therefore the support function of a triangular Bézier spline is given by a, possibly multi-valued, rational function defined over a partition of spherical triangles on the unit sphere.

The support function of a free-form surface is then approximated by the support function of the corresponding triangular Bézier spline approximation. In this way we can exploit this last SF representation to extract geometric information of the surface and to manipulate it for further modelling.

As an interesting application we show how to determine a *curvature dependent* surface sampling. The key idea is given by observing that a uniform point set on the unit sphere can be mapped through the *envelope operator* (defined from the SF) to a curvature dependent point set on the surface.

The remainder of the paper is divided into four sections. In the next section we review the main definitions and properties of support function representation of a surface. In Sect. 3 the approximation of general surfaces by quadratic ones is studied and the computation of the corresponding piecewise support functions is considered. Then Sect. 4 will be devoted to the computation of curvature-adaptive sampling points: the method is first presented in the univariate case and then the surface case is considered. Section 5 concludes the paper.

2. Preliminaries

Given a surface $\mathbf{x}(u,v) : \Omega \to \mathbb{R}^3 \ : (u,v) \mapsto \mathbf{x}(u,v)$, each point has an associated unit normal

$$\mathbf{N} \ : \ \Omega \to \mathbb{S}^2 \ : \ (u,v) \mapsto \mathbf{N}(u,v) \ , \tag{1}$$

where $\mathbb{S}^2$ is the unit sphere. The mapping $\mathbf{N}$, which depends smoothly on u, v and defines an orientation of the surface, is bijective provided that the surface does not have parabolic or singular points.

Let us consider now the *distance* between the tangent plane and the origin

$$H : \Omega \to \mathbb{R} \ : \ (u,v) \mapsto \mathbf{N}(u,v) \cdot \mathbf{x}(u,v) \ . \tag{2}$$

Then the *support function* is defined as the composition of the inverse map of $\mathbf{N}$ with the above function H

$$h : \mathbb{S}^2 \to \mathbb{R} \ : h = (\mathbf{N})^{-1} \circ H \ . \tag{3}$$

The support function assigns to each unit normal the distance between the corresponding tangent plane and the origin of the coordinate system.

If a support function $h : D \to \mathbb{R}$ is given, where $D \subseteq \mathbb{S}^2$, then the associated surface is obtained by computing the *envelope* of the tangent planes

$$\{\mathbf{p} : \ h(\mathbf{n}) = \mathbf{n} \cdot \mathbf{p}\}, \ \ \mathbf{n} \in D \ . \tag{4}$$

More precisely, for any $\mathbf{n} \in D$ we can compute the point on the envelope surface as

$$\mathbf{x}_h : \mathbf{n} \mapsto \mathbf{x}_h(\mathbf{n}) = h(\mathbf{n})\mathbf{n} + (\nabla_{\mathbb{S}^2} h)(\mathbf{n}), \tag{5}$$

where with $\nabla_{\mathbb{S}^2}$ we indicate the embedded intrinsic gradient with respect to the unit sphere. If h^* is the extension of h defined over all $\mathbb{R}^3$ then the

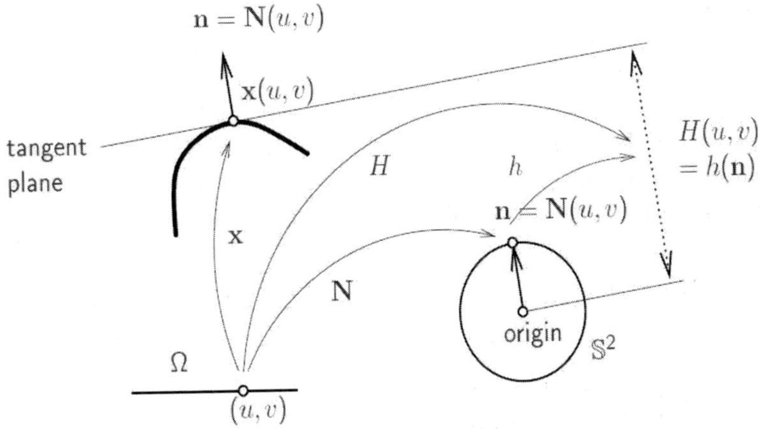

Fig. 1. A graphical scheme of the definition of the support function.

intrinsic gradient can be obtained by projecting the usual gradient into the tangent plane of the sphere,

$$(\nabla_{\mathbb{S}^2} h)(\mathbf{n}) = (\nabla h^*)(\mathbf{n}) - [(\nabla h^*)(\mathbf{n}) \cdot \mathbf{n}]\mathbf{n} \ . \tag{6}$$

In conclusion we can define the *envelope operator* $\mathcal{E}$ which associates a surface $\mathbf{x}_h$ to a support function h,

$$\mathcal{E} \ : \ C^1(\mathbb{S}^2, \mathbb{R}) \to C(\mathbb{S}^2, \mathbb{R}^3) \ : h \mapsto \mathbf{x}_h. \tag{7}$$

We may note that the envelope operator $\mathcal{E}$ is a linear mapping and defines an isomorphism between the linear spaces $C^1(D, \mathbb{R})$ and its images, where the addition in the image space is given by the so called convolution of the surfaces (for more details see Refs. 7,9).

3. Approximation of surfaces and their support functions

In general, given a quadratic surface patch

$$\mathbf{p}(u, v) = \frac{1}{2} a_{20} u^2 + a_{11} uv + \frac{1}{2} a_{02} v^2 + a_{10} u + a_{01} v + a_{00}, \tag{8}$$

its support function can be found by eliminating u, v from the following system of equations:

$$\begin{gathered} h(\mathbf{n}) = \mathbf{n} \cdot \mathbf{p}(u, v), \\ \mathbf{n} \cdot \mathbf{p}_u(u, v) = 0, \quad \mathbf{n} \cdot \mathbf{p}_v(u, v) = 0 \end{gathered} \tag{9}$$

The last two equations are linear in u, v, then we can make these parameters explicit with respect to $\mathbf{n} = (n_1, n_2, n_3)$ and substitute the result into in the first equation, finding the expression of the support function of the surface. By straightforward computation it is easy to see that we obtain a rational expression for h, which is given as the quotient of a cubic and a quadratic homogeneous polynomial.

In particular our attention will be focused on quadratic triangular Bézier surface patches

$$\mathbf{x}(u, v, w) = \sum_{i+j+k=2} B^2_{i,j,k}(u, v, w)\mathbf{b}_{i,j,k}, \qquad u, v, w \geq 0, \quad u + v + w = 1 \ . \tag{10}$$

where (u, v, w) are barycentric coordinates with respect to some domain triangle $\Delta \subset \mathbb{R}^2$, the basis functions $\frac{2}{i!j!k!}u^i v^j w^k$ are the bivariate Bernstein polynomials of degree 2 and the coefficient vectors $\mathbf{b}_{i,j,k}$ are called *control points*. This representation is very common in CAD applications for its simplicity and many useful geometric properties, for instance the surface patch is contained in the convex hull defined by the control points. An example is given in Fig. 2. For more details see for instance Ref. 1.

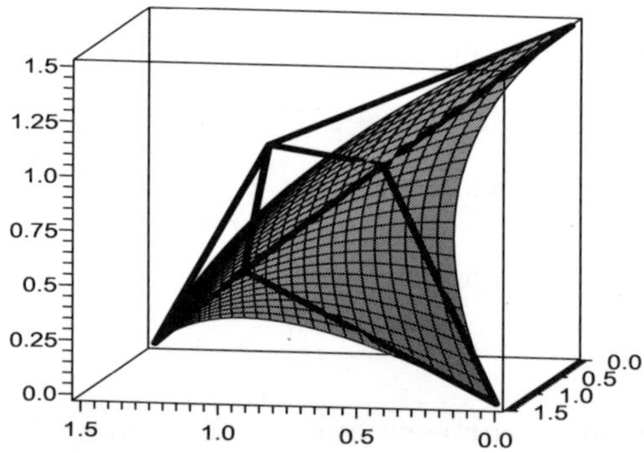

Fig. 2. An example of a quadratic triangular Bézier patch.

Recently it was found out that quadratic triangular Bézier splines also belong to the class of surfaces with odd rational support functions and therefore it can be proved that they belong to the family of surfaces which can be equipped with a linear field of normal vectors. This nice feature allows the exact computation of a rational parameterization of offsets and convolution surfaces.[9–12] A Bézier patch, by construction, is uniquely defined once its control points are assigned. A triangular quadratic patch is defined by 6 control points: the three vertices of the triangle and one additional point for each side (we have no inner control points in the quadratic case). Considering the patch of Fig. 2, its control points are given by

$$\begin{aligned} \mathbf{b}_{2,0,0} &= (2,0,0)^T, \mathbf{b}_{0,2,0} = (0,2,0)^T, \mathbf{b}_{0,0,2} = (0,0,2)^T, \\ \mathbf{b}_{1,1,0} &= (\tfrac{3}{2},0,0)^T, \mathbf{b}_{1,0,1} = (0,\tfrac{3}{2},0)^T, \mathbf{b}_{0,1,1} = (0,0,\tfrac{3}{2})^T, \end{aligned} \tag{11}$$

then the expression of the support function is given by

$$h(\mathbf{n}) = \frac{5(n_1^3+n_2^3+n_3^3)+4(n_1^2n_2+n_1^2n_3+n_1n_2^2+n_1n_3^2+n_3n_2^2+n_3^2n_2)+13n_1n_2n_3}{2({n_1}^2+3n_1n_2+{n_2}^2+3n_1n_3+3n_2n_3+{n_3}^2)}. \tag{12}$$

In the next paragraph we present a method to approximate general free-form surfaces by quadratic triangular Bézier splines.

3.1. *Quadratic spline approximation*

Given a free-form parametric surface we want to build a quadratic triangular Bézier spline of the form (10), which approximates it.

The construction is done component by component. For each component we may consider the C^1 quadratic spline quasi-interpolant described in Refs.[8,13] For the sake of completeness we report briefly the basic steps of the construction.

In the parameter domain Ω, that without loss of generality can be assumed equal to $[0,1]^2$, we consider a rectangular grid given by a set of mn isoparametric lines and then we endow it with the so-called criss-cross triangulation constructed taking all the diagonals of each subrectangle $\Omega_{i,j} = [\frac{i}{n}, \frac{i+1}{n}] \times [\frac{j}{m} \frac{j+1}{m}]$, $i = 0 \dots n-1$, $j = 0, \dots, m-1$.

Given a function f defined in Ω, we then consider the C^1 quadratic spline quasi-interpolant defined by

$$Qf = \sum_{i=0}^{n} \sum_{j=0}^{m} \mu_{ij}(f)\mathcal{B}_{ij}$$

where $\mathcal{B}_{ij}$ are the classical C^1 quadratic box splines obtained as translates of the Zwart-Powell element,[14] and $\mu_{ij}(f)$ are appropriate linear combinations of vertex values and/or centre values of adjacent sub-rectangles.[8,13] We note that various choices for the coefficients $\mu_{ij}(f)$ are possible, giving rise to different quasi-interpolants, sharing optimal properties.

We can see that these quasi-interpolant operators reproduce exactly the space of bivariate polynomials and produce optimal approximation order for smooth functions and their derivatives. More precisely, denoting with $k = max\{\frac{1}{n}, \frac{1}{m}\}$, with $\|\cdot\|_{\infty,\Omega} = \|\cdot\|_{\infty} =$ the supremum norm over Ω, and with $D^{\beta} = \frac{\partial^{|\beta|}}{\partial u^{\beta_1} \partial v^{\beta_2}}$, $|\beta| = \beta_1 + \beta_2$, we have[8,13]

$$\|f - Qf\|_{\infty} \leq C_0 k^3, \qquad \|D^{\beta}(f - Qf)\|_{\infty} \leq C_1 k^2 \text{ with } |\beta| = 1. \tag{13}$$

where the constants C_i depends solely on the function f.

In the case of parametric surfaces, we can obtain the above bounds for each component and taking the maximum norm we can conclude that the maximum difference between a surface $\mathbf{x}(u, v)$ and its approximant $\mathbf{p}(u, v)$ can be bounded by $\tilde{C}k^3$.

3.2. *Support function approximation*

Given a quadratic triangular Bézier spline, let us study now its support function.

Let us consider first a single triangular patch. In order to construct its support function we have to determine the region on the unit sphere which is the domain of the support function, that is the image of the mapping N introduced in (1). We should assume that the Bézier triangle does not contain any parabolic points. In Ref. 15 it was noted that the parabolic points of quadratic Bézier triangles determine curves which are images of straight lines in the parameter domain. Therefore in these cases it is sufficient to split the triangle along these lines, in order to exclude parabolic points.

Now, considering the points on the surface, the domain of the support function can be also seen as the image of the Gaussian map of the patch.

$$\mathbf{G} : M \in \mathbb{R}^3 \to \mathbb{S}^2 \ : \ \mathbf{p} \mapsto \mathbf{n_p} \tag{14}$$

With the help of a computer algebra system it is easy to see that the Gaussian image of a quadratic triangular patch is a spherical triangle with curved boundaries (conic sections). Such curves are obtained by the intersection of the sphere with three quadratic cones. If one of these cones is singular the spherical triangle may degenerate into a biangle. The support function will be then given eliminating u, v in (9) and solving them with respect to $\mathbf{n}$.

In the case of a quadratic spline the support function will be obtained collecting the support functions of all triangular patches constituting the spline. The resulting support function will be given by a, possibly multi-valued, piecewise C^0 rational function on the unit sphere, defined over a collection of curved spherical triangles.

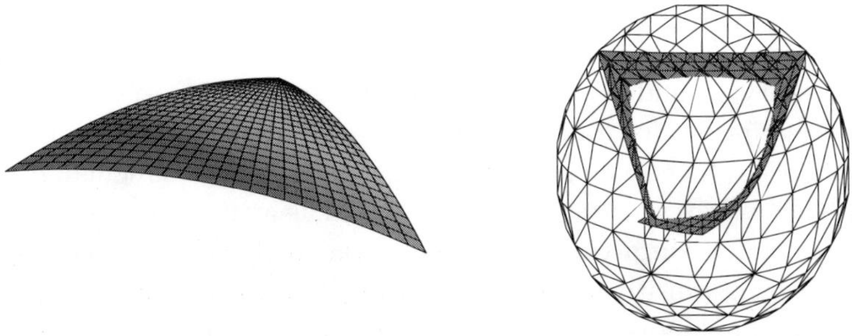

Fig. 3. A quadratic triangular patch and the domain of its support function on the unit sphere.

Remark 3.1. As we have said in the previous section, in order to have a well defined support function, we have to exclude singular points and parabolic ones. The last case can be checked considering for each triangular patch the sign of the principal curvatures. In particular we check the sign of the Gaussian curvature (achieved by computing the first and the second fundamental forms) and we classify the patches in elliptic and hyperbolic ones according to their curvature sign. Then we can treat them separately. The patches with some points of zero curvature can be subdivided till small (up to a fixed tolerance) regions containing isolated points or curves of parabolic points are detected According to the various applications these region will be handled differently.

Regarding the approximation order, from the error estimates (13), we can prove that the distance between the SF of a given surface h_x and that of its approximant h_p can be bounded by $\hat{C}k^3$.

4. Curvature-dependent sampling

In Computer Aided Geometric Design (CAGD), curves and surfaces have two standard representations: parametric and implicit. The parametric representation offers a number of advantages, e.g., simple techniques for display and for analyzing the geometric properties as well as fast generation of point meshes, fast visualization and interactive modelling. On the other hand, implicitly defined surfaces are better suited in many applications, for instance for the possibility of defining solids. Indeed the representation of geometric objects based on volumetric data structures guarantee e.g., fast surface interrogation or Boolean operations such as intersection and union. However, surface based algorithms like shape optimization (fairing) or freeform modelling often need a topological manifold representation where neighborhood information within the surface is explicitly available. Consequently, it is necessary to find effective conversion algorithms to generate explicit surface descriptions for the geometry which is implicitly defined by a volumetric data set. Since volume data is usually sampled on a regular grid with a given step width, we often observe severe alias artifacts at sharp features on the extracted surfaces. Then it is crucial to have a surface sampling which is feature sensitive and thus reduces these alias effects.

Surface sampling is used in many other applications such as Computer Graphics and Visualization (e.g.in biomedical problems), as well as subdivision and surface reconstruction (see for instance Refs. 16,17).

The aim of this section is to present a simple method to determine a surface sampling which is based on a local shape analysis. The main tool to achieve this is to use the support function representation of a surface.

4.1. *The curve case*

It is convenient to illustrate to idea first in the curve case. Given a quadratic spline it is always defined the corresponding support function. In this case the domain will be the unit circle. If we take a uniform set of points in the unit circle and we map back through the envelope operator defined by (7), we obtain a set of sampling points of the curve which is curvature dependent. This is basically due to the fact that the support function, being

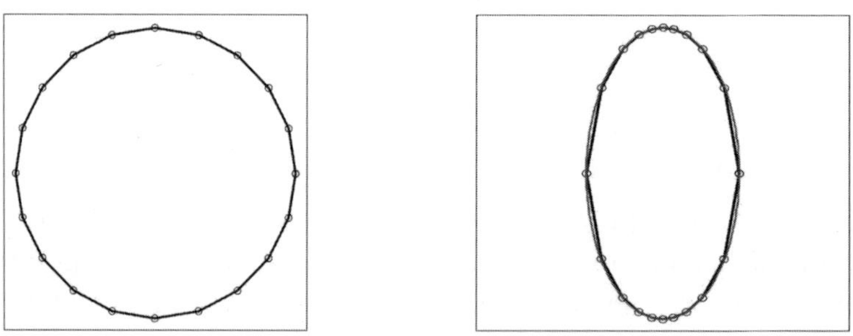

Fig. 4. A uniform point set on the unit circle is mapped in a curvature-dependent sampling on the curve.

defined in the Gaussian circle, by construction contains information about the curvature, see Fig. 4-5.

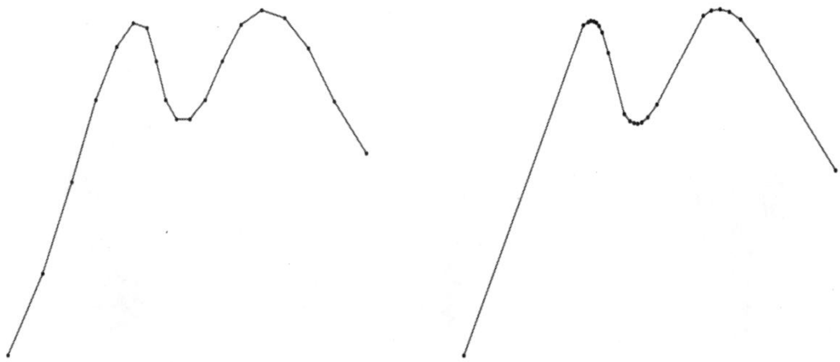

Fig. 5. Curve sampling. Left: the polygonal line connecting a uniformly sampled set of points in the curve. Right: the polygonal line connecting the curvature-dependent set.

The idea of sampling planar curves using an approach which takes into account curvature information was first investigated in Ref. 18, where the magnitude of the curvature signature function was studied.

4.2. *The surface case*

In the surface case we start from a uniformly distributed set of points into the unit sphere, for instance we can take the points coming from a uniform refinement of an icosahedron. In more detail having the quadratic triangular spline approximating a given surface, as we have seen in Sect. 3, the SF is given by a piecewise function defined over a partition of the unit sphere. Taking a point on the unit sphere if it belongs to the domain of the SF, it will be mapped by the envelope operator into a point of the surface. The number of sampling points depends by the number of points on the unit sphere. Figures 6-7 show two examples, where in the first one two sets of sampling points are constructed.

5. Conclusion

In this paper we have shown how the support function representation can be used effectively to construct a sampling set of a surface which is curvature dependent and therefore preserves its main features.

In general the SF may be not explicitly available for generic surfaces and therefore, as an intermediate step, a method to efficiently approximate a given surface with a triangular Bézier spline is presented. In this way we can use the SF of quadratic triangular splines which can be explicitly computed.

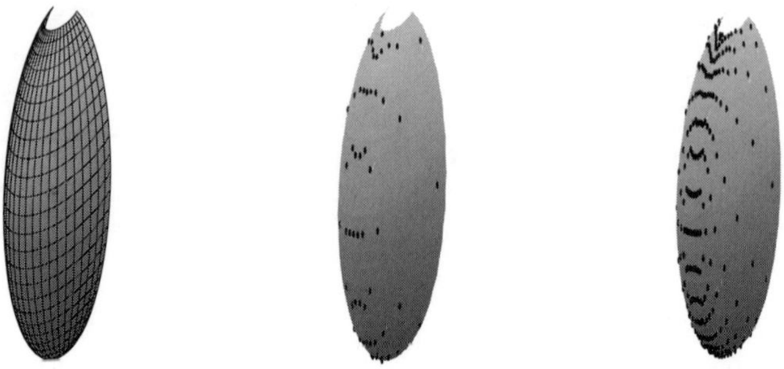

Fig. 6. Example of a part of an ellipsoid surface. Centre: a resulting sampling set. Right: a finer sampling.

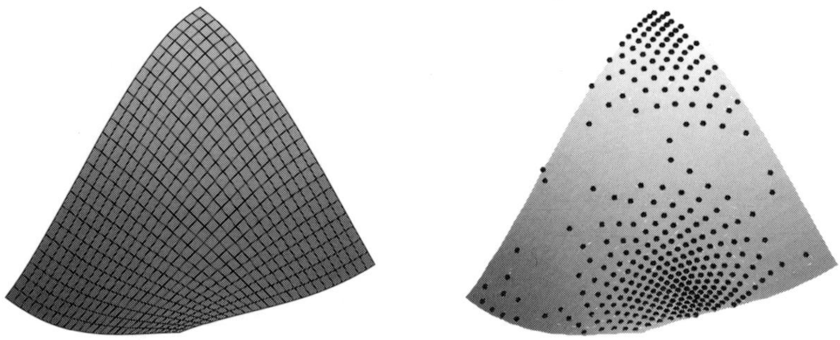

Fig. 7. Example of a non convex surface. Right: a resulting sampling set along with the approximating surface.

References

1. G. Farin, J. Hoschek and M.-S. Kim (eds.), *Hanbook of Computer Aided Geometric Design* (Elsevier Science, Amsterdam, 2002).
2. T. Bonnesen and W. Fenchel, *Theory of Convex Bodies* (BCS Associates, Moscow, Idaho, 1987).
3. R. Gardner, *Geometric Tomography* (Cambridge University Press, Cambridge, 1995).
4. M. Sabin, *A class of surfaces closed under five important geometric operations*, Technical Report No. VTO/MS/207, (British Aircraft Corp. 1974).
5. Z. Šír, J. Gravesen and B. Jüttler, *Computing convolution and Minkowski sums via support function representation*, in *Curve and Surface Design: Avignon 2006*, eds. P. Chenin, T. Lyche and L. L. Schumaker (Nashboro Press, Nashboro MN, 2007), pp. 244–253.
6. Z. Šír, J. Gravesen and B. Jüttler, Curves and surfaces represented by polynomial support functions, *Theoretical Comp. Science*, **392(1-3)**, 141–157 (2008).

7. J. Gravesen, B. Jüttler and Z. Šír, On rationally supported surfaces. *Comp. Aided Geom. Design* **25**, 320–331 (2008).
8. F. Foucher and P. Sablonnière, Approximating partial derivatives of first and second order by quadratic spline quasi-interpolants on uniform meshes, *Math. and Comput. Simulation* **77**, 202–208, (2008).
9. M. L. Sampoli, M. Peternell and B. Jüttler, Rational surfaces with linear normals and their convolution with rational surfaces, *Comp. Aided Geom. Design*, **23**, 179–192, (2006).
10. B. Jüttler and M. L. Sampoli, Hermite interpolation by piecewise polynomial surfaces with rational offsets, *Comp. Aided Geom. Design* **17**, 61–385 (2000).
11. M. Lávička and B. Bastl, Rational hypersurfaces with rational convolutions, *Comp. Aided Geom. Design*, **24**, 410–426, (2007).
12. M. Peternell and B. Odehnal, Convolution surfaces of quadratic triangular Bézier surfaces, *Comp. Aided Geom. Design*, **25**, 116–129, (2008).
13. C. Dagnino, P. Sablonnière, Error analysis for quadratic spline quasi-interpolants on non uniform criss-cross triangulations, *Prépublication IRMAR* 06-06, Rennes 2006.
14. C. de Boor, K. Höllig and S. Riemenschneider,*Box Splines* (Springer-Verlag, 1993).
15. B. Bastl, B. Jüttler, J. Kosinka and M. Lávička, Computing exact rational offsets of quadratic triangular Bezier surface patches, *Comp. Aided Design*, **40**, 197–209, (2008).
16. J. D. Boissonnat and S. Oudot, Provably good surface sampling and approximation, *ACM Intern. Conf. Proc. Series* **23** (Eurographics 2003), 9–18, (2003).
17. H. Pottmann, T. Steiner, M. Hofer, C. Haider, and A. Hanbury, The isophotic metric and its application to feature sensitive morphology on surfaces. In T. Pajdla and J. Matas editors, *Computer Vision — ECCV 2004, Part IV, Lecture Notes in Computer Science*, **3024**, 560–572, (2004).
18. T. Surazhsky and V. Surazhsky, Sampling planar curves using curvature-based shape analysis, in *Mathematical methods for curves and surfaces: Tromsø 2004*, eds. M. Dæhlen, K. Mørken and L. L. Schumaker (Nashboro Press, Nashboro MN, 2005), pp. 339–350.

THERMO-MECHANICAL MODELING OF GROUND DEFORMATION IN VOLCANIC AREAS

D. SCANDURA*[,1,2], A. BONACCORSO[1], G. CURRENTI[1], C. DEL NEGRO[1]

[1]*Istituto Nazionale di Geofisica e Vulcanologia, Sezione di Catania, Italy*
[2]*Dipartimento di Matematica e Informatica, Universit di Catania, Italy*
[*]*E-mail: scandura@ct.ingv.it*

Materials surrounding a long-lived magmatic source are significantly heated above the brittle-ductile transition and rocks no longer behave in a purely elastic manner. A thermo-mechanical numerical model is performed for evaluating the temperature dependency of the rheological behaviour. The inclusion of viscoelastic and elasto-plastic material around the volcanic source considerably reduces the magmatic pressure necessary to produce the observed surface deformation with respect to elastic model. The assumption on the medium rheology strongly affects the ground deformation and, hence, the estimate of source pressure, which is an important issue for improving the knowledge of the physics of volcano deformation and eruption hazard.

1. Introduction

In volcanic areas, the presence of heterogeneous materials and high temperatures affects the rheological behaviour of the Earth's crust that calls for considering the anelastic properties of the medium surrounding the magmatic sources.

Most volcano deformation models published to date assume that the Earth's crust behaves as a perfectly elastic solid. Over the last decades, elastic numerical models have contributed to asses how medium heterogeneity and topography can influence ground deformation especially near the volcano summit.[1–3] While elastic half space models fit a variety of crustal deformation data, this rheology in case of volcanic regions is an oversimplification. The elastic approximation is generally appropriate for small deformations of crustal materials with temperatures cooler than the brittle-ductile transition, about $300 \pm 500^0 C$ depending mainly on composition and strain rate. Materials surrounding a long-lived magmatic source are heated significantly above the brittle-ductile transition and rocks no longer behave in a purely elastic manner, but permanently deform because of the plastic deformation. Therefore, the thermal state of the volcanoes can greatly influence the surface deformation field, making the elastic approximation inappropriate to model the observed ground deformation.

In the present work, a mechanical numerical model is performed for evaluating, in terms of displacement field, the effects of pressure source of a spherical magma chamber in a volcanic edifice. We applied the 3D axi-symmetric Finite Element Method (FEM) to asses the role that rheology may play in computing deformation field. Particularly, we compared different simulations considering elastic, viscoelastic and elasto-plastic rheologies. In a second step we introduced the thermal problem and modified the properties of the rocks using the computed thermal profile. The effect of the thermal state on the deformation field was evaluated for different value of the brittle-ductile transition.

2. Numerical Model

We developed a 3D axi-symmetric Finite Element (FE) model to asses the role that rheology may play in computing deformation field. Numerical computations are carried out using the FE software COMSOL.

2.1. *Deformation model*

Our approach assumes radial symmetry about the source centre. To approximate an half-space, the axi-symmetric FEM is composed of $\sim$ 200000 triangular elements covering a region that extends 25 km horizontally from the source centre and 35km below the surface. The source is located at 4 km depth, and the radius of the spherical magma chamber is 0.7 km. We assumed free displacement values at the upper surface and zero displacement values at bottom and lateral boundaries. A step-like increase in pressure ($\Delta P = 320$MPa) is applied on the source wall.

We performed three different simulations considering: (i) elastic, (ii) viscoelastic and (iii) elasto-plastic rheology.

In the first case we considered a fully elastic half-space with Poisson ratio $\nu = 0.25$ and rigidity moduli $\mu = 30$GPa.

The second and third model was realized introducing a spherical shell (radius 1.7 km) surrounding the magmatic source. In the viscoelastic model it was supposed a viscoelastic behaviour inside the shell, an elastic behaviour outside it. It was assumed a generalized Maxwell rheology with Poisson ratio $\nu = 0.25$, rigidity moduli $\mu_1 = \mu_2 = 30$GPa, and viscosity

$\eta = 2{\cdot}10^{16}Pas$; a step-like increase in pressure ($\Delta P = 320$MPa) was applied at time t=0 on the source wall.

In the elasto-plastic model it was supposed an elasto-plastic behaviour inside the shell, while the remain domain was set as elastic. We implement the yield stress/strain laws considering an ideal plastic behaviour obeying to von Mises criterion; the yield strength of surrounding rocks is assumed as $\sigma_y = 15$MPa while the elastic parameters of the medium are those of models previously described. Figure 1 shows the comparison of ground uplift due to a pressure sourcc of 320 MPa considering elastic (dot line), viscoelastic (solid line) and elasto-plastic rheology (dashed line).

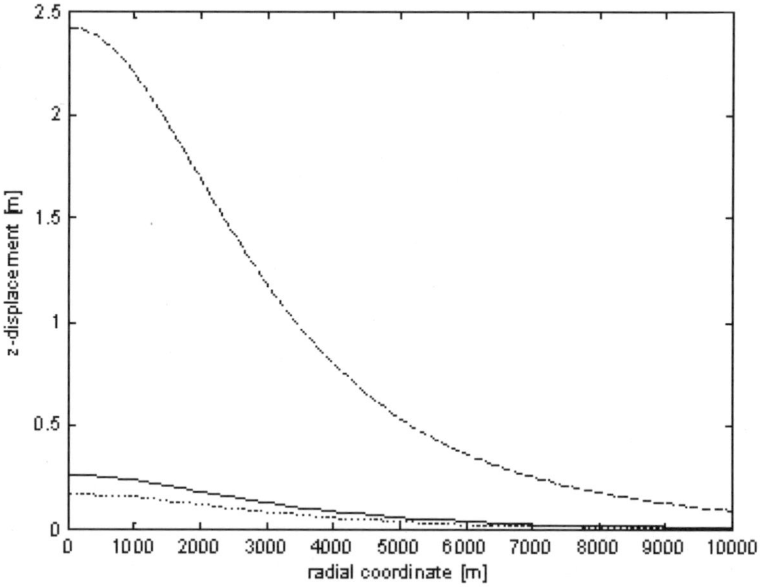

Fig. 1. : Comparison of ground uplift due to a pressure source of 320 MPa considering elastic (dot line), viscoelastic (solid line) and elasto-plastic rheology (dashed line). The radius of the source, located at a depth of 4km, is 0.7 km.

The maximum uplift reaches 0.17 m in the case of a fully elastic half-space, 0.27 m using a viscoelastic rheology and 2.4 m considering an elasto-plastic medium. Therefore, the numerical models that include an anelastic rheology enable to produce deformation comparable with those obtained from elastic model, requiring a significantly lower pressure. Particularly, the deformation in the viscoelastic solution are enhanced of about 1.6 in comparison with the elastic solution while, in the case of elasto-plastic model, the deformation field is about 14 times higher than those obtained from the fully elastic model. Therefore, to produce the same deformation the viscoelastic model requires a lower pressure changes ($\sim$ 200MPa) that is closer to the lithostatic load ($\sim$ 170MPa), but still higher than the crustal strength ($\sim$ 45MPa). In the case of elasto-plastic model, a non-linear dependence between pressure source and deformation field can be found (Fig. 2). Figure 2 shows that the maximum uplift of 0.17 m (elastic deformation) in the case of elasto-plastic model requires a pressure changes of $\sim$ 47MPa that is near to crustal strength.

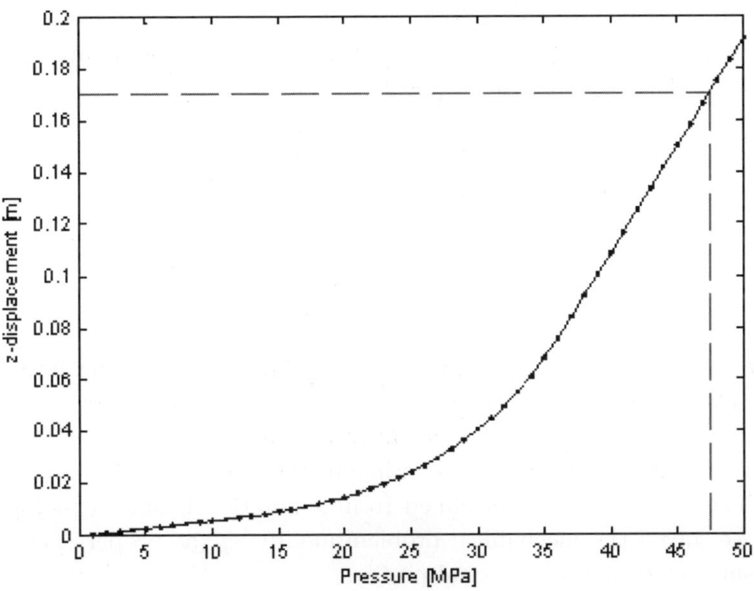

Fig. 2. : Non linear dependence between pressure source and ground uplift. Simulations were computed considering an ideal plastic medium with yield strength of $\sigma_y = 15$MPa.

2.2. *Thermal model*

The mechanical properties of the medium depend on the temperatures of the magma chamber and the surrounding rocks. If we assume a magma chamber maintained at a constant temperature, the temperature distribution around the magmatic source can be computed by solving a heat conduction equation. We developed the model in two steps solving separately: (i) the heat conduction equation to compute the temperature profile, and (ii) the mechanical problem to obtain the numerical solution of the deformation field.

To derive the temperature profile, we numerically solved the equations for heat conduction in a axial symmetric formulation

$$\nabla(k\nabla T) = -A \tag{1}$$

where $T = T(r, z)$ is the temperature field, r is the radial coordinate, z is the vertical coordinate, k is thermal conductivity, and $A(z) = A_s \exp(-z/b)$ is the volumetric crustal volumetric heat production, where A_s is the volumetric rate of heat production, and b is a characteristic depth of the order of 10 ± 5km. As boundary condition at the ground surface, we assumed that the surface is kept constant at atmospheric temperature, since the thermal conductivity of the air is much smaller than that of the ground. At bottom and lateral boundaries we assigned the geothermal temperature values, because they are far enough to not be affected by the magmatic source. We used the steady-state geothermal profile given by[4,5] :

$$T(z) = T_s + \left(\frac{q_m z}{k}\right) + \left(\frac{A_s b^2}{k}\right)(1 - e^{-z/b}) \tag{2}$$

where T_s is the surface temperature, q_m is the heat flow coming from the mantle. The temperature on the magma wall was set to $T_0 = 1500$K. Physically, this boundary condition is equivalent to stating that the magma walls act as heat sources, simulating a continuous refilling of the magma chamber.[6,7] The thermal profile obtained from simulation is shown in Fig. 3. Hence, we solved the mechanical problem modifying the properties of the rocks using the computed thermal profile.

In the viscoelastic model it was supposed a viscoelastic behaviour where the temperature is higher than a fixed threshold and an elastic behaviour where it is lower. Therefore, we modified the properties of the medium through the constitutive equations, allowing the element of the domain to behave elastically or viscoelastically in function of the temperature threshold. The solutions vary as a function of the threshold. Figure 4 shows the ground uplift as the temperature threshold increases from 600 K to 1000 K. When the threshold decreases, the thickness of the viscoelastic shell is wider and the deformation is enhanced.

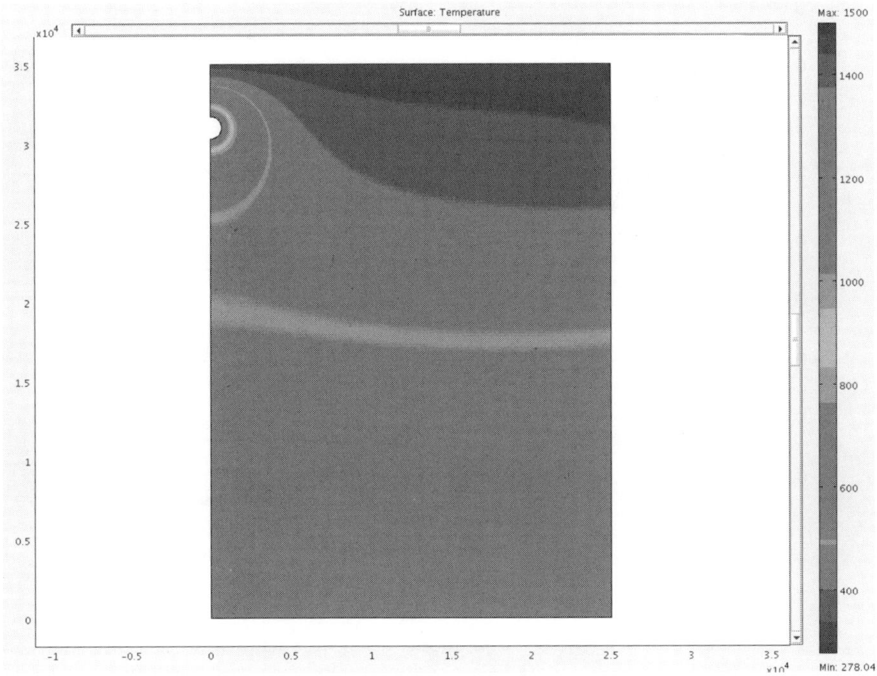

Fig. 3. : Temperature profile assuming a temperature of the magma wall of 1500 K

However, it is reasonable to assume that the viscosity around the magmatic source is not uniform but dependent on the temperature distribution. The model was then improved introducing an empirical relationship to relate the viscosity to the temperature. We estimated the medium viscosity surrounding the source region using the Arrhenius formula:

$$\eta = A_D \exp\left(\frac{E}{RT}\right) \tag{3}$$

where A_D is the Dorn parameter, E is the activation energy, R is the Boltzmann constant, and T the absolute temperature. Additional work is needed to define plausible values for the rheological parameters, and determine the extent to which these parameters vary in the region of high thermal gradients near the magma chamber. The values of the parameters used in the computations are summarized in Table 1.

We made a comparison between: (i) a model in which the viscoelastic shell thickness is fixed by a temperature threshold and (ii) a model in which the medium is fully viscoelastic. In the first model, the deformation increase

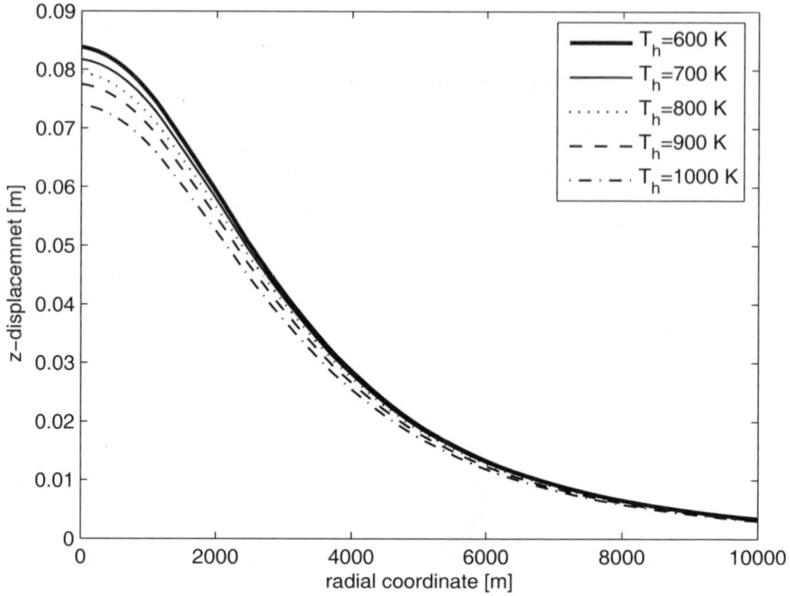

Fig. 4. : Ground uplift of the viscoelastic model when the shell thickness is dependent on the threshold temperature. Simulations are performed for threshold temperature ranging from 600 K to 1000 K.

Table 1. The best model parameters found by the GA.

Thermal parameters		
T_s	Surface temperature	$273K$
q_m	Heat flow	0.03mWm^{-2}
k	Thermal conductivity	$4\text{Wm}^{-1}K^{-1}$
A_s	Volumetric rate of heat production	$2.4710^{-6}\text{Wm}^{-3}$
b	Length scale for crustal radioactive decay	14.170km
Mechanical parameters		
A_D	Dorn parameter	10^9Pas
E	Activation energy	120kJ/mol
R	Boltzmann constant	8.314J/mol K

is not linearly proportional to the temperature threshold, but a saturation effect is observed (Fig. 5). The ground uplift of the second model does not differ too much from the first one when the temperature threshold is lower than 700 K (Fig. 6). Indeed, far away from the source, the temperature

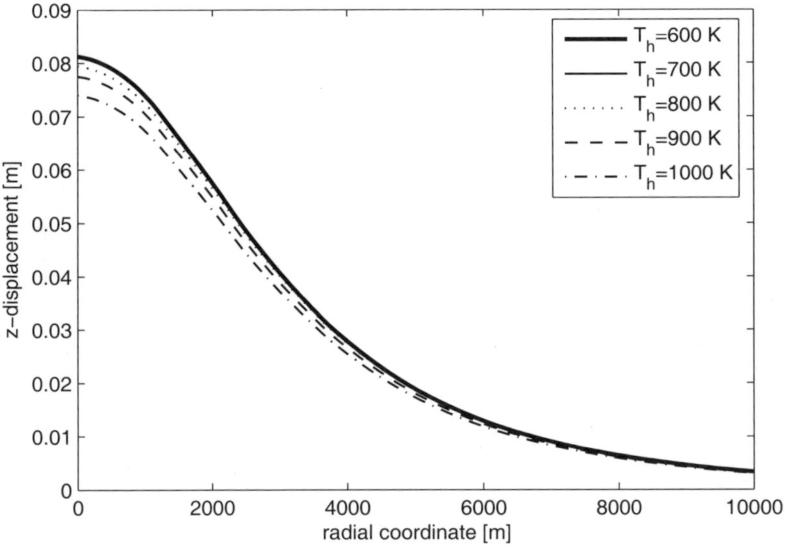

Fig. 5. : Ground uplift of the temperature-dependent viscoelastic model with the shell thickness dependent on the threshold temperature. Simulations are performed for threshold temperature ranging from 600 K to 1000 K.

decays, yielding a higher value of viscosity which makes the medium to behave as elastically.[8]

In the elasto-plastic model it was supposed an elasto-plastic behaviour where the temperature is higher than the threshold value, an elastic behaviour where it is lower. Figure 7 shows the ground uplift setting the threshold from 600 K to 1000 K. In the elasto-plastic model the transition temperature is found to be a very sensitive parameter. Indeed, as the threshold temperature decreases, the volume participating to the elasto-plastic flow increases and gives more contribution to the plastic deformation. A 100 K of variation in the transition temperature is sufficient to enhance the ground uplift from 0.18 m to 0.25 m.

2.3. *Discussion and conclusions*

A thermo-mechanical numerical model has been carried out to investigate the deformation caused by pressure changes within a magmatic source. The definition of elastic/anelastic rock properties strongly affects the solution and, especially in volcanic region, cannot disregard the thermal regime of

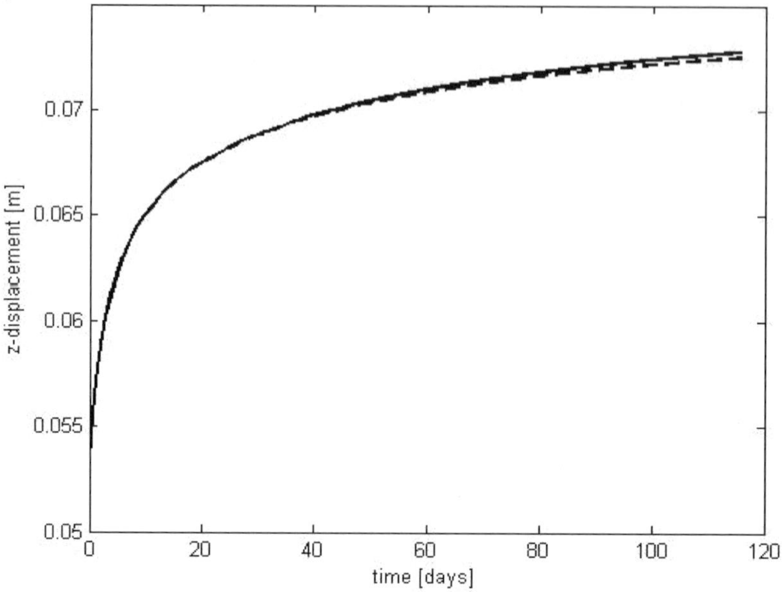

Fig. 6. : Comparison between the model in which the thickness of the shell is temperature dependent (dashed line) and the model in which all the medium is fully viscoelastic (solid line).

the crust. The rheology assumption influences the estimate of the magmatic pressure changes. Since the estimated pressures are significantly different from each other, the picture of the volcano state is completely altered. The thermo-mechanical model also evidences that the thermal state of the crust can play an important role in the deformation field: in the viscoelastic model the viscosity is dependent on the temperature distribution, in the elasto-plastic model the transition temperature is found to be a very sensitive parameter. Therefore, the definition of the rheology and of the thermal state becomes a key parameter in volcano hazard assessment.

References

1. V. Cayol and F. Cornet, *Geophys. Res. Lett.* **25**, 1979 (1998).
2. G. Currenti, C. Del Negro, G. Ganci and D. Scandura, *Phys. Earth Planet. Interiors* **168**, p. 8896 (2008).
3. C. Williams and G. Wadge, *J. Geophys. Res.* **105**, p. 1038120 (2000).
4. G. Ranalli, *Rheology of the Earth* (Chapman and Hall, London, 1995).

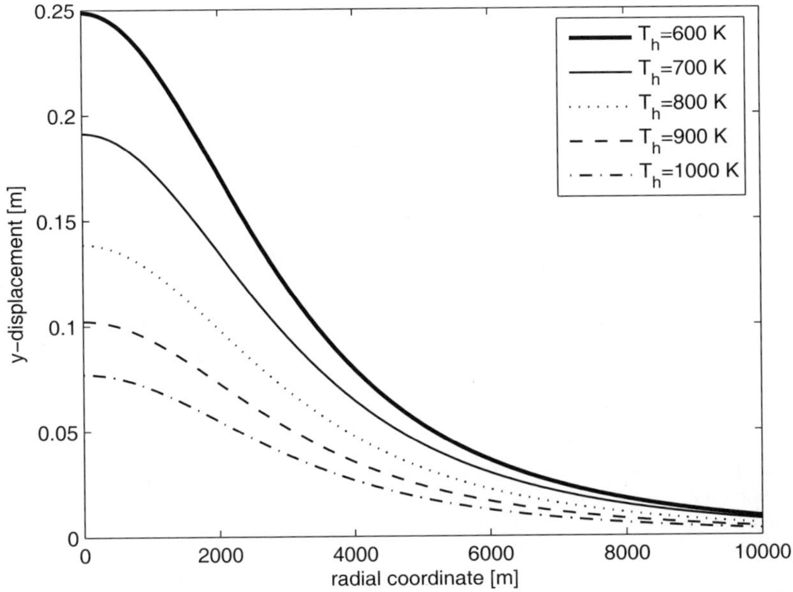

Fig. 7. : Ground uplift for the elasto-plastic model as the threshold temperature increases from 600 K to 1000 K.

5. D. Turcotte and G. Schubert, *of Continuum Physics to Geological Problems* (Wiley, New York, 1982).
6. M. Civetta, S. DAntonio, V. De Lorenzo and P. Gasparini, *Volcanol. Geotherm. Res.* **133**, p. 112 (2004).
7. P. Dragoni, H. P. and F. Mongelli, *Pure Appl. Geophys.* **150**, p. 181201 (1997).
8. C. Williams and R. Richardson, *Geophys. Res. Lett.* **25**, p. 15491552 (2000).

Qualitative Analysis for Walrasian Price Equilibrium Problem: Parametric Variational Approach

Francesco Scaramuzzino

Department of Mathematics
Messina University, Italy
fscaramuzzino@unime.it

This paper considers a qualitative analysis of the solution of the general economic equilibrium problem regulated by Walra's law. Some results relative to the existence, the uniqueness and the sensitivity analysis of the solution of the Walrasian price equilibrium problem have been collected. Such results have been applied in a particular parametric case. It has been shown a numerical application where the parameter represents the variation of the average annual inflation rate and for this numerical example it has been proved the existence of the solutions of the parametric variational inequality that describes the equilibrium conditions by means of the direct method.

Keywords: Walrasian price equilibrium problem, parametric variational inequality, parametric sensitivity analysis.

1. Introduction

In this paper the focus is on general economic equilibrium problems, in particular, a Walrasian price equilibrium or pure exchange equilibrium. This problem has been extensively studied in the economic literature dating to Walras (1874), see also Wald (1951), Debreu (1959), and Mas - Colell (1985); in particular many authors have been devoted to the study of the static general economic equilibrium problem regulated by Walra's law (see e.g. Arrow and Hann in Ref. 1; Border in Ref. 2 ; Dafermos in Ref. 5 , Dafermos and Nagurney in Ref. 3, Dafermos and Zhao in Ref. 6, Donato et al. in Ref. 9–11; Nagurney in Ref. 16; Nagurnay and Zhao in Ref. 17; Walker in Ref. 21; Walras in Ref. 22; Zhao in Ref. 24 and their bibliography. The aim of this paper is to provide a qualitative analysis of the solution of the static general economic equilibrium problem regulated by Walra's law in a parametric case. The equilibrium conditions that describe this pure exchange economic model are expressed in terms of a Parametric

Variational Inequality (PVI), for which existence and parametric sensitivity results are given. The case of PVI with a fixed constraint set is studied, for example, in Kyparisis[13], Robinson[18]. The case of PVI whose constraint set depends on a parameter has been considered, for example, in Tobin[20], Dafermos[4], Harker and Pang[12], Kyparisis[14], Yen[23]. In this paper the constraint set doesen't depend on a parameter but only the dependence of the excess demand function on a parameter has been considered. The interest of this parametric study resides in the fact that it is necessary to know how the solution changes under a suitable variation of the aggregate excess demand function to a parameter variation. The paper is organized as follows. In Sect. 2 it is introduced the Walrasian price equilibrium problem, for which existence, uniqueness and sensitivity results are given, in Sect. 3 it is introduced the parametric variational model with parametric sensitivity analysis result. Finally, in Sect. 4, in order to support the theoretical results, a numerical example has been provided: the Walrasian price equilibrium is solved by the PVI with the direct method given in Ref. 15.

2. Walrasian price equilibrium problem

Let us consider a pure exchange economy with l different commodities and with column price vector $p = (p_1, p_2, ..., p_l)^T$ taking value in $\mathbb{R}^l_+$. Let us denote the induced aggregate excess demand function by $z(p)$ and let us assume that is generally defined on a subcone C of $\mathbb{R}^l_+$ which contains the interior $\mathbb{R}^l_{++}$ of $\mathbb{R}^l_+$, that is, $\mathbb{R}^l_{++} \subset C \subset \mathbb{R}^l_+$. Hence, the possibility that the aggregate excess demand function may become unbounded when the price of a certain commodity vanishes is allowed; $z(p)$ is a row vector with components $z_1(p), z_2(p), ..., z_l(p)$, where: $z_i : C \to \mathbb{R}, \quad i = 1, ..., l$. Assuming that $z(p)$ will be homogeneous of degree zero in p on C, that is $z(\alpha p) = z(p)$ for all $p \in C, \alpha > 0$ and, as usual, let us assume that $z(p)$ satisfies Walras' law, that is:

$$\langle z(p), p \rangle = 0, \quad \forall p \in C.$$

Because of homogeneity, the prices may be normalized so that they take values in the simplex:

$$S^l = \left\{ p : p \in \mathbb{R}^l_+, \ \sum_{i=1}^{l} p_i = 1 \right\}$$

and, therefore, the aggregate excess demand function may be restricted to the intersection $D = S^l \cap C$. Let $S^l_+ = \{p : p > 0, p \in S^l\}$, $S^l_+ \subset D \subset S^l$

is noted. In a general economic equilibrium theory it is assumed that the function $z(p) : D \to \mathbb{R}^l_+$ is continuous and satisfies the Walra's law:

$$\langle z(p), p\rangle = \sum_{i=1}^{l} z_i(p) \cdot p_i = 0 \quad \forall p \in D.$$

The definition of a Walrasian equilibrium is now stated.

Definition 2.1. A price vector $p^* \in D$ is a Walrasian equilibrium price vector if $z(p^*) \leq 0$.

The following theorem establishes that the Walrasian price equilibrium can be characterized as a solution of a variational inequality.

Theorem 2.1. *A price vector $p^* \in D$ is a Walrasian equilibrium if and only if it satisfies the variational inequality:*

$$\langle z(p^*), p - p^*\rangle = \sum_{i=1}^{l} z_i(p^*) \cdot (p_i - p_i^*) \leq 0 \quad \forall p \in S^l. \tag{1}$$

2.1. *Existence results*

In order to give some results of existence of equilibria it is possible to adapt classical existence theorems to our variational inequality problem. Being $D = S^l \cap C$ a convex and bounded set, the following theorems provide the existence with or without the compactness assumption. If the aggregate excess demand function $z(p)$ is defined and is continuous on all of S^l, that is, $D = S^l$, the existence of at least one Walrasian equilibrium price vector in S^l follows immediately from Theorem 2.1 and from the following classic Theorem shown by Stampacchia in Ref. 19 and adapted in our case:

Theorem 2.2. *If D is a compact convex set and $z(p)$ is continuous on D, then the variational inequality admits at least a solution p^*.*

However, since D is not necessarily compact, Theorem 2.2 cannot be applied directly to the above variational inequality problem. Neverthless, this theorem may be still applied to deduce that $z(p)$ exhibits the needed behavior near the boundary of S^l, in particular, that at least some of the components of $z(p)$ become in a "wide" sense as p approaches points of the boundary of S^l that are not contained in D. Several existence proofs of this type can be found in Ref. 2. The result shown in Ref. 5 is now provided.

Theorem 2.3. *Let us assume that the aggregate excess demand function $z(p)$ satisfies the following assumption: If S^l/D is nonempty, then with any* *sequence* $\{p_n\}$ *in* S^l_+ *which converges to a point of* S^l/D *there is associated a point* $\bar{p} \in S^l_+$, *generally dependent on* $\{p_n\}$, *such that the sequence* $\langle z(p_n), \bar{p}\rangle$ *contains infinitely many positive terms. Then there exists a Walrasian equilibrium price vector* $p^* \in D$.

2.2. *Uniqueness result*

The uniqueness issue is now investigated; in particular, if the monotonicity assumption is now strengthened, the following result shown by Nagurney is obtained (see Theorem (9.4) of Ref. 16).

Theorem 2.4. *Let us assume that* $-z(p)$ *is strictly monotone on* D, *that is:*

$$\left\langle z\left(p^1\right) - z\left(p^2\right), p^1 - p^2\right\rangle < 0, \quad \forall\, p^1p^2 \in D, \quad p^1 \neq p^2.$$

Then there exists at most a single Walrasian price equilibrium vector $p^* \in D$.

2.3. *Sensitivity analysis*

A result related to the sensitivity of the solution price vector to changes in the data is now reported. Let us consider a pure exchange economy governed by the variational inequality (1). Let us assume that the aggregate excess demand function changes from $z(\cdot)$ to $z^*(\cdot)$, and let us establish the relation between the correspondent equilibria p and p^*. The following theorem, shown in Ref. 16, holds.

Theorem 2.5. *Let* $z(\cdot)$ *and* $z^*(\cdot)$ *denote two aggregate excess demand functions, and let* p *and* p^* *denote, respectively, their associated Walrasian equilibrium price vectors. Let us assume that* $-z(p)$ *satisfies the strong monotonicity assumption:*

$$\sum_{i=1}^{l}\left(-z_i\left(p^1\right) + z_i\left(p^2\right)\right)\cdot\left(p_i^1 - p_i^2\right) \geq \alpha\left\|p^1 - p^2\right\|^2, \quad \forall p^1, p^2 \in D, \ \alpha > 0.$$

Then

$$\|p^* - p\| \leq \frac{1}{\alpha}\left\|z^*\left(p^*\right) - z\left(p^*\right)\right\|.$$

3. Qualitative analysis: Parametric variational approach

In this section is provided the sensitivity analysis of the equilibrium price vector of a pure exchange economy when the aggregate excess demand function varies with a parameter λ. The economic means of the parameter λ, acting λ as any other external factor, is able to influence the agents' demand, e.g., increase/decrease of:

- inflation rate;
- production costs;
- consumer incom;
- competitive competition; etc.

Assumed that $\lambda \in \Lambda$, Λ is an open subset of $\mathbb{R}^k$. Let us consider the family of operators:

$$\{z(p,\lambda), p \in D, \lambda \in \Lambda\}.$$

where $z(p,\lambda)$ is the aggregate excess demand function defined on $D \times \Lambda \subseteq \mathbb{R}^l_+ \times \mathbb{R}^k$, taking values in $\mathbb{R}^l$, with components:

$$z_i(p,\lambda) : D \times \Lambda \to \mathbb{R} \quad \forall i = 1, ..., l.$$

Assuming that, for all λ in Λ the operators $z(p,\lambda)$ are continuous and satisfy Walra's law:

$$\langle z(p,\lambda), p\rangle = \sum_{i=1}^{l} z_i(p,\lambda) \cdot p_i = 0 \quad \forall p \in D.$$

The definition of a Walrasian equilibrium to a parameter variation is now stated.

Definition 3.1. A price vector $p^* \in D$ is a Walrasian equilibrium price vector if

$$z(p^*,\lambda) \leq 0 \;\; \forall \lambda \in \Lambda. \tag{2}$$

We observe that, since the market is regulated by Walra's law, it is possible to rewrite the equilibrium condition (2) in the following way:

Definition 3.2. A price vector $p^* \in D$ is a Walrasian equilibrium price vector if and only if for all $j = 1, ..., l$ and all $\lambda \in \Lambda$:

$$z^j(p^*,\lambda) \begin{cases} \leq 0 & \text{if} \quad (p^*)^j = 0 \\ = 0 & \text{if} \quad (p^*)^j > 0. \end{cases} \tag{3}$$

Condition (3) states that if the price of equilibrium of the goods j is equal to zero, then the aggregate excess demand $z^j(p^*, \lambda)$, related to the goods j, for all value of the parameter λ in Λ, must not be greater than zero, that is, the supply must be over the demand to varying of λ. If the price of equilibrium of the goods j is greater than zero, then $z^j(p^*, \lambda)$, related to the goods j, for all value of the parameter λ in Λ, must be equal to zero, that is, the demand and the supply must be equal to varying of λ.
The PVI problem is then given.
Let us determine $\bar{p} \in D$, satisfing:

$$\langle z(\bar{p}, \lambda), p - \bar{p} \rangle \leq 0 \quad \forall p \in D, \ \forall \lambda \in \Lambda, \tag{4}$$

which is equivalent to the parametric Walrasian equilibrium price condition.

Remark 3.1. It has been observed that in this parametric formulation, the aggregate excess demand function depends on the parameter λ, but the convex set over which the variational inequality problem is defined, does not depend on λ, that is, the constraints are not subject to a parameter perturbation.

Choosing λ in Λ, we have:

$$z(p, \lambda) = z_\lambda(p)$$

where $z_\lambda(\cdot) : D \to \mathbb{R}^l$, $D \subseteq \mathbb{R}^l$, then the PVI problem (4) is equivalent to determine $\bar{p} \in D$:

$$\langle z_\lambda(\bar{p}), p - \bar{p} \rangle \leq 0 \quad \forall p \in D, \tag{5}$$

that is equivalent to the variational problem (1).

3.1. *Existence and uniqueness results*

Assuming that $D = S^l$ and choosing λ in Λ, then the existence of a solution to a PVI problem (4) (or 5), follows from continuity of the function $z_\lambda(\cdot)$ and compactness of S^l thanks to the Theorem 2.2, but if the result $D \subseteq S^l$, since D is not necessarily compact, the existence of a solution to a PVI problem (4) follows from the Theorem 2.3. Moreover, if $z_\lambda(\cdot)$ is strictly monotone on D, thanks to the Theorem 2.4 the uniqueness of the solution $\bar{p} \in D$ is achieved.
Then, for all $\lambda \in \Lambda$ exist only one $\bar{p} \in D$ solution of (4), where:

$$\bar{p} : \Lambda \subseteq \mathbb{R}^k \to \mathbb{R}^l.$$

Considering $\lambda = \lambda^*$, let us suppose that for some $\lambda^* \in \Lambda$ the PVI (4) possesses an equilibrium solution $\bar{p}(\lambda^*) = \bar{p}^*$. Now, it is possible to effect the following qualitative analysis.

3.2. *Parametric sensitivity analysis*

It is now examined the sensitivity of the solution price vector to changes in the data, when the aggregate excess demand function varies with a parameter λ. To this aim, let us consider a pure exchange economy governed by the PVI (4). Let us assume that the following strong monotonicity condition on $-z(\cdot,\lambda)$ holds:

$$\sum_{i=1}^{l} \left(-z_i\left(p^1,\lambda\right) + z_i\left(p^2,\lambda\right)\right)\left(p_i^1 - p_i^2\right) \geq \alpha \left\|p^1 - p^2\right\|^2, \ \forall p^1, p^2 \in S^l, \ \forall \lambda \in \Lambda$$

where $\alpha > 0$. Let us assume that the aggregate excess demand function changes from $z(\cdot,\lambda)$ to $z^*(\cdot,\lambda')$ with $\lambda, \lambda' \in \Lambda$, and let us establish the relation between the correspondent equilibria $\tilde{p}$ and $\tilde{p}^*$, where $\tilde{p}$, $\tilde{p}^*$ are the equilibrium price vectors obtained, respectively, fixing, $\lambda = \tilde{\lambda}$ and $\lambda' = \tilde{\lambda}^*$. The following theorem establishes that small changes in the aggregate excess demand function, to varying of the parameter λ, lead to small changes in the Walrasian price equilibrium.

Theorem 3.1. *Let $z(\cdot,\lambda)$ and $z^*(\cdot,\lambda')$ be two excess demand functions, with $\lambda, \lambda' \in \Lambda$, and let $\tilde{p}$ and $\tilde{p}^*$ denote, respectively, their associated Walrasian price equilibrium vectors, obtained, fixing in Λ, respectively, $\lambda = \tilde{\lambda}$ and $\lambda' = \tilde{\lambda}^*$. Then, under strong monotonicity assumption, we have:*

$$\|\tilde{p}^* - \tilde{p}\| \leq \frac{1}{\alpha}\left\|z^*_{\tilde{\lambda}^*}(\tilde{p}^*) - z_{\tilde{\lambda}^*}(\tilde{p}^*)\right\|, \ \forall \tilde{\lambda}, \ \tilde{\lambda^*} \in \Lambda, \ \alpha > 0$$

and $\|\cdot\|$ is the euclidean norm.

Proof. *Let us start to prove the following result:*

$$\left\langle z^*_{\tilde{\lambda}^*}(\tilde{p}^*) - z_{\tilde{\lambda}}(\tilde{p}), \tilde{p}^* - \tilde{p}\right\rangle = \sum_{i=1}^{l}\left(\left(z^*_{\tilde{\lambda}^*}\right)_i(\tilde{p}^*) - \left(z_{\tilde{\lambda}}\right)_i(\tilde{p})\right)\cdot\left(\tilde{p}_i^* - \tilde{p}_i\right) \geq 0. \tag{6}$$

Fixing $\lambda = \tilde{\lambda}$, for some $\tilde{\lambda} \in \Lambda$ the PVI (4) possesses an equilibrium solution $\tilde{p} = p(\tilde{\lambda})$, then:

$$\left\langle z_{\tilde{\lambda}}(\tilde{p}), q - \tilde{p}\right\rangle \leq 0 \ \ \forall q \in S^l. \tag{7}$$

Fixing $\lambda' = \tilde{\lambda}^*$, *for some* $\tilde{\lambda}^* \in \Lambda$ *the PVI (4) possesses an equilibrium solution* $\tilde{p}^* = p(\tilde{\lambda}^*)$, *then:*

$$\left\langle z^*_{\tilde{\lambda}^*}(\tilde{p}^*), q - \tilde{p}^* \right\rangle \leq 0, \ \forall q \in S^l. \quad (8)$$

Letting $q = \tilde{p}^*$ *and* $q = \tilde{p}$, *respectively, and adding the two inequalities (7) and (8), one has that (6) hold. Then, the proof follows by the strong monotonicity condition and the Schwarz inequality and it is strictly close to that shown by Nagurney (see Ref. 16, Theorems 9.1 and 9.5).* □

4. Example

The Walrasian price equilibrium problem previously presented is now illustrated with a numerical example. Let us solve the PVI problem by means of the direct method of Ref. 15 (see also Ref. 7,8). Let us consider a pure exchange economy with: 2 goods, $j = 1, 2$, 2 agents, $a = 1, 2$, and 1 parameter $\lambda \in \Lambda$, where Λ is an open set in which the parameter λ takes values. It is assumed that the agent 1 and 2 have, respectively, the endowment vectors:

$$e_1 = \left(e_1^1, e_1^2\right) = (2, 0), \qquad e_2 = \left(e_2^1, e_2^2\right) = (1, 5).$$

In order to obtain the demand vectors the Cobb - Douglas utility function has been used, opportunely adapted to our parametric case, and it is as follows:

$$x_a^j(\lambda) = \frac{\beta_a^j(\lambda) \sum_{j=1}^{l} p^j e_a^j}{\sum_{j=1}^{l} p^j \beta_a^j(\lambda)}, \quad \forall j = 1, .., l, \ a = 1, .., l, \ \lambda \in \Lambda.$$

The demand by each agent depends on the function $\beta(\lambda) = \left(\beta_a^j(\lambda)\right)$, by means of which it is possible to control the demand in the market. Assuming that $\beta(\lambda) \in \mathbb{R}^{2\times 2}$, with:

$$\beta(\lambda) = \begin{bmatrix} \beta_1^1(\lambda) & \beta_1^2(\lambda) \\ \beta_2^1(\lambda) & \beta_2^2(\lambda) \end{bmatrix} = \begin{bmatrix} 1+\lambda & 2\lambda \\ 2+\lambda & 3\lambda \end{bmatrix}.$$

Consequently, the demand vectors are:

$$x_1(\lambda) = \left(x_1^1(\lambda), x_1^2(\lambda)\right) = \left(\frac{(1+\lambda)\,2p^1}{(1+\lambda)\,p^1 + 2\lambda p^2}, \frac{4\lambda p^1}{(1+\lambda)\,p^1 + 2\lambda p^2}\right),$$

$$x_2(\lambda) = \left(x_2^1(\lambda), x_2^2(\lambda)\right) = \left(\frac{(2+\lambda)\left(p^1 + 5p^2\right)}{(2+\lambda)\,p^1 + 3\lambda p^2}, \frac{3\lambda\left(p^1 + 5p^2\right)}{(2+\lambda)\,p^1 + 3\lambda p^2}\right),$$

$\forall j = 1,2,\ a = 1,2,\ \lambda \in \Lambda.$

Remark 4.1. The choice to use the Cobb - Douglas utility function is due to its particularly convenient properties (differentiability, quasi-concavity) and also thanks to the facility with which it is possible to deal with them analytically; for this reason, they are often used in the study of micro economic problems. Moreover, such functions have the same algebraical form of the Walrasian demand function associated with them.

Since in this economy there is only pure exchange, the excess demand function, related to the goods j, is so defined:

$$z^j(p,\lambda) = \sum_{a=1}^{2} x_a^j(\lambda) - \sum_{a=1}^{2} e_a^j, \quad \forall p \in S^2,\ j = 1,2,\ \lambda \in \Lambda,$$

and in our case it becomes:

$$z^1(p,\lambda) = \left(\frac{(1+\lambda)\,2p^1}{(1+\lambda)\,p^1 + 2\lambda p^2} + \frac{(2+\lambda)\left(p^1 + 5p^2\right)}{(2+\lambda)\,p^1 + 3\lambda p^2} - 3\right).$$

$$z^2(p,\lambda) = \left(\frac{4\lambda p^1}{(1+\lambda)\,p^1 + 2\lambda p^2} + \frac{3\lambda\left(p^1 + 5p^2\right)}{(2+\lambda)\,p^1 + 3\lambda p^2} - 5\right).$$

In order to find the Walrasian price equilibrium of this parametric economic problem, the PVI has been solved:

$$z^1(p^*,\lambda)\left(p^1 - (p^*)^1\right) + z^2(p^*,\lambda)\left(p^2 - (p^*)^2\right) \le 0,\ \forall p \in S^2,\ \lambda \in \Lambda.$$

that is:

$$\begin{aligned}&\left(\tfrac{(1+\lambda)2p^1}{(1+\lambda)p^1+2\lambda p^2} + \tfrac{(2+\lambda)\left(p^1+5p^2\right)}{(2+\lambda)p^1+3\lambda p^2} - 3\right)\cdot\left(p^1 - (p^*)^1\right) + \\ &+\left(\tfrac{4\lambda p^1}{(1+\lambda)p^1+2\lambda p^2} + \tfrac{3\lambda\left(p^1+5p^2\right)}{(2+\lambda)p^1+3\lambda p^2} - 5\right)\cdot\left(p^2 - (p^*)^2\right) \le 0\ \ \forall p \in S^2,\ \lambda \in \Lambda.\end{aligned}$$

By applying the direct method, it results that:

$$\begin{aligned}&\left(\tfrac{(1+\lambda)2p^1}{p^1-\lambda p^1+2\lambda} + \tfrac{(2+\lambda)\left(5-4p^1\right)}{2p^1-2\lambda p^1+3\lambda} - \tfrac{4\lambda p^1}{p^1-\lambda p^1+2\lambda} + \right. \\ &\left. - \tfrac{3\lambda\left(5-4p^1\right)}{2p^1-2\lambda p^1+3\lambda} + 2\right)\cdot\left(p^1 - (p^*)^1\right) \le 0,\ p^2 = 1 - p^1,\ 0 \le p^1 \le 1,\end{aligned} \tag{9}$$

and the exact solution to PVI (9) is obtained. Hence the Walrasian price equilibrium, to varying of λ, is given by:

$$\begin{cases} (p^*)^1 = \frac{2\lambda(2\lambda-5)}{(\lambda-1)(3\lambda-5)} \\ (p^*)^2 = \frac{2(\lambda-10)}{3(1-\lambda)(3\lambda-5)} - \frac{1}{3} \end{cases}$$

with $1-\sqrt{6}<\lambda<0 \vee \frac{5}{2}<\lambda<\sqrt{6}+1$. As application, it is supposed that Λ is the set of the average inflation rates calculated in the period: *December*2007 − *December*2008 (see Table 1), then the parameter λ represents the variation of the annual average inflation rate.

Table 1. Detail of average inflation concerning 2008: national indexes of prices (Source: ISTAT)

N.	Periods	Annual Inflation	Monthly Inflation	Average Inflation *partial value*	Indices
1	Dec 2006–2007	2.6%	0.3%	2.6%	[130.5][133.9]
2	Jan 2007-2008	2.9%	0.4%	2.8%	[130.6][134.4]
3	Feb 2007-2008	2.9%	0.3%	2.8%	[131][134.8]
4	Mar 2007-2008	3.3%	0.5%	2.9%	[131.2][135.5]
5	Apr 2007-2008	3.3%	0.2%	3.0%	[131.4][135.8]
6	May 2007-2008	3.6%	0.5%	3.1%	[131.8][136.5]
7	June 2007-2008	3.8%	0.4%	3.2%	[132.1][137.1]
8	July 2007-2008	4.1%	0.5%	3.3%	[132.4][137.8]
9	Aug 2007-2008	4.1%	0.1%	3.4%	[132.6][138]
10	Sep 2007-2008	3.8%	-0.3%	3.4%	[132.6][137.6]
11	Oct 2007-2008	3.5%	0.0%	3.4%	[133][137.6]
12	Nov 2007-2008	2.7%	-0.4%	3.4%	[133.5][137.1]
13	Dec 2007-2008	2.2%	-0.1%	3.3%	[133.9][136.9]
-	Annual Average Inflation	3.3%	——	——	——

Attributing to the parameter λ the values exposed in Table 1, by using MatLab computation and after a linear interpolation, the curves of equilibria to varying of λ are obtained as shown in Fig. 1 where the data are expressed in perceptual values.

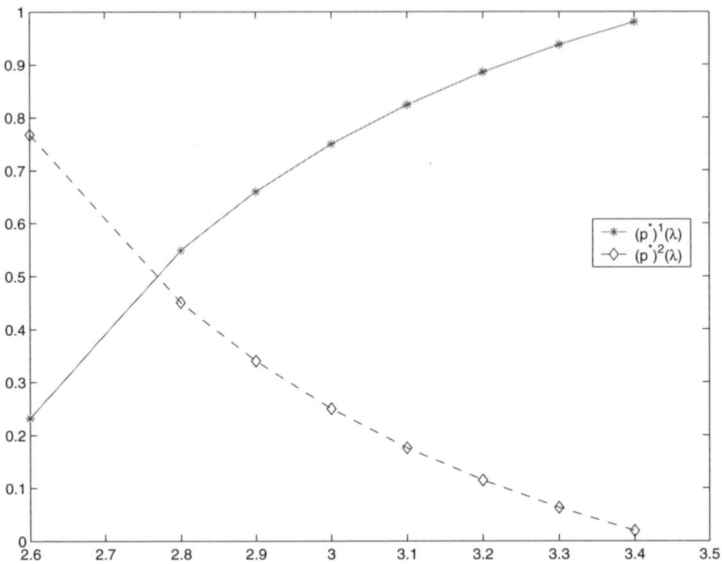

Fig. 1. Curves of equilibria

References

1. K.J. Arrow, F.H. Hahn: *General competitive analysis.* In: Bliss, C.J., Intriligator, M.D. (eds.) Advanced Textbooks in Economics (1991).
2. K.C. Border: *Fixed point theorems with application to economics and game theory.* Cambridge University Press, Cambridge (1985).
3. S. Dafermos, A. Nagurney: *Sensitivity analysis for the general spatial economic equilibrium problem.* Oper. Res. **32**, pp. 1069–1086 (1984).
4. S. Dafermnos: *Sensitivity analysis in variational inequalities.* Math. Oper. Res. **13**, pp. 421–434 (1988).
5. S. Dafermos: *Exchange price equilibria and variational inequalities.* Math. Programm. **46**, pp. 391–402 (1990).
6. S. Dafermos, L. Zhao: *General economic equilibrium and variational inequalities.* Oper. Res. Lett. **10**, pp. 369–376 (1991).
7. P. Daniele: *Evolutionary variational inequalities and economic models for demand-supply markets.* M3AS Math. Models Methods Appl. Sci. **4**(13), pp. 471–489 (2003).
8. P. Daniele, A. Maugeri: *The economic model for demand supply problems.* In: Giannessi, F., Maugeri, A., Pardalos, P. (eds.) Equilibrium Problems: Nonsmooth Optimization and Variational Inequality Models, pp. 59–69, Kluwer, The Netherlands (2001).

9. M.B. Donato, M. Milasi, C. Vitanza: *Duality theory for a Walrasian equilibrium problem.* Journal of nonlinear and convex analysis, **7**, n.3, pp. 393–404 (2006).
10. M.B. Donato, M. Milasi, C. Vitanza: *An existence result of a quasi-variational inequality associated to a Walrasian equilibirum problem.* J. Glob. Optim. (in press) (2007).
11. M.B. Donato, M. Milasi, C. Vitanza: *Quasi-variational inequality approach of a competitive economic equilibrium problem with utility function.* Mathematical models and methods in applied science, **18**, n.3, pp. 351–367 (2008).
12. P.T. Harker, J.S. Pang: *Finite-dimensional variational inequality and nonlinear complementarity problems: a survey of theory, algorithms, and applications.* Math. Programming **48**, pp. 161–220 (1990).
13. J. Kyparisis: *Perturbed solutions of variational problems over polyhedral sets.* Optim. Theory Appl. **57** pp. 295–305 (1988).
14. J. Kyparisis: *Sensitivity analysis for nonlinear programs and variational inequalities with nonunique multipliers.* Math. Oper. Res.**15**, pp.286–298 (1990).
15. A. Maugeri: *Convex programming, variational inequalities, and applications to the traffic equilibrium problem. In: Applied Mathematics and Optimization. Springer, Heidelberg(1987).*
16. A. Nagurney: *Network economics - a variational inequality approach.* 2nd edn. Kluwer, Dordrecht (1999).
17. A. Nagurney, L. Zhao: *Network formalism for pure exchange economic equilibria.* In: Du, D.Z., Pardalos, P.M. (eds.) Network Optimization Problems: Alghoritms, Complexity and Applications, pp. 363–386. World Scientific Press, Singapore (1993).
18. S.M. Robinson: *Generalized equations and their solutions, Part I: Basic theory.* Math. Programming Stud. **10**, pp. 128–141 (1979).
19. G. Stampacchia: *Variational inequalities.* In: Theory and Application of Monotone Operators (Proc. NATO Advanced Study Inst., Venice, 1968) Edizioni "Oderisi", Gubbio, pp. 101–192 (1969).
20. R.L. Tobin: *Sensitivity analysis for variational inequalities.* J. Optim. Theory Appl. **48**, pp.191–204 (1986).
21. D.A. Walker: *J. Polit. Econ.**94**(4).* pp. 753–773 (1987).
22. L. Walras: *Elements d' Economique Politique Pure.* Corbaz, Lausanne, Switzerland (1874).
23. N.D. Yen: *Holder continuity of solutions to a parametric variational inequality.* Appl. Math. Optim. **31**, pp. 245–255 (1995).
24. L. Zhao: *Variational inequalities in general equilibrium: analysis and computation.* Ph. D. Thesis, Division of Applied Mathematics, Brown University, Providence, Rhode Island, 1989, also appears as: LCDS 88-24, Lefschetz Center for Dynamical Systems, Brown University, Providence, Rhode Island (1988).

Predictive Numerical Models of Basin Evolution and Petroleum Generation

Anna Scotti
MOX, Mathematics Department, Politecnico di Milano, Milan, 20133, Italy
E-mail: anna.scotti@mail.polimi.it

Andrea Villa
Department of Mathematics, Università degli Studi di Milano, Milan, 20133, Italy
E-mail: andrea.villa@unimi.it

The present work deals with two aspects of the geological simulation of sedimentary basins: the modeling of their structural evolution and the hydrocarbon generation and expulsion. The sedimentary basin layers are modeled as stratified incompressible fluids and an original scheme is presented to solve this problem. For what concerns hydrocarbon generation the proposed numerical model includes differential retention effects and it is thus able to describe fractionation effects providing correct estimates of the expulsion time and the chemical composition of the expelled oil and gas.

Keywords: Numerical geology, multiphase fluids, multiphysics, level set

1. Introduction

The numerical simulation of geological features have had an increasing importance in the last decades. The applications of these models spread from the geothermal exploitation to the nuclear waste disposal and the petroleum exploitation. Our aim is to investigate the petroleum formation and migration: this kind of simulation requires the knowledge of pressure, temperature and porosity of the sedimentary layer where petroleum is formed. Therefore we have also developed a numerical tool that is capable to predict the geometrical deformation of the sedimentary layers along with the pressure and stress fields. At geologic time scales the sediments can be considered highly viscous incompressible fluids: the fluid-properties of the layers have a great variability. For instance the rock-salt is much weaker than the other sediments and has a great mobility. The salt structures reveal strong connections with the petroleum formation: in fact thanks to its low permeability the salt can create some traps for the hydrocarbons. Moreover the salt has a strong thermal conductivity and alters the temperature field and therefore it influences the chemical reactions that form the oil.

Hydrocarbon generation takes place inside the source rock, a layer of sediments rich in organic matter, when the rock is at the proper depth and temperature. The solid organic matter, kerogen, generates hydrocarbons that can in turn crack into lighter compounds. Hydrocarbons and water can be regarded as immiscible phases and they are expelled from the rock due to different driving forces (see[10]), the most relevant being compaction. Moreover there exist two types of retention processes: the generated hydrocarbons can be retained by kerogen with a solution process, or be adsorbed in the nanopores that are present in the organic matter and in the inorganic part of the source rock. These processes act as selective filters because the retained amount depends on the molecular characteristics of each compound. The numerical simulation of generation, retention and expulsion allows us, thanks to the structural simulation that provides the thermal history of the basin, to forecast the amount of generated hydrocarbons and the timing of expulsion, so that we can infer from the corresponding geologic scenario the existence of a possible reservoir. Finally, with a detailed reaction scheme and the inclusion of retention processes it is possible to compute the average chemical composition of expelled products.

In the first part of the paper we briefly describe a numerical tool capable to simulate the structural evolution of the basin while the second part is devoted to the construction of a simulation method for the formation and primary migration of hydrocarbons.

2. A stratified fluid model

The sediment layers can be modeled as highly viscous stratified fluids.[8] This mathematical model hides some numerical difficulties, namely the high jump of the viscosity coefficient between the layers and the necessity of tracking a large number of sedimentary layers that can experience large deformations. Sometimes also topological changes occur, therefore we need a tracking algorithm that is robust with respect to large modifications in the layers geometry. Let's introduce the geometric model of the sedimentary basin, see Figure 1. The domain $\Omega \in \mathbb{R}^3$ is divided into n_s subdomains ω_i (without overlapping regions), which represent different layers characterized by a specific value of density ρ_i and dynamic viscosity μ_i. On the external part of the basin we apply, at first, an homogeneous Dirichlet condition on the velocity field. Finally, we introduce the mathematical model that

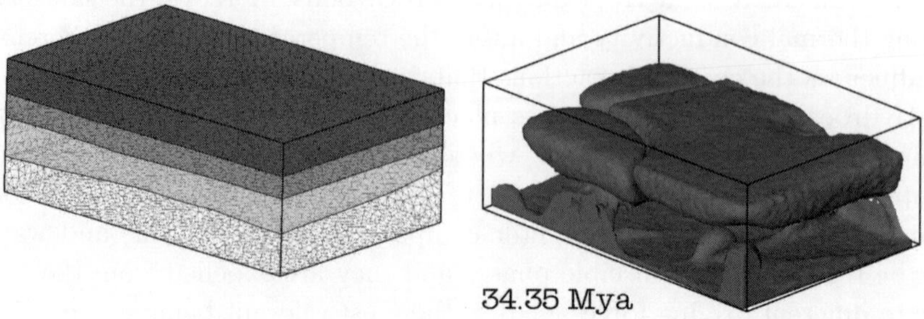

(a) (b)

Fig. 1. In Figure (a) an open three dimensional view of the sedimentary basin is displayed. The basin contains three horizons and four layers. In Figure (b) we show the evolution of the salt layer (the deepest, white layer in Figure (a)) at 34 Ma.

describes the geological evolution of the basin, modeled as a stratified fluid, in which the layers are immiscible and have constant properties:

$$\begin{cases} \overrightarrow{\nabla} \cdot \bar{\bar{\sigma}} - \overrightarrow{\nabla} P + \rho \overrightarrow{g} = 0, & \text{in } \Omega \times (0,T], \\ \overrightarrow{\nabla} \cdot \overrightarrow{V} = 0, & \text{in } \Omega \times (0,T], \\ \dfrac{\partial \rho}{\partial t} + \overrightarrow{V} \cdot \overrightarrow{\nabla} \rho = 0, \quad \dfrac{\partial \mu}{\partial t} + \overrightarrow{V} \cdot \overrightarrow{\nabla} \mu = 0, & \text{in } \Omega \times (0,T], \\ \rho = \rho^0, \ \mu = \mu^0, & \text{in } \Omega \times \{0\}, \\ \overrightarrow{V} = 0 & \text{on } \partial\Omega \times (0,T] \end{cases} \tag{1}$$

where the unknowns are the velocity and pressure fields, respectively $\overrightarrow{V}$ and P, and $\overrightarrow{g}$ is the gravitational acceleration. The density and viscosity fields are defined as:

$$\rho(\overrightarrow{X}) = \rho_i \text{ if } \overrightarrow{X} \in \omega_i \quad \mu(\overrightarrow{X}) = \mu_i \text{ if } \overrightarrow{X} \in \omega_i \tag{2}$$

We also assume, for now, a Newtonian law: $\bar{\bar{\sigma}} = \mu(\overrightarrow{\nabla}\overrightarrow{V} + (\overrightarrow{\nabla}\overrightarrow{V})^T)$. This relation may not seem representative enough of the rheological complexity of the sediments, however it is accepted in literature as a solid base model to study the Rayleigh–Taylor instabilities associated with diapirism.[12]

To solve (1) we must perform a numerical approximation of the problem. First of all let us introduce the following time discretization in the interval $[0,T]$: $[0, t^1, t^2, \ldots, T]$ with $\Delta t^n = t^{n+1} - t^n$. Then let us define:

$$\begin{gathered} \overrightarrow{V}^n(\overrightarrow{X}) = \overrightarrow{V}(t^n, \overrightarrow{X}), \quad P^n(\overrightarrow{X}) = P(t^n, \overrightarrow{X}), \quad \rho^n(\overrightarrow{X}) = \rho(t^n, \overrightarrow{X}), \\ \mu^n(\overrightarrow{X}) = \mu(t^n, \overrightarrow{X}), \quad \bar{\bar{\sigma}}^n(\overrightarrow{X}) = \bar{\bar{\sigma}}(t^n, \overrightarrow{X}). \end{gathered} \tag{3}$$

In order to solve the numerical problem we have implemented the following splitting algorithm used, for example, by:[7]

1° step: knowing ρ^n and μ^n, $\overrightarrow{V}^n$ and P^n can be computed solving the following Stokes problem:

$$\begin{cases} \overrightarrow{\nabla} \cdot \bar{\bar{\sigma}}^n - \overrightarrow{\nabla} P^n + \rho^n \overrightarrow{g} = 0, & \text{in}\, \Omega, \\ \overrightarrow{\nabla} \cdot \overrightarrow{V}^n = 0, & \text{in}\, \Omega, \\ \overrightarrow{V}^n = 0 & \text{on}\, \partial\Omega \times (0, T] \end{cases} \tag{4}$$

2° step: knowing $\overrightarrow{V}^n$ and P^n, ρ^{n+1} and μ^{n+1} can be computed solving the following hyperbolic equations:

$$\frac{\partial \rho}{\partial t} + \overrightarrow{\nabla} \cdot (\rho \overrightarrow{V}^n) = 0, \qquad \frac{\partial \mu}{\partial t} + \overrightarrow{\nabla} \cdot (\mu \overrightarrow{V}^n) = 0 \text{ In } [t^n, t^{n+1}] \times \Omega \tag{5}$$

Therefore the stratified fluid problem (1) can be split into a Stokes problem and a linear hyperbolic system. The Stokes problem is solved using an iterative Krilov-type preconditioned scheme. The hyperbolic problem is equivalent to the tracking of the fluid subdomains as we can always reconstruct a posteriori the density and viscosity fields. In particular introducing the characteristic functions of the subdomains:

$$\lambda_i(\overrightarrow{X}) = \begin{cases} 1, \text{ if } \overrightarrow{X} \in \omega_i, \\ 0, \text{ if } \overrightarrow{X} \notin \omega_i. \end{cases} \tag{6}$$

we can express the density and viscosity fields as:

$$\rho = \sum_{i=1}^{n_s} \lambda_i \rho_i, \qquad \mu = \sum_{i=1}^{n_s} \lambda_i \mu_i. \tag{7}$$

We can also express an evolution equation for the λ_i variables:

$$\begin{cases} \dfrac{\partial}{\partial t} \lambda_i + \overrightarrow{V}^n \cdot \overrightarrow{\nabla} \lambda_i = 0, \\ \lambda_i(0, \overrightarrow{X}) = \lambda_i^0(\overrightarrow{X}). \end{cases} \tag{8}$$

where λ_i^0 are the initial characteristic functions. We discretize equation (8) with a positive high-resolution finite volume scheme. Let $\mathcal{T}_h$ be a simplicial non structured grid with n_e elements e_r with $r = 1, \ldots, n_e$ and let us denote $\widehat{\mathcal{T}}_h$ as its dual grid constituted by $\widehat{n}_e$ elements $\widehat{e}_r$ with $r = 1, \ldots, \widehat{n}_e$. The discrete counterpart of λ_i is $\lambda_{h,i} \in \mathbb{V}^0$ where $\mathbb{V}^0 = \{\varphi_h \in L^2(\Omega) : \varphi_h|_{\widehat{e}_r} \in \mathbb{P}^0\}$. From λ_i we could also get some information about the interfaces between the subdomains by setting $\phi_{h,i} \in \mathbb{V}^1$, where $\mathbb{V}^1 = \{\varphi_h \in \mathbb{C}^1(\Omega) : \varphi|_{e_r} \in \mathbb{P}^1\}$, as the piecewise linear interpolant on $\mathcal{T}_h$ of $\lambda_{h,i}$. Then we get an approximation $\partial\omega_{h,i}$ of the boundary of ω_i using a level set interpretation of $\phi_{h,i}$, in fact:

$$\partial\omega_{h,i}(t) = \left\{ \overrightarrow{X} \in \Omega : \phi_{h,i}(t, \overrightarrow{X}) = \frac{1}{2} \right\}. \tag{9}$$

This new kind of level set method allows to track at the same time a large number of subdomains in place of the usual two fluid applications.[9,11] Moreover conservation properties of this level set method are much better than the original level set method - some results are shown in 2. Let us now

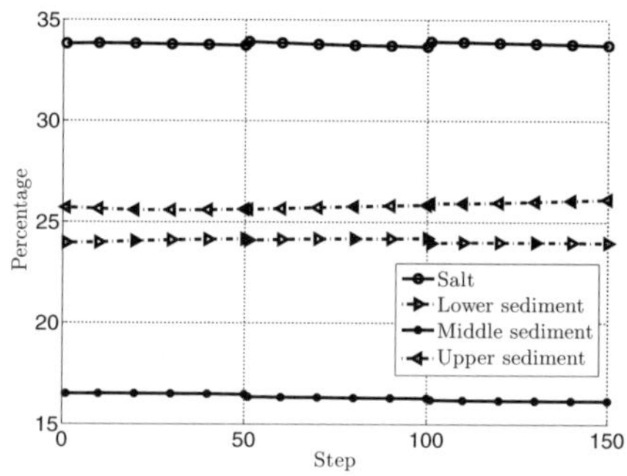

Fig. 2. Volume percentages of the four species. At steps 50 and 100 a re-initialization algorithm is applied.

introduce our test case, we consider four sedimentary layers shown in Figure 1, the deepest one is the salt layer. The overburden is heavier than the salt and a gravitational inversion occurs. We have tested the conservation properties of the algorithm: Figure 2 shows the percentage of volume of the four layers and it can be observed that the results are good in terms of volume conservation. An important feature in structural simulations is the computation of the stresses, indeed the stress field is very important for the wells excavation because the highly stressed regions near the diapir hide some technical difficulties during the drilling phase. An example of the stress field is shown in Figure 3. Another important issue of this kind of simulations is the need for an accurate description of the layers deformations and topological changes which we achieve thanks to the features of the tracking algorithm described above. The results of the structural simulation are the fundamental input data for the simulation of oil generation that will be the main topic of the following section.

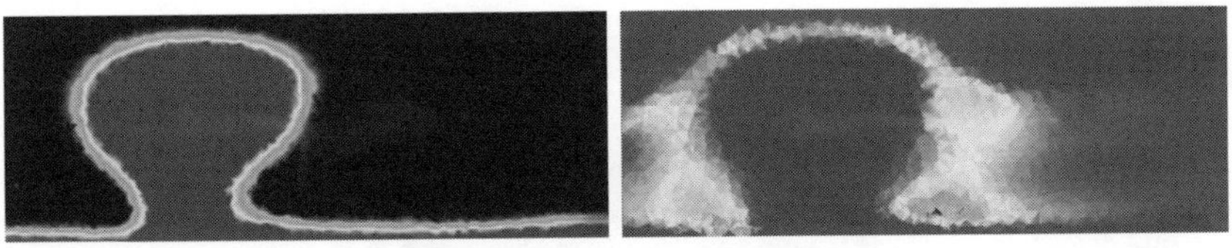

(a) (b)

Fig. 3. In Figure (a) a cross section of the basin is shown. In Figure (b) the corresponding stress field is shown, as we can see the stress is higher near the salt diapir.

3. Modeling generation and expulsion of hydrocarbons

The process of hydrocarbon generation and expulsion can be described by the flow chart in Figure 4: the organic matter generates hydrocarbons (primary generation), and these compounds undergo cracking reactions, i.e. the molecules break up to form lighter compounds, until they leave the source rock. The products of the chemical reactions can be retained by kerogen or nanoporosity: the retained amounts keep reacting, while the remaining is available for the expulsion process which can be modeled as a two-phase flow in a porous medium. The model we adopted for the flow

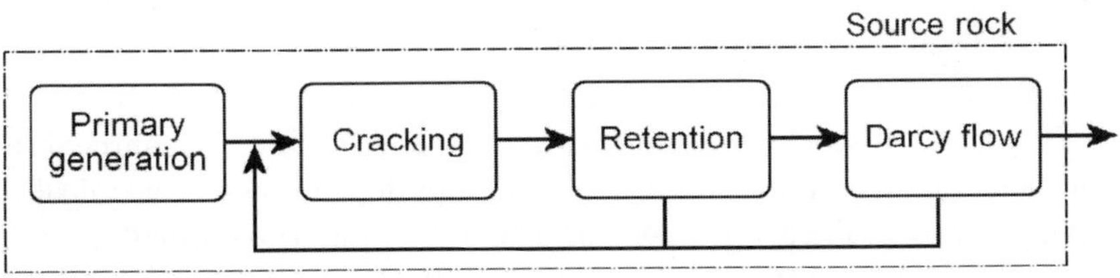

Fig. 4. Flow chart of the process.

in source rock is a 1D, two-phase Darcy model. Let z denote the vertical coordinate and let Σ be the projection of the rock layer on the horizontal plane: then for each point $(x, y) \in \Sigma$ we have that $z \in [0, h(x, y)]$ (see Figure 5). Let the subscript α take the values w and n which denote each of the two immiscible phases involved in the expulsion process, the wetting and

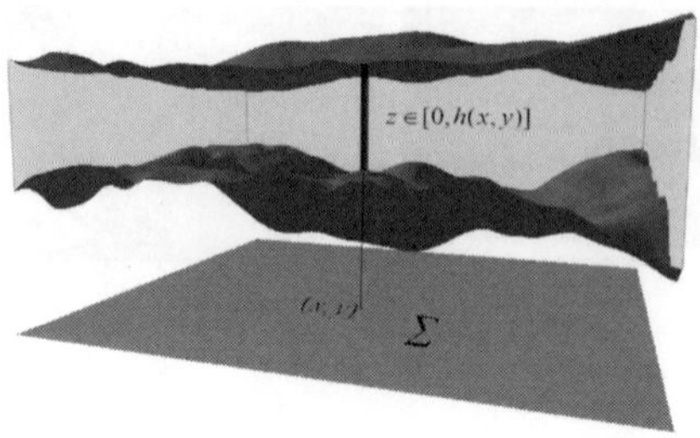

Fig. 5. Domain and its 1D approximation.

non-wetting phase. The model can be written as:

$$\frac{\partial(\Phi\rho_\alpha S_\alpha)}{\partial t} = -\frac{\partial(\rho_\alpha u_\alpha)}{\partial z} + \rho_\alpha q_\alpha \quad [0,h]\times(0,T], \tag{10}$$

$$u_\alpha = -\frac{k_{r\alpha}}{\mu_\alpha} K\left(\frac{\partial p_\alpha}{\partial z} - \rho_\alpha g\right) \quad [0,h]\times(0,T], \tag{11}$$

$$S_w + S_n = 1 \quad [0,h]\times(0,T], \tag{12}$$

$$p_n - p_w = p_c(S_w) \quad [0,h]\times(0,T], \tag{13}$$

where the six unknowns u_α, p_α, S_α, denote velocity, pressure and saturation of each phase. Densities ρ_α and viscosities μ_α are assumed for now to be constant, while the permeability tensor K depends on the porosity Φ. Finally $k_{r\alpha}$ are relative permeabilities and depend on S_w. The source terms q_α account for generation and destruction and depend on the chemical reactions as we will illustrate later on. The 1D assumption is motivated by the thickness of the source rock (only $\mathcal{O}(100m)$) that is small compared to the basin width ($\mathcal{O}(100km)$). The first two equations in (10) express respectively mass conservation and Darcy's law for the two phases while the two last algebraic equations close the model with the constraint that the saturations must sum to one and the definition of capillary pressure as a function of the water phase saturation only. The model can be written in the so called "global pressure formulation" (also known as "fractional flow formulation", see[1]) introducing an artificial variable p, the global pressure, and the total velocity $u = u_w + u_n$. Performing linear combinations of the equations in (10) we obtain the following system for the unknowns p, u, u_w

and S_w:

$$\frac{\partial(\Phi\rho_w S_w)}{\partial t} = \rho_w q_w - \frac{\partial(\rho_w u_w)}{\partial z} \tag{14}$$

$$\frac{\partial u}{\partial z} = q_w + q_n - \frac{\partial \Phi}{\partial t} \tag{15}$$

$$u = -\lambda K \left(\frac{\partial p}{\partial z} - G\right) \tag{16}$$

$$u_w = f_w u + \lambda_n f_w K(\frac{\partial p_c}{\partial z} + (\rho_w - \rho_n)g) \tag{17}$$

where $\lambda_\alpha = k_{r\alpha}/\mu_\alpha$ are the phase mobilities, $\lambda = \lambda_w + \lambda_n$ is the total mobility, $f_\alpha = \lambda_\alpha/\lambda$ is the fractional flow and $G = \frac{\lambda_w \rho_w \lambda_n \rho_n}{\lambda} g$ is a "modified gravity". This formulation results in a weaker coupling between the saturation equation and Darcy's law. In the global pressure formulation the pressure equation depends on saturation through the relative permeabilities, and the saturation equation depends on the phase velocity. The solution strategy relies on the IMPES (implicit pressure - explicit saturation) method to decouple the two equations. The procedure consists in two steps:

(1) computing pressure at time t^n from an elliptic equation whose coefficient depend on saturation at time t^n i.e.:

$$-\frac{\partial}{\partial z}\left(\lambda K \left(\frac{\partial p}{\partial z} - G\right)\right) = q_w + q_n - \frac{\partial \Phi}{\partial t} \tag{18}$$

(2) computing saturation at time t^{n+1} with the frozen velocity field (obtained from Darcy's law - step 1) using equation (14).

The saturation equation is a two-point degenerate parabolic equation: it can be shown that the coefficient of the diffusive term is zero both for $S_w = 0$ and $S_w = 1$.

$$\frac{\partial(\Phi\rho_w S_w)}{\partial t} = \rho_w q_w - \frac{\partial(\rho_w f_w u)}{\partial z} + \frac{\partial(\lambda_n f_w K \frac{\partial p_c}{\partial z})}{\partial z} + \frac{\partial(\lambda_n f_w K(\rho_w - \rho_n)g)}{\partial z} \tag{19}$$

Following the approach proposed by Dawson[4] an operator splitting technique was employed to solve (19). The first step consists in computing an intermediate solution S_w^* solving with Godunov method the following advection-reaction equation:

$$\frac{\partial(\Phi\rho_w S_w^*)}{\partial t} = \rho_w q_w - \frac{\partial(\rho_w f_w u)}{\partial z} + \frac{\partial(\lambda_n f_w K(\rho_w - \rho_n)g)}{\partial z} \tag{20}$$

and then computing the solution at time t^{n+1} using S_w^* as the initial condition for

$$\frac{\partial(\Phi \rho_w S_w^{n+1})}{\partial t} = \frac{\partial}{\partial z}\left(\lambda_n f_w K \frac{\partial p_c}{\partial z}\right). \tag{21}$$

In the problem of primary migration porosity is time dependent due to rock compaction (according to the so-called "phi-stress" relation) and to kerogen conversion. Moreover mass conservation must account, with the term q_o, for hydrocarbons generation (we assume that water does not participate in chemical reactions). Primary generation and cracking are described by a set of Ahrrenius reaction which arises a stiff system of ODEs in the form

$$\frac{d\mathbf{c}}{dt} = \mathbf{f}(\mathbf{c}, t) \tag{22}$$

where $\mathbf{c}$ is a vector containing the concentrations of the hydrocarbon species in the rock. Let us introduce the vector of expellable products ξ whose components are defined as follows

$$\begin{cases} \xi_i = c_i & \text{if } c_i \text{ is a fluid } - \text{ oil or gas} \\ \xi_i = 0 & \text{if } c_i \text{ is a solid.} \end{cases} \tag{23}$$

The source term q_o is thus

$$q_o = \rho_{\rm r} \sum_{i=1}^{N} \frac{d\xi_i}{dt} \frac{1}{\rho_i} \tag{24}$$

where ρ_i are the densities of the species and ρ_r is the density of the rock. The equations describing sets of chemical reactions are often difficult to solve numerically, mostly due to the stiffness that arises when different reactions rates coexist. Moreover, the numerical solution must reproduce two physical properties, that are analytical properties of the corresponding equations: conservativity and positivity. Different numerical methods were tested in order to compare their performances, including modified Patankar methods and second order implicit schemes (see[2,3]). The results of a synthetic multiphase and multicomponent case with two hydrocarbon species are shown in figures 6 and 7. Figure 6 shows the fractions of water and hydrocarbons filling pore space while figure 7 illustrates the evolution of the rock composition in terms of inorganic sediments, organic matter and pore space.

The model can also include retention effects that act as selective filters on the different hydrocarbon compounds according to the following

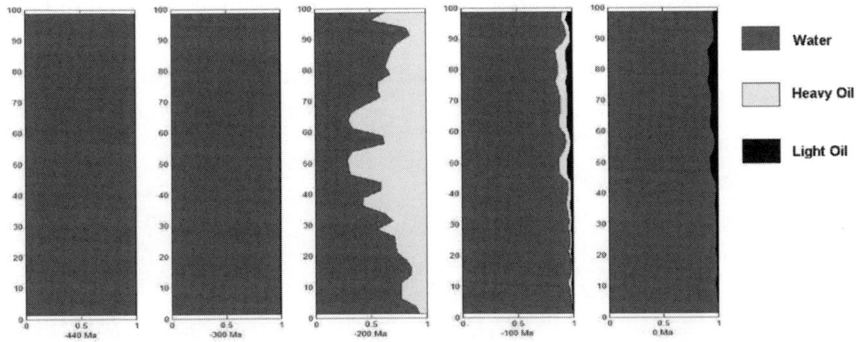

Fig. 6. Pore space filling: water and two oil species

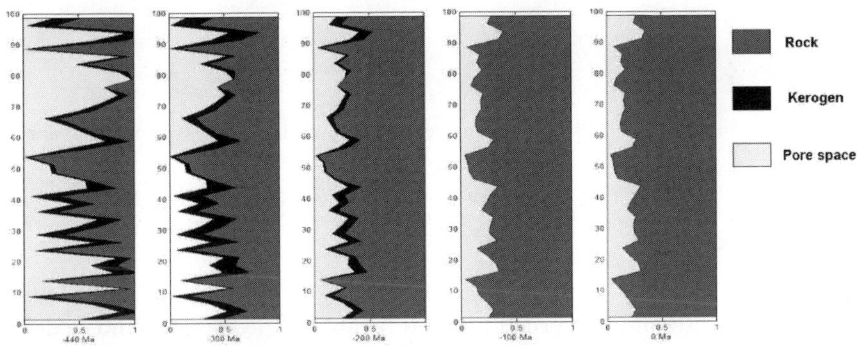

Fig. 7. Rock composition: pore space, kerogen and solid

equations

$$\begin{cases} \dfrac{\partial c_i}{\partial t} = f_i(\mathbf{c}, t) \quad \text{if } c_i \leq E_i \\ \dfrac{\partial c_i}{\partial t} = \dfrac{\partial E_i}{\partial t} \quad \text{if } c_i > E_i \\ \xi_i = \max(c_i - E_i, 0) \quad i = 1, ..., N \end{cases} \tag{25}$$

where the retention thresholds E_i, which express the maximum retainable amount for the i-th compound, are defined by experimental expressions and ξ_i are now re-defined as the amounts that exceed the thresholds and are thus available for the expulsion. The inclusion of these effects in the model results in a discontinuous forcing term for the ODE system that describes chemical reactions: step adaptivity and restarting are thus needed to deal with the discontinuity (see[5]). The importance of modelling differential retention is well highlighted by figure 8 which compares the cumulative expelled products of the same test case neglecting (a) and including (b) retention: it must be pointed out that the average chemical composition of expelled products in 8-(a) seems to be much more in agreement with experimental well data.

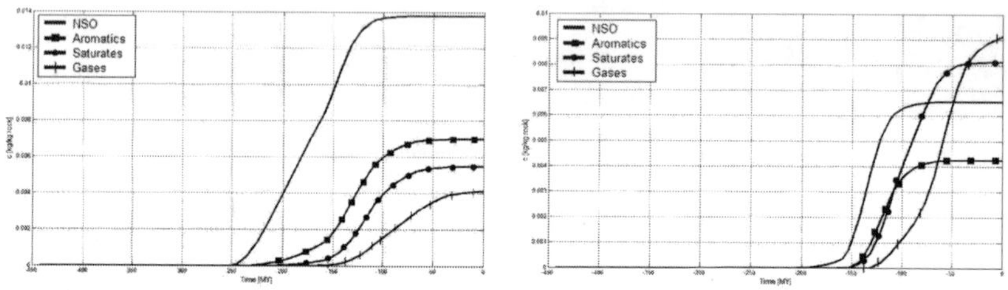

(a) (b)

Fig. 8. Cumulative expelled products neglecting (a) and including (b) retention effects.

4. Remarks

We have developed a tool for the structural modelling of the basin which, thanks to an original numerical method, is able to describe the large deformations of the sedimentary layers that occur in presence of salt structures with very good conservation properties. The results of structural simulations provide the input data for the simulation of petroleum generation and expulsion, modelled as a two-phase reactive flow in a porous medium. The solution strategy of the coupled model takes advantage of stepping and operator splitting techniques. The inclusion of a detailed description of the chemical reactions and retention processes lets us have a promising agreement between simulations and experimental data.

References

1. P.Bastian, *Numerical Computation of Multiphase Flow in Porous Media*, 1999.
2. E.Bertolazzi, Positive and conservative schemes for mass action kinetics, *Computers and Mathematics with Applications*, **32**, 6, (1996), pp. 29-43(15).
3. H. Burchard, E. Deleersnijder and A. Meister, A higher-order conservative Patankar-type discretisation for stiff systems of production-destruction equations, *Applied Numerical Mathematics*, **47**, 1, (2003), pp. 1-30(30).

4. C.Dawson, Godunov-mixed methods for advection-diffusion equations in multidimensions, *SIAM Journal on Numerical Analysis*, **30**, 5, pp. 1315-1332.
5. C. W. Gear and O. Østerby, Solving ordinary differential equations with discontinuities, *ACM Transactions on Mathematical software*, **10**, 1, (1984).
6. R. Huber and R. Helmig, Multiphase flow in heterogeneous porous media: a classical finite element method versus an implicit pressure-explicit saturation-based mixed finite element-finite volume approach, *Int. J. Numer. Meth. Fluids*, **29**, (1999).
7. P. Massimi, A. Quarteroni, and G. Scrofani, An adaptive finite element method for modeling salt diapirism, *M3AS* **16**, 4, (2006).
8. D. L. Turcotte and G. Schubert, *Geodynamics.* Cambridge University Press 2002.
9. S. Osher and R. P. Fedkiw, *Level Set Methods and Dynamic Implicit Surfaces.* Springer 2003.
10. X.Pang, Z.Jiang, S.Zuo and I.Lerche, Dynamics of hydrocarbon expulsion from shale source rock, *Energy exploration and exploitation*, **23**, 5, (2005) , pp. 333-355.
11. A. Sethian *Level Set Methods and Fast Marching Methods.* Cambridge University Press 1999.
12. A. Ismail-Zadeh, I. Tsepelev, C. Talbot, and A. Korotkii, Three-dimensional forward and backward modelling of diapirism: numerical approach and its applicability to the evolution of salt structures in the Pricaspian basin. *Tectonophysics* **387** (2004), pp. 81-103.

A NUMERICAL MODEL FOR BINARY FLUID MIXTURES

A. TIRIBOCCHI*, N. STELLA and G. GONNELLA
Dipartimento di Fisica, Università di Bari and INFN, Sezione di Bari,
Via Amendola 173, 70126 Bari, Italy
** E-mail:adriano.tiribocchi@ba.infn.it*
www.ba.infn.it

A. LAMURA
Istituto Applicazioni Calcolo, CNR,
Via Amendola 122/D, 70126 Bari, Italy
E-mail:a.lamura@ba.iac.cnr.it

In this paper a numerical approach for binary fluid mixtures is proposed. A lattice Boltzmann algorithm for the continuity and the Navier-Stokes equations is coupled to a finite-difference scheme for the convection-diffusion equation. A free-energy is used to derive the thermodynamic quantities related to the equilibrium properties of the system. Spurious velocities are reduced by using a general stencil scheme for discretizing spatial derivatives.

Keywords: Lattice Boltzmann method, finite differences, binary mixtures

1. Introduction

Lattice Boltzmann (LB) method is a widely used technique to simulate physical phenomena for fluid systems.[1] It consists of a phase-space discretization of the Boltzmann equation[2] and proved to be an excellent tool in several cases for simple and complex fluids.[3]

In this work we develop an hybrid approach to study the dynamical properties of a binary fluid mixture solving numerically the continuity, the Navier-Stokes and the convection-diffusion equations. In a previous model[4] these were solved by using two LB equations. The main inconvenient is that spurious (undesired) terms were present in the continuum limit of the hydrodynamic equations. We show that this drawback can be cured by using a LB method with a forcing term for the continuity and the Navier-Stokes equations and a finite-difference scheme for the convection-diffusion equation. This hybrid approach gives good numerical results in reducing spurious velocities when a general stencil is used to discretize spatial derivatives in the forcing term.

The paper is organized as follows. In Section 2 we describe the thermodynamic properties of a binary fluid mixture and present the LB method. Moreover, a description of the numerical scheme for the convection-diffusion equation is given. In Section 3 some numerical results are discussed and the Section 4 concludes this paper.

2. The model

The theoretical framework to describe the equilibrium properties of a binary fluid mixture is given by a Landau-type mean-field theory in which a free-energy is used to obtain thermodynamic quantities. A suitable form for the free-energy[5] is given by

$$\mathcal{F} = \int d\mathbf{r} \left[nT \ln n + \frac{a}{2}\varphi^2 + \frac{b}{4}\varphi^4 + \frac{\kappa}{2}(\nabla\varphi)^2 \right] \tag{1}$$

where n is the total density of the mixture, φ is the concentration difference between the two components and T is the temperature that is assumed constant. In Equation (1) the term depending on n is linked to the ideal gas pressure, the polynomial terms in φ describe the bulk properties of the fluid and the gradient term describes the interfacial ones.

When the two components coexist, the equilibrium values of concentration are given by $\varphi_{eq} = \pm\sqrt{(-a)/b}$. In this case the coefficient a must be negative. The equilibrium interface profile[6] is given by

$$\varphi(x) = \varphi_{eq} \tanh(\frac{2x}{\xi}), \tag{2}$$

where $\xi = 2\sqrt{2\kappa/(-a)}$ is the interface width and the surface tension is $\sigma = (2/3)\sqrt{2a^2\kappa/b}$.

From the free-energy (1) one can obtain the chemical potential

$$\mu = \frac{\delta\mathcal{F}}{\delta\varphi} = a\varphi + b\varphi^3 - \kappa\nabla^2\varphi \tag{3}$$

and the pressure tensor

$$P_{\alpha\beta} = p_0\delta_{\alpha\beta} + \kappa\partial_\alpha\varphi\partial_\beta\varphi. \tag{4}$$

In this last expression, p_0 is the diagonal part of the pressure tensor given by

$$p_0 = nT + \frac{a}{2}\varphi^2 + \frac{3b}{4}\varphi^4 - \kappa\varphi(\nabla^2\varphi) - \frac{\kappa}{2}(\nabla\varphi)^2, \tag{5}$$

while the second term is a concentration gradient contribution required to satisfy a general equilibrium condition for the pressure tensor $\partial_\alpha P_{\alpha\beta} = 0$.[7]

The dynamic behavior is described by the continuity and the Navier-Stokes equations that take into account, respectively, the conservation of mass and momentum, and by the convection-diffusion equation:[8]

$$\partial_t n + \partial_\alpha(n u_\alpha) = 0, \tag{6}$$

$$\partial_t(n u_\beta) + \partial_\alpha(n u_\alpha u_\beta) = -\partial_\alpha P_{\alpha\beta} + \partial_\alpha\{\eta(\partial_\alpha u_\beta + \partial_\beta u_\alpha - \frac{2\delta_{\alpha\beta}}{d}\partial_\gamma u_\gamma) + \zeta\delta_{\alpha\beta}\partial_\gamma u_\gamma\}, \tag{7}$$

$$\partial_t\varphi + \partial_\alpha(\varphi u_\alpha) = \Gamma\nabla^2\mu, \tag{8}$$

where u is the fluid velocity, η and ζ are the shear and the bulk viscosities, respectively, d is the dimensionality of the system and the constant Γ is a mobility coefficient.

To solve numerically these equations, we use a mixed technique that consists of a lattice Boltzmann scheme for Equations (6) and (7) and a finite-difference scheme for Equation (8).

2.1. *The Lattice Boltzmann scheme and the calculation of the forcing term*

The lattice Boltzmann scheme is defined on a square lattice of size $N \times N$ with first and second neighbors interactions. The horizontal and vertical links have length Δx, the diagonal ones $\sqrt{2}\Delta x$, where Δx is the space step. On each site $\mathbf{x}$ at each time t, a set of distribution functions $f_i(\mathbf{x}, t)$ is defined and each f_i is associated with a lattice velocity vector $\mathbf{e}_i$ that has modulus $|\mathbf{e}_i| = \frac{\Delta x}{\Delta t_{LB}} \equiv c$, being Δt_{LB} the LB time step, for $i = 1$ (east direction), 2 (north), 3 (west), 4 (south) and modulus $|\mathbf{e}_i| = \sqrt{2}c$ for $i = 5$ (north east), 6 (north west), 7 (south west), 8 (south east). A zero velocity vector $\mathbf{e}_0$ is also defined.

To simulate Equation (7) we use a LB approach with a forcing term. The evolution equation of the distribution functions is

$$f_i(\mathbf{x}+\mathbf{e}_i\Delta t_{LB}, t+\Delta t_{LB}) - f_i(\mathbf{x}, t) = -\frac{\Delta t_{LB}}{\tau}[f_i(\mathbf{x}, t) - f_i^{eq}(\mathbf{x}, t)] + \Delta t_{LB} F_i, \tag{9}$$

where F_i is the forcing term that must be determined, τ is a relaxation parameter, and $f_i^{eq}(\mathbf{x}, t)$ are the local equilibrium distribution functions. The total density n and the fluid momentum $n\mathbf{u}$ are given by

$$n = \sum_i f_i, \qquad n\mathbf{u} = \sum_i f_i \mathbf{e}_i + \frac{1}{2}\mathbf{F}\Delta t_{LB}, \tag{10}$$

where $\mathbf{F}$ is the force density acting on the fluid. The form of f_i^{eq} is chosen in such a way that the mass and the momentum are locally conserved at each collision step. These conditions are satisfied using the following relations

$$\sum_i f_i^{eq} = n, \qquad n\mathbf{u} = \sum_i f_i^{eq}\mathbf{e}_i. \tag{11}$$

To obtain the continuity and the Navier-Stokes equations in the continuum limit, a suitable choice for the equilibrium distribution functions is given by an expansion up to the second order in the fluid velocity $\mathbf{u}$[9]

$$f_i^{eq}(\mathbf{r},t) = \omega_i n\left[1 + \frac{\mathbf{e}_i \cdot \mathbf{u}}{c_s^2} + \frac{\mathbf{uu} : (\mathbf{e}_i\mathbf{e}_i - c_s^2\mathbf{I})}{2c_s^4}\right], \tag{12}$$

where $c_s = c/\sqrt{3}$ is the sound speed of the model and the weights ω_i are given by $\omega_0 = 4/9$, $\omega_i = 1/9$ for $i = 1-4$, $\omega_i = 1/36$ for $i = 5-8$. The expression chosen for f_i^{eq} also satisfies the relation

$$\sum_i f_i^{eq} e_{i\alpha} e_{i\beta} = n c_s^2 \delta_{\alpha\beta} + n u_\alpha u_\beta. \tag{13}$$

The forcing term F_i in Equation (9) is expressed as a second order expansion in the lattice velocity vectors[10]

$$F_i = \omega_i\left[A + \frac{\mathbf{B} \cdot \mathbf{e}_i}{c_s^2} + \frac{\mathbf{C} : (\mathbf{e}_i\mathbf{e}_i - c_s^2\mathbf{I})}{2c_s^4}\right]. \tag{14}$$

The coefficients A, $\mathbf{B}$ and $\mathbf{C}$ are related to the moments of the force by the following relations

$$\sum_i F_i = A, \qquad \sum_i F_i\mathbf{e}_i = \mathbf{B}, \qquad \sum_i F_i\mathbf{e}_i\mathbf{e}_i = c_s^2 A\mathbf{I} + \frac{1}{2}[\mathbf{C} + \mathbf{C}^T]. \tag{15}$$

and depend on the force density $\mathbf{F}$ which will be given later.

By using a Chapman-Enskog expansion at the second order in the Knudsen number ϵ and assuming the force term to be of the first order in ϵ,[11] the continuum limit is obtained. The choice

$$A = 0, \qquad \mathbf{B} = \left(1 - \frac{\Delta t_{LB}}{2\tau}\right)\mathbf{F}, \qquad \mathbf{C} = \left(1 - \frac{\Delta t_{LB}}{2\tau}\right)(\mathbf{uF} + \mathbf{Fu}) \tag{16}$$

results in the continuity and the Navier-Stokes equations in the following form

$$\partial_t(n u_\beta) + \partial_\alpha(n u_\alpha u_\beta) = -\partial_\beta(n c_s^2) + \partial_\alpha\{\eta(\partial_\alpha u_\beta + \partial_\beta u_\alpha)\} + F_\beta \tag{17}$$

are obtained. It is important to note that no spurious terms are present, except for a term of order u^3 that is neglected. This approximation is correct as far as the Mach number u/c_s is kept very small.[1] For this model the bulk viscosity is zero and the shear viscosity is

$$\eta = nc_s^2 \Delta t_{LB} \left(\frac{\tau}{\Delta t_{LB}} - \frac{1}{2} \right). \tag{18}$$

The final expression of the force term turns out to be

$$F_i = \left(1 - \frac{\Delta t_{LB}}{2\tau}\right) \omega_i \left[\frac{\mathbf{e}_i - \mathbf{u}^*}{c_s^2} + \frac{\mathbf{e}_i \cdot \mathbf{u}^*}{c_s^4} \mathbf{e}_i \right] \cdot \mathbf{F}, \tag{19}$$

being $\mathbf{F}$ the force density given by

$$\mathbf{F} = -\varphi \nabla \mu. \tag{20}$$

in the case of a binary mixture.

The force density (20) is computed by a general nine-point stencil representation of finite difference operators. This is done to ensure isotropy[12] and to reduce spurious velocities in the numerical solution.[13,14] The stencil representation for the first derivative and for the laplacian is

$$\partial_{Dx} = \frac{1}{\Delta x} \begin{bmatrix} -A & 0 & A \\ -B & 0 & B \\ -A & 0 & A \end{bmatrix} \tag{21}$$

$$\nabla_D^2 = \frac{1}{\Delta x^2} \begin{bmatrix} E & G & E \\ G & -4(E+G) & G \\ E & G & E \end{bmatrix} \tag{22}$$

with $2B + 4A = 1$ and $G + 2E = 1$ for consistency between continuous and discrete operators.[12] The subscript D stands for discrete operator. The central entry in the operators is the site in which the derivative is calculated, while the remaining entries refer to the other eight sites (first and second neighbors) around the central one. The parameters B and G are chosen to minimize the spurious velocities as shown later. The usual choice (UC) corresponding to the second order centered finite difference operators is $B = 1/2$ and $G = 1$. In the following we will compare this choice with the best one (BC) obtained numerically minimizing spurious velocities.

2.2. *Numerical implementation of the convection-diffusion equation*

To solve the convection-diffusion equation (8) we use a finite-difference scheme in which the time is discretized in time steps Δt_{FD} related to the lattice Boltzmann time steps by the relation $\Delta t_{LB} = p\Delta t_{FD}$, being p an integer. The order parameter $\varphi(\mathbf{x},t)$ is discretized on the same lattice used for the LB method. At discrete times $t^s = s\Delta t_{FD}$ $(s = 1, 2..)$ we define $\varphi^s_{l,m}$ as the value of φ calculated at time t^s on the lattice site (x_l, y_m). The numerical solution of the convection-diffusion equation comes out updating the value φ^s in two successive steps: $\varphi^s \to \varphi^{s+1/2} \to \varphi^{s+1}$. In the first step the convective part is integrated using an explicit Euler scheme[15] and calculating the derivative of the order parameter φ by using a first order backward scheme if $u > 0$ and a first order forward scheme if $u < 0$. In the second step the diffusive part is implemented using an explicit Euler scheme and discretizing the operator ∇^2, on the r.h.s. of the convection-diffusion equation, by using the form (22) with the usual choice of the parameter G. This scheme allows to have a good numerical stability.[16]

3. Results and discussion

We have simulated a system made of a circular liquid drop of radius

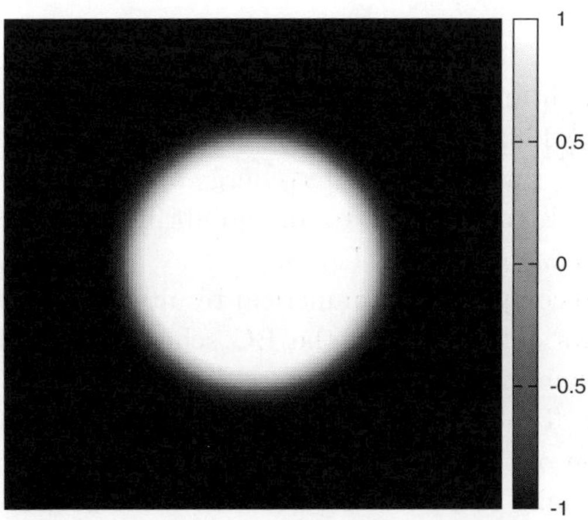

Fig. 1. Equilibrium pattern of the concentration φ of a liquid drop on a lattice of size $N = 128$ in the UC scheme.

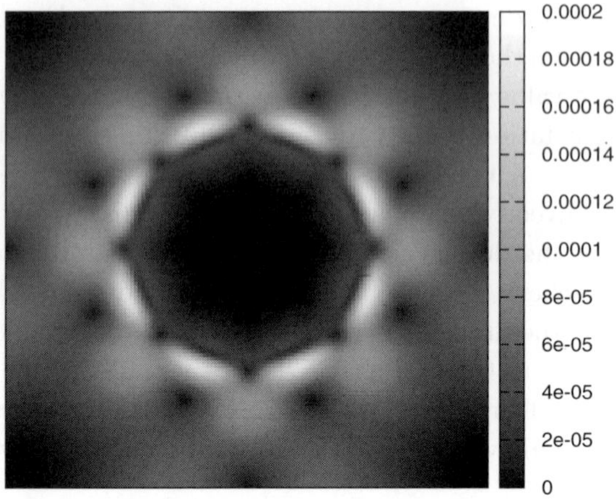

Fig. 2. Velocity field at equilibrium for the UC scheme with $\tau/\Delta t_{LB} = 1$. Velocities are measured in units $\Delta x/\Delta t_{LB}$.

$32\Delta x$ with an initial sharp interface placed at the center of a lattice of size $N = 128$. We have assumed $\Delta x = \Delta t_{FD} = \Delta t_{LB} = 1$, and the parameters a, b and κ in the free-energy are chosen so that $-a = b = 10^{-3}$, $\kappa = -3a$. This corresponds to an equilibrium interface of width $\xi = 5\Delta x$. Moreover $\tau/\Delta t_{LB} = 1$ and $\Gamma = 5$.

We have compared the numerical results obtained with the UC scheme and the ones obtained with the BC scheme. The equilibrium pattern in the UC scheme is shown in Figure 1, in which the drop has relaxed to equilibrium. The corresponding velocity field is reported in Figure 2. An investigation of the velocity pattern shows the existence of non-negligible values especially at interfaces. For this reason we have proposed the BC scheme.

The results for the BC case were obtained scanning different values of the parameters B and G to reduce the maximum value of the velocity on the whole lattice. The couple of values that minimize the velocity is $B = 0.3$ and $G = 2.5$. We found that this optimal choice holds over the range of

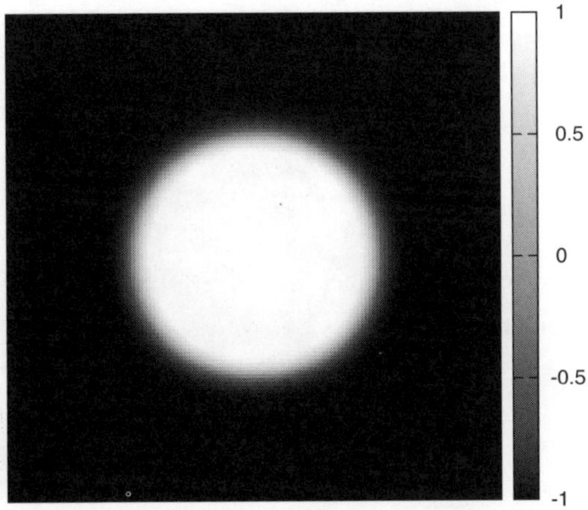

Fig. 3. Equilibrium pattern of the concentration φ of a liquid drop on a lattice of size $N = 128$ in the BC scheme.

values $[0.6; 10]$ of τ which is of physical interest. The equilibrium pattern and the velocity one for this case are shown in Figures 3-4. We can see that there is a decrease of a factor ~ 10 for the velocities when using the BC scheme respect to the UC scheme. Indeed, from Figure 2 and Figure 4, we can see that the maximum value of the velocity in the UC is $\sim 2 \cdot 10^{-4}$ while the one for the BC case is $\sim 2.5 \cdot 10^{-5}$ (velocities are measured in units $\Delta x/\Delta t_{LB}$).

4. Conclusions

In this work we have proposed a new method to study the dynamical properties of a binary fluid mixture that consists in solving the continuity and the Navier-Stokes equations by using a LB method with a forcing term coupled with a finite-difference scheme for the convection-diffusion equation.

The main result of this work is twofold. This mixed approach allows to obtain the correct form of Navier-Stokes equation without spurious terms. Moreover, in the BC scheme there is a decrease of the spurious velocities

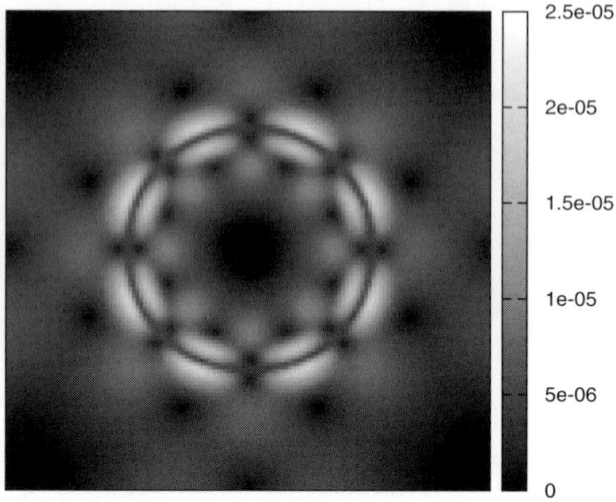

Fig. 4. Velocity field at equilibrium for the BC scheme with $\tau/\Delta t_{LB} = 1$. Velocities are measured in units $\Delta x/\Delta t_{LB}$.

when using a nine-point stencil scheme for discretizing space derivatives in the force term. The reduction with respect to the UC scheme is of about one order of magnitude. We have also probed the model over a wide range of values of τ finding a good stability for the same set of parameters of the BC scheme. Finally, we have shown that the time and space scales used for the convection-diffusion equation and the LB scheme are the same.

The analysis presented in this work opens the possibility of extending this approach to the case of thermal binary fluids, where the temperature is not assumed constant and is enslaved to the energy equation.

References

1. R. Benzi, S. Succi, and M. Vergassola, Phys. Rep. **222**, 145 (1992); S. Chen and G. D. Doolen, Annu. Rev. Fluid Mech. **30**, 329 (1998); S. Succi, *The Lattice Boltzmann Equation for Fluid Dynamics and Beyond* (Clarendon Press, Oxford, 2001).
2. D. A. Wolf-Gladrow, *Lattice Gas Cellular Automata and Lattice Boltzmann Models* (Springer verlag, Berlin, 2000).

3. B. Dünweg and A. J. C. Ladd, Adv. Polym. Sci. **221**, 89 (2009).
4. E. Orlandini, M. R. Swift and J. M. Yeomans, Europhys. Lett. **32**, 463 (1995); M. R. Swift, E. Orlandini, W. R. Osborn, and J. M. Yeomans, Phys. Rev. E **54**, 5041 (1996).
5. A. J. Bray, Adv. Phys. **43**, 357 (1994).
6. J. S. Rowlinson and B. Widom, *Molecular Theory of Capillarity* (Clarendon Press, Oxford, 1982).
7. R. Evans, Adv. Phys. **28**, 143 (1979).
8. S. R. De Groot and P. Mazur, *Non equilibrium Thermodynamics* (Dover Publications, New York, 1984).
9. Y. Qian, D. d'Humieres, and P. Lallemand, Europhys. Lett. **17**, 479 (1992).
10. A. J. C. Ladd and R. Verberg, J. Stat. Phys. **104**, 1191 (2001).
11. J. M. Buick and C. A. Greated, Phys. Rev. E **61**, 5307 (2000).
12. C. M. Pooley and K. Furtado, Phys. Rev. E **77**, 046702 (2008).
13. X. Shan, Phys. Rev. E **73**, 047701 (2006).
14. M. Sbragaglia, R. Benzi, L. Biferale, S. Succi, K. Sugiyama, and F. Toschi, Phys. Rev. E **75**, 026702 (2007).
15. J. C. Strikwerda, *Finite Difference Schemes and Partial Differential Equations* (Chapman & Hall, New York, 1989).
16. S.M. Fielding, Phys. Rev. E **77**, 021504 (2008)

Series on Advances in Mathematics for Applied Sciences

Aims and Scope

This Series reports on new developments in mathematical research relating to methods, qualitative and numerical analysis, mathematical modeling in the applied and the technological sciences. Contributions related to constitutive theories, fluid dynamics, kinetic and transport theories, solid mechanics, system theory and mathematical methods for the applications are welcomed.

This Series includes books, lecture notes, proceedings, collections of research papers. Monograph collections on specialized topics of current interest are particularly encouraged. Both the proceedings and monograph collections will generally be edited by a Guest editor.

High quality, novelty of the content and potential for the applications to modern problems in applied science will be the guidelines for the selection of the content of this series.

Instructions for Authors

Submission of proposals should be addressed to the editors-in-charge or to any member of the editorial board. In the latter, the authors should also notify the proposal to one of the editors-in-charge. Acceptance of books and lecture notes will generally be based on the description of the general content and scope of the book or lecture notes as well as on sample of the parts judged to be more significantly by the authors.

Acceptance of proceedings will be based on relevance of the topics and of the lecturers contributing to the volume.

Acceptance of monograph collections will be based on relevance of the subject and of the authors contributing to the volume.

Authors are urged, in order to avoid re-typing, not to begin the final preparation of the text until they received the publisher's guidelines. They will receive from World Scientific the instructions for preparing camera-ready manuscript.

SERIES ON ADVANCES IN MATHEMATICS FOR APPLIED SCIENCES

Published:*

Vol. 67 Computational Methods for PDE in Mechanics
by B. D'Acunto

Vol. 68 Differential Equations, Bifurcations, and Chaos in Economics
by W. B. Zhang

Vol. 69 Applied and Industrial Mathematics in Italy
eds. M. Primicerio, R. Spigler and V. Valente

Vol. 70 Multigroup Equations for the Description of the Particle Transport in Semiconductors
by M. Galler

Vol. 71 Dissipative Phase Transitions
eds. P. Colli, N. Kenmochi and J. Sprekels

Vol. 72 Advanced Mathematical and Computational Tools in Metrology VII
eds. P. Ciarlini et al.

Vol. 73 Introduction to Computational Neurobiology and Clustering
by B. Tirozzi, D. Bianchi and E. Ferraro

Vol. 74 Wavelet and Wave Analysis as Applied to Materials with Micro or Nanostructure
by C. Cattani and J. Rushchitsky

Vol. 75 Applied and Industrial Mathematics in Italy II
eds. V. Cutello et al.

Vol. 76 Geometric Control and Nonsmooth Analysis
eds. F. Ancona et al.

Vol. 77 Continuum Thermodynamics
by K. Wilmanski

Vol. 78 Advanced Mathematical and Computational Tools in Metrology and Testing
eds. F. Pavese et al.

Vol. 79 From Genetics to Mathematics
eds. M. Lachowicz and J. Miękisz

Vol. 80 Inelasticity of Materials: An Engineering Approach and a Practical Guide
by A. R. Srinivasa and S. M. Srinivasan

Vol. 82 Applied and Industrial Mathematics in Italy III
eds. E. De Bernardis, R. Spigler and V. Valente

*To view the complete list of the published volumes in the series, please visit:
http://www.worldscibooks.com/series/samas_series.shtml